Numerical Analysis 2000, Volume 7

Partial Differential Equations

Numerical Analysis 2000, Volume 7

Partial Differential Equations

Edited by

D. Sloan
University of Strathclyde
Dept. of Mathematics
Livingston Tower
26 Richmond St.
Glasgow G1 1XH
Scotland
UK

E. Süli
Oxford University
Computer Lab.
Wolfson Building
Parks Road
Oxford OX1 3QD
UK

S. Vandewalle
Katholieke Universiteit Leuven
Departement Computerwetenschappen
Celestijnenlaan 200A
Leuven (Haverlee) B-3001
Belgium

2001
ELSEVIER
Amsterdam - London - New York - Oxford - Paris - Shannon - Tokyo

ELSEVIER SCIENCE B.V.
Sara Burgerhartstraat 25
P.O. Box 211, 1000 AE Amsterdam, The Netherlands

First edition 2001

Library of Congress Cataloging in Publication Data
A catalog record from the Library of Congress has been applied for.

ISBN: 0 444 50616 0

♾ The paper used in this publication meets the requirements of ANSI/NISO Z39.48-1992 (Permanence of Paper).
Transferred to digital printing 2005

JOURNAL OF COMPUTATIONAL AND APPLIED MATHEMATICS

Volume 128, Numbers 1–2, 1 March 2001

Contents

PII: S0377-0427(01)00351-X

ELSEVIER

Journal of Computational and Applied Mathematics 128 (2001) ix–xi

JOURNAL OF COMPUTATIONAL AND APPLIED MATHEMATICS

www.elsevier.nl/locate/cam

Preface

Numerical Analysis 2000
Vol. VII: Partial Differential Equations

Over the second half of the 20th century the subject area loosely referred to as *numerical analysis of partial differential equations* (*PDEs*) has undergone unprecedented development. At its practical end, the vigorous growth and steady diversification of the field were stimulated by the demand for accurate and reliable tools for computational modelling in physical sciences and engineering, and by the rapid development of computer hardware and architecture. At the more theoretical end, the analytical insight into the underlying stability and accuracy properties of computational algorithms for PDEs was deepened by building upon recent progress in mathematical analysis and in the theory of PDEs.

To embark on a comprehensive review of the field of numerical analysis of partial differential equations within a single volume of this journal would have been an impossible task. Indeed, the 16 contributions included here, by some of the foremost world authorities in the subject, represent only a small sample of the major developments. We hope that these articles will, nevertheless, provide the reader with a stimulating glimpse into this diverse, exciting and important field.

The opening paper by *Thomée* reviews the history of numerical analysis of PDEs, starting with the 1928 paper by Courant, Friedrichs and Lewy on the solution of problems of mathematical physics by means of finite differences. This excellent survey takes the reader through the development of finite differences for elliptic problems from the 1930s, and the intense study of finite differences for general initial value problems during the 1950s and 1960s. The formulation of the concept of stability is explored in the Lax equivalence theorem and the Kreiss matrix lemmas. Reference is made to the introduction of the finite element method by structural engineers, and a description is given of the subsequent development and mathematical analysis of the finite element method with piecewise polynomial approximating functions. The penultimate section of Thomée's survey deals with 'other classes of approximation methods', and this covers methods such as collocation methods, spectral methods, finite volume methods and boundary integral methods. The final section is devoted to numerical linear algebra for elliptic problems.

The next three papers, by *Bialecki and Fairweather*, *Hesthaven and Gottlieb* and *Dahmen*, describe, respectively, spline collocation methods, spectral methods and wavelet methods. The work by Bialecki and Fairweather is a comprehensive overview of orthogonal spline collocation from its first appearance to the latest mathematical developments and applications. The emphasis throughout is on problems in two space dimensions. The paper by Hesthaven and Gottlieb presents a review of Fourier and Chebyshev pseudospectral methods for the solution of hyperbolic PDEs. Particular emphasis is placed on the treatment of boundaries, stability of time discretisations, treatment of non-smooth solu-

PII: S0377-0427(00)00508-2

tions and multidomain techniques. The paper gives a clear view of the advances that have been made over the last decade in solving hyperbolic problems by means of spectral methods, but it shows that many critical issues remain open. The paper by *Dahmen* reviews the recent rapid growth in the use of wavelet methods for PDEs. The author focuses on the use of adaptivity, where significant successes have recently been achieved. He describes the potential weaknesses of wavelet methods as well as the perceived strengths, thus giving a balanced view that should encourage the study of wavelet methods.

Aspects of finite element methods and adaptivity are dealt with in the three papers by *Cockburn*, *Rannacher* and *Suri*. The paper by Cockburn is concerned with the development and analysis of discontinuous Galerkin (DG) finite element methods for hyperbolic problems. It reviews the key properties of DG methods for nonlinear hyperbolic conservation laws from a novel viewpoint that stems from the observation that hyperbolic conservation laws are normally arrived at via model reduction, by elimination of dissipation terms. Rannacher's paper is a first-rate survey of duality-based a posteriori error estimation and mesh adaptivity for Galerkin finite element approximations of PDEs. The approach is illustrated for simple examples of linear and nonlinear PDEs, including also an optimal control problem. Several open questions are identified such as the efficient determination of the dual solution, especially in the presence of oscillatory solutions. The paper by Suri is a lucid overview of the relative merits of the *hp* and *p* versions of the finite element method over the *h* version. The work is presented in a non-technical manner by focusing on a class of problems concerned with linear elasticity posed on thin domains. This type of problem is of considerable practical interest and it generates a number of significant theoretical problems.

Iterative methods and multigrid techniques are reviewed in a paper by *Silvester*, *Elman*, *Kay and Wathen*, and in three papers by *Stüben*, *Wesseling and Oosterlee* and *Xu*. The paper by Silvester et al. outlines a new class of robust and efficient methods for solving linear algebraic systems that arise in the linearisation and operator splitting of the Navier–Stokes equations. A general preconditioning strategy is described that uses a multigrid V-cycle for the scalar convection–diffusion operator and a multigrid V-cycle for a pressure Poisson operator. This two-stage approach gives rise to a solver that is robust with respect to time-step-variation and for which the convergence rate is independent of the grid. The paper by Stüben gives a detailed overview of algebraic multigrid. This is a hierarchical and matrix-based approach to the solution of large, sparse, unstructured linear systems of equations. It may be applied to yield efficient solvers for elliptic PDEs discretised on unstructured grids. The author shows why this is likely to be an active and exciting area of research for several years in the new millennium. The paper by Wesseling and Oosterlee reviews geometric multigrid methods, with emphasis on applications in computational fluid dynamics (CFD). The paper is not an introduction to multigrid: it is more appropriately described as a refresher paper for practitioners who have some basic knowledge of multigrid methods and CFD. The authors point out that textbook multigrid efficiency cannot yet be achieved for all CFD problems and that the demands of engineering applications are focusing research in interesting new directions. Semi-coarsening, adaptivity and generalisation to unstructured grids are becoming more important. The paper by Xu presents an overview of methods for solving linear algebraic systems based on subspace corrections. The method is motivated by a discussion of the local behaviour of high-frequency components in the solution of an elliptic problem. Of novel interest is the demonstration that the method of subspace corrections is closely related to von Neumann's method of alternating projections. This raises the question as to whether certain error estimates for alternating directions that are available in the literature may be used to derive convergence estimates for multigrid and/or domain decomposition methods.

Moving finite element methods and moving mesh methods are presented, respectively, in the papers by *Baines* and *Huang and Russell.* The paper by Baines reviews recent advances in Galerkin and least-squares methods for solving first- and second-order PDEs with moving nodes in multidimensions. The methods use unstructured meshes and they minimise the norm of the residual of the PDE over both the computed solution and the nodal positions. The relationship between the moving finite element method and L_2 least-squares methods is discussed. The paper also describes moving finite volume and discrete l_2 least-squares methods. Huang and Russell review a class of moving mesh algorithms based upon a moving mesh partial differential equation (MMPDE). The authors are leading players in this research area, and the paper is largely a review of their own work in developing viable MMPDEs and efficient solution strategies.

The remaining three papers in this special issue are by *Budd and Piggott*, *Ewing and Wang* and *van der Houwen and Sommeijer*. The paper by Budd and Piggott on geometric integration is a survey of adaptive methods and scaling invariance for discretisations of ordinary and partial differential equations. The authors have succeeded in presenting a readable account of material that combines abstract concepts and practical scientific computing. Geometric integration is a new and rapidly growing area which deals with the derivation of numerical methods for differential equations that incorporate qualitative information in their structure. Qualitative features that may be present in PDEs might include symmetries, asymptotics, invariants or orderings and the objective is to take these properties into account in deriving discretisations. The paper by Ewing and Wang gives a brief summary of numerical methods for advection-dominated PDEs. Models arising in porous medium fluid flow are presented to motivate the study of the advection-dominated flows. The numerical methods reviewed are applicable not only to porous medium flow problems but second-order PDEs with dominant hyperbolic behaviour in general. The paper by van der Houwen and Sommeijer deals with approximate factorisation for time-dependent PDEs. The paper begins with some historical notes and it proceeds to present various approximate factorisation techniques. The objective is to show that the linear system arising from linearisation and discretisation of the PDE may be solved more efficiently if the coefficient matrix is replaced by an approximate factorisation based on splitting. The paper presents a number of new stability results obtained by the group at CWI Amsterdam for the resulting time integration methods.

We are grateful to the authors who contributed to the Special Issue and to all the referees who reviewed the submitted papers. We are also grateful to *Luc Wuytack* for inviting us to edit this Special Issue.

David M. Sloan
Department of Mathematics
University of Strathclyde
Livingstone Tower, 26 Richmond street
Glasgow, Scotland, G1 1XH, UK
E-mail address: dms@maths.strath.ac.uk

Endre Süli
University of Oxford, Oxford, UK

Stefan Vandewalle
Katholieke Universiteit Leuven
Leuven, Belgium

ELSEVIER

Journal of Computational and Applied Mathematics 128 (2001) 1–54

JOURNAL OF COMPUTATIONAL AND APPLIED MATHEMATICS

www.elsevier.nl/locate/cam

From finite differences to finite elements A short history of numerical analysis of partial differential equations

Vidar Thomée

Department of Mathematics, Chalmers University of Technology, S-412 96 Göteborg, Sweden

Received 30 April 1999; received in revised form 13 September 1999

Abstract

This is an account of the history of numerical analysis of partial differential equations, starting with the 1928 paper of Courant, Friedrichs, and Lewy, and proceeding with the development of first finite difference and then finite element methods. The emphasis is on mathematical aspects such as stability and convergence analysis. © 2001 Elsevier Science B.V. All rights reserved.

MSC: 01A60; 65-03; 65M10; 65N10; 65M60; 65N30

Keywords: History; Finite difference methods; Finite element methods

0. Introduction

This article is an attempt to give a personal account of the development of numerical analysis of partial differential equations. We begin with the introduction in the 1930s and further development of the finite difference method and then describe the subsequent appearence around 1960 and increasing role of the finite element method. Even though clearly some ideas may be traced back further, our starting point will be the fundamental theoretical paper by Courant, Friedrichs and Lewy (1928)[1] on the solution of problems of mathematical physics by means of finite differences. In this paper a discrete analogue of Dirichlet's principle was used to define an approximate solution by means of the five-point approximation of Laplace's equation, and convergence as the mesh width tends to zero was established by compactness. A finite difference approximation was also defined for the

E-mail address: thomee@math.chalmers.se (V. Thomée).

[1] We refer to original work with publication year; we sometimes also quote survey papers and textbooks which are numbered separately.

0377-0427/01/$ - see front matter
PII: S0377-0427(00)00507-0

wave equation, and the CFL stability condition was shown to be necessary for convergence; again compactness was used to demonstrate convergence. Since the purpose was to prove existence of solutions, no error estimates or convergence rates were derived. With its use of a variational principle for discretization and its discovery of the importance of mesh-ratio conditions in approximation of time-dependent problems this paper points forward and has had a great influence on numerical analysis of partial differential equations.

Error bounds for difference approximations of elliptic problems were first derived by Gerschgorin (1930) whose work was based on a discrete analogue of the maximum principle for Laplace's equation. This approach was actively pursued through the 1960s by, e.g., Collatz, Motzkin, Wasow, Bramble, and Hubbard, and various approximations of elliptic equations and associated boundary conditions were analyzed.

For time-dependent problems considerable progress in finite difference methods was made during the period of, and immediately following, the Second World War, when large-scale practical applications became possible with the aid of computers. A major role was played by the work of von Neumann, partly reported in O'Brien, Hyman and Kaplan (1951). For parabolic equations a highlight of the early theory was the important paper by John (1952). For mixed initial–boundary value problems the use of implicit methods was also established in this period by, e.g., Crank and Nicolson (1947). The finite difference theory for general initial value problems and parabolic problems then had an intense period of development during the 1950s and 1960s, when the concept of stability was explored in the Lax equivalence theorem and the Kreiss matrix lemmas, with further major contributions given by Douglas, Lees, Samarskii, Widlund and others. For hyperbolic equations, and particularly for nonlinear conservation laws, the finite difference method has continued to play a dominating role up until the present time, starting with work by, e.g., Friedrichs, Lax, and Wendroff.

Standard references on finite difference methods are the textbooks of Collatz [12], Forsythe and Wasow [14] and Richtmyer and Morton [28].

The idea of using a variational formulation of a boundary value problem for its numerical solution goes back to Lord Rayleigh (1894, 1896) and Ritz (1908), see, e.g., Kantorovich and Krylov [21]. In Ritz's approach the approximate solution was sought as a finite linear combination of functions such as, for instance, polynomials or trigonometrical polynomials. The use in this context of continuous piecewise linear approximating functions based on triangulations adapted to the geometry of the domain was proposed by Courant (1943) in a paper based on an address delivered to the American Mathematical Society in 1941. Even though this idea had appeared earlier, also in work by Courant himself (see Babuška [4]), this is often thought of as the starting point of the finite element method, but the further development and analysis of the method would occur much later. The idea to use an orthogonality condition rather than the minimization of a quadratic functional is attributed to Galerkin (1915); its use for time-dependent problems is sometimes referred to as the Faedo–Galerkin method, cf. Faedo (1949), or, when the orthogonality is with respect to a different space, as the Petrov–Galerkin or Bubnov–Galerkin method.

As a computational method the finite element method originated in the engineering literature, where in the mid 1950s structural engineers had connected the well established framework analysis with variational methods in continuum mechanics into a discretization method in which a structure is thought of as divided into elements with locally defined strains or stresses. Some of the pioneering work was done by Turner, Clough, Martin and Topp (1956) and Argyris [1] and the name of the finite element method appeared first in Clough (1960). The method was later applied to other classes of problems in continuum mechanics; a standard reference from the engineering literature is Zienkiewicz [43].

Independently of the engineering applications a number of papers appeared in the mathematical literature in the mid-1960s which were concerned with the construction and analysis of finite difference schemes by the Rayleigh–Ritz procedure with piecewise linear approximating functions, by, e.g., Oganesjan (1962, 1966), Friedrichs (1962), Céa (1964), Demjanovič (1964), Feng (1965), and Friedrichs and Keller (1966) (who considered the Neumann problem). Although, in fact, special cases of the finite element method, the methods studied were conceived as finite difference methods; they were referred to in the Russian literature as variational difference schemes.

In the period following this, the finite element method with piecewise polynomial approximating functions was analyzed mathematically in work such as Birkhoff, Schultz and Varga (1968), in which the theory of splines was brought to bear on the development, and Zlámal (1968), with the first stringent a priori error analysis of more complicated finite elements. The so called mixed finite element methods, which are based on variational formulations where, e.g., the solution of an elliptic equation and its gradient appear as separate variables and where the combined variable is a saddle point of a Lagrangian functional, were introduced in Brezzi (1974); such methods have many applications in fluid dynamical problems and for higher-order elliptic equations.

More recently, following Babuška (1976), Babuška and Rheinboldt (1978), much effort has been devoted to showing a posteriori error bounds which depend only on the data and the computed solution. Such error bounds can be applied to formulate adaptive algorithms which are of great importance in computational practice.

Comprehensive references for the analysis of the finite element method are Babuška and Aziz [5], Strang and Fix [34], Ciarlet [11], and Brenner and Scott [8].

Simultaneous with this development other classes of methods have arisen which are related to the above, and we will sketch four such classes: In a *collocation* method an approximation is sought in a finite element space by requiring the differential equation to be satisfied exactly at a finite number of collocation points, rather than by an orthogonality condition. In a *spectral* method one uses globally defined functions, such as eigenfunctions, rather than piecewise polynomials approximating functions, and the discrete solution may be determined by either orthogonality or collocation. A *finite volume* method applies to differential equations in divergence form. Integrating over an arbitrary volume and transforming the integral of the divergence into an integral of a flux over the boundary, the method is based on approximating such a boundary integral. In a *boundary integral* method a boundary value problem for a homogeneous elliptic equation in a d-dimensional domain is reduced to an integral equation on its $(d-1)$-dimensional boundary, which in turn can be solved by, e.g., the Galerkin finite element method or by collocation.

An important aspect of numerical analysis of partial differential equations is the numerical solution of the finite linear algebraic systems that are generated by the discrete equations. These are in general very large, but with sparse matrices, which makes iterative methods suitable. The development of convergence analysis for such methods has parallelled that of the error analysis sketched above. In the 1950s and 1960s particular attention was paid to systems associated with finite difference approximation of positive type of second-order elliptic equations, particularly the five-point scheme, and starting with the Jacobi and Gauss–Seidel methods techniques were developed such as the Frankel and Young successive overrelaxation and the Peaceman–Rachford (1955) alternating direction methods, as described in the influential book of Varga [39]. In later years systems with positive-definite matrices stemming from finite element methods have been solved first by the conjugate gradient method proposed by Hestenes and Stiefel (1952), and then making this more effective by preconditioning. The multigrid method was first introduced for finite difference methods in the 1960s by Fedorenko and Bahvalov and further developed by Brandt in the 1970s. For finite elements the multigrid method and the associated method of domain decomposition have been and are being intensely pursued by, e.g., Braess, Hackbusch, Bramble, and Widlund.

Many ideas and techniques are common to the finite difference and the finite element methods, and in some simple cases they coincide. Nevertheless, with its more systematic use of the variational approach, its greater geometric flexibility, and the way it more easily lends itself to error analysis, the finite element method has become the dominating approach both among numerical analysts and in applications. The growing need for understanding the partial differential equations modeling the physical problems has seen an increase in the use of mathematical theory and techniques, and has attracted the interest of many mathematicians. The computer revolution has made large-scale real-world problems accessible to simulation, and in later years the concept of computational mathematics has emerged with a somewhat broader scope than classical numerical analysis.

Our approach in this survey is to try to illustrate the ideas and concepts that have influenced the development, with as little technicalities as possible, by considering simple model situations. We emphasize the mathematical analysis of the discretization methods, involving stability and error estimates, rather than modeling and implementation issues. It is not our ambition to present the present state of the art but rather to describe the unfolding of the field. It is clear that it is not possible in the limited space available to do justice to all the many important contributions that have been made, and we apologize for omissions and inaccuracies; writing a survey such as this one is a humbling experience. In addition to references to papers which we have selected as important in the development we have quoted a number of books and survey papers where additional and more complete and detailed information can be found; for reason of space we have tried to limit the number of reference to any individual author.

Our presentation is divided into sections as follows. Section 1: The Courant–Friedrichs–Lewy paper; Section 2: Finite difference methods for elliptic problems; Section 3: Finite difference methods for initial value problems; Section 4: Finite difference methods for mixed initial–boundary value problems; Section 5: Finite element methods for elliptic problems; Section 6: Finite element methods for evolution equations; Section 7: Some other classes of approximation methods; and Section 8: Numerical linear algebra for elliptic problems.

1. The Courant–Friedrichs–Lewy paper

In this seminal paper from 1928 the authors considered difference approximations of both the Dirichlet problems for a second-order elliptic equation and the biharmonic equation, and of the initial–boundary value problem for a second-order hyperbolic equation, with a brief comment also about the model heat equation in one space variable. Their purpose was to derive existence results for the original problem by constructing finite-dimensional approximations of the solutions, for which the existence was clear, and then showing convergence as the dimension grows. Although the aim was not numerical, the ideas presented have played a fundamental role in numerical analysis. The paper appears in English translation, together with commentaries by Lax, Widlund, and Parter concerning its influence on the subsequent development, see the quotation in references.

The first part of the paper treats the Dirichlet problem for Laplace's equation,

$$-\Delta u = 0 \quad \text{in } \Omega, \text{ with } u = g \text{ on } \partial\Omega, \tag{1.1}$$

where $\Omega \subset \mathbb{R}^2$ is a domain with smooth boundary $\partial\Omega$. Recall that by Dirichlet's principle the solution minimizes $\int\int_\Omega |\nabla\varphi|^2\,dx$ over φ with $\varphi = g$ on $\partial\Omega$. For a discrete analogue, consider mesh-points $x_j = jh$, $j \in \mathbb{Z}^2$, and let Ω_h be the mesh-points in Ω for which the neighbors $x_{j\pm e_1}, x_{j\pm e_2}$ are in Ω ($e_1 = (1,0)$, $e_2 = (0,1)$), and let ω_h be those with at least one neighbor outside Ω. For $U_j = U(x_j)$ a mesh-function we introduce the forward and backward difference quotients

$$\partial_l U_j = (U_{j+e_l} - U_j)/h, \qquad \bar\partial_l U_j = (U_j - U_{j-e_l})/h, \quad l = 1,2. \tag{1.2}$$

By minimizing a sum of terms of the form $(\partial_1 U_j)^2 + (\partial_2 U_j)^2$ one finds a unique mesh-function U_j which satisfies

$$-\partial_1\bar\partial_1 U_j - \partial_2\bar\partial_2 U_j = 0 \quad \text{for } x_j \in \Omega_h, \text{ with } U_j = g(x_j) \text{ on } \omega_h; \tag{1.3}$$

the first equation is the well-known five-point approximation

$$4U_j - (U_{j+e_1} + U_{j-e_1} + U_{j+e_2} + U_{j-e_2}) = 0. \tag{1.4}$$

It is shown by compactness that the solution of (1.3) converges to a solution u of (1.1) when $h \to 0$. By the same method it is shown that on compact subsets of Ω difference quotients of U converge to the corresponding derivatives of u as $h \to 0$. Also included are a brief discussion of discrete Green's function representation of the solution of the inhomogeneous equation, of discretization of the eigenvalue problem, and of approximation of the solution of the biharmonic equation.

The second part of the paper is devoted to initial value problems for hyperbolic equations. In this case, in addition to the mesh-width h, a time step k is introduced and the discrete function values $U_j^n \approx u(x_j, t_n)$, with $t_n = nk, n \in \mathbb{Z}_+$. The authors first consider the model wave equation

$$u_{tt} - u_{xx} = 0 \quad \text{for } x \in \mathbb{R}^2,\ t \geqslant 0, \text{ with } u(\cdot,0),\ u_t(\cdot,0) \text{ given} \tag{1.5}$$

and the approximate problem, with obvious modification of notation (1.2),

$$\partial_t\bar\partial_t U_j^n - \partial_x\bar\partial_x U_j^n = 0 \quad \text{for } j \in \mathbb{Z},\ n \geqslant 1, \text{ with } U_j^n \text{ given for } n = 0,1.$$

When $k = h$ the equation may also be expressed as $U_j^{n+1} + U_j^{n-1} - U_{j+1}^n - U_{j-1}^n = 0$, and it follows at once that in this case the discrete solution at $(x,t) = (x_j, t_n)$ depends only on the initial data in the interval $(x-t, x+t)$. For a general time-step k the interval of dependence becomes $(x - t/\lambda, x + t/\lambda)$, where $\lambda = k/h$ is the mesh-ratio. Since the exact solution depends on data in $(x-t, x+t)$ it follows that if $\lambda > 1$, not enough information is used by the scheme, and hence a necessary condition for convergence of the discrete solution to the exact solution is that $\lambda \leqslant 1$; this is referred to as the Courant–Friedrichs–Lewy or CFL condition. By an energy argument it is shown that the appropriate sums over the mesh-points of positive quadratic forms in the discrete solution are bounded, and compactness is used to show convergence as $h \to 0$ when $\lambda = k/h$=constant $\leqslant 1$. The energy argument is a clever discrete analogue of an argument by Friedrichs and Lewy (1926): For the wave equation in (1.5) one may integrate the identity $0 = 2u_t(u_{tt} - u_{xx}) = (u_t^2 + u_x^2)_t - 2(u_x u_t)_x$ in x to show that $\int_R (u_t^2 + u_x^2)\,dx$ is independent of t, and thus bounded. The case of two spatial variables is also briefly discussed.

In an appendix brief discussions are included concerning a first-order hyperbolic equation, the model heat equation in one space variable, and of wave equations with lower-order terms.

2. Finite difference methods for elliptic problems

The error analysis of finite difference methods for elliptic problems started with the work of Gerschgorin (1930). In contrast to the treatment in Courant, Friedrichs and Lewy (1928) this work was based on a discrete version of the maximum principle. To describe this approach we begin with the model problem

$$-\Delta u = f \quad \text{in } \Omega, \quad \text{with } u = 0 \text{ on } \partial\Omega, \tag{2.1}$$

where we first assume Ω to be the square $\Omega=(0,1)\times(0,1)\subset\mathbb{R}^2$. For a finite difference approximation consider the mesh-points $x_j = jh$ with $h = 1/M$, $j \in \mathbb{Z}^2$ and the mesh-function $U_j = U(x_j)$. With the notation (1.2) we replace (2.1) by

$$-\Delta_h U_j := -\partial_1\bar{\partial}_1 U_j - \partial_2\bar{\partial}_2 U_j = f_j \quad \text{for } x_j \in \Omega, \; U_j = 0 \text{ for } x_j \in \partial\Omega. \tag{2.2}$$

This problem may be written in matrix form as $AU=F$, where A is a symmetric $(M-1)^2\times(M-1)^2$ matrix whose elements are $4, -1$, or 0, with 0 the most common occurrence, cf. (1.4).

For the analysis one first shows a discrete maximum principle: If U is such that $-\Delta_h U_j \leqslant 0$ ($\geqslant 0$) in Ω, then U takes its maximum (minimum) on $\partial\Omega$; note that $-\Delta_h U_j \leqslant 0$ is equivalent to $U_j \leqslant (U_{j+e_1} + U_{j-e_1} + U_{j+e_2} + U_{j-e_2})/4$. Letting $W(x) = \frac{1}{2} - |x - x_0|^2$ where $x_0 = (\frac{1}{2}, \frac{1}{2})$ we have $W(x) > 0$ in Ω, and applying the discrete maximum principle to the function $V_j = \pm U_j - \frac{1}{4}|\Delta_h U|_\Omega W_j$ one concludes easily that, for any mesh-function U on $\bar{\Omega}$,

$$|U|_{\bar{\Omega}} \leqslant |U|_{\partial\Omega} + C|\Delta_h U|_\Omega, \quad \text{where } |U|_S = \max_{x_j \in S} |U_j|.$$

Noting that the error $z_j = U_j - u(x_j)$ satisfies

$$-\Delta_h z_j = f_j + \Delta_h u(x_j) = (\Delta_h - \Delta)u(x_j) = \tau_j, \quad \text{with } |\tau_j| \leqslant Ch^2\|u\|_{C^4}, \tag{2.3}$$

one finds, since $z_j = 0$ on $\partial\Omega$, that

$$|U - u|_{\bar{\Omega}} = |z|_{\bar{\Omega}} \leqslant C|\tau|_\Omega \leqslant Ch^2\|u\|_{C^4}.$$

The above analysis uses the fact that all neighbors $x_{j\pm e_l}$ of the interior mesh-points $x_j \in \Omega$ are either interior mesh-points or belong to $\partial\Omega$. When the boundary is curved, however, there will be mesh-points in Ω which have neighbors outside $\bar{\Omega}$. If for such a mesh-point $x = x_j$ we take a point $b_{h,x} \in \partial\Omega$ with $|b_{h,x} - x| \leqslant h$ and set $U_j = u(b_{h,x}) = 0$, then it follows from Gerschgorin, loc. cit., that $|U - u|_{\bar{\Omega}} \leqslant Ch\|u\|_{C^3}$. To retain second-order accuracy Collatz (1933) proposed to use linear interpolation near the boundary: Assuming for simplicity that Ω is a convex plane domain with smooth boundary, we denote by Ω_h the mesh-points $x_j \in \Omega_h$ that are truly interior in the sense that all four neighbors of x_j are in $\bar{\Omega}$. (For the above case of a square, Ω_h simply consists of all mesh-points in Ω.) Let now $\tilde{\omega}_h$ be the mesh-points in Ω that are not in Ω_h. For $x_j \in \tilde{\omega}_h$ we may then find a neighbor $y = x_i \in \Omega_h \cup \tilde{\omega}_h$ such that the line through x and y cuts $\partial\Omega$ at a point $\bar{x}$ which is not a mesh-point. We denote by $\bar{\omega}_h$ the set of points $\bar{x} \in \partial\Omega$ thus associated with the points of $\tilde{\omega}_h$, and define for $x = x_j \in \tilde{\omega}_h$ the error in the linear interpolant

$$\ell_h u_j := u(x_j) - \alpha u(x_i) - (1-\alpha)u(\bar{x}), \quad \text{where } \alpha = \gamma/(1+\gamma) \leqslant \tfrac{1}{2} \text{ if } |x - \bar{x}| = \gamma h.$$

As $u=0$ on $\partial\Omega$ we now pose the problem

$$-\Delta_h U_j = f_j \quad \text{in } \Omega_h, \qquad \ell_h U_j = 0 \quad \text{in } \tilde{\omega}_h, \quad U(\bar{x}) = 0 \text{ on } \bar{\omega}_h$$

and since $\alpha \leqslant \frac{1}{2}$ it is not difficult to see that

$$|U|_{\Omega_h \cup \tilde{\omega}_h} \leqslant C(|U|_{\bar{\omega}_h} + |\ell_h U|_{\tilde{\omega}_h} + |\Delta_h U|_{\Omega}). \tag{2.4}$$

Using again (2.3) together with $|\ell_h z|_{\tilde{\omega}_h} \leqslant Ch^2\|u\|_{C^3}$, $z=0$ on $\bar{\omega}_h$, one finds that $|U-u|_{\Omega_h\cup\tilde{\omega}_h} \leqslant Ch^2\|u\|_{C^4}$. Another approximation near $\partial\Omega$ was proposed by Shortley and Weller (1938). For $x_j \in \tilde{\omega}_h$ it uses the points defined by the intersections of $\partial\Omega$ with the horizontal and vertical mesh-lines through x_j, which together with the neighbors that are in Ω form an irregular five-point star. This gives an approximation to $-\Delta$ which is only first-order accurate, but, using it in a boundary operator ℓ_h similarly to the Collatz interpolation error it will yield a second-order error bound.

Consider more generally the variable coefficient Dirichlet problem

$$Au := -\sum_{j,k=1}^{d} a_{jk}\frac{\partial^2 u}{\partial x_j \partial x_k} - \sum_{j=1}^{d} b_j \frac{\partial u}{\partial x_j} f \quad \text{in } \Omega \subset \mathbb{R}^d, \ u = g \text{ on } \partial\Omega, \tag{2.5}$$

where the matrix (a_{jk}) is uniformly positive definite in $\bar{\Omega}$, and a corresponding finite difference operator with finitely many terms of the form

$$A_h u(x) = -h^{-2}\sum_j a_j u(x - jh), \quad a_j = a_j(x,h), \ j \in \mathbb{Z}^2 \tag{2.6}$$

which is consistent with A so that $A_h u(x) \to Au(x)$ as $h \to 0$. Following Motzkin and Wasow (1953) such an operator is said to be of positive type if $a_j \geqslant 0$ for $j \neq 0$, with $a_0 < 0$.

For mesh-points x, let $N(x)$ be the convex hull of the set of neighbors of x defined by (2.6), i.e., the mesh-points $x - jh$ with $a_j(x,h) \neq 0$, and let Ω_h denote the set of mesh-points with $N(x) \not\subset \bar{\Omega}$. The remaining mesh-points in $\bar{\Omega}$ form the set $\tilde{\omega}_h$ of boundary mesh-points. We set $\bar{\Omega}_h = \Omega_h \cup \tilde{\omega}_h$. For $x \in \Omega_h$ we want to use the equation $A_h U(x) = M_h f(x)$ as an analogue of the differential equation in (2.5), where M_h is a linear operator approximating the identity operator I (in most cases $M_h = I$). For $x \in \tilde{\omega}_h$, $A_h U(x)$ is not defined by the values of U on $\bar{\Omega}_h$, and at such points we therefore want to choose an equation of the form

$$\ell_h u(x) := \sum_{x_j \in \bar{\Omega}} \tilde{a}_j u(x - jh) = m_h(g, f), \quad \tilde{a}_j = \tilde{a}_j(x,h),$$

where m_h is a suitable linear operator. The values of u at points in $\bar{\omega}_h$ will now be included in the right-hand side by $u = g$ on $\partial\Omega$. Together these equations form our difference scheme, and we say (see Bramble, Hubbard and Thomée (1969)) that this is of essentially positive type if A_h is of positive type and $\tilde{a}_0 = 1$, $\sum_{j\neq 0}|\tilde{a}_j| \leqslant \gamma < 1$ for $x \in \omega_h$. A discrete maximum principle shows that the analogue of (2.4) remains valid in this case (with Δ_h replaced by A_h and without the term $|U|_{\bar{\omega}_h}$). The scheme is said to be accurate of order q if $A_h u - M_h Au = O(h^q)$ on Ω_h, and $\ell_h u - m_h(u|_{\partial\Omega}, Au) = O(h^q)$ on $\tilde{\omega}_h$. Under somewhat more precise natural assumptions one may now conclude from (2.4) that $|U-u|_{\Omega_h} \leqslant Ch^q\|u\|_{C^{q+2}}$. Error bounds may also be expressed in terms of data, and a $O(h^q)$ error bound holds if $f \in C^s(\bar{\Omega})$, $g \in C^s(\partial\Omega)$ with $s > q$. For homogeneous boundary conditions this follows easily using the Schauder estimate $\|u\|_{C^{q+2}} \leqslant C\|f\|_{C^s}$. For inhomogeneous boundary conditions a more precise analysis may be based on a representation using a nonnegative discrete Green's function $G_{jl} = G(x_j, x_l)$,

$$U_j = h^d \sum_{x_l \in \Omega_h} G_{jl} A_h U_l + \sum_{x_l \in \omega_h} G_{jl} U_l \quad \text{for } x_j \in \bar{\Omega}_h,$$

where $h^d \sum_{x_l \in \Omega_h} G_{jl} \leqslant C$ and $\sum_{x_l \in \tilde{\omega}_h} G_{jl} \leqslant 1$, and where also a discrete analogue of the estimate $\int_{\Gamma_\delta} G(x,y)\,\mathrm{d}s \leqslant C\delta$ for the continuous problem holds, where $\Gamma_\delta = \{y \in \Omega;\ \mathrm{dist}(y, \partial\Omega) = \delta\}$. The latter is related to the important observation by Bahvalov (1959) that the regularity demands on the solution u of the continuous problem can be relaxed in some cases by essentially two derivatives at the boundary without loosing the convergence rate. For less regular data one can obtain correspondingly weaker convergence estimates: When $f \in C^s(\bar{\Omega})$, $g \in C^s(\partial\Omega)$ with $0 \leqslant s < q$, $O(h^s)$ order convergence may be shown by interpolation between C^s-spaces. The regularity demands of f may be further relaxed by choosing for M_h an averaging operator, see Tikhonov and Samarskii (1961); this paper also demonstrated how to construct finite difference approximations of elliptic equations with discontinuous coefficients by taking local harmonic averages. When the boundary itself is nonsmooth the exact solution may have singularities which make the above results not applicable. Laasonen (1957) showed that the presence of a corner with an accute inner angle does not affect the rate of convergence but if the angle is $\pi\alpha$ with $\alpha > 1$ he shows the weaker estimate $O(h^{1/\alpha - \varepsilon})$ for any $\varepsilon > 0$.

As an example of an operator of the form (2.6), consider the nine-point formula

$$-\Delta_h^{(9)} u(x) = \tfrac{1}{6} h^{-2} \left(20\, u(x) - 4 \sum_{|j|=1} u(x - jh) - \sum_{|j_1|=|j_2|=1} u(x - jh) \right).$$

With $M_h f = f + \frac{1}{12} h^2 \Delta_h f$ one finds $\Delta_h^{(9)} u + M_h \Delta u = O(h^4)$ and Bramble and Hubbard (1962) showed that the operator ℓ_h can be chosen so that the corresponding scheme is of essentially positive type and accurate of order $q = 4$. Further, Bramble and Hubbard (1963) constructed second-order accurate schemes of essentially positive type in the case of a general A ($d = 2$), also with mixed derivative terms. Here the neighbors of x may be several mesh-widths away from x. Related results were also obtained in special cases by Wasow (1952), Laasonen (1958), and Volkov (1966), see also Bramble, Hubbard and Thomée (1969).

We shall now turn to some schemes that approximate elliptic equations containing mixed derivatives but are not generally of essentially positive type. Assume now that A is in divergence form, $A = -\sum_{i,k=1}^{d} (\partial/\partial x_i)(a_{ik} \partial u/\partial x_k)$. To our above notation ∂_i, $\bar{\partial}_i$ of (1.2) we add the symmetric difference quotient $\hat{\partial}_i = (\partial_i + \bar{\partial}_i)/2$ and set, with $a_{ik}^{(h)} = a_{ik}(x + \frac{1}{2} h e_i)$,

$$A_h^{(1)} u = -\sum_{i,k} \bar{\partial}_i (a_{ik}^{(h)} \partial_k u), \qquad A_h^{(2)} u = -\sum_{i} \bar{\partial}_i (a_{ii}^{(h)} \partial_i u) - \sum_{i \neq k} \hat{\partial}_i (a_{ik} \hat{\partial}_k u).$$

These operators are obviously consistent with A and second-order accurate. Except in special cases the $A_h^{(l)}$ are not of positive type and the above analysis does not apply. Instead one may use energy arguments to derive error estimates in discrete l_2-norms, see Thomée (1964). For $x \in \tilde{\omega}_h$ let as above $b_{h,x} \in \partial\Omega$, and consider the discrete Dirichlet problem

$$A_h U = f \quad \text{on } \Omega_h, \qquad U = g(b_{h,x}) \quad \text{on } \tilde{\omega}_h, \quad \text{with } A_h = A_h^{(1)} \text{ or } A_h^{(2)}. \tag{2.7}$$

With $\mathscr{D}_h$ the mesh-functions which vanish outside Ω_h, we define for $U, V \in \mathscr{D}_h$

$$(U,V) = h^d \sum_j U_j V_j, \quad \|U\| = (U,U)^{1/2}, \quad \text{and} \quad \|U\|_1 = \|U\| + \sum_{j=1}^{d} \|\partial_i U\|.$$

By summation by parts one easily derives that $\|V\|_1^2 \leqslant C(A_h V, V)$ for $V \in \mathscr{D}_h$, and this shows at once the uniqueness and hence the existence of a solution of (2.7). When $\tilde{\omega}_h \subset \partial\Omega$, application to $U-u$, using the second-order accuracy of A_h shows that $\|U-u\|_1 \leqslant C(u)h^2$. When $\tilde{\omega}_h \not\subset \partial\Omega$, h^2 has to be replaced by $\sqrt{h}$, but with $\|\cdot\|_1'$ a slightly weaker norm it was shown in Bramble, Kellogg and Thomée (1968) that $\|U-u\| \leqslant \|U-u\|_1' \leqslant C(u)h$.

Consider now a constant coefficient finite difference operator of the form (2.6), which is consistent with A of (2.5). Introducing the symbol of A_h, the trigonometric polynomial $p(\xi) = \sum_j a_j e^{ij\xi}$, we say that A_h is elliptic if $p(\xi) \neq 0$ for $|\xi_l| \leqslant \pi$, $l = 1,2$, $\xi \neq 0$. For the five-point operator $-\Delta_h$ we have $p(\xi) = 4 - 2\cos\xi_1 - 2\cos\xi_2$ and $-\Delta_h$ is thus elliptic. For such operators A_h we have the following interior estimate by Thomée and Westergren (1968). Set $\|U\|_S = (h^d \sum_{x_j \in S} U_j^2)^{1/2}$ and $\|U\|_{k,S} = \left(\sum_{|\alpha| \leqslant k} \|\partial_h^\alpha U\|_S^2\right)^{1/2}$ where $\partial_h^\alpha = \partial_1^{\alpha_1} \cdots \partial_d^{\alpha_d}$, $|\alpha| = \alpha_1 + \cdots + \alpha_d$, and let $\bar{\Omega}_0 \subset \Omega_1 \subset \bar{\Omega}_1 \subset \Omega$. Then for any α and h small we have

$$|\partial_h^\alpha U|_{\Omega_0} \leqslant C(\|A_h U\|_{l,\Omega_1} + \|U\|_{\Omega_1}) \quad \text{if } l \geqslant |\alpha| + [d/2] - 1.$$

Thus, the finite difference quotients of the solution of the equation $A_h U = f$ may be bounded in Ω_0 by the difference quotients of f in a slightly larger domain Ω_1 plus a discrete l_2-norm of U in Ω_1. Assuming u is a solution of (2.1) this may be used to show that if A_h is accurate of order q and Q_h is a finite difference operator which approximates the differential operator Q to order q, then

$$|Q_h U - Qu|_{\Omega_0} \leqslant C(u)h^q + C\|U-u\|_{\Omega_1}. \tag{2.8}$$

Thus, if we already know that $U-u$ is of order $O(h^q)$ in the l_2-norm, then $Q_h u - Qu$ is of the same order in maximum norm in the interior of the domain.

Finite difference approximations for elliptic equations of higher order were studied by, e.g., Saulev (1957); the work in Thomée (1964) concerns also such equations.

3. Finite difference methods for initial value problems

In this section we sketch the development of the stability and convergence theory for finite difference methods applied to pure initial value problems. We first consider linear constant coefficient evolution equations and then specialize to parabolic and hyperbolic equations.

We begin with the initial value problem for a general linear constant coefficient scalar equation

$$u_t = P(D)u \quad \text{for } x \in \mathbb{R}^d,\ t > 0, \qquad u(\cdot,0) = v, \quad \text{where } P(\xi) = \sum_{|\alpha| \leqslant M} P_\alpha \xi^\alpha \tag{3.1}$$

with $u = u(x,t)$, $v = v(x)$, $\xi^\alpha = \xi_1^{\alpha_1} \cdots \xi_d^{\alpha_d}$, and $D = (\partial/\partial x_1, \ldots, \partial/\partial x_d)$. Such an initial value problem is said to be well-posed if it has a unique solution that depends continuously on the initial data, in some sense that has to be specified. For example, the one-dimensional wave equation $u_t = \rho u_x$ has the unique solution $u(x,t)=v(x+\rho t)$ and since $\|u(\cdot,t)\|_{L_p}=\|v\|_{L_p}$, this problem is well posed in L_p for $1\leqslant p\leqslant\infty$. Similarly, for the heat equation $u_t=u_{xx}$ we have

$$u(x,t)=\frac{1}{\sqrt{4\pi t}}\int_{-\infty}^{\infty}\mathrm{e}^{-(x-y)^2/4t}v(y)\,\mathrm{d}y$$

and $\|u(\cdot,t)\|_{L_p}\leqslant\|v\|_{L_p}$, $1\leqslant p\leqslant\infty$. More precisely, (3.1) is well-posed in L_p if $P(D)$ generates a semigroup $E(t)=\mathrm{e}^{tP(D)}$ in L_p which grows at most exponentially, so that the solution $u(t)=E(t)v$ satisfies $\|u(t)\|_{L_p}\leqslant C\mathrm{e}^{\kappa t}\|v\|_{L_p}$ for $t\geqslant 0$, for some κ. For $p=2$, which is the case we will concentrate on first, we see by Fourier transformation and Parseval's relation that this is equivalent to

$$|\mathrm{e}^{tP(\mathrm{i}\xi)}|\leqslant C\mathrm{e}^{\kappa t},\quad \forall\xi\in\mathbb{R}^d,\ t>0, \tag{3.2}$$

or $\operatorname{Re}P(\mathrm{i}\xi)\leqslant\kappa$ for $\xi\in\mathbb{R}^d$; if only the highest-order terms are present in $P(D)$, this is equivalent to $\operatorname{Re}P(\mathrm{i}\xi)\leqslant 0$ for $\xi\in\mathbb{R}^d$, in which case (3.2) holds with $\kappa=0$, so that $E(t)$ is uniformly bounded in L_2 for $t\geqslant 0$. In the above examples $P(\mathrm{i}\xi)=\mathrm{i}\rho\xi$ and $P(\mathrm{i}\xi)=-\xi^2$, respectively, and $\kappa=0$.

Generalizing to systems of the form (3.1) with u and v N-vectors and the P_α $N\times N$ matrices, the condition for well-posedness in L_2 is again (3.2) where now $|\cdot|$ denotes any matrix norm. Here it is clear that a necessary condition for (3.2) is that $\operatorname{Re}\lambda_j(\xi)\leqslant\kappa$ for $\xi\in\mathbb{R}^d$, for any eigenvalue $\lambda_j(\xi)$ of $P(\mathrm{i}\xi)$, and if $P(\mathrm{i}\xi)$ is a normal matrix this is also sufficient. Necessary and sufficient conditions for (3.2) were given by Kreiss (1959b), and depend on the following lemma. Here, for any $N\times N$ matrix A with eigenvalues $\{\lambda_j\}_{j=1}^N$ we set $\Lambda(A)=\max_j\operatorname{Re}\lambda_j$ and $\operatorname{Re}A=(A+A^*)/2$, where A^* is the adjoint matrix. For Hermitian matrices $A\leqslant B$ means $Au\cdot u\leqslant Bu\cdot u$ for all $u\in\mathbb{R}^N$. With this notation, (3.2) holds if and only if the set $\mathscr{F}=\{P(\mathrm{i}\xi)-\kappa;\,\xi\in\mathbb{R}^d\}$ satisfies the conditions of the following lemma:

Let $\mathscr{F}$ be a set of $N\times N$ matrices. Then the following conditions are equivalent:

(i) $|\mathrm{e}^{tA}|\leqslant C$ for $A\in\mathscr{F}$, $t\geqslant 0$, for some $C>0$;

(ii) $\Lambda(A)\leqslant 0$ *and* $\operatorname{Re}(z\,|R(A;z)|)\leqslant C$ *for* $A\in\mathscr{F}$, $\operatorname{Re}z>0$, *for some* $C>0$;

(iii) $\Lambda(A)\leqslant 0$ *for* $A\in\mathscr{F}$ *and there exist* C_1 *and* C_2 *and for each* $A\in\mathscr{F}$ *a matrix* $S=S(A)$ *such that* $\max(|S|,|S^{-1}|)\leqslant C_1$ *and such that* $SAS^{-1}=B=(b_{jk})$ *is a triangular matrix with off-diagonal elements satisfying* $|b_{jk}|\leqslant C_2\min(|\operatorname{Re}\lambda_j|,|\operatorname{Re}\lambda_k|)$ *where* $\lambda_j=b_{jj}$;

(iv) *Let* $0\leqslant\gamma<1$. *There exists* $C>0$ *such that for each* $A\in\mathscr{F}$ *there is a Hermitian matrix* $H=H(A)$ *with* $C^{-1}I\leqslant H\leqslant CI$ *and* $\operatorname{Re}(HA)\leqslant\gamma\Lambda(A)H\leqslant 0$.

Equations of higher order in the time variable such as the wave equation $u_{tt}=u_{xx}$ may be written in system form (3.1) by introducing the successive time derivatives of u as dependent variables, and is therefore covered by the above discussion.

For the approximate solution of the initial value problem (3.1), let h and k be (small) positive numbers. We want to approximate the solution at time level $t_n=nk$ by U^n for $n\geqslant 0$ where $U^0=v$ and $U^{n+1}=E_kU^n$ for $n\geqslant 0$, and where E_k is an operator of the form

$$E_kv(x)=\sum_j a_j(h)v(x-jh),\quad \lambda=k/h^M=\text{constant} \tag{3.3}$$

with summation over a finite set of multi-indices $j=(j_1,\ldots,j_d)\in\mathbb{Z}^d$; such operators are called explicit. The purpose is to choose E_k so that E_k^nv approximates $u(t_n)=E(t_n)v=E(k)^nv$. In numerical applications we would apply (3.3) only for $x=x_l=lh$, $l\in\mathbb{Z}^d$, but for convenience in the analysis we shall think of $E_k v$ as defined for all x. As an example, for the heat equation $u_t = u_{xx}$ the simplest such operator is obtained by replacing derivatives by finite difference quotients, $\partial_t U_j^n = \partial_x \bar{\partial}_x U^n$. Solving for U^{n+1} we see that this defines $E_k v(x) = \lambda v(x+h) + (1-2\lambda)v(x) + \lambda v(x-h)$, $\lambda = k/h^2$. We shall consider further examples below.

We say that E_k is consistent with (3.1) if $u(x,t+k) = E_k u(x,t) + \mathrm{o}(k)$ when $k \to 0$, for sufficiently smooth solutions $u(x,t)$ of (3.1), and accurate of order r if the term $\mathrm{o}(k)$ may be replaced by $k\mathrm{O}(h^r)$. By Taylor series expansion around (x,t) these conditions are seen to be equivalent to algebraic conditions between the coefficients $a_j(h)$. In our above example, $u(x,t+k) - E_k u(x,t) = ku_t + \frac{1}{2}k^2 u_{tt} - \lambda h^2 u_{xx} - \frac{1}{12}\lambda h^4 u_{xxxx} = k\mathrm{O}(h^2)$ when $u_t = u_{xx}$, so that $r = 2$.

For the purpose of showing convergence of $E_k^n v$ to $E(t_n)v$ and to derive an error bound one needs some stability property of the operator E_k^n: This operator is said to be stable in L_2 if for any $T > 0$ there are constants C and κ such that $\|E_k^n v\| \leqslant C\mathrm{e}^{\kappa nk}\|v\|$ for $v \in L_2$, $n \geqslant 0$, where $\|\cdot\| = \|\cdot\|_{L_2}$. If this holds, and if (3.1) is well posed in L_2 and E_k is accurate of order r, then it follows from the identity $E_k^n - E(t_n) = \sum_{j=0}^{n-1} E_k^{n-1-j}(E_k - E(k))E(t_j)$ that, with $\|\cdot\|_s = \|\cdot\|_{H^s(\mathbb{R}^d)}$,

$$\|(E_k^n - E(t_n))v\| \leqslant C \sum_{j=0}^{n-1} kh^r \|E(t_j)v\|_{M+r} \leqslant C_T h^r \|v\|_{M+r} \quad \text{for } t_n \leqslant T, \tag{3.4}$$

where we have also used the fact that spatial derivatives commute with $E(t)$.

The sufficiency of stability for convergence of the solution of the discrete problem to the solution of the continuous initial value problem was shown in particular cases in many places, e.g., Courant, Friedrichs and Lewy (1928), O'Brien, Hyman and Kaplan (1951), Douglas (1956). It was observed by Lax and Richtmyer (1959) that stability is actually a necessary condition for convergence to hold for all $v \in L_2$; the general Banach space formulation of stability as a necessary and sufficient condition for convergence is known as the Lax equivalence theorem, see Richtmyer and Morton [28]. We note that for individual, sufficiently regular v convergence may hold without stability; for an early interesting example with analytic initial data and highly unstable difference operator, see Dahlquist (1954). Without stability, however, roundoff errors will then overshadow the theoretical solution in actual computation.

We shall see that the characterization of stability of finite difference operators (3.3) is parallel to that of well-posedness of (3.1). Introducing the trigonometric polynomial $E_k(\xi) = \sum_j a_j(h)\mathrm{e}^{\mathrm{i}jh\xi}$, the symbol of E_k, Fourier transformation shows that E_k is stable in L_2 if and only if, cf. (3.2),

$$|E_k(\xi)^n| \leqslant C\mathrm{e}^{\kappa nk}, \quad \forall \xi \in \mathbb{R}^d, \ n \geqslant 0.$$

In the scalar case this is equivalent to $|E_k(\xi)| \leqslant 1 + Ck$ for $\xi \in \mathbb{R}^d$ and small k, or $|E_k(\xi)| \leqslant 1$ for $\xi \in \mathbb{R}^d$ when the coefficients of E_k are independent of h, as is normally the case when no lower-order terms occur in $P(D)$. For our above example $E_k(\xi) = 1 - 2\lambda + 2\lambda\cos h\xi$ and stability holds if and only if $\lambda = k/h^2 \leqslant \frac{1}{2}$. For the equation $u_t = \rho u_x$ and the scheme $\partial_t U_j^n = \hat{\partial}_x U_j^n$ we have $E_k(\xi) = 1 + \mathrm{i}\rho\lambda \sin h\xi$ and this method is therefore seen to be unstable for any choice of $\lambda = k/h$.

Necessary for stability in the matrix case is the von Neumann condition

$$\rho(E_k(\xi)) \leqslant 1 + Ck, \quad \forall \xi \in \mathbb{R}^d, \tag{3.5}$$

where $\rho(A) = \max_j |\lambda_j|$ is the spectral radius of A, and for normal matrices $E_k(\xi)$ this is also sufficient. This covers the scalar case discussed above. Necessary and sufficient conditions are given in the following discrete version of the above Kreiss matrix lemma, see Kreiss (1962), where we denote $|u|_H=(Hu,u)^{1/2}$ and $|A|_H=\sup_{u\neq 0}|Au|_H/|u|_H$ for H positive definite:

Let $\mathscr{F}$ *be a set of* $N\times N$ *matrices. Then the following conditions are equivalent*:

(i) $|A^n|\leqslant C$ *for* $A\in\mathscr{F}$, $n\geqslant 0$, *for some* $C>0$;

(ii) $\rho(A)\leqslant 1$ *and* $(|z|-1)|R(A;z)|\leqslant C$ *for* $A\in\mathscr{F}$, $|z|>1$, *with* $C>0$;

(iii) $\rho(A)\leqslant 1$ *for* $A\in\mathscr{F}$ *and there are* C_1 and C_2 *and for each* $A\in\mathscr{F}$ *a matrix* $S=S(A)$ *such that* $\max(|S|,|S^{-1}|)\leqslant C_1$ *and such that* $SAS^{-1}=(b_{jk})$ *is triangular with off-diagonal elements satisfying* $|b_{jk}|\leqslant C_2\min(1-|\lambda_j|,1-|\lambda_k|)$ *where* $\lambda_j=b_{jj}$;

(iv) *let* $0\leqslant\gamma<1$. *There exists* $C>0$ *and for each* $A\in\mathscr{F}$ *a Hermitian matrix* $H=H(A)$ *with* $C^{-1}I\leqslant H\leqslant CI$ *and* $|A|_H\leqslant 1-\gamma+\gamma\rho(A)$.

Application shows that if E_k is stable in L_2, then there is a κ such that $\mathscr{F}=\{\mathrm{e}^{-\kappa k}E_k(\xi);\ k\leqslant k_0,\ \xi\in\mathbb{R}^d\}$ satisfies conditions (i)–(iv). On the other hand, if one of these conditions holds for some κ, then E_k is stable in L_2.

Other related sufficient conditions were given in, e.g., Kato (1960), where it was shown that if the range of an $N\times N$ matrix is in the unit disc, i.e., if $|Av\cdot v|\leqslant 1$ for $|v|\leqslant 1$, then $|A^n v\cdot v|\leqslant 1$, and hence, by taking real and imaginary parts, $|A^n|\leqslant 2$ for $n\geqslant 1$.

Using the above characterizations one can show that a necessary and sufficient condition for the existence of an L_2-stable operator which is consistent with (3.1) is that (3.1) is well-posed in L_2, see Kreiss (1959a). It was also proved by Wendroff (1968) that for initial value problems which are well-posed in L_2 one may construct L_2-stable difference operators with arbitrarily high-order of accuracy.

It may be shown that von Neumann's condition (3.5) is equivalent to growth of at most polynomial order of the solution operator in L_2, or $\|E_k^n v\|_{L_2}\leqslant Cn^q\|v\|_{L_2}$ for $t_n\leqslant T$, for some $q\geqslant 0$. This was used by Forsythe and Wasow [14] and Ryabenkii and Filippov [29] as a definition of stability.

For variable coefficients it was shown by Strang (1965) that if the initial-value problem for the equation $u_t=\sum_{|\alpha|\leqslant M}P_\alpha(x)D^\alpha u$ is well-posed in L_2, then the one for the equation without lower-order terms and with coefficients fixed at $x\in\mathbb{R}^d$ is also well-posed, and a similar statement holds for the stability of the finite difference scheme $E_k v(x)=\sum_j a_j(x,h)v(x-jh)$. However, Kreiss (1962) showed that well-posedness and stability with frozen coefficients is neither necessary nor sufficient for well-posedness and stability of a general variable coefficient problem. We shall return below to variable coefficients for parabolic and hyperbolic equations.

We now consider the special case when system (3.1) is parabolic, and begin by quoting the fundamental paper of John (1952) in which maximum-norm stability was shown for finite difference schemes for second-order parabolic equations in one space variable. For simplicity, we restrict our presentation to the model problem

$$u_t=u_{xx}\quad\text{for } x\in R,\ t>0,\quad \text{with } u(\cdot,0)=v\quad\text{in } \mathbb{R} \tag{3.6}$$

and a corresponding finite difference approximation of the form (3.3) with $a_j(h)=a_j$ independent of h. Setting $a(\xi)=\sum_j a_j\mathrm{e}^{-\mathrm{i}j\xi}=E_k(h^{-1}\xi)$ one may write

$$U^n(x)=\sum_j a_{nj}v(x-jh),\quad \text{where } a_{nj}=\frac{1}{2\pi}\int_{-\pi}^{\pi}\mathrm{e}^{-\mathrm{i}j\xi}a(\xi)^n\,\mathrm{d}\xi. \tag{3.7}$$

Here the von Neumann condition reads $|a(\xi)|\leqslant 1$, and is necessary and sufficient for stability in L_2. To show maximum-norm stability we need to estimate the a_{nj} in (3.7). It is easily seen that if the difference scheme is consistent with (3.6) then $a(\xi)=e^{-\lambda\xi^2}+o(\xi^2)$ as $\xi\to 0$, and if we assume that $|a(\xi)|<1$ for $0<|\xi|\leqslant\pi$ it follows that

$$|a(\xi)|\leqslant e^{-c\xi^2} \quad \text{for } |\xi|\leqslant\pi, \text{ with } c>0. \tag{3.8}$$

One then finds at once from (3.7) that $|a_{nj}|\leqslant Cn^{-1/2}$, and integration by parts twice, using $|a'(\xi)|\leqslant C|\xi|$, shows $|a_{nj}|\leqslant Cn^{1/2}j^{-2}$. Thus

$$\sum_j |a_{nj}|\leqslant C\sum_{j\leqslant n^{1/2}} n^{-1/2}+\sum_{j>n^{1/2}} n^{1/2}j^{-2}\leqslant C,$$

so that $\|U^n\|_\infty\leqslant C\|v\|_\infty$ by (3.7) where $\|\cdot\|_\infty=\|\cdot\|_{L_\infty}$. We remark that for our simple example above we have $|E_k v(x)|\leqslant(\lambda+|1-2\lambda|+\lambda)\|v\|_\infty=\|v\|_\infty$ for $\lambda\leqslant\frac{1}{2}$, which trivially yields maximum-norm stability.

In the general constant coefficient case system (3.1) is said to be parabolic in Petrowski's sense if $\Lambda(P(\xi))\leqslant-\delta|\xi|^M+C$ for $\xi\in\mathbb{R}^d$, and using (iii) of the Kreiss lemma one shows easily that the corresponding initial value problem is well-posed in L_2. For well-posedness in L_∞ we may write

$$u(x,t)=(E(t)v)(x)=\int_{\mathbb{R}^d} G(x-y,t)v(y)\,dy$$

where, cf., e.g., [15], with $D^\alpha=D_x^\alpha$,

$$|D^\alpha G(x,t)|\leqslant C_T t^{-(|\alpha|+d)/M} e^{(-\delta(|x|^M/t)^{1/(M-1)})} \quad \text{for } 0<t\leqslant T. \tag{3.9}$$

This implies that $\|D^\alpha u(\cdot,t)\|_\infty\leqslant Ct^{-|\alpha|/M}\|v\|_\infty$, so that the solution, in addition to being bounded, is smooth for positive time even if the initial data are only bounded. Consider now a difference operator of the form (3.3) which is consistent with (3.1). Generalizing from (3.8) we define this operator to be parabolic in the sense of John if, for some positive δ and C,

$$\rho(E_k(h^{-1}\xi))\leqslant e^{-\delta|\xi|^M}+Ck, \quad \text{for } \xi\in Q=\{\xi;\ |\xi_j|\leqslant\pi,\ j=1,\dots,d\},$$

such schemes always exist when (3.1) is parabolic in Petrowski's sense. Extending the results of John (1952) to this situation Aronson (1963) and Widlund (1966) showed that if we write $U^n(x)=E_k^n v(x)=\sum_j a_{nj}(h)v(x-jh)$, then, denoting difference quotients corresponding to D^α by ∂_h^α, we have, cf. (3.9),

$$|\partial_h^\alpha a_{nj}(h)|\leqslant Ch^d t_n^{-(|\alpha|+d)/M} e^{(-c(|j|^M/n)^{1/(M-1)})} \quad \text{for } t_n\leqslant T,$$

which implies that $\|\partial_h^\alpha E_k^n v\|_\infty\leqslant Ct_n^{-|\alpha|/M}\|v\|_\infty$. In the work quoted also multistep methods and variable coefficients were treated.

From estimates of these types follow also convergence estimates such as, if D_h^α is a difference operator consistent with D^α and accurate of order r,

$$\|D_h^\alpha U^n-D^\alpha u(t_n)\|_\infty\leqslant Ch^r\|v\|_{W_\infty^{r+|\alpha|}} \quad \text{for } t_n\leqslant T.$$

Note in the convergence estimate for $\alpha=0$ that, as a result of the smoothing property for parabolic equations, less regularity is required of initial data than for the general well-posed initial value problem, cf. (3.4). For even less regular initial data lower convergence rates have to be expected; by interpolation between the W_∞^s spaces one may show (see Peetre and Thomée (1967)), e.g.,

$$\|U^n-u(t_n)\|_\infty\leqslant Ch^s\|v\|_{W_\infty^s} \quad \text{for } 0\leqslant s\leqslant r,\ t_n\leqslant T.$$

We remark that for nonsmooth initial data v it is possible to make a preliminary smoothing of v to recover full accuracy for t bounded away from zero: It was shown in Kreiss, Thomée and Widlund (1970) that there exists a smoothing operator of the form $M_h v = \Phi_h * v$, where $\Phi_h(x) = h^{-d}\Phi(h^{-1}x)$, with Φ an appropriate function, such that if $M_h v$ is chosen as initial data for the difference scheme, then

$$\|U^n - u(t_n)\|_\infty = \|E_h^n M_h v - E(t_n)v\|_\infty \leqslant Ch^r t_n^{-r/M} \|v\|_\infty.$$

Let us also note that from a known convergence rate for the difference scheme for fixed initial data v conclusions may be drawn about the regularity of v. For example, under the above assumptions, assume that for some p,s with $1 \leqslant p \leqslant \infty, 0 < s \leqslant r$, we know that $\|U^n - u(t_n)\|_{L_p} \leqslant Ch^s$ for $t_n \leqslant T$. Then v belongs to the Besov space $B_p^{s,\infty}$ ($\approx W_p^s$), and, if $s > r$ then $v = 0$. Such inverse results were given in, e.g., Hedstrom (1968), Löfström (1970), and Thomée and Wahlbin (1974).

We now turn our attention to hyperbolic equations, and consider first systems with constant real coefficients

$$u_t = \sum_{j=1}^{d} A_j u_{x_j} \quad \text{for } t > 0, \quad \text{with } u(0) = v. \tag{3.10}$$

Such a system is said to be strongly hyperbolic if it is well-posed in L_2, cf. Strang (1967). With $P(\xi) = \sum_{j=1}^{d} A_j \xi_j$ this holds if and only if for each $\xi \in \mathbb{R}^d$ there exists a nonsingular matrix $S(\xi)$, uniformly bounded together with its inverse, such that $S(\xi)P(\xi)S(\xi)^{-1}$ is a diagonal matrix with real elements. When the A_j are symmetric this holds with $S(\xi)$ orthogonal; the system is then said to be symmetric hyperbolic. The condition is also satisfied when the eigenvalues of $P(\xi)$ are real and distinct for $\xi \neq 0$; in this case the system is called strictly hyperbolic.

One important feature of hyperbolic systems is that the value $u(x,t)$ of the solution at (x,t) only depends on the initial values on a compact set $K(x,t)$, the smallest closed set such that $u(x,t) = 0$ when v vanishes in a neighborhood of $K(x,t)$. The convex hull of $K(x,t)$ may be described in terms of $P(\xi)$: if $K = K(0,1)$ we have for its support function $\phi(\xi) = \sup_{x \in K} x\xi = \lambda_{\max}(P(\xi))$.

Consider now the system (3.10) and a corresponding finite difference operator of the form (3.3) (with $M = 1$). Here we introduce the domain of dependence $\tilde{K}(x,t)$ of E_k as the smallest closed set such that $E_k^n v(x) = 0$ for all n,k with $nk = t$ when v vanishes in a neighborhood of $\tilde{K}(x,t)$. Corresponding to the above the support function of $\tilde{K}(x,t)$ satisfies $\tilde{\phi}(\xi) = \lambda^{-1}\max_{a_j \neq 0} j\xi$. Since clearly convergence, and hence stability, demands that $\tilde{K} = \tilde{K}(0,1)$ contains the continuous domain of dependence $K = K(0,1)$ it is necessary for stability that $\lambda_{\max}(P(\xi)) \leqslant \lambda^{-1}\max_{a_j \neq 0} j\xi$; this is the CFL-condition, cf. Section 1. In particular, $\min_{a_j \neq 0} j_l \leqslant \lambda\lambda_{\min}(A_l) \leqslant \lambda\lambda_{\max}(A_l) \leqslant \max_{a_j \neq 0} j_l$. For the equation $u_t = \rho u_x$ and a difference operator of the form (3.3) using only $j = -1, 0, 1$, this means $\lambda|\rho| \leqslant 1$: In this case $u(0,1) = v(\rho)$ so that $K = \{\rho\}$, and the condition is thus $\rho \in \tilde{K} = [-\lambda^{-1}, \lambda^{-1}]$.

We shall now give some sufficient conditions for stability. We first quote from Friedrichs (1954) that if a_j are symmetric and positive semidefinite, with $\sum_j a_j = I$, then $|E_k(h^{-1}\xi)| = |\sum_j a_j e^{ij\xi}| \leqslant 1$ so that the scheme is L_2-stable. As an example, the first order accurate Friedrichs operator

$$E_k v(x) = \frac{1}{2}\sum_{j=1}^{d}((d^{-1}I + \lambda A_j)v(x + he_j) + (d^{-1}I - \lambda A_j)v(x - he_j)) \tag{3.11}$$

is L_2-stable if $0 < \lambda \leqslant (d\rho(A_j))^{-1}$. It was observed by Lax (1961) that this criterion is of limited value in applications because it cannot in general be combined with accuracy higher than first order.

The necessary and sufficient conditions for L_2-stability of the Kreiss stability lemma are of the nature that the necessary von Neumann condition (3.5) has to be supplemented by conditions assuring that the eigenvalues of $E_k(\xi)$ sufficiently well describe the growth behavior of $E_k(\xi)^n$. We now quote some such criteria which utilize relations between the behavior of the eigenvalues of $E_k(\xi)$ for small ξ and the accuracy of E_k. In Kreiss (1964) E_k is defined to be dissipative of order v, v even, if $\rho(\tilde{E}_k(h^{-1}\xi)) \leqslant 1 - \delta|\xi|^v$ for $\xi \in Q$ with $\delta > 0$, and it is shown that if E_k is consistent with the strongly hyperbolic system (3.10), accurate of order $v-1$, and dissipative of order v, then it is L_2-stable. Further, it was shown by Parlett (1966) that if the system is symmetric hyperbolic, and E_k is dissipative of order v it suffices that it is accurate of order $v-2$, and by Yamaguti (1967) that if the system is strictly hyperbolic, then dissipativity of some order v is sufficient. For strictly hyperbolic systems Yamaguti also showed that if the eigenvalues of $E_k(\xi)$ are distinct for $Q \ni \xi \neq 0$, then von Neumann's condition is sufficient for stability in L_2.

For $d=1$ (with $A = A_1$) the well-known Lax–Wendroff (1960) scheme

$$E_k v(x) = \tfrac{1}{2}(\lambda^2 A^2 + \lambda A)v(x+h) + (1 - \lambda^2 A^2)v(x) + \tfrac{1}{2}(\lambda^2 A^2 - \lambda A)v(x-h)$$

is L_2-stable for $\lambda\rho(A) \leqslant 1$ if the system approximated is strongly hyperbolic. L_2-stable finite difference schemes of arbitrarily high order for such systems were constructed in, e.g., Strang (1962).

It is also possible to analyze multistep methods, e.g., by rewriting them as single-step systems. A popular stable such method is the leapfrog method $\hat{\partial}_t U_j^n = \hat{\partial}_x U_j^n$ (or $U_j^{n+1} = U_j^{n-1} + \lambda\rho(U_{j+1}^n - U_{j-1}^n)$). The eigenvalues $\tau_1(\xi), \tau_2(\xi)$ appearing in the stability analysis then satisfy the characteristic equation $\tau - \tau^{-1} - 2\lambda\rho \sin\xi = 0$, and we find that the von Neumann condition, $|\tau_j(\xi)| \leqslant 1$ for $\xi \in \mathbb{R}$, $j = 1,2$, is satisfied if $\lambda\rho \leqslant 1$, and using the Kreiss lemma one can show that L_2-stability holds for $\lambda\rho < 1$.

As an example for $d = 2$ we quote the operator introduced in Lax and Wendroff (1964) defined by

$$\begin{aligned} E_k(h^{-1}\xi) = {} & I + \mathrm{i}\lambda(A_1 \sin\xi_1 + A_2 \sin\xi_2) \\ & - \lambda^2(A_1^2(1-\cos\xi_1) + \tfrac{1}{2}(A_1A_2 + A_2A_1)\sin\xi_1 \sin\xi_2 + A_2^2(1-\cos\xi_2)). \end{aligned}$$

Using the above stability criterion of Kato they proved stability for $\lambda|A_j| \leqslant 1/\sqrt{8}$ in the symmetric hyperbolic case.

Consider now a variable coefficient symmetric hyperbolic system

$$u_t = \sum_{j=1}^{d} A_j(x) u_{x_j}, \quad \text{where } A_j(x)^* = A_j(x). \tag{3.12}$$

Using an energy argument Friedrichs (1954) showed that the corresponding initial value problem is well-posed; the boundedness follows at once by noting that after multiplication by u, integration over $x \in \mathbb{R}^d$, and integration by parts,

$$\frac{\mathrm{d}}{\mathrm{d}t}\|u(t)\|^2 = -\int_{\mathbb{R}^d} \left(\sum_j \partial A_j/\partial x_j u, u \right) \mathrm{d}x \leqslant 2\kappa \|u(t)\|^2.$$

For finite difference approximations $E_k v(x) = \sum_j a_j(x) v(x - jh)$ which are consistent with (3.12) various sufficient conditions are available. Kreiss, et al. (1970), studied difference operators which are dissipative of order v, i.e., such that $\rho(E(x,\xi)) \leqslant 1 - c|\xi|^v$ for $x \in \mathbb{R}^d$, $\xi \in Q$, with $c > 0$, where $E(x,\xi) = \sum_j a_j(x) \mathrm{e}^{-\mathrm{i}j\xi}$. As in the constant coefficient case it was shown that E_k is stable in L_2 if the $a_j(x)$ are symmetric and if E_k is accurate of order $v-1$ and dissipative of order v; if (3.12) is strictly hyperbolic, accuracy of order $v-2$ suffices. The proofs are based on the Kreiss stability lemma and a perturbation argument.

In an important paper it was proved by Lax and Nirenberg (1966) that if E_k is consistent with (3.12) and $|E(x,\xi)| \leqslant 1$ then E_k is strongly stable with respect to the L_2-norm, i.e., $\|E_k\| \leqslant 1 + Ck$. In particular, if the $a_j(x)$ are symmetric positive-definite stability follows by the result of Friedrichs quoted earlier for constant coefficients. Further, if E_k is consistent with (3.12) and $|E(x,\xi)v \cdot v| \leqslant 1$ for $|v| \leqslant 1$, then E_k is L_2-stable.

As mentioned above, higher-order equations in time may be expressed as systems of the form (3.1), and finite difference methods may be based on such formulations. In Section 1 we described an example of a difference approximation of a second-order wave equation in its original form.

We turn to estimates in L_p norms with $p \neq 2$. For the case when the equation in (3.10) is symmetric hyperbolic it was shown by Brenner (1966) that the initial value problem is well-posed in $L_p, p \neq 2$, if and only if the matrices A_j commute. This is equivalent to the simultaneous diagonalizability of these matrices, which in turn means that by introducing a new dependent variable it can be transformed to a system in which all the A_j are diagonal, so that the system consists of N uncoupled scalar equations.

Since stability of finite difference approximations can only occur for well-posed problems it is therefore natural to consider the scalar equations $u_t = \rho u_x$ with ρ real, and corresponding finite difference operators (3.3). With $a(\xi)$ as before we recall that such an operator is stable in L_2 if and only if $|a(\xi)| \leqslant 1$ for $\xi \in \mathbb{R}$. Necessary and sufficient conditions for stability in $L_p, p \neq 2$, as well as rates of growth of $\|E_k^n\|_{L_p}$ in the nonstable case have been given, e.g., in Brenner and Thomée (1970), cf. also references in Brenner, Thomée and Wahlbin [9]. If $|a(\xi)| < 1$ for $0 < |\xi| \leqslant \pi$ and $a(0) = 1$ the condition is that $a(\xi) = e^{i\alpha\xi - \beta\xi^v(1+o(1))}$ as $\xi \to 0$, with α real, $\operatorname{Re}\beta > 0$, v even. Thus E_k is stable in L_p, $p \neq 2$, if and only if there is an even number v such that E_k is dissipative of order v and accurate of order $v-1$. As an example of an operator which is stable in L_2 but not in L_p, $p \neq 2$, we may take the Lax–Wendroff operator introduced above (with A replaced by ρ). For $0 < \lambda|\rho| < 1$, this operator is stable in L_2, is dissipative of order 4, but accurate only of order 2. It may be proved that for this operator $cn^{1/8} \leqslant \|E_k^n\|_\infty \leqslant Cn^{1/8}$ with $c > 0$, which shows a weak instability in the maximum-norm.

An area where finite difference techniques continue to flourish and to form an active research field is for nonlinear fluid flow problems. Consider the scalar nonlinear conservation law

$$u_t + F(u)_x = 0 \quad \text{for } x \in \mathbb{R},\ t > 0, \text{ with } u(\cdot,0) = v \text{ on } \mathbb{R}, \tag{3.13}$$

where $F(u)$ is a nonlinear function of u, often strictly convex, so that $F''(u) > 0$. The classical example is Burger's equation $u_t + u\,u_x = 0$, where $F(u) = u^2/2$. Discontinuous solutions may arise even when v is smooth, and one therefore needs to consider weak solutions. Such solutions are not necessarily unique, and to select the unique physically relevant solution one has to require so called entropy conditions. This solution may also be obtained as the limit as $\varepsilon \to 0$ of the diffusion equation with εu_{xx} replacing 0 on the right in (3.13).

One important class of methods for (3.13) are finite difference schemes in conservation form

$$U_j^{n+1} = U_j^n - \lambda(H_{j+1/2}^n - H_{j-1/2}^n) \quad \text{for } n \geqslant 0, \text{ with } U_j^0 = v(jh), \tag{3.14}$$

where $H^n_{j+1/2} = H(U^n_j, U^n_{j+1})$ is a numerical flux function which has to satisfy $H(V,V) = F(V)$ for consistency. Here stability for the linearized equation is neither necessary nor sufficient for the nonlinear equation in the presence of discontinuities and much effort has been devoted to the design of numerical flux functions with good properties. When the right-hand side of (3.14) is increasing in $U^n_{j+l}, l = -1, 0, 1$, the scheme is said to be monotone (corresponding to positive coefficients in the linear case) and such schemes converge for fixed λ to the entropy solution as $k \to 0$, but are at most first-order accurate. Examples are the Lax–Friedrichs scheme which generalizes (3.11), the scheme of Courant, Isaacson and Rees (1952), which is one sided (upwinding), and the Engquist–Osher scheme which pays special attention to changes in sign of the characteristic direction. Godunov's method replaces U^n by a piecewise constant function, solves the corresponding problem exactly from t_n to t_{n+1} and defines U^{n+1} by an averaging process.

Higher-order methods with good properties are also available, and are often constructed with an added artificial diffusion term which depends on the solution, or so-called total variation diminishing (TVD) schemes. Early work in the area was Lax and Wendroff (1960), and more recent contributions have been given by Engquist, Harten, Kuznetsov, MacCormack, Osher, Roe, Yee; for overviews and generalizations to systems and higher dimension, see Le veque [23] and Godlewski and Raviart [16].

4. Finite differences for mixed initial–boundary value problems

The pure initial value problem discussed in Section 3 is often not adequate to model a given physical situation and one needs to consider instead a problem whose solution is required to satisfy the differential equation in a bounded spatial domain $\Omega \subset \mathbb{R}^d$, as well as boundary conditions on $\partial\Omega$ for positive time, and to take on given initial values. For such problems the theory of finite difference methods is less complete and satisfactory. In the same way as for the stationary problem treated in Section 2 one reason for this is that for $d \geqslant 2$ only very special domains may be well represented by mesh-domains, and even when $d = 1$ the transition between the finite difference approximation in the interior and the boundary conditions may be complex both to define and to analyze. Again there are three standard approaches to the analysis, namely methods based on maximum principles, energy arguments, and spectral representation. We illustrate this first for parabolic and then for hyperbolic equations.

As a model problem for parabolic equations we shall consider the one-dimensional heat equation in $\Omega = (0,1)$,

$$u_t = u_{xx} \quad \text{in } \Omega, \quad u = 0 \quad \text{on } \partial\Omega = \{0,1\}, \quad \text{for } t \geqslant 0, \quad u(\cdot,0) = v \text{ in } \Omega. \tag{4.1}$$

For the approximate solution we introduce mesh-points $x_j = jh$, where $h = 1/M$, and time levels $t_n = nk$, where k is the time step, and denote the approximate solution at (x_j, t_n) by U^n_j. As for the pure initial value problem we may then approximate (4.1) by means of the explicit forward Euler difference scheme

$$\partial_t U^n_j = \partial_x \bar{\partial}_x U^n_j \quad \text{in } \Omega, \text{ with } U^{n+1}_j = 0 \text{ on } \partial\Omega, \quad \text{for } n \geqslant 0$$

with $U^0_j = V_j = v(x_j)$ in Ω. For U^{n-1} given this defines U^n through

$$U^n_j = \lambda(U^{n-1}_{j-1} + U^{n-1}_{j+1}) + (1 - 2\lambda)U^{n-1}_j, \quad 0 < j < M, \quad U^n_j = 0, \ j = 0, M.$$

For $\lambda = k/h^2 \leqslant 1/2$ the coefficients are nonnegative and their sum is 1 so that we conclude that $|U^{n+1}| \leqslant |U^n|$ where $|U| = \max_{x_j \in \Omega} |U_j|$. It follows that $|U^n| \leqslant |V|$, and the scheme is thus stable in maximum norm. Under these assumptions one shows as for the pure initial value problem that $|U^n - u(t_n)| \leqslant C(u)(h^2 + k) \leqslant C(u)h^2$. It is easy to see that $k \leqslant h^2/2$ is also a necessary condition for stability.

To avoid to have to impose the quite restrictive condition $\lambda \leqslant \frac{1}{2}$, Laasonen (1949) proposed the implicit backward Euler scheme

$$\bar{\partial}_t U_j^n = \partial_x \bar{\partial}_x U_j^n \quad \text{in } \Omega, \text{ with } U_j^n = 0 \text{ on } \partial\Omega, \quad \text{for } n \geqslant 1$$

with $U_j^0 = v(x_j)$ as above. For U^{n-1} given one now needs to solve the linear system

$$(1 + 2\lambda)U_j^n - \lambda(U_{j-1}^n + U_{j+1}^n) = U_j^{n-1}, \; 0 < j < M, \quad U_j^n = 0, \; j = 0, M.$$

This method is stable in maximum norm without any restrictions on k and h. In fact, we find at once, for suitable k,

$$|U^n| = |U_k^n| \leqslant \frac{2\lambda}{1 + 2\lambda}|U^n| + \frac{1}{1 + 2\lambda}|U^{n-1}|$$

and hence $|U^n| \leqslant |U^{n-1}| \leqslant |V|$. Here, the error is of order $O(h^2 + k)$.

Although the backward Euler method is unconditionally stable, it is only first-order accurate in time. A second-order accurate method was proposed by Crank and Nicolson (1947) which uses the equation

$$\bar{\partial}_t U_j^n = \partial_x \bar{\partial}_x U_j^{n-1/2}, \quad 0 < j \leqslant M, \text{ where } U_j^{n-1/2} = \tfrac{1}{2}(U_j^n + U_j^{n-1}). \tag{4.2}$$

The above approach will now show $|U^n| \leqslant |U^{n-1}|$ only if $\lambda \leqslant 1$.

For this problem, however, the energy method may be used to show unconditional stability in l_2-type norms: With $(V, W) = h\sum_{j=1}^{M-1} V_j W_j$ and $\|V\| = (V, V)^{1/2}$ one finds, upon multiplication of (4.2) by $2U_j^{n-1/2}$, summation, and summation by parts,

$$\bar{\partial}_t \|U^n\|^2 = 2(\bar{\partial}_t U^n, U^{n-1/2}) = -2h^{-2} \sum_{j=1}^{M} (U_j^{n-1/2} - U_{j-1}^{n-1/2})^2 \leqslant 0,$$

which shows $\|U^n\| \leqslant \|U^{n-1}\|$, i.e., stability in l_2 holds for any $\lambda > 0$. In the standard way this yields the error estimate $\|U^n - u(t_n)\| \leqslant C(u)(h^2 + k^2)$; a corresponding estimate in a discrete H^1-norm may be obtained similarly, and this also yields a maximum-norm estimate of the same order by using a discrete Sobolev inequality. The energy approach was developed by, e.g., Kreiss (1959a), Lees (1960) and Samarskii (1961).

Stability in l_2 may also be deduced by spectral analysis, as observed in O'Brien, Hyman and Kaplan (1951). We illustrate this for the Crank–Nicolson method: Representing the mesh-functions vanishing at $x = 0$ and 1 in terms of the eigenfunctions φ_p of $\partial_x \bar{\partial}_x$, the vectors with components $\varphi_{pj} = \sqrt{2} \sin \pi pjh$, $0 \leqslant j \leqslant M$, and eigenvalues $\mu_p = -2h^{-2}(1 - \cos \pi ph)$, one finds

$$U^n = \sum_{p=1}^{M-1} (V, \varphi_p) E(\pi ph)^n \varphi_p \quad \text{where } E(\xi) = \frac{1 - \lambda(1 - \cos \xi)}{1 + \lambda(1 - \cos \xi)}$$

and $\|U^n\| \leqslant \|V\|$ follows by Parseval's relation since $|E(\xi)| \leqslant 1$ for $\xi \in \mathbb{R}$, $\lambda > 0$. This is analogous to the Fourier analysis of Section 3 for pure initial value problems.

We remark that although the maximum-principle-type argument for stability for the Crank–Nicolson method requires $\lambda \leqslant 1$, it was shown by Serdjukova (1964) using Fourier analysis that the maximum-norm bound $|U^n| \leqslant 23|V|$ holds. Precise convergence analyses in maximum-norm for initial data with low regularity were carried out in, e.g., Juncosa and Young (1957).

For the pure initial value problem, difference schemes or arbitrarily high order of accuracy can be constructed by including the appropriate number of terms in a difference operator of the form $U_j^{n+1} = \sum_{l=-q}^{s} a_l U_{j-l}^n$. For application to the mixed initial–boundary value problem (4.1) such a formula would require additional equations for U_j^n near $x = 0$ or $x = 1$ when $s \geqslant 2$ or $q \geqslant 2$. For the semiinfinite interval $(0, \infty)$, Strang (1964) showed that with $s = 1$ any order of accuracy may be achieved together with stability by choosing an "unbalanced" operator with $q \geqslant 2$. The stability of schemes with additional boundary conditions has been analyzed in the parabolic case by Varah (1970) and Osher (1972) using the GKS-technique which we briefly describe for hyperbolic equations below.

We note that the above methods may be written as $U^{n+1} = E_k U^n$, where E_k acts in different spaces $\mathcal{N}_k$ of vectors $V = (V_0, \ldots, V_M)^{\mathrm{T}}$ with $V_0 = V_M = 0$, where M depends on k. In order to deal with the stability problem in such situations, Godunov and Ryabenkii (1963) introduced a concept of spectrum of a family of operators $\{E_k\}$, with E_k defined in a normed space $\mathcal{N}_k$ with norm $\|\cdot\|_k$, k small. The spectrum $\sigma(\{E_k\})$ is defined as the complex numbers z such that for any $\varepsilon > 0$ and sufficiently small k there is a $U_k \in \mathcal{N}_k, U_k \neq 0$, such that $\|E_k U_k - z U_k\|_k \leqslant \varepsilon \|U_k\|_k$, and the following variant of von Neumann's criterion holds: If $\{E_k\}$ is stable in the sense that $\|E_k^n V\|_k \leqslant C\|V\|_k$ for $t_n \leqslant T$, with C independent of k, then $\sigma(\{E_k\}) \subset \{z; |z| \leqslant 1\}$. It was demonstrated that the spectrum of a family such as one of the above is the union of three sets, one corresponding to the pure initial value problem and one to each of the one-sided boundary value problems for the differential equation in $\{x \geqslant 0, t \geqslant 0\}$ and in $\{x \leqslant 1, t \geqslant 0\}$, with boundary conditions given at $x = 0$ and 1, respectively. For instance, in the example of the explicit method, $\sigma(\{E_k\})$ equals the set of eigenvalues of the operator $\tilde{E}_k$ associated with the pure initial value problem, which is easily shown to be equal to the interval $[1 - 4\lambda, 1]$, and hence $\lambda \leqslant \frac{1}{2}$ is a necessary condition for stability. The proof of the equality between these sets is nontrivial as the eigenfunctions of $\tilde{E}_k$ do not satisfy the boundary conditions. Using instead a boundary condition of the form $u_0 - \gamma u_1 = 0$ at $x = 0$, will result in instability for certain choices of γ.

We now turn to the two-dimensional model problem in the square $\Omega = (0,1)^2$,

$$u_t = \Delta u \quad \text{in } \Omega, \qquad u = 0 \text{ on } \partial\Omega, \qquad \text{for } t > 0, \text{ with } u(\cdot, 0) = v \text{ in } \Omega. \tag{4.3}$$

Again with $h = 1/M$ we use mesh-points $x_j = jh$, now with $j \in \mathbb{Z}^2$. We consider the above methods collectively as the ϑ-method, with $0 \leqslant \vartheta \leqslant 1$,

$$\bar{\partial}_t U_j^n = \Delta_h(\vartheta U_j^n + (1 - \vartheta) U_j^{n-1}) \quad \text{in } \Omega, \quad U_j^n = 0 \text{ on } \partial\Omega,$$

where $\vartheta = 0, 1$, and $\frac{1}{2}$ for the forward and backward Euler methods and the Crank–Nicolson method, respectively. The above stability and error analysis carries over to this case; the ϑ-method is unconditionally stable in l_2 for $\frac{1}{2} \leqslant \vartheta \leqslant 1$ whereas for $0 \leqslant \vartheta < \frac{1}{2}$ the mesh-ratio condition $\lambda(1 - 2\vartheta) \leqslant \frac{1}{4}$ has to be imposed.

For the model problem (4.3) we consider also the alternating direction implicit (ADI) scheme of Peaceman and Rachford (1955). Noting that the Crank–Nicolson scheme requires the solution at time t_n of the two-dimensional elliptic problem $(I - \frac{1}{2}k\Delta_h)U^n = (I + \frac{1}{2}k\Delta_h)U^{n-1}$, the purpose of the

ADI method is to reduce the computational work by solving only one-dimensional problems. This is done by introducing the intermediate value $U^{n-1/2}$ for the solution at $t_{n-1/2}=t_n-k/2$ by the equations

$$\frac{U^{n-1/2}-U^{n-1}}{k/2}=\partial_1\bar{\partial}_1U^{n-1/2}+\partial_2\bar{\partial}_2U^{n-1},$$

$$\frac{U^{n}-U^{n-1/2}}{k/2}=\partial_1\bar{\partial}_1U^{n-1/2}+\partial_2\bar{\partial}_2U^{n}.$$

Elimination of $U^{n-1/2}$ gives, since the various operators commute,

$$U^n=E_kU^{n-1}=(I-\tfrac{k}{2}\partial_1\bar{\partial}_1)^{-1}(I+\tfrac{k}{2}\partial_1\bar{\partial}_1)(I-\tfrac{k}{2}\partial_2\bar{\partial}_2)^{-1}(I+\tfrac{k}{2}\partial_2\bar{\partial}_2)U^{n-1}.$$

By either energy arguments or by the spectral method one sees easily that this method is stable in the discrete l_2-norm and since it is second-order accurate in both space and time one finds $\|U^n-u(t_n)\|\leqslant C(u)(h^2+k^2)$. We may also express the definition of E_k (using a different $U^{n-1/2}$) by

$$U^{n-j/2}=(I-\tfrac{k}{2}\partial_j\bar{\partial}_j)^{-1}(I+\tfrac{k}{2}\partial_j\bar{\partial}_j)U^{n-(j-1)/2}\quad\text{for } j=0,1.$$

In this form, which generalizes in an obvious way to more dimensions, it is referred to as a fractional step method and depends on the splitting of the operator Δ_h into $\partial_1\bar{\partial}_1$ and $\partial_2\bar{\partial}_2$. This has been a very active area of research during the 1960s, with contributions by, e.g., Douglas, Kellogg, Temam, Wachspress, Dyakonov, Samarskii, Marchuk, and Yanenko, see the survey article by Marchuk [25].

In the same way as for the elliptic problems studied in Section 2, complications arise when the boundary mesh-points do not fall exactly on $\partial\Omega$, which is the case in the presence of a curved boundary. Using, e.g., linear interpolation or Shortley–Weller-type approximations of the Laplacian one may show $O(k+h^2)$ error estimates for the backward Euler method by means of a discrete maximum-principle, and one may also use energy arguments, with the crudest boundary approximation, to show $O(k+h^{1/2})$ error estimates, see Thomée [36,37].

We now turn to hyperbolic equations and consider first the spatially one-dimensional wave equation

$$u_{tt}=u_{xx}\quad\text{in }\Omega=(0,1),\text{ with } u(\cdot,0),\ u_t(\cdot,0)\text{ given.}$$

Here one may define the ϑ-method, $\vartheta\in[0,1]$, which is a special case of a family of schemes studied in Newmark (1959), by

$$\partial_t\bar{\partial}_tU_j^n=\partial_x\bar{\partial}_x((1-\vartheta)U_j^n+\tfrac{1}{2}\vartheta(U_j^{n+1}+U_j^{n-1}))\tag{4.4}$$

with $U_0^n=U_M^n=0$ for $n\geqslant 0$ and U_j^0 and U_j^1 given. This scheme is unconditionally stable for $\frac{1}{2}\leqslant\vartheta\leqslant 1$, and for $0\leqslant\vartheta<\frac{1}{2}$ it is stable if $\lambda^2=k^2/h^2<1/(1-2\vartheta)$; for $\vartheta=0$ we recognize the explicit scheme of Section 1, which can easily be shown to be unstable for $\lambda=1$, see Raviart and Thomas [27].

As a simple first-order hyperbolic model problem we consider, with $\rho>0$,

$$u_t=\rho u_x\quad\text{in }\Omega=(0,1),\quad u(1,t)=0,\quad\text{for } t>0,\text{ with } u(\cdot,0)=v\text{ in }\Omega.$$

Note that since the solution is constant along the characteristics $x+\rho t=$constant no boundary condition is needed at $x=0$.

Consider first the "upwind" scheme, see Courant, Isaacson and Rees (1952),

$$\partial_tU_j^n=\rho\partial_xU_j^n,\quad j=0,\dots,M-1,\qquad U_M^{n+1}=0,\quad\text{for } n\geqslant 0,\text{ with } U_j^0=v(x_j),$$

which may be written in explicit form as

$$U_j^{n+1} = (1-\lambda\rho)U_j^n + \lambda\rho U_{j+1}^n, \quad 0 \leqslant j < M, \quad U_M^{n+1} = 0.$$

When $\lambda\rho \leqslant 1$ the method is stable in maximum norm; this condition may be expressed by saying that the characteristic traced back from (x_j, t_{n+1}) cuts $t = t_n$ in $[x_j, x_{j+1}]$, which we recognize as the CFL condition. By the lack of symmetry it is only first-order accurate. For $\rho\lambda > 1$ one can use instead the Carlson scheme (see [28])

$$\partial_t U_j^n = \rho\bar{\partial}_x U_j^{n+1}, \quad 1 \leqslant j \leqslant M, \qquad U_M^{n+1} = 0, \text{ for } n \geqslant 0, \text{ with } U_j^0 = v(x_j),$$

which determines U_j^{n+1} for decreasing j by $U_{j-1}^{n+1} = (1-(\rho\lambda)^{-1})U_j^{n+1} + (\rho\lambda)^{-1}U_j^n$. Again this method is maximum-norm stable, but only first-order accurate.

A second-order method is the box scheme of Wendroff (1960),

$$\partial_t U_{j-1/2}^n = \rho\bar{\partial}_x U_j^{n+1/2}, \quad 1 \leqslant j \leqslant M, \qquad U_M^{n+1} = 0, \quad U_{j-1/2} = \tfrac{1}{2}(U_j + U_{j-1}).$$

With U^n and U_M^{n+1} given this determines U_{j-1}^{n+1} for decreasing j as a combination of U_j^{n+1}, U_{j-1}^n, and U_j^n. Stability in l_2 may be shown by an energy argument.

We end this section with two examples where the finite difference operators used in the interior of Ω require modification near the boundary or additional artificial boundary conditions. In our first example we describe a special case of an energy argument proposed by Kreiss and Scherer (1974), see also Gustafsson, Kreiss and Oliger [18]. We consider the initial–boundary value problem, with $\rho > 0$,

$$u_t - \rho u_x = f \quad \text{in } \Omega, \text{ with } u(1,t) = 0, \quad \text{for } t \geqslant 0, \; u(\cdot,0) = v \text{ in } \Omega.$$

Assume that we want to apply the six-point Crank–Nicolson equation

$$\bar{\partial}_t U_j^n - \rho\hat{\partial}_x U_j^{n-1/2} = f_j^{n-1/2} \quad \text{for } 1 \leqslant j \leqslant M-1. \tag{4.5}$$

At the right endpoint of Ω we set $U_M^n = 0$. For $x = 0$ the value of u is not given, and we therefore use the one-sided equation $\bar{\partial}_t U_0^n - \rho\partial_x U_0^{n-1/2} = f_0^{n-1/2}$. With the obvious definition of the composite difference operator D_x we may write $\bar{\partial}_t U_j^n - \rho D_x U_j^{n-1/2} = f_j^{n-1/2}$ for $0 \leqslant j < M$. Introducing temporarily the inner product $(U,V) = \frac{1}{2}hU_0V_0 + h\sum_{j=1}^{M-1} U_jV_j$ we have $(D_xU, U) = -\frac{1}{2}U_0^2$ if $U_M = 0$, which yields $(\bar{\partial}_t U^n, U^{n-1/2}) + \frac{1}{2}\rho(U_0^{n-1/2})^2 = (f^{n-1/2}, U^{n-1/2})$. Together with the inequality $hf_0U_0 \leqslant \rho U_0^2 + Ch^2f_0^2$ this easily shows the stability estimate

$$\|U^n\|^2 \leqslant C\|U^0\|^2 + Ck\sum_{l=0}^{n-1} \|\tilde{f}^{l+1/2}\|^2, \quad \text{where } \tilde{f} = (hf_0, f_1, \ldots, f_{M-1}).$$

Note that the choice of the term with $j = 0$ in $(\cdot,\cdot)$ is essential for the argument.

Applying this to the error $U - u$, with the truncation error τ for f, and observing that $\tau_j = O(h^2)$ for $j \geqslant 1$, $\tau_0 = O(h)$, we find $\|U - u\| = O(h^2)$. This approach was also used in the reference quoted to construct higher-order schemes. We note that the modification of (4.5) for $j = 0$ may also be interpreted as using (4.5) for $j = 0$, and adding the boundary condition $h^2\partial_x\bar{\partial}_x U_0^n = U_1^n - 2U_0^n + U_{-1}^n = 0$.

We finally give an example of the stability analysis based on the use of discrete Laplace transforms developed by Kreiss (1968), Gustafsson, Kreiss and Sundström (1972), the so-called GKS-theory. Consider the initial–boundary value problem, again with $\rho > 0$,

$$u_t = \rho u_x \quad \text{for } x > 0, \; t > 0, \quad \text{with } u(x,0) = v(x) \text{ for } x \geqslant 0.$$

Assume that we want to use the leapfrog scheme

$$\hat{\partial}_t U_j^n = \rho \hat{\partial}_x U_j^n \quad \text{for } j \geqslant 1, n \geqslant 1, \text{ with } U_j^0,\ U_j^1 \text{ given for } j \geqslant 0, \tag{4.6}$$

where we assume $\rho\lambda = \rho k/h < 1$ so that the corresponding scheme for the pure initial value problem is stable. Again an additional boundary condition is required for $j=0$ in order to apply the equation at $j=1$; following Strikwerda [35] we illustrate the theory by sketching the argument for stability in choosing the extrapolation $U_0^n = U_1^{n-1}$ for $n \geqslant 1$.

By subtracting a solution of the pure initial value problem one is reduced to assuming that $U_j^0 = U_j^1 = 0$ for $j \geqslant 0$, but then has to impose the inhomogeneous boundary condition $U_0^{n+1} = U_1^n + \beta^n$ for $n \geqslant 0$, and to show for this problem the stability estimate

$$\|U\|_\kappa \leqslant C_\kappa |\beta|_\kappa, \quad \text{with } \|U\|_\kappa^2 = k \sum_{n=0}^\infty \mathrm{e}^{-2\kappa t_n} h \sum_{j=0}^\infty |U_j^n|^2, \quad |\beta|_\kappa^2 = k \sum_{n=0}^\infty \mathrm{e}^{-2\kappa t_n} |\beta^n|^2.$$

Note that κ is a parameter allowing for a certain exponential growth in time.

Applying discrete Laplace transforms in time to (4.6) one finds that the transformed solution $\tilde{U}_j(z)$ satisfies

$$(z - z^{-1})\tilde{U}_j = \lambda\rho(\tilde{U}_{j+1} - \tilde{U}_{j-1}), \quad \text{where } \tilde{U}(z) = k \sum_{n=0}^\infty z^{-n} U^n,$$

which is referred to as the resolvent equation. It is a second-order difference equation in j, and provided the two roots $\tau_1(z), \tau_2(z)$ of its characteristic equation $\lambda\rho(\tau - \tau^{-1}) = z - z^{-1}$ are distinct, the general solution is $\tilde{U}_j(z) = c_1(z)\tau_1(z)^j + c_2(z)\tau_2(z)^j$. It follows from the stability of the scheme for the initial value problem that for $|z| > 1$, with the proper ordering, $|\tau_1(z)| < 1$ and $|\tau_2(z)| > 1$. In fact, if this were not so and since $\tau_1(z)\tau_2(z) = 1$, we have $\tau_{1,2}(z) = \mathrm{e}^{\pm \mathrm{i}\xi}$ for some ξ, and z is therefore a solution of the characteristic equation $z - z^{-1} - 2\mathrm{i}\lambda\rho \sin\xi = 0$ of the leapfrog scheme for the pure initial value problem. By von Neumann's condition we therefore have $|z| \leqslant 1$ which contradicts our assumption. Since we want $\tilde{U}_j$ to be in $l_2(Z_+)$ we must have $c_2(z) = 0$, and taking the Laplace transform also at $j = 0$, we find $c_1(z)(z - \tau_1(z)) = z\tilde{\beta}(z)$, and thus $\tilde{U}_j(z) = z\tilde{\beta}(z)\tau_1(z)^j/(z - \tau_1(z))$. With $z = \mathrm{e}^{sk}$, $s = \kappa + \mathrm{i}\eta$ we obtain using Parseval's relation

$$\|U\|_\kappa^2 = kh \sum_j \int_{-\pi/k}^{\pi/k} |\tilde{U}_j(z)|^2 \,\mathrm{d}\eta = h \int_{-\pi/k}^{\pi/k} \frac{|z|^2 |\tilde{\beta}(z)|^2}{|z - \tau_1(z)|^2 (1 - |\tau_1(z)|^2)} \,\mathrm{d}\eta.$$

By studying the behavior of $\tau_1(z)$ one may show that, $|z - \tau_1(z)| \geqslant c$ and $1 - |\tau_1(z)|^2 \geqslant 1 - |\tau_1(z)| \geqslant c(|z| - 1) = c(\mathrm{e}^{\kappa k} - 1) \geqslant c\kappa k$, with $c > 0$. Hence,

$$\|U\|_\kappa^2 \leqslant Ch(k\kappa)^{-1} \int_{-\pi/k}^{\pi/k} |\tilde{\beta}(z)|^2 \,\mathrm{d}\eta = C(\lambda\kappa)^{-1} |\beta|_\kappa^2,$$

which shows that the method is stable. Using similar arguments it is possible to show that the alternative extrapolation defined by $U_0^n = U_1^n$ is unstable.

5. Finite element methods for elliptic problems

In this section we summarize the basic definitions, properties, and successive development of the finite element method for elliptic problems. As a model problem we consider Dirichlet's problem for Poisson's equation in a domain $\Omega \subset \mathbb{R}^d$,

$$-\Delta u = f \quad \text{in } \Omega, \text{ with } u = 0 \text{ on } \partial\Omega. \tag{5.1}$$

The standard finite element method uses a variational formulation to define an approximate solution u_h of (5.1) in a finite-dimensional linear space S_h, normally consisting of continuous, piecewise polynomial functions on some partition of Ω: By Dirichlet's principle the solution u of (5.1) may be characterized as the function which minimizes $J(v) = \|\nabla v\|^2 - 2(f, v)$ over $H_0^1 = H_0^1(\Omega)$, where $(\cdot,\cdot)$ and $\|\cdot\|$ are the standard inner product and norm in $L_2 = L_2(\Omega)$. The Euler equation for this minimization problem is

$$(\nabla u, \nabla \varphi) = (f, \varphi), \quad \forall \varphi \in H_0^1; \tag{5.2}$$

this weak or variational form of (5.1) may also be derived by multiplying the elliptic equation in (5.1) by $\varphi \in H_0^1$, integrating over Ω, and applying Green's formula in the left-hand side. The standard finite element method assumes $S_h \subset H_0^1$ and defines the approximate solution u_h as the minimizer of $J(v)$ over S_h, or, equivalently,

$$(\nabla u_h, \nabla \chi) = (f, \chi), \quad \forall \chi \in S_h. \tag{5.3}$$

In terms of a basis $\{\Phi_j\}_{j=1}^{N_h}$ for S_h, our discrete problem (5.3) may be stated in matrix form as $A\alpha = \tilde{f}$, where A is the matrix with elements $a_{jk} = (\nabla\Phi_j, \nabla\Phi_k)$ (the stiffness matrix), $\tilde{f}$ the vector with entries $f_j = (f, \Phi_j)$, and α the vector of unknown coefficients α_j in $u_h = \sum_{j=1}^{N_h} \alpha_j \Phi_j$. Here A is a Gram matrix and thus, in particular, positive definite and invertible, so that (5.3) has a unique solution. From (5.2) and (5.3) follows that $(\nabla(u_h - u), \nabla\chi) = 0$ for $\chi \in S_h$, that is, u_h is the orthogonal projection of u onto S_h with respect to the Dirichlet inner product $(\nabla v, \nabla w)$.

We recall that defining u_h as the minimizer of $J(\chi)$ is referred to as the Ritz method, and using instead (5.3), which is suitable also for nonsymmetric differential equations, as Galerkin's method. Some further historical remarks are collected in the introduction to this paper.

For the purpose of error analysis we briefly consider the approximation in S_h of smooth functions in Ω which vanish on $\partial\Omega$. We first exemplify by the Courant elements in a convex plane domain Ω. For such a domain, let $\mathscr{T}_h$ denote a partition into disjoint triangles τ such that no vertex of any triangle lies on the interior of a side of another triangle and such that the union of the triangles determine a polygonal domain $\Omega_h \subset \Omega$ with boundary vertices on $\partial\Omega$. Let h denote the maximal length of the sides of the triangles of $\mathscr{T}_h$, and assume that the angles of the $\mathscr{T}_h$ are bounded below by a positive constant, independently of h. Let now S_h denote the continuous functions on the closure $\bar{\Omega}$ of Ω which are linear in each triangle of $\mathscr{T}_h$ and which vanish outside Ω_h. With $\{P_j\}_{j=1}^{N_h}$ the interior vertices of $\mathscr{T}_h$, a function in S_h is then uniquely determined by its values at the points P_j and thus $\dim(S_h) = N_h$. Let Φ_j be the "pyramid" function in S_h which takes the value 1 at P_j but vanishes at the other vertices; these functions form a basis for S_h. A given smooth function v on Ω which vanishes on $\partial\Omega$ may now be approximated by, e.g., its interpolant $I_h v = \sum_{j=1}^{N_h} v(P_j)\Phi_j \in S_h$, which agrees with v at the interior vertices, and one may show

$$\|I_h v - v\| \leqslant Ch^2\|v\|_2 \quad \text{and} \quad \|\nabla(I_h v - v)\| \leqslant Ch\|v\|_2, \quad \text{for } v \in H^2 \cap H_0^1, \tag{5.4}$$

where $\|\cdot\|_r$ denotes the norm in the Sobolev space $H^r = H^r(\Omega)$.

More generally we consider the case when $\Omega\subset\mathbb{R}^d$ and $\{S_h\}$ is a family of finite-dimensional subspaces of H_0^1 such that, for some integer $r\geqslant 2$,

$$\inf_{\chi\in S_h}\{\|v-\chi\|+h\|\nabla(v-\chi)\|\}\leqslant Ch^s\|v\|_s, \quad \text{for } 1\leqslant s\leqslant r,\ v\in H^s\cap H_0^1. \tag{5.5}$$

The spaces S_h are thought of as consisting of piecewise polynomials of degree at most $r-1$ on a partition $\mathscr{T}_h$ of Ω, and bound (5.5) shown by exhibiting a $\chi=I_hu$ where $I_h: H^r\cap H_0^1\to S_h$ is an interpolation type operator, see Zlámal (1968). The proof often involves the lemma of Bramble and Hilbert (1970):

Let $D\subset\mathbb{R}^d$ and assume that F is a bounded linear functional on $H^r(D)$ which vanishes for all polynomials of degree $<r$. Then $|F(u)|\leqslant C\sum_{|\alpha|=r}\|D^\alpha u\|_{L_2(D)}$.

To use this to show (5.4), e.g., one considers the difference I_hu-u on an individual $\tau\in\mathscr{T}_h$, transforms this to a unit size reference triangle $\hat{\tau}$, invokes the Bramble–Hilbert lemma with $D=\hat{\tau}$, noting that I_hu-u vanishes for linear functions, and transforms back to τ, using the fact that the bound for $|F(u)|$ in the lemma only contains the highest-order derivatives. In this example $\Omega_h\neq\Omega$ but the width of $\Omega\setminus\Omega_h$ is of order $O(h^2)$ and the contribution from this set is bounded appropriately. When $\partial\Omega$ is curved and $r>2$, however, there are difficulties in the construction and analysis of such operators I_h near the boundary; we shall return to this problem below. When Ω is polygonal and $\Omega_h=\Omega$, the Bramble–Hilbert argument for (5.5) may be used also for $r>2$, but in this case the solution of (5.1) will not normally have the regularity required. For comprehensive accounts of various choices of partitions $\mathscr{T}_h$ and finite element spaces S_h we refer to, e.g., Ciarlet [11], and Brenner and Scott [8].

We return to the finite element equation (5.3) using Courant elements. One way of triangulating $\Omega\subset\mathbb{R}^2$ is to start with the three families of straight lines $x_1=lh$, $x_2=lh$, $x_1+x_2=lh$, $l\in\mathbb{Z}$. The triangles thus formed may be used in the interior of Ω and then supplemented by other triangles near $\partial\Omega$ to form a triangulation $\mathscr{T}_h$ with the desired properties. With the notation (1.2) the equation corresponding to an interior vertex $x_j=jh$, $j\in\mathbb{Z}^2$ then takes the form

$$-\partial_1\bar{\partial}_1U_j-\partial_2\bar{\partial}_2U_j=h^{-2}(f,\Phi_j), \quad \text{where } U_j=u_h(x_j). \tag{5.6}$$

We recognize this as essentially the five-point finite difference equation (2.2), but with the right-hand side $f_j=f(x_j)$ replaced by an average of f over a neighborhood of x_j. Taking $f(x_j)$ may be considered as a quadrature rule for the right-hand side of (5.6). Recall that such averages were proposed also for finite difference methods.

Whereas a finite difference method may be obtained by replacing derivatives by finite differences, with some ad hoc modification near the boundary, the basic finite element method thus uses a variational formulation in a way that automatically accomodates the boundary conditions. We recall that the error analysis for the finite difference method uses a local estimate for the truncation error, together with some stability property, such as a discrete maximum principle. The finite element error analysis, as we shall now see, is based directly on the variational formulation and is global in nature. The difficulties in the construction of finite difference equations near the boundary are even greater for Neumann-type boundary conditions, whereas in the variational approach these are natural boundary conditions which do not have to be imposed on the approximating functions.

Under assumption (5.5) we now demonstrate the optimal order error estimate

$$\|u_h-u\|+h\|\nabla(u_h-u)\|\leqslant Ch^s\|u\|_s \quad \text{for } 1\leqslant s\leqslant r. \tag{5.7}$$

Starting with the error in the gradient we note that since u_h is the orthogonal projection of u onto S_h with respect to $(\nabla v, \nabla w)$, we have, by (5.5),

$$\|\nabla(u_h - u)\| = \inf_{\chi\in S_h} \|\nabla(\chi - u)\| \leqslant Ch^{s-1}\|u\|_s \quad \text{for } 1\leqslant s\leqslant r \tag{5.8}$$

for linear finite elements this was observed in Oganesjan (1963). For the L_2-error we apply a duality argument by Aubin (1967) and Nitsche (1968): Let φ be arbitrary in L_2, take $\psi \in H^2 \cap H_0^1$ as the solution of

$$-\Delta\psi = \varphi \quad \text{in } \Omega, \quad \text{with } \psi = 0 \text{ on } \partial\Omega, \tag{5.9}$$

and recall the elliptic regularity inequality $\|\psi\|_2 \leqslant C\|\Delta\psi\| = C\|\varphi\|$. We then have for the error $e = u_h - u$, for any $\chi \in S_h$,

$$(e,\varphi) = -(e,\Delta\psi) = (\nabla e, \nabla\psi) = (\nabla e, \nabla(\psi - \chi)) \leqslant \|\nabla e\|\,\|\nabla(\psi - \chi)\| \tag{5.10}$$

and hence, using (5.8) and (5.5) with $s = 2$, the desired result follows from

$$(e,\varphi) \leqslant (Ch^{s-1}\|u\|_s)(Ch\|\psi\|_2) \leqslant Ch^s\|u\|_s\|\varphi\|.$$

In the case of a more general, not necessarily symmetric, elliptic equation, and an approximation by Galerkin's method, the estimate for the gradient may be obtained by application with $V = H_0^1$ of the lemma of Céa (1964):

Let V be a Hilbert space with norm $|\cdot|$ and let $A(u,v)$ be a continuous bilinear form on V such that $|A(u,v)| \leqslant M\,|u|\,|v|$ and $A(u,u) \geqslant \mu|u|^2$, $\mu > 0$. For F a continuous linear functional on V, consider the equation

$$A(u,\varphi) = F(\varphi), \quad \forall\varphi \in V. \tag{5.11}$$

Let $S_h \subset V$ and let *$u_h \in S_h$ be the solution of $A(u_h,\chi) = F(\chi)$ for $\chi \in S_h$. Then $|u_h - u| \leqslant M\mu^{-1}\inf_{\chi\in S_h}|\chi - u|$.*

Since $A(u_h - u, \chi) = 0$ for $\chi \in S_h$ this follows at once from

$$\mu|u_h - u|^2 \leqslant A(u_h - u, u_h - u) = A(u_h - u, \chi - u) \leqslant M|u_h - u|\,|\chi - u|.$$

Note that the problem (5.11) has a unique solution in V by the Lax–Milgram lemma.

We remark that the finite element error estimate for, e.g., the Courant elements, will require the solution to have two derivatives, whereas four derivatives were needed in the five-point finite difference method. This advantage of finite elements stems from the use of averages and disappears when a quadrature rule is used.

The error analysis given above assumed the approximation property (5.5) for some $r \geqslant 2$. The most natural example of such a family in a plane domain Ω would be to take for S_h the continuous piecewise polynomials of degree at most $r-1$ on a triangulation $\mathcal{T}_h$ of Ω of the type described above, which vanish on $\partial\Omega$. However, for $r > 2$ and in the case of a domain with curved boundary, it is then not possible, in general, to satisfy the homogeneous boundary conditions exactly, and the above analysis therefore does not apply. One method to deal with this difficulty is to consider elements near $\partial\Omega$ that are polynomial maps of a reference triangle $\hat{\tau}$, so called isoparametric elements, such that these elements define a domain Ω_h which well approximates Ω, and to use the corresponding maps of polynomials on $\hat{\tau}$ as approximating functions. Such finite element spaces were proposed by Argyris and by Fried, Ergatoudis, Irons, and Zienkiewicz, and Felipa and Clough, and analyzed in, e.g., Ciarlet and Raviart (1972), and other types of curved finite elements were considered by, e.g., Zlámal and Scott, see Ciarlet [11].

Another example of how to deal with the boundary condition is provided by the following method proposed by Nitsche (1971), again in a plane domain Ω. It uses a family $\mathscr{T}_h$ of triangulations which is quasi-uniform in the sense that area $(\tau) \geqslant ch^2$ for $\tau \in \mathscr{T}_h$, with $c > 0$ independent of h. In this case certain inverse inequalities hold, such as $\|\nabla\chi\| \leqslant Ch^{-1}\|\chi\|$ for $\chi \in S_h$; this follows at once from the corresponding result for each $\tau \in \mathscr{T}_h$, for which it is shown by transformation to a fixed reference triangle and using the fact that all norms on a finite-dimensional space are equivalent, see, e.g., [11] With $\langle\cdot,\cdot\rangle$ the L_2-inner product on $\partial\Omega$, the solution of (5.1) satisfies, for $\chi \in S_h$,

$$N_\gamma\,(u,\chi) := (\nabla u, \nabla\chi) - \left\langle \frac{\partial u}{\partial n}, \chi \right\rangle - \left\langle u, \frac{\partial \chi}{\partial n} \right\rangle + \gamma h^{-1}\langle u,\chi\rangle = -(\Delta u,\chi) = (f,\chi).$$

Using inverse and trace inequalities, the bilinear form $N_\gamma(\cdot,\cdot)$ is seen to be positive definite on S_h for γ fixed and sufficiently large, and we may therefore pose the discrete problem $N_\gamma(u_h,\chi) = (f,\chi)$ for $\chi \in S_h$. Nitsche showed

$$\|u_h - u\| + h\|\nabla(u_h - u)\| + h^{1/2}\|u_h\|_{L_2(\partial\Omega)} \leqslant Ch^r\|u\|_r.$$

The bound for the third term expresses that u_h almost vanishes on $\partial\Omega$.

Other examples of methods used to deal with curved boundaries for which $S_h \not\subset H_0^1$ include a method of Babuška (1973) with Lagrangian multipliers, the method of interpolated boundary conditions by Berger, Scott and Strang (1972), Scott (1975), and an approach by Bramble, Dupont and Thomée (1972) and Dupont (1974) where the finite element method is based on an approximating polygonal domain with a correction built into the boundary conditions.

In some situations one may want to use finite element spaces S_h defined by piecewise polynomial approximating functions on a partition $\mathscr{T}_h$ of Ω which are not continuous across interelement boundaries, so called nonconforming elements. Assuming Ω polygonal so that it is exactly a union of elements τ, one may introduce a discrete bilinear form by $D_h(\psi,\chi) = \sum_{\tau\in\mathscr{T}_h}(\nabla\psi,\nabla\chi)_\tau$. Provided S_h is such that $\|\chi\|_{1,h} = D_h(\chi,\chi)^{1/2}$ is a norm on S_h, a unique nonconforming finite element solution u_h of (5.1) is now defined by $D_h(u_h,\chi) = (f,\chi)$ for $\chi \in S_h$, and it was shown in Strang (1972) that

$$\|u_h - u\|_{1,h} \leqslant C \inf_{\chi\in S_h} \|u - \chi\|_{1,h} + C \sup_{\chi\in S_h} \frac{|D_h(u,\chi) - (f,\chi)|}{\|\chi\|_{1,h}}. \tag{5.12}$$

As an example, consider an axes parallel rectangular domain, partitioned into smaller such rectangles with longest edge $\leqslant h$, and let S_h be piecewise quadratics which are continuous at the corners of the partition. Then $\|\cdot\|_{1,h}$ is a norm on S_h. In Wilson's rectangle, the six parameters involved on each small rectangle are determined by the values at the corners plus the (constant) values of $\partial^2\chi/\partial x_l^2$, $l = 1,2$. The functions in S_h are not in $C(\bar\Omega)$ but using (5.12) one may still show $\|u_h - u\|_{1,h} \leqslant C(u)h$.

The analysis above assumes that all inner products are calculated exactly. An analysis where quadrature errors are permitted was also worked out by Strang (1972). For instance, if (f,χ) is replaced by a quadrature formula $(f,\chi)_h$, a term of the form $C\sup_{\chi\in S_h}|(f,\chi) - (f,\chi)_h|/\|\nabla\chi\|$ has to be added to the bound for $\|\nabla(u_h - u)\|$. For example, if the quadrature formula is exact on each element for constants and if $f \in W_q^1(\Omega)$ with $q > 2$, then the $O(h)$ error for $\|\nabla(u_h - u)\|$ is maintained. The situations when curved boundaries, nonconforming elements, or quadrature errors occur, so that the basic assumptions of the variational formulation are not satisfied, are referred to in Strang (1972), as variational crimes.

Because of the variational formulation of Galerkin's method, the natural error estimates are expressed in L_2-based norms. In the maximum-norm it was shown by Natterer (1975), Nitsche (1975), and Scott (1976), see Schatz and Wahlbin (1982), that, for piecewise linear approximating functions on a quasi-uniform family $\mathscr{T}_h$ in a plane domain Ω, we have

$$\|u_h - u\|_{L_\infty} \leqslant Ch^2 \log(1/h)\|u\|_{W^2_\infty}, \qquad \|\nabla(u_h - u)\|_{L_\infty} \leqslant Ch\|u\|_{W^2_\infty}.$$

For polygonal domains and with piecewise polynomials of degree $r-1>1$,

$$\|u_h - u\|_{L_\infty} + h\|\nabla(u_h - u)\|_{L_\infty} \leqslant Ch^r\|u\|_{W^r_\infty},$$

but Haverkamp (1984) has proved that the above factor $\log(1/h)$ for piecewise linears may not be removed, even though it is not needed when estimating $I_h u - u$.

We shall now consider a finite element method for our model problem (5.1) which is based on a so called mixed formulation of this problem. Here the gradient of the solution u is introduced as a separate dependent variable whose approximation is sought in a different finite element space than the solution itself. This may be done in such a way that ∇u may be approximated to the same order of accuracy as u. With ∇u as a separate variable, (5.1) may thus be formulated

$$-\operatorname{div}\sigma = f \quad \text{in } \Omega, \quad \sigma = \nabla u \text{ in } \Omega, \text{ with } u = 0 \text{ on } \partial\Omega. \tag{5.13}$$

With $H=\{\omega=(\omega_1,\omega_2)\in L_2\times L_2;\ \operatorname{div}\omega\in L_2\}$ we note that the solution $(u,\sigma)\in L_2\times H$ also solves the variational problem

$$(\operatorname{div}\sigma,\varphi)+(f,\varphi)=0, \quad \forall\varphi\in L_2, \quad (\sigma,\omega)+(u,\operatorname{div}\omega)=0, \ \forall\omega\in H, \tag{5.14}$$

where the $(\cdot,\cdot)$ denote the appropriate L_2 inner products, and a smooth solution of (5.14) satisfies (5.13). Setting $L(v,\mu)=\frac{1}{2}\|\mu\|^2+(\operatorname{div}\mu+f,v)$ the solution (u,σ) of (5.13) may also be characterized as the saddle-point satisfying

$$L(v,\sigma)\leqslant L(u,\sigma)\leqslant L(u,\mu), \quad \forall v\in L_2,\ \mu\in H \tag{5.15}$$

and the key to the existence of a solution is the inequality

$$\inf_{v\in L_2}\sup_{\mu\in H}\frac{(v,\operatorname{div}\mu)}{\|v\|\,\|\mu\|_H}\geqslant c>0, \quad \text{where } \|\mu\|_H^2=\|\mu\|^2+\|\operatorname{div}\mu\|^2. \tag{5.16}$$

With S_h and H_h certain finite-dimensional subspaces of L_2 and H we shall consider the discrete analogue of (5.14) to find $(u_h,\sigma_h)\in S_h\times H_h$ such that

$$(\operatorname{div}\sigma_h,\chi)+(f,\chi)=0, \quad \forall\chi\in S_h, \ (\sigma_h,\psi)+(u_h,\operatorname{div}\psi)=0, \ \forall\psi\in H_h. \tag{5.17}$$

As in the continuous case this problem is equivalent to the discrete analogue of the saddle-point problem (5.15), and in order for this discrete problem to have a solution with the desired properties the choice of combinations $S_h\times H_h$ has to be such that the analogue of (5.16) holds, in this context referred to as the Babuška–Brezzi *inf–sup* condition (Babuška (1971); Brezzi (1974)).

One family of pairs of spaces which satisfy the *inf–sup* condition was introduced in Raviart and Thomas (1977); the first-order accurate pair of this family is as follows: With $\mathscr{T}_h$ a quasi-uniform family of triangulation of Ω, which we assume here to be polygonal, we set $S_h=\{\chi\in L_2;\chi|_\tau$ linear, $\forall\tau\in\mathscr{T}_h\}$, with no continuity required across inter-element boundaries. We then define

$H_h = \{\psi = (\psi_1, \psi_2) \in H; \psi|_\tau \in H(\tau), \quad \forall \tau \in \mathcal{T}_h\}$, where $H(\tau)$ denotes affine maps of quadratics on a reference triangle $\hat{\tau}$ of the form $(l_1(\xi) + \alpha\xi_1(\xi_1 + \xi_2), l_2(\xi) + \beta\xi_2(\xi_1 + \xi_2))$, with $l_1(\xi), l_2(\xi)$ linear, $\alpha, \beta \in R$. This space thus consists of piecewise quadratics on the triangulation $\mathcal{T}_h$ which are of the specific form implied by the definition of $H(\tau)$, and $\dim H(\tau) = 8$. As degrees of freedom for H_h one may use the values of $\psi \cdot n$ at two points on each side of τ (6 conditions) and in addition the mean values of ψ_1 and ψ_2 over τ (2 conditions). We note that the condition $\psi \in H$ in the definition of H_h requires that $\operatorname{div} \psi \in L_2$, which is equivalent to the continuity of $\chi \cdot n$ across inter-element boundaries. For the solutions of (5.17) and (5.13) holds

$$\|u_h - u\| \leqslant Ch^2 \|u\|_2 \quad \text{and} \quad \|\sigma_h - \sigma\| \leqslant Ch^s \|u\|_{s+1}, \quad s = 1,2$$

and correspondingly higher-order estimates were derived for higher-order Raviart–Thomas elements.

We now turn to negative norm estimates and superconvergence. Recalling the error estimate (5.7) which holds for the model problem under the approximation assumption (5.5), we shall now see that for $r > 2$, the duality argument used to show the L_2-norm estimate yields an error estimate in a negative order norm. Introducing such negative norms by $\|v\|_{-s} = \sup_{\varphi \in H^s} (v, \varphi)/\|\varphi\|_s$ for $s \geqslant 0$, the error in u_h satisfies

$$\|u_h - u\|_{-s} \leqslant Ch^{q+s} \|u\|_q, \quad \text{for } 0 \leqslant s \leqslant r-2, \ 1 \leqslant q \leqslant r. \tag{5.18}$$

In particular, $\|u_h - u\|_{-(r-2)} \leqslant Ch^{2r-2} \|u\|_r$. Since $2r - 2 > r$ for $r > 2$ the power of h in this estimate is higher than in the standard $O(h^r)$ error estimate in the L_2-norm. To show (5.18), we use the solution ψ of (5.9) and recall that $\|\psi\|_{s+2} \leqslant C\|\varphi\|_s$. This time (5.10) yields, for $0 \leqslant s \leqslant r-2$,

$$|(e,\varphi)| \leqslant \|\nabla e\| \inf_{\chi \in S_h} \|\nabla(\psi - \chi)\| \leqslant \|\nabla e\| (Ch^{s+1} \|\psi\|_{s+2}) \leqslant Ch^{s+1} \|\nabla e\| \, \|\varphi\|_s.$$

By (5.8) this gives $|(e,\varphi)| \leqslant Ch^{q+s} \|u\|_q \|\varphi\|_s$ for $\varphi \in H^s$, which shows (5.18). As an application of (5.18), assume we want to evaluate the integral $F(u) = \int_\Omega u\psi \, dx = (u, \psi)$, where u is the solution of (5.1) and $\psi \in H^{r-2}$. Then for the obvious approximation $F(u_h) = (u_h, \psi)$ we find the superconvergent order error estimate

$$|F(u_h) - F(u)| = |(u_h - u, \psi)| \leqslant \|u_h - u\|_{-(r-2)} \|\psi\|_{r-2} \leqslant Ch^{2r-2} \|u\|_r \, \|\psi\|_{r-2}.$$

One more example of these ideas is provided by Douglas and Dupont (1974a), which concerns superconvergent nodal approximation in the two-point boundary value problem

$$Au = -\frac{d}{dx}\left(a\frac{du}{dx}\right) + a_0 u = f \quad \text{in } I = (0,1), \quad \text{with } u(0) = u(1) = 0. \tag{5.19}$$

Defining the partition $0 = x_0 < x_1 < \cdots < x_M = 1$, with $x_{i+1} - x_i \leqslant h$, we set

$$S_h = \{\chi \in C(\bar{I}); \ \chi|_{I_i} \in \Pi_{r-1}, \ 1 \leqslant i \leqslant M; \ \chi(0) = \chi(1) = 0\},$$

where $I_i = (x_{i-1}, x_i)$. Clearly this family satisfies our assumption (5.5). The finite element solution is now defined by $A(u_h, \chi) = (f, \chi)$ for $\chi \in S_h$, where $A(v,w) = (av_x, w_x) + (a_0 v, w)$, and the error estimate (5.18) holds.

Let $g = g^{\bar{x}}$ denote the Green's function of the two-point boundary value problem (5.19) with singularity at the partition point $\bar{x}$, which we now consider fixed, so that $w(\bar{x}) = A(w,g)$ for $w \in H_0^1 = H_0^1(I)$. Applied to the error $e = u_h - u$, and using the orthogonality of e to S_h with respect to $A(\cdot,\cdot)$, we find $e(\bar{x}) = A(e,g) = A(e, g-\chi)$ for $\chi \in S_h$, and hence that

$$|e(\bar{x})| \leqslant C\|e\|_1 \inf_{\chi \in S_h} \|g - \chi\|_1 \leqslant Ch^{r-1} \|u\|_r \inf_{\chi \in S_h} \|g - \chi\|_1.$$

Although $g^{\bar{x}}$ is not a smooth function at $\bar{x}$ it may still be approximated well by a function in S_h since it is smooth except at $\bar{x}$ and the discontinuity of the derivative at $\bar{x}$ can be accommodated in S_h. In particular, we have

$$\inf_{\chi\in S_h} \|g-\chi\|_1 \leqslant Ch^{r-1}(\|g\|_{H^r((0,\bar{x}))} + \|g\|_{H^r((\bar{x},1))}) \leqslant Ch^{r-1},$$

so that $|e(\bar{x})| \leqslant Ch^{2r-2}\|u\|_r$. Note that for $A = -\mathrm{d}^2/\mathrm{d}x^2$ the Green's function $g^{\bar{x}}$ is linear outside $\bar{x}$ and so $g^{\bar{x}} \in S_h$. We may then conclude that $e(\bar{x}) = 0$, which is a degenerate case.

We now touch on some superconvergent order estimates for the gradient in the two-dimensional model problem (5.1) using piecewise linear approximations for S_h in (5.3). It was shown in Oganesjan and Ruhovec (1969) that if the triangulations $\mathcal{T}_h$ are uniform then $\|\nabla(u_h - I_h u)\|_{L_2(\Omega_h)} \leqslant Ch^2\|u\|_{H^3}$, where as above I_h denotes the interpolant into S_h. This implies that at the midpoints of the edges of $\mathcal{T}_h$ the average of ∇u_h from the two adjacent triangles is a $O(h^2)$ approximation to ∇u in a discrete l_2 sense. Such results have been improved to maximum-norm estimates and to triangulations that are perturbations in various ways of uniform triangulations by Chen, Lin, Xu, Zhou, Zhu, and others, and the approximation at other points than midpoints of edges has also been studied, see, e.g., references in Křížek and Neittaanmäki [22] or Wahlbin [41]. We remark that for uniform, axes parallel triangulations it follows from (5.6) that finite differences may be used as in (2.8) to approximate both the gradient and higher order derivatives to order $O(h^2)$ in the interior of Ω.

All error estimates quoted above are a priori error estimates in that they depend on certain norms of the exact solution of the problem. In principle, these norms could be bounded in terms of norms of the data of the problem, but generally such bounds would be rather crude. During the last decades so-called a posteriori error estimates have been developed which depend directly on the computed solution, and on the data. Such estimates may be applied to include an adaptive aspect in the solution method, by detecting areas in a computational domain where the error is larger than elsewhere, and using this information to refine the mesh locally to reduce the error by an additional computation. Pioneering work is Babuška (1976) and Babuška and Rheinboldt (1978); for a recent survey, see Verfürth [40].

We illustrate this approach for the two-dimensional problem (5.1) in a polygonal domain Ω, using piecewise linear finite element approximations. With $\{\Phi_j\}$ the basis of pyramid functions we define Ω_j by $\bar{\Omega}_j = \operatorname{supp}\Phi_j$. Given the finite element solution $u_h \in S_h$ of (5.3), we now consider the local error equation

$$-\Delta w_j = f \quad \text{in } \Omega_j, \quad \text{with } w_j = u_h \text{ on } \partial\Omega_j.$$

It is then proved in Babuška and Babuška and Rheinboldt (1978), that, with c and C positive constants which depend on geometrical properties of the triangulations $\mathcal{T}_h$,

$$c\sum_j \eta_j^2 \leqslant \|\nabla(u_h - u)\|^2 \leqslant C\sum_j \eta_j^2, \quad \text{where } \eta_j = \|\nabla(w_j - u_h)\|.$$

The error in ∇u_h is thus bounded both above and below in terms of the local quantities η_j, which can be approximately determined. It is argued that a triangulation for which the quantities η_j are of essentially the same size gives a small error in ∇u_h, and this therefore suggests an adaptive strategy for the solution of (5.1).

Another approach was taken in Eriksson and Johnson (1991), showing an a posteriori error estimate of the form

$$\|\nabla(u_h-u)\|\leqslant C\left(\left(\sum_\tau h_\tau^2\|f\|_{L_2(\tau)}^2\right)^{1/2}+\left(\sum_\gamma h_\gamma^2\left|\left[\frac{\partial u_h}{\partial n}\right]_\gamma\right|\right)^{1/2}\right), \tag{5.20}$$

where the γ with length h_γ are the edges of $\mathscr{T}_h$ and $[\,\cdot\,]_\gamma$ denotes the jump across γ. Under a certain assumption on the local variation of h_τ, which is weaker than quasi-uniformity the a priori estimate

$$\|\nabla(u_h-u)\|\leqslant C\left(\sum_\tau h_\tau^2\|u\|_{H^2(\tau)}^2\right)^{1/2} \tag{5.21}$$

is also derived. Together with (5.20) this may be used to justify an adaptive scheme with a given tolerance and an essentially minimal number of triangles. Analogous a posteriori and a priori bounds are demonstrated for $\|u_h-u\|$ and, in Eriksson (1994), also in maximum norm, where the analogue of (5.21) reads $\|\nabla(u_h-u)\|_{L_\infty}\leqslant C\max_\tau(h_\tau\|u\|_{W_\infty^2(\tau)})$.

Superconvergence of the error in the gradient has been used in, e.g., Zienkiewicz and Zhu (1992) to derive a posteriori error bounds for adaptive purposes.

We finally mention the p- and h–p-versions of the finite element method: So far it has been assumed that the approximating subspaces S_h are piecewise polynomial spaces of a fixed degree based on partitions $\mathscr{T}_h$ with $\max_{\tau\in\mathscr{T}_h}\operatorname{diam}(\tau)\leqslant h$, and higher accuracy is achieved by refining the partition. An alternative approach proposed in Babuška, Szabó and Katz (1981) is to fix the mesh and then let the degree of the polynomials grow. The two approaches are referred to as the h-version and the p-version of the finite element method, respectively. A combination of the two methods, the h–p-version has been studied in Babuška and Dorr (1981). For more material about the p- and h–p-methods, see Babuška and Suri (1990).

6. Finite element methods for evolution equations

This section is concerned with the application of the finite element method to time dependent problems. We begin with the model heat equation and discuss then the wave equation and finally some simple first-order hyperbolic model problems.

We consider thus first the approximate solution of the parabolic problem

$$u_t-\Delta u=f(t)\quad\text{in }\Omega,\ \text{with } u=0 \text{ on } \partial\Omega,\ t>0,\ u(\cdot,0)=v \text{ in } \Omega, \tag{6.1}$$

in a finite-dimensional space S_h belonging to a family satisfying (5.5). As a first step we discretize this problem in the spatial variable by writing it in variational form and defining $u_h=u_h(\cdot,t)\in S_h$ for $t\geqslant 0$ by

$$(u_{h,t},\chi)+(\nabla u_h,\nabla\chi)=(f(t),\chi),\quad \forall\chi\in S_h,\ t>0,\ u_h(0)=v_h\approx v. \tag{6.2}$$

With respect to a basis $\{\Phi_j\}_{j=1}^{N_h}$ of S_h this may be written as a system of ordinary differential equations $B\alpha'+A\alpha=\tilde{f}$ where A is the stiffness matrix introduced in Section 5 and where $B=(b_{jk})$, $b_{jk}=(\Phi_j,\Phi_k)$, is referred to as the mass matrix. A fully discrete time-stepping scheme may then be obtained by discretization of this system in time, using, e.g., the single-step ϑ-method:

With k the time step, $t_n = nk$, $\bar{\partial}_t U^n = (U^n - U^{n-1})/k$, and with $\vartheta \in [0,1]$, the approximation $U^n \in S_h$ of $u(t_n)$ for $n \geqslant 1$ is then defined by

$$(\bar{\partial}_t U^n, \chi) + (\nabla(\vartheta U^n + (1-\vartheta)U^{n-1}), \nabla\chi) = (f^{n-1+\vartheta}, \chi), \quad \forall \chi \in S_h,\ n \geqslant 1 \tag{6.3}$$

with $f^s = f(sk)$ and U^0 given. For $\vartheta = 0$ and 1 these are the forward and backward Euler methods and for $\vartheta = \frac{1}{2}$ the Crank–Nicolson method. Note that the forward Euler method is not explicit because the matrix B is nondiagonal.

For the semidiscrete problem (6.2) Douglas and Dupont (1970) showed that

$$\|u_h(t) - u(t)\| + \left(\int_0^t \|\nabla(u_h - u)\|^2 \,\mathrm{d}s\right)^{1/2} \leqslant \|v_h - v\| + C(u)h^{r-1}. \tag{6.4}$$

For a proof we note that the error $e = u_h - u$ satisfies $(e_t, \chi) + (\nabla e, \nabla \chi) = 0$ for $\chi \in S_h, t > 0$, and hence $(e_t, e) + (\nabla e, \nabla e) = (e_t, \chi - u) + (\nabla e, \nabla(\chi - u))$, from which the result follows after integration and with the appropriate choice of χ. Because of the contribution from $\|\nabla(\chi - u)\|$ on the right, (6.4) is of suboptimal order in L_2-norm. In this regard the estimate was improved by Wheeler (1973) to

$$\|u_h(t) - u(t)\| \leqslant \|v_h - v\| + Ch^r\left(\|v\|_r + \int_0^t \|u_t\|_r \,\mathrm{d}s\right) \quad \text{for } t > 0. \tag{6.5}$$

This was done by introducing the elliptic or Ritz projection, the orthogonal projection $R_h : H_0^1 \to S_h$ with respect to the Dirichlet inner product, thus defined by $(\nabla(R_h u - u), \nabla\chi) = 0$ for $\chi \in S_h$, and writing $e = (u_h - R_h u) + (R_h u - u) = \theta + \rho$. Here, by the error estimate (5.7) for the elliptic problem, $\|\rho(t)\| \leqslant Ch^r \|u(t)\|_r$, which is bounded as desired, and one also finds $(\theta_t, \chi) + (\nabla\theta, \nabla\chi) = -(\rho_t, \chi)$ for $\chi \in S_h$, $t > 0$. Choosing $\chi = \theta$ and integrating this yields $\|\theta(t)\| \leqslant \|\theta(0)\| + \int_0^t \|\rho_t\| \,\mathrm{d}s$ which is easily bounded as desired. In particular, for $v_h \in S_h$ suitably chosen, this shows an optimal order error estimate in L_2.

Defining the discrete Laplacian $\Delta_h : S_h \to S_h$ by $-(\Delta_h \psi, \chi) = (\nabla\psi, \nabla\chi)\ \forall \psi, \chi \in S_h$ and using the L_2-projection P_h onto S_h the above equation for θ may be written as $\theta_t - \Delta_h \theta = -P_h \rho_t$, and, with $E_h(t) = e^{\Delta_h t}$ the solution operator of (6.2) with $f = 0$, we find by Duhamel's principle that $\theta(t) = E_h(t)\theta(0) + \int_0^t E_h(t-s)P_h\rho_t(s)\,\mathrm{d}s$. An obvious energy argument shows the stability property $\|E_h(t)v_h\| \leqslant \|v_h\|$, which again gives the above bound for θ. The error estimate for the semidiscrete problem thus follows from the stability of $E_h(t)$ together with error estimates for the elliptic problem; for finite difference methods stability was similarly combined with a bound for the truncation error.

The use of the elliptic projection also yields an estimate of superconvergent order for $\nabla\theta$. In fact, by choosing this time $\chi = \theta_t$ in the variational equation for θ, we find after integration and simple estimates that $\|\nabla\theta(t)\| \leqslant C(u)h^r$ if $v_h = R_h v$. For piecewise linears ($r = 2$) this may be combined with the superconvergent second-order estimate for $\nabla(R_h u - I_h u)$ quoted in Section 5 to bound $\nabla(u_h - I_h u)$, with similar consequences as in the elliptic case, see Thomée, Xu and Zhang (1989).

Estimates for the fully discrete ϑ-method (6.3) were also shown in Douglas and Dupont (1970) and Wheeler (1973). The contribution from the time discretization that has to be added to (6.5) at $t = t_n$ is then $Ck \int_0^{t_n} \|u_{tt}\| \,\mathrm{d}s$, with a stability condition $k \leqslant \gamma_\vartheta h^2$ for $0 \leqslant \vartheta < \frac{1}{2}$, and $Ck^2 \int_0^{t_n} (\|u_{ttt}\| + \|\Delta u_{tt}\|)\,\mathrm{d}s$ for $\vartheta = \frac{1}{2}$.

The ϑ-method for the homogeneous equation may be defined by $U^n = E_{kh}^n v_h$ where $E_{kh} = r(-k\Delta_h)$, $r(\lambda) = (1+(1-\vartheta)\lambda)/(1+\vartheta\lambda)$. Two-level schemes using more general rational functions $r(\lambda)$ of arbitrary order of accuracy were constructed in Baker, Bramble and Thomée (1977), under the stability assumption $|r(\lambda)| \leqslant 1$ for $\lambda \in \sigma(-k\Delta_h)$. Stable two-level time-stepping methods for the inhomogeneous equation of arbitrary order of accuracy may be constructed in the form $U^{n+1} = r(-k\Delta_h)U^n + k\sum_{j=1}^m q_j(-k\Delta_h)f(t_n+\tau_j k)$ where $r(\lambda)$ and the $q_j(\lambda)$ are rational functions, with, e.g., the backward Euler method included for $m=1$, $\tau_1=0$, $r(\lambda)=q_1(\lambda)=1/(1+\lambda)$, see Brenner, Crouzeix and Thomée (1982). Stable multistep time discretization schemes of accuracy of order $q \leqslant 6$ have also been derived by Le Roux (1979) and others, see Thomée [38].

The regularity requirements needed for optimal order convergence in some of the above error estimates make it natural to enquire about error estimates under weaker regularity assumptions on data or on the solution. To illustrate this we now consider the solution of the homogeneous equation, i.e., (6.1) with $f=0$, and recall that the solution of this problem is smooth for $t>0$ even if v is only in L_2, say, and satisfies $\|u(t)\|_s \leqslant Ct^{-s/2}\|v\|$. Similarly, for the semidiscrete solution, $\|\Delta_h^{s/2}u_h(t)\| \leqslant Ct^{-s/2}\|v_h\|$, and using this one may show the nonsmooth data error estimate

$$\|u_h(t)-u(t)\| \leqslant Ch^r t^{-r/2}\|v\| \quad \text{for } t>0 \quad \text{if } v_h = P_h v,$$

so that optimal order $O(h^r)$ convergence holds for $t>0$, without any regularity assumptions on v. The corresponding result for the backward Euler method reads

$$\|U^n - u(t_n)\| \leqslant C(h^r t_n^{-r/2} + kt_n^{-1})\|v\|, \quad \text{for } n \geqslant 1 \text{ if } v_h = P_h v.$$

Results of this type were shown by spectral representation in, e.g., Blair (1970), Helfrich (1974), and later by energy methods, permitting also time-dependent coefficients, in Huang and Thomée (1981), Luskin and Rannacher (1982), Sammon (1983a,b), see [38]. For stable fully discrete approximations of the form $U^n = E_{kh}^n v_h$ with $E_{kh} = r(-k\Delta_h)$ one then has to require $|r(\infty)| < 1$, see Baker, Bramble and Thomée (1977). The Crank–Nicolson method lacks this smoothing property, but Rannacher (1984) showed that using this method with two initial steps of the backward Euler method, one has $\|U^n - u(t_n)\| \leqslant C(h^r t_n^{-r/2} + k^2 t_n^{-2})\|v\|$ for $n \geqslant 1$.

The methods quoted in Section 5 for handling the difficulty of incorporating homogeneous Dirichlet boundary conditions in the approximating spaces S_h have been carried over from the elliptic to the semidiscrete parabolic case in Bramble, Schatz, Thomée and Wahlbin (1977). This may be accomplished by replacing the gradient term in (6.2) and (6.3) by more general bilinear forms such as $N_\gamma(\cdot,\cdot)$ described there, or by using other approximations of the Laplacian than the above Δ_h. Within this framework negative norm estimates and superconvergence results were derived in Thomée (1980). We also quote Johnson and Thomée (1981) where the mixed method discussed in Section 5 is applied to (6.1).

In the fully discrete schemes discussed above, Galerkin's method was applied in space but a finite-difference-type method was used in time. We shall now describe an approach which uses a Galerkin-type method also in time, the discontinuous Galerkin time-stepping method. This method was introduced and analyzed in Lesaint and Raviart (1974) and Jamet (1978), and generalized in the case of ordinary differential equations in Delfour, Hager and Trochu (1981). In the present context it was studied in Eriksson, Johnson and Thomée (1985). With a not necessarily uniform partition of $[0,\infty)$ into intervals $J_n=[t_{n-1},t_n)$, $n \geqslant 1$, let $\mathscr{S}_h=\{X=X(x,t); X|_{J_n}=\sum_{j=0}^{q-1}\chi_j t^j, \chi_j \in S_h\}$, where S_h are finite element spaces satisfying (5.5). Since the elements $X \in \mathscr{S}_h$ are not required to be continuous at the t_j we set $U^n_\pm = U(t_n \pm 0)$ and $[U^{n-1}] = U^{n-1}_+ - U^{n-1}_-$. The discontinuous Galerkin method may then be stated: With $U^0_- = v_h$ given, find $U \in \mathcal{S}_h$ such that, for $n \geqslant 1$,

$$\int_{J_n} [(U_t, X) + (\nabla U, \nabla X)]\,\mathrm{d}s + ([U^{n-1}], X^{n-1}_+) = \int_{J_n} (f, X)\,\mathrm{d}s, \quad \forall X \in \mathcal{S}_h. \tag{6.6}$$

In the piecewise constant case, $q = 0$, this may be written, with $k_n = t_n - t_{n-1}$,

$$(\bar\partial_{t_n} U^n, \chi) + (\nabla U^n, \nabla \chi) = \left(k_n^{-1} \int_{J_n} f\,\mathrm{d}s, \chi\right), \quad \bar\partial_{t_n} U^n = (U^n - U^{n-1})/k_n, \tag{6.7}$$

this reduces to the standard backward Euler method when the average of f is replaced by $f(t_n)$. It was shown in Eriksson and Johnson (1991) that for the error in the time discretization in (6.7)

$$\|U^n - u_h(t_n)\| \leqslant C\ell_n \max_{j \leqslant n} (k_j \|u_{h,t}\|_{J_j}), \quad \text{where } \ell_n = (1 + \log(t_n/k_n))^{1/2},$$

with u_h the solution of the semidiscrete problem (6.2) and $\|\varphi\|_{J_j} = \sup_{t \in J_j} \|\varphi(t)\|$.

For $q = 1$ the method requires the determination on J_n of $U(t)$ of the form $U(t) = U^{n-1}_+ + (t - t_{n-1})/k_n V_n$ with $U^{n-1}_+, V_n \in S_h$, and such that (6.6) holds, which gives a 2×2 system for these elements in S_h. In this case we have

$$\|U - u_h\|_{J_n} \leqslant C\ell_n \max_{j \leqslant n} (k_j^2 \|u_{h,tt}\|_{J_j}), \quad \|U^n_- - u_h(t_n)\| \leqslant C\ell_n \max_{j \leqslant n} (k_j^3 \|\Delta_h u_{h,tt}\|_{J_j}),$$

thus with third-order superconvergence at the nodal points. For the total error in the fully discrete scheme, with, e.g., piecewise linear elements in space, one has

$$\|U^n_- - u(t_n)\| \leqslant C\ell_n \max_{j \leqslant n} (k_j^3 \|\Delta u_{tt}\|_{J_j} + h^2 \|u\|_{2,J_j}), \quad \|u\|_{2,J_j} = \sup_{t \in J_j} \|u(t)\|_2.$$

All our error estimates so far have been a priori error estimates, expressed in terms of the unknown exact solution of our parabolic problem. We close by mentioning briefly some a posteriori estimates by Eriksson and Johnson (1991), based on an idea of Lippold (1991), where the error bounds are expressed in terms of the data and the computed solution. For $q = 0$, $r = 2$ such an estimate is

$$\|U^n_- - u(t_n)\| \leqslant C\ell_n \max_{j \leqslant n} ((h^2 + k_j)\|f\|_{J_j} + k_j \|\bar\partial_t U^j\| + h^2 \|U^j\|_{2,h}), \tag{6.8}$$

where $\|\cdot\|_{2,h}$ is a discrete H^2-norm defined by $\|U\|_{2,h} = (\sum_\gamma |[\partial U/\partial n]_\gamma|^2)^{1/2}$, with γ denoting the edges of $\mathcal{T}_h$ and $[\,\cdot\,]_\gamma$ the jumps across γ. Error bounds are also available for $q = 1$, and the estimates generalize to variable h.

Estimates such as (6.8) may be used to design adaptive schemes in which the time step is successively chosen so that the error is bounded by a given tolerance. The earlier a priori estimates are then needed to show that such a procedure will end in a finite number of steps, cf. Eriksson and Johnson (1992). This approach was further developed in a sequence of paper by Eriksson and Johnson, see the survey paper Eriksson, Estep, Hansbo and Johnson [13]. A Petrov–Galerkin method with continuous in time approximations was studied by Aziz and Monk (1989).

We now briefly consider the question of maximum-norm stability for the finite element scheme. For the solution operator $E(t)$ of the homogeneous equation in (6.1) ($f = 0$) the maximum-principle shows at once that $\|E(t)v\|_{L_\infty} \leqslant \|v\|_{L_\infty}$, and the smoothing estimate $\|\Delta^{s/2} E(t)v\|_{L_\infty} \leqslant Ct^{-s/2}\|v\|_{L_\infty}$ also holds for $s > 0$. However, considering the case $r = d = 2$ one can easily see (cf. [38]) that the maximum principle does not apply for the semidiscrete finite element analogue. This is in contrast to the corresponding finite difference method and is related to the fact that the mass matrix B is nondiagonal. In this regard, it was shown by Fujii (1973) that if B is replaced by a diagonal matrix whose diagonal elements are the row sums of B, and if all angles of the triangulation are nonobtuse, then the maximum principle holds and hence $\|u_h(t)\|_{L_\infty} \leqslant \|v_h\|_{L_\infty}$ for $t>0$. This method is called the lumped mass method and can also be defined by

$$(u_{h,t},\chi)_h + (\nabla u_h, \nabla \chi) = 0, \quad \forall \chi \in S_h,\ t>0,$$

where the first term has been obtained by replacing the first term in (6.2) by using the simple quadrature expression $Q_{h\tau}(u_{h,t}\chi)$ on each τ, where $Q_{h\tau}(\varphi) = \frac{1}{3}\,\mathrm{area}(\tau)\sum_{j=1}^{3}\varphi(P_j)$ with P_j the vertices of τ.

Even though the maximum principle does not hold for (6.2), it was shown in Schatz, Thomée and Wahlbin (1980) that, for $d=r=2$ and quasi-uniform $\mathscr{T}_h$,

$$\|E_h(t)v_h\|_{L_\infty} + t\|E_h'(t)v_h\|_{L_\infty} \leqslant C\ell_h\|v_h\|_{L_\infty}, \quad \text{for } t>0, \quad \ell_h = \log(1/h).$$

The proof uses a weighted norm technique to estimate a discrete fundamental solution. For $d=1,2,3$ and $r\geqslant 4$, Nitsche and Wheeler (1981–82) subsequently proved stability without the factor ℓ_h, and this and the corresponding smoothing result were shown for $d=1$ and $r\geqslant 2$ in Crouzeix, Larsson and Thomée (1994). Recently, logarithm-free stability and smoothness bounds have been shown for general d and r, first for Neumann boundary conditions in Schatz, Thomée and Wahlbin (1998), and then for Dirichlet boundary conditions in Thomée and Wahlbin (1998). We note that the combination of stability and smoothing shows that the semigroup $E_h(t)$ is analytic, and via a resolvent estimate for its infinitesimal generator Δ_h this may be used to derive stability estimates also for fully discrete approximations of the form $U^n = r(-k\Delta_h)^n v_h$ where $r(z)$ is a rational function with the appropriate stability and consistency properties, see Palencia (1992) and Crouzeix, Larsson, Piskarev and Thomée (1993). Other maximum-norm error bounds have been given in the literature by, e.g., Dobrowolski (1978), Nitsche (1979), and Rannacher (1991).

We now turn to hyperbolic equations and begin with a brief discussion of semidiscrete and fully discrete finite element schemes for the initial–boundary value problem for the wave equation

$$u_{tt} - \Delta u = f \quad \text{in } \Omega \subset \mathbb{R}^d, \text{ with } u=0 \text{ on } \partial\Omega, \text{ for } t>0$$

$$u(\cdot,0) = v, \quad u_t(\cdot,0) = w \quad \text{in } \Omega.$$

Assuming as usual that $S_h \subset H_0^1$ satisfies (5.5), the semidiscrete analogue of our problem is to find $u_h(t) \in S_h$ for $t\geqslant 0$ from

$$(u_{h,tt},\chi) + (\nabla u_h, \nabla\chi) = (f,\chi) \quad \forall \chi \in S_h,\ t>0, \quad \text{with } u_h(0) = v_h,\ u_{h,t}(0) = w_h.$$

Similarly to the parabolic case this problem may be written in matrix form, this time as $B\alpha'' + A\alpha = \tilde{f}$ for $t>0$, with $\alpha(0)$ and $\alpha'(0)$ given, where B and A are the mass and stiffness matrices.

Analogously to the analysis in the parabolic case it was shown in Dupont (1973a), with a certain improvement in Baker (1976), that under natural regularity assumptions and with appropriate choices of v_h and w_h,

$$\|u_h(t) - u(t)\| + h\|\nabla(u_h(t) - u(t))\| \leqslant C(u)h^r.$$

One possible fully discrete method for the wave equation is, cf. the case $\vartheta = \frac{1}{2}$ of the Newmark-type method (4.4),

$$(\partial_t\bar{\partial}_t U^n, \chi) + (\nabla(\tfrac{1}{4}U^{n+1} + \tfrac{1}{2}U^n + \tfrac{1}{4}U^{n-1}), \nabla\chi) = (f(t_n),\chi), \quad \forall \chi \in S_h,\ n\geqslant 1,$$

where U^0 and U^1 are given approximations of $u(0)=v$ and $u(k)$, respectively. Setting $U^{n+1/2}=(U^n+U^{n+1})/2$ one shows for the homogeneous equation ($f=0$) that the energy $\|\partial_t U^n\|^2+\|\nabla U^{n+1/2}\|^2$ is conserved for $n\geqslant 0$, and, also in the general case, that $\|U^{n+1/2}-u(t_n+\frac{1}{2}k)\|=O(h^r+k^2)$ for appropriate initial values U^0 and U^1, u sufficiently regular, and t_n bounded. Although the error is then estimated at the points $t_n+\frac{1}{2}k$ it is easy to derive approximations also at the points t_n. For the homogeneous equation more general time-stepping schemes based on rational functions of the discrete Laplacian Δ_h were studied in Baker and Bramble (1979), where the second-order wave equation was written as a first-order system.

We proceed with some results for first-order hyperbolic equations, and begin with the periodic model problem

$$u_t+u_x=f \quad \text{for } x\in\mathbb{R},\ t>0, \ \text{ with } u(\cdot,0)=v \text{ on } \mathbb{R}, \tag{6.9}$$

where f, v, and the solution sought are 1-periodic in x; the L_2 inner products and norms used below are based on intervals of length 1.

To define an approximate solution, let $S_h\subset C^k(\mathbb{R})$, with $k\geqslant 0$, denote 1-periodic splines of order r (i.e., piecewise polynomials of degree $r-1$) based on a partition with maximal interval length h. The standard Galerkin method for (6.9) is then to find $u_h(t)\in S_h$ for $t\geqslant 0$ such that

$$(u_{h,t},\chi)+(u_{h,x},\chi)=(f,\chi), \quad \forall\chi\in S_h,\ t>0, \ \text{ with } u_h(0)=v_h.$$

The equation may again be written in the form $B\alpha'+A\alpha=\tilde{f}$ where as usual B is the mass matrix but where the matrix A now has elements $a_{jk}=(\Phi_j',\Phi_k)$ and is skew-symmetric.

We first establish the simple error estimate, cf. Swartz and Wendroff (1969),

$$\|u_h(t)-u(t)\|\leqslant\|v_h-v\|+C(u)h^{r-1}, \quad \text{for } t\geqslant 0. \tag{6.10}$$

For this, we use an interpolation operator Q_h into S_h which commutes with time differentiation and is such that $\|Q_hv-v\|+h\|(Q_hv-v)_x\|\leqslant Ch^r\|v\|_r$. It remains to bound $\theta=u_h-Q_hu$, which satisfies $(\theta_t,\chi)+(\theta_x,\chi)=-(\rho_t+\rho_x,\chi)$ for $\chi\in S_h$. Setting $\chi=\theta$ and observing that $(\theta_x,\theta)=0$ by periodicity, we conclude

$$\frac{1}{2}\frac{\mathrm{d}}{\mathrm{d}t}\|\theta\|^2=\|\theta\|\frac{\mathrm{d}}{\mathrm{d}t}\|\theta\|\leqslant(\|\rho_t\|+\|\rho_x\|)\|\theta\|\leqslant C(u)(h^r+h^{r-1})\|\theta\|, \tag{6.11}$$

which shows $\|\theta(t)\|\leqslant\|\theta(0)\|+C(u)h^{r-1}\leqslant\|v_h-v\|+C(u)h^{r-1}$, and yields (6.10).

We observe that estimate (6.10) is of nonoptimal order $O(h^{r-1})$, because the first derivative of the error in Q_hu occurs on the right side of (6.11). For special cases more accurate results are known. For example, for the homogeneous equation, with S_h consisting of smooth splines ($k=r-2$) on a uniform partition, the last term in (6.10) may be replaced by the optimal order term $C(u)h^r$. In this case superconvergence takes place at the nodes in the sense that $\|u_h(t)-I_hu(t)\|\leqslant C(u)h^{2r}$ if $v_h=I_hv$, where I_hv denotes the interpolant of v in S_h. This follows from Fourier arguments, see Thomée (1973), after observing that the Galerkin method may be interpreted as a finite difference method for the coefficients with respect to a basis for S_h. This was generalized to variable coefficients in Thomée and Wendroff (1974). It was shown by Dupont (1973b), however, that the improvement to optimal order is not always possible. In fact, if S_h is defined by a uniform partition with $r=4$, $k=1$ (Hermite cubics), and if v is a nonconstant analytic 1-periodic function and $v_h\in S_h$ is arbitrary, then $\sup_{t\in(0,t^*)}\|u_h(t)-u(t)\|\geqslant ch^3$, $c>0$, for any $t^*>0$.

Leaving the standard Galerkin method, it was shown in Wahlbin (1974) that with S_h defined on a uniform partition and $0 \leqslant k \leqslant r-2$, optimal order convergence holds for the Petrov–Galerkin method

$$(u_{h,t} + u_{h,x}, \chi + h\chi_x) = (f, \chi + h\chi_x), \quad \forall \chi \in S_h, \ t > 0, \ \text{with } u_h(0) = v_h.$$

In Dendy (1974) and Baker (1975) nonstandard variational schemes with optimal order convergence without requiring uniform meshes were exhibited for the initial–boundary value problem

$$u_t + u_x = f \quad \text{for } x \in I = (0,1), \quad u(0,t) = 0, \ \text{for } t \geqslant 0, \ \text{with } u(\cdot, 0) = v.$$

We now quote some space–time methods for this initial–boundary value problem for $(x,t) \in \Omega = I \times J\ (J = (0,T))$, and introduce the characteristic directional derivative $Du = u_t + u_x$. Let $\mathscr{T}_h = \{\tau\}$ be a quasi-uniform triangulation of Ω with max diam$(\tau) = h$ and let $\mathscr{S}_h = \{\chi \in C(\bar{\Omega}); \chi|_\tau \in \Pi_{r-1} : \chi = 0 \text{ on } \partial\Omega^-\}$, where $\partial\Omega^-$ is the inflow boundary $(\{0\} \times J) \cup (I \times \{0\})$. With $((v,w)) = \int_0^T \int_0^1 v\, w\, \mathrm{d}x\, \mathrm{d}t$, the standard Galerkin method in this context is then to find $u_h \in \mathscr{S}_h$ such that

$$((Du_h, \chi)) = ((f, \chi)), \quad \forall \chi \in \mathscr{S}_h. \tag{6.12}$$

Standard arguments show as above the nonoptimal order error estimate

$$\|u_h - u\| + \|u_h - u\|_{L_2(\partial\Omega^+)} \leqslant Ch^{r-1}\|u\|_r, \tag{6.13}$$

where $\partial\Omega^+ = \partial\Omega \backslash \partial\Omega^-$ is the outflow boundary. This method does not work well in the case of discontinuous solutions. To stabilize the scheme one could consider an artificial dissipation in the form of an additional term $h((\nabla u_h, \nabla \chi))$ on the left in (6.12), but such a method would be at most first-order accurate.

The so-called streamline diffusion method was introduced by Hughes and Brooks (1979) and analyzed in Johnson and Pitkäranta (1986). It consists in substituting $\chi + hD\chi$ for the test function χ in (6.12), and (6.13) then holds with h^{r-1} replaced by $h^{r-1/2}$. In the discontinuous Galerkin method studied by Lesaint and Raviart (1974), and Johnson and Pitkäranta (1986), cf. the corresponding method for parabolic equations introduced above, one determines $u_h \in \mathscr{S}_h = \{\chi \in L_2(\Omega); \chi|_\tau \in \Pi_{r-1}; \chi = 0 \text{ on } \partial\Omega^-\}$, thus without requiring $u_h \in C(\bar{\Omega})$, from

$$((Du_h, \chi))_\tau - \int_{\partial\tau^-} [u_h]_{\partial\tau}(n_t + n_x)\, \mathrm{d}s = ((f, \chi))_\tau, \quad \forall \chi \in \Pi_{r-1}, \ \tau \in \mathscr{T}_h,$$

where $((\cdot,\cdot))_\tau$ denotes restriction to τ. This method also satisfies (6.12), with $h^{r-1/2}$ instead of h^{r-1}, and here the error in Du_h is of optimal order $O(h^{r-1})$.

Winther (1981) investigated a Petrov–Galerkin method defining u_h in continuous, piecewise Π_{r-1} spaces $\mathscr{S}_h$ based on rectangles $\tau = I_j \times J_l$, where I_j and J_l are partitions of I and J, by the equation $((Du_h, \chi))_\tau = ((f, \chi))_\tau$ for all $\chi \in \Pi_{r-2}$ and all τ, and proved optimal order $O(h^r)$ convergence. This method coincides with the cell vertex finite-volume method, and is associated with the Wendroff box scheme, see Morton [26].

These types of approaches have been developed also for advection dominated diffusion problems, see, e.g., Johnson, Nävert and Pitkäranta (1984), Hughes, Franca and Mallet (1987) and to non-linear conservation laws, Johnson and Szepessy (1987), Szepessy (1989). For such time-dependent problems Pirroneau (1982), Douglas and Russel (1982), and others have also analyzed the so-called characteristic Galerkin method in which one adopts a Lagrangian point of view in the time stepping, following an approximate characteristic defined by the advection term, in combination with a finite element approximation in the diffusive term.

7. Some other classes of approximation methods

Methods other than finite difference and finite element methods, but often closely related to these, have also been developed, and in this section we sketch briefly four such classes of methods, namely collocation methods, spectral methods, finite volume methods, and boundary element methods.

In a *collocation method* one seeks an approximate solution of a differential equation in a finite-dimensional space of sufficiently regular functions by requiring that the equation is satisfied exactly at a finite number of points. Such a procedure for parabolic equations in one space variable was analyzed by Douglas (1972), Douglas and Dupont (1974b); we describe it for the model problem

$$u_t = u_{xx} \quad \text{in } I=(0,1), \qquad u(0,t)=u(1,t)=0 \text{ for } t>0, \text{ with } u(\cdot,0)=v \text{ in } I.$$

Setting $h=1/M$, $x_j=jh$, $j=0,\dots,M$, and $I_j=(x_{j-1},x_j)$, we introduce the piecewise polynomial space $S_h=\{\chi\in C^1(\bar I); v|_{I_j}\in\Pi_{r-1},\ v(0)=v(1)=0\}$, with $r\geqslant 4$. Letting ξ_i, $i=1,\dots,r-2$, be the Gaussian points in $(0,1)$, the zeros of the Lagrange polynomial P_{r-2}, we define the collocation points $\xi_{ji}=x_{j-1}+h\xi_i$ in I_j, and pose the spatially semidiscrete problem to find $u_h\in S_h$ such that

$$u_{h,t}(\xi_{ji},t)=u_{h,xx}(\xi_{ji},t), \quad \text{for } j=1,\dots,M,\ i=1,\dots,r-2,\ t>0,$$

with $u_h(\cdot,0)=v_h$ an approximation of v. This method may be considered as a Galerkin method using a discrete inner product based on a Gauss quadrature rule. For v_h appropriately chosen one may then show the global error estimate

$$\|u_h(t)-u(t)\|_{L_\infty}\leqslant Ch^r\left(\max_{s\leqslant t}\|u(s)\|_{r+2}+\left(\int_0^t\|u_t\|_{r+2}^2\,\mathrm{d}s\right)^{1/2}\right).$$

Further, for $r>4$, and with a more refined choice of initial approximation v_h, superconvergence takes place at the nodes,

$$|u_h(x_j,t)-u(x_j,t)|\leqslant C_T h^{2r-4}\sup_{s\leqslant t}\sum_{p+2q\leqslant 2r-1}\|u^{(q)}(s)\|_p \quad \text{for } t\leqslant T.$$

We note the more stringent regularity requirements than for the Galerkin methods discussed in Section 6. These results carry over to fully discrete methods using both finite difference approximations and collocation in time.

For a two-point boundary value problem, results of a similar nature were derived by de Boor and Swartz (1973).

Spectral methods are in many ways similar to Galerkin/collocation methods. The main difference is in the choice of finite-dimensional approximating spaces. We begin by considering an evolution equation in a Hilbert space framework.

Let thus H be a Hilbert space with inner product $(\cdot,\cdot)$ and norm $\|\cdot\|$, and assume L is a nonnegative operator defined in $D(L)\subset H$, so that $(Lu,u)\geqslant 0$. Consider the initial value problem

$$u_t+Lu=f \quad \text{for } t>0, \text{ with } u(0)=v. \tag{7.1}$$

Let now $\{\varphi_j\}_{j=1}^\infty\subset H$ be a sequence of linearly independent functions in $D(L)$ which span H and set $S_N=\operatorname{span}\{\varphi_j\}_{j=1}^N$. We define a "spatially" semidiscrete approximation $u_N=u_N(t)\in S_N$ of (7.1) by

$$(u_{N,t},\chi)+(Lu_N,\chi)=(f,\chi)\quad \forall\chi\in S_N,\ t\geqslant 0, \quad \text{with } u_N(0)=v_N. \tag{7.2}$$

Introducing the orthogonal projection $P_N : H \to S_N$ we may write (7.2) as

$$u_{N,t} + L_N u_N = f_N := P_N f, \quad \text{for } t \geqslant 0, \text{ where } L_N = P_N L P_N.$$

Clearly $(L_N \chi, \chi) = (L P_N \chi, P_N \chi) \geqslant 0$. With $u_N(t) = \sum_{j=1}^N \alpha_j(t)\varphi_j$, this equation may be written $B\alpha' + A\alpha = \tilde{f}$ for $t \geqslant 0$, where the elements of the matrices A and B are $(L\varphi_i, \varphi_j)$ and (φ_i, φ_j), respectively. Clearly B is a Gram matrix and so positive definite.

As a simple example, let $L = -(\mathrm{d}/\mathrm{d}x)^2$ on $I = (0,1)$ and $H = L_2(I)$, $D(L) = H^2 \cap H_0^1$, and let the $\varphi_j(x) = c_j \sin \pi j x$ be the normalized eigenfunctions of L. Then $B = I, A$ is positive definite and P_N is simply the truncation of the Fourier series, $P_N v = \sum_{j=1}^N (v, \varphi_j)\varphi_j$, with $L_N = \sum_{j=1}^N (\pi j)^2 (v, \varphi_j)\varphi_j = P_N L v$.

We note that the error $e_N = u_N - u$ satisfies

$$e_{N,t} + L_N e_N = f_N - f + (L_N - L)u \quad \text{for } t > 0, \quad e_N(0) = v_N - v$$

and hence, since $E_N(t) = \mathrm{e}^{-L_N t}$ is bounded,

$$\|e_N\| \leqslant \|v_N - v\| + \int_0^t (\|(P_N - I)f\| + \|(L_N - L)u\|)\,\mathrm{d}s. \tag{7.3}$$

It follows that the error is small with $v_N - v$, $(P_N - I)f$, and $(L_N - L)u$.

In our above example we see that if $v_N = P_N v$, and if the Fourier series for v, f, and Lu converge, then the error is small. In particular, the convergence is of order $\mathrm{O}(N^{-r})$ for any r provided the solution is sufficiently regular.

Another way to define a semidiscrete numerical method employing the space S_N of our example is to make S_N a Hilbert space with the inner product $(v,w)_N = h\sum_{j=0}^{N-1} v(x_j)w(x_j)$ where $x_j = j/(N-1)$. This gives rise to a projection P_N defined by $P_N u(x_j) = u(x_j), j = 0,\ldots,N-1$, and the semidiscrete equation (7.2) now becomes the collocation equation

$$u_{N,t}(x_j,t) + Lu_N(x_j,t) = f(x_j,t) \quad \text{for } j = 0,\ldots,N-1,\ t \geqslant 0.$$

This is also referred to as a pseudospectral method and the error estimate (7.3) will be valid in the discrete norm corresponding to $(\cdot,\cdot)_N$.

Spectral and pseudospectral methods using the above sinusoidal basis functions are particularly useful for periodic problems. For initial–boundary value problems for hyperbolic equations basis functions related to Chebyshev and Lagrange polynomials are sometimes useful. Such methods are successfully applied in fluid dynamics calculations. Spectral methods have been studied since the 1970s, see Gottlieb and Orszag [17], Canuto, Hussaini, Quarteroni and Zhang [10], Boyd [6], and references therein.

We now turn to the *finite volume method* which we exemplify for the model problem

$$-\Delta u = f \text{ in } \Omega, \text{ with } u = 0 \text{ on } \partial\Omega, \tag{7.4}$$

where Ω is a convex polygonal domain in $\mathbb{R}^2$. The basis for this approach is the observation that for any $V \subset \Omega$ we have by Green's formula that

$$\int_{\partial V} \frac{\partial u}{\partial n}\,\mathrm{d}s = \int_V f\,\mathrm{d}x. \tag{7.5}$$

Let now $\mathscr{T}_h = \{\tau_j\}_{j=1}^{N_h}$ be a triangulation of Ω and consider (7.5) with $V = \tau_j, j = 1,\ldots,N_h$. Let Q_j be the center of the circumscribed circle of τ_j. If τ_i has an edge γ_{ji} in common with τ_j, then $Q_i - Q_j$ is orthogonal to γ_{ji}, and $\partial u/\partial n$ in (7.5) may be approximated by the difference quotient

$(u(Q_i) - u(Q_j))/|Q_i - Q_j|$. This produces a finite difference scheme on the nonuniform mesh $\{Q_j\}$; for the boundary triangles one may use the boundary values in (7.4). Writing the discrete problem as $AU = F$ the matrix A is symmetric positive definite, and the solution satisifies a discrete maximum principle. When the $\mathscr{T}_h$ is quasi-uniform (and such that the Q_j are on $\partial\Omega$) one has $\|U - u\|_{1,h} \leqslant Ch^{s-1}\|u\|_s$ for $s = 2$ in a certain discrete H^1-norm, and, under an additional symmetry assumption on $\mathscr{T}_h$, also for $s = 3$. This method may be described as cell centered and goes back to Tikhonov and Samarskii (1961) in the case of rectangular meshes; for further developments, see Samarskii, Lazarov and Maharov [30]. For such meshes it was used in Varga [39] to construct finite difference schemes.

An associated method is the following vertex centered method, also referred to as the finite volume element method: Let $S_h \subset H_0^1$ be the piecewise linear finite element space defined by $\mathscr{T}_h$. The straight lines connecting a vertex of $\tau \in \mathscr{T}_h$ with the midpoint of the opposite edge intersect at the barycenter of τ and divide τ into six triangles. Let $B_{j,\tau}$ be the union of the two of these which have P_j as a vertex. For each interior vertex P_j we let B_j be the union of the corresponding $B_{j,\tau}$, and let $\bar{S}_h$ denote the associated piecewise constant functions. Motivated by (7.5) we then pose the Petrov-Galerkin method to find $u_h \in S_h$ such that

$$\bar{A}(u_h, \psi) := \sum_j \psi_j \int_{\partial B_j} \frac{\partial u_h}{\partial n}\, \mathrm{d}s = (f, \psi) \quad \forall \psi \in \bar{S}_h, \tag{7.6}$$

this may also be thought of as a finite difference scheme on the irregular mesh $\{P_j\}$. The B_j are referred to as control volumes; they were called mesh regions in Mac Neal (1953). Associating with $\chi \in S_h$ the function $\bar{\chi} \in \bar{S}_h$ which agrees with χ at the vertices of $\mathscr{T}_h$ one finds that $\bar{A}(\psi, \bar{\chi}) = A(\psi, \chi)$ so that (7.6) may be written $A(u_h, \chi) = (f, \bar{\chi})$ for $\chi \in S_h$. In particular, (7.6) has a unique solution, and using the Babuška–Brezzi *inf–sup* condition it was shown in Bank and Rose (1987) that the standard error estimate $\|u_h - u\|_1 \leqslant Ch\|u\|_2$ holds for this method.

Finite volume methods are useful for operators in divergence form and have also been applied to time-dependent conservation laws, see Heinrich [20], Morton [26]. For the model heat equation the vertex centered method is similar to the lumped mass finite element method.

In a *boundary integral method* a boundary value problem for a homogeneous partial differential equation in a domain Ω is reformulated as an integral equation over the boundary $\partial\Omega$. This equation may then be used as a basis for numerical approximation. We shall illustrate this approach for the model problem

$$\Delta u = 0 \quad \text{in } \Omega \subset \mathbb{R}^d, \text{ with } u = g \text{ on } \partial\Omega, \tag{7.7}$$

$\partial\Omega$ smooth. To pose the boundary integral equation, let $\Gamma(x) = -(2\pi)^{-1}\log|x|$ for $d = 2$ and $\Gamma(x) = c_d|x|^{-d+2}$ for $d > 2$ be the fundamental solution of the Laplacian in $\mathbb{R}^d$. For any u with $\Delta u = 0$ on $\partial\Omega$ we have by Green's formula

$$u(x) = \int_{\partial\Omega} \Gamma(x-y)\frac{\partial u}{\partial n_y}\, \mathrm{d}s_y - \int_{\partial\Omega} \frac{\partial \Gamma}{\partial n_y}(x-y)u(y)\, \mathrm{d}s_y, \quad x \in \Omega. \tag{7.8}$$

With x on $\partial\Omega$ the integrals on the right define the single- and double-layer potentials $V\partial u/\partial n$ and Wu (note that $K(x,y) = (\partial\Gamma/\partial n_y)(x-y) = O(|x-y|^{-(d-2)})$ for $x, y \in \partial\Omega$). For $x \in \Omega$ approaching $\partial\Omega$ the two integrals tend to $V\partial u/\partial n$ and $\frac{1}{2}u + Wu$, respectively, so that (7.8) yields $\frac{1}{2}u = V\partial u/\partial n + Wu$. With $u = g$ on $\partial\Omega$ this is a Fredholm integral equation of the first kind to determine $\partial u/\partial n$ on $\partial\Omega$, which inserted into (7.8) together with $u = g$ on $\partial\Omega$ gives the solution of (7.7).

Instead of this direct method one may use the indirect method of assuming that the solution of (7.8) may be represented as a potential of a function on $\partial\Omega$, so that

$$u(x)=\int_{\partial\Omega}\Gamma(x-y)v(y)\,\mathrm{d}s_y \quad \text{or } u(x)=\int_{\partial\Omega}\frac{\partial\Gamma}{\partial n_y}(x-y)w(y)\,\mathrm{d}s_y, \quad x\in\Omega.$$

With V and W as above, if such functions v and w exist, they satisfy the first and second kind Fredholm integral equations

$$Vv=g \quad \text{and} \quad \frac{1}{2}w+Ww=g. \tag{7.9}$$

Writing $H^s=H^s(\partial\Omega)$, V and W are pseudodifferential operators of order -1, bounded operators $H^s\to H^{s+1}$, in particular compact on H^s; for $d=2$ the kernel of W is actually smooth. The first kind equation is uniquely solvable provided a certain measure, the transfinite diameter $\delta_{\partial\Omega}$ of $\partial\Omega$, is such that $\delta_{\partial\Omega}\neq 1$, and the second kind equation in (7.8) always has a unique solution. Similar reformulations may be used also for Neumann boundary conditions, for a large number of other problems involving elliptic-type equations, and for exterior problems; in fact, this approach to the numerical solution is particularly useful in the latter cases.

The use of boundary integral equations, particularly of the second kind, to study boundary value problems for elliptic equations has a long history, and includes work of Neumann and Fredholm. We shall not dwell on this here but refer to, e.g., Atkinson [2].

In the boundary element method (BEM) one determines the approximate solution in a piecewise polynomial finite element-type space of a boundary integral formulation such as the above using the Galerkin or the collocation method.

The numerical solution of second kind equations by projection methods, which include both Galerkin and collocation methods, were studied in an abstract Banach space setting in the important paper by Kantorovich (1948), and their convergence was established under the appropriate assumptions on the projection operator involved. Consider, e.g., the second kind equation in (7.9) (with $d=2$) in $C(\partial\Omega)$ with the maximum norm $|\cdot|_\infty$, and let $S_h\subset C(\partial\Omega)$ be finite dimensional. With $P_h: C(\partial\Omega)\to S_h$ a projection operator, the corresponding discrete problem is $\frac{1}{2}w_h+P_hWw_h=P_hg$, and if $|P_hW-W|_\infty\to 0$ in operator norm one may show that $|w_h-w|_\infty\leqslant C|P_hw-w|_\infty$, so that the discrete solution converges as fast as the projection as $h\to 0$.

The collocation method may also be combined with quadrature, as suggested in Nyström (1930): as an example we may use, e.g., the composite trapezoidal rule on a uniform partition, and then apply collocation at the nodal points so that, with $\partial\Omega=\{x(s);\ 0\leqslant s\leqslant l\}$ and $K(s,t)$ the kernel of W, the discrete solution is

$$w_h(x(s))=2g(x(s))-2h\sum_{j=1}^{N_h}K(s,s_j)w_j \quad \text{for } 0\leqslant s\leqslant l,$$

where the w_j are determined by setting $w_h(x(s_i))=w_i$ for $i=1,\dots,N_h$. It is not difficult to see that since the trapezoidal rule is infinitely accurate for smooth periodic functions we have $|w_h-w|_\infty=O(h^r)$ for any $r>0$.

For the second kind equation in (7.9), using Galerkin's method and a finite dimensional subspace S_h of $L_2(\partial\Omega)$, the discrete approximation $w_h\in S_h$ to w is determined from

$$\tfrac{1}{2}\langle w_h,\chi\rangle+\langle Ww_h,\chi\rangle=\langle g,\chi\rangle, \quad \forall\chi\in S_h, \quad \text{where } \langle\cdot,\cdot\rangle=(\cdot,\cdot)_{L_2(\partial\Omega)}.$$

Writing $|\cdot|_s$ for the norm in H^s, one has $|w_h - w|_0 \leqslant C_r(u)h^r$ if S_h is accurate of order $O(h^r)$, and by a duality argument one may show the superconvergent order negative norm estimate $|w_h - w|_{-r} \leqslant C_r(u)h^{2r}$, see Sloan and Thomée (1985); using an iteration argument by Sloan (1976) this may be used, in principle, to define an approximate solution $\tilde{w}_h$ with $|\tilde{w}_h - w|_0 = O(h^{2r})$.

After early work of Wendland (1968) and Nedelec and Planchard (1973) the study of the Galerkin finite element approach using first kind equations was pursued from the mid 1970s by Hsiao, Le Roux, Nedelec, Stephan, Wendland, and others, see the surveys in Atkinson [2], Sloan [31], and Wendland [42]. Within this framework we consider the numerical solution of the first kind equation in (7.9) with $d = 2$ in the finite-dimensional space S_h of periodic smoothest splines of order r, i.e., $S_h \subset C^{r-2}$ consists of piecewise polynomials in Π_{r-1}. Our discrete problem is then to find $v_h \in S_h$ such that

$$\langle Vv_h, \chi \rangle = \langle g, \chi \rangle, \quad \forall \chi \in S_h.$$

It was shown in the work quoted above that the bilinear form $\langle Vv, w \rangle$ associated with $V : H^{-1/2} \to H^{1/2}$ is symmetric, bounded, and coercive in $H^{-1/2}$, i.e.,

$$\langle Vv, w \rangle = \langle v, Vw \rangle \leqslant C|v|_{-1/2}|w|_{-1/2} \quad \text{and} \quad \langle Vv, v \rangle \geqslant c|v|_{-1/2}^2, \text{ with } c > 0.$$

An application of Céa's lemma and approximation properties of S_h then show

$$|v_h - v|_{-1/2} \leqslant C \inf_{\chi \in S_h} |\chi - v|_{-1/2} \leqslant Ch^{r+1/2}|v|_r$$

and an Aubin–Nitsche-type duality argument first used by Hsiao and Wendland (1981) implies $|v_h - v|_{-r-1} \leqslant Ch^{2r+1}|v|_r$. For x an interior point of Ω we therefore find for $u_h = Vv_h$ that $|u_h(x) - u(x)| \leqslant C_x|v_h - v|_{-r-1} \leqslant Ch^{2r+1}$, since $\Gamma(x-y)$ is smooth when $y \neq x$.

Expressed in terms of a basis $\{\Phi_j\}$ of S_h this problem may be written in matrix form as $A\alpha = \tilde{g}$ where A is symmetric positive definite. However, although the dimension of A has been reduced by the reduction of the original two-dimensional problem to a one-dimensional one, in contrast to the finite element method for a differential equation problem, the matrix A is now not sparse. We also note that the elements $\langle V\Phi_i, \Phi_j \rangle$ require two integrations, one in forming $V\Phi_i$ and one in forming the inner product.

In order to reduce this work the collocation method has again been considered by Arnold and Wendland (1983); here v_h is determined from $Vv_h(x(s_j)) = g(x(s_j))$ at N_h quadrature points s_j in $[0, l]$, where $N_h = \dim S_h$. Applied to our above model problem this method, using smoothest splines of even order r, has a lower order of maximal convergence rate, $O(h^{r+1})$ rather than $O(h^{2r+1})$; if r is odd and the mesh uniform Saranen (1988) has shown $O(h^{r+1})$ in $\|\cdot\|_{-2}$. A further step in the development is the qualocation method proposed by Sloan (1988), which is a Petrov–Galerkin method, thus with different trial and test spaces. For S_h the smoothest splines of order r on a uniform mesh (so that Fourier analysis may be applied) and with the quadrature rule suitably chosen, negative norm estimates of order $O(h^{r+3})$ for even r and $O(h^{r+4})$ for odd r may be shown.

In the vast literature on the numerical boundary integral methods much attention has been paid to the complications arising when our above regularity assumptions fail to be satisfied, such as for domains with corners in which case V and W are not compact.

8. Numerical linear algebra for elliptic problems

Both finite difference and finite element methods for elliptic problems such as (2.1) lead to linear algebraic systems

$$AU = F, \tag{8.1}$$

where A is a nonsingular matrix. When Ω is a d-dimensional domain, using either finite differences or finite elements based on quasi-uniform triangulations, the dimension N of the corresponding finite-dimensional problem is of order $O(h^{-d})$, where h is the mesh-width, and for $d>1$ direct solution by Gauss elimination is normally not feasible as this method requires $O(N^3)=O(h^{-3d})$ algebraic operations. Except in special cases one therefore turns to iterative methods. In this section we summarize the historical development of such methods.

As a basic iterative method we consider the Picard method

$$U^{n+1} = U^n - \tau(AU^n - F) \quad \text{for } n \geqslant 0, \text{ with } U^0 \text{ given}, \tag{8.2}$$

where τ is a positive parameter. With U the exact solution of (8.1) we have

$$U^n - U = R(U^{n-1} - U) = \cdots = R^n(U^0 - U), \quad \text{where } R = I - \tau A$$

and hence the rate of convergence of the method depends on $\|R^n\|$ where $\|\cdot\|$ is the matrix norm subordinate to the Euclidean norm in R^N. When A is symmetric positive definite (SPD) we have $\|R^n\| = \rho^n$ where $\rho = \rho(R) = \max_i|1-\tau\lambda_i|$ denotes the spectral radius of R, and (8.2) converges if $\rho<1$. The optimal choice is $\tau=2/(\lambda_1+\lambda_N)$, which gives $\rho=(\kappa-1)/(\kappa+1)$, where $\kappa=\kappa(A)=\lambda_N/\lambda_1$ is the condition number or A; note, however, that this choice of τ requires knowledge of λ_1 and λ_N which is not normally at hand. In applications to second-order elliptic problems one often has $\kappa=O(h^{-2})$ so that $\rho\leqslant 1-ch^2$ with $c>0$. Hence with the optimal choice of τ the number of iterations required to reduce the error to a small $\varepsilon>0$ is of order $O(h^{-2}|\log\varepsilon|)$. Since each iteration uses $O(h^{-d})$ operations in the application of $I-\tau A$ this shows that the total number of operations needed to reduce the error to a given tolerance is of order $O(h^{-d-2})$, which is smaller than for the direct solution when $d\geqslant 2$.

The early more refined methods were designed for finite difference methods of positive type for second-order elliptic equations, particularly the five-point operator (2.2). The corresponding matrix may then be written $A=D-E-F$ where D is diagonal and E and F are (elementwise) nonnegative and strictly lower and upper triangular. The analysis was often based on the Perron–Frobenius theory of positive matrices. A commonly used property is diagonal dominance: $A=(a_{ij})$ is diagonally dominant if $\sum_{j\neq i}|a_{ij}|\leqslant|a_{ii}|, =1\ldots,N$, irreducibly diagonally dominant if it is also irreducible, so that (8.1) cannot be written as two lower-order systems, and strictly diagonally dominant if there is strict inequality for at least one i. Examples are the Jacobi (after Jacobi (1845)) and Gauss–Seidel (Gauss (1823); Seidel (1874) or Liebmann (1918)) methods which are defined by

$$U^{n+1} = U^n - B(AU^n - F) = RU^n + BF, \quad \text{with } R = I - BA, \tag{8.3}$$

in which $B=B_J=D^{-1}$ or $B=B_{GS}=(D-E)^{-1}$ with $R_J=D^{-1}(E+F)$ and $R_{GS}=(D-E)^{-1}F$, respectively. In the application to the model problem (2.1) in the unit square, using the five-point operator, the equations may be normalized so that $D=4I$ and the application of R_J simply means that the new value at any interior mesh-point x_j is obtained by replacing it by the average of the old values at the four neighboring points $x_{j\pm e_l}$. The Gauss–Seidel method also takes averages, but with the mesh-points taken in a given order, and successively uses the values already obtained in forming the averages. The methods were referred to in Geiringer (1949) as the methods of simultaneous and successive displacements, respectively. For the model problem one may easily determine the eigenvalues and eigenvectors of A and show that with $h = 1/M$ one has $\rho(R_{\mathrm{J}}) = \cos \pi h = 1 - \frac{1}{2}\pi^2 h^2 + \mathrm{O}(h^4)$ and $\rho(R_{\mathrm{GS}}) = \rho(R_{\mathrm{J}})^2 = 1 - \pi^2 h^2 + \mathrm{O}(h^4)$ so that the number of iterates needed to reduce the error to ε is of the orders $2h^{-2}\pi^2|\log \varepsilon|$ and $h^{-2}\pi^2|\log \varepsilon|$. The Gauss–Seidel method thus requires about half as many iterations as the Jacobi method.

If A is irreducibly diagonally dominant then $\rho(R_{\mathrm{J}}) < 1$ and $\rho(R_{\mathrm{GS}}) < 1$ so that both methods converge, see Geiringer (1949); for A strictly diagonally dominant this was shown in Collatz (1942). Further, Stein and Rosenberg (1948) showed that if D is positive and E and F nonnegative and $\rho(B_{\mathrm{J}}) < 1$ then $\rho(B_{\mathrm{GS}}) < \rho(B_{\mathrm{J}})$, i.e., the Gauss–Seidel method converges faster than the Jacobi method.

Forming the averages in the Jacobi and Gauss–Seidel methods may be thought as relaxation; in the early work by Gauss and Seidel this was not done in a cyclic order as described above, and which is convenient on computers, but according to the size of the residual or other criteria, see Southwell [33] and Fox (1948). It turns out that one may obtain better results than those described above by overrelaxation, i.e., choosing $B_\omega = (D - \omega E)^{-1}$ and $R_\omega = (D - \omega E)^{-1}((1-\omega)E + F)$ with $\omega > 1$. These methods were first studied by Frankel (1950) in the case of the model problem, and Young (1950, 1954) in more general cases of matrices satisfying his property $\mathscr{A}$, which holds for a large class of difference approximations of elliptic problems in general domains. Frankel proved that for the model problem the optimal choice of the parameter is $\omega_{\mathrm{opt}} = 2/(1 + \sqrt{1-\rho^2})$ where $\rho = \rho(B_{\mathrm{J}}) = \cos \pi h$, i.e., $\omega_{\mathrm{opt}} = 2/(1 + \sin \pi h) = 2 - 2\pi h + \mathrm{O}(h^2)$, and that correspondingly $\rho(R_{\omega_{\mathrm{opt}}}) = \omega_{\mathrm{opt}} - 1 = 1 - 2\pi h + \mathrm{O}(h^2)$. The number of iterations required is thus then of order $\mathrm{O}(h^{-1})$, which is significally smaller than for the above methods. It was shown by Kahan (1958), also for nonsymmetric A, that $\rho(R_\omega) \geqslant |\omega - 1|$ so that convergence can only occur for $0 < \omega < 2$. On the other hand, Ostrowski (1954) showed that if A is SPD, then $\rho(R_\omega) < 1$ if and only if $0 < \omega < 2$.

We consider again an iterative method of the form (8.3) with $\rho(R) < 1$, and introduce now the new sequence $V^n = \sum_{j=0}^n \beta_{nj} U^j$ where the β_{nj} are real. Setting $p_n(\lambda) = \sum_{j=0}^n \beta_{nj}\lambda^j$, and assuming $p_n(1) = \sum_{j=0}^n \beta_{nj} = 1$ for $n \geqslant 0$, we obtain easily $V^n - U = p_n(R)(U^0 - U)$. For V^n to converge fast to U one wants to choose the β_{nj} in such a way that $\rho(p_n(R))$ becomes small with n. By the Cayley–Hamilton theorem $p_n(R) = 0$ if p_n is the characteristic polynomial of R, and hence $V^n = U$ if $n \geqslant N$, but this is a prohibitively large number of iterations. For $n < N$ we have by the spectral mapping theorem that $\rho(p_n(R)) = \max_{\lambda \in \sigma(R)} |p_n(\lambda)|$. In particular, if R is symmetric and $\rho = \rho(R)$, a simple calculation shows that, taking the maximum instead over $[-\rho, \rho] \supset \sigma(R)$, the optimal polynomial is $p_n(\lambda) = T_n(\lambda/\rho)/T_n(1/\rho)$ where T_n is the nth Chebyshev polynomial, and the corresponding value of $\rho(p_n(R))$ is bounded by

$$T_n(1/\rho)^{-1} = 2\left(\left(\frac{1+\sqrt{1-\rho^2}}{\rho}\right)^n + \left(\frac{1+\sqrt{1-\rho^2}}{\rho}\right)^{-n}\right)^{-1} \leqslant 2\left(\frac{\rho}{1+\sqrt{1-\rho^2}}\right)^n.$$

For the model problem using the Gauss–Seidel basic iteration we have as above $\rho = 1 - \pi^2 h^2 + \mathrm{O}(h^4)$ and we find that the average error reduction factor per iteration step in our present method is bounded by $1 - \sqrt{2}\pi h + \mathrm{O}(h^2)$, which is of the same order of magnitude as for SOR. The use of the sequence

V^n instead of the U^n was called linear acceleration by Forsythe (1953) and is sometimes attributed to Richardson (1910); in [39] it is referred to as a semiiterative method.

We now describe the Peaceman–Rachford alternating direction implicit iterative method for the model problem (2.1) on the unit square, using the five-point discrete elliptic equation with $h=1/M$. In this case we may write $A=H+V$ where H and V correspond to the horizontal and vertical difference operators $-h^2\partial_1\bar{\partial}_1$ and $-h^2\partial_2\bar{\partial}_2$. Note that H and V are positive definite and commute. Introducing an acceleration parameter τ and an intermediate value $U^{n+1/2}$ we may consider the scheme defining U^{n+1} from U^n by

$$(\tau+H)U^{n+1/2}=(\tau-V)U^n+F, \quad (\tau+V)U^{n+1}=(\tau-H)U^{n+1/2}+F,$$

or after elimination, with G_τ appropriate and using that H and V commute,

$$U^{n+1}=R_\tau U^n+G_\tau, \quad \text{where } R_\tau=(\tau I-H)\,(\tau I+H)^{-1}(\tau I-V)\,(\tau I+V)^{-1}.$$

The error satisfies $U^n-U=R_\tau^n(U^0-U)$, and with μ_i the (common) eigenvalues of H and V, $\|R_\tau\|\leqslant \max_i|(\tau-\mu_i)/(\tau+\mu_i)|^2<1$, and it is easy to see that the maximum occurs for $i=1$ or M. With $\mu_1=4\sin^2(\frac{1}{2}\pi h)$, $\mu_M=4\cos^2(\frac{1}{2}\pi h)$ the optimal τ is $\tau_{\text{opt}}=(\mu_1\mu_M)^{1/2}$ with the maximum for $i=1$, so that, with $\kappa=\kappa(H)=\kappa(V)=\mu_M/\mu_1$,

$$\|R\|_{\tau_{\text{opt}}}\leqslant\left(\frac{(\mu_1\mu_M)^{1/2}-\mu_1}{(\mu_1\mu_M)^{1/2}+\mu_1}\right)^{1/2}=\frac{\kappa^{1/2}-1}{\kappa^{1/2}+1}=1-\pi h+O(h^2).$$

This again shows the same order of convergence rate as for SOR.

A more efficient procedure is obtained by using varying acceleration parameters τ_j, $j=1,2,\ldots$, corresponding to the n step error reduction matrix $\tilde{R}_n=\prod_{j=1}^n R_{\tau_j}$. It can be shown that the τ_j can be chosen cyclically with period m in such a way that $m\approx c\log\kappa\approx c\log(1/h)$ and

$$\|\tilde{R}_m\|^{1/m}=\max_{1\leqslant i\leqslant M}\left(\prod_{j=0}^{m-1}\left|\frac{\tau_j-\mu_i}{\tau_j+\mu_i}\right|\right)^{2/m}\leqslant 1-c(\log(1/h))^{-1}, \quad c>0.$$

The analysis indicated depends strongly on the fact that H and V commute, which only happens for rectangles and constant coefficients, but the method may be defined and shown convergent for more general cases, see Birkhoff and Varga (1959). We remark that these iterative schemes may often be associated with time-stepping methods for parabolic problems and that our discussion in Section 4 of fractional step and splitting methods are relevant also in the present context. For a comprehensive account of the above methods for solving systems associated with finite difference methods, including historical remarks, see Varga [39].

We now turn to the development of iterative methods for systems mainly associated with the emergence of the finite element method. We begin by describing the conjugate gradient method by Hestenes and Stiefel (1952), and assume that A is SPD. Considering the iterative method

$$U^{n+1}=(I-\tau_n A)U^n+\tau_n F \quad \text{for } n\geqslant 0, \text{ with } U^0=0,$$

we find at once that, for any choice of the parameters τ_j, U^n belongs to the Krylov space $K_n(A;F)=\text{span}\{F, AF, \dots, A^{n-1}F\}$. The conjugate gradient method defines these parameters so that U^n is the best approximation of U in $K_n(A;F)$ with respect to the norm defined by $|U|=(AU,U)^{1/2}$, i.e., as the orthogonal projection of U onto $K_n(A;F)$ with respect to the inner product (AV, W). By our above discussion it follows that, with $\kappa=\kappa(A)$ the condition number of A,

$$|U^n - U| \leqslant (T_n(1/\rho))^{-1}|U| \leqslant 2\left(\frac{\kappa^{1/2}-1}{\kappa^{1/2}+1}\right)^n |U|. \tag{8.4}$$

The computation of U^n can be done by a two-term recurrence relation, for instance in the following form using the residuals $r^n = F - AU^n$ and the auxiliary vectors $q^n \in K_{n+1}(A;F)$, orthogonal to $K_n(A;F)$,

$$U^{n+1} = U^n + \frac{(r^n, q^n)}{(Aq^n, q^n)}q^n, \quad q^{n+1} = r^{n+1} - \frac{(Ar^{n+1}, q^n)}{(Aq^n, q^n)}q^n, \quad U^0 = 0, \; q^0 = F.$$

In the preconditioned conjugate gradient (PCG) method the conjugate gradient method is applied to Eq. (8.1) after multiplication by some easy to determine SPD approximation B of A^{-1} and using the inner product $(B^{-1}V, W)$; we note that BA is SPD with respect to this inner product. The error estimate (8.4) is now valid with $\kappa=\kappa(BA)$; B would be chosen so that this condition number is smaller than $\kappa(A)$. For the recursion formulas the only difference is that now $r^n = B(F - AU^n)$ and $q^0 = BF$. An early application of PCG to partial differential equations is Wachspress (1963) and it is systematically presented in Marchuk [24] and Axelsson and Barker [3], where reference to other work can be found.

One way of defining a preconditioner is by means of the multigrid method. This method is based on the observation that large components of the errors are associated with low frequencies in a spectral representation. The basic idea is then to work in a systematic way with a sequence of triangulations and reduce the low-frequency errors on coarse triangulations, which corresponds to small size problems, and higher frequency residual errors on finer triangulations by a smoothing operator, such as a step of the Jacobi method, which is relatively inexpensive.

One common situation is as follows: Assuming Ω is a plane polygonal domain we first perform a coarse triangulation of Ω. Each of the triangles is then divided into four similar triangles, and this process is repeated, which after a finite number M of steps leads to a fine triangulation with each of the original triangles devided into 4^M small triangles. Going from one level of fineness to the next the procedure may be described in three steps: (1) presmoothing on the finer triangulation, (2) correction on the coarser triangulation by solving a residual equation, (3) postsmoothing on the finer triangulation. This procedure is then used recursively between the levels of the refinement leading to, e.g., the V- or W-cycle algorithms. It turns out that under some assumptions the error reduction matrix R corresponding to one sweep of the algorithm satisfies $\|R\| \leqslant \rho < 1$, with ρ independent of M, i.e., of h, and that the number of operations is of order $O(N)$ where $N = O(h^{-2})$ is the dimension of the matrix associated with the finest triangulation.

The multigrid method was first introduced for finite difference methods in the 1960s by Fedorenko (1964) and Bahvalov (1966) and further developed and advocated by Brandt in the 1970s, see, e.g., Brandt (1977). For finite elements it has been intensely pursued by, e.g., Braess and Hackbusch, Bramble and Pasciak, Mandel, McCormick and Bank; for overviews with further references, see Hackbusch [19], Bramble [7].

A class of iterative methods that have attracted a lot of attention recently is the so-called domain decomposition methods. These assume that the domain Ω in which we want to solve our elliptic problem may be decomposed into subdomains Ω_j, $j=1,\dots,M$, which could overlap. The idea is to reduce the boundary value problem on Ω into problems on each of the Ω_j, which are then coupled by their values on the intersections. The problems on the Ω_j could be solved independently on parallel processors. This is particularly efficient when the individual problems may be solved very fast, e.g., by fast transform methods. Such a case is provided by the model problem (2.1) on the unit square which may be solved directly by using the discrete Fourier transform, defined by $\hat{F}_m=\sum_j F^j e^{-2\pi i\, mjh}$. In fact, we then have $(-\Delta_h U)\hat{}_m = 2\pi^2|m|^2\hat{U}_m$ and hence $\hat{U}_m=(2\pi^2|m|^2)^{-1}\hat{F}_m$ so that by the inverse discrete Fourier transform $U^j=\sum_m(2\pi^2|m|^2)^{-1}\hat{F}_m e^{2\pi i\, mj h}$. Using the fast Fourier transform both $\hat{F}_m$ and U^j may be calculated in $O(N\log N)$ operations.

The domain decomposition methods go back to the alternating procedure by Schwarz (1869), in which $\Omega=\Omega_1\cup\Omega_2$. Considering the Dirichlet problem (2.1) on Ω one defines a sequence $\{u^k\}$ starting with a given u^0 vanishing on $\partial\Omega$, by

$$-\Delta u^{2k+1}=f \quad \text{in } \Omega_1, \text{ with } u^{2k+1}=u^{2k} \text{ on } \partial\Omega_1\cap\Omega_2,\ u^{2k+1}=0 \text{ on } \partial\Omega_1\cap\partial\Omega,$$

$$-\Delta u^{2k+2}=f \quad \text{in } \Omega_2, \text{ with } u^{2k+2}=u^{2k+1} \text{ on } \partial\Omega_2\cap\Omega_1,\ u^{2k+2}=0 \text{ on } \partial\Omega_2\cap\partial\Omega$$

and this procedure can be combined with numerical solution by, e.g., finite elements. A major step in the analysis of this so-called multiplicative form of the Schwarz alternating procedure was taken by Lions (1988). A modification referred to as the additive form was first studied by Matsokin and Nepomnyashchikh (1985) and Dryja and Widlund (1987).

The following alternative approach may be pursued when Ω_1 and Ω_2 are disjoint but with a common interface $\partial\Omega_1\cap\partial\Omega_2$: If u_j denotes the solution in $\Omega_j, j=1,2$, transmission conditions $u_1=u_2$, $\partial u_1/\partial n=\partial u_2/\partial n$ have to be satisfied on the interface. One method is then to reduce the problem to an integral-type equation on the interface and use this as a basis of an iterative method. For a survey of domain decomposition techniques, see Smith, Bjørstad and Gropp [32].

References

Arnold, D.N., Wendland, W.L., 1983. On the asymptotic convergence of collocation methods. Math. Comp. 41, 349–381.

Aronson, D.G., 1963. The stability of finite difference approximations to second order linear parabolic differential equations. Duke Math. J. 30, 117–128.

Aubin, J.P., 1967. Behavior of the error of the approximate solution of boundary value problems for linear elliptic operators by Galerkin's and finite difference methods. Ann. Scuola Norm. Sup. Pisa Cl. Sci. 21, 599–637.

Aziz, A.K., Monk, P., 1989. Continuous finite elements in space and time for the heat equation. Math. Comp. 52, 255–274.

Babuška, I., 1971. Error bounds for finite element method. Numer. Math. 16, 322–333.

Babuška, I., 1973. The finite element method with Lagrangian multipliers. Numer. Math. 20, 179–192.

Babuška, I., 1976. The selfadaptive approach in the finite element method. In: Whiteman, J.R. (Ed.), The Mathematics of Finite Elements and Applications, II. Academic Press, New York, pp. 125–142.

Babuška, I., Dorr, M.R., 1981. Error estimates for the combined h and p-version of the finite element method. Numer. Math. 37, 257–277.

Babuška, I., Rheinboldt, W.C., 1978. Error estimates for adaptive finite element computations. SIAM J. Numer. Anal. 15, 736–754.

Babuška, I., Suri, M., 1990. *The p- and h-p versions of the finite element method, an overview* In: Canuto, C., Quarteroni, A., (Eds.), Spectral and High Order Methods for Partial Differential Equations. North-Holland, Amsterdam, pp. 5–26.

Babuška, I., Szabó, B.A., Katz, I.N., 1981. The p-version of the finite element method. SIAM J. Numer. Anal. 18, 512–545.

Bahvalov, N.S., 1959. Numerical solution of the Dirichlet problem for Laplace's equation. Vestnik Moskov. Univ. Ser. Math. Meh. Astronom. Fiz. Him. 3, 171–195.

Bahvalov, N.S., 1966. On the convergence of a relaxation method with natural constraints on the elliptic operator. USSR Comput. Math. and Math. Phys. 6, 101–135.

Baker, G.A., 1975. A finite element method for first order hyperbolic equations. Math. Comp. 20, 995–1006.

Baker, G.A., 1976. Error estimates for finite element methods for second order hyperbolic equations. SIAM J. Numer. Anal. 13, 564–576.

Baker, G.A., Bramble, J.H., 1979. Semidiscrete and single step fully discrete approximations for second order hyperbolic equations. RAIRO Anal. Numér. 13, 75–100.

Baker, G.A., Bramble, J.H., Thomée, V., 1977. Single step Galerkin approximations for parabolic problems. Math. Comp. 31, 818–847.

Bank, R.E., Rose, D.J., 1987. Some error estimates for the box method. SIAM J. Numer. Anal. 24, 777–787.

Berger, A.E., Scott, L.R., Strang, G., 1972. *Approximate boundary conditions in the finite element method.* Symposia Mathematica, Vol. 10. Academic Press, New York, 295–313.

Birkhoff, G., Schultz, M.H., Varga, R.S., 1968. *Piecewise Hermite interpolation in one and two variables with applications to partial differential equations.* Numer. Math., 232–256.

Blair, J., 1970. Approximate solution of elliptic and parabolic boundary value problems. Ph.D. Thesis. University of California, Berkeley.

Bramble, J.H., Hilbert, S.R., 1970. Estimation of linear functionals on Sobolev spaces with application to Fourier transforms and spline interpolation. SIAM J. Numer. Anal. 7, 113–124.

Bramble, J.H., Hubbard, B.E., 1962. On the formulation of finite difference analogues of the Dirichlet problem for Poisson's equation. Numer. Math. 4, 313–327.

Bramble, J. H., Hubbard, B.E., 1963. A theorem on error estimation for finite difference analogues of the Dirichlet problem for elliptic equations. Contrib. Differential Equations 2, 319–340.

Bramble, J.H., Dupont, T., Thomée, V., 1972. Projection methods for Dirichlet's problem in approximating polygonal domains with boundary value corrections. Math. Comp. 26, 869–879.

Bramble, J.H., Hubbard, B.E., Thomée, V., 1969. Convergence estimates for essentially positive type discrete Dirichlet problems. Math. Comp. 23, 695–710.

Bramble, J.H., Kellogg, R.B., Thomée, V., 1968. On the rate of convergence of some difference schemes for second order elliptic equations. Nordisk Tidskr. Inform. Behandling (BIT) 8, 154–173.

Bramble, J.H., Schatz, A.H., Thomée, V., Wahlbin, L.B., 1977. Some convergence estimates for semidiscrete Galerkin type approximations for parabolic equations. SIAM J. Numer. Anal. 14, 218–241.

Brandt, A., 1977. Multi-level adaptive solution to boundary-value problems. Math. Comp. 31, 333–390.

Brenner, P., 1966. The Cauchy problem for symmetric hyperbolic systems in L_p. Math. Scand. 19, 27–37.

Brenner, P., Thomée, V., 1970. Stability and convergence rates in L_p for certain difference schemes. Math. Scand. 27, 5–23.

Brenner, P., Crouzeix, M., Thomée, V., 1982. Single step methods for inhomogeneous linear differential equations in Banach space. RAIRO Anal. Numér. 16, 5–26.

Brezzi, F., 1974. On the existence, uniqueness and approximation of saddle-point problems arising from Lagrangian mulitpliers. RAIRO Anal. Numér. 8, 129–151.

Céa, J., 1964. Approximation variationelle des problèmes aux limites. Ann. Inst. Fourier. (Grenoble) 14, 345–444.

Clough, R.W., 1960. *The finite element method in plane stress analysis.* Proceedings of Second ASCE Conference on Electronic Computation, Vol. 8, Pittsburg, Pennsylvania pp. 345–378.

Collatz, L., 1933. Bemerkungen zur Fehlerabschätzung für das Differenzenverfahren bei partiellen Differentialgleichungen. Z. Angew. Math. Mech. 13, 56–57.

Collatz, L., 1942. Fehlerabschätzung für das Iterationsverfahren zur Auflösung linearer Gleichungssysteme. Z. Angew. Math. Mech. 22, 357–361.

Courant, R., 1943. Variational methods for the solution of problems of equilibrium and vibrations. Bull. Amer. Math. Soc. 49, 1–23.

Courant, R., Friedrichs, K.O., Lewy, H., 1928. *Über die partiellen Differenzengleichungen der Mathematischen Physik.* Math. Ann. 100, 32–74 (English translation, with commentaries by Lax, P.B., Widlund, O.B., Parter, S.V., in IBM J. Res. Develop. 11 (1967)).

Courant, R., Isaacson, E., Rees, M., 1952. On the solution of nonlinear hyperbolic differential equations by finite differences. Comm. Pure Appl. Math. 5, 243–255.

Crank, J., Nicolson, P., 1947. A practical method for numerical integration of solution of partial differential equations of heat-conduction type. Proc. Cambridge Philos. Soc. 43, 50–67.

Crouzeix, M., Larsson, S., Thomée, V., 1994. Resolvent estimates for elliptic finite element operators in one dimension. Math. Comp. 63, 121–140.

Crouzeix, M., Larsson, S., Piskarev, S., Thomée, V., 1993. The stability of rational approximations of analytic semigroups. BIT 33, 74–84.

Dahlquist, G., 1954. Convergence and stability for a hyperbolic difference equation with analytic initial-values. Math. Scand. 2, 91–102.

de Boor, D., Swartz, B., 1973. Collocation at Gaussian points. SIAM J. Numer. Anal. 10, 582–606.

Delfour, M., Hager, W., Trochu, F., 1981. Discontinuous Galerkin method for ordinary differential equations. Math. Comp. 36, 453–473.

Demjanovič, Ju.K., 1964. The net method for some problems in mathematical physics. Dokl. Akad. Nauk SSSR 159, 250–253.

Dendy, J.E., 1974. Two methods of Galerkin type achieving optimum L^2 rates of convergence for first order hyperbolics. SIAM J. Numer. Anal. 11, 637–653.

Dobrowolski, M., 1978. L^∞-convergence of linear finite element approximations to quasi-linear initial boundary value problems. RAIRO Anal. Numér. 12, 247–266.

Douglas Jr. J., 1956. On the relation between stability and convergence in the numerical solution of linear parabolic and hyperbolic differential equations. J. Soc. Indust. Appl. Math. 4, 20–37.

Douglas Jr. J., 1972. A superconvergence result for the approximate solution of the heat equation by a collocation method. In: Aziz, A.K. (Ed.), The Mathematical Foundations of the Finite Element Method with Applications to Partial Differential Equations. Academic Press, New York, pp. 475–490.

Douglas Jr. J., Dupont, T., 1970. Galerkin methods for parabolic equations. SIAM J. Numer. Anal. 7, 575–626.

Douglas Jr. J., Dupont, T., 1974a. Galerkin approximations for the two point boundary problem using continuous, piecewise polynomial spaces. Numer. Math. 22, 99–109.

Douglas Jr. J., Dupont, T., 1974b. Collocation Methods for Parabolic Equations in a Single Space Variable. Lecture Notes in Mathematics, Vol. 385. Springer, Berlin.

Douglas Jr. J., Russel, T.F., 1982. Numerical methods for convection-dominated diffusion problems based on combining the method of characteristics with finite element or finite difference procedures. SIAM J. Numer. Anal. 19, 871–885.

Dryja, M., Widlund, O.B., 1987. An additive variant of the Schwarz alternating method for the case of many subregions. Technical Report, Department of Computer Science, Courant Institute of Mathematical Sciences New York University.

Dupont, T., 1973a. L^2-estimates for Galerkin methods for second order hyperbolic equations. SIAM J. Numer. Anal. 10, 880–889.

Dupont, T., 1973b. Galerkin methods for first order hyperbolics: an example. SIAM J. Numer. Anal. 10, 890–899.

Dupont, T., 1974. L_2 error estimates for projection methods for parabolic equations in approximating domains. In: C. de Boor, Mathematical Aspects of Finite Elements in Partial Differential Equations. Academic Press, New York, pp. 313-352.

Eriksson, K., 1994. An adaptive finite element method with efficient maximum norm error control for elliptic problems. Math. Models Methods Appl. Sci. 4, 313–329.

Eriksson, K., Johnson, C., 1991. Adaptive finite element methods for parabolic problems. I: A linear model problem. SIAM J. Numer. Anal. 28, 43–77.

Eriksson, K., Johnson, C., Thomée, V., 1985. A discontinuous in time Galerkin method for parabolic type problems. RAIRO Math.Anal. 19, 611–643.

Faedo, S., 1949. Un nuovo metodo de l'analisi esistenziale e quantitative dei problemi di propagazione. Ann. Scuola Norm. Sup Pisa Cl. Sci. 1 (3), 1–40.

Fedorenko, R., 1964. The speed of convergence of one iterative process. USSR Comput. Math. Math. Phys. 4, 227–235.

Feng, K., 1965. Finite difference schemes based on variational principles. Appl. Math. Comput. Math. 2, 238–262.

Fox, L., 1948. A short account of relaxation methods. Quart. J. Mech. Appl. Math. 1, 253–280.

Frankel, S.P., 1950. Convergence rates of iterative treatments of partial differential equations. Math. Tables Aids Comput. 4, 65–75.

Friedrichs, K.O., 1954. Symmetric hyperbolic linear differential equations. Comm. Pure Appl. Math 7, 345–392.

Friedrichs, K.O., 1962. A finite difference scheme for the Neumann and Dirichlet problem. Report No NYO-9760, AEC Comp. and Appl. Math. Center, Courant Inst., Math. Sci., New York Univ.

Friedrichs, K.O., Keller, H.B., 1966. A finite difference scheme for generalized Neumann problems. Numerical Solution of Partial Differential Equations. Academic Press, New York. pp. 1-19.

Friedrichs, K.O., Lewy, H., 1926. Über die Eindeutigkeit und das Abhängigkeitsgebiet der Lösungen beim Anfangswertproblem linearer hyperbolischer Differentialgleichungen. Math. Ann. 98, 192–204.

Fujii, H., 1973. Some remarks on finite element analysis of time-dependent field problems. Theory and Practice in Finite Element Structural Analysis. University of Tokyo, pp. 91–106.

Galerkin, B.G., 1915. Rods and plates. Series occurring in various questions concerning the elastic equilibrium of rods and plates. Vestnik Insz. 19, 897–908.

Gauss, C.F., 1823. Brief an Gerling. Werke 9, 278–281.

Geiringer, H., 1949. On the solution of systems of linear equations by certain iterative methods. Reissner Anniversary Volume. J.W. Edwards, Ann Arbor, Mi., pp. 365–393.

Gerschgorin, S., 1930. Fehlerabschätzung für das Differenzenverfahren zur Lösung partieller Differentialgleichungen. Z. Angew. Math. Mech. 10, 373–382.

Godunov, S.K., Ryabenkii, V.S., 1963. Special criteria of stability of boundary-value problems for non-selfadjoint difference equations. Uspekhi Mat. Nauk 18, 3–14.

Gustafsson, B., Kreiss, H.O., Sundström, A., 1972. Stability theory for difference approximations of mixed initial boundary value problems. Math. Comp. 26, 649–686.

Haverkamp, R., 1984. Eine Aussage zur L_∞-Stabilität und zur genauen Konvergenzordnung der H_0^1-Projektionen. Numer. Math. 44, 393–405.

Hedstrom, G.W., 1968. The rate of convergence of some difference schemes. SIAM J. Numer. Anal. 5, 363–406.

Helfrich, H.-P., 1974. Fehlerabschätzungen für das Galerkinverfahren zur Lösung von Evolutions-gleichungen. Manuscripta Math. 13, 219–235.

Hestenes, M., Stiefel, E., 1952. Methods of conjugate gradients for solving linear systems. J. Res. Nat. Bur. Standards 49, 409–436.

Hsiao, G., Wendland, W., 1981. The Aubin-Nitsche lemma for integral equations. J. Integral Equations 3, 299–315.

Hughes, T.J.R., Brooks, A., 1979. A multidimensional upwind scheme with no crosswind diffusion. In: Hughes, T.J.R. (Ed.), Finite Element Methods for Convection Dominated Flows, Vol. 34. American Society of Mechanical Engineers, New York, pp. 19–35.

Hughes, T.J.R., Franca, L.P., Mallet, M., 1987. A new finite element formulation for computational fluid dynamics: VI. Convergence analysis of the generalized SUPG formulation for linear time dependent multi-dimensional advective-diffusive systems. Comput. Meth. Appl. Mech. Eng. 63, 97–112.

Jamet, P., 1978. Galerkin-type approximations which are discontinuous in time for parabolic equations in a variable domain. SIAM J. Numer. Anal. 15, 912–928.

John, F., 1952. On integration of parabolic equations by difference methods. Comm. Pure Appl. Math. 5, 155–211.

Johnson, C., Pitkäranta, J., 1986. An analysis of the discontinuous Galerkin method for a scalar hyperbolic equation. Math. Comp. 46, 1–26.

Johnson, C., Szepessy, A., 1987. On the convergence of a finite element method for a nonlinear hyperbolic conservation law. Math. Comp. 49, 427–444.

Johnson, C., Thomée, V., 1981. Error estimates for some mixed finite element methods for parabolic type problems. RAIRO Modél. Math. Anal. Numér. 15, 41–78.

Johnson, C., Nävert, U., Pitkäranta, J., 1984. Finite element methods for linear hyperbolic problems. Comput. Methods Appl. Mech. Eng. 45, 285–312.

Juncosa, M.L., Young, D.M., 1957. On the Crank-Nicolson procedure for solving parabolic partial differential equations. Proc. Cambridge Philos. Soc. 53, 448–461.

Kantorovich, L., 1948. Functional analysis and applied mathematics. Uspehi Mat. Nauk 3, 89–185.

Kato, T., 1960. Estimation of iterated matrices, with application to the von Neumann condition. Numer. Math. 2, 22–29.

Kreiss, H.O., 1959a. Über die Lösung des Cauchyproblems für lineare partielle Differenzialgleichungen mit Hilfe von Differenzengleichungen. Acta Math. 101, 179–199.

Kreiss, H.O., 1959b. Über Matrizen die beschränkte Halbgruppen erzeugen, Math. Scand. 7, 72–80.

Kreiss, H.O., 1962. Über die Stabilitätsdefinition für Differenzengleichungen die partielle Differentialgleichungen approximieren. Nordisk Tidskr. Inform.-Behandling 2, 153–181.

Kreiss, H.O., 1964. On difference approximations of dissipative type for hyperbolic differential equations. Comm. Pure Appl. Math. 17, 335–353.

Kreiss, H.O., 1968. Stability theory for difference approximations of mixed initial boundary value problems. Math. Comp. 22, 703–714.

Kreiss, H.O., Scherer, G., 1974. Finite Element and Finite Difference Methods for Hyperbolic Partial Differential Equations. Academic Press, New York, 195–212.

Kreiss, H.O., Thomée, V., Widlund, O.B., 1970. Smoothing of initial data and rates of convergence for parabolic difference equations. Comm. Pure Appl. Math. 23, 241–259.

Laasonen, P., 1949. Über eine Methode zur Lösung der Wärmeleitungsgleichung. Acta Math. 81, 309–317.

Laasonen, P., 1958. On the solution of Poisson's difference equation. J. Assoc. Comput. Mach. 5, 370–382.

Lax, P.D., 1961. On the stability of difference approximations to solutions of hyperbolic differential equations. Comm. Pure Appl. Math. 14, 497–520.

Lax, P.D., Nirenberg, L., 1966. On stability for difference schemes: a sharp form of Gårding's inequality. Comm. Pure Appl. Math. 19, 473–492.

Lax, P.D., Richtmyer, R.D., 1959. Survey of the stability of linear finite difference equations. Comm. Pure Appl. Math. 9, 473–492.

Lax, P.D., Wendroff, B., 1960. Systems of conservation laws. Comm. Pure Appl. Math. 13, 217–237.

Lax, P.D., Wendroff, B., 1964. Difference schemes for hyperbolic equations with high order of accuracy. Comm. Pure Appl. Math. 17, 381–398.

Lees, M., 1960. A priori estimates for the solution of difference approximations to parabolic differential equations. Duke Math. J. 27, 297–311.

Le Roux, M.N., 1979. Semidiscretization in time for parabolic problems. Math. Comp. 33, 919–931.

Lesaint, P., Raviart, P.-A., 1974. On the finite element method for solving the neutron transport equation. In: de Boor, C. (Ed.), Mathematical Aspects of Finite Elements in Partial Differential Equations. Academic Press, New York, pp. 89–123.

Liebmann, H., 1918. Die angenäherte Ermittlung harmonischer Funktionen und konformer Abbildung (nach Ideen von Boltzmann und Jacobi). Sitzungsberichte der Beyer. Akad. Wiss., Math.-Phys. Kl. 47, 385–416.

Lions, P.-L., 1988. On the Schwarz alternating method I. In: Glowinski, R., Golub, G.H., Meurant, G.A., Périaux, J. (Eds.), First International Symposium on Domain Decomosition Methods. SIAM, Philadelphia, pp. 1–42.

Lippold, G., 1991. Error estimates and step-size control for the approximate solution of a first order evolution equation. RAIRO Modél. Math. Anal. Numér. 25, 111–128.

Löfström, J., 1970. Besov spaces in the theory of approximation. Ann. Mat. Pura Appl. 85, 93–184.

Luskin, M., Rannacher, R., 1982. On the smoothing property of the Crank-Nicolson scheme. Appl. Anal. 14, 117–135.

Mac Neal, R.H., 1953. An asymmetric finite difference network. Quart. Appl. Math. 1, 295–310.

Matsokin, A.M., Nepomnyashchikh, S.V., 1985. A Schwarz alternating method in a subspace. Izv. VUZ Mat. 29 (10), 61–66.

Motzkin, T.S., Wasow, W., 1953. On the approximation of linear elliptic differential equations by difference equations with positive coefficients. J. Math. Phys. 31, 253–259.

Natterer, F., 1975. Über die punktweise Konvergenz finiter Elemente. Numer. Math. 25, 67–77.

Nedelec, J.C., Planchard, J., 1973. Une méthode variationelle d'élements finis pour la résolution numérique d'un problème extérieur dans R^3. RAIRO, Anal. Numér. 7, 105–129.

Newmark, N.M., 1959. A method of computation for structural dynamics. Proc. ASCE, J. Eng. Mech. (EM3) 85, 67–94.

Nitsche, J., 1968. Ein Kriterium für die Quasioptimalität des Ritzschen Verfahrens. Numer. Math. 11, 346–348.

Nitsche, J., 1971. Über ein Variationsprinzip zur Lösung von Dirichlet-problemen bei Verwendung von Teilräumen, die keinen Randbedingungen unterworfen sind. Abh. Math. Semin. Univ. Hamb. 36, 9–15.

Nitsche, J.A., 1975. L_∞-convergence of finite element approximation. Second Conference on Finite Elements. Rennes, France.

Nitsche, J.A., 1979. L_∞-convergence of finite element Galerkin approximations for parabolic problems. RAIRO Anal. Numér. 13, 31–54.

Nitsche, J.A., Wheeler, M.F., 1981–82. L_∞-boundedness of the finite element Galerkin operator for parabolic problems. Numer. Funct. Anal. Optim. 4, 325–353.

Nyström, E., 1930. Über die praktische Auflösung von Integralgleichungen mit Anwendungen auf Randwertaufgaben. Acta Math. 54, 185–204.

O'Brien, G.G., Hyman, M.A., Kaplan, S., 1951. A study of the numerical solution of partial differential equations. J. Math. Phys. 29, 223–251.

Oganesjan, L.A., 1963. Numerical Solution of Plates. Lensovnarchoz. CBTI, Sbornik.

Oganesjan, L.A., 1966. Convergence of difference schemes in case of improved approximation of the boundary. Ž. Vyčisl. Mat. Mat. Fiz. 6, 1029–1042.

Oganesjan, L.A., Ruhovec, L.A., 1969. An investigation of the rate of convergence of variational-difference schemes for second order elliptic equations in a two-dimensional region with smooth boundary. Zh. Vychisl. Mat. Mat. Fiz. 9, 1102–1120.

Osher, S., 1972. Stability of parabolic difference approximations to certain mixed initial boundary value problems. Math. Comp. 26, 13–39.

Ostrowski, A.M., 1954. On the linear iteration procedures for symmetric matrices. Rend. Mat. Appl. 14, 140–163.

Palencia, C., 1992. A stability result for sectorial operators in Banach spaces. SIAM J. Numer. Anal. 30, 1373–1384.

Parlett, B., 1966. Accuracy and dissipation in difference schemes. Comm. Pure Appl. Math. 19, 111–123.

Peaceman, D.W., Rachford Jr. H.H., 1955. The numerical solution of parabolic and elliptic differential equations. J. Soc. Indust. Appl. Math. 3, 28–41.

Peetre, J., Thomée, V., 1967. On the rate of convergence for discrete initial-value problems. Math. Scand. 21, 159–176.

Pirroneau, O., 1982. On the transport-diffusion algorithm and its applications to the Navier-Stokes equations. Numer. Math. 38, 309–332.

Rannacher, R., 1984. Finite element solution of diffusion problems with irregular data. Numer. Math. 43, 309–327.

Rannacher, R., 1991. L^∞-stability and asymptotic error expansion for parabolic finite element equations. Bonner Math. Schriften 228.

Raviart, P., Thomas, J., 1977. A mixed finite element method for 2nd order elliptic problems. Proceedings of the Symposium on the Mathematical Aspects of the Finite Element Method, Vol. 606, Rome, December, 1975. Springer Lecture Notes in Mathematics, Springer, Berlin, pp. 292–315.

Lord Rayleigh, Theory of Sound (1894, 1896), Vols. I, II., Macmillan, London.

Ritz, W., 1908. Über eine neue Methode zur Lösing gewisser Variationsprobleme der mathematischen Physik. J. Reine Angew. Math. 135, 1–61.

Saranen, J., 1988. The convergence of even degree spline collocation solution for potential problems in smooth domains of the plane. Numer. Math. 53, 299–314.

Samarskii, A.A., 1961. A priori estimates for the solution of the difference analogue of a parabolic differential equation. Zh. Vychisl. Mat. Mat. Fiz, 1, 441–460 (in Russian) (English translation in U.S.S.R. Comput. Math. Math. Phys. 1 487–512.)

Sammon, P., 1983a. Convergence estimates for semidiscrete parabolic equation approximations. SIAM J. Numer. Anal. 19, 68–92.

Sammon, P., 1983b. Fully discrete approximation methods for parabolic problems with nonsmooth initial data. SIAM J. Numer. Anal. 20, 437–470.

Saulev, V.K., 1957. On a class of elliptic equations solvable by the method of finite-differences. Vycisl. Mat. 1, 81–86.

Schatz, A.H., Wahlbin, L.B., 1982. On the quasi-optimality in L_∞ of the $H^\circ 1$-projection into finite element spaces. Math. Comp. 38, 1–22.

Schatz, A.H., Thomée, V., Wahlbin, L.B., 1980. Maximum-norm stability and error estimates in parabolic finite element equations. Comm. Pure Appl. Math. 33, 265–304.

Schatz, A.H., Thomée, V., Wahlbin, L.B., 1998. Stability, analyticity, and almost best approximation in maximum-norm for parabolic finite element equations. Comm. Pure Appl. Math. 51, 1349–1385.

Schwarz, H.A., 1869. Über einige Abbildungsaufgaben. J. Reine Angew. Math 70, 105–120.

Scott, L.R., 1975. Interpolated boundary conditions in the finite element method. SIAM J. Numer. Anal. 12, 404–427.

Scott, L.R., 1976. Optimal L^∞ estimates for the finite element method. Math. Comp. 30, 681–697.

Seidel, L., 1874. Über ein Verfahren, die Gleichungen, auf welche die Methode der kleinsten Quadrate führt, sowie lineäre Gleichungen überhaupt, durch successive Annäherung aufzulösen. Abh. Math.-Phys. Kl., Bayerische Akad. Wiss. München 11, 81–108.

Serdjukova, S.J., 1964. *The uniform stability with respect to the initial data of a sixpoint symmetrical scheme for the heat conduction equation.* Numerical Methods for the Solution of Differential Equations and Quadrature Formulae. Nauka, Moscow, pp. 212–216.

Shortley, G.H., Weller, R., 1938. The numerical solution of Laplace's equation. J. Appl. Phys. 9, 334–344.

Sloan, I.H., 1976. Improvement by iteration for compact operator equations. Math. Comp. 30, 758–764.

Sloan, I.H., 1988. A quadrature-based approach to improving the collocation method. Numer. Math. 54, 41–56.

Sloan, I.H., Thomée, V., 1985. Superconvergence of the Galerkin iterates for integral equations of the second kind. J. Integral Equations 9, 1–23.

Stein, P., Rosenberg, R.L., 1948. On the solution of linear simultaneous equations by iteration. J. London Math. Soc. 23, 111–118.

Strang, G., 1962. Trigonometric polynomials and difference methods of maximum accuracy. J. Math. Phys. 41, 147–154.

Strang, G., 1964. Unbalanced polynomials and difference methods for mixed problems. SIAM J. Numer. Anal. 2, 46–51.

Strang, G., 1965. Necessary and insufficient conditions for well-posed Cauchy problems. J. Differential Equations 2, 107–114.

Strang, G., 1967. On strong hyperbolicity. J. Math. Kyoto Univ. 6, 397–417.

Strang, G., 1972. Variational crimes in the finite element method. In: Aziz, A.K. (Ed.), The Mathematical Foundations of the Finite Element Method with Applications to Partial Differential Equations. Academic Press, New York, pp. 689–710.

Swartz, B., Wendroff, B., 1969. Generalized finite-difference schemes. Math. Comp. 23, 37–49.

Szepessy, A., 1989. Convergence of a shock-capturing streamline diffusion finite element method for a scalar conservation law in two space dimensions. Math. Comp. 53, 527–545.

Tikhonov, A.N., Samarskii, A.A., 1961. Homogeneous difference schemes. Zh. Vychisl. Mat. Mat. Fiz. 1, 5–63, (U.S.S.R. Comput. Math. Math. Phys. 1, 5–67).

Thomée, V., 1964. Elliptic difference equations and Dirichlet's problem. Contrib. Differential Equations 3, 301–324.

Thomée, V., 1973. Convergence estimates for semi-discrete Galerkin methods for initial-value problems. Numerische, insbesondere approximationstheoretische Behandlung von Funktionalgleichungen, Vol. 333, Springer Lecture Notes in Mathematics. Springer, Berlin, pp. 243–262.

Thomée, V., 1980. Negative norm estimates and superconvergence in Galerkin methods for parabolic problems. Math. Comp. 34, 93–113.

Thomée, V., Wahlbin, L.B., 1974. Convergence rates of parabolic difference schemes for non-smooth data. Math. Comp. 28, 1–13.

Thomée, V., Wahlbin, L.B., 1998. Stability and analyticity in maximum-norm for simplicial Lagrange finite element semidiscretizations of parabolic equations with Dirichlet boundary conditions. Preprint no. 1998-44, Department of Mathematics, Göteborg.

Thomée, V., Wendroff, B., 1974. Convergence estimates for Galerkin methods for variable coefficient initial value problems. SIAM J. Numer. Anal. 1, 1059–1068.

Thomée, V., Westergren, B., 1968. Elliptic difference equations and interior regularity. Numer. Math. 11, 196–210.

Thomée, V., Xu, J.C., Zhang, N.Y., 1989. Superconvergence of the gradient in piecewise linear finite-element approximation to a parabolic problem. SIAM J. Numer. Anal. 26, 553–573.

Turner, M.J., Clough, R.W., Martin, H.C., Topp, L., 1956. Stiffness and deflection analysis of complex structures. J. Aero. Sci. 23, 805–823.

Varah, J.M., 1970. Maximum norm stability of difference approximations to the mixed initial boundary value problem for the heat equation. Math. Comp. 24, 31–44.

Volkov, E.A., 1966. *Obtaining an error estimate for a numerical solution of the Dirichlet problem in terms of known quantities.* Z. Vyčisl. Mat. Mat. Fiz. 6 (4)(suppl)., 5–17.

Wahlbin, L.B., 1974. A dissipative Galerkin method applied to some quasilinear hyperbolic equations. RAIRO Anal. Numér. 8, 109–117.

Wasow, W., 1952. On the truncation error in the solution of Laplace's equation by finite differences. J. Res. Nat. Bur. Standards 48, 345–348.

Wachspress, E., 1963. Extended application of alternating direction implicit iteration model problem theory. SIAM J. Appl. Math. 11, 994–1016.

Wendland, W., 1968. Die Behandlung von Randwertaufgaben im R_3 mit Hilfe von Einfach- und Doppelschichtpotentialen. Numer. Math. 11, 380–404.

Wendroff, B., 1960. On central difference equations for hyperbolic systems. J. Soc. Indust. Appl. Math. 8, 549–555.

Wendroff, B., 1968. Well-posed problems and stable difference operators. SIAM J. Numer. Anal. 5, 71–82.

Wheeler, M.F., 1973. A priori L_2 error estimates for Galerkin approximations to parabolic partial differential equations. SIAM J. Numer. Anal. 10, 723–759.

Widlund, O.B., 1966. Stability of parabolic difference schemes in the maximum-norm. Numer. Math. 8, 186–202.

Winther, R., 1981. A stable finite element method for initial-boundary value problems for first order hyperbolic equations. Math. Comp. 36, 65–86.

Yamaguti, M., 1967. Some remarks on the Lax-Wendroff finite-difference scheme for nonsymmetric hyperbolic systems. Math. Comp. 21, 611–619.

Young, Jr. D.M., Iterative methods for solving partial differential equations of elliptic type Thesis, Harvard University, 1950, reprinted in `http://www-sccm.stanford.edu/pub/sccm/`.

Young, D.M., 1954. Iterative methods for soliving partial difference equations of ellipic type. Trans. Amer. Math. Soc. 76, 92–111.

Zienkiewicz, O.C., Zhu, J.Z., 1992. The superconvergent patch recovery and a posteriori error estimates. Internat. J. Numer. Methods Eng. 33, 1331–1382.

Zlámal, M., 1968. On the finite element method. Numer. Math. 12, 394–409.

Survey articles and books

[1] J.H. Argyris, Energy Theorems and Structural Analysis. A Generalised Discourse with Applications on Energy Principles of Structural Analysis Including the Effects of Temperature and Non-linear Stress-Strain Relations, Butterworths, London, 1960.

[2] K.E. Atkinson, The Numerical Solution of Integral Equations of the Second Kind, Cambridge University Press, Cambridge, UK, 1997.

[3] O. Axelsson, V.A. Barker, Finite Element Solution of Boundary Value Problems, Academic Press, 1984.

[4] I. Babuška, Courant element: before and after, in: M. Křížek, P. Neittaanmäki, R. Stenberg (Eds.), Finite Element Methods: Fifty Years of the Courant Element, Decker, New York, 1994, pp. 37–51.

[5] I. Babuška, A.K. Aziz, Survey lectures on the mathematical foundation of the finite element method, in: A.K. Aziz (Ed.), The Mathematical Foundations of the Finite Element Method with Applications to Partial Differential Equations, Academic Press, New York, 1972, pp. 5–359.

[6] J.P. Boyd, Chebyshev & Fourier Spectral Methods, Lecture Notes in Engineering, Vol. 49, Springer, Berlin, 1989.

[7] J.H. Bramble, Multigrid Methods, Pitman Research Notes in Mathematics, Longman, London, 1993.

[8] S.C. Brenner, L.R. Scott, The Mathematical Theory of Finite Element Methods, Springer, New York, 1994.

[9] P. Brenner, V. Thomée, L.B. Wahlbin, in: Besov Spaces and Applications to Difference Methods for Initial Value Problems, Springer Lecture Notes in Mathematics, Vol. 434, Springer, Berlin, 1975.

[10] C. Canuto, Y.M. Hussaini, A. Quarteroni, T.A. Zang, Spectral Methods in Fluid Dynamics, Springer, Berlin, 1988.

[11] P.G. Ciarlet, The Finite Element Method for Elliptic Problems, North-Holland, Amsterdam, 1978.

[12] L. Collatz, Numerische Behandlung von Differentialgbeichungen, Springer, Berlin, 1955.

[13] K. Eriksson, D. Estep, P. Hansbo, C. Johnson, Introduction to adaptive methods for differential equations, Acta Numer. 4 (1995) 105–158.

[14] G.E. Forsythe, W.R. Wasow, Finite Difference Methods for Partial Differential Equations, Wiley, New York, 1960.

[15] A. Friedman, Partial Differential Equations of Parabolic Type, Prentice-Hall, Englewood Cliffs, NJ, 1964.

[16] E. Godlewski, P.-A. Raviart, Hyperbolic Systems of Conservation Laws, Ellipse, Paris, 1991.

[17] D. Gottlieb, S.A. Orszag, Numerical Analysis of Spectral Methods: Theory and Applications, SIAM, Philadelphia, 1977.

[18] B. Gustafsson, H.-O. Kreiss, J. Oliger, Time Dependent Problems and Difference Methods, Wiley, New York, 1995.

[19] W. Hackbusch, in: Multi-Grid Methods and Applications, Springer Series in Computational Mathematics, Vol. 4, Springer, New York, 1985.
[20] B. Heinrich, Finite Difference Methods on Irregular Networks, Akademie-Verlag, Berlin, 1987.
[21] L.V. Kantorovich, V.I. Krylov, Approximate Methods in Higher Analysis, Interscience, New York, 1958.
[22] M. Križek, P. Neittaanmäki, On superconvergence techniques, Acta Appl. Math. 9 (1987) 175–198.
[23] R.J. Le Veque, Numerical Methods for Conservation Laws, Birkhäuser, Basel, 1990.
[24] G.I. Marchuk, Methods of Numerical Mathematics, Springer, New York, 1975.
[25] G.I. Marchuk, Splitting and alternating direction methods, in: P.G. Ciarlet, J.L. Lions (Eds.), Handbook of Numerical Analysis, Vol. I. Finite Difference Methods (Part 1), North-Holland, Amsterdam, 1990, pp. 197–462.
[26] K.W. Morton, Numerical Solution of Convection-Diffusion Problems, Chapman & Hall, London, 1996.
[27] P.A. Raviart, J.M. Thomas, Introduction à l'analyse numérique des équations aux dérivées partielles, Masson, Paris, 1983.
[28] R.D. Richtmyer, K.W. Morton, Difference Methods for Initial-Value Problems, Interscience, New York, 1967.
[29] V.S. Ryabenkii, A.F. Filippov, Über die Stabilität von Differenzengleichungen, Deutscher Verlag der Wissenschaften, Berlin, 1960.
[30] A.A. Samarskii, R.D. Lazarov, V.L. Makarov, Difference Schemes for Differential Equations with Generalized Solutions, Vysshaya Shkola, Moscow, 1987 (in Russian).
[31] I.H. Sloan, Error analysis of boundary integral methods, Acta Numer. 1 (1992) 287–339.
[32] B.F. Smith, P.E. Bjørstad, W.D. Gropp, Domain Decomposition: Parallel Multilevel Methods for Elliptic Partial Differential Equations, Cambridge University Press, Cambridge, 1996,.
[33] R.V. Southwell, Relaxation Methods in Theoretical Physics, I, II, Clarendon Press, Oxford, 1946, 1956.
[34] G. Strang, G.J. Fix, An Analysis of the Finite Element Method, Prentice-Hall, Englewood Cliffs, NJ, 1973.
[35] J.C. Strikwerda, Finite Difference Schemes and Partial Differential Equations, Chapman & Hall, New York, 1990.
[36] V. Thomée, Stability theory for partial difference operators, SIAM Rev. 11 (1969) 152–195.
[37] V. Thomée, Finite difference methods for linear parabolic equations, in: P.G. Ciarlet, J.L. Lions (Eds.), Handbook of Numerical Analysis, Vol. I. Finite Difference Methods (Part 1), North-Holland, Amsterdam, 1990, pp. 5–196.
[38] V. Thomée, Galerkin Finite Element Methods for Parabolic Problems, Springer Series in Computational Mathematics, Vol. 25, Springer, Berlin, 1997.
[39] R.S. Varga, Matrix Iterative Analysis, Prentice-Hall, Englewood Cliffs, NJ, 1962.
[40] R. Verfürth, A Review of A Posteriori Error Estimation and Adaptive Mesh-Refinement Techniques, Wiley, New York, Teubner, Stuttgart, 1996.
[41] L.B. Wahlbin, Superconvergence in Galerkin Finite Element Methods, Lecture Notes in Mathematics, Vol. 1605, Springer, Berlin, 1995.
[42] W.L. Wendland, Boundary element methods for elliptic problems, in: A.H. Schatz, V. Thomée, W.L. Wendland (Eds.), Mathematical Theory of Finite and Boundary Element Methods, Birkhäuser, Basel, 1990.
[43] O.C. Zienkiewicz, The Finite Element Method in Engineering Science, McGraw-Hill, London, 1977.

ELSEVIER

Journal of Computational and Applied Mathematics 128 (2001) 55–82

JOURNAL OF COMPUTATIONAL AND APPLIED MATHEMATICS

www.elsevier.nl/locate/cam

Orthogonal spline collocation methods for partial differential equations ☆

B. Bialecki, G. Fairweather *

Department of Mathematical and Computer Sciences, Colorado School of Mines, Golden, CO 80401-1887, USA

Received 4 February 2000; received in revised form 24 March 2000

Abstract

This paper provides an overview of the formulation, analysis and implementation of orthogonal spline collocation (OSC), also known as spline collocation at Gauss points, for the numerical solution of partial differential equations in two space variables. Advances in the OSC theory for elliptic boundary value problems are discussed, and direct and iterative methods for the solution of the OSC equations examined. The use of OSC methods in the solution of initial–boundary value problems for parabolic, hyperbolic and Schrödinger-type systems is described, with emphasis on alternating direction implicit methods. The OSC solution of parabolic and hyperbolic partial integro-differential equations is also mentioned. Finally, recent applications of a second spline collocation method, modified spline collocation, are outlined.

1. Introduction

1.1. Preliminaries

In [70], Fairweather and Meade provided a comprehensive survey of spline collocation methods for the numerical solution of differential equations through early 1989. The emphasis in that paper is on various collocation methods, primarily smoothest spline collocation, modified spline collocation and orthogonal spline collocation (OSC) methods, for boundary value problems (BVPs) for ordinary differential equations (ODEs). Over the past decade, considerable advances have been made in the formulation, analysis and application of spline collocation methods, especially OSC for partial differential equations (PDEs). In this paper, we review applications of OSC (also called spline collocation at Gauss points) to elliptic, parabolic, hyperbolic and Schrödinger-type PDEs, as well

☆ This work was supported in part by National Science Foundation grant DMS-9805827.

* Corresponding author.

E-mail address: gfairwea@mines.edu (G. Fairweather).

0377-0427/01/$ - see front matter

PII: S0377-0427(00)00509-4

as to parabolic and hyperbolic partial integro-differential equations. The emphasis throughout is on problems in two space variables.

A brief outline of the paper is as follows. In Section 2, OSC for linear two-point BVPs for ODEs is described. OSC for such problems was first analyzed in the seminal paper of deBoor and Swartz [32], which laid the foundation for the formulation and analysis of OSC methods for a wide variety of problems and the development of software packages for their solution; see [70]. In Section 2, we also describe the continuous-time OSC method and the discrete-time Crank–Nicolson OSC method for linear parabolic initial–boundary value problems (IBVPs) in one space variable and outline applications of OSC to other equations in one space variable such as Schrödinger-type equations. OSC methods for linear and nonlinear parabolic problems in one space variable were first formulated and analyzed in [59,61–63], and Cerutti and Parter [43] tied together the results of [32] and those of Douglas and Dupont. Following the approach of Douglas and Dupont, Houstis [83] considered OSC for nonlinear second-order hyperbolic problems.

Section 3 is devoted to elliptic BVPs. The development of the convergence theory of OSC for various types of elliptic problems is examined, followed by an overview of direct and iterative methods for solving the OSC equations. Several of these methods reduce to the OSC solution of linear BVPs for ODEs. The OSC solution of biharmonic problems is also described and the section closes with a discussion of recent work on domain decomposition OSC methods.

Section 4 concerns IBVPs problems for parabolic and hyperbolic equations and Schrödinger-type systems with emphasis on the formulation and analysis of alternating direction implicit (ADI) methods. The ADI methods considered are based on the two space variable Crank–Nicolson OSC method, and reduce to independent sets of one space variable problems of the type considered in Section 2. Spline collocation methods for certain types of partial integro-differential equations are also discussed. In Section 5, we give a brief synopsis of modified spline collocation methods for elliptic BVPs.

In the remainder of this section, we introduce notation that is used throughout the paper.

1.2. Notation

We denote the unit interval $(0,1)$ by I and the unit square $I \times I$ by Ω. Let π be a partition given by

$$\pi\colon\ 0 = x^{(0)} < x^{(1)} < \cdots < x^{(N)} = 1.$$

Let

$$\mathcal{M}_r(\pi) = \{v \in C^1(\bar{I})\colon\ v|_{[x^{(i-1)},x^{(i)}]} \in P_r, i = 1,2,\ldots,N\},$$

where P_r denotes the set of all polynomials of degree $\leqslant r$. Also let

$$\mathcal{M}_r^0(\pi) = \{v \in \mathcal{M}_r(\pi)\colon\ v(0) = v(1) = 0\}.$$

Note that

$$\dim \mathcal{M}_r^0(\pi) \equiv M = N(r-1), \quad \dim \mathcal{M}_r(\pi) = M + 2.$$

When $r = 3$, the spaces $\mathcal{M}_r(\pi)$ $(\mathcal{M}_r^0(\pi))$ and $\mathcal{M}_r(\pi) \otimes \mathcal{M}_r(\pi)$ $(\mathcal{M}_r^0(\pi) \otimes \mathcal{M}_r^0(\pi))$ are commonly known as piecewise Hermite cubics and piecewise Hermite bicubics, respectively.

Let $\{\sigma_k\}_{k=1}^{r-1}$ be the nodes of the $(r-1)$-point Gauss–Legendre quadrature rule on I, and let the Gauss points in I be defined by

$$\xi_{(i-1)(r-1)+k} = x^{(i-1)} + h_i\sigma_k, \quad k = 1,2,\ldots,r-1, \quad i = 1,\ldots,N, \tag{1}$$

where $h_i = x^{(i)} - x^{(i-1)}$. We set $h = \max_i h_i$.

By optimal order estimates, we mean bounds on the error which are $O(h^{r+1-j})$ in the H^j norm, $j = 0,1,2$, and $O(h^{r+1})$ in the L^∞ norm.

2. Problems in one space variable

2.1. Boundary value problems for ordinary differential equations

In this section, we briefly discuss OSC for the two-point BVP

$$\mathscr{L}u \equiv -a(x)u'' + b(x)u' + c(x)u = f(x), \quad x \in I, \tag{2}$$

$$\mathscr{B}_0u(0) \equiv \alpha_0u(0) + \beta_0u'(0) = g_0, \qquad \mathscr{B}_1u(1) \equiv \alpha_1u(1) + \beta_1u'(1) = g_1, \tag{3}$$

where α_i, β_i, and g_i, $i = 0,1$, are given constants. The OSC method for solving (2)–(3) consists in finding $u_h \in \mathscr{M}_r(\pi)$, $r \geqslant 3$, such that

$$\mathscr{L}u_h(\xi_j) = f(\xi_j), \quad j = 1,2,\ldots,M, \qquad \mathscr{B}_0u_h(0) = g_0, \quad \mathscr{B}_1u_h(1) = g_1, \tag{4}$$

where $\{\xi_j\}_{j=1}^M$ are the *collocation points* given by (1). If $\{\phi_j\}_{j=1}^{M+2}$ is a basis for $\mathscr{M}_r(\pi)$, we may write

$$u_h(x) = \sum_{j=1}^{M+2} u_j\phi_j(x)$$

and hence the collocation equations (4) reduce to a system of linear algebraic equations for the coefficients $\{u_j\}_{j=1}^{M+2}$. If the basis is of Hermite type, B-splines or monomial basis functions, the coefficient matrices are *almost block diagonal* (ABD) [31]. For example, if $\boldsymbol{u} = [u_1,u_2,\ldots,u_{M+2}]^{\mathrm{T}}$, and $\boldsymbol{f} = [g_0, f(\xi_1), f(\xi_2),\ldots, f(\xi_{M-1}), f(\xi_M), g_1]^{\mathrm{T}}$, then, for Hermite-type or B-spline bases with standard orderings, the collocation equations have the form $A\boldsymbol{u} = \boldsymbol{f}$, where A has the ABD structure

$$\begin{bmatrix} D_0 & & & & & \\ W_{11} & W_{12} & W_{13} & & & \\ & & W_{21} & W_{22} & W_{23} & \\ & & & & \ddots & \\ & & & & W_{N1} & W_{N2} & W_{N3} \\ & & & & & & D_1 \end{bmatrix}. \tag{5}$$

The 1×2 matrices $D_0 = [\alpha_0 \;\; \beta_0]$, $D_1 = [\alpha_1 \;\; \beta_1]$ arise from the boundary conditions, and the matrices $W_{i1} \in \mathbb{R}^{(r-1)\times 2}$, $W_{i2} \in \mathbb{R}^{(r-1)\times(r-3)}$, $W_{i3} \in \mathbb{R}^{(r-1)\times 2}$ come from the collocation equations on the ith subinterval. Such systems are commonly solved using the package *colrow* [56,57]. The packages *abdpack* and *abbpack* [108–110] are designed to solve the systems arising when monomial bases are employed, in which case the ABD structure is quite different from that in (5). The solution procedures implemented in these packages are all variants of Gaussian elimination with partial pivoting. A review of methods for solving ABD systems is given in [3].

2.2. Parabolic problems in one space variable

Consider the parabolic IBVP

$$
\begin{aligned}
&u_t + \mathscr{L}u = f(x,t), \quad (x,t) \in I \times (0,T],\\
&\mathscr{B}_0 u(0,t) \equiv \alpha_0 u(0,t) + \beta_0 u_x(0,t) = g_0(t), \quad t \in (0,T],\\
&\mathscr{B}_1 u(1,t) \equiv \alpha_1 u(1,t) + \beta_1 u_x(1,t) = g_1(t), \quad t \in (0,T],\\
&u(x,0) = u_0(x), \quad x \in \bar{I},
\end{aligned}
$$

where

$$\mathscr{L}u = -a(x,t)u_{xx} + b(x,t)u_x + c(x,t)u. \tag{6}$$

The continuous-time OSC approximation is a differentiable map $u_h : [0,T] \to \mathscr{M}_r(\pi)$ such that

$$
\begin{aligned}
&[(u_h)_t + \mathscr{L}u_h](\xi_i,t) = f(\xi_i,t), \quad i = 1,2,\ldots,M, \ t \in (0,T],\\
&\mathscr{B}_0 u_h(0,t) = g_0(t), \qquad \mathscr{B}_1 u_h(1,t) = g_1(t), \quad t \in (0,T],
\end{aligned} \tag{7}
$$

where $u_h(\cdot,0) \in \mathscr{M}_r(\pi)$ is determined by approximating the initial condition using either Hermite or Gauss interpolation. With $u_h(x,t) = \sum_{j=1}^{M+2} u_j(t)\phi_j(x)$, where $\{\phi_j\}_{j=1}^{M+2}$ is a basis for $\mathscr{M}_r(\pi)$, (7) is an initial value problem for a first-order system of ODEs. This system can be written as

$$B\boldsymbol{u}_h'(t) + A(t)\boldsymbol{u}_h(t) = \boldsymbol{F}(t), \quad t \in (0,T], \quad \boldsymbol{u}(0) \text{ prescribed}, \tag{8}$$

where B and $A(t)$ are both ABD matrices of the same structure.

A commonly used discrete-time OSC method for solving (7) is the Crank–Nicolson OSC method [63]. This method consists in finding $u_h^k \in \mathscr{M}_r(\pi)$, $k=1,\ldots,K$, which satisfies the boundary conditions and, for $k = 0,\ldots,K-1$,

$$\left[\frac{u_h^{k+1} - u_h^k}{\Delta t} + \mathscr{L}^{k+1/2} u_h^{k+1/2}\right](\xi_m) = f(\xi_m, t_{k+1/2}), \quad m = 1,2,\ldots,M,$$

where

$$K\Delta t = T, \quad t_{k+1/2} = (k+1/2)\Delta t, \quad u_h^{k+1/2} = (u_h^k + u_h^{k+1})/2$$

and $\mathscr{L}^{k+1/2}$ is the operator $\mathscr{L}$ of (6) with $t = t_{k+1/2}$. In matrix–vector form, this method can be written as

$$[B + \tfrac{1}{2}\Delta t A(t_{k+1/2})]\boldsymbol{u}_h^{k+1} = [B - \tfrac{1}{2}\Delta t A(t_{k+1/2})]\boldsymbol{u}_h^k + \Delta t \boldsymbol{F}(t_{k+1/2}),$$

which is essentially the trapezoidal method for (8). Thus, with a standard basis, an ABD system must be solved at each time step.

Several time-dependent problems in one space variable (with homogeneous Dirichlet boundary conditions) have been solved using a method of lines (MOL) approach using OSC with monomial bases for the spatial discretization, which results in an initial value problem for a system of differential algebraic equations (DAEs). In a series of papers, Robinson and Fairweather [124–126] adopted this approach in the solution of the cubic Schrödinger equation

$$\mathrm{i}u_t + u_{xx} + q|u|^2 u = 0, \quad (x,t) \in I \times (0,T],$$

where $i^2 = -1$ and q is a given positive constant, and in the so-called two-dimensional parabolic equation of Tappert

$$u_t = \frac{i}{2}k_0[n^2(x,t) - 1 + iv(x,t)]u + \frac{i}{2k_0}u_{xx}, \quad (x,t) \in I \times (0,T],$$

where $n(x,t)$ and $v(x,t)$ are given functions. In each case, the DAEs are solved using the package D02NNF from the NAG Library. In [126], an optimal order L^2 estimate of the error in the semidiscrete (continuous-time) approximation at each time level is derived. In [124,125], it is shown that the use of monomial bases is particularly convenient in problems in layered media. In [122,123], this work is extended to the Schrödinger equation with general power nonlinearity

$$iu_t + u_{xx} + q|u|^{p-1}u = 0, \quad (x,t) \in I \times (0,T]$$

and to the generalized nonlinear Schrödinger equation

$$iu_t + u_{xx} + q_c|u|^2u + q_q|u|^4u + iq_m(|u|^2)_xu + iq_u|u|^2u_x = 0, \quad (x,t) \in I \times (0,T],$$

where q_c, q_q, q_m and q_u are real constants.

In [112], the MOL OSC approach is used in the solution of the Rosenau equation

$$u_t + u_{xxxxt} = f(u)_x, \quad (x,t) \in I \times (0,T],$$

where

$$f(u) = \sum_{i=1}^{n} \frac{c_i u^{p_i+1}}{p_i + 1}$$

with $c_i \in \mathbb{R}$ and p_i a nonnegative integer. This equation is first reformulated as a system by introducing the function

$$v = -u_{xx}$$

to obtain

$$u_t - v_{xxt} = f(u)_x, \quad v + u_{xx} = 0, \quad (x,t) \in I \times (0,T].$$

Here the DAEs are solved using an implicit Runge–Kutta method. Optimal order L^2 and L^∞ error estimates are obtained. The same approach is adopted and similar error estimates obtained in [111] for the Kuramoto–Sivashinsky equation

$$u_t + vu_{xxxx} + u_{xx} + uu_x = 0, \quad (x,t) \in I \times (0,T],$$

where v is a positive constant.

Several software packages have been developed for solving systems of nonlinear IBVPs in one space variable using an MOL OSC approach. The first, *pdecol* [107], uses B-spline bases and solves the resulting linear systems using the code *solveblok* [31]. The code *epdcol* [92] is a variant of *pdecol* in which *solveblok* is replaced by *colrow* [57]. In the MOL code based on OSC with monomial bases described in [116], the linear algebraic systems are solved using *abdpack* [110].

3. Elliptic boundary value problems

3.1. Introduction

In this section, we discuss OSC for the Dirichlet BVP

$$\begin{aligned} &Lu = f(x), \quad x = (x_1, x_2) \in \Omega, \\ &u(x) = g(x), \quad x \in \partial\Omega, \end{aligned} \tag{9}$$

where the linear second-order partial differential operator L is in the divergence form

$$Lu = -\sum_{i=1}^{2} (a_i(x) u_{x_i})_{x_i} + \sum_{i=1}^{2} b_i(x) u_{x_i} + c(x) u \tag{10}$$

with a_1, a_2 satisfying

$$0 < a_{\min} \leqslant a_i(x), \quad x \in \Omega, \; i = 1, 2, \tag{11}$$

or in the nondivergence form

$$Lu = -\sum_{i,j=1}^{2} a_{ij}(x) u_{x_i x_j} + \sum_{i=1}^{2} b_i(x) u_{x_i} + c(x) u, \quad a_{12} = a_{21} \tag{12}$$

with a_{ij} satisfying

$$a_{\min} \sum_{i=1}^{2} \eta_i^2 \leqslant \sum_{i,j=1}^{2} a_{ij}(x) \eta_i \eta_j, \quad x \in \Omega, \; \eta_1, \eta_2 \in \mathbb{R}, \; a_{\min} > 0. \tag{13}$$

Although the principal part of L is elliptic in the sense of (11) or (13), the operator L itself could be indefinite. It should be noted that, while the divergence form (10) is more natural for finite element Galerkin methods, the nondivergence form (12) is more appropriate for spline collocation methods (cf., for example, [90], where modified spline collocation is considered). The BVP most frequently considered in the literature is Poisson's equation with homogeneous Dirichlet boundary conditions, viz.,

$$\begin{aligned} &-\Delta u = f(x), \quad x \in \Omega, \\ &u(x) = 0, \quad x \in \partial\Omega, \end{aligned} \tag{14}$$

where Δ denotes the Laplacian.

The OSC problem for (9) consists in finding $u_h \in \mathcal{M}_r(\pi) \otimes \mathcal{M}_r(\pi)$, $r \geqslant 3$, such that

$$\begin{aligned} &L u_h(\xi_m, \xi_n) = f(\xi_m, \xi_n), \quad m, n = 1, \ldots, M, \\ &u_h(x) = \tilde{g}(x), \quad x \in \partial\Omega, \end{aligned} \tag{15}$$

where, on each side of $\partial\Omega$, $\tilde{g}$ is an approximation to g in $\mathcal{M}_r(\pi)$.

3.2. Convergence theory

Prenter and Russell [120] considered (15) for $r = 3$, $g = 0$, and L of (10) with $b_1 = b_2 = 0$ and $c \geqslant 0$. Assuming the existence of the OSC solution and uniform boundedness of partial derivatives of certain divided difference quotients, they derived optimal order H^1 and L^2 error estimates. For the same r and L, but nonhomogeneous boundary conditions, Bialecki and Cai [15] chose $\tilde{g}$ to be the piecewise Hermite cubic interpolant of g or the piecewise cubic interpolant of g at the boundary collocation points, that is, the Gauss interpolant. In both cases, they used energy inequalities to prove existence and uniqueness of the OSC solution for sufficiently small meshsize h, and superconvergence properties of the piecewise Hermite bicubic interpolant of u to derive an optimal order H^1 error estimate.

Percell and Wheeler [117] studied (15) for $r \geqslant 3$, $g = 0$, and L of (10) with $a_1 = a_2$. They proved existence and uniqueness of the OSC solution for sufficiently small h and derived optimal order H^1 and L^2 error estimates. The assumption $a_1 = a_2$ is essential in their approach, which reduces the analysis to that for the Laplacian Δ. Dillery [58] extended the results of Percell and Wheeler to $g \neq 0$ and $a_1 \neq a_2$.

Bialecki [11] analyzed (15) for $g = 0$, and L as in (12) in the case $r \geqslant 3$. He established existence and uniqueness of the OSC solution for sufficiently small h and derived the optimal order H^2, H^1, and L^2 error estimates by proving H^2 stability of the OSC problem using a transformation of Bernstein and Nitsche's trick, and by bounding the truncation error using superconvergence properties of an interpolant of u in $\mathcal{M}_r^0(\pi) \otimes \mathcal{M}_r^0(\pi)$. The results of [11] extend and generalize all previous theoretical OSC results for two-dimensional BVPs, which do not consider mixed partial derivatives and provided no H^2 convergence analysis.

Houstis [84] considered (15) for $r = 3$, $g = 0$, and two cases of L. In the first case, L is given by (12), and in the second case L is that of (10) with $b_1 = b_2 = 0$ and $c \geqslant 0$. Using a Green's function approach (cf. [32] for two-point BVPs), he derived an optimal order L^∞ error estimate in the first case, and an optimal order L^2 error estimate in the second. However, it appears that his analysis is based on the unrealistic assumption that a partial derivative of the corresponding Green's function be uniformly bounded.

It was proved in [32] that OSC possesses superconvergence properties for two-point BVPs. Specifically, if the exact solution of the problem is sufficiently smooth and if the OSC solution $u_h \in \mathcal{M}_r(\pi)$, $r \geqslant 3$, then, at the partition nodes, the values of the OSC solution and the values of its first derivative approximate the corresponding values of the exact solution with error $O(h^{2r-2})$. For BVPs on rectangles, the same rate of convergence in the values of the OSC solution and the values of its first partial derivatives at the partition nodes was first observed numerically in [22] for $r = 3$, and in [21] for $r > 3$ and has since been observed in all applications of OSC to elliptic BVPs and IBVPs which we have examined. For (14) with $r = 3$ and a uniform partition, Bialecki [12] proved the fourth-order convergence rate in the first-order partial derivatives at the partition nodes. The approach used in [12] is a combination of a discrete Fourier method and a discrete maximum principle applied in the two different coordinate directions.

Grabarski [74] considered OSC with $r = 3$ for the solution of a nonlinear problem comprising (14) with $f(x)$ replaced with $f(x, u)$. He proved the existence and uniqueness of the OSC solution using Browder's theorem and derived an optimal order H^1 error estimate using a superconvergence property of the piecewise Hermite bicubic interpolant of u.

Aitbayev and Bialecki [1] analyzed OSC with $r = 3$ for the nonlinear problem (9) with $g = 0$ and

$$Lu = -\sum_{i,j=1}^{2} a_{ij}(x, u) u_{x_i x_j} + \sum_{i=1}^{2} b_i(x, u) u_{x_i} + c(x, u) u,$$

where a_{ij} satisfy the ellipticity condition

$$a_{\min}\sum_{i=1}^{2}\eta_i^2\leqslant\sum_{i,j=1}^{2}a_{ij}(x,s)\eta_i\eta_j,\quad x\in\Omega,\ s\in\mathbb{R},\ \eta_1,\eta_2\in\mathbb{R},\ a_{\min}>0.$$

For sufficiently small h, they proved existence and uniqueness of the OSC solution, and established optimal order H^2 and H^1 error bounds. The approach of [1] is based on showing consistency and stability of the nonlinear OSC problem. Consistency is proved using superconvergence properties of the piecewise Hermite bicubic interpolant of u and stability is established using Banach's lemma. Then existence, uniqueness and error bounds are obtained using general results similar to those of [94]. Newton's method is analyzed for the solution of the resulting nonlinear OSC problem.

For the solution of second-order linear elliptic boundary value problems with differential operators of the form (12), Houstis et al. [86] described three OSC algorithms which use piecewise Hermite bicubics. The first, GENCOL, is for problems on general two-dimensional domains, the second for problems on rectangular domains with general linear boundary conditions, and the third, INTCOL, for problems on rectangular domains with uncoupled boundary conditions. FORTRAN implementations of these algorithms are given in [87,88]; see also [121]. In [100], the algorithm INTCOL is extended to regions whose sides are parallel to the axes. No convergence analysis of the algorithms is provided. However, in [115], Mu and Rice present an experimental study of the algorithm GENCOL on a large sample of problems in nonrectangular regions which demonstrates that the rate of convergence of the OSC solution at the nodes is fourth order.

3.3. Direct methods for linear OSC problems

Consider the special case of (9) with $g=0$ and $Lu=-u_{x_1x_1}+L_2u$, where

$$L_2u=-a_2(x_2)u_{x_2x_2}+b_2(x_2)u_{x_2}+c_2(x_2)u, \tag{16}$$

which includes (14) in polar, cylindrical, and spherical coordinate systems. In recent years, several matrix decomposition algorithms have been developed for the fast solution of the linear algebraic systems arising when finite difference, finite element Galerkin and spectral methods are applied to problems of this type. These methods, which are described in [20], depend on knowledge of the eigensystem of the second derivative operator subject to certain boundary conditions. In this section, we first describe an OSC matrix decomposition algorithm developed in [6,22,23,68] for the case in which $r=3$ and the partition of Ω is uniform at least in the x_1 direction. Then we formulate a matrix decomposition algorithm for $r\geqslant 3$ for the solution of more general elliptic problems [21].

Let $\{\phi_n\}_{n=1}^{M}$ be a basis for $\mathcal{M}_3^0(\pi)$, and write the piecewise Hermite bicubic approximation in the form

$$u_h(x_1,x_2)=\sum_{m=1}^{M}\sum_{n=1}^{M}u_{m,n}\phi_m(x_1)\phi_n(x_2).$$

If $\boldsymbol{u}=[u_{1,1},\ldots,u_{1,M},\ldots,u_{M,1},\ldots,u_{M,M}]^{\mathrm{T}}$, then the collocation equations (15) can be written as

$$(A_1\otimes B+B\otimes A_2)\boldsymbol{u}=\boldsymbol{f}, \tag{17}$$

where $\boldsymbol{f}=[f_{1,1},\ldots,f_{1,M},\ldots,f_{M,1},\ldots,f_{M,M}]^{\mathrm{T}}$, $f_{m,n}=f(\xi_m,\xi_n)$,

$$A_1=(-\phi_n''(\xi_m))_{m,n=1}^{M},\quad A_2=(L_2\phi_n(\xi_m))_{m,n=1}^{M},\quad B=(\phi_n(\xi_m))_{m,n=1}^{M}. \tag{18}$$

In [22], real nonsingular matrices $\Lambda = \mathrm{diag}(\lambda_j)_{j=1}^{M}$ and Z, a matrix of sines and cosines, are determined such that

$$Z^{\mathrm{T}}B^{\mathrm{T}}A_1Z = \Lambda, \quad Z^{\mathrm{T}}B^{\mathrm{T}}BZ = I. \tag{19}$$

System (17) can then be written in the form

$$(Z^{\mathrm{T}}B^{\mathrm{T}} \otimes I)(A_1 \otimes B + B \otimes A_2)(Z \otimes I)(Z^{-1} \otimes I)\boldsymbol{u} = (Z^{\mathrm{T}}B^{\mathrm{T}} \otimes I)\boldsymbol{f},$$

which becomes, on using (19),

$$(\Lambda \otimes B + I \otimes A_2)(Z^{-1} \otimes I)\boldsymbol{u} = (Z^{\mathrm{T}}B^{\mathrm{T}} \otimes I)\boldsymbol{f}.$$

Hence the matrix decomposition algorithm for solving (17) takes the following form:

Algorithm A

1. Compute $\boldsymbol{g} = (Z^{\mathrm{T}}B^{\mathrm{T}} \otimes I)\boldsymbol{f}$.
2. Solve $(\Lambda \otimes B + I \otimes A_2)\boldsymbol{v} = \boldsymbol{g}$.
3. Compute $\boldsymbol{u} = (Z \otimes I)\boldsymbol{v}$.

In step 1, matrix–vector multiplications involving the matrix B^{T} require a total of $O(N^2)$ arithmetic operations. FFT routines can be used to perform multiplications by the matrix Z^{T} in step 1 and by the matrix Z in step 3. The corresponding cost of each of these steps is $O(N^2 \log N)$ operations. Step 2 consists of solving M almost block diagonal linear systems of order M, the coefficient matrix of the jth system being of the form $A_2 + \lambda_j B$. The use of *colrow* [57] to solve these systems requires $O(N^2)$ operations. Thus the total cost of the algorithm is $O(N^2 \log N)$ operations. Note that step 2 is equivalent to solving the OSC problem for a two-point BVP of the form

$$-v'' + \lambda_j v = f(x), \quad x \in I, \quad v(0) = v(1) = 0.$$

Algorithm A is Algorithm II of [22]. Algorithm I of that paper describes an OSC matrix decomposition procedure for (14) in which the linear system (17) is diagonalized by applying FFTs in both variables. This algorithm costs twice as much as Algorithm A for (14).

Sun and Zamani [137] also developed a matrix decomposition algorithm for solving the OSC equations (15) for (14). Their algorithm is based on the fact that the eigenvalues of the matrix $B^{-1}A_1$ are real and distinct [64] and hence there exists a real nonsingular matrix Q such that $B^{-1}A_1 = Q\Lambda Q^{-1}$. Sun and Zamani's algorithm, which also requires $O(N^2 \log N)$ operations, appears to be more complicated and less efficient than Algorithm A, which hinges on the existence of a real nonsingular matrix Z satisfying (19). In particular, the utilization of the second equation in (19) distinguishes Algorithm A and makes it more straightforward.

Algorithm A can be generalized to problems in which, on the sides $x_2 = 0, 1$ of $\partial\Omega$, u satisfies either the Robin boundary conditions

$$\alpha_0 u(x_1, 0) + \beta_0 u_{x_2}(x_1, 0) = g_0(x_1), \quad \alpha_1 u(x_1, 1) + \beta_1 u_{x_2}(x_1, 1) = g_1(x_1), \quad x_1 \in \bar{I},$$

where α_i, β_i, $i = 0, 1$, are constants, or the periodic boundary conditions

$$u(x_1, 0) = u(x_1, 1), \quad u_{x_2}(x_1, 0) = u_{x_2}(x_1, 1), \quad x_1 \in \bar{I}.$$

On the sides $x_1 = 0,1$ of $\partial\Omega$, u may be subject to either Dirichlet conditions, Neumann conditions, mixed Dirichlet–Neumann boundary conditions or periodic boundary conditions. Details are given in [6,23,68]. Other extensions have been considered, to problems in three dimensions [119] and to OSC with higher degree piecewise polynomials [134].

For the case of Poisson's equation with pure Neumann or pure periodic boundary conditions, Bialecki and Remington [28] formulated a matrix decomposition approach for determining the least-squares solution of the singular OSC equations when $r = 3$.

In [127], an eigenvalue analysis is presented for spline collocation differentiation matrices corresponding to periodic boundary conditions. In particular, the circulant structure of piecewise Hermite cubic matrices is used to develop a matrix decomposition FFT algorithm for the OSC solution of a general second-order PDE with constant coefficients. The proposed algorithm, whose cost is $O(N^2 \log N)$, requires the use of complex arithmetic. An eigenvalue analysis for Dirichlet and Neumann boundary conditions and arbitrary collocation points is presented in [135].

In [13], Algorithm A was extended to the OSC solution of the polar form of the Poisson's equation on a disk

$$r^{-1}(ru_r)_r + r^{-2}u_{\theta\theta} = f(r,\theta), \quad (r,\theta) \in (0,1) \times (0,2\pi), \tag{20}$$

subject to Dirichlet or Neumann boundary conditions. The new algorithm is remarkably simple, due, in part, to a new treatment of the boundary condition on the side of the rectangle corresponding to the center of the disk. For Dirichlet boundary conditions, the OSC solution is obtained as the superposition of two OSC solutions on the rectangle. (The superposition approach was also used in [138] for a finite difference scheme.) The first OSC solution is obtained using the FFT matrix decomposition method of [6] which is based on the knowledge of the eigensystem for the OSC discretization of the second derivative operator subject to periodic boundary conditions. A simple analytical formula is derived for the second OSC solution (this is not the case for the finite difference scheme of [138]). For Neumann boundary conditions, the corresponding OSC problem is singular. In this case, the matrix decomposition method is modified to obtain an OSC approximation corresponding to the particular continuous solution with a specified value at the center of the disk. Each algorithm requires $O(N^2 \log N)$ operations. While the numerical results demonstrate fourth-order accuracy of the OSC solution and third-order accuracy of its partial derivatives at the nodes, the analysis of the OSC scheme is an open question.

Sun [132] considered the piecewise Hermite bicubic OSC solution of (20) on an annulus and on a disk. In the case of the annulus, Sun's FFT matrix decomposition algorithm is based on that of [137]. For the disk, the approach of [138] is used to derive an additional equation corresponding to the center of the disk. In contrast to [13], this equation is not solved independently of the rest of the problem. A convergence analysis for piecewise Hermite bicubic OSC solution of (20) on an annulus or a disk has yet to be derived.

For the case $r \geqslant 3$, we now describe an algorithm for determining OSC approximations to solutions of BVPs of the form

$$(L_1 + L_2)u = f(x), \quad x \in \Omega,$$
$$u(x) = 0, \quad x \in \partial\Omega,$$

where

$$L_1 u = -a_1(x_1)u_{x_1x_1} + c_1(x_1)u$$

and L_2 is given by (16), with $a_i \geqslant a_{\min} > 0$, $c_i \geqslant 0$, $i = 1,2$. The collocation equations can again be written in the form (17) with A_2 and B as in (18) and $A_1 = (L_1\phi_n(\xi_m))_{m,n=1}^M$. As before, these matrices are ABD for the usual choices of bases of $\mathscr{M}_r^0(\pi)$.

Now, let $W = \operatorname{diag}(h_1w_1, h_1w_2, \ldots, h_1w_{r-1}, \ldots, h_Nw_1, h_Nw_2, \ldots, h_Nw_{r-1})$, where $\{w_i\}_{i=1}^{r-1}$ are the weights of the $(r-1)$-point Gauss–Legendre quadrature rule on I. For v defined on I, let $D(v) = \operatorname{diag}(v(\xi_1), v(\xi_2), \ldots, v(\xi_M))$. If

$$F = B^{\mathrm{T}}WD(1/a_1)B, \quad G = B^{\mathrm{T}}WD(1/a_1)A_1, \tag{21}$$

then F is symmetric and positive definite, and G is symmetric. Hence, there exist a real $\Lambda = \operatorname{diag}(\lambda_j)_{j=1}^M$ and a real nonsingular Z such that

$$Z^{\mathrm{T}}GZ = \Lambda, \quad Z^{\mathrm{T}}FZ = I, \tag{22}$$

[73]. By (21), the matrices Λ and Z can be computed by using the decomposition $F = HH^{\mathrm{T}}$, where $H = B^{\mathrm{T}}[WD(1/a_1)]^{1/2}$, and solving the symmetric eigenproblem for

$$C = H^{-1}GH^{-\mathrm{T}} = [WD(1/a_1)]^{1/2}A_1B^{-1}[WD(1/a_1)]^{-1/2}$$

to obtain

$$Q^{\mathrm{T}}CQ = \Lambda \tag{23}$$

with Q orthogonal. If $Z = B^{-1}[WD(1/a_1)]^{-1/2}Q$, then Λ and Z satisfy (22). Thus,

$$[Z^{\mathrm{T}}B^{\mathrm{T}}WD(1/a_1) \otimes I](A_1 \otimes B + B \otimes A_2)(Z \otimes I) = \Lambda \otimes B + I \otimes A_2,$$

which leads to the following matrix decomposition algorithm, Algorithm II of [21]:

Algorithm B

1. Determine Λ and Q satisfying (23).
2. Compute $\boldsymbol{g} = (Q^{\mathrm{T}}[WD(1/a_1)]^{1/2} \otimes I)\boldsymbol{f}$.
3. Solve $(\Lambda \otimes B + I \otimes A_2)\boldsymbol{v} = \boldsymbol{g}$.
4. Compute $\boldsymbol{u} = (B^{-1}[W_1D_1(1/a_1)]^{-1/2}Q \otimes I)\boldsymbol{v}$.

Steps 1, 3, and 4 each involve solving M independent ABD systems which are all of order M. In Step 1, C can be determined efficiently from $B^{\mathrm{T}}[WD(1/a_1)]^{1/2}C = A_1^{\mathrm{T}}[WD(1/a_1)]^{1/2}$. Computing the columns of C requires solving ABD systems with the same coefficient matrix, the transpose of the ABD matrix $[WD(1/a_1)]^{1/2}B$. This ABD matrix is factored once and the columns of C determined. This factored form is also used in Step 4. In Step 3, the ABD matrices have the form $A_2 + \lambda_jB$, $j = 1,2,\ldots,M$. Assuming that on average only two steps of the QR algorithm are required per eigenvalue when solving the symmetric, tridiagonal eigenvalue problem corresponding to (23), the total cost of this algorithm is $O(r^3N^3 + r^4N^2)$.

Matrix decomposition algorithms similar to those of [21] are described in [118] for the OSC solution of Poisson's equation in three dimensions. The authors claim that these algorithms are competitive with FFT-based methods since the cost of solving one-dimensional collocation eigenvalue problems is low compared to the total cost.

Bialecki [9] developed cyclic reduction and Fourier analysis-cyclic reduction methods for the solution of the linear systems which arise when OSC with piecewise Hermite bicubics is applied to (14). On a uniform partition, the cyclic reduction and Fourier analysis-cyclic reduction methods require $O(N^2 \log N)$ and $O(N^2 \log(\log N))$ arithmetic operations, respectively.

3.4. Iterative methods for linear OSC problems

The OSC problem (15) with $r \geqslant 3$, $g = 0$ and separable $L = L_1 + L_2$, where

$$L_i u = -(a_i(x_i) u_{x_i})_{x_i} + c_i(x_i) u, \quad i = 1, 2 \tag{24}$$

or

$$L_i u = -a_i(x_i) u_{x_i x_i} + c_i(x_i) u, \quad i = 1, 2, \tag{25}$$

$a_i \geqslant a_{\min} > 0$, $c_i \geqslant 0$, can be solved by an alternating direction implicit (ADI) method. The matrix–vector form of this method is defined as follows: given $\boldsymbol{u}^{(0)}$, for $k = 0, 1, \ldots,$ compute $\boldsymbol{u}^{(k+1)}$ from

$$\begin{aligned}
[(A_1 + \gamma_{k+1}^{(1)} B) \otimes B]\boldsymbol{u}^{(k+1/2)} &= \boldsymbol{f} - [B \otimes (A_2 - \gamma_{k+1}^{(1)} B)]\boldsymbol{u}^{(k)}, \\
[B \otimes (A_2 + \gamma_{k+1}^{(2)} B)]\boldsymbol{u}^{(k+1)} &= \boldsymbol{f} - [(A_1 - \gamma_{k+1}^{(2)} B) \otimes B]\boldsymbol{u}^{(k+1/2)},
\end{aligned}$$

where, for $i=1,2$, $A_i = (L_i \phi_n(\xi_m))_{m,n=1}^{M}$, and $B = (\phi_n(\xi_m))_{m,n=1}^{M}$ as before, and $\gamma_{k+1}^{(1)}$, $\gamma_{k+1}^{(2)}$ are acceleration parameters. Introducing

$$\boldsymbol{v}^{(k)} = (B \otimes I)\boldsymbol{u}^{(k)}, \quad \boldsymbol{v}^{(k+1/2)} = (I \otimes B)\boldsymbol{u}^{(k+1/2)},$$

we obtain

$$\begin{aligned}
[(A_1 + \gamma_{k+1}^{(1)} B) \otimes I]\boldsymbol{v}^{(k+1/2)} &= \boldsymbol{f} - [I \otimes (A_2 - \gamma_{k+1}^{(1)} B)]\boldsymbol{v}^{(k)}, \\
[I \otimes (A_2 + \gamma_{k+1}^{(2)} B)]\boldsymbol{v}^{(k+1)} &= \boldsymbol{f} - [(A_1 - \gamma_{k+1}^{(2)} B) \otimes I]\boldsymbol{v}^{(k+1/2)},
\end{aligned}$$

which, at each iteration step, avoids unnecessary multiplications by B, and the solution of systems with coefficient matrix B. Note that each step of the ADI method requires the solution of ABD linear systems similar to those arising in step 3 of Algorithm B.

Dyksen [64] used the ADI method for (14) with $r = 3$, and a uniform partition. The eigenvalues (taken in increasing order) of the OSC eigenvalue problem corresponding to $-v'' = \lambda v$, $v(0) = v(1) = 0$, are used as the acceleration parameters. The cost of the resulting algorithm with $2N$ ADI iterations is $O(N^3)$. Based on numerical results, Dyksen claims that reasonable accuracy is achieved with fewer than $2N$ ADI iterations. Numerical results are included to demonstrate that the same acceleration parameters work well for more general operators L_i. In [65], Dyksen considered the solution of certain elliptic problems in three space variables in a rectangular parallelepiped using Hermite bicubic OSC to discretize in x_1 and x_2 and symmetric finite differences in x_3. The resulting equations were solved using an ADI approach.

Cooper and Prenter [53] analyzed the ADI method for L_i of (24) with $r = 3$, and a nonuniform partition. They proved convergence of the ADI method with arbitrary positive acceleration parameters but did not determine an optimal sequence of such parameters. Generalizations of this approach are discussed in [50,51].

For L_i of (25), $r \geqslant 3$, and a nonuniform partition, Bialecki [7] showed that, with Jordan's selection of the acceleration parameters, the cost of the ADI method is $O(N^2 \log^2 N)$ for obtaining an approximation to the OSC solution within the accuracy of the truncation error.

For non-separable, positive-definite, self-adjoint and nonself-adjoint L of the form (10), Richardson and minimal residual preconditioned iterative methods were presented in [10] for solving (15) with $r = 3$. In these methods, the OSC discretization of $-\Delta$, $-\Delta_h$, is used as a preconditioner since it is shown that the OSC discretization of L, L_h, and $-\Delta_h$ are spectrally equivalent on the space $\mathscr{M}_3^0(\pi) \otimes \mathscr{M}_3^0(\pi)$ with respect to the collocation inner product, that is, the discrete inner product defined by the collocation points. At each iteration step, OSC problems with $-\Delta_h$ are solved using Algorithm A. As an alternative, for self-adjoint, positive definite L of (10), in [18] the general theory of additive and multiplicative Schwarz methods is used to develop multilevel methods for (15). In these methods, variable coefficient additive and multiplicative preconditioners for L_h are, respectively, sums and products of "one-dimensional" operators which are defined using the energy collocation inner product generated by the operator $(L_h^* + L_h)/2$, where L_h^* is the operator adjoint to L_h with respect to the collocation inner product. This work includes implementations of additive and multiplicative preconditioners. Numerical tests show that multiplicative preconditioners are faster than additive preconditioners.

For L of the form (10) with $a_1 = a_2 = 1$, and u subject to a combination of Dirichlet, Neumann and Robin's boundary conditions, an application of a bilinear finite element preconditioner is considered in [98] for the solution of the corresponding OSC problem on a uniform partition. The finite element matrix $\tilde{L}_N$ associated with the self-adjoint and positive-definite operator $\tilde{L}u = -\Delta u + \tilde{c}(x)u$ is used to precondition the OSC matrix L_N corresponding to $r = 3$. Bounds and clustering results are obtained for the singular values of the preconditioned matrix $\tilde{L}_N^{-1} L_N$. This approach is used in [102] to solve the systems arising in the OSC solution of problems of linear elasticity, and is extended to quasiuniform partitions in [97].

The application of finite difference preconditioners to the solution of (15) is discussed in [136] and [95]. The theory of [136] appears to be applicable only for the case $g = 0$, and $Lu = -\Delta u + cu$, with c a positive constant, for $r = 3$ and a uniform partition. It should be noted that, in this special case, as well as in the case when c is a function of either x_1 or x_2, (15) can be solved by Algorithm A. Even if c is a function of both x_1 and x_2 and nonnegative, (15) can be solved very efficiently by the preconditioned conjugate gradient (PCG) method with $-\Delta_h$ as a preconditioner. In [95], the same operator L as in [98] (but with u subject to homogeneous Dirichlet boundary conditions) is preconditioned by a finite difference operator. Again bounds and clustering results are obtained for the singular values of the preconditioned matrix.

Some researchers [79,101,133] have investigated applications of classical iterative methods, such as Jacobi, Gauss-Seidel or SOR, to the solution of (15) for (14) for $r = 3$ and a uniform partition. In [131], an application of these methods was considered for $r = 3$, a nonuniform partition, and $L = L_1 + L_2$ with L_i of (25) and $g = 0$. As in the case of the finite difference discretization, classical iterative methods applied to these specialized problems are not as efficient as Algorithm A or the ADI method; see [129].

In [2], iterative methods are developed and implemented for the solution of (15) with L of the form (12). The PCG method is applied, with Δ_h^2 as a preconditioner, to the solution of the normal problem

$$L_h^* L_h u_h = L_h^* f_h.$$

Using an H^2 stability result of [11], it is proved that the convergence rate of this method is independent of h. On a uniform partition, the preconditioned OSC problem is solved with cost $O(N^2 \log N)$ by a modification of Algorithm A.

Lai et al. [100] considered the application of iterative methods (SOR and CG types, and GMRES) for the solution of piecewise Hermite bicubic OSC problems on regions with sides parallel to the axes.

3.5. Biharmonic problems

A common approach to solving the biharmonic equation

$$\Delta^2 u = f(x), \quad x \in \Omega$$

is to use the splitting principle in which an auxiliary function $v = \Delta u$ is introduced and the biharmonic equation rewritten in the form

$$-\Delta u + v = 0, \quad -\Delta v = -f(x), \quad x \in \Omega. \tag{26}$$

In the context of the finite element Galerkin method, this approach is known as the mixed method of Ciarlet and Raviart [48].

Using this approach, Lou et al. [106] derived existence, uniqueness and convergence results for piecewise Hermite bicubic OSC methods and developed implementations of these methods for the solution of three biharmonic problems. The boundary conditions for the first problem comprise $u = g_1$ and $\Delta u = g_2$ on $\partial\Omega$, and the problem becomes one of solving two nonhomogeneous Dirichlet problems for Poisson's equation. The resulting linear systems can be solved effectively with cost $O(N^2 \log N)$ on a uniform partition using Algorithm A. In this case, optimal order H^j error estimates, $j = 0, 1, 2$, are derived. In the second problem, the boundary condition in the first problem on the horizontal sides of $\partial\Omega$, $\Delta u = g_2$, is replaced by the condition $u_{x_2} = g_3$. Optimal order H^1 and H^2 error estimates are derived and a variant of Algorithm A is formulated for the solution of the corresponding algebraic problem. This algorithm also has cost $O(N^2 \log N)$ on a uniform partition. The third problem is the biharmonic Dirichlet problem,

$$\begin{gathered} -\Delta u + v = 0, \quad -\Delta v = -f(x), \quad x \in \Omega, \\ u = g_1(x), \quad u_n = g_2(x), \quad x \in \partial\Omega, \end{gathered} \tag{27}$$

where the subscript n denotes the outward normal derivative. Again optimal order H^1 and H^2 error estimates are derived. In this case, the OSC linear system is rather complicated and is solved by a direct method which is based on the capacitance matrix technique with the second biharmonic problem as the auxiliary problem. On a uniform partition, the total cost of the capacitance matrix method for computing the OSC solution is $O(N^3)$ since the capacitance system is first formed explicitly and then solved by Gauss elimination. Results of some numerical experiments are presented which, in particular, demonstrate the fourth-order accuracy of the approximations and the superconvergence of the derivative approximations at the mesh points.

Piecewise Hermite bicubic OSC methods for the biharmonic Dirichlet problem (27) were considered by Cooper and Prenter [52] who proposed an ADI OSC method for the discretization of the BVP and the solution of the discrete problem, and by Sun [130] who presented an algorithm, the cost of which is $O(N^3 \log N)$. Numerical results show that these methods produce approximations which are fourth-order accurate at the nodes, but no rigorous proof of this result is provided in [52] or [130].

In [14], Bialecki developed a very efficient Schur complement method for obtaining the piecewise Hermite bicubic OSC solution to the biharmonic Dirichlet problem (27). In this approach, which is similar to that of [29] for finite differences, the OSC biharmonic Dirichlet problem is reduced to a Schur complement system involving the approximation to v on the vertical sides of $\partial\Omega$ and to an auxiliary OSC problem for a related biharmonic problem with v, instead of u_n, specified on the two vertical sides of $\partial\Omega$. The Schur complement system with a symmetric and positive definite matrix is solved by the PCG method with a preconditioner obtained from the OSC problem for a related biharmonic problem with v, instead of u_n, specified on the two horizontal sides of $\partial\Omega$. On a uniform partition, the cost of solving the preconditioned system and the cost of multiplying the Schur complement matrix by a vector are $O(N^2)$ each. With the number of PCG iterations proportional to $\log N$, the cost of solving the Schur complement system is $O(N^2 \log N)$. The solution of the auxiliary OSC problem is obtained using a variant of Algorithm A at a cost of $O(N^2 \log N)$. Thus the total cost of solving the OSC biharmonic Dirichlet problem is $O(N^2 \log N)$, which is essentially the same as that of Bjørstad's algorithm [29] for solving the second-order finite difference biharmonic Dirichlet problem. Numerical results indicate that the L^2, H^1, and H^2 errors in the OSC approximations to u and v are of optimal order. Convergence at the nodes is fourth order for the approximations to u, v, and their first-order derivatives.

3.6. Domain decomposition

Bialecki [8] used a domain decomposition approach to develop a fast solver for the piecewise Hermite bicubic OSC solution of (14). The square Ω is divided into parallel strips and the OSC solution is first obtained on the interfaces by solving a collection of independent tridiagonal linear systems. Algorithm A is then used to compute the OSC solution on each strip. Assuming that the strips have the same width and that their number is proportional to $N/\log N$, the cost of the domain decomposition solver is $O(N^2 \log(\log N))$.

Mateescu et al. [113] considered linear systems arising from piecewise Hermite bicubic collocation applied to general linear two-dimensional second-order elliptic PDEs on Ω with mixed boundary conditions. They constructed an efficient, parallel preconditioner for the GMRES(k) method. The focus in [113] is on rectangular domains decomposed in one dimension.

For the same problem as in [8], Bialecki and Dillery [17] analyzed the convergence rates of two Schwarz alternating methods. In the first method, Ω is divided into two overlapping subrectangles, while three overlapping subrectangles are used in the second method. Fourier analysis is used to obtain explicit formulas for the convergence factors by which the H^1 norm of the error is reduced in one iteration of the Schwarz methods. It is shown numerically that while these factors depend on the size of the overlap, they are independent of h. Using a convex function argument, Kim and Kim [96] bounded theoretically the convergence factors of [17] by quantities that depend only on the way that Ω is divided into the overlapping subrectangles.

In [16], an overlapping domain decomposition method was considered for the solution of the piecewise Hermite bicubic OSC problem corresponding to (14). The square is divided into overlapping squares and the additive Schwarz, conjugate gradient method involves solving independent OSC problems using Algorithm A.

Lai et al. [99] considered a generalized Schwarz splitting method for solving elliptic BVPs with interface conditions that depend on a parameter that might differ in each overlapping region. The method is coupled with the piecewise Hermite bicubic collocation discretization to solve the corresponding BVP in each subdomain. The main objective of [99] is the mathematical analysis of the iterative solution of the so-called enhanced generalized Schwarz splitting collocation equation corresponding to (14).

Bialecki and Dryja [19] considered the piecewise Hermite bicubic OSC solution of

$$\Delta u = f \text{ in } \hat{\Omega}, \quad u = 0 \text{ on } \partial\hat{\Omega}, \tag{28}$$

where $\hat{\Omega}$ is the L-shaped region given by

$$\hat{\Omega} = (0,2)\times(0,1)\cup(0,1)\times(1,2). \tag{29}$$

The region $\hat{\Omega}$ is partitioned into three nonoverlapping squares with two interfaces. On each square, the approximate solution is a piecewise Hermite bicubic that satisfies Poisson's equation at the collocation points in the subregion. The approximate solution is continuous throughout the region and its normal derivatives are equal at the collocation points on the interfaces, but continuity of the normal derivatives across the interfaces is not guaranteed. The solution of the collocation problem is first reduced to finding the approximate solution on the interfaces. The discrete Steklov–Poincaré operator corresponding to the interfaces is self-adjoint and positive definite with respect to the discrete inner product associated with the collocation points on the interfaces. The approximate solution on the interfaces is computed using the PCG method with the preconditioner obtained from two discrete Steklov–Poincaré operators corresponding to two pairs of the adjacent squares. Once the solution of the discrete Steklov–Poincaré equation is obtained, the collocation solutions on subregions are computed using Algorithm A. On a uniform partition, the total cost of the algorithm is $O(N^2 \log N)$, where the number of unknowns is proportional to N^2.

4. Time-dependent problems

4.1. Parabolic and hyperbolic problems

In this section, we discuss OSC for the IBVP

$$\begin{aligned} &u_t + (L_1 + L_2)u = f(x,t), \quad (x,t)\in\Omega_T \equiv \Omega\times(0,T],\\ &u(x,0) = g_1(x), \quad x\in\bar{\Omega},\\ &u(x,t) = g_2(x,t), \quad (x,t)\in\partial\Omega\times(0,T] \end{aligned} \tag{30}$$

and the second-order hyperbolic IBVP

$$\begin{aligned} &u_{tt} + (L_1 + L_2)u = f(x,t), \quad (x,t)\in\Omega_T,\\ &u(x,0) = g_1(x), \quad u_t(x,0) = g_2(x), \quad x\in\bar{\Omega},\\ &u(x,t) = g_3(x,t), \quad (x,t)\in\partial\Omega\times(0,T], \end{aligned} \tag{31}$$

where the second-order differential operators L_1 and L_2 are of the form

$$L_1 u = -(a_1(x,t)u_{x_1})_{x_1} + b_1(x,t)u_{x_1} + c(x,t)u, \quad L_2 u = -(a_2(x,t)u_{x_2})_{x_2} + b_2(x,t)u_{x_2} \tag{32}$$

or

$$L_1u = -a_1(x,t)u_{x_1x_1} + b_1(x,t)u_{x_1} + c(x,t)u, \quad L_2u = -a_2(x,t)u_{x_2x_2} + b_2(x,t)u_{x_2} \tag{33}$$

with a_1, a_2 satisfying

$$0 < a_{\min} \leqslant a_i(x,t) \leqslant a_{\max}, \quad (x,t) \in \Omega_T, \quad i = 1,2. \tag{34}$$

For $r \geqslant 3$, Greenwell-Yanik and Fairweather [76] analyzed continuous-time OSC methods and Crank–Nicolson schemes for nonlinear problems of the form (30) and (31) with $f(x,t,u)$ in place of $f(x,t)$, $g_2=0$, L_1 and L_2 of (33) with $a_1=a_2$, and $c=0$. Using the OSC solution of the corresponding elliptic problem as a comparison function, they derived optimal order L^2 error estimates.

For $r = 3$, Grabarski [75] considered the continuous-time OSC method for a nonlinear problem, (30) with $f(x,t,u)$ in place of $f(x,t)$, $g_2 = 0$, L_1 and L_2 of (33) with $a_1 = a_2$, $b_1 = b_2 = c = 0$. He derived an optimal order H^1 error estimate using a superconvergence property of the piecewise Hermite bicubic interpolant of u.

ADI OSC methods, without convergence analysis, have been used over the last 20 years to solve certain parabolic problems (see, for example, [5,33,36–42,54,80–82]). In general, the ADI Crank–Nicolson scheme for (30) consists in finding $u_h^k \in \mathcal{M}_r(\pi) \otimes \mathcal{M}_r(\pi)$, $k = 1,\dots,K$, such that, for $k = 0,\dots,K-1$,

$$\left[\frac{u_h^{k+1/2} - u_h^k}{\Delta t/2} + L_1^{k+1/2}u_h^{k+1/2} + L_2^{k+1/2}u_h^k\right](\xi_m,\xi_n) = f(\xi_m,\xi_n,t_{k+1/2}), \quad m,n = 1,\dots,M,$$

$$\left[\frac{u_h^{k+1} - u_h^{k+1/2}}{\Delta t/2} + L_1^{k+1/2}u_h^{k+1/2} + L_2^{k+1/2}u_h^{k+1}\right](\xi_m,\xi_n) = f(\xi_m,\xi_n,t_{k+1/2}), \quad m,n = 1,\dots,M, \tag{35}$$

where $L_1^{k+1/2}$ and $L_2^{k+1/2}$ are the differential operators L_1 and L_2 with $t = t_{k+1/2}$, respectively, $u_h^0 \in \mathcal{M}_r(\pi) \otimes \mathcal{M}_r(\pi)$, $u_h^k|_{\partial\Omega}$, $k = 1,\dots,K$, are prescribed by approximating the initial and boundary conditions of (30) using either piecewise Hermite or Gauss interpolants, and for each $n = 1,\dots,M$, $u_h^{k+1/2}(\cdot,\xi_n) \in \mathcal{M}_r(\pi)$ satisfies

$$u_h^{k+1/2}(\alpha,\xi_n) = [(1/2)(u_h^{k+1} + u_h^k) + (\Delta t/4)L_2^{k+1/2}(u_h^{k+1} - u_h^k)](\alpha,\xi_n), \quad \alpha = 0,1.$$

For $r \geqslant 3$, Fernandes and Fairweather [72] analyzed (35) for the heat equation with homogeneous Dirichlet boundary conditions, which is the special case of (30) with $L_iu = -u_{x_ix_i}$, $i=1,2$, and $g_2=0$. They proved second-order accuracy in time and optimal order accuracy in space in the L^2 and H^1 norms.

For $r=3$, Bialecki and Fernandes [24] considered two- and three-level Laplace-modified (LM) and ADI LM schemes for the solution of (30) with L_1 and L_2 of (32). Also in [24], a Crank–Nicolson ADI OSC scheme (35), with $r = 3$, $L_2^{k+1/2}$ replaced by L_2^k in the first equation and by L_2^{k+1} in the second equation, was considered for L_1 and L_2 of (32) with $b_1 = b_2 = c = 0$. The stability proof of the scheme hinges on the fact that the operators L_1 and L_2 are nonnegative definite with respect to the collocation inner product. The derived error estimate shows that the scheme is second-order accurate in time and third-order accurate in space in a norm which is stronger than the L^2 norm but weaker than the H^1 norm.

In [25], scheme (35) was considered for $r = 3$, and L_1 and L_2 of (33). It was shown that the scheme is second-order accurate in time and of optimal accuracy in space in the H^1 norm. The analysis in [25] can be easily extended to the case $r \geqslant 3$. A new efficient implementation of the scheme is presented and tested on a sample problem for accuracy and convergence rates in various norms. Earlier implementations of ADI OSC schemes are based on determining, at each time level, a two-dimensional approximation defined on Ω. In the new implementation, at each time level, one-dimensional approximations are determined along horizontal and vertical lines passing through Gauss points and the two-dimensional approximation on Ω is determined only when desired. Note that the two-level, parameter-free ADI OSC scheme does not have a finite element Galerkin counterpart. The method of [60] of comparable accuracy is the three-level ADI LM scheme requiring the selection of a stability parameter.

A nonlinear parabolic IBVP on a rectangular polygon with variable coefficient Robin boundary conditions is considered in [27]. An approximation to the solution at a desired time value is obtained using an ADI extrapolated Crank–Nicolson scheme in which OSC with $r \geqslant 3$ is used for spatial discretization. For rectangular and L-shaped regions, an efficient B-spline implementation of the scheme is described and numerical results are presented which demonstrate the accuracy and convergence rates in various norms.

Fernandes and Fairweather [72] considered the wave equation with homogeneous Dirichlet boundary conditions, which is the special case of the hyperbolic problem (31) with $L_i u = -u_{x_i x_i}$, $i = 1,2$, and $g_1 = 0$. First, the wave equation is rewritten as a system of two equations in two unknown functions. Then an ADI OSC scheme with $r \geqslant 3$ is developed and its second-order accuracy in time and optimal order accuracy in space in the L^2 and H^1 norms are derived.

In [71], two schemes are formulated and analyzed for the approximate solution of (31) with L_1 and L_2 of (32). OSC with $r = 3$ is used for the spatial discretization, and the resulting system of ODEs in the time variable is discretized using perturbations of standard finite difference procedures to produce LM and ADI LM schemes. It is shown that these schemes are unconditionally stable, and of optimal order accuracy in time and of optimal order accuracy in space in the H^1 norm, provided that, in each scheme, the LM stability parameter is chosen appropriately. The algebraic problems to which these schemes lead are also described and numerical results are presented for an implementation of the ADI LM scheme to demonstrate the accuracy and rate of convergence of the method.

In [26], the approximate solution of (31) with L_1 and L_2 of (33) is considered. The new ADI OSC scheme consists in finding $u_h^k \in \mathcal{M}_3(\pi) \otimes \mathcal{M}_3(\pi)$, $k = 2, \ldots, K$, such that, for $k = 1, \ldots, K-1$,

$$[I + \tfrac{1}{2}(\Delta t)^2 L_1^k]\tilde{u}_h^k(\xi_m, \xi_n) = f(\xi_m, \xi_n, t_k) + 2(\Delta t)^{-2} u_h^k(\xi_m, \xi_n), \quad m,n = 1, \ldots, M,$$
$$[I + \tfrac{1}{2}(\Delta t)^2 L_2^k](u_h^{k+1} + u_h^{k-1})(\xi_m, \xi_n) = (\Delta t)^2 \tilde{u}_h^k(\xi_m, \xi_n), \quad m,n = 1, \ldots, M, \tag{36}$$

where L_1^k and L_2^k are the differential operators of (33) with $t = t_k$, and $u_h^0, u_h^1 \in \mathcal{M}_3(\pi) \otimes \mathcal{M}_3(\pi)$, $u_h^k|_{\partial\Omega}$, $k = 2, \ldots, M$, are prescribed by approximating the initial and boundary conditions of (31) using either piecewise Hermite or Gauss interpolants. In the first equation of (36), for each $n = 1, \ldots, M$, $\tilde{u}_h^k(\cdot, \xi_n) \in \mathcal{M}_3(\pi)$ satisfies

$$\tilde{u}_h^k(\alpha, \xi_n) = (\Delta t)^{-2}[I + \tfrac{1}{2}(\Delta t)^2 L_2^k](u_h^{k+1} + u_h^{k-1})(\alpha, \xi_n), \quad \alpha = 0, 1.$$

It is shown in [26] that scheme (36) is second-order accurate in time and of optimal order accuracy in space in the H^1 norm. An efficient implementation of the scheme is similar to that for the scheme of [25] and involves representing u_h^k in terms of basis functions with respect to x_2 alone while $\tilde{u}_h^k$

is represented in terms of basis functions with respect to x_1 only. It is interesting to note that, for variable coefficient hyperbolic problems, the parameter-free ADI OSC scheme (36) does not have a finite element Galerkin counterpart.

In [66], an OSC method is considered for the solution of PDEs that arise in investigating invariant tori for dynamical systems.

4.2. Schrödinger-type problems

In [103], Crank–Nicolson and ADI OSC schemes based on the Crank–Nicolson approach are formulated and analyzed for the approximate solution of the linear Schrödinger problem

$$\begin{aligned} &\psi_t - \mathrm{i}\Delta\psi + \mathrm{i}\sigma(x,t)\psi = f(x,t), \quad (x,t) \in \Omega_T, \\ &\psi(x,0) = \psi^0(x), \quad x \in \Omega, \\ &\psi(x,t) = 0, \quad (x,t) \in \partial\Omega \times (0,T], \end{aligned} \tag{37}$$

where σ is a real function, while ψ, ψ^0 and f are complex valued. A problem of this type of current interest is the so-called parabolic wave equation which arises in wave propagation problems in underwater acoustics; see, for example, [128]. The functions ψ, f and ψ^0 are written as $\psi_1 + \mathrm{i}\psi_2$, $f_1 + \mathrm{i}f_2$ and $\psi_1^0 + \mathrm{i}\psi_2^0$, respectively. Taking real and imaginary parts of (37) then yields

$$\begin{aligned} &\boldsymbol{u}_t + S(-\Delta + \sigma(x,t))\boldsymbol{u} = \boldsymbol{F}(x,t), \quad (x,t) \in \Omega_T, \\ &\boldsymbol{u}(x,0) = \boldsymbol{u}^0(x), \quad x \in \Omega, \\ &\boldsymbol{u}(x,t) = \boldsymbol{0}, \quad (x,t) \in \partial\Omega \times (0,T], \end{aligned} \tag{38}$$

where $\boldsymbol{u} = [\psi_1 \; \psi_2]^{\mathrm{T}}$, $\boldsymbol{F} = [f_1 \; f_2]^{\mathrm{T}}$, and $\boldsymbol{u}^0 = [\psi_1^0 \; \psi_2^0]^{\mathrm{T}}$ are real-valued vector functions, and

$$S = \begin{bmatrix} 0 & -1 \\ 1 & 0 \end{bmatrix}.$$

Hence, (38), and thus (37), is not parabolic but a Schrödinger-type system of partial differential equations. OSC with $r \geqslant 3$ is used for the spatial discretization of (38). The resulting system of ODEs in the time variable is discretized using the trapezoidal rule to produce the Crank–Nicolson OSC scheme, which is then perturbed to obtain an ADI OSC scheme of the form (35). The stability of these schemes is examined, and it is shown that they are second-order accurate in time and of optimal order accuracy in space in the H^1 and L^2 norms. Numerical results are presented which confirm the analysis.

Li et al. [104] considered the OSC solution of the following problem governing the transverse vibrations of a thin square plate clamped at its edges:

$$\begin{aligned} &u_{tt} + \Delta^2 u = f(x,t), \quad (x,t) \in \Omega_T, \\ &u(x,0) = g_0(x), \quad u_t(x,0) = g_1(x), \quad x \in \Omega, \\ &u(x,t) = 0, \quad u_n(x,t) = 0, \quad (x,t) \in \partial\Omega \times (0,T]. \end{aligned} \tag{39}$$

With $u_1 = u_t$, and $u_2 = \Delta u$, and $\boldsymbol{U} = [u_1 \ u_2]^{\mathrm{T}}$, $\boldsymbol{F} = [f \ 0]^{\mathrm{T}}$, and $\boldsymbol{G} = [g_1 \ \Delta g_0]^{\mathrm{T}}$, this problem can be reformulated as the Schrödinger-type problem

$$\begin{aligned} &\boldsymbol{U}_t - S\Delta \boldsymbol{U} = \boldsymbol{F}, \quad (x,t) \in \Omega_T, \\ &\boldsymbol{U}(x,0) = \boldsymbol{G}(x), \quad x \in \Omega, \\ &u_1(x,t) = (u_1)_n(x,t) = 0, \quad (x,t) \in \partial\Omega \times (0,T]. \end{aligned} \tag{40}$$

An approximate solution of (39) is computed by first determining an approximation to the solution U of (40) using the Crank–Nicolson OSC scheme with $r=3$. The existence, uniqueness and stability of this scheme are analyzed and it is shown to be second-order accurate in time and of optimal order accuracy in space in the H^1 and H^2 norms. An approximation to the solution u of (39) with these approximation properties is determined by integrating the differential equation $u_t = u_1$ using the trapezoidal method with u_1 replaced by its Crank–Nicolson OSC approximation.

Similar results also hold for vibration problems with other boundary conditions, specifically, "hinged" boundary conditions,

$$u(x,t) = 0, \quad \Delta u(x,t) = 0, \quad (x,t) \in \partial\Omega \times (0,T]$$

and conditions in which the vertical sides are hinged and the horizontal sides are clamped:

$$\begin{aligned} &u(x,t) = 0, \quad (x,t) \in \partial\Omega \times (0,T], \\ &\Delta u(x,t) = 0, \quad (x,t) \in \partial\Omega_1 \times (0,T], \\ &u_n(x,t) = 0, \quad (x,t) \in \partial\Omega_2 \times (0,T], \end{aligned}$$

where $\partial\Omega_1 = \{(\alpha, x_2)\colon \alpha = 0, 1, \ 0 \leqslant x_2 \leqslant 1\}$ and $\partial\Omega_2 = \{(x_1, \alpha)\colon 0 \leqslant x_1 \leqslant 1, \ \alpha = 0, 1\}$. For these boundary conditions, it is also possible to formulate and analyze Crank–Nicolson ADI OSC methods which cost $O(N^2)$ per time step. Details are given in [105] for piecewise polynomials of arbitrary degree. In the case of (39), no ADI method has been found and, for the case $r = 3$, to solve the linear systems arising at each time step of the Crank–Nicolson scheme, a capacitance matrix method, which was used effectively in [106] for the solution of Dirichlet biharmonic problems, is employed. The cost per time step of this method is $O(N^2 \log N)$. Results of numerical experiments are presented which confirm the theoretical analysis and also indicate that the L^2 norm of the error is of optimal order for the first and third choices of boundary conditions, a result which has yet to be proved analytically.

4.3. *Partial integro-differential equations*

In various fields of engineering and physics, systems which are functions of space and time are often described by PDEs. In some situations, such a formulation may not accurately model the physical system because, while describing the system as a function at a given time, it fails to take into account the effect of past history. Particularly in such fields as heat transfer, nuclear reactor dynamics, and viscoelasticity, there is often a need to reflect the effects of the "memory" of the system. This has resulted in the inclusion of an integral term in the basic PDE yielding a partial integro-differential equation (PIDE). For example, consider PIDEs of the form

$$u_t = \mu\Delta u + \int_0^t a(t-s)\Delta u(x,s)\,\mathrm{d}s + f(x,t), \quad (x,t) \in \Omega_T, \tag{41}$$

which arise in several areas such as heat flow in materials with memory, and in linear viscoelastic problems. When $\mu > 0$, the equation is parabolic; in the case where $\mu = 0$, if the memory function a is differentiable and $a(0) > 0$, the equation is then hyperbolic because (41) can be differentiated with respect to t to give

$$u_{tt} = a(0)\Delta u + \int_0^t a'(t-s)\Delta u(x,s)\,\mathrm{d}s + f_t(x,t), \quad (x,t) \in \Omega_T.$$

However, if a has a "strong" singularity at the origin, that is, $\lim_{t\to 0} a(t) = \infty$, Eq. (41) can only be regarded as intermediate between parabolic and hyperbolic.

In recent years, considerable attention has been devoted to the development and analysis of finite element methods, particularly finite element Galerkin methods, for the solution of parabolic and hyperbolic PIDEs; see, for example, [114] and references cited therein. Much of this work has attempted to carry over to PIDEs standard results for PDEs. OSC methods for PIDEs were first considered by Yanik and Fairweather [143] who formulated and analyzed discrete-time OSC methods for parabolic and hyperbolic PIDEs in one space variable of the form

$$c(x,t,u)\frac{\partial^j u}{\partial t^j} - u_{xx} = \int_0^t f(x,t,s,u(x,s),u_x)\,\mathrm{d}s, \quad (x,t) \in (0,1)\times(0,T],$$

where $j = 1,2$, respectively. In the case of two space variables, techniques developed in [76] can be used to derive optimal order estimates for fully discrete approximations to equations of the form

$$c(x,t)u_t - \Delta u = \int_0^t f(x,t,s,u(x,s))\,\mathrm{d}s, \quad (x,t) \in \Omega_T.$$

However this analysis does not extend to spline collocation methods applied to the PIDE (41).

Fairweather [67] examined two types of spline collocation methods, OSC and modified cubic spline collocation (see Section 5) for (41) in the special case when $\mu = 0$. The error analyses presented in [67] are based on the assumption that the memory function a is "positive", that is,

$$\int_0^t v(\sigma)\int_0^\sigma a(\sigma - s)v(s)\,\mathrm{d}s\,\mathrm{d}\sigma \geqslant 0 \tag{42}$$

for every $v \in C[0,T]$ and for every $t \in [0,T]$. This condition is guaranteed by easily checked sign conditions on the function a and its derivatives, namely,

$$a \in C^2[0,T], \quad (-1)^k a^{(k)}(t) \geqslant 0, \quad k = 0,1,2, \quad a' \not\equiv 0.$$

In the hyperbolic case (μ=0), condition (42) excludes a large class of memory functions. By employing a different approach involving the Laplace transform with respect to time, Yan and Fairweather [142] gave a complete analysis of the stability and convergence of the continuous-time OSC method for PIDEs of the form (41) which is more general than that presented in [67]. Their convergence results hold under much weaker conditions on the function a than (42). For example, a can be a C^2 function with only the additional condition $a(0) > 0$. Moreover, a may be singular, for example, $a(t) = \mathrm{e}^{-\beta t}t^{-\alpha}$ with $0 < \alpha < 1$ and $\beta \in R$.

5. Modified spline collocation methods

The first spline collocation method proposed for the solution of two-point boundary value problems of the form (2)–(3) was the nodal cubic spline method. In this method, one seeks an approximate solution in the space of cubic splines which satisfies the differential equation at the nodes of the partition π and the boundary conditions. De Boor [30] proved that this procedure is second-order accurate and no better, which led researchers to seek cubic spline collocation methods of optimal accuracy, namely, fourth order. Such a method, defined only on a uniform partition, was developed independently by Archer [4] and Daniel and Swartz [55]. In this method, often called modified cubic spline collocation, either the nodal cubic spline collocation solution is improved using a deferred correction-type procedure (the two-step method) or a fourth-order approximation is determined directly by collocating a high-order perturbation of the differential equation (the one-step method). Similar approaches were adopted in [85] in the development of optimal quadratic spline collocation methods for problems of the form (2)–(3) and in [91] for quintic spline collocation methods for general linear fourth-order two-point boundary value problems.

Modified spline collocation methods have also been derived for the solution of elliptic boundary value problems on rectangles; Houstis et al. [89,90] considered both approaches using bicubic splines, and Christara [45] has examined the use of biquadratics. Various direct and iterative methods for the solution of the resulting linear systems were considered for the cubic case in two and three space variables in [77,78,139], and in the biquadratic case in [44,46,47]. In several cases, analysis is carried out only for Helmholtz problems with constant coefficients and homogeneous Dirichlet boundary conditions.

Recently, for Poisson and Helmholtz equations, matrix decomposition algorithms have been developed for both the biquadratic and bicubic cases. In [49], the two-step method was considered for biquadratic splines. In this matrix decomposition approach, the coefficient matrix is diagonalized using FFTs as in Algorithm I of [22]. In [69], the two-step bicubic method was used for various combinations of boundary conditions but a matrix decomposition procedure for the one-step method has been derived only for Dirichlet problems.

A practical advantage of modified spline collocation over OSC is that, for a given partition, there are fewer unknowns with the same degree of piecewise polynomials, thereby reducing the size of the linear systems. However, it does require a uniform partition over the whole of the domain and is only applicable for low degree splines.

It should be noted that modified spline collocation methods may have considerable potential in computational physics, for example, where traditional spline collocation seems to be gaining in popularity over finite element Galerkin methods; see, for example, [34,35,93,140,141].

Acknowledgements

The authors wish to thank Apostolos Hadjidimos, Sang Dong Kim, Yu-Ling Lai, Amiya Pani, and Mark Robinson for providing information for this paper, and also Patrick Keast for his careful reading of the manuscript and many useful suggestions. The comments of the referees were also exceedingly helpful.

References

[1] A. Aitbayev, B. Bialecki, Orthogonal spline collocation for quasilinear elliptic Dirichlet boundary value problems, SIAM J. Numer. Anal., to appear.

[2] A. Aitbayev, B. Bialecki, A preconditioned conjugate gradient method for orthogonal spline collocation problems for linear, non-selfadjoint, indefinite boundary value problems, preprint.

[3] P. Amodio, J.R. Cash, G. Roussos, R.W. Wright, G. Fairweather, I. Gladwell, G.L. Kraut, M. Paprzycki, Almost block diagonal linear systems: sequential and parallel solution techniques, and applications, Numer. Linear Algebra Appl. 7 (2000) 275–317.

[4] D. Archer, An $O(h^4)$ cubic spline collocation method for quasilinear parabolic equations, SIAM J. Numer. Anal. 14 (1977) 620–637.

[5] V.K. Bangia, C. Bennett, A. Reynolds, R. Raghavan, G. Thomas, Alternating direction collocation methods for simulating reservoir performance, Paper SPE 7414, 53rd SPE Fall Technical Conference and Exhibition, Houston, Texas, 1978.

[6] K.R. Bennett, Parallel collocation methods for boundary value problems, Ph.D. Thesis, University of Kentucky, 1991.

[7] B. Bialecki, An alternating direction implicit method for orthogonal spline collocation linear systems, Numer. Math. 59 (1991) 413–429.

[8] B. Bialecki, A fast domain decomposition Poisson solver on a rectangle for Hermite bicubic orthogonal spline collocation, SIAM J. Numer. Anal. 30 (1993) 425–434.

[9] B. Bialecki, Cyclic reduction and FACR methods for piecewise Hermite bicubic orthogonal spline collocation, Numer. Algorithms 8 (1994) 167–184.

[10] B. Bialecki, Preconditioned Richardson and minimal residual iterative methods for piecewise Hermite bicubic orthogonal spline collocation equations, SIAM J. Sci. Comput. 15 (1994) 668–680.

[11] B. Bialecki, Convergence analysis of orthogonal spline collocation for elliptic boundary value problems, SIAM J. Numer. Anal. 35 (1998) 617–631.

[12] B. Bialecki, Superconvergence of orthogonal spline collocation in the solution of Poisson's equation, Numer. Methods Partial Differential Equations 15 (1999) 285–303.

[13] B. Bialecki, Piecewise Hermite bicubic orthogonal spline collocation for Poisson's equation on a disk, preprint.

[14] B. Bialecki, A fast solver for the orthogonal spline collocation solution of the biharmonic Dirichlet problem on rectangles, preprint.

[15] B. Bialecki, X.-C. Cai, H^1-norm error estimates for piecewise Hermite bicubic orthogonal spline collocation schemes for elliptic boundary value problems, SIAM J. Numer. Anal. 31 (1994) 1128–1146.

[16] B. Bialecki, X.-C. Cai, M. Dryja, G. Fairweather, An additive Schwarz algorithm for piecewise Hermite bicubic orthogonal spline collocation, Proceedings of the Sixth International Conference on Domain Decomposition Methods in Science and Engineering, Contemporary Mathematics, Vol. 157, American Mathematical Society, Providence, RI, 1994, pp. 237–244.

[17] B. Bialecki, D.S. Dillery, Fourier analysis of Schwarz alternating methods for piecewise Hermite bicubic orthogonal spline collocation, BIT 33 (1993) 634–646.

[18] B. Bialecki, M. Dryja, Multilevel additive and multiplicative methods for orthogonal spline collocation problems, Numer. Math. 77 (1997) 35–58.

[19] B. Bialecki, M. Dryja, A domain decomposition method for orthogonal spline collocation, preprint.

[20] B. Bialecki, G. Fairweather, Matrix decomposition algorithms for separable elliptic boundary value problems in two dimensions, J. Comput. Appl. Math. 46 (1993) 369–386.

[21] B. Bialecki, G. Fairweather, Matrix decomposition algorithms in orthogonal spline collocation for separable elliptic boundary value problems, SIAM J. Sci. Comput. 16 (1995) 330–347.

[22] B. Bialecki, G. Fairweather, K.R. Bennett, Fast direct solvers for piecewise Hermite bicubic orthogonal spline collocation equations, SIAM J. Numer. Anal. 29 (1992) 156–173.

[23] B. Bialecki, G. Fairweather, K.A. Remington, Fourier methods for piecewise Hermite bicubic orthogonal spline collocation, East–West J. Numer. Math. 2 (1994) 1–20.

[24] B. Bialecki, R.I. Fernandes, Orthogonal spline collocation Laplace-modified and alternating-direction collocation methods for parabolic problems on rectangles, Math. Comp. 60 (1993) 545–573.

[25] B. Bialecki, R.I. Fernandes, An orthogonal spline collocation alternating direction implicit Crank–Nicolson method for linear parabolic problems on rectangles, SIAM J. Numer. Anal. 36 (1999) 1414–1434.

[26] B. Bialecki, R.I. Fernandes, An orthogonal spline collocation alternating direction implicit method for linear second order hyperbolic problems on rectangles, preprint.

[27] B. Bialecki, R.I. Fernandes, An orthogonal spline collocation alternating direction implicit Crank–Nicolson method for nonlinear parabolic problems on rectangular polygons, preprint.

[28] B. Bialecki, K.A. Remington, Fourier matrix decomposition method for least squares solution of singular Neumann and periodic Hermite bicubic collocation problems, SIAM J. Sci. Comput. 16 (1995) 431–451.

[29] P.E. Bjørstad, Fast numerical solution of the biharmonic Dirichlet problem on rectangles, SIAM J. Numer. Anal. 20 (1983) 59–71.

[30] C. de Boor, The method of projections as applied to the numerical solution of two point boundary value problems using cubic splines, Ph.D. Thesis, University of Michigan, Ann Arbor, Michigan, 1966.

[31] C. de Boor, A Practical Guide to Splines, Applied Mathematical Sciences, Vol. 27, Springer, New York, 1978.

[32] C. de Boor, B. Swartz, Collocation at Gauss points, SIAM J. Numer. Anal. 10 (1973) 582–606.

[33] J.F. Botha, M. Celia, The alternating direction collocation approximation, Proceedings of the Eighth South African Symposium on Numerical Mathematics, Durban, South Africa, July 1982, pp. 13–26.

[34] C. Bottcher, M.R. Strayer, The basis spline method and associated techniques, in: C. Bottcher, M.R. Strayer, J.B. McGrory (Eds.), Computational Atomic and Nuclear Physics, World Scientific Co., Singapore, 1990, pp. 217–240.

[35] C. Bottcher, M.R. Strayer, Spline methods for conservation equations, in: D. Lee, A.R. Robinson, R. Vichnevetsky (Eds.), Computational Acoustics, Vol. 2, Elsevier Science Publishers, Amsterdam, 1993, pp. 317–338.

[36] M.A. Celia, L.R. Ahuja, G.F. Pinder, Orthogonal collocation and alternating-direction methods for unsaturated flow, Adv. Water Resour. 10 (1987) 178–187.

[37] M.A. Celia, G.F. Pinder, Collocation solution of the transport equation using a locally enhanced alternating direction formulation, in: H. Kardestuncer (Ed.), Unification of Finite Element Methods, Elsevier, New York, 1984, pp. 303–320.

[38] M.A. Celia, G.F. Pinder, An analysis of alternating-direction methods for parabolic equations, Numer. Methods Partial Differential Equations 1 (1985) 57–70.

[39] M.A. Celia, G.F. Pinder, Generalized alternating-direction collocation methods for parabolic equations. I. Spatially varying coefficients, Numer. Methods Partial Differential Equations 6 (1990) 193–214.

[40] M.A. Celia, G.F. Pinder, Generalized alternating-direction collocation methods for parabolic equations. II. Transport equations with application to seawater intrusion problems, Numer. Methods Partial Differential Equations 6 (1990) 215–230.

[41] M.A. Celia, G.F. Pinder, Generalized alternating-direction collocation methods for parabolic equations. III. Nonrectangular domains, Numer. Methods Partial Differential Equations 6 (1990) 231–243.

[42] M.A. Celia, G.F. Pinder, L.J. Hayes, Alternating-direction collocation solution to the transport equation, in: S.Y. Wang et al. (Eds.), Proceedings of the Third International Conference on Finite Elements in Water Resources, University of Mississippi, Oxford, MS, 1980, pp. 3.36–3.48.

[43] J.H. Cerutti, S.V. Parter, Collocation methods for parabolic partial differential equations in one space dimension, Numer. Math. 26 (1976) 227–254.

[44] C.C. Christara, Schur complement preconditioned conjugate gradient methods for spline collocation equations, Comput. Architecture News 18 (1990) 108–120.

[45] C.C. Christara, Quadratic spline collocation methods for elliptic partial differential equations, BIT 34 (1994) 33–61.

[46] C.C. Christara, Parallel solvers for spline collocation equations, Adv. Eng. Software 27 (1996) 71–89.

[47] C.C. Christara, B. Smith, Multigrid and multilevel methods for quadratic spline collocation, BIT 37 (1997) 781–803.

[48] P.G. Ciarlet, P.A. Raviart, A mixed finite element method for the biharmonic equation, in: C. de Boor (Ed.), Mathematical Aspects of Finite Elements in Partial Differential Equations, Academic Press, New York, 1974, pp. 125–145.

[49] A. Constas, Fast Fourier transform solvers for quadratic spline collocation, M.Sc. Thesis, Department of Computer Science, University of Toronto, 1996.

[50] K.D. Cooper, Domain-embedding alternating direction method for linear elliptic equations on irregular regions using collocation, Numer. Methods Partial Differential Equations 9 (1993) 93–106.

[51] K.D. Cooper, K.M. McArthur, P.M. Prenter, Alternating direction collocation for irregular regions, Numer. Methods Partial Differential Equations 12 (1996) 147–159.

[52] K.D. Cooper, P.M. Prenter, A coupled double splitting ADI scheme for first biharmonic using collocation, Numer. Methods Partial Differential Equations 6 (1990) 321–333.

[53] K.D. Cooper, P.M. Prenter, Alternating direction collocation for separable elliptic partial differential equations, SIAM J. Numer. Anal. 28 (1991) 711–727.

[54] M.C. Curran, M.B. Allen III, Parallel computing for solute transport models via alternating direction collocation, Adv. Water Resour. 13 (1990) 70–75.

[55] J.W. Daniel, B.K. Swartz, Extrapolated collocation for two-point boundary value problems using cubic splines, J. Inst. Math. Appl. 16 (1975) 161–174.

[56] J.C. Diaz, G. Fairweather, P. Keast, FORTRAN packages for solving certain almost block diagonal linear systems by modified alternate row and column elimination, ACM Trans. Math. Software 9 (1983) 358–375.

[57] J.C. Diaz, G. Fairweather, P. Keast, Algorithm 603 COLROW and ARCECO: FORTRAN packages for solving certain almost block diagonal linear systems by modified alternate row and column elimination, ACM Trans. Math. Software 9 (1983) 376–380.

[58] D.S. Dillery, High order orthogonal spline collocation schemes for elliptic and parabolic problems, Ph.D. Thesis, University of Kentucky, 1994.

[59] J. Douglas Jr., A superconvergence result for the approximate solution of the heat equation by a collocation method, in: A.K. Aziz (Ed.), Mathematical Foundations of the Finite Element Method with Applications to Partial Differential Equations, Academic Press, New York, 1972, pp. 475–490.

[60] J. Douglas Jr., T. Dupont, Alternating direction Galerkin methods on rectangles, in: B. Hubbard (Ed.), Numerical Solution of Partial Differential Equations – II, Academic Press, New York, 1971, pp. 133–214.

[61] J. Douglas Jr., T. Dupont, A finite element collocation method for the heat equation, Ist. Nazionale Alta Mat. Symp. Math. 10 (1972) 403–410.

[62] J. Douglas Jr., T. Dupont, A finite element collocation method for quasilinear parabolic equations, Math. Comp. 27 (1973) 17–28.

[63] J. Douglas Jr., T. Dupont, in: Collocation Methods for Parabolic Equations in a Single Space Variable, Lecture Notes in Mathematics, Vol. 385, Springer, New York, 1974.

[64] W.R. Dyksen, Tensor product generalized ADI methods for separable elliptic problems, SIAM J. Numer. Anal. 24 (1987) 59–76.

[65] W.R. Dyksen, A tensor product generalized ADI method for the method of planes, Numer. Methods Partial Differential Equations 4 (1988) 283–300.

[66] K.D. Edoh, R.D. Russell, W. Sun, Computation of invariant tori by orthogonal collocation, Appl. Numer. Math. 32 (2000) 273–289.

[67] G. Fairweather, Spline collocation methods for a class of hyperbolic partial integro-differential equations, SIAM J. Numer. Anal. 31 (1994) 444–460.

[68] G. Fairweather, K.R. Bennett, B. Bialecki, Parallel matrix decomposition algorithms for separable elliptic boundary value problems, in: B.J. Noye, B.R. Benjamin, L.H. Colgan (Eds.), Computational Techniques and Applications: CTAC-91, Proceedings of the 1991 International Conference on Computational Techniques and Applications Adelaide, South Australia, July 1991, Computational Mathematics Group, Australian Mathematical Society, 1992, pp. 63–74.

[69] G. Fairweather, A. Karageorghis, Matrix decomposition algorithms for modified cubic spline collocation for separable elliptic boundary value problems. I. Helmholtz problems, Technical Report 01/2000, Department of Mathematics and Statistics, University of Cyprus, 2000.

[70] G. Fairweather, D. Meade, A survey of spline collocation methods for the numerical solution of differential equations, in: J.C. Diaz (Ed.), Mathematics for Large Scale Computing, Lecture Notes in Pure and Applied Mathematics, Vol. 120, Marcel Dekker, New York, 1989, pp. 297–341.

[71] R.I. Fernandes, Efficient orthogonal spline collocation methods for solving linear second order hyperbolic problems on rectangles, Numer. Math. 77 (1997) 223–241.

[72] R.I. Fernandes, G. Fairweather, Analysis of alternating direction collocation methods for parabolic and hyperbolic problems in two space variables, Numer. Methods Partial Differential Equations 9 (1993) 191–211.

[73] G.H. Golub, C.F. Van Loan, Matrix Computations, 3rd Edition, The Johns Hopkins University Press, Baltimore, Maryland, 1996.

[74] A. Grabarski, The collocation method for the quasi-linear elliptic equation of second order, Demonstratio Math. 29 (1986) 431–447.

[75] A. Grabarski, The collocation method for a quasi-linear parabolic equation in two space variables, Demonstratio Math. 29 (1986) 831–839.

[76] C.E. Greenwell-Yanik, G. Fairweather, Analyses of spline collocation methods form parabolic and hyperbolic problems in two space variables, SIAM J. Numer. Anal. 23 (1986) 282–296.

[77] A. Hadjidimos, E.N. Houstis, J.R. Rice, E. Vavalis, Iterative line cubic spline collocation methods for elliptic partial differential equations in several dimensions, SIAM J. Sci. Statist. Comput. 14 (1993) 715–734.

[78] A. Hadjidimos, E.N. Houstis, J.R. Rice, E. Vavalis, Analysis of iterative line spline collocation methods for elliptic partial differential equations, SIAM J. Matrix Anal. Appl. 21 (1999) 508–521.

[79] A. Hadjidimos, Y.G. Saridakis, Modified successive overrelaxation (MSOR) and equivalent 2-step iterative methods for collocation matrices, J. Comput. Appl. Math. 42 (1992) 375–393.

[80] L.J. Hayes, An alternating-direction collocation method for finite element approximation on rectangles, Comput. Math. Appl. 6 (1980) 45–50.

[81] L.J. Hayes, A comparison of alternating-direction collocation methods for the transport equation, in: T.J.R. Hughes et al. (Eds.), New Concepts in Finite Element Analysis, AMS-Vol. 44, American Society of Mechanical Engineers, New York, 1981, pp. 169–177.

[82] L.J. Hayes, G.F. Pinder, M. Celia, Alternating-direction collocation for rectangular regions, Comput. Methods Appl. Mech. Eng. 27 (1981) 265–277.

[83] E.N. Houstis, Application of method of collocation on lines for solving nonlinear hyperbolic problems, Math. Comp. 31 (1977) 443–456.

[84] E.N. Houstis, Collocation methods for linear elliptic problems, BIT 18 (1978) 301–310.

[85] E.N. Houstis, C.C. Christara, J.R. Rice, Quadratic-spline collocation methods for two-point boundary value problems, Internat. J. Numer. Methods Eng. 26 (1988) 935–952.

[86] E.N. Houstis, W.F. Mitchell, J.R. Rice, Collocation software for second-order elliptic partial differential equations, ACM Trans. Math. Software 11 (1985) 379–412.

[87] E.N. Houstis, W.F. Mitchell, J.R. Rice, Algorithm GENCOL: Collocation on general domains with bicubic Hermite polynomials, ACM Trans. Math. Software 11 (1985) 413–415.

[88] E.N. Houstis, W.F. Mitchell, J.R. Rice, Algorithms INTCOL and HERMCOL: collocation on rectangular domains with bicubic Hermite polynomials, ACM Trans. Math. Software 11 (1985) 416–418.

[89] E.N. Houstis, J.R. Rice, E.A. Vavalis, Parallelization of a new class of cubic spline collocation methods, in: R. Vichnevetsky, R.S. Stepleman (Eds.), Advances in Computer Methods for Partial Differential Equations VI, Publ. IMACS, 1987, pp. 167–174.

[90] E.N. Houstis, E.A. Vavalis, J.R. Rice, Convergence of $O(h^4)$ cubic spline collocation methods for elliptic partial differential equations, SIAM J. Numer. Anal. 25 (1988) 54–74.

[91] M. Irodotou-Ellina, E.N. Houstis, An $O(h^6)$ quintic spline collocation method for fourth order two-point boundary value problems, BIT 28 (1988) 288–301.

[92] P. Keast, P.H. Muir, Algorithm 688: EPDCOL: a more efficient PDECOL code, ACM Trans. Math. Software 17 (1991) 153–166.

[93] D.R. Kegley Jr., V.E. Oberacker, M.R. Strayer, A.S. Umar, J.C. Wells, Basis spline collocation method for solving the Schrödinger equation in axillary symmetric systems, J. Comput. Phys. 128 (1996) 197–208.

[94] H.B. Keller, Approximate methods for nonlinear problems with application to two-point boundary value problems, Math. Comp. 29 (1975) 464–474.

[95] H.O. Kim, S.D. Kim, Y.H. Lee, Finite difference preconditioning cubic spline collocation method of elliptic equations, Numer. Math. 77 (1997) 83–103.

[96] S. Kim, S. Kim, Estimating convergence factors of Schwarz algorithms for orthogonal spline collocation method, Bull. Korean Math. Soc. 36 (1999) 363–370.

[97] S.D. Kim, Preconditioning cubic spline collocation discretization by P_1 finite element method on an irregular mesh, Kyungpook Math. J. 35 (1996) 619–631.

[98] S.D. Kim, S.V. Parter, Preconditioning cubic spline collocation discretizations of elliptic equations, Numer. Math. 72 (1995) 39–72.

[99] Y.-L. Lai, A. Hadjidimos, E.N. Houstis, A generalized Schwarz splitting method based on Hermite collocation for elliptic boundary value problems, Appl. Numer. Math. 21 (1996) 265–290.

[100] Y.-L. Lai, A. Hadjidimos, E.N. Houstis, J.R. Rice, General interior Hermite collocation methods for second order elliptic partial differential equations, Appl. Numer. Math. 16 (1994) 183–200.

[101] Y.-L. Lai, A. Hadjidimos, E.N. Houstis, J.R. Rice, On the iterative solution of Hermite collocation equations, SIAM J. Matrix Anal. Appl. 16 (1995) 254–277.

[102] Y.H. Lee, Numerical solution of linear elasticity by preconditioning cubic spline collocation, Comm. Korean Math. Soc. 11 (1996) 867–880.

[103] B. Li, G. Fairweather, B. Bialecki, Discrete-time orthogonal spline collocation methods for Schrödinger equations in two space variables, SIAM J. Numer. Anal. 35 (1998) 453–477.

[104] B. Li, G. Fairweather, B. Bialecki, A Crank–Nicolson orthogonal spline collocation method for vibration problems, Appl. Numer. Math. 33 (2000) 299–306.

[105] B. Li, G. Fairweather, B. Bialecki, Discrete-time orthogonal spline collocation methods for vibration problems, Technical Report 00-05, Dept. of Mathematical and Computer Sciences, Colorado School of Mines, 2000.

[106] Z.-M. Lou, B. Bialecki, G. Fairweather, Orthogonal spline collocation methods for biharmonic problems, Numer. Math. 80 (1998) 267–303.

[107] N.K. Madsen, R.F. Sincovec, Algorithm 540. PDECOL: general collocation software for partial differential equations, ACM Trans. Math. Software 5 (1979) 326–351.

[108] F. Majaess, P. Keast, G. Fairweather, Packages for solving almost block diagonal linear systems arising in spline collocation at Gaussian points with monomial basis functions, in: J.C. Mason, M.G. Cox (Eds.), Scientific Software Systems, Chapman & Hall, London, 1990, pp. 47–58.

[109] F. Majaess, P. Keast, G. Fairweather, The solution of almost block diagonal linear systems arising in spline collocation at Gaussian points with monomial basis functions, ACM Trans. Math. Software 18 (1992) 193–204.

[110] F. Majaess, P. Keast, G. Fairweather, K.R. Bennett, Algorithm 704: ABDPACK and ABBPACK – FORTRAN programs for the solution of almost block diagonal linear systems arising in spline collocation at Gaussian points with monomial basis functions, ACM Trans. Math. Software 18 (1992) 205–210.

[111] A.V. Manickam, K.M. Moudgalya, A.K. Pani, Second-order splitting combined with orthogonal cubic spline collocation method for the Kuramoto-Sivashinsky equation, Comput. Math. Appl. 35 (1998) 5–25.

[112] S.A.V. Manickam, A.K. Pani, S.K. Chung, A second-order splitting combined with orthogonal cubic spline collocation method for the Rosenau equation, Numer. Methods Partial Differential Equations 14 (1998) 695–716.

[113] G. Mateescu, C.J. Ribbens, L.T. Watson, A domain decomposition algorithm for collocation problems, preprint.

[114] W. McLean, V. Thomée, Numerical solution of an evolution equation with a positive type memory term, J. Austral. Math. Soc. Ser. B 35 (1993) 23–70.

[115] M. Mu, J.R. Rice, An experimental performance analysis of the rate of convergence of collocation on general domains, Numer. Methods Partial Differential Equations 5 (1989) 45–52.

[116] T.B. Nokonechny, P. Keast, P.H. Muir, A method of lines package, based on monomial spline collocation, for systems of one-dimensional parabolic differential equations, in: D.F. Griffiths, G.A. Watson (Eds.), Numerical Analysis, A. R. Mitchell 75th Birthday Volume, World Scientific, Singapore, 1996, pp. 207–223.

[117] P. Percell, M.F. Wheeler, A C^1 finite element collocation method for elliptic equations, SIAM J. Numer. Anal. 12 (1980) 605–622.

[118] L. Plagne, J.-Y. Berthou, Tensorial basis spline collocation method for Poisson's equation, J. Comput. Phys. 157 (2000) 419–440.

[119] R. Pozo, K. Remington, Fast three-dimensional elliptic solvers on distributed network clusters, in: G.R. Joubert et al. (Eds.), Parallel Computing: Trends and Applications, Elsevier, Amsterdam, 1994, pp. 201–208.

[120] P.M. Prenter, R.D. Russell, Orthogonal collocation for elliptic partial differential equations, SIAM J. Numer. Anal. 13 (1976) 923–939.

[121] J.R. Rice, R.F. Boisvert, Solving Elliptic Problems using ELLPACK, Springer, New York, 1985.

[122] M.P. Robinson, Orthogonal spline collocation solution of nonlinear Schrödinger equations, Mathematics of Computation 1943–1993: a Half-Century of Computational Mathematics, Vancouver, BC, 1993, Proceedings of Symposium on Applied Mathematics, Vol. 48, American Mathematical Society, Providence RI, 1994, pp. 355–360.

[123] M.P. Robinson, The solution of nonlinear Schrödinger equations using orthogonal spline collocation, Comput. Math. Appl. 33 (1997) 39–57. [Corrigendum: Comput. Math. Appl. 35 (1998) 151.]

[124] M.P. Robinson, G. Fairweather, Orthogonal cubic spline collocation solution of underwater acoustic wave propagation problems, J. Comput. Acoust. 1 (1993) 355–370.

[125] M.P. Robinson, G. Fairweather, An orthogonal spline collocation method for the numerical solution of underwater acoustic wave propagation problems, in: D. Lee, A.R. Robinson, R. Vichnevetsky (Eds.), Computational Acoustics, Vol. 2, North-Holland, Amsterdam, 1993, pp. 339–353.

[126] M.P. Robinson, G. Fairweather, Orthogonal spline collocation methods for Schrödinger-type equations in one space variable, Numer. Math. 68 (1994) 355–376.

[127] R.D. Russell, W. Sun, Spline collocation differentiation matrices, SIAM J. Numer. Anal. 34 (1997) 2274–2287.

[128] F. Saied, M.J. Holst, Multigrid methods for computational acoustics on vector and parallel computers, in: R.L. Lau, D. Lee, A.R. Robinson (Eds.), Computational Ocean Acoustics, Vol. 1, North-Holland, Amsterdam, 1993, pp. 71–80.

[129] A.A. Samarskii, E.S. Nikolaev, Numerical Methods for Grid Equations. Vol. II: Iterative Methods, Birkhäuser, Basel, 1989.

[130] W. Sun, Orthogonal collocation solution of biharmonic equations, Internat J. Comput. Math. 49 (1993) 221–232.

[131] W. Sun, Iterative algorithms for solving Hermite bicubic collocation equations, SIAM J. Sci. Comput. 16 (1995) 720–737.

[132] W. Sun, A higher order direct method for solving Poisson's equation on a disc, Numer. Math. 70 (1995) 501–506.

[133] W. Sun, Block iterative algorithms for solving Hermite bicubic collocation equations, SIAM J. Numer. Anal. 33 (1996) 589–601.

[134] W. Sun, Fast algorithms for high-order spline collocation systems, Numer. Math. 81 (1998) 143–160.

[135] W. Sun, Spectral analysis of Hermite cubic spline collocation systems, SIAM J. Numer. Anal. 36 (1999) 1962–1975.

[136] W. Sun, W. Huang, R.D. Russell, Finite difference preconditioning for solving orthogonal collocation for boundary value problems, SIAM J. Numer. Anal. 33 (1996) 2268–2285.

[137] W. Sun, N.G. Zamani, A fast algorithm for solving the tensor product collocation equations, J. Franklin Inst. 326 (1989) 295–307.

[138] P.N. Swarztrauber, R.A. Sweet, The direct solution of the discrete Poisson equation on a disk, SIAM J. Numer. Anal. 10 (1973) 900–907.

[139] P. Tsompanopoulou, E. Vavalis, ADI methods for cubic spline collocation discretizations of elliptic PDEs, SIAM J. Sci. Comput. 19 (1998) 341–363.

[140] A.S. Umar, Three-dimensional HF and TDHF calculations with the basis-spline collocation technique, in: C. Bottcher, M.R. Strayer, J.B. McGrory (Eds.), Computational Atomic and Nuclear Physics, World Scientific Co., Singapore, 1990, pp. 377–390.

[141] A.S. Umar, J. Wu, M.R. Strayer, C. Bottcher, Basis-spline collocation method for the lattice solution of boundary value problems, J. Comput. Phys. 93 (1991) 426–448.

[142] Y. Yan, G. Fairweather, Orthogonal spline collocation methods for some partial integrodifferential equations, SIAM J. Numer. Anal. 29 (1992) 755–768.

[143] E.G. Yanik, G. Fairweather, Finite element methods for parabolic and hyperbolic partial integro-differential equations, Nonlinear Anal. 12 (1988) 785–809.

ELSEVIER

Journal of Computational and Applied Mathematics 128 (2001) 83–131

JOURNAL OF COMPUTATIONAL AND APPLIED MATHEMATICS

www.elsevier.nl/locate/cam

Spectral methods for hyperbolic problems ☆

D. Gottlieb, J.S. Hesthaven *

Division of Applied Mathematics, Brown University, Box F, Providence, RI 02912, USA

Received 9 February 2000

Abstract

We review the current state of Fourier and Chebyshev collocation methods for the solution of hyperbolic problems with an eye to basic questions of accuracy and stability of the numerical approximations. Throughout the discussion we emphasize recent developments in the area such as spectral penalty methods, the use of filters, the resolution of the Gibbs phenomenon, and issues related to the solution of nonlinear conservations laws such as conservation and convergence. We also include a brief discussion on the formulation of multi-domain methods for hyperbolic problems, and conclude with a few examples of the application of pseudospectral/collocation methods for solving nontrivial systems of conservation laws. © 2001 Elsevier Science B.V. All rights reserved.

Keywords: Spectral; Pseudospectral; Collocation; Penalty methods; Discontinuous solutions; Gibbs phenomenon; Stability; Filtering; Vanishing viscosity; Multi-domain methods

1. Introduction

The theory, implementation and application of spectral and pseudospectral methods for the solution of partial differential equations has traditionally been centered around problems with a certain amount of inherent regularity of the solutions, e.g., elliptic/parabolic problems. Among many examples, the application that is perhaps most responsible for the widespread use of spectral methods is the accurate and efficient solution of the incompressible Navier–Stokes equations [10].

On the other hand, the application of spectral and pseudospectral methods for the solution of hyperbolic problems, and in particular nonlinear conservation laws which are prone to develop discontinuous solutions in finite time, has traditionally been viewed as problematic. Indeed, with a few noticeable exceptions, very little work was done to adapt spectral and pseudospectral methods to the

☆ This work has been supported by DARPA/AFOSR grant F49620-96-1-0426, by DOE grant DE-FG02-95ER25239, and by NSF grant ASC-9504002.

* Corresponding author.

E-mail addresses: dig@cfm.brown.edu (D. Gottlieb), jan_hesthaven@brown.edu (J.S. Hesthaven).

PII: S0377-0427(00)00510-0

solution of such important classes of problems as the equations of gas-dynamics and electromagnetics until the late 1980s and early 1990s.

The reasons for the perceived difficulty are several. Contrary to parabolic and elliptic problems, there is no physical dissipation inherent in the hyperbolic problem. This again implies that even minor errors and under resolved phenomena can cause the scheme to become unstable, i.e., the question of stability of the spectral approximations tends to be even more critical than for other types of problems. Perhaps the most important reason, however, for the slow acceptance of the use of high-order methods in general and pseudospectral methods in particular for solving hyperbolic conservation laws can be found in the appearance of the Gibbs phenomenon as finite-time discontinuities develop in the solution. Left alone, the nonlinear mixing of the Gibbs oscillations with the approximate solution will eventually cause the scheme to become unstable. Moreover, even if stability is maintained sufficiently long, the computed solution appears to be only first-order accurate in which case the use of a high-order method is questionable. More fundamental issues of conservation and the ability of the scheme to compute the correct entropy solution to conservation laws have also caused considerable concern among practitioners and theoreticians alike.

While many of these issues are genuine and requires careful attention they are not causing the pseudospectral approach to fail if applied correctly. This was hinted to in early work around 1980 [74,66,36] where the first numerical solution of problems with discontinuous solutions and general nonlinear conservation laws were presented. It is, however, mainly within the last decade that many of the most significant advances has been made to establish the soundness of the pseudospectral approach for such problems, often confirming that the superior behavior of these methods for smooth problems carries over to problems involving nonsmooth solutions.

It is the central components of these more recent developments that we shall review in the following. We do not attempt to be complete in the treatment and discussion. We do hope, however, that the review will offer enough information to allow the reader to venture deeper into the fascinating theory and the application of spectral and pseudospectral methods to the solution of hyperbolic conservations laws. It is a topic that has challenged researchers in the last decades of this century and is certain to continue to do so in the next decades.

What remains of this review is organized as follows. In Section 2 we introduce the spectral and pseudospectral approximation of spatial derivatives using Fourier and Chebyshev series approximations of the function. We highlight the duality between the modal representation, exploiting quadratures to approximate inner products, and the nodal formulation, using Lagrange interpolation polynomials, and emphasize how this duality suggests two computationally different but mathematically equivalent formulations of spectral/pseudospectral methods. Section 3 offers a brief overview of the relevant approximation results for spectral and pseudospectral Fourier and Chebyshev expansions of smooth functions. We subsequently discuss the behavior of such expansions for nonsmooth functions, the introduction of the Gibbs phenomenon and recent results on how to improve on the convergence of the approximation away from the discontinuity by the use of filters. We also discuss the complete resolution of the Gibbs phenomenon through the reconstruction of an exponentially convergent approximation to a piecewise smooth function using the information in the global Fourier or Chebyshev expansions only. This sets the stage for Section 4 in which we introduce collocation approximations to hyperbolic problems in general. While we briefly discuss the issues related to Fourier approximations, the emphasis is on the formulation of Chebyshev approximations and techniques to enforce the prescribed boundary conditions. Strongly as well as weakly enforced boundary conditions are addressed and it is shown, among other things, that the discontinuous Galerkin method is a special case of a much larger class of methods. The flexibility of this general approach, known as spectral penalty methods, is illustrated through a few examples. The critical question of stability is considered in Section 5 where we review the stability of Fourier and Chebyshev pseudospectral approximations of linear and variable coefficient hyperbolic problems. We return to the formulation of the penalty methods in more detail and address the question of stability for linear systems of equations, before we briefly attend to the stability of linear problems with nonsmooth initial conditions and conclude with a brief discussion on issues related to fully discrete stability. The introduction of nonlinearities introduces a number of new complications, discussed in Section 6, such as the impact of the Gibbs phenomenon on the stability of the approximation, the use of stabilization by filtering and conservation properties of the pseudospectral approximations. We briefly review recent results on spectrally vanishing viscosity and its relation to filtering techniques. Section 7 is devoted to a brief overview of multi-domain techniques that allow for solving hyperbolic problems in geometrically complex domains or allows the use of a spatially varying resolution. This last development is critical in enabling the use of pseudospectral methods for the solution of hyperbolic conservations laws for realistic problems as we shall illustrate through a few examples in the concluding Section 8.

2. Modes and nodes

Embedded in all numerical schemes for solving partial differential equations (PDEs) lies an assumption about the behavior of the solution to the PDE as reflected in a choice of how to represent the approximate solution.

For spectral methods the traditional approach has been to assume that the solution, $u(x,t)$, can be expressed as a series of smooth basis functions of the form

$$\mathscr{P}_N u(x,t) = \sum_{n=0}^{N} \hat{u}_n(t)\phi_n(x), \tag{1}$$

where the projection, $\mathscr{P}_N u(x,t)$, of the solution is assumed to approximate $u(x,t)$ well in some appropriate norm as N approaches infinity. Hence, the approximate solution, $\mathscr{P}_N u(x,t) \in \mathsf{P}_N$, at all times where the space, $\mathsf{P}_N \in \{\phi_n(x)\}_{n=0}^{N}$, is spanned by smooth basis-functions, $\phi_n(x)$, which we assume form an L^2-complete basis. For the sake of computational efficiency this basis is typically chosen to be orthogonal in a weighted inner-product although this is not a requirement. However, the actual choice of the basis, $\phi_n(x)$, and the way in which the expansion coefficients, $\hat{u}_n(t)$, are computed offers a number of alternative methods which we shall discuss further in the following.

Leaving these details open for a minute it is clear that, under the assumption of Eq. (1), the approximation of a spatial derivative can be expressed in two different ways:

$$\frac{\partial u(x,t)}{\partial x} \simeq \frac{\partial \mathscr{P}_N u(x,t)}{\partial x} = \sum_{n=0}^{N} \hat{u}_n(t)\frac{\mathrm{d}\phi_n(x)}{\mathrm{d}x} = \sum_{n=0}^{N} \hat{u}'_n(t)\phi_n(x), \tag{2}$$

where the first formulation simply involves derivatives of the smooth basis function, $\phi_n(x)$, while the second expression involves the direct expansion of the spatial derivative itself.

Rather than expressing the unknown solution in terms of a basis as in Eq. (1), one could choose to introduce a grid and assume that the solution can be expressed as a global interpolation polynomial

$$\mathscr{I}_N u(x,t) = \sum_{j=0}^{N} u(x_j,t) l_j(x), \tag{3}$$

where $l_j(x)$ represents the Lagrange interpolation polynomial based on the grid points, x_j. Here we have introduced the notation $\mathscr{I}_N u(x,t)$ to reflect the interpolation property, i.e., $\mathscr{I}_N u(x_j,t) = u(x_j,t)$. In this setting spatial derivatives are approximated by

$$\frac{\partial u(x,t)}{\partial x} \simeq \frac{\partial \mathscr{I}_N u(x,t)}{\partial x} = \sum_{j=0}^{N} u(x_j,t) \frac{\mathrm{d} l_j(x)}{\mathrm{d}x}. \tag{4}$$

As we shall see shortly, the approximation, $\mathscr{P}_N u$, in Eqs. (1)–(2) and the interpolation, $\mathscr{I}_N u$, in Eqs. (3)–(4) are closely related and provides a duality which is pivotal in the formulation, analysis and implementation of efficient spectral methods for the solution of hyperbolic problems.

2.1. Modal expansions and derivatives

If one considers the solution of periodic problems it is natural to express the unknown solution as a Fourier series

$$\mathscr{P}_N u(x) = \sum_{n=-N}^{N} \hat{u}_n \exp(inx) \tag{5}$$

with $\phi_n(x) = \exp(inx)$ in Eq. (1). Here and in the following we suppress the explicit time dependency of $u(x,t)$ for simplicity.

The expansion coefficients are obtained directly as

$$\hat{u}_n = \frac{1}{2\pi}(u, \exp(inx))_{L^2[0,2\pi]} = \frac{1}{2\pi}\int_0^{2\pi} u(x)\exp(-inx)\,\mathrm{d}x, \tag{6}$$

through the orthogonality of the basis in the inner product

$$(f,g)_{L^2[0,2\pi]} = \int_0^{2\pi} f\bar{g}\,\mathrm{d}x, \quad \|f\|^2_{L^2[0,2\pi]} = \int_0^{2\pi} |f|^2\,\mathrm{d}x$$

with the associated norm $\|\cdot\|_{L^2[0,2\pi]}$.

The simplicity of the Fourier series makes it straightforward to approximate the spatial derivative

$$\frac{\mathrm{d}^p u(x)}{\mathrm{d}x^p} \simeq \frac{\mathrm{d}^p \mathscr{P}_N u(x)}{\mathrm{d}x^p} = \sum_{n=-N}^{N} (in)^p \hat{u}_n \phi_n(x) = \sum_{n=-N}^{N} \hat{u}_n^{(p)} \phi_n(x),$$

i.e., $\hat{u}_n^{(p)} = (in)^p \hat{u}_n$, for the approximation of an arbitrary derivative of a function given by its Fourier coefficients.

The computation of the Fourier coefficients, $\hat{u}_n$, poses a problem as one cannot, in general, evaluate the integrals in Eq. (6). The natural solution is to introduce a quadrature approximation to Eq. (6) of the form

$$\tilde{u}_n = \frac{1}{2N\tilde{c}_n} \sum_{j=0}^{2N-1} u(x_j) \exp(-inx_j), \tag{7}$$

where $\tilde{c}_N = \tilde{c}_{-N} = 2$ and $\tilde{c}_n = 1$ otherwise. We recognize this as the trapezoidal rule with the equidistant grid

$$x_j = \frac{2\pi}{2N} j, \quad j \in [0, 2N-1]$$

As N increases one hopes that $\tilde{u}_n$ is a good representation of $\hat{u}_n$. To quantify this, we can express $\tilde{u}_n$ using $\hat{u}_n$ as

$$\tilde{c}_n \tilde{u}_n = \hat{u}_n + \sum_{\substack{m=-\infty \\ m \neq 0}}^{m=\infty} \hat{u}_{n+2Nm},$$

where the second term is termed the aliasing error. In particular, if $u(x)$ is bandlimited such that $\hat{u}_{n+2Nm} = 0$ for $|m| > 0$, Eq. (7) is exact.

While the use of trigonometric polynomials is natural for the approximation of periodic problems, an alternative has to be sought when nonperiodic problems are being considered.

A natural basis for the approximation of functions on a finite interval, normalized for convenience to $x \in [-1, 1]$, employs the Chebyshev polynomials, $T_n(x)$, and an approximating expansion of the form

$$\mathscr{P}_N u(x) = \sum_{n=0}^{N} \hat{u}_n \cos(n \arccos x) = \sum_{n=0}^{N} \hat{u}_n T_n(x), \tag{8}$$

i.e., $\phi_n(x) = T_n(x) = \cos(n \arccos x)$ in Eq. (1). The continuous expansion coefficients, $\hat{u}_n$, are found by exploiting the weighted L^2-orthogonality of $T_n(x)$ in the inner product

$$(f, g)_{L^2_w[-1,1]} = \int_{-1}^{1} fg \frac{1}{\sqrt{1-x^2}} \, dx, \quad \|f\|^2_{L^2_w[-1,1]} = \int_{-1}^{1} |f|^2 \frac{1}{\sqrt{1-x^2}} \, dx$$

with the associated norm, $\|\cdot\|_{L^2_w[-1,1]}$. With this, one immediately recovers

$$\hat{u}_n = \frac{2}{\pi c_n} (u, T_n)_{L^2_w[-1,1]}, \tag{9}$$

where $c_0 = 2$ and $c_n = 1$ otherwise.

Given the series for $u(x)$ in Eq. (8) we need, as for the Fourier series, to recover an approximation to the spatial derivative of $u(x)$. This involves the expression of the derivative of the basis in terms of the basis itself. Utilizing the recursion

$$T_n(x) = -\frac{1}{2(n-1)} T'_{n-1}(x) + \frac{1}{2(n+1)} T'_{n+1}(x),$$

we recover

$$T'_n(x) = 2n \sum_{\substack{p=0 \\ p+n \text{ odd}}}^{n-1} \frac{T_p(x)}{c_p}.$$

Hence, the expansion coefficients for the spatial derivative is recovered by matching terms in Eq. (2) to obtain

$$c_n\hat{u}'_n = 2 \sum_{\substack{p=n+1 \\ p+n \text{ odd}}}^{N} p\hat{u}_p.$$

Similar types of expressions can be derived to express higher derivatives. Contrary to the Fourier series, however, the computation of $\hat{u}'_n$ involves global spectral information which makes the straightforward formulation computationally inefficient. The resolution lies in realizing that for all finite expansions the coefficients can be recovered through a backward recursion of the form

$$c_{n-1}\hat{u}'_{n-1} = 2n\hat{u}_n + \hat{u}'_{n+1},$$

keeping in mind that $\hat{u}'_{N+1} = \hat{u}'_N = 0$ due to the nature of the finite approximation. This reduces the computation of a derivative to a linear process.

As for the Fourier expansion the evaluation of the continuous inner product, Eq. (9), is a source of considerable problems. The classical solution lies in the introduction of a Gauss quadrature of the form

$$\tilde{u}_n = \frac{2}{\tilde{c}_n\pi}\sum_{j=0}^{N} u(x_j)T_n(x_j)w_j,$$

as an approximation to the inner product with $N+1$ being the number of grid points in the Gauss quadrature. Among several possible choices, the Chebyshev–Gauss–Lobatto quadrature with

$$x_j = -\cos\left(\frac{\pi}{N}j\right), \quad w_j = \frac{\pi}{\tilde{c}_j N}, \quad \tilde{c}_j = \begin{cases} 2, & j = 0, N, \\ 1, & j = 1,\dots,N-1 \end{cases} \tag{10}$$

is the most popular as it includes the endpoints of the finite interval among the grid points. This is clearly an advantage if one needs to impose boundary conditions. With this approximation, which suffers from aliasing similar to the Fourier series, approximations to derivatives can be recovered as if the continuous expansion coefficients were being used.

2.2. Nodal methods and differentiation matrices

As the use of the modal expansions for all practical purposes requires the introduction of a finite grid one may question the need to consider special basis functions at all. Indeed, given a specific nodal set, x_j, we can construct a global interpolation

$$\mathcal{I}_N u(x) = \sum_{j=0}^{N} u(x_j)l_j(x),$$

where the Lagrange interpolating polynomials, $l_j(x)$, takes the form

$$l_j(x) = \frac{q(x)}{(x-x_j)q'(x_j)}, \quad q(x) = \prod_{j=0}^{N}(x-x_j).$$

Clearly, if the x_j's are distinct, $l_j(x)$ is uniquely determined as an Nth-order polynomial specified at $N+1$ points and we can approximate derivatives of $u(x)$ directly as in Eq. (4). In particular, if we restrict our attention to the approximation of the derivative of $u(x)$ at the grid points, x_j, we have

$$\left.\frac{\mathrm{d}u}{\mathrm{d}x}\right|_{x_i} \simeq \left.\frac{\mathrm{d}\mathscr{I}_N u}{\mathrm{d}x}\right|_{x_i} = \sum_{j=0}^{N} u(x_j)\left.\frac{\mathrm{d}l_j}{\mathrm{d}x}\right|_{x_i} = \sum_{j=0}^{N} u(x_j) D_{ij},$$

where D_{ij} is recognized as a differentiation matrix similar in spirit to a traditional finite difference approximation to spatial derivatives. One should keep in mind, however, that the global nature of the interpolation implies that the differentiation matrix is full.

The specification of the nodes uniquely defines the interpolation polynomials and, hence, the differentiation matrices. Indeed, if we simply take an equidistant grid for $x\in[0,2\pi[$ of the form

$$x_j = \frac{2\pi}{2N} j, \quad j\in[0,2N-1]$$

and assume that the interpolating polynomial itself it 2π-periodic we recover

$$\mathscr{I}_N u(x) = \sum_{j=0}^{2N-1} u(x_j) g_j(x) = \sum_{j=0}^{2N-1} u(x_j)\frac{1}{2N}\sin[N(x-x_j)]\cot\left[\frac{1}{2}(x-x_j)\right] \tag{11}$$

with the entries of the differentiation matrix being

$$D_{ij} = \left.\frac{\mathrm{d}g_j}{\mathrm{d}x}\right|_{x_i} = \begin{cases} \frac{1}{2}(-1)^{i+j}\cot\left[\frac{x_i - x_j}{2}\right], & i\neq j, \\ 0, & i=j. \end{cases} \tag{12}$$

Among other things, we see that the differentiation matrix is an anti-symmetric, circulant Toeplitz matrix. It is interesting to note that Eq. (12) essentially can be obtained as the limit of an infinite order central finite difference stencil under the assumption of periodicity [25].

Turning to the interpolation of nonperiodic functions on finite intervals it is well known [56,26] that one must abandon the equidistant grid to avoid the Runge-phenomenon and the divergence of the Lagrange polynomials and choose a grid that clusters quadratically as

$$x_j \sim -1 + c\left(\frac{j}{N}\right)^2,$$

close to the endpoints. Such nodal sets are plentiful, among them all the zeros of the classical orthogonal polynomials and their derivatives. Indeed, if we consider the set of nodes given as

$$x_j = -\cos\left(\frac{\pi}{N} j\right),$$

which are the roots of the polynomial, $(1-x^2)T_N'(x)$, and recognized as the Chebyshev–Gauss–Lobatto quadrature nodes, Eq. (10), we recover

$$\mathscr{I}_N u(x) = \sum_{j=0}^{N} u(x_j) h_j(x) = \sum_{j=0}^{N} u(x_j)\frac{(-1)^{N+1+j}(1-x^2)T_N'(x)}{\tilde{c}_j N^2 (x-x_j)}, \tag{13}$$

where $\tilde{c}_0 = \tilde{c}_N = 2$ and $\tilde{c}_j = 1$ otherwise. The corresponding differentiation matrix has the entries

$$D_{ij} = \left.\frac{\mathrm{d}h_j}{\mathrm{d}x}\right|_{x_i} = \begin{cases} -\dfrac{2N^2+1}{6}, & i=j=0, \\ \dfrac{\tilde{c}_i}{\tilde{c}_j}\dfrac{(-1)^{i+j+N}}{x_i - x_j}, & i \neq j, \\ -\dfrac{x_i}{2(1-x_i^2)}, & 0 < i = j < N, \\ \dfrac{2N^2+1}{6}, & i=j=N. \end{cases} \tag{14}$$

It is easy to see that $D_{ij} = -D_{N-i,N-j}$, i.e., the differentiation matrix is centro-antisymmetric as a consequence of the reflection symmetry of the nodal set. Moreover, one can show that D is nilpotent.

2.3. The duality between modes and nodes

While there is a great deal of flexibility in the choice of the quadrature rules used to compute the discrete expansion coefficients in the modal expansions, and similar freedom in choosing a nodal set on which to base the Lagrange interpolation polynomials, particular choices are awarded by a deeper insight.

Consider, as an example, the modal expansion, Eq. (5), with the expansion coefficients approximated as in Eq. (7). Inserting the latter directly into the former yields

$$\begin{aligned} \mathscr{P}_N u(x) &= \sum_{n=-N}^{N} \left[\frac{1}{2N\tilde{c}_n} \sum_{j=0}^{2N-1} u(x_j) \exp(-inx_j) \right] \exp(inx) \\ &= \sum_{j=0}^{2N-1} u(x_j) \left[\frac{1}{2N} \sum_{n=-N}^{N} \frac{1}{\tilde{c}_n} \exp(in(x-x_j)) \right] \\ &= \sum_{j=0}^{2N-1} u(x_j) \frac{1}{2N} \sin[N(x-x_j)] \cot\left[\frac{1}{2}(x-x_j)\right], \end{aligned}$$

which we recognize as the periodic interpolation polynomial based on the equidistant grid, Eq. (11). In other words, we have that

$$\mathscr{I}_N u(x) = \sum_{n=-N}^{N} \tilde{u}_n \exp(inx) = \sum_{j=0}^{2N-1} u(x_j) g_j(x),$$

provided the expansion coefficients are approximated by the trapezoidal rule as in Eq. (7). This particular combination of grid points and quadrature rules results in two mathematically equivalent, but computationally very different, ways of expressing the interpolation and hence the computation of spatial derivatives.

In a similar fashion one can show that as a consequence of the Christoffel–Darboux identity for orthogonal polynomials [81] we have

$$\mathscr{I}_N u(x) = \sum_{n=0}^{N} \tilde{u}_n T_n(x) = \sum_{j=0}^{N} u(x_j) h_j(x)$$

with $h_j(x)$ being given in Eq. (13), provided only that the Chebyshev–Gauss–Lobatto quadrature is used to approximate the expansion coefficients, $\tilde{u}_n$, in the modal expansion, Eq. (8). Similar results can be obtained if one chooses a different Gauss quadrature and the corresponding nodes as the grid for the interpolation polynomials.

It is important to appreciate that one need not choose the special quadratures used here or the special nodal sets for the interpolation polynomials to obtain robust and stable spectral schemes. Doing so, however, provides a duality in the formulation that have major advantages in the analysis as well as in the practical implementation of methods for solving hyperbolic problems.

3. Approximation results

In attempting to understand the quality of the computed approximate solutions we need to consider the behavior of the finite-order expansions given in Eqs. (1) and (3) as N increases.

While we shall focus the attention on the behavior of the polynomial interpolation, Eq. (3), we shall also find it useful understand the purely modal expansion, Eq. (1), as the difference between the two is a measure of the aliasing error.

We have chosen to split the subsequent discussion of the approximation results into that of problems possessing a minimum amount of smoothness and the approximation of truly discontinuous functions. Many more results on the approximation of functions using spectral expansions can be found in [10,27,4,6] and references therein.

3.1. Smooth problems

Although the approximations based on Fourier and Chebyshev series are closely related it is instructive to split the discussion of the two as the latter provides an example of results for the much broader class of approximations based on orthogonal polynomials.

3.1.1. Fourier expansions

We begin by considering the truncated continuous Fourier expansion

$$\mathscr{P}_N u(x) = \sum_{n=-N}^{N} \hat{u}_n \exp(inx)\,dx, \quad \hat{u}_n = \frac{1}{2\pi}\int_0^{2\pi} u(x)\exp(-inx)\,dx \tag{15}$$

for which it is clear that

$$\|u - \mathscr{P}_N u\|^2_{L^2[0,2\pi]} = 2\pi \sum_{|n|>N} |\hat{u}_n|^2,$$

as a direct consequence of Parsevals identity for Fourier series. The truncation error depends solely on the decay of $\hat{u}_n$ which behaves as

$$|\hat{u}_n| = \frac{1}{2\pi n^q}\left|\int_0^{2\pi} u^{(q)}(x)\exp(-inx)\,\mathrm{d}x\right|,$$

provided $u(x) \in C^{(q-1)}[0, 2\pi]$, i.e., $u^{(q)} \in L^2[0, 2\pi]$ and $u(x)$ as well as its first $(q-1)$-derivatives are 2π-periodic. Hence, the truncation error is directly related to the smoothness of $u(x)$ as

$$\|u - \mathscr{P}_N u\|_{L^2[0,2\pi]} \leqslant C(q)N^{-q}\|u^{(q)}\|_{L^2[0,2\pi]}. \tag{16}$$

In the event that $u(x)$ is analytic one recovers the remarkable property that [83]

$$\|u - \mathscr{P}_N u\|_{L^2[0,2\pi]} \leqslant C(q)N^{-q}\|u^{(q)}\|_{L^2[0,2\pi]}$$

$$\sim C(q)\frac{q!}{N^q}\|u\|_{L^2[0,2\pi]} \sim C(q)e^{-cN}\|u\|_{L^2[0,2\pi]},$$

known as spectral accuracy or spectral convergence. This is indeed the basic property that has given name to spectral methods.

The diagonality of the modal differentiation operator implies that truncation and differentiation commutes, i.e.,

$$\mathscr{P}_N\frac{\mathrm{d}^q u}{\mathrm{d}x^q} = \frac{\mathrm{d}^q}{\mathrm{d}x^q}\mathscr{P}_N u$$

for $u(x) \in C^{(q-1)}[0, 2\pi]$, and by repeatedly applying Eq. (16) we obtain

$$\|u - \mathscr{P}_N u\|_{W^p[0,2\pi]} \leqslant C(p,q)N^{p-q}\|u\|_{W^q[0,2\pi]},$$

provided only that $0 \leqslant p \leqslant q$. Here, we have introduced the Sobolev norm

$$\|u\|^2_{W^q[0,2\pi]} = \sum_{s=0}^{q}\|u^{(s)}\|^2_{L^2[0,2\pi]}$$

to measure the error on the spatial derivative. Clearly, as long as u is sufficiently smooth and periodic, the error decays rapidly for increasing number of terms, N, and we recover spectral convergence for all spatial derivatives if $u(x) \in C^{\infty}[0, 2\pi]$.

These results address the mean convergence, while the pointwise convergence is a harder but often more useful measure. For the truncated Fourier series one recovers [10]

$$\|u - \mathscr{P}_N u\|_{L^{\infty}[0,2\pi]} \leqslant C(q)(1 + \log N)N^{-q}\|u^{(q)}\|_{L^{\infty}[0,2\pi]},$$

where we have introduced the familiar $L^{\infty}[0, 2\pi]$ norm to measure the maximum pointwise error. This result provides an early indication that we may experience problems if $u(x)$ is only piecewise smooth.

Before addressing this, however, let us consider the behavior of the discrete expansion, Eq. (3), as N increases. As discussed in Section 2.1, the continuous and the discrete expansion coefficients are related as

$$\tilde{c}_n\tilde{u}_n = \hat{u}_n + \sum_{\substack{m=-\infty \\ m \neq 0}}^{m=\infty} \hat{u}_{n+2Nm},$$

provided $u(x) \in W^{1/2}[0, 2\pi]$. The aliasing error, however, has been shown to be of the same order as the truncation error [66], obtained directly from Eq. (16). Thus, as for the continuous expansion, we recover

$$\|u - \mathscr{I}_N u\|_{L^2[0,2\pi]} \leqslant C(q) N^{-q} \|u^{(q)}\|_{L^2[0,2\pi]}$$

and exponential convergence in cases where $u(x) \in C^\infty[0, 2\pi]$ [83]. Numerical evidence for this was first given in [75,25].

Contrary to the continuous expansion, however, truncation and differentiation does not commute

$$\mathscr{I}_N \frac{du}{dx} \neq \frac{d}{dx} \mathscr{I}_N u$$

as a consequence of the aliasing error. Nevertheless, if $u(x)$ is at least continuous, the difference is bounded as [76,83]

$$\left\| \mathscr{I}_N \frac{du}{dx} - \frac{d}{dx} \mathscr{I}_N u \right\|_{L^2[0,2\pi]} \leqslant C(q) N^{1-q} \|u^{(q)}\|_{L^2[0,2\pi]}, \tag{17}$$

provided $u(x) \in W^q[0, 2\pi]$, $q > \frac{1}{2}$. This suggests that the estimate

$$\|u - \mathscr{I}_N u\|_{W^p[0,2\pi]} \leqslant C(p, q) N^{p-q} \|u\|_{W^q[0,2\pi]}$$

for $0 \leqslant p \leqslant q$ provides a good bound on the expected accuracy of the approximation of spatial derivatives using the discrete expansions.

An estimate of the pointwise error is made difficult by the influence of the aliasing error. A bound on the pointwise error is given as [57]

$$\|u - \mathscr{I}_N u\|_{L^\infty[0,2\pi]} \leqslant C(q) \log(N) N^{-q} \|u^{(q)}\|_{L^\infty[0,2\pi]},$$

provided $u(x) \in C^q[0, 2\pi]$, $q > 0$.

3.1.2. Chebyshev expansions

Turning to the truncated Chebyshev expansion we have

$$\mathscr{P}_N u(x) = \sum_{n=0}^{N} \hat{u}_n T_n(x), \quad \hat{u}_n = \frac{2}{\pi c_n} \int_{-1}^{1} u(x) T_n(x) \frac{1}{\sqrt{1-x^2}} \, dx. \tag{18}$$

A direct consequence of Bessels equality for orthogonal expansions is

$$\|u - \mathscr{P}_N u\|^2_{L^2_w[-1,1]} = \frac{2}{\pi} \sum_{|n|>N} |\hat{u}_n|^2.$$

Hence, as for the Fourier series, the truncation error depends solely on the decay of $\hat{u}_n$ as

$$|\hat{u}_n| = \frac{2}{\pi c_n n^{2q}} \left| \int_{-1}^{-1} \left[\sqrt{1-x^2} \frac{d}{dx} \right]^{2q} u(x) T_n(x) \frac{1}{\sqrt{1-x^2}} \, dx \right|,$$

provided $u(x) \in C^{2q-1}[-1, 1]$. This implies that the truncation error is directly related to the smoothness of $u(x)$ as

$$\|u - \mathscr{P}_N u\|_{L^2_w[-1,1]} \leqslant C(q) N^{-q} \left\| \left[\sqrt{1-x^2} \frac{d}{dx} \right]^q u \right\|_{L^2_w[-1,1]} \leqslant C N^{-q} \|u\|_{W^q_w[-1,1]}, \tag{19}$$

where we, as for the Fourier series, have introduced the weighted Sobolev norm

$$\|u\|^2_{W^q_w[-1,1]} = \sum_{s=0}^{q} \|u^{(s)}\|^2_{L^2_w[-1,1]}$$

as a measure of the regularity of $u(x)$. As for the Fourier approximation, the truncation error decays faster than any algebraic order of N if $u(x) \in C^\infty[-1,1]$ [83]. This exponential convergence, however, is achieved with no conditions on $u(x)$ at the boundaries, emphasizing the usefulness of Chebyshev expansions for the approximation of problems defined on finite, nonperiodic domains.

Contrary to the Fourier series, however, even the continuous expansions do not permit commutation of truncation and differentiation without introducing an error as [12]

$$\left\| \mathscr{P}_N \frac{\mathrm{d}u}{\mathrm{d}x} - \frac{\mathrm{d}}{\mathrm{d}x}\mathscr{P}_N u \right\|_{L^2_w[-1,1]} \leqslant C(q) N^{3/2-q} \|u\|_{W^q_w[-1,1]}$$

for $q > 1$. Thus, the estimate

$$\|u - \mathscr{P}_N u\|_{W^p_w[-1,1]} \leqslant C(p,q) N^{2p-q-1/2} \|u\|_{W^q_w[-1,1]}$$

and $u(x) \in C^{(q-1)}[-1,1]$, $q \geqslant 1$, provides a bound on the accuracy we can expect for the Chebyshev approximation of spatial derivatives of smooth functions.

Let us finally consider the properties of the truncated discrete expansion, or interpolation, of the form

$$\mathscr{I}_N u(x) = \sum_{n=0}^{N} \tilde{u}_n T_n(x), \quad \tilde{u}_n = \frac{2}{\tilde{c}_n \pi} \sum_{j=0}^{N} u(x_j) T_n(x_j) w_j,$$

where the grid points, x_j, and the weights, w_j, of the Chebyshev–Gauss–Lobatto quadrature rule are given in Eq. (10). As for the Fourier interpolation, the grid introduces aliasing errors as reflected in the connection between the discrete and the continuous expansion coefficients

$$\tilde{c}_n \tilde{u}_n = \hat{u}_n + \sum_{\substack{m=-\infty \\ m \neq 0}}^{m=\infty} \hat{u}_{n+2Nm},$$

provided $u(x) \in W^{1/2}_w[-1,1]$, i.e., the aliasing term takes the exact same form as for the Fourier series. Recalling that the Chebyshev series can be written as a cosine series, this is only natural and one recovers a bound on the interpolation error [10,5]

$$\|u - \mathscr{I}_N u\|_{L^2_w[-1,1]} \leqslant C(q) N^{-q} \|u\|_{W^q_w[-1,1]}$$

for $q \geqslant 1$. Hence, the aliasing error is of the same order as the truncation error and the interpolation maintains exponential convergence if $u(x) \in C^\infty[-1,1]$.

Exploiting the close connection between the Fourier basis and the Chebyshev basis one can show that [10]

$$\left\| \mathscr{I}_N \frac{\mathrm{d}u}{\mathrm{d}x} - \frac{\mathrm{d}}{\mathrm{d}x}\mathscr{I}_N u \right\|_{L^2_w[-1,1]} \leqslant C(q) N^{2-q} \|u\|_{W^q_w[-1,1]}$$

for $q \geqslant 1$ being a special case of the general estimate [11]

$$\|u - \mathscr{I}_N u\|_{W_w^p[-1,1]} \leqslant C(p,q) N^{2p-q} \|u\|_{W_w^q[-1,1]}$$

for $0 \leqslant p \leqslant q$. This provides a bound on the accuracy we can expect for a discrete Chebyshev approximation of a spatial derivative.

The pointwise error is given as [11]

$$\|u - \mathscr{I}_N u\|_{L^\infty[-1,1]} \leqslant C(q) N^{1/2-q} \|u\|_{W_w^q[-1,1]}$$

when $q \geqslant 1$.

3.2. Nonsmooth problems

The main conclusion to draw from the previous section is that if the solution possesses significant regularity we can expect the spectral expansion to be highly efficient for the representation of the solution and its spatial derivatives. In other words, only relatively few terms are needed in the expansion to produce a very accurate approximation.

Considering problems with only limited regularity, however, the picture is considerably more complex and the results given above tell us little about the accuracy of the approximation of such solutions. In particular, if the solution is only piecewise smooth the results discussed so far ensure mean convergence only while the question of pointwise convergence remains open.

It is by now a classical result that the Fourier series, Eq. (15), in the neighborhood of a point of discontinuity, x_0, behaves as [39]

$$\mathscr{P}_N u\left(x_0 + \frac{2z}{2N+1}\right) \sim \frac{1}{2}[u(x_0^+) + u(x_0^-)] + \frac{1}{\pi}[u(x_0^+) - u(x_0^-)]\ \mathrm{Si}(z),$$

where z is a constant and $\mathrm{Si}(z)$ signifies the sine integral. Away from the point of discontinuity, x_0, we recover linear pointwise convergence as $\mathrm{Si}(z) \simeq \pi/2$ for z large. Close to the point of discontinuity, however, we observe that for any fixed value of z, pointwise convergence is lost regardless of the value of N. We recognize this nonuniform convergence and complete loss of pointwise convergence as the celebrated Gibbs phenomenon. Moreover, the oscillatory behavior of the sine integral is recognized as the familiar Gibbs oscillations which are high frequency and appear as being noise.

While the use of the Chebyshev expansion, Eq. (18), eliminates the Gibbs phenomenon at the boundaries of the domain, the problem remains in the interior of the domain on the form [39]

$$\mathscr{P}_N u\left(x_0 + \frac{2z}{\sqrt{1-x_0^2}(2N+1)}\right) \sim \frac{1}{2}[u(x_0^+) + u(x_0^-)] + \frac{1}{\pi}[u(x_0^+) - u(x_0^-)]\ \mathrm{Si}(z),$$

i.e., a nonuniform convergence of the expansion close to the point of discontinuity.

With the situation being even more complex and the details partially unresolved for the discrete expansions, the Gibbs phenomenon and the loss of fast global convergence is often perceived as an argument against the use of spectral expansions for the representation of piecewise smooth functions and, ultimately, against the use of spectral methods for the solution of problems with discontinuous solutions.

As we shall discuss in the following, however, recent results allow us to dramatically improve on this situation and even completely overcome the Gibbs oscillations to recover an exponential accurate approximation to a piecewise analytic function based on the information contained in the global expansion coefficients.

3.2.1. Enhanced convergence by the use of filters

One manifestation of the slow and nonuniform convergence of $\mathcal{I}_N u$ for a piecewise smooth functions is found in the linear decay of the global expansion coefficients, $\tilde{u}_n$. This realization also suggests that one could attempt to modify the global expansion coefficients to enhance the convergence rate of the spectral approximation. The critical question to address naturally is exactly how one should modify the expansion to ensure enhanced convergence to the correct solution.

Let us consider the filtered approximation, $\mathcal{F}_N u_N(x)$, of the form

$$\mathcal{F}_N u_N(x) = \sum_{n=-N}^{N} \sigma\left(\frac{n}{N}\right) \tilde{u}_n \exp(inx), \tag{20}$$

where $\tilde{u}_n$ signifies the discrete expansion coefficients of $u_N(x,t)$ and $\sigma(\eta)$ is a real filter function with the following properties [88]:

$$\sigma(\eta) = \begin{cases} \sigma(-\eta), \\ \sigma(0) = 1, \\ \sigma^{(q)}(0) = 0, & 1 \leqslant q \leqslant 2p-1, \\ \sigma(\eta) = 0, & |\eta| \geqslant 1. \end{cases} \tag{21}$$

If $\sigma(\eta)$ has at least $2p-1$ continuous derivatives, $\sigma(\eta)$ is termed a filter of order $2p$.

As the filter is nothing more than a lowpass filter, it is not surprising that the filtered function converges faster than the unfiltered filtered original expansion. To understand exactly how the filtering modifies the convergence rate, let us assume that $u(x)$ is piecewise $C^{2p}[0,2\pi]$ with one discontinuity located at $x=\xi$. Let us furthermore assume that the filter is of order $2p$. Then the pointwise error of the filtered approximation is given as [88,42]

$$|u(x) - \mathcal{F}_N u_N(x)| \leqslant C \frac{1}{N^{2p-1} d(x,\xi)^{2p-1}} K(u) + C \frac{\sqrt{N}}{N^{2p}} \|u^{(2p)}\|_{L_B^2[0,2\pi]},$$

where $d(x,\xi)$ measures the distance from x to the point of discontinuity, ξ, $K(u)$ is uniformly bounded away from the discontinuity and a function of $u(x)$ only. Also, $\|\cdot\|_{L_B^2[0,2\pi]}$ signifies the broken $L^2[0,2\pi]$-norm.

While the details of the proof of this result are quite technical and can be found in [88,42], the interpretation of the result is simple, and perhaps somewhat surprising. It states that the convergence rate of the filtered approximation is determined solely by the order, $2p$, of the filter, $\sigma(\eta)$, and the regularity of the function, $u(x)$, away from the point of discontinuity. In particular, if the function, $u(x)$, is piecewise analytic and the order of the filter increases with N, one recovers an exponentially accurate approximation to the unfiltered function everywhere except very close to the discontinuity [88,42].

The actual choice of the filter function, $\sigma(\eta)$, is one of great variety and numerous alternatives are discussed in [10,42]. A particularly popular one is the exponential filter [74]

$$\sigma(\eta) = \exp(-\alpha\eta^{2p}),$$

which satisfies all the conditions in Eq. (21) except that of being zero for $|\eta| \geqslant 1$. However, by choosing $\alpha = -\ln \varepsilon_M$, with ε_M representing the machine accuracy, $\sigma(\eta)$ vanishes for all practical purposes when $|\eta|$ exceeds one and the exponential filter allows for the recovery of a piecewise exponentially accurate representation of a piecewise analytic function away from the point of discontinuity.

When applying the filter it is worth noticing that, as for differentiation, the spectral filtering, Eq. (20), has a dual formulation, expressed as a matrix operation, of the form

$$\mathscr{F}_N u_N(x_i) = \sum_{j=0}^{2N-1} F_{ij} u_N(x_j), \quad F_{ij} = \frac{1}{N} \sum_{n=0}^{N} \frac{1}{c_n} \sigma\left(\frac{n}{N}\right) \cos(n(x_i - x_j)),$$

where $c_0 = c_N = 2$ and $c_n = 1$ otherwise.

The use of a filter in a Chebyshev collocation method is similar to that of the Fourier approximation, i.e., it can be expressed as

$$\mathscr{F}_N u_N(x) = \sum_{n=0}^{N} \sigma\left(\frac{n}{N}\right) \tilde{u}_n T_n(x),$$

or on its dual form

$$\mathscr{F}_N u_N(x_i) = \sum_{j=0}^{N} F_{ij} u_N(x_j), \quad F_{ij} = \frac{2}{N c_j} \sum_{n=0}^{N} \frac{1}{c_n} \sigma\left(\frac{n}{N}\right) T_n(x_i) T_n(x_j).$$

While the use of spectral filtering as discussed in the above remains the most popular way of enhancing the convergence rate, an alternative approach can be realized by observing that the Gibbs phenomenon is also a manifestation of the global nature of the interpolating polynomial. In other words, one could attempt to improve on the quality of the approximation by localizing the approximation close to the point of the discontinuity.

This approach, known as physical space filtering, operates directly on the interpolating polynomials rather than the expansion coefficients. To illustrate the basic idea consider the filtered Fourier approximation

$$\mathscr{F}_N u_N(x) = \sum_{j=0}^{2N-1} u(x_j) l_j^{\sigma}(x),$$

where x_j are the usual equidistant grid and the filtered Lagrange interpolation polynomial takes the form

$$l_j^{\sigma}(x) = \frac{1}{2N} \sum_{n=-N}^{N} \sigma\left(\frac{n}{N}\right) \frac{1}{\tilde{c}_n} \exp(in(x - x_j)).$$

Clearly, if $\sigma(\eta) = 1$ we recover the Fourier interpolation polynomial given in Eq. (11). However, we do not need to maintain the close connection to the trigonometric interpolation polynomials, as expressed in $l_j^{\sigma}(x)$, but can choose to use any reasonable kernel, $\psi(x, x_j)$, that approximates a delta function as $x - x_j$ approaches zero. Following [44,42] we can exemplify this idea by considering

$$\psi(x, x_j) = \rho(\xi(x, x_j)) \frac{1}{2\varepsilon} \sum_{n=-N}^{N} \exp(in\xi(x, x_j)),$$

where

$$\xi(x,x_j) = \frac{x_j - x}{\varepsilon}.$$

We assume that $u(x) \in C^{2p}[x_j - \varepsilon, x_j + \varepsilon]$ and $\rho(\xi(x,x_j))$ controls the amounts of localization as

$$\rho(0) = 1, \quad \rho(\xi(x,x_j)) = 0 \quad \text{for } |\xi(x,x_j)| \geqslant 1,$$

i.e., the kernel vanished outside of the symmetric interval $[x_j - \varepsilon, x_j + \varepsilon]$. Note also that $(1/2\varepsilon)\sum_{n=-N}^{N} \exp(in\xi(x,x_j))$ is an approximation to a x_j-centered delta function.

In this setting it can be shown [44,42] that the order of the filter and the regularity of the function away from the point of discontinuity solely determines the convergence rate of the filtered approximation. Exponential convergence can be recovered everywhere except very close to the point of discontinuity as measured through ε. The need to specify the size, ε, of the symmetric interval remains a practical obstacle to the use of the physical space filters.

While the use of filters can have a dramatic impact on the quality of the convergence of a global approximation to a discontinuous function, such techniques are unable to improve on the quality of the approximation as one approaches the point of discontinuity. Moreover, filtering generally treats the Gibbs oscillations as a source of noise and attempts to remove it. However, as has been speculated in the past [67], and recently shown rigorously, the Gibbs oscillations are not noise but rather contain sufficient information to reconstruct an exponentially convergent approximation everywhere provided only that the location of the discontinuity is known, i.e., the Gibbs phenomenon can be overcome completely.

3.2.2. Resolving the Gibbs phenomenon

In the following, we outline the key elements of a general theory that establishes the possibility of recovering a piecewise exponentially convergent series to a piecewise analytic function, $u(x) \in L^2_w[-1,1]$, having knowledge of the global expansion coefficients and the position of the discontinuities only.

The basic element of this new approach is the identification of a new basis with very special properties and, subsequently, the expansion of the slowly convergent truncated global expansion in this new basis. Provided this new basis satisfies certain conditions, the new expansion has the remarkable property that it is exponentially convergent to the original piecewise analytic function even though it uses information from the slowly convergent global expansion.

As previously we assume that

$$\mathscr{P}_N u(x) = \sum_{n=0}^{N} \hat{u}_n \phi_n(x), \quad \hat{u}_n = \frac{1}{\gamma_n}(u, \phi_n)_w$$

with $\gamma_n = \|\phi_n\|_w^2$. Note in particular that this also contains the Fourier case provided

$$\phi_n(x) = \exp\left[\mathrm{i}\left(n - \frac{N}{2}\right)\pi x\right].$$

Let us also assume that there exists an interval $[a,b] \subset [-1,1]$ in which $u(x)$ is analytic and, furthermore, that the original truncated expansion is pointwise convergent in all of $[-1,1]$ with the exception of a finite number of points. We shall introduce the scaled variable

$$\xi(x) = -1 + 2\frac{x-a}{b-a}.$$

Clearly, $\xi : [a,b] \to [-1,1]$.

We define a new basis, $\psi_n^\lambda(\xi)$, which is orthogonal in the weighted inner product, $(\cdot,\cdot)_w^\lambda$ where λ signifies that the weight, $w(x)$, may depend on λ, i.e.,

$$(\psi_k^\lambda, \psi_n^\lambda)_w^\lambda = \|\psi_n^\lambda\|^2_{L^2_w[-1,1]}\delta_{kn} = \gamma_n^\lambda \delta_{kn}.$$

Furthermore, we require that if $v(\xi)$ is analytic then

$$\mathscr{P}_\lambda v(\xi) = \sum_{n=0}^{\lambda} \frac{1}{\gamma_n^\lambda}(v, \psi_n^\lambda)_w^\lambda \psi_n^\lambda(\xi)$$

is pointwise exponentially convergent as λ increases, i.e.,

$$\|v - \mathscr{P}_\lambda v\|_{L^\infty[-1,1]} \leqslant C\mathrm{e}^{-c\lambda}$$

with $c > 0$. This is similar to the case of classical expansions discussed in Section 3.1.

A final condition, however, sets this basis apart and is central in order to overcome the Gibbs phenomenon. We shall require that there exists a number $\beta < 1$, such that for $\lambda = \beta N$ we have

$$\left|\frac{1}{\gamma_n^\lambda}(\phi_k(x(\xi)), \psi_n^\lambda(\xi))_w^\lambda\right| \|\psi_n^\lambda\|_{L^\infty[-1,1]} \leqslant \left(\frac{\alpha N}{k}\right)^\lambda \tag{22}$$

for $k > N$, $n \leqslant \lambda$ and $\alpha < 1$. The interpretation of this condition is that the projection of the high modes of ϕ_k onto the basis, ψ_n^λ, is exponentially small in the interval, $\xi \in [-1,1]$. In other words, by reexpanding the slowly decaying ϕ_k-based global expansion in the local ψ_n^λ-basis an exponentially accurate local approximation is recovered. We shall term this latter, and crucial, condition on ψ_n^λ the Gibbs condition to emphasize its close connection to the resolution of the Gibbs phenomenon.

Provided only that the ψ_n^λ-basis, which we term the Gibbs complementary basis, is complete we recover the key result

$$\left\| u(x) - \sum_{n=0}^{\lambda} \frac{1}{\gamma_n^\lambda}(\mathscr{P}_N u, \psi_n^\lambda)_w^\lambda \psi_n^\lambda(\xi(x)) \right\|_{L^\infty[a,b]} \leqslant C\exp(-cN),$$

where $\lambda = \beta N$ and $u(x)$ is analytic in the interval $[a,b]$.

In other words, if a Gibbs complementary basis exists it is possible to reconstruct a piecewise exponentially convergent approximation to a piecewise analytic function from the information contained in the original very slowly converging global approximation using only knowledge about the location of the points of discontinuity. Hence, the impact of the Gibbs phenomenon can be overcome.

A constructive approach to the identification of the complementary basis is currently unknown. The existence of such a basis, however, has been established by carefully examining the properties of the basis

$$\psi_n^\lambda(\xi) = C_n^\lambda(\xi),$$

where $C_n^{\lambda}(\xi)$ represent the Gegenbauer polynomials, also known as the symmetric Jacobi polynomials or ultraspherical polynomials [81]. It is well known that the polynomials are orthogonal in the inner product

$$(f,g)_w^{\lambda} = \int_{-1}^{1} f(x)g(x)(1-x^2)^{\lambda-1/2}\,dx$$

and that

$$\gamma_n^{\lambda} = (\psi_n^{\lambda}, \psi_n^{\lambda})_w^{\lambda} = \sqrt{\pi}\frac{\Gamma(n+2\lambda)}{n!\Gamma(2\lambda)}\frac{\Gamma(\lambda+\frac{1}{2})}{(n+\lambda)\Gamma(\lambda)}.$$

The spectral convergence of Gegenbauer expansions of analytic functions is natural as the polynomials appears as eigensolutions to a singular Sturm–Liouville problem. Hence, to establish that the Gegenbauer polynomials provide an example of a Gibbs complementary basis we need to verify the Gibbs condition, Eq. (22).

If we consider the Fourier basis

$$\phi_n(x) = \exp(in\pi x),$$

it must be established that

$$\left|\frac{1}{\gamma_n^{\lambda}}(\phi_k, \psi_n^{\lambda})_w^{\lambda}\right| \leqslant \left(\frac{\alpha N}{k}\right)^{\lambda}$$

for $k > N$, $0 < \alpha < 1$, and $n \leqslant \beta N = \lambda$.

For the Fourier basis the inner product allows an exact evaluation

$$\frac{1}{\gamma_n^{\lambda}}(\phi_k, \psi_n^{\lambda})_w^{\lambda} = \mathrm{i}^n\Gamma(\lambda)\left(\frac{2}{\pi k\varepsilon}\right)^{\lambda}(n+\lambda)J_{n+\lambda}(\pi\varepsilon k)$$

with $J_\nu(x)$ being the Bessel function and $\varepsilon = b - a$ measures the width of the interval. Using the properties of the Bessel function and the Stirling formula for the asymptotic of the Γ-function, the Gibbs condition is satisfied if [42]

$$\beta = \frac{2\pi\varepsilon}{27}.$$

This establishes the existence of a Gibbs complementary basis to the Fourier basis.

For the Chebyshev case with

$$\phi_n(x) = T_n(x),$$

it can again be shown that the Gegenbauer polynomials provides an example of a Gibbs complementary basis although the proof is considerably more complicated as the means by which the inner product is bounded is recursive [42,43]. Nevertheless, the existence of a Gibbs complementary basis for the Chebyshev basis has been established. This paves the way for postprocessing of global approximations of piecewise smooth problems to recover a piecewise exponentially accurate approximation. Whether the Gegenbauer basis is the optimal choice as the Gibbs complementary basis for the Fourier and Chebyshev basis remains an open questions.

We have focused on the resolution of the Gibbs phenomenon for continuous spectral expansions which, unfortunately, have little practical importance. A similar discussion and analysis for the discrete expansion is complicated by the introduction of the aliasing error.

If one constructs the special interpolating polynomial

$$v(\xi) = \mathscr{I}_N[(1-\xi^2)^{\lambda-1/2}u(x(\xi))],$$

the proof that the Gegenbauer polynomials remain a Gibbs complementary basis for the Fourier and Chebyshev polynomials has be completed [41]. However, experience shows that the straightforward approach in which one considers the expansion

$$\mathscr{I}_\lambda v(\xi) = \sum_{n=0}^{\lambda} \tilde{v}_n^\lambda C_n^\lambda(\xi)$$

with the discrete expansion coefficients being approximated by a Gaussian quadrature rule as

$$\tilde{v}_n^\lambda = \frac{1}{\tilde{\gamma}_n^\lambda}\sum_{j=0}^{\lambda} v(\xi_j)w_j C_n^\lambda(\xi_j) = \frac{1}{\tilde{\gamma}_n^\lambda}\sum_{j=0}^{\lambda} \mathscr{I}_N u(x(\xi_j))w_i C_n^\lambda(\xi_j),$$

works well although a proof for this remains unknown.

The reconstruction of piecewise smooth solutions to conservation laws as a postprocessing technique has been exploited in [18,20,31]. Other applications of the reconstruction technique can be found in [89,90] and a two-dimensional version is discussed in [30].

It is worth mentioning in passing that alternatives to the identification of the Gibbs complementary basis for the reconstruction of piecewise analytic functions are known. These techniques all exploit the idea, originally proposed in [36], that by knowing or computing the location and size of the discontinuities, one can subtract these to recover a function with enhanced regularity. This was originally used as a postprocessing technique only [36] but later used as part of a time-dependent solution [8]. A high-order version of this approach, accounting also for discontinuities in the derivatives of the solution, has recently been developed [23,3] and tested on linear hyperbolic problems [22].

4. Collocation approximations of hyperbolic problems

Let us now turn the attention towards the actual solution of hyperbolic problems. Prominent examples of such problems include Maxwells equations from electromagnetics, the Euler equations from gas dynamics and the equations of elasticity. However, for the sake of simplicity we shall concentrate on methods for the scalar conservation law of the type

$$\frac{\partial u}{\partial t} + \frac{\partial f(u)}{\partial x} = 0,$$

subject to appropriate boundary and initial conditions. As a prominent special case we shall devote much attention to the variable coefficient linear wave problem

$$\frac{\partial u}{\partial t} + a(x)\frac{\partial u}{\partial x} = 0, \tag{23}$$

where $a > 0$ implies a rightward propagating wave and $a < 0$ corresponds to a leftward propagating wave.

4.1. Fourier methods

Restricting the attention to problems of a purely periodic character, i.e.,

$$\frac{\partial u}{\partial t}+\frac{\partial f(u)}{\partial x}=0, \tag{24}$$

$$u(0,t)=u(2\pi,t),$$

$$u(x,0)=g(t),$$

it is only natural to seek numerical solutions, $u_N(x,t)$, expressed in terms of the Fourier series. If we require that the solution, $u_N(x,t)$, satisfies Eq. (24) in a collocation sense, $u_N(x,t)$ is a trigonometric polynomial that satisfies the equation

$$\left.\frac{\partial u_N}{\partial t}\right|_{x_j}+\left.\frac{\partial \mathcal{I}_N f(u_N)}{\partial x}\right|_{x_j}=0$$

at the grid points, x_j. It is worth while emphasizing that in solving the conservation law, we encounter three types of solutions. The exact solution, $u(x,t)$, will generally not be available. However, when solving the partial differential equation we conjecture that the computable numerical solution, $u_N(x,t)$, is very close to the interpolation of the exact solution, $\mathcal{I}_N u(x,t)$. Due to aliasing errors and effects of nonlinearities the two solutions are generally not equivalent although for well resolved problems it is a reasonable assumption. A complete analysis of these aspects involves the derivation of the error equation which is a complex task, even for simple equations.

Consider the simple wave equation, Eq. (23), for which the Fourier collocation scheme is given as

$$\left.\frac{\partial u_N}{\partial t}\right|_{x_j}+a(x_j)\left.\frac{\partial u_N}{\partial x}\right|_{x_j}=0,$$

where we have left the derivative on symbolic form to emphasize the two mathematically equivalent, but computationally different ways of computing this operation.

In the same simple fashion, the Fourier collocation approximation to Burgers equation

$$\frac{\partial u}{\partial t}+\frac{1}{2}\frac{\partial u^2}{\partial x}=0$$

is obtained by seeking the approximate solution, $u_N(x,t)$, such that

$$\left.\frac{\partial u_N}{\partial t}\right|_{x_j}+\left.\frac{1}{2}\frac{\partial}{\partial x}\mathcal{I}_N u_N^2\right|_{x_j}=0. \tag{25}$$

Note that while the partial differential equation has the equivalent formulation

$$\frac{\partial u}{\partial t}+u\frac{\partial u}{\partial x}=0,$$

the corresponding nonconservative Fourier approximation

$$\left.\frac{\partial u_N}{\partial t}\right|_{x_j}+u_N(x_j)\left.\frac{\partial u_N}{\partial x}\right|_{x_j}=0$$

is not equivalent to Eq. (25) and will in general yield a different results due to the aliasing errors and the mixing of these through the nonlinear term.

4.2. Chebyshev methods

Let us now consider the more general initial–boundary value problem

$$\frac{\partial u}{\partial t}+\frac{\partial f(u)}{\partial x}=0, \tag{26}$$

$$u(x,0)=g(t)$$

posed on a finite domain which we take to be $[-1,1]$ without loss of generality. For the problem to be wellposed, we must specify boundary conditions of the form

$$\alpha u(-1,t)=f^{-}(t), \quad \beta u(1,t)=f^{+}(t),$$

where the specification of α and β is closely related to the flux function, e.g., if

$$x\frac{\partial f}{\partial u}<0,$$

at the boundary, information is incoming and a boundary condition must be given. For a system of equations, the equivalent condition is posed through the characteristic variables, i.e., characteristic waves entering the computational domain must be specified and, hence, require a boundary condition to ensure wellposedness of the problem.

What separates the polynomial collocation approximation from the trigonometric schemes discussed in Section 4.1 is the need to impose the boundary conditions in such a way that we restrict the numerical solutions, $u_N(x,t)$, to those obeying the boundary conditions. The details of how this is done leads to different schemes.

4.2.1. Strongly imposed boundary conditions

In the classic approach one requires that the boundary conditions are imposed strongly, i.e., exactly. Hence, we shall seek a polynomial, $u_N(x,t)$, that satisfies Eq. (26) in a collocation sense at all the interior grid points, x_j, as

$$\left.\frac{\partial u_N}{\partial t}\right|_{x_j}+\left.\frac{\partial \mathcal{I}_N f(u_N)}{\partial x}\right|_{x_j}=0,$$

while the boundary conditions are imposed exactly

$$\alpha u_N(-1,t)=f^{-}(t), \quad \beta u_N(1,t)=f^{+}(t).$$

If we again consider the wave equation, Eq. (23), the Chebyshev collocation scheme becomes

$$\left.\frac{\partial u_N}{\partial t}\right|_{x_j}+a(x_j)\left.\frac{\partial u_N}{\partial x}\right|_{x_j}=0$$

at all interior grid points, i.e., for $a>0$, $j\in[1,N]$, while $u_N(x_0,t)=f^{-}(t)$.

In the same spirit the conservative Chebyshev collocation approximation to Burgers equation becomes

$$\left.\frac{\partial u_N}{\partial t}\right|_{x_j}+\left.\frac{1}{2}\frac{\partial}{\partial x}\mathcal{I}_N u_N^2\right|_{x_j}=0,$$

which, if subjected to pure Dirichlet boundary conditions, are computed under the constraint

$$u_N(-1,t)=f^{-}(t), \quad u_N(1,t)=f^{+}(t).$$

4.2.2. Weakly imposed boundary conditions

The conceptual leap that leads one to consider other ways of imposing boundary conditions is the observation that it is sufficient to impose the boundary conditions to the order of the scheme, i.e., weakly, such that only in the limit of infinite order is the boundary condition is enforced exactly.

This simple idea, put forward in the context of spectral methods in [11] in a weak formulation and in [28,29] for the strong formulation considered here, has recently been developed further into a flexible and very general technique to impose boundary conditions in pseudospectral approximations to a variety of problems [19,52,13,49,53,50,55].

In this setting, one seeks a polynomial solution, $u_N(x,t)$, to Eq. (26) satisfying

$$\frac{\partial u_N}{\partial t} + \mathcal{I}_N \frac{\partial \mathcal{I}_N f(u_N)}{\partial x} = -\tau^- \alpha Q^-(x)[u_N(-1,t) - \tilde{f}^-(t)] + \tau^+ \beta Q^+(x)[u_N(1,t) - \tilde{f}^+(t)], \qquad (27)$$

where we have introduced the polynomials, $Q^{\pm}(x) \in \mathsf{P}_N$, and the scalars, $\tau^{\pm}$.

To complete the scheme we must specify how the equation is to be satisfied which in most cases amounts to a choice between a Galerkin or a collocation approach. Moreover, we must choose $Q^{\pm}(x)$ and an approach by which to specify the scalar parameters, $\tau^{\pm}$. While the latter choice usually is dictated by requiring semi-discrete stability, the former choice of $Q^{\pm}(x)$ is associated with a great deal of freedom.

Before we discuss this in more detail, let us briefly introduce the discrete expansions based on Legendre polynomials, $P_n(x)$, as

$$\mathcal{I}_N u(x) = \sum_{n=0}^{N} \tilde{u}_n P_n(x), \quad \tilde{u}_n = \frac{1}{\tilde{\gamma}_n} \sum_{j=0}^{N} u(x_j) P_n(x_j) w_j,$$

where the quadrature to compute $\tilde{u}_n$ is based on the Legendre–Gauss–Lobatto points, i.e., the zeros of $(1-x^2)P_N'(x)$, or the Gauss points, i.e., the zeros of $P_{N+1}(x)$. The weights, w_i, of the quadrature naturally depend on the choice of the grid points [27]. We recall that the summation is exact provided $f(x)$ is a polynomial of degree at most $2N+1$ if the Gauss quadrature is used and exact for a polynomial of order at most $2N-1$ if the Gauss–Lobatto quadrature is used.

Consider the approximation to the constant coefficient wave equation, Eq. (23),

$$\frac{\partial u_N}{\partial t} + a\frac{\partial u_N}{\partial x} = -\tau^- a Q^-(x)[u_N(-1,t) - f(t)],$$

where $u_N(x,t)$ is based on the Legendre–Gauss–Lobatto points. A viable choice of $Q^-(x)$ is

$$Q^-(x) = \frac{(1-x)P_N'(x)}{2P_N'(-1)} = \begin{cases} 1, & x=-1, \\ 0, & x = x_j \neq -1, \end{cases}$$

where x_j refers to the Legendre–Gauss–Lobatto points. By requesting that the equation be satisfied in a collocation sense, we recover the scheme

$$\left.\frac{\partial u_N}{\partial t}\right|_{x_j} + a\left.\frac{\partial u_N}{\partial x}\right|_{x_j} = -a\frac{N(N+1)}{4}\frac{(1-x_j)P_N'(x_j)}{2P_N'(-1)}[u_N(-1,t) - f(t)],$$

which we shall show in Section 5.2 to be asymptotically stable. Although the boundary condition is imposed only weakly, the approximation is clearly consistent, i.e., if $u_N(x,t) = u(x,t)$ the penalty term vanishes identically.

To illustrate the flexibility of the weakly imposed boundary conditions, let us again consider

$$\frac{\partial u_N}{\partial t} + a\frac{\partial u_N}{\partial x} = -\tau^- a Q^-(x)[u_N(-1,t) - f(t)],$$

where $Q^-(x)$ is as above, but $u_N(x,t)$ is based on the Chebyshev–Gauss–Lobatto grid, Eq. (10). In this case, $Q^-(x)$ is different from zero at all the interior Chebyshev–Gauss–Lobatto grid points and the boundary term reflects a global correction. Nevertheless, if we require that the wave equation be satisfied at the Legendre–Gauss–Lobatto nodes we recover the scheme

$$\left.\frac{\partial u_N}{\partial t}\right|_{x_j} + a\left.\frac{\partial u_N}{\partial x}\right|_{x_j} = -a\frac{N(N+1)}{4}\frac{(1-x_j)P_N'(x_j)}{2P_N'(-1)}[u_N(-1,t) - f(t)],$$

where x_j are the Chebyshev–Gauss–Lobatto nodes. In other words, we have constructed an asymptotically stable Legendre–Gauss–Lobatto collocation method using the Chebyshev–Gauss–Lobatto grid for the approximation. This method, known as the Chebyshev–Legendre method [19], provides an example of a scheme where the equation is satisfied at points different from those on which the approximation is based. This example also shows that there is nothing special about the quadrature points in terms of stability. Indeed, we can construct a stable scheme on any set of grid points [13]. In terms of accuracy, however, the use of very special families of grid points is crucial.

As a final example, let us consider the general conservation law, Eq. (24), subject to boundary conditions at $x = \pm 1$, and assume that the polynomial solution, $u_N(x,t)$, is based on the Legendre–Gauss nodes, i.e.,

$$u_N(x,t) = \sum_{n=0}^{N} \tilde{u}_n P_n(x) = \sum_{j=0}^{N} u(x_j) l_j(x) = \sum_{j=0}^{N} u(x_j)\frac{P_{N+1}(x)}{(x-x_j)P_{N+1}'(x_j)},$$

where x_j signifies the Legendre–Gauss nodes.

We now request that Eq. (27) be satisfied in the following Galerkin-like way [53]:

$$\begin{aligned}\int_{-1}^{1}\left(\frac{\partial u_N}{\partial t} + \frac{\partial f_N}{\partial x}\right) l_i(x)\,\mathrm{d}x &= \oint_{-1}^{1} \tau(x) l_i(x)[f_N(x,t) - g(x,t)]\,\mathrm{d}s \\ &= -\tau^- l_i(-1)[f_N(-1,t) - g^-(t)] \\ &\quad +\tau^+ l_i(1)[f_N(1,t) - g^+(t)],\end{aligned}$$

where we have abused the notation a bit to make the multi-dimensional generalization more obvious. Here the boundary integral enforces the boundary conditions and we have introduced

$$f_N(x) = \mathcal{I}_N f(u_N) = \sum_{j=0}^{N} f_j l_j(x)$$

with $f_j = \mathcal{I}_N f(u_N(x_j))$. After integration by parts once, we recover

$$\begin{aligned}&\sum_{j=0}^{N} M_{ij}\left.\frac{\mathrm{d}u_N}{\mathrm{d}t}\right|_{x_j} - \sum_{j=0}^{N} S_{ij} f_j + f_N(1) l_i(1) - f_N(-1) l_i(-1) \\ &\quad = -\tau^- l_i(-1)[f_N(-1) - g^-(t)] + \tau^+ l_i(1)[f_N(1) - g^+(t)]\end{aligned}$$

with the mass-matrix, M, and the stiffness matrix, S, having the entries

$$M_{ij} = (l_i, l_j)_{L^2_w[-1,1]}, \quad S_{ij} = \left(\frac{\mathrm{d}l_i}{\mathrm{d}x}, l_j\right)_{L^2_w[-1,1]}$$

and the inner product is the usual unweighted inner product, i.e., $w(x) = 1$.

Exploiting the exactness of the Gauss quadrature and the fact that $l_i(x)$ are based on the Gauss-nodes we recover

$$M_{ij} = \begin{cases} w_i, & i = j, \\ 0, & i \neq j, \end{cases} \quad S_{ij} = \left.\frac{\mathrm{d}l_i}{\mathrm{d}x}\right|_{x_j} w_j = D_{ji} w_j,$$

where w_j are the Gauss–Legendre quadrature weights [27] and D_{ji} represents, in the spirit of Section 2.2, the entries of the differentiation matrix based on the Gauss–Legendre grid point. This results in the collocation scheme

$$\left.\frac{\mathrm{d}u_N}{\mathrm{d}t}\right|_{x_i} - \sum_{j=0}^{N} D_{ji} f_j \frac{w_j}{w_i} + f_N(1)\frac{l_i(1)}{w_i} - f_N(-1)\frac{l_i(-1)}{w_i}$$

$$= -\tau^- \frac{l_i(-1)}{w_i}[f_N(-1) - g^-(t)] + \tau^+ \frac{l_i(1)}{w_i}[f_N(1) - g^+(t)] \tag{28}$$

at the grid points, x_i. Taking $\tau^\pm = 1$ we recover

$$\left.\frac{\mathrm{d}u_N}{\mathrm{d}t}\right|_{x_j} - \sum_{j=0}^{N} D_{ji} f_j \frac{w_j}{w_i} + g^+(t)\frac{l_i(1)}{w_i} - g^-(t)\frac{l_i(-1)}{w_i} = 0,$$

which one recognizes as the collocation form of the discontinuous Galerkin method [16,63], i.e., by taking $\tau^\pm$ to unity, one ensures that the scheme is conservative. Nevertheless, the discontinuous Galerkin method is only a special case of a much larger family of schemes with weakly imposed boundary conditions. The advantage in this realization lies in the flexibility of choosing $\tau^\pm$. In particular, if conservation is unnecessary, as for linear or smooth nonlinear problems, one can simply require asymptotic stability of Eq. (28) as [53]

$$\frac{1}{2}\frac{\mathrm{d}}{\mathrm{d}t}\|u_N\|^2_{L^2_w[-1,1]} \leqslant 0 \rightarrow \tau^\pm \geqslant \frac{1}{2},$$

i.e., by sacrificing conservation we may lower $\tau^\pm$ while maintaining stability. As discussed in [52], a lower value of $\tau^\pm$ typically allows for increasing the discretely stable time-step when using explicit time-stepping.

5. Stability results for hyperbolic problems

To establish stability for the collocation schemes one traditionally either exploits the structure in the differentiation matrices, e.g., the Fourier differentiation matrices are antisymmetric, or utilizes the exactness of the quadrature rules to go from the semi-discrete formulation to the continuous formulation.

These techniques all lead to semi-discrete energy-estimates of the form

$$\|u_N(t)\| \leqslant K\mathrm{e}^{\alpha t}\|u_N(0)\|,$$

assuming homogeneous boundary conditions without loss of generality. Clearly, $\alpha \leqslant 0$ implies asymptotic stability.

5.1. Stability of the Fourier collocation method

Consider the variable coefficient linear wave problem, Eq. (23), subject to periodic boundary conditions for which the Fourier collocation approximation becomes

$$\frac{d}{dt}\boldsymbol{u} + AD\boldsymbol{u} = 0, \tag{29}$$

where $\boldsymbol{u} = [u_N(x_0), \ldots, u_N(x_{2N-1})]^T$ represents the solution vector, D is the Fourier differentiation matrix, Eq. (12), and $A_{jj} = a(x_j)$ is diagonal.

Let us define the discrete inner product and L^2-equivalent norm as

$$[f, g]_N = \frac{\pi}{N} \sum_{j=0}^{2N-1} f(x_j) g(x_j), \quad \|f\|_N^2 = [f, f]_N.$$

If we initially assume that $|a(x)| > 0$ [75,65,25,39,76], it is easy to see that for $\boldsymbol{v} = A^{-1/2}\boldsymbol{u}$, we recover

$$\frac{d}{dt}\boldsymbol{v} + A^{1/2} D A^{1/2} \boldsymbol{v} = 0,$$

such that

$$\frac{1}{2}\frac{d}{dt}\|v_N\|_N^2 = \frac{1}{2}\frac{d}{dt}\boldsymbol{u}^T A^{-1}\boldsymbol{u} = \frac{1}{2}\frac{d}{dt}\boldsymbol{u}^T H \boldsymbol{u} = \frac{1}{2}\frac{d}{dt}\|u_N\|_H^2 = 0,$$

since $A^{1/2} D A^{1/2}$ is antisymmetric. Here we have introduced the usual notation for the L^2-equivalent H-norm, $\|\cdot\|_H$ [25], also known as the elliptic norm.

For the general case where $a(x)$ changes sign within the computational domain, the situation is more complex. The straightforward way to guarantee stability is to consider the skew-symmetric form [65]

$$\frac{\partial u}{\partial t} + \frac{1}{2}\frac{\partial a(x) u}{\partial x} + \frac{1}{2} a(x) \frac{\partial u}{\partial x} - \frac{1}{2} a_x(x) u(x) = 0 \tag{30}$$

with the discrete form

$$\left.\frac{\partial u_N}{\partial t}\right|_{x_j} + \frac{1}{2} \left.\frac{\partial \mathcal{I}_N a(x) u_N}{\partial x}\right|_{x_j} + \frac{1}{2} a(x_j) \left.\frac{\partial u_N}{\partial x}\right|_{x_j} - \frac{1}{2} a_x(x_j) u_N(x_j) = 0.$$

Stability follows since

$$\frac{1}{2}\frac{d}{dt}\|u_N\|_N^2 \leqslant \frac{1}{2} \max_{x \in [0, 2\pi]} |a_x(x)| \, \|u_N\|_N^2.$$

The disadvantage of the skew-symmetric formulation, however, clearly lies in a doubling of the computational work.

The question of stability of the simple formulation, Eq. (29), for general $a(x)$ has remained an open question until very recently, although partial results has been known for a while [40,86]. The difficulty in resolving this issue is associated with the development of very steep spatial gradients which, for a fixed resolution, eventually introduce significant aliasing that affect the stability. A testament to this observation is the trivial stability of the Fourier–Galerkin approximation [39,84]. By carefully examining the interplay between aliasing, resolution, and stability, it has recently been shown [34] that the Fourier approximation is only algebraically stable [39], i.e.,

$$\|u_N(t)\|_N \leqslant C(t)N\|u_N(0)\|_N, \tag{31}$$

or weakly unstable. However, the weak instability spreads from the high modes through aliasing and results in at most an $\mathcal{O}(N)$ amplification of the Fourier components of the solution. In other words, for well-resolved computations where these aliasing components are very small the computation will appear stable for all practical purposes.

As an example of one of the few nonlinear cases for which stability can be established, recall Burgers equation

$$\frac{\partial u}{\partial t} + \frac{1}{2}\frac{\partial u^2}{\partial x} = 0. \tag{32}$$

A stable approximation is obtained by considering the skew-symmetric form [82]

$$\frac{\partial u}{\partial t} + \frac{1}{3}\frac{\partial u^2}{\partial x} + \frac{1}{3}u\frac{\partial u}{\partial x} = 0,$$

from which stability of the collocation approximation

$$\left.\frac{\partial u_N}{\partial t}\right|_{x_j} + \frac{1}{3}\left.\frac{\partial}{\partial x}\mathcal{I}_N u_N^2\right|_{x_j} + \frac{1}{3}u_N(x_j)\left.\frac{\partial u_N}{\partial x}\right|_{x_j} = 0$$

follows directly from the exactness of the quadrature.

5.2. Stability of Chebyshev collocation method

Establishing stability of the Chebyshev collocation approximations is considerably more challenging than for the Fourier collocation approximation. To expose the sources of this difficulty, let us consider the simple wave equation, Eq. (23), with $a(x) = 1$ and subject to the conditions

$$u(x,0) = g(x), \quad u(-1,t) = 0.$$

A Chebyshev collocation scheme based on the Gauss–Lobatto nodes yields

$$\frac{\mathrm{d}}{\mathrm{d}t}\boldsymbol{u} = -\tilde{D}\boldsymbol{u}. \tag{33}$$

Here $\boldsymbol{u}(t) = [u_N(-1,t),\ldots,u_N(x_j,t),\ldots,u_N(1,t)]^{\mathrm{T}}$ represents the grid vector at the Gauss–Lobatto nodes, x_j, and the matrix $\tilde{D}$ represents the Chebyshev differentiation matrix, Eq. (14), modified to enforce the boundary condition strongly, i.e., by introducing zeros in the first row and column.

The strongly enforced boundary condition introduces the first main obstacle as the delicate structure of the differentiation matrix is destroyed by this modification, leaving us with the quadrature formula in trying to establish stability. The straightforward quadrature formula, however, is closely related to the weighted inner product, $(f,g)_{L^2_w[-1,1]}$, in which the Chebyshev polynomials are orthogonal. The norm associated with this inner product is, unfortunately, not uniformly equivalent to the usual L^2-norm [39]. This loss of equivalence eliminates the straightforward use of the quadrature rules in the quest to establish stability as the corresponding norm is too weak. Thus, the two central techniques utilized for the Fourier methods are not directly applicable to the case of the Chebyshev collocation methods. It is worth mentioning that the situation for the Legendre approximations is considerably better as the Legendre polynomials are orthogonal in the unweighted $L^2[-1,1]$ inner product.

It seems natural to attempt to construct a new inner product and associated norm, uniformly equivalent to L^2, and subsequently establish stability in this norm. This is exactly the approach that was taken in [35,37,46] where Eq. (23) is considered and the following inner product was introduced [39]:

$$(f,g)_{L^2_{\tilde{w}}[-1,1]} = \int_{-1}^{1} f(x)g(x)\frac{1-x}{\sqrt{1-x^2}}\,\mathrm{d}x \tag{34}$$

with the associated quadrature rule

$$\int_{-1}^{1} f(x)\frac{1-x}{\sqrt{1-x^2}}\,\mathrm{d}x = \sum_{j=0}^{N-1} f(x_j)\tilde{w}_j,$$

which is exact for $f(x) \in \mathrm{P}_{2N-2}$. The weights, $\tilde{w}_j$, are given in [35,46] where it is also shown that $L^2_{\tilde{w}[-1,1]}$ is equivalent to $L^2_{w[-1,1]}$.

We note that the quadrature sum does not include the outflow boundary at $x=1$. To utilize the above quadrature we introduce the grid vector, $\boldsymbol{v}(t) = [u_N(-1,t),\ldots,u_N(x_j,t),\ldots,u_N(x_{N-1},t)]^{\mathrm{T}}$, containing the first N components of u_N. The evolution of $\boldsymbol{v}$ is simply described

$$\frac{\mathrm{d}}{\mathrm{d}t}\boldsymbol{v} = -\hat{D}\boldsymbol{v}, \tag{35}$$

where $\hat{D}$ signifies $\tilde{D}$, Eq. (33), with the last row and column removed.

Stability of $\boldsymbol{v}$ in Eq. (35) follows from the exactness of the quadrature as

$$\frac{1}{2}\frac{\mathrm{d}}{\mathrm{d}t}\sum_{j=0}^{N-1} v_N^2(x_j)\tilde{w}_j \leqslant 0.$$

The exact relation between this result and the stability of $\boldsymbol{u}$ in Eq. (33) is nontrivial and we refer to [35] where stability is established by directly relating $\boldsymbol{u}$ and $\boldsymbol{v}$.

The more general variable coefficient problem, Eq. (23), with $a(x)$ being smooth can be addressed using a similar approach. In particular, if $a(x)$ is smooth and uniformly bounded away from zero stability is established in the elliptic norm [35]

$$\frac{1}{2}\frac{\mathrm{d}}{\mathrm{d}t}\sum_{j=0}^{N-1} v_N^2(x_j)\frac{\tilde{w}_j}{a(x_j)} \leqslant 0.$$

For the more general case of $a(x)$ changing sign the only known results are based on the skew-symmetric form [11], Eq. (30), although numerical experiments suggest that the straightforward Chebyshev collocation approximation of the wave equation with a variable coefficient behaves much as the Fourier approximation discussed above, i.e., if the solution is well resolved, the approximation is stable [39,35].

The extension of these results to a hyperbolic system of equations with constant coefficients is discussed in [37,38]. Stability of the scalar problem in combination with a dissipative boundary operator is shown to be sufficient to guarantee algebraic stability of the Chebyshev approximation to a hyperbolic system of equations.

What we have discussed so far can be viewed as the traditional approach to stability, i.e., one formulates a meaningful approximation to the hyperbolic problem and subsequently attempts to establish stability of the scheme. As we have experienced this approach may well lead to very significant technical difficulties and it is worth while looking for an alternative approach.

Rather than first proposing an approximation and then attempting to establish stability it would seem natural to ensure stability as part of the construction. If we recall the idea of enforcing the boundary conditions only weakly, discussed in some detail in Section 4.2.2, we realize that this approach provides an example of just such a constructive approach to stability.

Let us again consider the general variable coefficient problem, Eq. (23), subject to appropriate boundary conditions. If we first assume that a is a constant, and recall the Chebyshev–Legendre approximation [19] discussed in Section 4.2.2 we have

$$\left.\frac{\partial u_N}{\partial t}\right|_{x_j} + a\left.\frac{\partial u_N}{\partial x}\right|_{x_j} = -a\tau\frac{(1-x_j)P'_N(x_j)}{2P'_N(-1)}[u_N(-1,t)-f(t)].$$

To establish stability, we exploit that both sides can be represented as an Nth-order polynomial specified at $N+1$ grid points, i.e., it is unique. We can thus read it at the Legendre–Gauss–Lobatto point, y_j, multiply from the left with $u_N(y_j)w_j$ with w_j being the weight associated with the Legendre–Gauss–Lobatto quadrature [27], and sum over all the nodes to obtain

$$\frac{1}{2}\frac{\mathrm{d}}{\mathrm{d}t}\sum_{j=0}^{N}u_N^2(y_j,t)w_j = -\frac{1}{2}a[u_N^2(1,t)-u_N^2(-1,t)] - \tau\omega_0 a u_N^2(-1,t),$$

from which we recover L^2 stability provided only that

$$\tau \geqslant \frac{1}{2\omega_0} = \frac{N(N+1)}{4}$$

as mentioned in Section 4.2.2. A direct proof of stability for the Chebyshev approximation using a Chebyshev penalty term is given in [17]. These results generalize directly to the case of a variable coefficient with a constant sign by introducing an elliptic norm and the general case can be addressed by writing the problem on skew-symmetric form, Eq. (30).

Let us finally consider the strictly hyperbolic system

$$\frac{\partial U}{\partial t} + A\frac{\partial U}{\partial x} = 0,$$

where $U = [u_1(x,t),\ldots,u_M(x,t)]^{\mathrm{T}}$ represent the statevector and A is an $M\times M$ matrix which we without loss of generality take to be diagonal. If we split A into A^- and A^+, corresponding to the the negative and positive entries of A, respectively, the scheme is given as

$$\begin{aligned}\frac{\mathrm{d}}{\mathrm{d}t}\boldsymbol{U} + (A\otimes D)\boldsymbol{U} = {}& -\tau^-(A^+\otimes Q^-)(I_M\otimes I_N[\boldsymbol{U}(-1,t)-\boldsymbol{g}^-(t)]) \\ & +\tau^+(A^-\otimes Q^+)(I_M\otimes I_N[\boldsymbol{U}(+1,t)-\boldsymbol{g}^+(t)]),\end{aligned}$$

where $\otimes$ signifies the Kronecker product, the grid vector $\boldsymbol{U}$ is ordered by the component as $\boldsymbol{U}=[u_1(-1,t),\ldots,u_1(1,t),u_2(-1,t),\ldots,u_M(1,t)]^{\mathrm{T}}$ and we have introduced the two diagonal matrices

$$Q^-_{jj}=\frac{(1-x_j)P'_N(x_j)}{2P'_N(-1)},\quad Q^+_{jj}=\frac{(1+x_j)P'_N(x_j)}{2P'_N(1)}, \tag{36}$$

while I_L signifies the order L identity matrix. The boundary conditions are represented by the vectors $\boldsymbol{g}^-(t)$ and $\boldsymbol{g}^+(t)$ accounting for the left and right boundaries, respectively.

The stability of this approximation follows directly by choosing $\tau^{\pm}$ as for the scalar case. This illustrates well how this approach lends it self to the formulation of stable spectral approximations of even very complex systems of equations [52,54,49,50].

For cases with variable coefficients, the situation is unchanged as long as the coefficients vary smoothly and the frozen coefficient problem is stable [64]. The same is true for the nonlinear case with smooth solutions. For problems with discontinuities, however, the general question of stability remains a significant challenge that we shall discuss further in the following section.

5.3. Stability of problems with nonsmooth initial conditions

Prior to that, however, let us briefly consider a situation where one solves Eq. (23) with discontinuous initial conditions. The question to raise is whether we can expect anything meaningful from such a solution due to the appearance of the Gibbs phenomenon and its potential impact on the time-dependent solution.

The simplest case of a constant coefficient problem clearly poses no problem as there is no means by which the aliasing errors can be redistributed. Let us therefore focus the attention on the linear variable coefficient problem [1]

$$\frac{\partial u}{\partial t}+\frac{1}{2}a(x)\frac{\partial u}{\partial x}+\frac{1}{2}\frac{\partial au}{\partial x}-\frac{1}{2}a_x(x)u=\frac{\partial u}{\partial t}+\mathscr{L}u=0. \tag{37}$$

Note that we have written $\mathscr{L}$ on skew-symmetric form to avoid the instabilities discussed above. We wish to solve this problem subject to periodic boundary conditions and with a discontinuous initial condition

$$u(x,t)=u_0(x).$$

Approximating the initial condition introduces the Gibbs phenomenon and the variable coefficient enables the mixing of the aliasing error with the solution itself. One could speculate that this process eventually could destroy the accuracy of the computed solution.

To understand this scenario let us introduce the dual problem

$$\frac{\partial v}{\partial t}+\frac{1}{2}a(x)\frac{\partial v}{\partial x}+\frac{1}{2}\frac{\partial av}{\partial x}-\frac{1}{2}a_x(x)v=\frac{\partial v}{\partial t}-\mathscr{L}^*v=0, \tag{38}$$

where $(\mathscr{L}u,v)_{L^2[0,2\pi]}=(u,\mathscr{L}^*v)_{L^2[0,2\pi]}$. Contrary to Eq. (37), we assume that Eq. (38) has the smooth initial condition

$$v(x,0)=v_0(x).$$

An immediate consequence of the structure of Eqs. (37)–(38) is that

$$\frac{\mathrm{d}}{\mathrm{d}t}(u,v)_{L^2[0,2\pi]}=(\mathscr{L}u,v)_{L^2[0,2\pi]}-(u,\mathscr{L}^*v)_{L^2[0,2\pi]}=0$$

implying that

$$(u(t), v(t))_{L^2[0,2\pi]} = (u(0), v(0))_{L^2[0,2\pi]}. \tag{39}$$

If we consider the pseudospectral Fourier approximations of Eqs. (37)–(38)

$$\frac{\partial u_N}{\partial t} + \mathscr{L}_N u_N = 0, \quad \frac{\partial v_N}{\partial t} - \mathscr{L}_N^* v_N = 0,$$

where $u_N \in \mathsf{P}_N$ and $v_N \in \mathsf{P}_N$ represent the polynomial solutions and $\mathscr{L}_N = \mathscr{I}_N \mathscr{L} \mathscr{I}_N$ and $\mathscr{L}_N^* = \mathscr{I}_N \mathscr{L}^* \mathscr{I}_N$, respectively, we recover

$$\frac{\pi}{N} \sum_{j=0}^{2N-1} u_N(x_j, t) v_N(x_j, t) = \frac{\pi}{N} \sum_{j=0}^{2N-1} u_N(x_j, 0) v_N(x_j, 0) \tag{40}$$

as a consequence of the skew-symmetry of $\mathscr{L}_N$ and $\mathscr{L}_N^*$.

A deceptive element of the pseudospectral approximation of the initial conditions is that it hides the oscillations, i.e., if we look at $u_N(x_j, 0)$ it appears perfectly smooth. To reinforce the oscillatory behavior of $u_N(0)$ we preprocess the initial conditions, u_0 and v_0, such that they are the pseudospectral representation of the Galerkin representation, i.e., the truncated continuous expansion, of u_0 and v_0. For the latter, this will have little impact as v_0 is smooth. However, the Galerkin representation of u_0 is oscillatory and this will be reflected in $u_N(x_j, 0)$. Since $u_N(0)$ and $v_N(0)$ are both Nth-order trigonometric polynomials Eq. (40) implies

$$\frac{\pi}{N} \sum_{j=0}^{2N-1} u_N(x_j, t) v_N(x_j, t) = (u_N(0), v_N(0))_{L^2[0,2\pi]},$$

by the quadrature. Exploiting the smoothness of v_0 it is straightforward to show that [1]

$$(u_0, v_0)_{L^2[0,2\pi]} = (u_N(0), v_N(0))_{L^2[0,2\pi]} + CN^{-q} \| v_0^{(q)} \|_{L^2[0,2\pi]}.$$

In combination with Eqs. (39) and (40) this yields

$$\frac{\pi}{N} \sum_{j=0}^{2N-1} u_N(x_j, t) v_N(x_j, t) = (u(t), v(t))_{L^2[0,2\pi]} + CN^{-q} \| v_0^{(q)} \|_{L^2[0,2\pi]}.$$

Further assuming that the approximation of the dual problem is stable, which is supported by the discussion in Section 5.1 and the assumption that v_0 is smooth, we have convergence as

$$\| v(t) - v_N(t) \|_{L^2[0,2\pi]} \leqslant CN^{1-q} \| v^{(q)} \|_{L^2[0,2\pi]}.$$

We can therefore replace v_N with v, thereby introducing an error of the order of the scheme, to obtain

$$\frac{\pi}{N} \sum_{j=0}^{2N-1} u_N(x_j, t) v(x_j, t) = (u(t), v(t))_{L^2[0,2\pi]} + \varepsilon,$$

where ε is exponentially small if $v(x,t)$ is analytic. In other words, $u_N(x,t)$ approximates $u(x,t)$ weakly to within spectral accuracy and the stability of the problem with smooth initial conditions is sufficient to guarantee stability of the problem with nonsmooth initial conditions.

Moreover, there exists a smooth function, $v(x,t)$, that allows one to extract highly accurate information about $u(x,t)$ from $u_N(x,t)$ even after propagation in time and the accumulated effects of the aliasing error. This justifies the use of the Gibbs reconstruction techniques discussed in Section 3.2 as a postprocessing technique after propagating the oscillatory initial conditions.

5.4. Aspects of fully discrete stability

The results summarized in the past sections on semi-discrete stability provide a necessary foundation for understanding the behavior of the fully discrete approximation, i.e., an approximation in which also the temporal dimension is discretized. This last, yet essential step, in constructing a fully discrete scheme introduces a number of additional complications.

A thorough discussion of temporal integration techniques and their properties is well beyond the scope of this review and we shall focus the attention on the widely used Runge–Kutta methods. Numerous alternative techniques, explicit as well as implicit, are discussed in [39,10].

Let us consider the problem, Eq. (23), on the generic form, Eq. (37), and assume that $a(x)$ is uniformly bounded away from zero to avoid unnecessary complications. In this case, the fully discrete s-stage Runge–Kutta scheme

$$u_N^{n+1} = \sum_{k=0}^{s} \frac{(\Delta t \mathscr{L}_N)^k}{k!} = \mathscr{P}(\Delta t \mathscr{L}_N) u_N^n,$$

advances the solution, u_N^n, from $t = n\Delta t$ to $t = (n+1)\Delta t$. The central issue to address is which value of Δt ensures that this is a stable process in the sense that $\|\mathscr{P}(\Delta t \mathscr{L}_N)^n\|$ remains bounded for all n.

If we first consider the Fourier collocation approximation to Eq. (23) we have

$$\mathscr{L}_N = AD,$$

in the notation of Section 5.1. Recall that D, Eq. (12), is skew-symmetric which immediately implies that $D = S\Lambda S^{\mathrm{T}}$ with $\|S\| = \|S^{\mathrm{T}}\| = 1$ and [65]

$$\Lambda = \mathrm{diag}[-\mathrm{i}(N-1), -\mathrm{i}(N-2), \ldots, -\mathrm{i}, 0, 0, \mathrm{i}, \ldots, \mathrm{i}(N-2), \mathrm{i}(N-1)],$$

such that $\|D\| = (N-1)$. We note the double zero eigenvalue which appears as a result of having $2N+1$ modes but only $2N$ nodes in the expansion. The double eigenvalues is not degenerate and is thus not introducing any problems.

Given the purely imaginary spectrum, this yields a necessary and sufficient condition for fully discrete stability on the form

$$\Delta t \leqslant \frac{C_{\Delta t}}{a_\infty (N-1)}, \quad a_\infty = \max_{x \in [0, 2\pi]} |a(x)|, \tag{41}$$

and $C_{\Delta t} = \sqrt{3}$ for the third-order third-stage Runge–Kutta method while it is $C_{\Delta t} = \sqrt{8}$ for the fourth-order fourth-stage Runge–Kutta method.

While the von Neumann stability analysis suffices to establish both necessary and sufficient conditions for discrete stability of the Fourier method it fails to provide more than necessary conditions for the Chebyshev case. This is caused by $\hat{D}$, Eq. (14) modified to account for boundary conditions, being nonnormal, i.e., even if $\hat{D}$ can be diagonalized as $\hat{D} = S\Lambda S^{-1}$ we cannot in general ensure that $\|S\|\|S^{-1}\|$ remains bounded for $N \to \infty$.

As for the semi-discrete case discussed in Section 5.2 the difficulty in establishing rigorous stability results lies in the need to identify the right norm. If we restrict the attention to Eq. (23) in the simplest case of $a(x) = a > 0$ it is natural to employ the norm used in Eq. (34) and consider the stability of the slightly changed problem

$$\frac{\mathrm{d}}{\mathrm{d}t} \boldsymbol{v} + aD\boldsymbol{v} = 0, \tag{42}$$

where $\boldsymbol{v}(t) = [u_N(-1,t),\ldots,u_N(x_j,t),\ldots,u_N(x_{N-1},t)]^{\mathrm{T}}$, contains the first N components of u_N and $\hat{D}$ is discussed in relation with Eq. (35).

Using a first-order one-stage Runge–Kutta scheme, also known as the forward Euler method, a necessary and sufficient condition for fully discrete stability of the Chebyshev collocation approximation of Eq. (42) has been obtained for Eq. (35) of the form [45]

$$\Delta t a \left(N^2 + \frac{2}{\Delta x_{\min}} \right) \leqslant \frac{1}{4},$$

where the first term, N^2, is associated with the Chebyshev basis itself, while

$$\Delta x_{\min} = \min(1 + x_0, 1 - x_N),$$

reflects a dependency on the minimum grid size. Although one can only conjecture that the discrete stability remains valid for the pseudospectral Chebyshev–Gauss–Lobatto approximation, Eq. (29), of $\boldsymbol{u}$ rather than $\boldsymbol{v}$ it is interesting to note that

$$\Delta x_{\min} \simeq \frac{\pi^2}{2} N^{-2}.$$

Hence, we recover the well-known empirical stability condition [39]

$$\Delta t \leqslant \frac{C}{aN^2},$$

where C is of order one. Similar results have been established for higher-order Runge–Kutta schemes in [68].

To associate this limit with the clustering of the grid, however, is a deceptive path. To see this consider Eq. (23) with $a(x) > 0$, in which case the stability condition becomes [45]

$$\Delta t \left(a_\infty N^2 + 2 \max_{x_j} \frac{a(x_j)}{1 - x_j} \right) \leqslant \frac{1}{4}.$$

Clearly, in the event where $a(x_j)$ approaches zero faster than N^{-2} it is the first term, $a_\infty N^2$, rather than the minimum grid size that controls the time step.

The effect of the nonnormality on the performance of the pseudospectral approximations is discussed in detail in [87] for both Legendre and Chebyshev approximations. Attempts to extend the applicability of the von Neumann analysis to problems with nonnormal operators has been discussed in the context of spectral methods in [78,79] where it is advocated that one considers the pseudo-spectrum rather than the simpler eigenspectrum of D to properly understand the fully discrete stability of the approximation.

6. Convergence results for nonlinear hyperbolic problems

The return to the general nonlinear problem

$$\frac{\partial u}{\partial t} + \frac{\partial f(u)}{\partial x} = 0 \tag{43}$$

introduces additional issues and new problems which are not relevant or of less importance for the variable coefficient problem discussed in Section 5.

One of the central difficulties in using spectral methods for the solution of nonlinear conservation laws lies in the potential development of nonsmooth solutions in finite time even for problems with very smooth initial conditions. As we have discussed previously, this introduces the Gibbs phenomenon which, through the nonlinearity, interacts with the solution. What we shall discuss in the following is the impact this has on the performance of the numerical approximation and techniques that allow us to recover accurate and physically meaningful solutions to the conservation laws even when the Gibbs oscillations are apparent.

6.1. Stability by the use of filters

Maintaining stability of the numerical approximation becomes increasingly hard as the discontinuity develops and generates energy with higher and higher frequency content. This process, amplified by the nonlinear mixing of the Gibbs oscillations and the numerical solution, eventually renders the scheme unstable.

Understanding the source of the stability problem, i.e., accumulation of high-frequency energy, also suggests a possible solution by introducing a dissipative mechanism that continuously remove these high-frequency components.

A classical way to accomplish this is to modify the original problem by adding artificial dissipation as

$$\frac{\partial u}{\partial t} + \frac{\partial f(u)}{\partial x} = \varepsilon(-1)^{p+1}\frac{\partial^{2p} u}{\partial x^{2p}}.$$

A direct implementation of this, however, may be costly and could introduce additional stiffness which would limit the stable time-step. We shall hence seek a different approach to achieve a similar effect.

In Section 3.2.1 we discussed the use of low pass filtering to improve on the convergence rate of the global approximation away from the point of discontinuity. This was achieved by modifying the numerical solution, $u_N(x,t)$, through the use of a spectral filter as

$$\mathscr{F}_N u_N(x,t) = \sum_{n=-N}^{N} \sigma\left(\frac{n}{N}\right) \tilde{u}_n(t)\exp(inx). \tag{44}$$

To understand the impact of using the filter at regular intervals as a stabilizing mechanism, a procedure first proposed in [74,66], let us consider the exponential filter

$$\sigma(\eta) = \exp(-\alpha\eta^{2p}).$$

As discussed in Section 3.2.1 this filter allows for a dramatic improvement in the accuracy of the approximation away from points of discontinuity.

To appreciate its impact on stability, consider the generic initial value problem

$$\frac{\partial u}{\partial t} = \mathscr{L}u$$

with the pseudospectral Fourier approximation

$$\frac{\mathrm{d}}{\mathrm{d}t}\boldsymbol{u} = \mathscr{L}_N\boldsymbol{u}.$$

Advancing the solution from $t=0$ to Δt followed by the filtering is conveniently expressed as

$$\boldsymbol{u}(\Delta t)=\mathscr{F}_N \exp(\mathscr{L}_N\Delta t)\boldsymbol{u}(0).$$

If we first assume that $\mathscr{L}_N$ represents the constant coefficient hyperbolic problem, $\mathscr{L}=a(\partial/\partial x)$, we recover that

$$\tilde{u}_n(\Delta t)=\exp(-\alpha\eta^{2p}+a(\mathrm{i}k)\Delta t)\tilde{u}_n(0), \tag{45}$$

i.e., we are in fact computing the solution to the modified problem

$$\frac{\partial u}{\partial t}=a\frac{\partial u}{\partial x}-\alpha\frac{(-1)^p}{\Delta t N^{2p}}\frac{\partial^{2p}u}{\partial x^{2p}}.$$

The effect of the filter is thus equivalent to the classical approach of adding a small dissipative term to the original equation, but the process of adding the dissipation is very simple as discussed in Section 3.2.1. Note in particular that ΔtN essentially represents the *CFL* condition, Eq. (41), and hence is of order one.

For a general $\mathscr{L}$, e.g., with a variable coefficient or of a nonlinear form, in which case $\mathscr{F}_N$ and $\mathscr{L}_N$ no longer commute, the modified equation being solved takes the form

$$\frac{\partial u}{\partial t}=\mathscr{L}u-\alpha\frac{(-1)^p}{\Delta t N^{2p}}\frac{\partial^{2p}u}{\partial x^{2p}}+\mathscr{O}(\Delta t^2),$$

by viewing the application of the filter as an operator splitting problem [7,20].

With this in mind it is not surprising that using a filter has a stabilizing effect. Moreover, we observe that if p increases with N the modification caused by the filter vanishes spectrally as N increases. These loose arguments for the stabilizing effect of filtering have been put on firm ground for problem with smooth and nonsmooth initial data [74,66,83] for the Fourier approximation to the general variable coefficient problem, Eq. (23). These results, however, are typically derived under the assumption that $\sigma(\eta)$ is of polynomial form. While such filtering indeed stabilizes the approximation it also reduces the global accuracy of the scheme [74,88,42]. Let us therefore briefly consider the stabilizing effect of the exponential filter in the pseudospectral Fourier approximation of Eq. (23), known to be weakly unstable as discussed in Section 5.1, Eq. (31).

Consider the filtered approximation of the form

$$\frac{\partial u_N}{\partial t}+\mathscr{I}_N\left(a(x)\frac{\partial u_N}{\partial x}\right)=\varepsilon_N(-1)^{p+1}\frac{\partial^{2p}u_N}{\partial x^{2p}}, \tag{46}$$

where the superviscosity term on the right can be implemented through a filter and

$$\varepsilon_N=\frac{\alpha}{\Delta t N^{2p}}.$$

To establish stability, let us rewrite Eq. (46) as

$$\frac{\partial u_N}{\partial t}+\mathscr{N}_1u_N+\mathscr{N}_2u_N+\mathscr{N}_3u_N=\varepsilon_N(-1)^{p+1}\frac{\partial^{2p}u_N}{\partial x^{2p}},$$

where

$$\mathscr{N}_1u_N=\frac{1}{2}\frac{\partial}{\partial x}\mathscr{I}_Na(x)u_N+\frac{1}{2}\mathscr{I}_N\left(a(x)\frac{\partial u_N}{\partial x}\right)$$

is the skew-symmetric form of the operator, Eq. (30),

$$\mathscr{N}_2 u_N = \frac{1}{2}\mathscr{I}_N\left(a(x)\frac{\partial u_N}{\partial x}\right) - \frac{1}{2}\mathscr{I}_N\frac{\partial a(x)u_N}{\partial x}$$

and

$$\mathscr{N}_3 u_N = \frac{1}{2}\mathscr{I}_N\frac{\partial a(x)u_N}{\partial x} - \frac{1}{2}\frac{\partial}{\partial x}\mathscr{I}_N a(x)u_N.$$

To establish stability, consider

$$\frac{1}{2}\frac{\mathrm{d}}{\mathrm{d}t}\|u_N\|_N^2 = -[u_N, \mathscr{N}_1 u_N]_N - [u_N, \mathscr{N}_2 u_N]_N$$

$$-[u_N, \mathscr{N}_3 u_N]_N + \left[u_N, \varepsilon_N(-1)^{p+1}\frac{\partial^{2p}u_N}{\partial x^{2p}}\right]_N.$$

Clearly $[u_N, \mathscr{N}_1 u_N]_N = 0$ due to the skew-symmetry of $\mathscr{N}_1 u_N$ and by inspection we can bound

$$[u_N, \mathscr{N}_2 u_N]_N \leqslant \tfrac{1}{2}\max_{x\in[0,\,2\pi]}|a_x(x)|\,\|u_N\|_N^2.$$

It is indeed the term associated with $\mathscr{N}_3 u_N$ that is the troublemaker. To appreciate this, simply note that if $\mathscr{P}_N$ was used rather than $\mathscr{I}_N$ such that differentiation and truncation commute, the term would vanishes identically and the scheme would be stable. To bound this term, we can use that

$$[u_N, \mathscr{N}_3 u_N]_N \leqslant C(\|u_N\|_N^2 + \|\mathscr{N}_3 u_N\|_N^2).$$

Noting that $\|\mathscr{N}_3 u_N\|^2_{L^2[0,2\pi]}$ is nothing more than the commutation error and that $\|\cdot\|_N$ is L^2-equivalent, we can borrow the result of Eq. (17) to obtain

$$[u_N, \mathscr{N}_3 u_N]_N \leqslant C(\|u_N\|^2_{L^2[0,2\pi]} + N^{2-2p}\|u_N^{(p)}\|^2_{L^2[0,2\pi]}),$$

where C depends on $a(x)$ and its first p derivatives. If we finally note that

$$\left[u_N, \varepsilon_N(-1)^{p+1}\frac{\partial^{2p}u_N}{\partial x^{2p}}\right]_N = -\varepsilon_N\|u_N^{(p)}\|_N^2,$$

it is clear that we can always choose $\varepsilon_N = AN^{2-2p}$ and A sufficiently large to ensure stability. In other words, using an exponential filter is sufficient to stabilize the Fourier approximation of Eq. (23).

There is one central difference in the effect of using the filter in the Fourier and the Chebyshev approximation. In the latter, the modified equation takes the form

$$\frac{\partial u}{\partial t} = \mathscr{L}u - \alpha\frac{(-1)^p}{\Delta t N^{2p}}\left[\sqrt{1-x^2}\frac{\partial}{\partial x}\right]^{2p} u + \mathcal{O}(\Delta t^2). \tag{47}$$

Hence, while the filtering continues to introduce dissipation, it is spatially varying. In particular, it vanishes as one approaches the boundaries of the domain. In computations with moving discontinuities this may be a source of problems since the stabilization decreases as the discontinuity approaches the boundaries of the computational domain.

6.2. Spectrally vanishing viscosity and entropy solutions

The foundation of a convergence theory for spectral approximations to conservation laws has been laid in [85,72,15] for the periodic case and subsequently extended in [73] to the Legendre approximation and recently to the Chebyshev–Legendre scheme in [70,71].

To appreciate the basic elements of this convergence theory let us first restrict ourselves to the periodic case. For the discrete approximation to Eq. (43) we must add a dissipative term that is strong enough to stabilize the approximation, yet small enough not to ruin the spectral accuracy of the scheme. In [85,72] the following spectral viscosity method was considered

$$\frac{\partial u_N}{\partial t} + \frac{\partial}{\partial x}\mathscr{P}_N(f(u_N)) = \varepsilon_N(-1)^{p+1}\frac{\partial^p}{\partial x^p}\left[Q_m(x,t)\frac{\partial^p u_N}{\partial x^p}\right], \tag{48}$$

where

$$\frac{\partial^p}{\partial x^p}\left[Q_m(x,t)\frac{\partial^p u_N}{\partial x^p}\right] = \sum_{m<|n|\leqslant N}(\mathrm{i}k)^{2p}\hat{Q}_n\hat{u}_n\exp(\mathrm{i}nx).$$

To ensure that stability is maintained m should not be taken too big. On the other hand, taking m too small will impact the accuracy in a negative way. An acceptable compromise seems to be

$$m \sim N^\theta, \quad \theta < \frac{2p-1}{2p}.$$

Moreover, the smoothing factors, $\hat{Q}_n$, should only be activated for high modes as

$$\hat{Q}_n = 1 - \left(\frac{m}{|n|}\right)^{(2p-1)/\theta}$$

for $|n| > m$ and $\hat{Q}_n = 1$ otherwise. Finally, we shall assume that the amplitude of the viscosity is small as

$$\varepsilon_N \sim \frac{C}{N^{2p-1}}.$$

Under these assumptions, one can prove for $p = 1$ that the solution is bounded in $L^\infty[0,2\pi]$ and obtain the estimate

$$\|u_N\|_{L^2[0,2\pi]} + \sqrt{\varepsilon_N}\left\|\frac{\partial u_N}{\partial x}\right\|_{L^2_{loc}} \leqslant C.$$

Convergence to the correct entropy solution then follows from compensated compactness arguments [85,72].

To realize the close connection between the spectral viscosity method and the use of filters discussed in Section 6.1, consider the simple case where $f(u) = au$. In this case, the solution to Eq. (48) is given as

$$\hat{u}_n(t) = \exp(\mathrm{i}nat - \varepsilon_N n^2\hat{Q}_n)\hat{u}_n(0), \quad |n| > m,$$

which is equivalent to the effect of the filtering discussed in Section 6.1. Note that the direct application of the vanishing viscosity term in Eq. (48) amounts to $2p$ spatial derivatives while filtering as discussed in Section 3.2.1 can be done at little or no additional cost.

For $p \neq 1$ a bound on the $L^\infty[0,2\pi]$ is no longer known. However, experience suggests that it is better to filter from the first mode but to employ a slower decay of the expansion coefficients, corresponding to taking $p > 1$. This yields the superviscosity method in which one solves

$$\frac{\partial u_N}{\partial t} + \frac{\partial}{\partial x}\mathscr{P}_N f(u_N) = \varepsilon_N(-1)^{p+1}\frac{\partial^{2p} u_N}{\partial x^{2p}},$$

which we recognize from Eq. (46) as being equivalent to that obtained when using a high-order exponential filter.

The vanishing viscosity approximation to Eq. (43) using a Chebyshev collocation approach takes the form

$$\frac{\partial u_N}{\partial t} + \frac{\partial}{\partial x}\mathscr{I}_N f(u_N) = \varepsilon_N(-1)^{p+1}\left[\sqrt{1-x^2}\frac{\partial}{\partial x}\right]^{2p} u_N + \mathscr{B}u_N,$$

where again

$$\varepsilon_N \sim \frac{C}{N^{2p-1}},$$

and p grows with N [73]. Here the boundary operator, $\mathscr{B}u_N$, may vanish or it may take values as

$$\mathscr{B}u_N = -\tau^- \frac{(1-x)T_N'(x)}{2T_N'(-1)}(u_N(-1,t) - g^-) + \tau^+ \frac{(1+x)T_N'(x)}{2T_N'(1)}(u_N(1,t) - g^+),$$

which we recognize as the weakly imposed penalty terms discussed in Sections 4.2 and 5.2. Note again that the vanishing viscosity term is equivalent to that obtained from the analysis of the effect of spectral space filtering, Eq. (47). Similar results can be obtained for the Legendre approximation and for the Chebyshev–Legendre method for which convergence has been proven [70,71], using arguments similar to those in [85,72], for $p = 1$ as well as for $p > 1$.

6.3. Conservation

It is natural to question whether the introduction of the artificial Gibbs oscillations has any impact on the basic physical properties described by the conservation law, e.g., mass conservation and the speed by which discontinuities propagate.

To come to an understanding of this, assume a spatially periodic problem and consider the pseudospectral Fourier scheme

$$\frac{\mathrm{d}}{\mathrm{d}t}\boldsymbol{u} + D\boldsymbol{f} = 0,$$

where $\boldsymbol{u} = [u_N(0,t),\ldots,u_N(x_{2N-1},t)]^{\mathrm{T}}$ represent the grid vector and the interpolation of the flux is given as $\boldsymbol{f} = [\mathscr{I}_N f(u_N(0,t),t),\ldots,\mathscr{I}_N f(u_N(x_{2N-1},t),t)]^{\mathrm{T}}$.

The first thing to note is that

$$\int_0^{2\pi} u_N(x,t)\,\mathrm{d}x = \int_0^{2\pi} u_N(x,0)\,\mathrm{d}x$$

as an immediate consequence of the accuracy of the trapezoidal rule and the assumption of periodicity. Hence, the approximation conserves the 'mass' of the interpolation of the initial conditions.

Let us introduce a smooth periodic test function, $\psi(x,t)$, with the corresponding grid vector, $\boldsymbol{\psi}=[\psi_N(x_0,t),\ldots,\psi_N(x_{2N-1},t)]^{\mathrm{T}}$. The test function, $\psi(x,t)$, is assumed to vanish at large t. If we consider [36]

$$\boldsymbol{\psi}^{\mathrm{T}}\left(\frac{\mathrm{d}}{\mathrm{d}t}\boldsymbol{u}+D\boldsymbol{f}\right)=0$$

and utilize the accuracy of the trapezoidal rule we recover

$$\int_0^{2\pi}\left[\psi_N(x,t)\frac{\partial u_N(x,t)}{\partial t}-\frac{\partial\psi_N(x,t)}{\partial x}\mathscr{I}_N f(u_N(x,t),t)\right]\mathrm{d}x=0,$$

after integration by parts which is permitted if the solution, $u_N(x,t)$, is bounded. This implicitly assumes that the numerical approximation itself is stable which generally implies that a vanishing viscosity term is to be added, potentially through the use of a filter.

Integrating over time and by parts once more, we recover the result

$$\int_0^{\infty}\int_0^{2\pi}\left[u_N(x,t)\frac{\partial\psi_N(x,t)}{\partial t}+\frac{\partial\psi_N(x,t)}{\partial x}\mathscr{I}_N f(u_N(x,t),t)\right]\mathrm{d}x\,\mathrm{d}t$$
$$+\int_0^{2\pi}\psi_N(x,0)u_N(x,0)\,\mathrm{d}x=0.$$

Thus, for $N\to\infty$ the solution, $u_N(x,t)$, is a weak solution to the conservation law. This again implies that the limit solution satisfies the Rankine–Hugoniot conditions which guarantees that shocks propagate at the right speed to within the order of the scheme. Results similar to these have also been obtained for the Chebyshev approximation to the conservation law [36].

To appreciate that the addition of the vanishing viscosity has no impact on the conservation of the scheme, consider the Legendre superviscosity case [73] and let $\psi(x,t)$ be a test function in $C^3[-1,1]$ that vanishes at the endpoints. Taking $\psi_{N-1}(x,t)=\mathscr{I}_{N-1}\psi(x,t)$, then clearly $\psi_{N-1}(x,t)\to\psi(x,t)$, $(\psi_{N-1})_x(x,t)\to\psi_x(x,t)$, and $(\psi_{N-1})_t(x,t)\to\psi_t(x,t)$ uniformly in N.

Since $\psi_{N-1}(x)$ is a polynomial that vanishes at the boundaries we have

$$\int_{-1}^{1}(1+x)P_N'(x)\psi_{N-1}(x,t)\,\mathrm{d}x=0,\qquad\int_{-1}^{1}(1-x)P_N'(x)\psi_{N-1}(x,t)\,\mathrm{d}x=0.$$

Moreover, integration by parts yields that

$$\lim_{N\to\infty}\frac{\varepsilon_N(-1)^p}{N^{2p-1}}\int_{-1}^{1}\psi_{N-1}(x,t)\left[\frac{\partial}{\partial x}(1-x^2)\frac{\partial}{\partial x}\right]^{2p}u_N(x,t)\,\mathrm{d}x=0.$$

Hence, the superviscosity term does not cause any problems and one can show that

$$-\int_0^{\mathrm{T}}\int_{-1}^{1}\left(u_N(x,t)\frac{\partial\psi_{N-1}(x,t)}{\partial t}+\mathscr{I}_N f(u_N(x,t))\frac{\partial\psi_{N-1}(x,t)}{\partial x}\right)\mathrm{d}x\,\mathrm{d}t$$
$$-\int_{-1}^{1}u_N(x,0)\psi_{N-1}(x,0)\,\mathrm{d}x=0.$$

The main conclusion of this is that if $u_N(x,t)$ is a solution to the Legendre collocation approximation at the Gauss–Lobatto points and if $u_N(x,t)$ converges almost everywhere to a function $u(x,t)$, then

$u(x,t)$ is a weak solution to Eq. (43). The technical details of this proof can be found in [14] where also a similar result for the Chebyshev superviscosity approximation is given.

The theory of convergence of spectral methods equipped with spectral viscosity or superviscosity is limited to the scalar case as discussed in Section 6.2. For the system case a more limited result can be obtained, stating that if the solution converges to a bounded solution, it converges to the correct weak solution.

7. Multi-domain methods

As a final technique, playing a pivotal role in making many of the techniques discussed previously amenable to the solution of problems of interest to scientists and engineers, let us briefly discuss multi-domain methods for hyperbolic problems.

The original motivation for the introduction of multi-domain methods can be found in the restrictions that the fixed grids, required to ensure the high spatial accuracy, impose. This fixed grid makes it difficult to utilize adaptivity and, for multi-dimensional problems, to address problems in complex geometries. Moreover, the use of global spectral expansions makes it difficult to achieve a high parallel efficiency on contemporary parallel computers.

Many of these concerns can be overcome if one splits the computational domain into a number of geometrically simple building blocks, e.g., squares and cubes, and then employs tensor-product forms of the simple one-dimensional approximations as the basis of an element by element approximation. While this technique opens up for the use of a highly nonuniform resolution and the ability to model problems in geometrically complex domains, it also introduces the need to connect the many local solutions in an accurate, stable, and efficient manner to reconstruct the global solution.

7.1. Patching techniques

The patching of the local solutions in a way consistent with the nature of the hyperbolic problem can be performed in at least two different yet related ways. Borrowing the terminology introduced in [59], we shall refer to these two different methods as the differential and the correctional method, respectively.

To expose the differences between the two methods, let us consider the two domain scalar problem

$$\frac{\partial u}{\partial t} + \frac{\partial f(u)}{\partial x} = 0, \quad x \in [-1,0], \tag{49}$$

$$\frac{\partial v}{\partial t} + \frac{\partial f(v)}{\partial x} = 0, \quad x \in [0,1].$$

To recover the global solution $U = [u,v]$ under the constraint that $u(0,t) = v(0,t)$, the central issue is how one decides which of the multiple solutions at $x = 0$ takes preference and hence determines the evolution of $u(0,t)$ and $v(0,t)$.

Provided that the initial conditions are consistent with the continuity condition it will clearly remain continuous if we ensure that $u_t(0,t) = v_t(0,t)$. This approach, known as the differential method, involves the exchange of information between the two domains to ensure that the flux of

$u(0,t)$ and $v(0,t)$ are identical throughout the computation. There are, however, several ways to do so.

In the original work [58], the solution is assumed to be smooth and one introduces the flux derivative

$$\lambda = \left.\frac{\partial f}{\partial u}\right|_{u(0,t)} = \left.\frac{\partial f}{\partial v}\right|_{v(0,t)}$$

and requires that u and v be updated at $x=0$ as

$$\left.\frac{\partial u}{\partial t}\right|_{x=0} = \left.\frac{\partial v}{\partial t}\right|_{x=0} = -\frac{1}{2}(\lambda + |\lambda|)\left.\frac{\partial u}{\partial x}\right|_{x=0} - \frac{1}{2}(\lambda - |\lambda|)\left.\frac{\partial v}{\partial x}\right|_{x=0}.$$

This can be recognized as nothing else than pure upwinding. The extension to systems of equations employs the characteristic form of the system and the multi-dimensional case is treated by dimensional splitting.

An alternative formulation, based on the weakly imposed boundary conditions discussed in Section 4.2.2 and introduced in [48,19,50], takes the form

$$\left.\frac{\partial u}{\partial t}\right|_{x=0} + \left.\frac{\partial f(u)}{\partial x}\right|_{x=0} = -\tau\frac{|\lambda - |\lambda||}{2}(u(0,t) - v(0,t)),$$

$$\left.\frac{\partial v}{\partial t}\right|_{x=0} + \left.\frac{\partial f(v)}{\partial x}\right|_{x=0} = -\tau\frac{|\lambda + |\lambda||}{2}(v(0,t) - u(0,t)),$$

which again amounts to upwinding, although on a weak form. The advantage of this latter formulation is that it allows for establishing stability and it makes the enforcement of very complex interface conditions simple. The extension to systems employs the characteristic variables and is discussed in detail in [48,50] while the multi-dimensional case is treated in [49]. Similar developments for methods employing multi-variate polynomials [53,55] or a purely modal basis [69,91,93] defined on triangles and tetrahedra has recently been developed, pawing the way for the formulation of stable spectral methods for the solution of hyperbolic conservation laws using a fully unstructured grid.

Rather than correcting the local temporal derivative to ensure continuity of the flux across the interface one could choose to modify the solution itself. This observation provides the basic foundation for correctional methods in which both u and v is advanced everywhere within the each domain, leading to a multiplicity of solutions at $x=0$. For the specific case discussed here, the correctional approach amounts to

$$u(0,t) = v(0,t) = \begin{cases} u(0,t) & \text{if } \lambda \geqslant 0, \\ v(0,t) & \text{if } \lambda < 0, \end{cases}$$

which we again recognize as upwinding. The system case is treated similarly by exploiting the characteristic variables. As for the differential methods, the use of the characteristics implicitly assumes a minimum degree of smoothness of the solution. However, as no information about the spatial derivatives are passed between domains, the correctional method imposes no constraints on the smoothness of the grid.

The main appeal of the correctional method lies in its simplicity and robustness and it has been utilized to formulate very general multi-domain method for problems in gas dynamics [59,60], in acoustics and elasticity [2], and electromagnetic [95,96,51,97].

Note that both methods employ the local flux Jacobian, λ, which implicitly requires a certain amount of smoothness of the solution at the interface. A differential method overcoming this can be realized by borrowing a few ideas from classical finite volume methods.

Consider the cell-averaged formulation

$$\frac{\mathrm{d}\bar{u}_j}{\mathrm{d}t} + \frac{f(u(x_{j+1/2})) - f(u(x_{j-1/2}))}{\Delta x_j} = 0,$$

where

$$\Delta x_j = x_{i+1/2} - x_{i-1/2}, \quad \bar{u}_j = \frac{1}{\Delta x_j} \int_{x_{j-1/2}}^{x_{j+1/2}} u(s)\,\mathrm{d}s.$$

Here $x_{j\pm 1/2}$ signifies the Chebyshev–Gauss–Lobatto grid and x_j refers to the interlaced Chebyshev–Gauss grid. No assumptions are made about the smoothness of the flux and since each individual cell requires reconstruction, the patching of the subdomains is achieved by flux-splitting techniques known from finite volume methods. This approach was first proposed in [8] for Fourier methods and subsequently in [9] for the Chebyshev approximation and has the advantage of being conservative by construction. The averaging and reconstruction procedure, which can be done in an essentially nonoscillatory way, is essential for the accuracy and stability of the scheme and several alternatives, exploiting a similar framework, has been proposed in [80,32,33].

The use of a staggered grid, collocating the solution u at the Gauss grid and the fluxes, $f(u)$, at the Gauss–Lobatto grid, has the additional advantage of allowing for the formulation of multi-dimensional multi-domain methods with no grid points at the vertices of the elements. This approach, introduced in [62,61], has been developed for smooth problems and eliminates complications associated with the treatment of vertices in multi-domain methods [60].

Alternative differential patching methods has been discussed in [77,10] where the patching is achieved by the use of compatibility conditions.

7.2. Conservation properties of multi-domain schemes

The important question of the conservation properties of multi-domain schemes is discussed in [14] in which the following polynomial approximation to Eq. (49) is considered:

$$\begin{aligned}\frac{\partial u_N}{\partial t} + \frac{\partial}{\partial x}\mathscr{I}_N f(u_N) &= \tau_1 Q_{\mathrm{I}}^{+}(x)[f^{+}(u_N(0,t)) - f^{+}(v_N(0,t))] \\ &\quad + \tau_2 Q_{\mathrm{I}}^{+}(x)[f^{-}(u_N(0,t)) - f^{-}(v_N(0,t))] + \mathrm{SV}(u_N),\\ \frac{\partial v_N}{\partial t} + \frac{\partial}{\partial x}\mathscr{I}_N f(v_N) &= -\tau_3 Q_{\mathrm{II}}^{-}(x)[f^{+}(v_N(0,t)) - f^{+}(u_N(0,t))] \\ &\quad - \tau_4 Q_{\mathrm{II}}^{-}(x)[f^{-}(v_N(0,t)) - f^{-}(u_N(0,t))] + \mathrm{SV}(v_N),\end{aligned}$$

where $Q^{\pm}(x)$ are polynomials given in Eq. (36), i.e., they vanish at all collocation points except $x=\pm 1$. Furthermore, $\mathrm{SV}(u_N)$ and $\mathrm{SV}(v_N)$ represent the vanishing viscosity terms, or filtering, required to stabilize the nonlinear problem as discussed in Sections 6.1 and 6.2, while $f=f^{+}+f^{-}$ signifies a splitting into the upwind and downwind components of the flux.

To establish conservation of the approximation, consider a test function $\psi(x)$ and denote by ψ_{I} and ψ_{II} its restriction to the first and second domain respectively. We can assume that ψ_{I} and ψ_{II} are polynomials of order $N-1$ and that ψ_{I} vanishes at $x=-1$ of the first domain while ψ^{II} vanishes at $x=1$ of the second domain, but not at $x=0$.

Repeated integration by parts using the fact that $\mathrm{SV}(u_N)$ vanishes at the boundaries of each domain yields

$$\int_{-1}^{0} \psi_{\mathrm{I}}(x)\,\mathrm{SV}(u_N)\,\mathrm{d}x = (-1)^P \int_{-1}^{0} u_N\,\mathrm{SV}(\psi_{\mathrm{I}})\,\mathrm{d}x$$

which tends to zero with increasing N. A similar result can be obtained for the second domain.

Consider now

$$\int_{-1}^{0} \psi^{\mathrm{I}}(x)\frac{\partial u_N}{\partial t}\,\mathrm{d}x + \int_{0}^{1} \psi^{\mathrm{II}}(x)\frac{\partial v_N}{\partial t}\,\mathrm{d}x.$$

To recover that u_N and v_N are weak solutions to Eq. (49), i.e., the above integral vanishes, we must require

$$\tau_1+\tau_3=1, \quad \tau_2+\tau_4=1. \tag{50}$$

However, for linear stability, as we discussed in Section 5.2, one can show that

$$\tau_1 \geqslant \tfrac{1}{2}, \quad \tau_2 \leqslant \tfrac{1}{2},$$

$$\tau_3 \leqslant \tfrac{1}{2}, \quad \tau_4 \geqslant \tfrac{1}{2},$$

together with Eq. (50), are necessary and sufficient conditions to guarantee stability.

This leaves us with a set of conditions under which to design stable and conservative scheme. In particular, if we choose to do pure upwinding at the interfaces by specifying

$$\tau_1=\tau_4=1, \quad \tau_2=\tau_3=0,$$

we essentially recover the discontinuous Galerkin method discussed in Section 4.2.

An appealing alternative appears by considering

$$\tau_1=\tau_2=\tau_3=\tau_4=\tfrac{1}{2},$$

which yields a marginally stable and conservative scheme of the form

$$\frac{\partial u_N}{\partial t} + \frac{\partial}{\partial x}\mathcal{I}_N f(u_N) = \frac{1}{2}Q_{\mathrm{I}}^{+}(x)[f(u_N(0,t)) - f(v_N(0,t))] + \mathrm{SV}(u_N),$$

$$\frac{\partial v_N}{\partial t} + \frac{\partial}{\partial x}\mathcal{I}_N f(v_N) = -\frac{1}{2}Q_{\mathrm{II}}^{-}(x)[f(v_N(0,t)) - f(u_N(0,t))] + \mathrm{SV}(v_N),$$

i.e., the interface boundary conditions are imposed on the fluxes f rather than on the split fluxes f^+ and f^-.

7.3. Computational efficiency

An interesting question pertaining to the use of multi-domain methods is how one decides how many elements and what resolution to use within each element. In pragmatic terms, what we are interested in is to identify the optimal combination of the order of the polynomial, N, and the number of elements, K, needed to solve a particular problem to within a maximum error using minimum computational resources.

On one hand, it is the high order of the interpolation polynomial that results in the very accurate approximation. On the other hand, the computation of derivatives generally scales as $\mathcal{O}(N^2)$ while the total work scales only linearly with the number of elements. To develop guidelines for choosing the optimal N and K, consider a one-dimensional wave problem with a smooth solution. Assume that the approximation error, $E(N,K)$, scales as

$$E(N,K) \propto \left(\frac{\pi k}{KN}\right)^N,$$

where k is the maximum wavenumber in the solution, i.e., it is proportional to the inverse of the minimum spatial length scale. We shall require that the maximum absolute error, E, is bounded as $E \leqslant \exp(-\gamma)$, and estimate the computational work as

$$W(N,K) = c_1 KN^2 + c_2 KN,$$

where c_1 and c_2 are problem specific constants. Minimizing the work subject to the error constraint yields the optimal values

$$N_{\text{opt}} = \gamma, \quad K_{\text{opt}} = \frac{\pi k}{N_{\text{opt}}} \exp\left(\frac{\gamma}{N_{\text{opt}}}\right).$$

One observes that high accuracy, i.e., γ large, should be achieved by using a large number of modes, N, and not, as one could expect, by employing many subdomains each with a low number of modes. For very smooth and regular functions, where k is small, or if only moderate accuracy is required, the use of many domains may not be the optimal method of choice. On the other hand, if the function exhibits strongly localized phenomena, i.e., k is large, one should introduce many domains to minimize the computational burden. While these arguments are loose, they indicate that an optimal choice of N and K for most problems seems to be a few larger subdomains, each with a reasonable number of modes to maintain an acceptable spatial accuracy.

These results have been confirmed in computational experiments in [48,49,47] indicating that $N = 8$–16 is reasonable for two-dimensional problems and $N = 4$–8 is reasonable for three-dimensional problems. If this results in insufficient resolution one should generally increase the number of domains rather than the resolution. Similar conclusions have been reached for the analysis of spectral multi-domain methods in a parallel setting [24].

8. A few applications and concluding remarks

Less than a decade ago, the formulation and implementation of robust, accurate and efficient spectral methods for the solution of conservation laws was considered an extremely challenging task. This was partly due to problems of a more theoretical character but partly due also to many practical concerns introduced by the appearance of discontinuous solutions and the need to accurately model the long time behavior of hyperbolic problems in complex geometries.

To illustrate the impact of many of the recent developments discussed in this review, let us conclude by presenting a few contemporary examples of the use of spectral methods for the solution

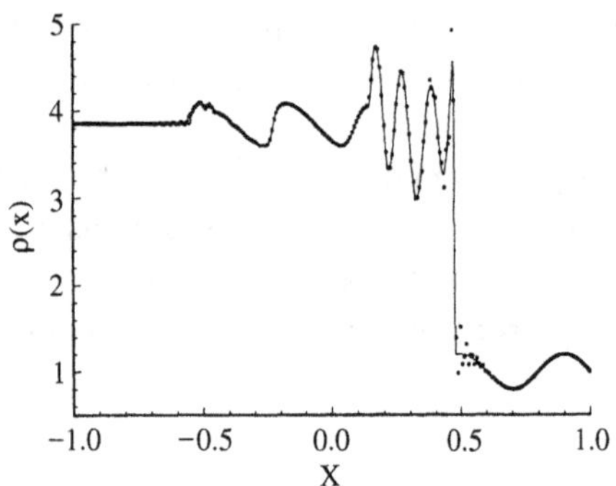

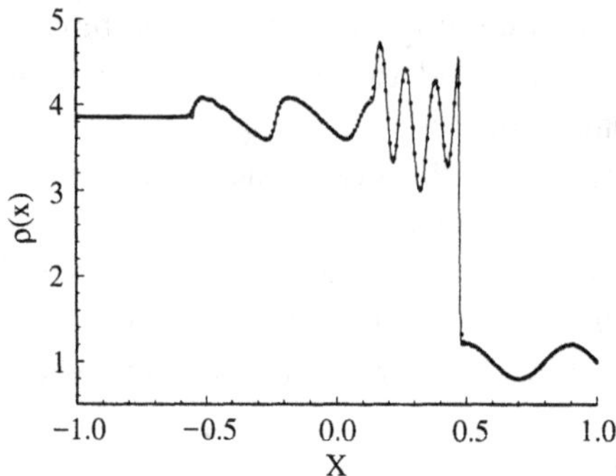

Fig. 1. The solution of the compressible Euler equations for a Mach 3 shock interacting with an entropy wave. On the left is shown the density at $t = 1.8$ computed by using a Chebyshev collocation method with 256 modes and a stabilizing filter. On the right is shown the solution after Gegenbauer reconstruction, removing the effect of the Gibbs phenomenon. The solid line represents the solution computed with a high-order ENO scheme [18,20].

of conservation laws. These few examples all use extensive parts of what we have discussed here and it would indeed have been difficult, if at all possible, to complete the computations without these recent developments.

As a first example, consider the solution of the one-dimensional Euler equations for a compressible gas. We consider the situation where a Mach 3 shock interacts with an entropy wave, producing a strongly oscillatory moving shock [18,20]. Fig. 1 shows the density at $t = 1.8$ computed using a Chebyshev collocation method and compared to the solution obtained using a high-order ENO scheme. A high-order exponential filter is used to stabilize the solution which exhibits strong Gibbs oscillations. Using the Gegenbauer reconstruction technique discussed in Section 3.2, one can recover the nonoscillatory solution with a sharp shock front as is also illustrated in Fig. 1, and a global solution that is essentially identical to that computed using the ENO scheme.

As a second more realistic problem, consider that of shock-induced combustion in which a strong shock, propagating in a oxygen atmosphere, impinges on one or several hydrogen jets, igniting the combustion by compressional heating. This is a problem of great interest to designers of jet engines but also of great difficulty due to the very rich dynamics, the strong shock and the development of very sharp interfaces and flame fronts.

This problem has been studied intensively in [18,20] from which also Fig. 2 is taken. The setting is a strong shock propagating through two aligned hydrogen jets, causing combustion and strong mixing of the fluid. The computation is performed using a two-dimensional Chebyshev collocation method with filtering, a simplified model for the combustion processes and high-order time-stepping. The accuracy as well as the efficiency of the spectral code for combustion problems with strong shocks has been confirmed by detailed comparisons with high-order ENO schemes [21,20].

As a final example, emphasizing geometric complexity rather than shock-induced complications, consider the problem of time-domain electromagnetic scattering by a geometrically very complex metallic scatterer, exemplified in this case by an F-15 fighter (see Fig. 3). The solution of the vectorial Maxwells equations for a problem of such complexity is achieved through the use of novel high-order multi-variate spectral methods defined on tetrahedra [55] with a flexible, stable, and efficient implementation of the penalty method as a discontinuous Galerkin formulation [92,94].

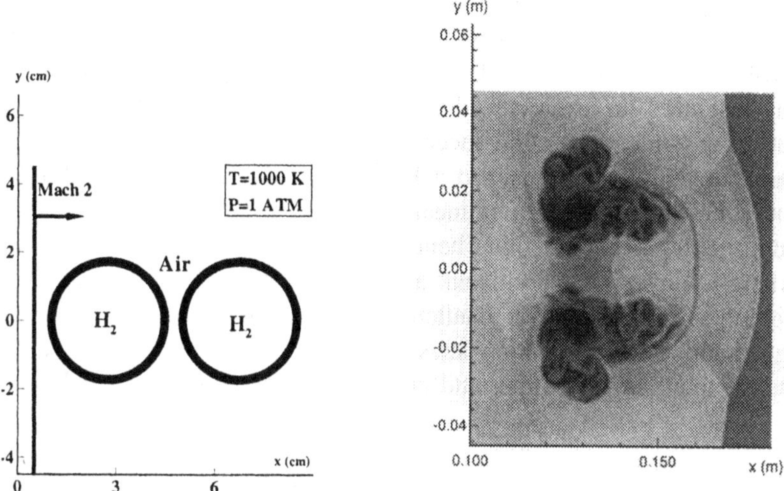

Fig. 2. Example of shock-induced combustion computed using a stabilized Chebyshev collocation method. On the left is shown the initial two-dimensional configuration of two hydrogen jets in an oxygen atmosphere. The left shows a snapshot of the density shortly after the strong shock has passed the jets, showing the very rich dynamics of the mixing process as well as the very complex shock structure [20].

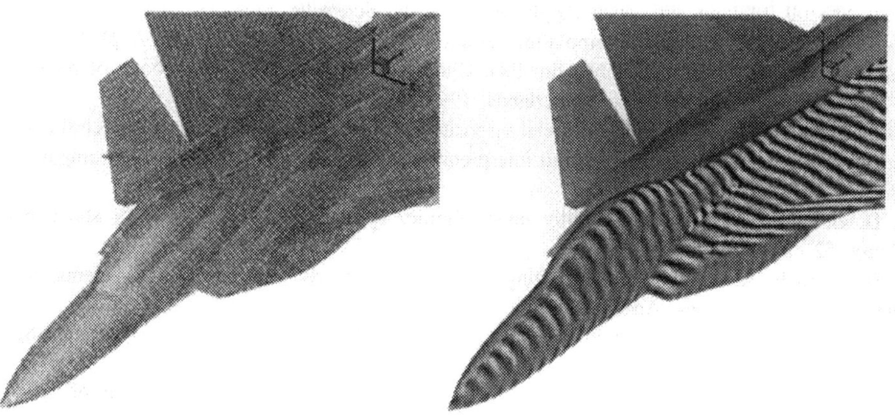

Fig. 3. Application of a fully unstructured spectral multi-domain method to the solution of electromagnetic scattering from an F-15 fighter. The frequency of the incoming plane wave is 600 MHz. On the left is shown a part of the triangulated surface grid and on the right is shown one of the magnetic field components on the surface of the plane. The computation is performed with fourth-order elements and approximately 120.000 tetrahedra to fill the computational volume.

This framework allows for the use of existing unstructured grid technology from finite elements to achieve similar geometric flexibility while it dramatically improves on the accuracy and paves the way for completing reliable simulations of very large scattering and penetration problems. Although the particular example here is for the solution of Maxwells equations, the general computational framework is amenable to the solution of general systems of conservation laws.

While the developments of spectral methods during the last decade have been very significant, a number of critical issues remains open. On the theoretical side, many issues related to stability, even for linear problems, remains open. The results are naturally even more sparse for nonlinear problems. In many instances the experienced user of spectral methods can do much more than can be justified — and often with remarkable success.

Spectral collocation methods have reached a level of sophistication and maturity where it allows for the accurate and efficient solution of nonlinear problems with strong discontinuities using only one domain. For smooth problems, one the other hand, the development of multi-domain formulations has reached a level where it allows for the efficient and accurate solution of problems of almost arbitrary geometric complexity. One of the great challenges of the immediate future lies in understanding how to do strongly nonlinear conservation laws in complex geometries in a stable manner using a spectrally accurate, geometrically flexible, and computationally efficient formulation.

References

[1] S. Abarbanel, D. Gottlieb, E. Tadmor, Spectral methods for discontinuous problems, in: K.W. Morton, M.J. Baines (Eds.), Numerical methods for fluid dynamics II, Clarendon Press, Oxford, 1986, pp. 129–153.

[2] Ø. Andreassen, I. Lie, Simulation of acoustical and elastic waves and interaction, J. Acoust. Soc. Am. 95 (1994) 171–186.

[3] N.S. Banerjee, J. Geer, Exponential approximations using fourier series partial sums, ICASE Report No. 97-56, NASA Langley Research Center, VA, 1997.

[4] G. Ben-Yu, Spectral Methods and Their Applications, World Scientific, Singapore, 1998.

[5] C. Bernardi, Y. Maday, Polynomial interpolation results in sobolev spaces, J. Comput. Appl. Math. 43 (1992) 53–80.

[6] C. Bernardi, Y. Maday, Spectral methods, in: P.G. Ciarlet, J.L. Lions (Eds.), Handbook of Numerical Analysis V, Elsevier Sciences, North-Holland, The Netherlands, 1999.

[7] J.P. Boyd, Two comments on filtering (artificial viscosity) for Chebyshev and Legendre spectral and spectral element methods: preserving boundary conditions and interpretation of the filter as a diffusion, J. Comput. Phys. 143 (1998) 283–288.

[8] W. Cai, D. Gottlieb, C.W. Shu, Essentially nonoscillatory spectral fourier methods for shock wave calculations, Math. Comp. 52 (1989) 389–410.

[9] W. Cai, D. Gottlieb, A. Harten, Cell averaging Chebyshev methods for hyperbolic problems, in: Computers and Mathematics with Applications, Academic Press, New York, 1990.

[10] C. Canuto, M.Y. Hussaini, A. Quarteroni, T.A. Zang, Spectral Methods in Fluid Dynamics, Springer Series in Computational Physics, Springer, New York, 1988.

[11] C. Canuto, A. Quarteroni, Error estimates for spectral and pseudospectral approximations of hyperbolic equations, SIAM J. Numer. Anal. 19 (1982) 629–642.

[12] C. Canuto, A. Quarteroni, Approximation results for orthogonal polynomials in sobolev spaces, Math. Comput. 38 (1982) 67–86.

[13] M.H. Carpenter, D. Gottlieb, Spectral methods on arbitrary grids, J. Comput. Phys. 129 (1996) 74–86.

[14] M.H. Carpenter, D. Gottlieb, C.W, Shu, On the Law-Wendroff theorem for spectral methods, (2000), in preparation.

[15] G.Q. Chen, Q. Du, E. Tadmor, Spectral viscosity approximations to multidimensional scalar conservation laws, Math. Comp. 61 (1993) 629–643.

[16] B. Cockburn, C.W. Shu, Discontinuous Galerkin methods for convection-dominated problems, SIAM Rev. (2000), submitted for publication.

[17] L. Dettori, B. Yang, On the Chebyshev penalty method for parabolic and hyperbolic equations, M^2AN 30 (1996) 907–920.

[18] W.S. Don, Numerical study of pseudospectral methods in shock wave applications, J. Comput. Phys. 110 (1994) 103–111.

[19] W.S. Don, D. Gottlieb, The Chebyshev–Legendre method: implementing Legendre methods on Chebyshev points, SIAM J. Numer. Anal. 31 (1994) 1519–1534.
[20] W.S. Don, D. Gottlieb, Spectral simulation of supersonic reactive flows, SIAM J. Numer. Anal. 35 (1998) 2370–2384.
[21] W.S. Don, C.B. Quillen, Numerical simulation of reactive flow, Part I: resolution, J. Comput. Phys. 122 (1995) 244–265.
[22] K.S. Eckhoff, On discontinuous solutions of hyperbolic equations, Comput. Methods Appl. Mech. Eng. 116 (1994) 103–112.
[23] K.S. Eckhoff, Accurate reconstructions of functions of finite regularity from truncated series expansions, Math. Comput. 64 (1995) 671–690.
[24] P. Fischer, D. Gottlieb, On the optimal number of subdomains for hyperbolic problems on parallel computers, Int. J. Supercomput. Appl. High Perform. Comput. 11 (1997) 65–76.
[25] B. Fornberg, On a Fourier method for the integration of hyperbolic problems, SIAM J. Numer. Anal. 12 (1975) 509–528.
[26] B. Fornberg, A Practical Guide to Pseudospectral Methods, Cambridge University Press, Cambridge, UK, 1996.
[27] D. Funaro, in: Polynomial Approximation of Differential Equations, Lecture Notes in Physics, Vol. 8, Springer, Berlin, 1992.
[28] D. Funaro, D. Gottlieb, A new method of imposing boundary conditions in pseudospectral approximations of hyperbolic equations, Math. Comp. 51 (1988) 599–613.
[29] D. Funaro, D. Gottlieb, Convergence results for pseudospectral approximations of hyperbolic systems by a penalty-type boundary treatment, Math. Comp. 57 (1991) 585–596.
[30] A. Gelb, D. Gottlieb, The resolution of the Gibbs phenomenon for spliced functions in one and two dimensions, Comput. Math. Appl. 33 (1997) 35–58.
[31] A. Gelb, E. Tadmor, Enhanced spectral viscosity approximations for conservation laws, Appl. Numer. Math. 33 (2000) 3–21.
[32] J.G. Giannakouros, G.E. Karniadakis, Spectral element-FCT method for the compressible Euler equations, J. Comput. Phys. 115 (1994) 65–85.
[33] J.G. Giannakouros, D. Sidilkover, G.E. Karniadakis, Spectral element-FCT method for the one- and two-dimensional compressible Euler equations, Comput. Methods Appl. Mech. Eng. 116 (1994) 113–121.
[34] J. Goodman, T. Hou, E. Tadmor, On the stability of the unsmoothed fourier method for hyperbolic equations, Numer. Math. 67 (1994) 93–129.
[35] D. Gottlieb, The stability of pseudospectral chebyshev methods, Math. Comput. 36 (1981) 107–118.
[36] D. Gottlieb, L. Lustman, S.A. Orszag, Spectral calculations of one-dimensional inviscid compressible flows, SIAM J. Sci. Comput. 2 (1981) 296–310.
[37] D. Gottlieb, L. Lustman, E. Tadmor, Stability analysis of spectral methods for hyperbolic initial–boundary value systems, SIAM J. Numer. Anal. 24 (1987) 241–256.
[38] D. Gottlieb, L. Lustman, E. Tadmor, Convergence of spectral methods for hyperbolic initial–boundary value systems, SIAM J. Numer. Anal. 24 (1987) 532–537.
[39] D. Gottlieb, S.A. Orszag, Numerical Analysis of Spectral Methods: Theory and Applications, CBMS-NSF, Vol. 26, SIAM, Philadelphia, PA, 1977.
[40] D. Gottlieb, S.A. Orszag, E. Turkel, Stability of pseudospectral and finite-difference methods for variable coefficient problems, Math. Comput. 37 (1981) 293–305.
[41] D. Gottlieb, C.W. Shu, On the gibbs phenomenon V: recovering exponential accuracy from collocation point values of a piecewise analytic function, Numer. Math. 71 (1995) 511–526.
[42] D. Gottlieb, C.W. Shu, On the gibbs phenomenon and its resolution, SIAM Review 39 (1997) 644–668.
[43] D. Gottlieb, C.W. Shu, A general theory for the resolution of the gibbs phenomenon, in: Tricomi's Ideas and Contemporary Applied Mathematics, National Italian Academy of Science, 1997.
[44] D. Gottlieb, E. Tadmor, Recovering pointwise values of discontinuous data with spectral accuracy, in: Progress and Supercomputing in Computational Fluid Dynamics, Birkhäuser, Boston, 1984, pp. 357–375.
[45] D. Gottlieb, E. Tadmor, The CFL condition for spectral approximations to hyperbolic initial-value problems, Math. Comput. 56 (1991) 565–588.
[46] D. Gottlieb, E. Turkel, Spectral Methods for Time-Dependent Partial Differential Equations, Lecture Notes in Mathematics, Vol. 1127, Springer, Berlin, 1983, pp. 115–155.

[47] D. Gottlieb, C.E. Wasberg, Optimal strategy in domain decomposition spectral methods for wave-like phenomena, SIAM J. Sci. Comput. (1999), to appear.

[48] J.S. Hesthaven, A stable penalty method for the compressible Navier–Stokes equations: II, one-dimensional domain decomposition schemes, SIAM J. Sci. Comput. 18 (1997) 658–685.

[49] J.S. Hesthaven, A stable penalty method for the compressible Navier–stokes equations: III, multidimensional domain decomposition schemes, SIAM J. Sci. Comput. 20 (1999) 62–93.

[50] J.S. Hesthaven, Spectral penalty methods, Appl. Numer. Math. 33 (2000) 23–41.

[51] J.S. Hesthaven, P.G. Dinesen, J.P. Lynov, Spectral collocation time-domain modeling of diffractive optical elements, J. Comput. Phys. 155 (1999) 287–306.

[52] J.S. Hesthaven, D. Gottlieb, A stable penalty method for the compressible Navier–Stokes equations, I, open boundary conditions, SIAM J. Sci. Comput. 17 (1996) 579–612.

[53] J.S. Hesthaven, D. Gottlieb, Stable spectral methods for conservation laws on triangles with unstructured grids, Comput. Methods Appl. Mech. Eng. 175 (1999) 361–381.

[54] J.S. Hesthaven, J. Juul Rasmussen, L. Bergé, J. Wyller, Numerical studies of localized wave fields govenered by the Raman-extended derivative nonlinear Schrödinger equation, J. Phys. A: Math. Gen. 30 (1997) 8207–8224.

[55] J.S. Hesthaven, C.H. Teng, Stable spectral methods on tetrahedral elements, SIAM J. Sci. Comput. (2000), to appear.

[56] E. Isaacson, H.B. Keller, Analysis of Numerical Methods, Dover Publishing, New York, 1966.

[57] D. Jackson, The Theory of Approximation, American Mathematical Society, Colloquim Publication, Vol. 11, Providence, RI, 1930.

[58] D.A. Kopriva, A spectral multidomain method for the solution of hyperbolic systems, Appl. Numer. Math. 2 (1986) 221–241.

[59] D.A. Kopriva, Computation of hyperbolic equations on complicated domains with patched and overset chebyshev grids, SIAM J. Sci. Statist. Comput. 10 (1989) 120–132.

[60] D.A. Kopriva, Multidomain spectral solution of the Euler gas-dynamics equations, J. Comput. Phys. 96 (1991) 428–450.

[61] D.A. Kopriva, A conservative staggered-grid Chebyshev multidomain method for compressible flows, II, a semi-structured method, J. Comput. Phys. 128 (1996) 475–488.

[62] D.A. Kopriva, J.H. Kolias, A conservative staggered-grid Chebyshev multidomain method for compressible flows, II, a semi-structured method, J. Comput. Phys. 125 (1996) 244–261.

[63] D.A. Kopriva, S.L. Woodruff, M.Y. Hussaini, Discontinuous spectral element approximation of Maxwell's equations, in: B. Cockburn, G.E. Karniadakis, C.W. Shu (Eds.), Discontinuous Galerkin Methods: Theory, Computation and Applications, Lecture Notes in Computational Science and Engineering, Vol. 11, Springer, New York, 2000, pp. 355–362.

[64] H.O. Kreiss, J. Lorenz, Initial-Boundary Value Problems and the Navier–Stokes Equations, Series in Pure and Applied Mathematics, Academic Press, San Diego, 1989.

[65] H.O. Kreiss, J. Oliger, Comparison of accurate methods for the integration of hyperbolic problems, Tellus 24 (1972) 199–215.

[66] H.O. Kreiss, J. Oliger, Stability of the Fourier method, SIAM J. Numer. Anal. 16 (1979) 421–433.

[67] P.D. Lax (Ed.), Accuracy and resolution in the computation of solutions of linear and nonlinear equations, in: Proceedings of Recent Advances in Numerical Analysis, University of Wisconsin, Academic Press, New York, 1978, pp. 107–117.

[68] D. Levy, E. Tadmor, From semi-discrete to fully-discrete: stability of Runge–Kutta schemes by the energy method, SIAM Rev. 40 (1998) 40–73.

[69] I. Lomtev, C.B. Quillen, G.E. Karniadakis, Spectral/hp methods for viscous compressible flows on unstructured 2D meshes, J. Comput. Phys. 144 (1998) 325–357.

[70] H. Ma, Chebyshev–Legendre super spectral viscosity method for nonlinear conservation laws, SIAM J. Numer. Anal. 35 (1998) 869–892.

[71] H. Ma, Chebyshev–Legendre spectral viscosity method for nonlinear conservation laws, SIAM J. Numer. Anal. 35 (1998) 893–908.

[72] Y. Maday, E. Tadmor, Analysis of the spectral vanishing viscosity method for periodic conservation laws, SIAM J. Numer. Anal. 26 (1989) 854–870.

[73] Y. Maday, S.M. Ould Kaper, E. Tadmor, Legendre pseudospectral viscosity method for nonlinear conservation laws, SIAM J. Numer. Anal. 30 (1993) 321–342.

[74] A. Majda, J. McDonough, S. Osher, The Fourier method for nonsmooth initial data, Math. Comp. 32 (1978) 1041–1081.

[75] S.A. Orszag, Comparison of pseudospectral and spectral approximation, Stud. Appl. Math. 51 (1972) 253–259.

[76] J.E. Pasciak, Spectral and pseudospectral methods for advection equations, Math. Comp. 35 (1980) 1081–1092.

[77] A. Quarteroni, Domain decomposition methods for systems of conservation laws: spectral collocation approximations, SIAM J. Sci. Statist. Comput. 11 (1990) 1029–1052.

[78] S.C. Reddy, L.N. Trefethen, Lax-stability of fully discrete spectral methods via stability regions and pseudo-eigenvalues, Comput. Methods Appl. Mech. Eng. 80 (1990) 147–164.

[79] S.C. Reddy, L.N. Trefethen, Stability of the method of lines, Numer. Math. 62 (1992) 235–267.

[80] D. Sidilkover, G.E. Karniadakis, Non-oscillatory spectral element Chebyshev method for shock wave calculations, J. Comput. Phys. 107 (1993) 10–22.

[81] G. Szegő, in: Orthogonal polynomials, 4th Edition, American Mathematical Society, Vol. 23, Colloquim Publication, Providence, RI, 1975.

[82] E. Tadmor, Skew-selfadjoint form for systems of conservation laws, J. Math. Anal. Appl. 103 (1984) 428–442.

[83] E. Tadmor, The exponential accuracy of fourier and Chebyshev differencing methods, SIAM Rev. 23 (1986) 1–10.

[84] E. Tadmor, Stability analysis of finite-difference, pseudospectral, and Fourier–Galerkin approximations for time-dependent problems, SIAM Rev. 29 (1987) 525–555.

[85] E. Tadmor, Convergence of spectral methods for nonlinear conservation laws, SIAM J. Numer. Anal. 26 (1989) 30–44.

[86] H. Tal-Ezer, Ph.D. Thesis, Tel Aviv University, 1983.

[87] L.N. Trefethen, M.R. Trummer, An instability phenomenon in spectral methods, SIAM J. Numer. Anal. 24 (1987) 1008–1023.

[88] H. Vandeven, Family of spectral filters for discontinuous problems, J. Sci. Comput. 8 (1991) 159–192.

[89] L. Vozovoi, M. Israeli, A. Averbuch, Analysis and application of Fourier–Gegenbauer method to stiff differential equations, SIAM J. Numer. Anal. 33 (1996) 1844–1863.

[90] L. Vozovoi, A. Weill, M. Israeli, Spectrally accurate solution of nonperiodic differential equations by the Fourier–Gegenbauer method, SIAM J. Numer. Anal. 34 (1997) 1451–1471.

[91] T. Warburton, G.E. Karniadakis, A discontinuous Galerkin method for the viscous MHD equations, J. Comput. Phys. 152 (1999) 608–641.

[92] T. Warburton, Application of the discontinuous Galerkin method to Maxwell's equations using unstructured polymorphic hp-finite elements, in: B. Cockburn, G.E. Karniadakis, C.W. Shu (Eds.), Discontinuous Galerkin Methods: Theory, Computation and Applications. Lecture Notes in Computational Science and Engineering, Vol. 11, Springer, New York, 2000, pp. 451–458.

[93] T. Warburton, I. Lomtev, Y. Du, S. Sherwin, G.E. Karniadakis, Galerkin and discontinuous Galerkin spectral/hp methods, Comput. Methods Appl. Mech. Eng. 175 (1999) 343–359.

[94] T. Warburton, J.S. Hesthaven, Stable unstructured grid scheme for Maxwells equations, J. Comput. Phys. (2000), submitted for publication.

[95] B. Yang, D. Gottlieb, J.S. Hesthaven, Spectral simulations of electromagnetic wave scattering, J. Comput. Phys. 134 (1997) 216–230.

[96] B. Yang, J.S. Hesthaven, A pseudospectral method for time-domain computation of electromagnetic scattering by bodies of revolution, IEEE Trans. Antennas and Propagation 47 (1999) 132–141.

[97] B. Yang, J.S. Hesthaven, Multidomain pseudospectral computation of Maxwells equations in 3-D general curvilinear coordinates, Appl. Numer. Math. 33 (2000) 281–289.

ELSEVIER

Journal of Computational and Applied Mathematics 128 (2001) 133–185

JOURNAL OF COMPUTATIONAL AND APPLIED MATHEMATICS

www.elsevier.nl/locate/cam

Wavelet methods for PDEs — some recent developments

Wolfgang Dahmen[1]

Institut für Geometrie und Praktische Mathematik, RWTH Aachen, Templergraben 55, 52056 Aachen, Germany

Received 5 July 1999; received in revised form 7 December 1999

Abstract

This paper is concerned with recent developments of wavelet schemes for the numerical treatment of operator equations with special emphasis on two issues: adaptive solution concepts and nontrivial domain geometries. After describing a general multiresolution framework the key features of wavelet bases are highlighted, namely locality, norm equivalences and cancellation properties. Assuming first that wavelet bases with these properties are available on the relevant problem domains, the relevance of these features for a wide class of stationary problems is explained in subsequent sections. The main issues are preconditioning and the efficient (adaptive) application of wavelet representations of the involved operators. We indicate then how these ingredients combined with concepts from nonlinear or best N-term approximation culminate in an adaptive wavelet scheme for elliptic selfadjoint problems covering boundary value problems as well as boundary integral equations. These schemes can be shown to exhibit convergence rates that are in a certain sense asymptotically optimal. We conclude this section with some brief remarks on data structures and implementation, interrelations with regularity in a certain scale of Besov spaces and strategies of extending such schemes to unsymmetric or indefinite problems. We address then the adaptive evaluation of nonlinear functionals of wavelet expansions as a central task arising in connection with nonlinear problems. Wavelet bases on nontrivial domains are discussed next. The main issues are the development of Fourier free construction principles and criteria for the validity of norm equivalences. Finally, we indicate possible combinations of wavelet concepts with conventional discretizations such as finite element or finite volume schemes in connection with convection dominated and hyperbolic problems. © 2001 Elsevier Science B.V. All rights reserved.

MSC: 41A25; 41A46; 46A35; 46B03; 65F35; 65N12; 65N55; 65R20; 65Y20

Keywords: Multiresolution; Wavelet bases; Norm equivalences; Cancellation properties; Preconditioning; Multiscale transformations; Best N-term approximation; Fast matrix/vector multiplication; Adaptive solvers; Convergence rates; Besov regularity; Nonlinear functionals of wavelet expansions; Wavelet bases on domains; Fourier free criteria; Stable completions; Direct and inverse estimates; Domain decomposition; Convection dominated problems; Hyperbolic problems

[1] WWW: http://www.igpm.rwth-aachen.de/~dahmen/

E-mail address: dahmen@igpm.rwth-aachen.de (W. Dahmen).

0377-0427/01/$ - see front matter © 2001 Elsevier Science B.V. All rights reserved.
PII: S0377-0427(00)00511-2

1. Introduction

Quoting from [67] the concept of wavelets can be viewed as a synthesis of ideas which originated during the last 30 years from engineering (subband coding), physics (coherent states, renormalization group) and pure mathematics (Calderón–Zygmund operators). In addition the construction of spline bases for Banach spaces can be viewed as a further relevant source from pure mathematics [30]. It is understandable that this multitude of origins appeal to scientists in many different areas. Consequently, the current research landscape concerned with wavelet ideas is extremely varied. Here is an incomplete list of related active topics: *construction* of wavelets; wavelets as tools in interdisciplinary *applications*; wavelets as *modeling tools*; wavelets as *analysis tools*; multiscale *geometry* representation; wavelets in *statistics*; wavelets in *numerical analysis* and large-scale computation.

The main focus of this article will be on recent developments in the last area. Rather than trying to give an exhaustive account of the state of the art I would like to bring out some mechanisms which are in my opinion important for the application of wavelets to operator equations. To accomplish this I found it necessary to address some of the pivotal issues in more detail than others. Nevertheless, such a selected 'zoom in' supported by an extensive list of references should provide a sound footing for conveying also a good idea about many other related branches that will only be briefly touched upon. Of course, the selection of material is biased by my personal experience and therefore is not meant to reflect any objective measure of importance. The paper is organized around two essential issues namely *adaptivity* and the development of concepts for coping with a major obstruction in this context namely *practically relevant domain geometries.*

Being able to solve a *given* large scale discrete problem in optimal time is one important step. In this regard a major breakthrough has been caused by multi-grid methods. A further, sometimes perhaps even more dramatic reduction of complexity may result from *adapting* in addition the discretization to the individual application at hand. This means to realize a desired overall accuracy at the expense of a possibly small number of involved degrees of freedom.

A central theme in subsequent discussions is to indicate why wavelet concepts appear to be particularly promising in this regard. In fact, the common attraction of wavelets in all the above listed areas hinges on their principal ability of organizing the decomposition of an object into components characterized by different length scales. This calls for locally adapting the level of resolution and has long been praised as a promising perspective. As intuitive as this might be a quantitative realization of such ideas faces severe obstructions. This note is to reflect some of the ideas that have been driving these developments. A more detailed account of a much wider scope of related topics can be found in [31,49].

The material is organized as follows. After collecting some preliminaries in Section 2, Section 3 is devoted to a discussion of two central features which are fundamental for wavelet concepts in numerical analysis namely *cancellation properties* and *isomorphisms* between function and sequence spaces. Assuming at this point that wavelet bases with these properties are indeed available (even for nontrivial domains) some consequences of these features are described in subsequent sections. Section 4 addresses *preconditioning* for elliptic problems as well as the connection with *nonlinear* or *best N-term approximation.* Section 5 is concerned with the *efficiency* of matrix/vector multiplications when working in wavelet coordinates. In particular, some implications pertaining to the treatment of boundary integral equations are outlined. It is indicated in Section 6 how these concepts enter the analysis of an adaptive scheme for a wide class of elliptic operator equations. Since these rather recent results reflect in my opinion the potential of wavelet methodologies quite well and since the involved tools are probably not very familiar in the numerical analysis community a somewhat more detailed discussion of this topic is perhaps instructive. In Section 7 a basic ingredient for the treatment of *nonlinear* problems is discussed namely the adaptive computation of nonlinear functionals of multiscale expansions. In particular, a recent fast evaluation scheme is sketched. Section 8 is concerned with the practical realization of those properties of wavelet bases that have been identified before as being crucial. This covers "Fourier free" criteria for norm equivalences as well as several construction principles such as *stable completions* or the *lifting scheme*. It is then indicated in Section 9 how to use these concepts for the concrete construction of wavelet bases on bounded domains. In particular, this covers an approach that leads in a natural way to convergent domain decomposition schemes for singular integral equations. Some remarks on finite element based wavelet bases conclude this section. Finally, we briefly touch in Section 10 upon recent applications of adaptive wavelet concepts within conventional finite element or finite volume discretization for convection diffusion equations, respectively hyperbolic conservation laws.

For convenience we will write $a \lesssim b$ to express that a can be bounded by some constant multiple of b uniformly in all parameters on which a and b might depend as long as the precise value of such constants does not matter. Likewise $a \sim b$ means $a \lesssim b$ and $b \lesssim a$.

2. Some preliminary comments

2.1. About the general spirit

To motivate the subsequent discussions consider an admittedly oversimplified example namely the representation of a real number x by means of decimal or more generally p-adic expansions $x=\sum_{j=l}^{-\infty} d_j b^j$, where b is the *base*, $b=10$, say. An irrational number is an *array* of *infinitely* many *digits*. Depending on the context such a number may only be needed up to a certain accuracy saying how many digits are required. So in a sense one may identify the number with an *algorithm* that produces the required number of digits. It is therefore natural to ask: Is there anything similar for *functions*? Of course, all classical expansions of Laurent, Taylor or Fourier type reflect exactly this point of view. Weierstraß was only willing to call an object a function if it could be in principle computed through a countable number of basic arithmetic operations based on such an expansion. A wavelet basis $\Psi=\{\psi_I\colon I \in \mathscr{J}\}$ (where $\mathscr{J}$ is a suitable index set to be commented on later in more detail) is in this sense just another collection of *basis* functions in terms of which a given function f can be expanded

$$f = \sum_{I \in \mathscr{J}} d_I(f)\psi_I. \tag{2.1}$$

Such an expansion associates with f the array $\boldsymbol{d} = \{d_I(f)\}_{I \in \mathscr{J}}$ of digits. There are a few points though by which a wavelet expansion differs from more classical expansions. First of all, for instance, a Taylor expansion puts strong demands on the regularity of f such as analyticity, while (2.1) is typically valid for much larger classes of functions such as square integrable ones. By this we mean that the series on the right-hand side of (2.1) converges in the corresponding norm. More importantly, the digits d_I convey very detailed information on f. To explain this, a word on the indices $I \in \mathscr{J}$

is in order. Each I comprises information of different type such as *scale* and *spatial location*. For instance, classical wavelets on the real line are generated by *scaling* and translating a single function ψ, i.e., $\psi_{j,k} = 2^{j/2}\psi(2^j \cdot -k)$. The simplest example is to take the box function $\phi := \chi_{[0,1)}$ to form the *Haar wavelet* $\psi := (\phi(2\cdot) - \phi(2\cdot -1))/\sqrt{2}$. Obviously ψ is orthogonal to ϕ. It is easy to see that the collection of functions $\{\phi(\cdot - k): k \in \mathbb{Z}\} \cup \{2^{j/2}\psi(2^j \cdot -k): k,j \in \mathbb{Z}\}$ forms an *orthonormal* basis for $L_2(\mathbb{R})$. Likewise $\{\phi\} \cup \{2^{j/2}\psi(2^j \cdot -k): k = 0,\ldots,2^j - 1, j = 0,1,\ldots\}$ is an orthonormal basis of $L_2([0,1])$. In this case one has $(j,k) \leftrightarrow I$. Note that the whole basis consists in both examples of two groups of functions namely the box functions on a fixed *coarsest level* and the set of 'true' wavelets on all successively higher levels. When expanding a function in such a basis the box functions are needed to recover the constant part while the rest encodes further *updates*. Such a coarsest level is usually needed when dealing with bounded domains. When employing isotropic scaling by powers of two, wavelets on $\mathbb{R}^d$ are obtained by scaling and translating $2^d - 1$ fixed functions $2^{jd/2}\psi_e(2^j \cdot -k)$, $e \in \{0,1\}^d \backslash \{\mathbf{0}\}$ so that then $I \leftrightarrow (j,k,e)$. For more general domains one has to relax the strict interrelation through translation but the nature of the indices will stay essentially the same as will be explained later for corresponding examples. At this point it suffices to denote by $j = |I|$ the *scale* j and by k the *location* k of the wavelet ψ_I. This means that ψ_I is associated with *detail information* of length scale $2^{-|I|}$ around a spatial location encoded by k. In particular, when the wavelets have compact support the diameter of the support of a wavelet on level $|I|$ behaves like

$$\operatorname{diam}\operatorname{supp}\psi_I \sim 2^{-|I|}. \tag{2.2}$$

Thus when the coefficient $|d_I(f)|$ is large this indicates that near $k(I)$ the function f has significant components of length scale $2^{-|I|}$.

Another characteristic feature of wavelets is that certain *weighted sequence norms* of the coefficients in a wavelet expansion are *equivalent* to certain function space norms of the corresponding functions. Again the Haar basis serves as the simplest example where the function space is L_2 and the Euclidean norm is used to measure the coefficients. This implies, in particular, that discarding small wavelet coefficients, results in small perturbations of the function in that norm. In a finite difference or finite element context adaptivity is typically associated with *local mesh refinement* where the refinement criteria are extracted from foregoing calculations usually with the aid of a posteriori error bounds or indicators. In the case of wavelet analysis the concept is somewhat different. In contrast to *approximating* the solution of a given operator equation on some mesh (of fixed highest resolution) wavelet-based schemes aim at determining its *representation* with respect to a basis. This seems to be a minor distinction of primarily philosophical nature but turns out to suggest a rather different way of thinking. In fact, adaptivity then means to track during the solution process only those coefficients in the unknown array $\boldsymbol{d}(f)$ in (2.1) that are *most significant* for approximating f with possibly few degrees of freedom.

2.2. Fourier techniques

Part of the fascination about wavelets stems from the fact that a conceptually fairly simple mathematical object offers at least an extremely promising potential in several important application areas. The *technical simplicity* and the practicality of this object, however, is by no means independent of its domain of definition. The development of wavelets has been largely influenced by *Fourier techniques* and harmonic analysis concepts, quite in agreement with the nature of the primary application areas. Fourier techniques not only account for the elegance of constructions but also for the analysis of applications [34,67]. It is therefore quite natural that first attempts of expanding the application of wavelet techniques to other areas such as numerical analysis tried to exploit the advantages of Fourier concepts. To a get a quick impression of the potential suggests to consider first model problems that help minimizing severe technical obstructions. Consequently periodic problems have largely served as test beds for such investigations. As much as one might argue about the practicability of such settings these investigations have been of pioneering importance. For instance, Beylkin et al. [15] has initiated important developments centering on *matrix compression* and fast processing of certain densely populated matrices. Likewise the concept of *vaguelette* draws on ideas from the theory of Calderón–Zygmund operators [87,110]. Putting more weight on the actual *inversion* of operators complements traditional thinking in numerical analysis, see e.g. [5,16,87,92,49] for a more detailed exposition and further references. Among other things these developments have led to interesting applications, for instance, in physical modeling concerning the simulation of coherent states in 2D turbulent flows or combustion [8,18,102]. Again for details the reader is referred to the original literature.

2.3. The curse of geometry

Nevertheless, the periodic setting poses severe limitations because it excludes, for instance, domain induced singularities or turbulent behavior due to no slip boundary conditions. However, a number of obstructions arise when leaving the periodic realm. First of all the essential features of wavelets closely tied to Fourier concepts have to be re-identified in a correspondingly more flexible context. New construction principles are needed. Thus, the systematic development of conceptual substitutes for Fourier techniques is of pivotal importance and will be a recurrent theme in subsequent discussions, see e.g. [26,46,48,49,108,109] for contributions in this spirit. We will outline next a multi-resolution framework which is flexible enough to cover later applications of varying nature.

2.4. Multi-resolution

Recall that a convenient framework for constructing and analyzing wavelets on $\mathbb{R}^d$ is the concept of *multi-resolution approximation* formalized in [93]. Here is a somewhat generalized variant of multi-resolution that will work later for more general domain geometries. To this end, consider an ascending sequence $\mathscr{S}$ of closed nested spaces $S_j, j \in \mathbb{N}_0$, $S_j \subset S_{j+1}$ whose union is dense in a given Banach space B, say. A particularly important case arises when $B = H$ is a Hilbert space with inner product $\langle \cdot,\cdot \rangle$ and norm $\|\cdot\|_H := \langle \cdot,\cdot \rangle^{1/2}$. In the classical multi-resolution setting as well as in all cases of practical interest the spaces S_j are spanned by what will be called a *single-scale basis* $\Phi_j = \{\phi_{j,k}: k \in \mathscr{I}_j\}$, i.e., $S_j = \operatorname{span} \Phi_j =: S(\Phi_j)$. In view of the intended applications, we consider here only finite-dimensional spaces S_j. Moreover, we are only interested in local bases, i.e., as in (2.2)

$$\operatorname{diam} \operatorname{supp} \phi_{j,k} \sim 2^{-j}. \tag{2.3}$$

Nestedness of the S_j implies then that the $\phi_{j,k}$ satisfy a *two-scale* or *refinement relation*

$$\phi_{j,k} = \sum_{l \in \mathscr{I}_{j+1}} m_{j,l,k} \phi_{j+1,l}, \quad k \in \mathscr{I}_j. \tag{2.4}$$

Collecting the *filter* or *mask coefficients* $m_{j,l,k}$ in the $\mathscr{I}_{j+1} \times \mathscr{I}_j$-matrix $\boldsymbol{M}_{j,0} = (m_{j,l,k})_{l \in \mathscr{I}_{j+1}, k \in \mathscr{I}_j}$, (2.4) is conveniently expresses as a matrix/vector relation

$$\Phi_j^{\mathrm{T}} = \Phi_{j+1}^{\mathrm{T}} \boldsymbol{M}_{j,0}. \tag{2.5}$$

The simplest example has been described above when $H = L_2([0,1])$ and the S_j are all piecewise constants on $[0,1]$ with mesh size 2^{-j}. Here $\phi_{j,k} = 2^{j/2}\chi_{2^{-j}[k,k+1)}$. A further typical example is $H = H_0^1(\Omega)$ where Ω is a polygonal domain in $\mathbb{R}^2$ which is partitioned into a collection Δ_0 of triangles. This means that any two triangles have either empty intersection or share a common edge or vertex. Let Δ_{j+1} be obtained by decomposing each triangle in Δ_j into four congruent sub-triangles and let S_j be the space of globally continuous piecewise affine functions with respect to Δ_j. In this case one has $m_{j,l,k} = \phi_{j,k}(l)/|\phi_{j+1,l}(l)|$.

The next step is to describe S_{j+1} as an *update* of the coarser space S_j. If $\mathscr{Q} = \{Q_j\}_{j \in \mathbb{N}_0}$ is a uniformly H-bounded sequence of projectors onto the spaces S_j the *telescoping expansion*

$$v = \sum_{j=0}^{\infty} (Q_j - Q_{j-1})v, \quad (Q_{-1} := 0) \tag{2.6}$$

converges strongly to v. The summands $(Q_j - Q_{j-1})v$ reflect *updates* of the coarse approximations $Q_{j-1}v$ of v. Wavelets (on level $j-1$) are basis functions that represent the *detail*

$$(Q_j - Q_{j-1})v = \sum_{|I|=j-1} d_I(v)\psi_I. \tag{2.7}$$

In the above example of the Haar wavelets the Q_j are just orthogonal projectors which, in particular, satisfy

$$Q_l Q_j = Q_l, \quad l \leqslant j. \tag{2.8}$$

Note that, in general, (2.8) means that $Q_j - Q_{j-1}$ is also a projector and that

$$W_j := (Q_{j+1} - Q_j)H \tag{2.9}$$

is a direct summand

$$S_{j+1} = S_j \oplus W_j. \tag{2.10}$$

Thus, the spaces W_j form a multilevel splitting of H and the union $\Psi = \Phi_0 \cup_j \Psi_j$ of the *complement bases* $\Psi_j := \{\psi_I \colon |I| = j\}$ and the coarse level generator basis Φ_0 is a candidate for a wavelet basis. This is trivially the case in the example of the Haar basis. Less obvious specifications such as, for instance, piecewise affine functions on triangulations will be revisited later, see also [49].

At this point we will continue collecting some further general ingredients for later use. Note that decomposition (2.6) and (2.7) suggest a *levelwise* organization of the indices $I \in \mathscr{J}$. Truncating expansion (2.6) simply means to cut off all frequencies above a certain level. It will therefore be convenient to denote by $\mathscr{J}_j = \mathscr{I}_0 \cup \{I \colon 0 \leqslant |I| < j\}$ the index set corresponding to the finite multiscale basis $\Psi_{\mathscr{J}_j} := \Phi_0 \cup_{l=0}^{j-1} \Psi_j$. Thus $S_j = S(\Phi_j) = S(\Psi_{\mathscr{J}_j})$. Associating with any I, $|I| = j$, a mesh size 2^{-j}, the space S_j corresponds to a *uniform* mesh and this setting will be referred to as *uniform refinement of level j*.

The complement bases Ψ_j are called *H-stable* if

$$\|\boldsymbol{d}_j\|_{\ell_2} \sim \|\boldsymbol{d}_j^{\mathrm{T}} \Psi_j\|_H. \tag{2.11}$$

Note that the complement basis functions for W_j in (2.9) must be linear combinations of the elements in Φ_{j+1}. Hence there must exist a $(\#\mathcal{I}_{j+1} \times (\#\mathcal{I}_{j+1} - \#\mathcal{I}_j))$-matrix $\mathbf{M}_{j,1}$ such that

$$\Psi_j^{\mathrm{T}} = \Phi_{j+1}^{\mathrm{T}} \mathbf{M}_{j,1}. \tag{2.12}$$

It is easy to see that $\Phi_j \cup \Psi_j$ is indeed a basis for $S(\Phi_{j+1}) = S_{j+1}$ if and only if the composed matrix $\mathbf{M}_j = (\mathbf{M}_{j,0}, \mathbf{M}_{j,1})$, where $\mathbf{M}_{j,0}$ is the refinement matrix from (2.5), is nonsingular. Thus, the search for a complement basis is equivalent to a *matrix completion* problem and many properties of such bases can be expressed in terms of matrices. In fact, the bases $\Phi_j \cup \Psi_j$ are uniformly L_2-stable if and only if

$$\mathrm{cond}_2(\mathbf{M}_j) := \|\mathbf{M}_j\| \|\mathbf{M}_j^{-1}\| \lesssim 1, \tag{2.13}$$

where $\|\cdot\|$ denotes here the spectral norm. In this case $\mathbf{M}_{j,1}$ is called a *stable completion* of $\mathbf{M}_{j,0}$ [26].

It is convenient to block the inverse $\mathbf{G}_j = \mathbf{M}_j^{-1}$ as $\mathbf{G}_j = \binom{\mathbf{G}_{j,0}}{\mathbf{G}_{j,1}}$. Since by (2.5) and (2.12), $(\Phi_j^{\mathrm{T}}, \Psi_j^{\mathrm{T}}) = \Phi_{j+1}^{\mathrm{T}} \mathbf{M}_j$ and $\mathbf{M}_j \mathbf{G}_j = \mathrm{id}$ one has $\Phi_j^{\mathrm{T}} \mathbf{c}_j + \Psi_j^{\mathrm{T}} \mathbf{d}_j = \Phi_{j+1}^{\mathrm{T}}(\mathbf{M}_{j,0}\mathbf{c}_j + \mathbf{M}_{j,1}\mathbf{d}_j)$. Hence the transformation

$$\mathbf{T}_J : \mathbf{d}^J := (\mathbf{c}_0^{\mathrm{T}}, \mathbf{d}_0^{\mathrm{T}}, \ldots, \mathbf{d}_{J-1}^{\mathrm{T}})^{\mathrm{T}} \to \mathbf{c}_J \tag{2.14}$$

that takes the array $\mathbf{d}^J$ of *multiscale coefficients* of an element $v_J \in S_J = S(\Phi_J)$ into its *single-scale coefficients* $\mathbf{c}_J$ is realized by a successive application of the steps

$$(\mathbf{c}_j, \mathbf{d}_j) \to \mathbf{c}_{j+1} := \mathbf{M}_{j,0}\mathbf{c}_j + \mathbf{M}_{j,1}\mathbf{d}_j. \tag{2.15}$$

Likewise since a function $v_{j+1} = \Phi_{j+1}^{\mathrm{T}} \mathbf{c}_{j+1} \in S_{j+1}$ can be written as $v_{j+1} = (\Phi_j^{\mathrm{T}}, \Psi_j^{\mathrm{T}}) \mathbf{G}_j \mathbf{c}_{j+1} = \Phi_j^{\mathrm{T}} \mathbf{G}_{j,0} \mathbf{c}_{j+1} + \Psi_j^{\mathrm{T}} \mathbf{G}_{j,1} \mathbf{c}_{j+1}$, the inverse transformation $\mathbf{T}_J^{-1}$ taking single scale into multiscale coefficients is given by a successive application of

$$\mathbf{c}_{j+1} \to (\mathbf{c}_j := \mathbf{G}_{j,0}\mathbf{c}_{j+1}, \mathbf{d}_j := \mathbf{G}_{j,1}\mathbf{c}_{j+1}). \tag{2.16}$$

Under the locality assumptions (2.2) and (2.3) the matrices $\mathbf{M}_j$ are sparse in the sense that the number of nonzero entries in each row and column of $\mathbf{M}_j$ remains uniformly bounded in j. Thus a geometric series argument shows that the execution of $\mathbf{T}_J$ requires $\mathcal{O}(N_J)$ operations where $N_J = \#\Phi_J$. Hence for both transformations $\mathbf{T}_J$ and $\mathbf{T}_J^{-1}$ to be efficient also the inverse $\mathbf{G}_j$ should be sparse. This is known to be rarely the case and indicates that the construction of suitable wavelets is a delicate task.

When Ψ is an orthonormal basis the transformations $\mathbf{T}_J$ are orthogonal and hence well-conditioned. In general, one can show the following fact, see e.g. [46].

Remark 2.1. The $\mathbf{T}_J$ have uniformly bounded condition numbers if and only if Ψ is a Riesz basis in L_2, i.e., each function $v \in L_2$ possesses a unique expansion $v = \mathbf{d}^{\mathrm{T}}\Psi$ and $\|\mathbf{d}\|_{\ell_2} \sim \|\mathbf{d}^{\mathrm{T}}\Psi\|_{L_2}$.

3. The key features

This section is devoted to collecting those specific properties of bases Ψ that are essential for the numerical treatment of operator equations.

3.1. Cancellation property

The Haar wavelet is orthogonal to constants. Thus, integrating a function against the Haar wavelet annihilates the constant part of the function. In more sophisticated examples wavelets are arranged to be *orthogonal to polynomials* up to a certain order. The wavelets are then said to have a corresponding order of *vanishing moments*. This concept is very convenient whenever dealing with domains where polynomials are well defined. Thinking of functions living on more general manifolds such as boundary surfaces it is preferable to express the *cancellation property* of wavelets by estimates of the form

$$|\langle f,\psi_I\rangle| \lesssim 2^{-|I|(\tilde{m}+d/2)}|f|_{W^{\infty,\tilde{m}}(\mathrm{supp}\,\psi_I)}, \tag{3.1}$$

where d is the spatial dimension and the integer $\tilde{m}$ signifies the strength of the cancellation property. In fact, $\tilde{m}$ will later be seen to be the approximation order of the range of the adjoints Q_j' of the projectors in (2.6) above. A simple Taylor expansion argument shows that (3.1) is implied by $\tilde{m}$th-order vanishing moments. Thus $\tilde{m}=1$ for Haar wavelets.

3.2. Norm equivalences

An important property of classical wavelets on $\mathbb{R}^d$ is that they provide a particularly tight relationship between the function and its coefficient sequence. For instance, since the Haar basis is orthonormal the ℓ_2-norm of the coefficients *equals* the L_2-norm of the function. It is of paramount importance to preserve properties of this type for multiscale bases defined on relevant domain geometries as indicated already by Remark 2.1.

To describe these features in more concrete terms note first that orthonormality is actually a strong property. In many practical situations it interferes with locality and is hard to realize. Moreover, in many applications it turns even out *not* to be optimal because different energy inner products may be relevant. A suitable more flexible format can be described first again for a Hilbert space H with norm $\|\cdot\|_H$. A collection $\Psi\subset H$ is called H-*stable* if and only if every $v\in H$ has a unique expansion $v=\sum_{I\in\mathcal{J}} d_I(v)\psi_I$ such that for some fixed positive weight coefficients D_I

$$c_1\left(\sum_{I\in\mathcal{J}} D_I^2|d_I(v)|^2\right)^{1/2} \leqslant \left\|\sum_{I\in\mathcal{J}} d_I(v)\psi_I\right\|_H \leqslant c_2\left(\sum_{I\in\mathcal{J}} D_I^2|d_I(v)|^2\right)^{1/2}, \tag{3.2}$$

where the constants c_1,c_2 are independent of v. It will be convenient to view Ψ and the corresponding array $\boldsymbol{d}(v)=\{d_I(v)\}_{I\in\mathcal{J}}$ as *vectors* to write briefly $\boldsymbol{d}(v)^{\mathrm{T}}\Psi:=\sum_{I\in\mathcal{J}} d_I(v)\psi_I$. Likewise, collecting the weights D_I in the (infinite) *diagonal matrix* $\boldsymbol{D}:=(D_I\delta_{I,I'})_{I,I'\in\mathcal{J}}$, (3.2) can be rewritten as

$$\|\boldsymbol{D}\boldsymbol{d}(v)\|_{\ell_2(\mathcal{J})}\sim\|\boldsymbol{d}^{\mathrm{T}}(v)\Psi\|_H.$$

In other words, the *scaled* collection $\boldsymbol{D}^{-1}\Psi$ is a *Riesz basis* for H. c_1,c_2 are called Riesz constants and the ratio c_2/c_1 is sometimes referred to as the *condition* of the Riesz basis. When H is the energy space for an elliptic variational problem the condition of the Riesz basis will be seen later to determine the performance of *preconditioners* for corresponding discrete problems.

Of course, in the case of the Haar basis on the interval $(0,1)$, say, we have $H=L_2((0,1))$, $\boldsymbol{D}=\mathrm{id}$, $c_1=c_2=1$. One easily concludes with the aid of the Riesz representation theorem that for any given duality pairing $\langle\cdot,\cdot\rangle$ for H and its *dual* H' there exists a collection $\tilde{\Psi} \subset H'$ such that $\langle \psi_I, \tilde{\psi}_{I'}\rangle = \delta_{I,I'}$, $I, I' \in \mathscr{J}$, and $\boldsymbol{D}\tilde{\Psi}$ is a Riesz basis in H'. In fact, a duality argument shows that (3.2) implies

$$c_2^{-1}\|\boldsymbol{D}^{-1}\boldsymbol{d}\|_{\ell_2(\mathscr{J})} \leqslant \|\boldsymbol{d}^{\mathrm{T}}\tilde{\Psi}\|_{H'} \leqslant c_1^{-1}\|\boldsymbol{D}^{-1}\boldsymbol{d}\|_{\ell_2(\mathscr{J})}, \tag{3.3}$$

which will later turn out to be important for residual estimates and the evaluation of 'negative' Sobolev norms. Defining for any countable collections $\Phi \subset H, \Theta \subset H'$ by $\langle \Phi, \Theta\rangle$ the corresponding *matrix* $(\langle \phi, \theta\rangle)_{\phi\in\Phi, \theta\in\Theta}$, the above *biorthogonality* relation between the ψ_I and $\tilde{\psi}_I$ is expressed as

$$\langle \Psi, \tilde{\Psi}\rangle = \mathrm{id}. \tag{3.4}$$

One typically has relations of type (3.2) not only for a *single* space H but even for a whole *scale* of spaces which are continuously embedded in some reference space like L_2. Specifically, denoting as usual by $H^s(\Omega)$ ($H^0(\Omega) = L_2(\Omega)$) the classical Sobolev space of (real positive) order $s > 0$ on some d-dimensional domain or manifold Ω, the relevant spaces are of the form $H^s := H^s(\Omega) \cap H_0$, where H_0 (is either equal to $H^s(\Omega)$ or) is some closed subspace of $H^s(\Omega)$ determined, for instance, by homogeneous boundary conditions. The dual $(H^s)'$ of H^s will briefly be denoted by H^{-s} endowed as usual with the norm

$$\|v\|_{H^{-s}} := \sup_{w\in H^s} \frac{\langle v, w\rangle}{\|w\|_{H^s}}. \tag{3.5}$$

Note that the dual will therefore depend on the constraints imposed by the subspace H_0. One is then interested in the validity of (3.2) for $H = H^s$ in some range of s depending on the choice of bases $\Psi, \tilde{\Psi}$. The weights arising in this case can be chosen as

$$D_I := 2^{s|I|}. \tag{3.6}$$

Recall that usually the actual Sobolev norm plays only an auxiliary role in a variational problem. What matters more is a given *energy norm* of the form

$$\|\cdot\|^2 := a(\cdot,\cdot), \quad a(v,w) := \langle v, \mathscr{L}w\rangle, \tag{3.7}$$

where $\mathscr{L}$ is some symmetric positive-definite (integral or differential) operator. It is often the case that $a(\cdot,\cdot)$ is H^s-*elliptic*, i.e., $\|\cdot\|$ is equivalent to $\|\cdot\|_{H^s}$ and (3.2) holds with $\boldsymbol{D}$ from (3.6). This means that $\|2^{-s|I|}\psi_I\|_{H^s} \sim 1$. However, this relation depends on the ellipticity constant. For instance, for $\mathscr{L}u = -\varepsilon\Delta u + au$ with $\varepsilon, a > 0$ these constants depend on the coefficients ε, a. In this case one can show that when Ψ satisfies (3.2) for $H = L_2$, $\boldsymbol{D} = \mathrm{id}$ and $H = H^1$, $\boldsymbol{D}$ as in (3.6)

$$a(v,v) \sim \|\boldsymbol{D}\boldsymbol{d}\|_{\ell_2(\mathscr{J})} \quad \text{for } v = \boldsymbol{d}^{\mathrm{T}}\Psi, \quad D_I := \sqrt{\max\{a, \varepsilon 2^{2|I|}\}} \tag{3.8}$$

with constants *independent* of the parameters ε, a in $\mathscr{L}$. Therefore taking

$$\boldsymbol{D} := \mathrm{diag}\,\langle \Psi, \mathscr{L}\Psi\rangle \tag{3.9}$$

one still obtains an equivalence of type (3.2) with $\|\cdot\|_H$ replaced by the energy norm (3.7). Since, by definition (3.9) of $\boldsymbol{D}$, the scaled basis $\boldsymbol{D}^{-1}\Psi$ is now normalized in the energy norm one can expect a better condition c_2/c_1 of this basis, a fact that has been confirmed by the numerical experiments in [94].

Classical examples for (3.2) when $\Omega = \mathbb{R}^d$ and Ψ is an orthonormal wavelet basis can be found in [67,95], see [34] for biorthogonal bases on $\mathbb{R}$.

It is important to note that relations like (3.2) are by no means confined to Hilbert spaces but extend to other smoothness spaces such as *Besov spaces.* These latter spaces can be defined for all relevant types of domains with the aid of the L_p*-modulus of continuity* $\omega_m(v,t,L_p(\Omega)) = \sup_{|h|\leqslant t}\|\Delta_h v\|_{L_p(\Omega_{h,m})}$ [71]. Here $\Delta_m v$ denotes the mth-order forward difference of v in the direction h and $\Omega_{h,m} := \{x\in\Omega\colon x+lh\in\Omega, 0\leqslant l\leqslant m\}$. For $0<q\leqslant\infty$, $0<p\leqslant\infty$, $0\leqslant s<m$, the space $B^s_q(L_p(\Omega))$ consists of those elements in $L_p(\Omega)$ for which

$$\|v\|_{B^s_q(L_p(\Omega))} := \left(\|v\|^q_{L_p(\Omega)} + \sum_{j=0}^{\infty} 2^{sjq}\omega_m(v,2^{-j},L_p(\Omega))^q\right)^{1/q} \tag{3.10}$$

is finite. Thus s reflects again the smoothness which is measured now in L_p while q is some fine tuning parameter. Note that we do admit $p<1$ which will be important later for the analysis of adaptive schemes. In this case L_p is only a quasi-Banach space since the triangle inequality holds only up to a constant. For a detailed discussion of these spaces in the context of wavelet analysis the reader is referred to [31,68,70]. Besov spaces are *interpolation spaces* and so are the spaces induced by weighted sequence norms of type (3.2) or (3.10). Keeping in mind that wavelets are usually normalized to have L_2-norm of unit order one readily checks that $\|\psi_I\|_{L_p} \sim 2^{d(|I|/2-|I|/p)}$. Using interpolation arguments then yields under the above assumptions on the multi-resolution spaces

$$\|v\|_{B^s_q(L_p(\Omega))} \sim \left(\sum_j 2^{jq(s+d/2-d/p)}\|\langle v,\tilde{\Psi}_j\rangle\|^q_{\ell_p}\right)^{1/q}, \tag{3.11}$$

see e.g. [31,47,68].

One should note that a significant portion of these key features of wavelet bases are shared by earlier constructions of *spline bases* for Banach spaces on compact domains or manifolds [28–30]. In particular, such spline systems give rise to isomorphisms of the form (3.11) between Besov and sequence spaces [28]. We will return to this point later in Section 9.3.

A particularly interesting case of (3.11) arises when the regularity s and the parameters q,p are *coupled* by

$$\frac{1}{\tau} = \frac{s}{d} + \frac{1}{2}. \tag{3.12}$$

Specializing (3.11) to this case gives

$$\|\boldsymbol{d}^{\mathrm{T}}\Psi\|_{B^s_\tau(L_\tau(\Omega))} \sim \|\boldsymbol{d}\|_{\ell_\tau}, \tag{3.13}$$

which will be used later. Note that s large implies $\tau<1$. Moreover, (3.12) is the limiting line for Sobolev embedding, i.e., the $B^s_\tau(L_\tau)$ are just still embedded in L_2 [31,68,70].

The main consequences of the above two main features are roughly concerned with the following issues. The cancellation property will be seen to account for *efficient matrix/vector multiplication.* The norm equivalences guarantee optimal *preconditioning*. Moreover, they allow us to make use of concepts like *best N-term approximation* in sequence spaces. A combination of all three components will facilitate a rigorous analysis of adaptive schemes. We will postpone the question of constructing bases with the above properties for nontrivial domain geometries to Sections 8 and 9. Instead, we will discuss first the relevance and implications of the above features.

4. Well-posedness in Euclidean metric

In this section several consequences of the norm equivalences (3.2) or (3.11) are highlighted. The significance of these properties is illustrated best for the following abstract setting.

4.1. A class of elliptic problems

Suppose that the linear operator $\mathcal{L}$ is an *isomorphism* from a Hilbert space H into its dual H', i.e., there exist constants $c_{\mathcal{L}}, C_{\mathcal{L}}$ such that

$$c_{\mathcal{L}}\|v\|_H \leqslant \|\mathcal{L}v\|_{H'} \leqslant C_{\mathcal{L}}\|v\|_H, \quad v \in H. \tag{4.1}$$

The problem to find for a given $f \in H'$ a $u \in H$ such that

$$\mathcal{L}u = f. \tag{4.2}$$

has therefore a unique solution.

Simple examples are (i) $H = H^1_{0,\Gamma_D}(\Omega)$ with $\mathcal{L}v = -\mathrm{div}(\boldsymbol{a}\nabla v) + cv$, $c \geqslant 0$, $\boldsymbol{x}^{\mathrm{T}}\boldsymbol{a}\boldsymbol{x} > 0$, $\boldsymbol{x} \in \bar{\Omega}$, where $H^1_{0,\Gamma_D}(\Omega)$ denotes the closure of all C^∞ functions on Ω that vanish on $\Gamma_D \subseteq \partial\Omega$; (ii) $H = H^2_0(\Omega)$ with $\mathcal{L} = \Delta^2$. Aside from such boundary value problems a second important class of problems concerns *boundary integral equations* which arise from reformulating, for instance, an *exterior domain* problem originally posed on $\mathbb{R}^d \backslash \Omega$, where Ω is a compact domain, as a *singular integral equation* on the boundary $\Gamma := \partial\Omega$. It is well known that there are several ways to do that exploiting in different ways the fundamental solution for the Laplace operator ($d \geqslant 3$) $\mathscr{E}(\boldsymbol{x},\boldsymbol{y}) = (4\pi|\boldsymbol{x} - \boldsymbol{y}|)^{-1}(-(1/2\pi)\log|\boldsymbol{x} - \boldsymbol{y}|$ in case $d = 2$). The relevant boundary integral operators on Γ are the *single-layer potential* $\mathscr{V}\sigma(\boldsymbol{x}) := \int_\Gamma \mathscr{E}(\boldsymbol{x},\boldsymbol{y})\sigma(\boldsymbol{y})\,\mathrm{d}s_y$, the *double-layer potential* $\mathscr{K}\sigma(\boldsymbol{x}) := \int_\Gamma (\partial_{n_y}\mathscr{E}(\boldsymbol{x},\boldsymbol{y}))\sigma(\boldsymbol{y})\,\mathrm{d}s_y$, its adjoint $\mathscr{K}'\sigma(\boldsymbol{x}) = \int_\Gamma (\partial_{n_x}\mathscr{E}(\boldsymbol{x},\boldsymbol{y}))\sigma(\boldsymbol{y})\,\mathrm{d}s_y$ and the *hyper-singular operator* $\mathscr{W}\sigma(\boldsymbol{x}) := -\partial_{n_x}\int_\Gamma (\partial_{n_y}\mathscr{E}(\boldsymbol{x},\boldsymbol{y}))\sigma(\boldsymbol{y})\,\mathrm{d}s_y$, $\boldsymbol{x} \in \Gamma$, see e.g. [75].

It is well known that $u(\boldsymbol{x}) = u^-(\boldsymbol{x})$ for $\boldsymbol{x} \in \Gamma$, where $u^-(\boldsymbol{x})$ is the limit $\boldsymbol{y} \to \boldsymbol{x} \in \Gamma$, $\boldsymbol{y} \in \Omega$, satisfies (for $d = 3$) the relations

$$\begin{aligned} u(\boldsymbol{x}) &= (\tfrac{1}{2}\mathrm{id} + \mathscr{K})u(\boldsymbol{x}) - \mathscr{V}\partial_n u(\boldsymbol{x}), \quad \boldsymbol{x} \in \Gamma, \\ \partial_n u(\boldsymbol{x}) &= -\mathscr{W}u(\boldsymbol{x}) + (\tfrac{1}{2}\mathrm{id} - \mathscr{K}')\partial_n u(\boldsymbol{x}), \quad \boldsymbol{x} \in \Gamma, \end{aligned} \tag{4.3}$$

which can now be used to express u as a solution of an integral equation depending on the type of given boundary data. Thus one obtains equations of the type (4.2) where, for instance, it is known that (4.1) holds with $H = L_2(\Gamma) = H'$ for $\mathcal{L} = \mathscr{K} - \frac{1}{2}\mathrm{id}$, $H = H^{-1/2}(\Gamma), H' = H^{1/2}(\Gamma)$ for $\mathcal{L} = \mathscr{V}$ and $H = H^{1/2}(\Gamma), H' = H^{-1/2}(\Gamma)$ for the hyper-singular operator.

The scope of problems covered by (4.1) is actually considerably larger and covers also (indefinite) *systems* of operator equations such as the Stokes problem. This issue will be addressed later in more detail, see Section 6.9. At this point we remark that whenever the operator $\mathcal{L}$ in any of the above cases possesses a *global* Schwartz kernel

$$(\mathcal{L}v)(\boldsymbol{x}) = \int_\Omega K(\boldsymbol{x},\boldsymbol{y})v(\boldsymbol{y})\,\mathrm{d}\boldsymbol{y}, \tag{4.4}$$

where K is smooth except for $\boldsymbol{x} = \boldsymbol{y}$, then K satisfies

$$|\partial_x^\alpha \partial_y^\beta K(\boldsymbol{x},\boldsymbol{y})| \lesssim \mathrm{dist}(\boldsymbol{x},\boldsymbol{y})^{-(d+\rho+|\alpha|+|\beta|)}. \tag{4.5}$$

Let us point out next several consequences of the above cancellation properties and norm equivalences for problems of the above type.

4.2. Preconditioning

Given a dual pair $\Psi, \tilde{\Psi}$ of wavelet bases the operator equation (4.2) can be rewritten as an *infinite* system of linear equations. In fact, making the ansatz $u = \mathbf{d}^{\mathrm{T}}\Psi$, whenever $\Theta = \{\theta_I : I \in \mathscr{J}\}$ is also a basis (and hence total over H), u satisfies (4.2) if and only if $\langle \mathscr{L}u - f, \theta\rangle = 0$, $\theta \in \Theta$, which is equivalent to $\langle \Theta, \mathscr{L}\Psi\rangle \mathbf{d} = \langle \Theta, f\rangle$. Of course, $\Theta = \Psi$ is a possible choice. In addition, we will introduce a particular scaling suggested by the norm equivalence (3.2). In fact, suppose that Θ also satisfies (3.2) with a (possibly different) diagonal matrix $\hat{\mathbf{D}}$ and constants $\hat{c}_1, \hat{c}_2$. This leads to the equivalent system $\hat{\mathbf{D}}^{-1}\langle \Theta, \mathscr{L}\Psi\rangle \mathbf{D}^{-1}\mathbf{D}\mathbf{d} = \hat{\mathbf{D}}^{-1}\langle \Theta, f\rangle$. The first central result can then be formulated as follows.

Theorem 4.1. *The function* $u = \mathbf{d}^{\mathrm{T}}\Psi$ *solves* (4.2) *if and only if* $\mathbf{u} := \mathbf{D}\mathbf{d}$ *solves*

$$\mathbf{A}\mathbf{u} = \mathbf{f}, \tag{4.6}$$

where

$$\mathbf{A} := \hat{\mathbf{D}}^{-1}\langle \Theta, \mathscr{L}\Psi\rangle \mathbf{D}^{-1}, \quad \mathbf{f} := \hat{\mathbf{D}}^{-1}\langle \Theta, f\rangle. \tag{4.7}$$

Moreover, when (4.1) *and* (3.2) *hold then*

$$\|\mathbf{A}\|_{\ell_2(\mathscr{J})}\|\mathbf{A}^{-1}\|_{\ell_2(\mathscr{J})} \leqslant \frac{C_{\mathscr{L}}\tilde{c}_2\hat{c}_2}{c_{\mathscr{L}}\tilde{c}_1\hat{c}_1}. \tag{4.8}$$

Once brought into clear focus the proof is rather simple. In fact, by (3.2), (3.3) and (4.1), one has for any $v = \mathbf{d}^{\mathrm{T}}\Psi \in H$

$$\begin{aligned}\|\mathbf{D}\mathbf{d}\|_{\ell_2(\mathscr{J})} &\leqslant \tilde{c}_1^{-1}\|v\|_H \leqslant c_{\mathscr{L}}^{-1}\tilde{c}_1^{-1}\|\mathscr{L}v\|_{H'} \leqslant c_{\mathscr{L}}^{-1}\tilde{c}_1^{-1}\hat{c}_1^{-1}\|\hat{\mathbf{D}}^{-1}\langle \Theta, \mathscr{L}v\rangle\|_{\ell_2(\mathscr{J})} \\ &= c_{\mathscr{L}}^{-1}\tilde{c}_1^{-1}\hat{c}_1^{-1}\|\hat{\mathbf{D}}^{-1}\langle \Theta, \mathscr{L}\Psi\rangle \mathbf{D}^{-1}\mathbf{D}\mathbf{d}\|_{\ell_2(\mathscr{J})} = c_{\mathscr{L}}^{-1}\tilde{c}_1^{-1}\hat{c}_1^{-1}\|\mathbf{A}\mathbf{D}\mathbf{d}\|_{\ell_2(\mathscr{J})}.\end{aligned} \tag{4.9}$$

Combining this with completely analogous estimates from below confirms (4.8). □

In essence Theorem 4.1 says that due to the norm equivalences of the form (3.2) a diagonal scaling in wavelet coordinates undoes a possible regularity shift caused by the operator $\mathscr{L}$ and turns the original problem into one which is *well-posed in Euclidean metric*. Eq. (4.8) shows how the condition depends on the Riesz constants of the bases and the ellipticity constants in (4.1).

So far we are still dealing with the *infinite-dimensional* original operator equation. Numerical approximations are obtained by considering *finite sections* of the matrix $\mathbf{A}$. Specifically, let $\mathscr{I} \subset \mathscr{J}$ be *any finite* subset and let $\Psi_{\mathscr{I}} := \{\psi_I : I \in \mathscr{I}\}$, $S(\Psi_{\mathscr{I}}) := \operatorname{span}\Psi_{\mathscr{I}}$. Finding $u_{\mathscr{I}} \in S(\Psi_{\mathscr{I}})$ such that

$$\langle \theta, \mathscr{L}u_{\mathscr{I}}\rangle = \langle \theta, f\rangle, \quad \theta \in \Theta_{\mathscr{I}}, \tag{4.10}$$

corresponds to a *Petrov–Galerkin scheme* with trial space $S(\Psi_{\mathscr{I}})$ and test space $S(\Theta_{\mathscr{I}})$. In the sequel, the subscript $\mathscr{I}$ is to indicate that all indices are restricted to $\mathscr{I}$. Obviously, $u_{\mathscr{I}} = \mathbf{d}_{\mathscr{I}}^{\mathrm{T}}\Psi_{\mathscr{I}}$ solves (4.10) if and only if $\mathbf{u}_{\mathscr{I}} := \mathbf{D}_{\mathscr{I}}\mathbf{d}_{\mathscr{I}}$ solves

$$\mathbf{A}_{\mathscr{I}}\mathbf{u}_{\mathscr{I}} = \mathbf{f}_{\mathscr{I}}. \tag{4.11}$$

All it takes now to estimate the condition numbers of the matrices $A_{\mathcal{J}}$ is to invoke the stability of the Petrov–Galerkin scheme. The Petrov–Galerkin scheme is called *H-stable* if

$$\check{c}_{\mathcal{L}}\|v\|_H \leqslant \|\langle \Theta_{\mathcal{J}}, \mathcal{L}v\rangle \tilde{\Theta}_{\mathcal{J}}\|_{H'} \leqslant \check{C}_{\mathcal{L}}\|v\|_H, \quad v \in S(\Psi_{\mathcal{J}}), \tag{4.12}$$

where again $\Theta, \tilde{\Theta}$ are supposed to form a dual pair of bases with $\tilde{\Theta} \in H'$. The same estimates as in (4.9) with $c_{\mathcal{L}}$ replaced by $\check{c}_{\mathcal{L}}$ and (4.1) replaced by (4.12) yields the following fact.

Theorem 4.2. *If in addition to the assumptions in Theorem* 4.1 *the Petrov–Galerkin scheme* (4.10) *is H-stable one has*

$$\|A_{\mathcal{J}}\|_{\ell_2(\mathcal{J})}\|A_{\mathcal{J}}^{-1}\|_{\ell_2(\mathcal{J})} \leqslant \frac{\check{C}_{\mathcal{L}}\tilde{c}_2\hat{c}_2}{\check{c}_{\mathcal{L}}\tilde{c}_1\hat{c}_1}. \tag{4.13}$$

If $\mathcal{L}$ is symmetric positive definite and $\Theta = \Psi$ the corresponding classical Galerkin discretizations are automatically H-stable. In this case, the uniform boundedness of the condition numbers ensures that *conjugate gradient* iterations on systems (4.11) are *asymptotically optimal.* This means that in connection with *nested iteration* discretization error accuracy of the discrete systems is accomplished at the expense of a uniformly bounded number of iterations on the system of highest resolution. In fact, for second-order elliptic boundary value problems on planar domains and orthonormal wavelet bases the conclusion (4.13) already appears in [80]. Under much more relaxed conditions concerning the bases and the type of boundary value problems (4.13) was obtained in [52], see also [14] for periodic problems. The extraction of the two essential ingredients namely stability of the discretization and norm equivalences has been perhaps brought into clear focus for the first time in [61] in order to identify the essential requirements on wavelet bases for a much wider class of equations. It is therefore important to note that in the above form (in contrast to Schwarz schemes) $\mathcal{L}$ neither need to be symmetric nor must have positive order. Of course, lacking symmetry it is not so clear beforehand what estimate (4.13) means for the complexity of a solution process. However, it will be seen later that (4.13) will be essential in combination with least-squares formulations.

Drawing on the theory of function spaces closely related concepts were already used in [96] to prove optimality of the BPX-preconditioner [21] for finite element discretizations, see [52] for the case of adaptively refined triangulations. In fact, for symmetric problems of positive order the use of explicit bases can be avoided. It suffices to resort to the weaker notion of *frame* to realize optimal preconditioners. A suitable general framework is offered by *Schwarz schemes* covering domain decomposition and multilevel splittings [77,97,114]. For the relation to wavelet preconditioners see [49].

4.3. Best N-term approximation

Aside from preconditioning the fact that the norm equivalences allow one to transform the original problem into a well-posed one in $\ell_2(\mathcal{J})$ has further important consequences. Suppose again that $\boldsymbol{u}$ is the solution of (4.6), or equivalently that $u = \boldsymbol{u}^{\mathrm{T}}\boldsymbol{D}^{-1}\Psi$ solves the continuous problem (4.2), one is interested in finding a possibly good approximation with possibly few coefficients. The smallest possible error is realized by the *best N-term approximation* defined as

$$\sigma_{N,H}(v) := \inf_{v_I : I \in \mathcal{J}, \#\mathcal{J} \leqslant N} \left\| v - \sum_{I \in \mathcal{J}} v_I D_I^{-1} \psi_I \right\|_H . \tag{4.14}$$

Since by (3.2),

$$\|\boldsymbol{v}\|_{\ell_2(\mathcal{J})} \sim \|\boldsymbol{v}^{\mathrm{T}} \boldsymbol{D}^{-1} \Psi\|_H, \tag{4.15}$$

$\sigma_{N,H}(u)$ is *asymptotically* the same as the error of *best N-term approximation* in $\ell_2(\mathcal{J})$

$$\sigma_{N,\ell_2(\mathcal{J})}(\boldsymbol{u}) := \inf_{\boldsymbol{w}, \# \mathrm{supp}\, \boldsymbol{w} \leqslant N} \|\boldsymbol{u} - \boldsymbol{w}\|_{\ell_2(\mathcal{J})}. \tag{4.16}$$

The latter notion is well understood, see e.g. [68]. Obviously, $\sigma_{N,\ell_2(\mathcal{J})}(\boldsymbol{u})$ is realized by retaining the N *largest* coefficients in $\boldsymbol{u}$ which at this point are, of course, unknown. Nevertheless, before trying to track them numerically when $\boldsymbol{u}$ is only given *implicitly* as the solution of an infinite system of the form (4.6), it is important to understand how this error behaves.

To this end, denote for any $\boldsymbol{v} \in \ell_2(\mathcal{J})$ by $\boldsymbol{v}^* = \{v_{I_l}\}_{l \in \mathbb{N}}$ its *decreasing rearrangement* in the sense that $|v_{I_l}| \geqslant |v_{I_{l+1}}|$ and let

$$\Lambda(\boldsymbol{v}, N) := \{I_l \colon l = 1, \ldots, N\}, \quad \boldsymbol{v}_N := \boldsymbol{v}|_{\Lambda(\boldsymbol{v},N)}. \tag{4.17}$$

It is clear that $\boldsymbol{v}_N$ is a best N-term approximation of $\boldsymbol{v}$.

In particular, it will be important to characterize the sequences in $\ell_2(\mathcal{J})$ whose best N-term approximation behaves like N^{-s} for some $s > 0$. The following facts are well known [32,68,70]. Let for $0 < \tau < 2$

$$|\boldsymbol{v}|_{\ell_\tau^w(\mathcal{J})} := \sup_{n \in \mathbb{N}} n^{1/\tau} |v_n^*|, \quad \|\boldsymbol{v}\|_{\ell_\tau^w(\mathcal{J})} := \|\boldsymbol{v}\|_{\ell_2(\mathcal{J})} + |\boldsymbol{v}|_{\ell_\tau^w(\mathcal{J})}. \tag{4.18}$$

It is easy to see that

$$\|\boldsymbol{v}\|_{\ell_\tau^w(\mathcal{J})} \leqslant 2 \|\boldsymbol{v}\|_{\ell_\tau}, \tag{4.19}$$

so that by Jensen's inequality, in particular, $\ell_\tau^w(\mathcal{J}) \subset \ell_2(\mathcal{J})$.

Proposition 4.3. *Let*

$$\frac{1}{\tau} = s + \frac{1}{2}, \tag{4.20}$$

then

$$\boldsymbol{v} \in \ell_\tau^w(\mathcal{J}) \Leftrightarrow \|\boldsymbol{v} - \boldsymbol{v}_N\|_{\ell_2(\mathcal{J})} \lesssim N^{-s} \|\boldsymbol{v}\|_{\ell_\tau^w(\mathcal{J})}. \tag{4.21}$$

The following remark shows why convergence rates of the form N^{-s} are of interest.

Remark 4.4. Note that by (4.19), $\boldsymbol{u} \in \ell_\tau$ implies $\boldsymbol{u} \in \ell_\tau^w(\mathcal{J})$. Using (3.13), one can show that for s and τ related through (4.20) $\boldsymbol{u} \in \ell_\tau$ means for $D_I = 2^{t|I|}$ that $u \in B_\tau^{sd+t}(L_\tau)$.

The exact relation between the rate of best N-term approximation (in the energy norm) and a certain Besov regularity can be stated as follows. We follow [41] and suppose again that $H = H^t$. Let $\gamma > 0$ denote the sup of all α such that $\Psi \subset H^\alpha$. Then the following holds [41].

Proposition 4.5. *Assume that* $\alpha, t < \gamma$ *and let for* $t \leqslant \alpha$

$$\frac{1}{\tau^*} := \frac{\alpha - t}{d} + \frac{1}{2}. \tag{4.22}$$

Then one has

$$\sum_{n=1}^{\infty} (N^{(\alpha-t)/d} \sigma_{N,H^t}(g))^{\tau^*} < \infty \tag{4.23}$$

if and only if $g \in B^{\alpha}_{\tau^*}(L_{\tau^*}(\Omega))$.

Of course, (4.23) implies that the best N-term approximation in H^t (and hence the near best N-term approximation with respect to the energy norm) $\sigma_{N,H^t}(g)$ decays at least like $N^{-(\alpha-t)/d}$, provided that g is in $B^{\alpha}_{\tau^*}(L_{\tau^*}(\Omega))$. Note that (4.22) means that $B^{\alpha}_{\tau^*}(L_{\tau^*}(\Omega))$ is just embedded in H^t but need not have any excess Sobolev regularity beyond the energy space.

Results of the above type hold for other L_p norms and play an important role in *nonlinear approximation* and applications to image compression, denoising and encoding, see [31,33,68,69,71]. In the above form they are fundamental for the understanding and interpretation of adaptive schemes as will be explained later.

5. Near sparsity of matrix representations

So far we have seen that scaled wavelet representations lead for a variety of operator equations to well-posed problems in ℓ_2. Moreover, some mechanisms pertaining to the underlying norm equivalences have been outlined. On the other hand, uniformly bounded condition numbers just by themselves do not ensure yet efficient solution processes. This requires more specific information on the matrices $\boldsymbol{A}$ in (4.6).

5.1. Vaguelettes

In this regard, one concept is to *adapt* the choice of the test basis Θ to the problem in the following way. Denoting by $\mathcal{L}'$ the adjoint of $\mathcal{L}$, i.e., $\langle \mathcal{L}v, w\rangle = \langle v, \mathcal{L}'w\rangle$, let $\Theta := (\mathcal{L}^{-1})'\tilde{\Psi}$, $\tilde{\Theta} := \mathcal{L}\Psi$. Clearly biorthogonality of Ψ and $\tilde{\Psi}$ (3.4) implies then $\langle \Theta, \tilde{\Theta}\rangle = \mathrm{id}$. Hence $\langle \Theta, f\rangle = \langle \Theta, \mathcal{L}\Psi\rangle \boldsymbol{d} = \boldsymbol{d}$ so that the solution u is given by

$$u = \langle f, \Theta\rangle \Psi. \tag{5.1}$$

In this case the above Petrov–Galerkin scheme simply consists of truncating the expansion (5.1). The collection Θ consists of so-called *vaguelettes* [110]. Of course, determining these vaguelettes requires the inversion of $\mathcal{L}'$ for each wavelet. When $\mathcal{L}$ is a constant coefficient elliptic differential operator on the torus this inversion can be done once and for all in the Fourier domain which yields an optimal method. For applications of these concepts to combustion and turbulence analysis see e.g. [18,73]. Although this will not work nearly as well for variable coefficients and other domain geometries it is possible to use this concept for *preconditioning* in such more general situations, see e.g. [87].

5.2. Change of bases

Let us focus now on the case $\Theta=\Psi$ which corresponds to Galerkin discretizations. The exploitation of the bounded condition numbers for an iterative solution of the discrete systems (4.11) now hinges on the efficiency of corresponding matrix/vector multiplications. To sketch the main developments in this regard let us first assume that $\mathscr{L}$ is a differential operator. Due to the fact that the inner products $\langle\psi_I,\mathscr{L}\psi_{I'}\rangle$ involve wavelets on different scales and hence with different support sizes (see (2.2)) entails that the matrices $\boldsymbol{A}_{\mathscr{I}}$ from (4.11) (with $\Theta=\Psi$) are not as sparse as typical finite element stiffness matrices. In fact, it is not hard to see that in the case of *uniform refinements* up to level J the order of nonvanishing entries is of the order $J2^{dJ}$. However, under assumption (2.3), the matrix $\boldsymbol{B}_J:=\langle\Phi_J,\mathscr{L}\Phi_J\rangle$ will be sparse, i.e., has the order of $N_J=\dim S_J$ nonvanishing entries. It is easy to check that the matrices $\boldsymbol{B}_J$ and $\boldsymbol{A}_J:=\boldsymbol{A}_{\mathscr{I}_J}$ are related by

$$\boldsymbol{A}_J=\boldsymbol{D}_J^{-1}\boldsymbol{T}_J^{\mathrm{T}}\boldsymbol{B}_J\boldsymbol{T}_J\boldsymbol{D}_J^{-1}, \tag{5.2}$$

where $\boldsymbol{T}_J$ is the multiscale transformation from (2.15). The so-called *change-of-bases preconditioner* consists now of solving the standard sparse Galerkin system $\boldsymbol{B}_J\boldsymbol{c}_J=\boldsymbol{f}_J:=\langle\Phi_J,f\rangle$ by conceptually applying the conjugate gradient method to the well-conditioned matrix $\boldsymbol{A}_J$. However, the less sparse matrix $\boldsymbol{A}_J$ need *not* be assembled. In fact, as pointed out before, the application of $\boldsymbol{T}_J$ to a vector $\boldsymbol{d}_J$ can be carried out in $\mathcal{O}(N_J)$ operations. Therefore, due to the sparseness of $\boldsymbol{B}_J$ the multiplication $\boldsymbol{A}_J\boldsymbol{d}_J$ can be carried out in $\mathcal{O}(N_J)$ steps and thus ensures *asymptotic optimality* of a wavelet preconditioner [52] for such uniformly refined trial spaces. The *hierarchical basis preconditioner* in [115] is of this type and very efficient due to the extreme sparseness of the matrices $\boldsymbol{M}_{j,1}$ in this case. However, it fails to provide uniformly bounded condition numbers (4.13) since (3.2) is not valid for hierarchical bases and $H=H_0^1(\Omega)$. The scheme in [111] aims at retaining this efficiency while improving the stability of the hierarchical bases, see Section 10.1.

Numerical experiences: In general, for more sophisticated wavelets with better stability properties the cost of the multiscale transformations, although asymptotically being of the right order, increases due to the larger supports of the wavelets and correspondingly higher density of the multiscale transformation matrices. In their straightforward realization this renders such a wavelet scheme somewhat less efficient than the above-mentioned closely related Schwarz schemes, at least for such simple uniform refinements and when the diagonal matrices contain the standard weights $2^{-t|I|}$, see e.g [90,91]. This has motivated successful attempts to construct wavelets with possibly small support in a finite element context [106,107]. Moreover, the stable bases are usually derived by modifying the simple hierarchical bases through the concept of *stable completions*, see Section 8.2. This allows one to factor each step (2.15) in the multiscale transformation retrieving much of the original efficiency of the hierarchical bases [66,107], see (8.12) below. Furthermore, since (suitably scaled) wavelets form stable bases for $H_0^1(\Omega)$ as well as for $L_2(\Omega)$ the change-of-bases preconditioners are typically more robust for operators of the type $\mathscr{L}v=-\Delta v+qv$ with respect to varying the parameter q than for instance the BPX scheme. Moreover, recent results in [94], indicate that scalings like (3.9) yield much smaller condition numbers which make the scheme competitive again, recall (3.8), (3.9).

Recently, wavelet solvers for problems of the type $-\Delta u+qu=f$ on L-shaped domains based on the above change-of-basis preconditioner have been implemented and tested in [24,113,112]. While [24] contains 3D realizations the results in [113] show that such schemes are particularly efficient in combination with *nested iteration*. In fact, to achieve discretization error accuracy 10–12 conjugate gradient iterations on each level turn out to suffice independently of the coefficient q.

Nevertheless, the main drawback of this strategy is that the multiscale transformation is best suited for index sets $\mathcal{I}_J$ corresponding to uniform refinements. Thus, if an adaptive solution process yields an approximate solution with significant indices in a very *lacunary* index set $\mathcal{I}$ where by no means all indices of each level $< J$ are present, $\#\mathcal{I}$ could be much smaller than $\dim S_J$ where $J-1 := \max\{|I|: I \in \mathcal{I}\}$. Then the complexity of multiscale transformation $\boldsymbol{T}_J$ is much higher than $\#\mathcal{I}$ which means that previously gained efficiency is wasted.

Moreover, when dealing with global operators the above approach is completely useless because $\boldsymbol{B}_J$ is densely populated with significant entries essentially everywhere.

5.3. Nonstandard representation

An alternative strategy of organizing a fast approximate application of a given operator to the array of wavelet coefficients is based on the so-called *nonstandard representation* [5,15,16]. Denoting by $Q_j v := \langle v, \tilde{\Phi}_j \rangle \Phi_j$ the canonical projector onto S_j, the nonstandard representation of $\mathcal{L}$ results from truncating the telescoping expansion

$$\mathcal{L} = \sum_{j=0}^{\infty} (Q'_{j+1} \mathcal{L} Q_{j+1} - Q'_j \mathcal{L} Q_j) + Q'_0 \mathcal{L} Q_0.$$

The Application of $\boldsymbol{A}_J$ to an array $\boldsymbol{u}_J$ is then reduced to the application of enlarged arrays $\hat{\boldsymbol{d}}_J$ involving the wavelet coefficients $\boldsymbol{d}_j$ as well as scaling function coefficients $\boldsymbol{c}_j$ on all levels $j < J$ to a larger matrix $\hat{\boldsymbol{A}}_J$. This latter matrix $\hat{\boldsymbol{A}}_J$ is *not* a matrix representation of $\mathcal{L}$ in a strict sense but corresponds to the representation of an element from S_J with respect to a *redundant* spanning set. The advantage is that $\hat{\boldsymbol{A}}_J$ consists of blocks whose entries in contrast to $\boldsymbol{A}_J$ are inner products of functions belonging to a *single* level only, i.e., the mixing of different levels is avoided. Moreover, for periodic problems these blocks are circulants so that even when dealing with integral operators fast Fourier transforms can be used for the calculation of $\hat{\boldsymbol{A}}_J \hat{\boldsymbol{d}}_J$.

This last advantage disappears for non-periodic problems. In this case one has to exploit the fact that many entries of $\hat{\boldsymbol{A}}_J$ are actually small and to some extent negligible, see e.g. [16]. In fact, the blocks in $\hat{\boldsymbol{A}}_J$ are of the form $\langle \Phi_j, \mathcal{L}\Psi_j \rangle$, $\langle \Psi_j, \mathcal{L}\Phi_j \rangle$ or $\langle \Psi_j, \mathcal{L}\Psi_j \rangle$, i.e., at least one factor in the inner products is a wavelet so that the cancellation property (3.1) can be used to confirm a certain decay. However, since in the *standard representation* $\boldsymbol{A}_J$ from (4.6) except for the few functions on the coarsest level only wavelets appear in the inner products this *compression* effect is stronger there. This will be made more precise next.

5.4. Decay estimates

Suppose that $\mathcal{L}$ is either a differential operator or of the form (4.4) satisfying (4.5). Using the norm equivalences (3.2) and continuity properties of $\mathcal{L}$, one can show that in the case $H = H'$ the entries of the scaled matrix $\boldsymbol{A}$ from (4.1) (for $\Theta = \Psi$) exhibit the following decay:

$$2^{-(|\lambda'|+|\lambda|)t} |\langle \mathcal{L}\psi_{I'}, \psi_I \rangle| \lesssim \frac{2^{-\||I|-|I'|\|\sigma}}{(1+d(I,I'))^{d+2\tilde{m}+2t}}, \tag{5.3}$$

where $d(I,I') := 2^{\min(|I|,|I'|)}\mathrm{dist}(\mathrm{supp}\,\psi_I, \mathrm{supp}\,\psi_{I'})$, [61,43,100]. Here $\sigma > d/2$ depends on the *regularity* of the wavelets. Hence the entries of $\mathbf{A}$ have a scalewise decay whose strength depends on the regularity of Ψ and a spatial decay determined by the order $\tilde{m}$ of the cancellation properties, or equivalently, by the approximation order of the dual multi-resolution $\tilde{\mathscr{S}}$, see Remark 8.4 below. The appearance of $2\tilde{m}$ in the exponent reflects that the cancellation properties of both wavelets ψ_I and $\psi_{I'}$ have been exploited, see e.g. [66,100].

Matrices satisfying (5.3) represent Calderón–Zygmund operators. In fact, $\sigma > d/2$ already implies that the matrix defines a bounded map on ℓ_2 which in the present context has been inferred in Theorem 4.1 from the properties of $\mathscr{L}$. For sufficiently large σ the class of such matrices forms an algebra. The question when such a matrix has an inverse in the same class is much more delicate, see e.g. [110] for a more detailed discussion.

5.5. Pseudo-differential and boundary integral equations

To exploit decay estimates of type (5.3) for the efficient numerical treatment of certain singular integral equations has been proposed first in the pioneering paper [15]. It has been shown there that for periodic univariate operators of order zero ($t=0$) it suffices to retain the order of $N\log N$ nonzero entries in the (standard and nonstandard) wavelet representation while preserving accuracy ε. This is often referred to as *matrix compression* and has since initiated numerous further investigations.

To understand a central issue of these investigations one should note that in spite of their startling nature the results in [15] are not quite satisfactory yet for the following reason. Roughly speaking, the accuracy ε is kept fixed while the size N of the linear systems increases. Of course, the only reason for increasing the number N of unknowns is to *increase* accuracy which clearly requires relating ε and N. First, systematic investigation of *asymptotic* properties of matrix compression for a general class of *elliptic* pseudo-differential equations were presented in [59,60] first for the periodic setting and in [61,98–100] for equations defined on *closed manifolds*. A central point of view taken in [61] was to identify those properties of wavelet bases for a given problem that facilitate *asymptotically optimal* compression and solution schemes. The main result can roughly be summarized as follows, see [98–100] for zero order operators.

Theorem 5.1. *Assume that* $\mathscr{L}$ *has the form* (4.4), (4.5) *and satisfies* (4.1) *for* $H=H^t$. *Furthermore assume that* $\Psi,\tilde{\Psi}$ *is a dual pair of* (*local*) *wavelet bases inducing norm equivalences of the form* (3.2) *for* $H = H^s, D_I = s^{s|I|}$, *in the range* $s \in (-\tilde{\gamma},\gamma)$ (*see also* (8.5) *below*) *and having cancellation properties* (3.1) *of order* $\tilde{m}$. *Let* $\mathbf{A}$ *be the preconditioned standard matrix representation of* $\mathscr{L}$ *given by* (4.7) *with* $\Theta=\Psi$, $\mathbf{D}=\hat{\mathbf{D}}=(2^{t|I|}\delta_{I,I'})_{I,I'\in\mathscr{I}}$. *As before let* $\mathbf{A}_J=\mathbf{A}_{\mathscr{I}_J}$ *be the corresponding stiffness matrix for the trial spaces* S_J *which have approximation order m, see* (8.3) *below. Finally, suppose that the parameters* $\gamma,\tilde{\gamma},m,\tilde{m}$ *signifying the bases* Ψ *and* $\tilde{\Psi}$ *satisfy*

$$\gamma > t, \quad \tilde{\gamma} > -t, \quad \tilde{m} > m - 2t. \tag{5.4}$$

Then there exists a compressed matrix $\mathbf{A}_J^c$ *with the following properties*:

(i) *The number of nonvanishing entries of* $\mathbf{A}_J^c$ *is of the order* $N_J(\log N_J)^b$ *uniformly in J for some fixed positive number b where* $N_J=\dim S_J$.

(ii) *The matrices* A_J^c *have still uniformly bounded condition numbers*

$$\mathrm{cond}_2(A_J^c) \lesssim 1. \tag{5.5}$$

(iii) *If the solution* $u = \mathbf{u}^T \mathbf{D}^{-1}\Psi$ *of* (4.2) *respectively* (4.6) *belongs to* H^s *then the solution* $\mathbf{u}_J^c$ *of the compressed system* $A_J^c \mathbf{u}_J^c = \mathbf{f}_{\mathscr{I}_J}$ *has asymptotically optimal accuracy, i.e.* $u_J^c = (\mathbf{u}_J^c)^T \mathbf{D}_{\mathscr{I}_J}^{-1} \Psi_{\mathscr{I}_J}$ *satisfies*

$$\|u - u_J^c\|_{H^\tau} \lesssim 2^{J(\tau - s)} \|u\|_{H^s}, \quad J \in \mathbb{N}, \tag{5.6}$$

for

$$-m + 2t \leqslant \tau < \gamma, \quad \tau \leqslant s, \ t \leqslant s \leqslant m. \tag{5.7}$$

In particular, when $u \in H^m$ *optimal order* $2^{-J2(m-t)}$ *is retained.*

In contrast to [15] operators of order $2t \neq 0$ are covered. The proof is based on a sophisticated *level-dependent* thresholding strategy in combination with preconditioning (4.8), see [59,61] for details. Note that for operators of order less than or equal to zero the above result ensures optimality only when the order $\tilde{m}$ of cancellation properties is *larger* than the order m of accuracy of the trial spaces. This rules out orthogonal wavelets in this case and stresses the need for the more flexible concept of biorthogonal bases.

First numerical tests for two-dimensional polyhedral closed surfaces are given in [51] for a collocation scheme. In [99] a corresponding analysis is given for discontinuous *multi-wavelets*. The compression in (i) above has been improved in [103] by incorporating a *second compression* depending on the distance of high-scale wavelets from the *singular support* of an overlapping lower-scale wavelet. This allows one to remove the log-factor and to realize compressed matrices even with only $\mathcal{O}(N_J)$ nonvanishing entries.

The practical exploitation of the above results now leaves two tasks:

(I) Find bases that satisfy the requirements in Theorem 5.1, in particular (5.4).
(II) Compute the compressed matrix A_J^c at a computational expense that stays possibly close to the order N_J of nonvanishing entries.

Clearly, if (I) and (II) can be realized at least for positive-definite symmetric problems the above theorem ensures also optimal $\mathcal{O}(N_J)$ solution complexity.

We will comment on (I) later. Task (II) poses a serious difficulty. In order to preserve optimal complexity roughly $\mathcal{O}(1)$ operations are to be spent on average per nonvanishing entry while retaining discretization error accuracy. It would be straightforward to realize that for entries of the type $\langle \phi_{J,k}, \mathscr{L}\phi_{J,k'} \rangle$ but this would require computing N_J^2 such entries. Therefore one has to work directly with the wavelet representation. However, then one has to compute entries involving wavelets of different scales, in particular those from the coarsest scale. A naive application of quadrature rules would require already a computational expense of order N_J to ensure sufficient accuracy for a single entry involving coarse scale basis functions. The key to overcoming this difficulty is to balance the quadrature accuracy with the decay of the entries. A first fully discrete scheme for zero-order operators is given in [99,86]. In this case Haar wavelets are employed that are suitable for zero-order operators and can conveniently be realized for patchwise defined boundary surfaces. They are almost optimal in the above sense and a similar complexity as above is confirmed. The computation of the compressed stiffness matrices is accomplished via a suitable Gaussian quadrature scheme of variable order. In [85] numerical experiments are reported treating systems up to a million unknowns. A similar strategy of adaptive quadrature is proposed in [103] which is combined with second compression and realizes full asymptotic complexity.

5.6. *Fast approximate matrix/vector multiplication*

The approach described in the previous section can be advanced in the following two respects. Firstly, instead of working only with full levelwise index sets $\mathcal{I}_J$ one can consider *adaptively* generated highly *lacunary* sets $\mathcal{I}$. Secondly, instead of exploiting only compressibility of $\boldsymbol{A}$ also the possible near sparseness of the arrays of wavelet coefficients can be taken into account. This culminates in a new different type of fast *matrix/vector multiplication* based on the following somewhat different notion of compressibility proposed in [32]. The rest of this section follows the development in [32]. The (infinite) matrix $\boldsymbol{C}$ belongs to the class $\mathscr{C}_s$ if there exists a positive summable sequence $(\alpha_j)_{j\geqslant 0}$ and for every $j\geqslant 0$ there exists a matrix $\boldsymbol{C}_j$ with at most $2^j\alpha_j$ nonzero entries per row and column such that

$$\|\boldsymbol{C}_j - \boldsymbol{C}\| \lesssim \alpha_j 2^{-sj}. \tag{5.8}$$

Decay properties of the type (5.3) turn out to imply such compressibility in a certain range.

Proposition 5.2. *Let*

$$s^* := \min\left\{\frac{\sigma}{d} - \frac{1}{2}, \frac{2t+2\tilde{m}}{d}\right\}, \tag{5.9}$$

where $\sigma, t, \tilde{m}$ are the parameters from (5.3). *Then for every $s < s^*$ the matrix $\boldsymbol{A}$ from* (4.7) *belongs to $\mathscr{C}_s$.*

In order to exploit the sparseness of $\boldsymbol{A}$ and a given array $\boldsymbol{v}$ let $\boldsymbol{v}_{[j]} := \boldsymbol{v}_{2^j}$ denote the best N-term approximation of $\boldsymbol{v}\in\ell_2(\mathcal{I})$ for $N = 2^j$ and define

$$\boldsymbol{w}_j := \boldsymbol{A}_j\boldsymbol{v}_{[0]} + \boldsymbol{A}_{j-1}(\boldsymbol{v}_{[1]} - \boldsymbol{v}_{[0]}) + \cdots + \boldsymbol{A}_0(\boldsymbol{v}_{[j]} - \boldsymbol{v}_{[j-1]}). \tag{5.10}$$

Then $\boldsymbol{w}_j$ is an approximation to the product $\boldsymbol{A}\boldsymbol{v}$. The estimation of the accuracy of this approximation relies on the characterization of best N-term approximation from Section 4.3.

Proposition 5.3. *If $\boldsymbol{A}\in\mathscr{C}_s$ then, whenever $\boldsymbol{v}\in\ell^w_\tau(\mathcal{I})$ for $1/\tau = s + \frac{1}{2}$, one has*

$$\|\boldsymbol{A}\boldsymbol{v} - \boldsymbol{w}_j\|_{\ell_2(\mathcal{I})} \lesssim 2^{-sj}\|\boldsymbol{v}\|_{\ell^w_\tau(\mathcal{I})}. \tag{5.11}$$

Moreover, $\boldsymbol{A}$ is bounded on $\ell^w_\tau(\mathcal{I})$, i.e.,

$$\|\boldsymbol{A}\boldsymbol{v}\|_{\ell^w_\tau(\mathcal{I})} \lesssim \|\boldsymbol{v}\|_{\ell^w_\tau(\mathcal{I})}. \tag{5.12}$$

In fact, it follows from the definition of $\mathscr{C}_s$ that the support of $\boldsymbol{w}_j$ is of the order 2^j so that (5.12) follows from Proposition 4.3. Moreover, (5.11) says that the computational work $\boldsymbol{CW}(\eta)$ needed to realize an approximation $\boldsymbol{w}_\eta$ to $\boldsymbol{A}\boldsymbol{v}$ such that $\|\boldsymbol{A}\boldsymbol{v} - \boldsymbol{w}_\eta\|_{\ell_2(\mathcal{I})}\leqslant\eta$ is of the order

$$\boldsymbol{CW}(\eta) \sim \#\mathrm{supp}\,\boldsymbol{w}_\eta \lesssim \eta^{-1/s}\|\boldsymbol{v}\|^{1/s}_{\ell^w_\tau(\mathcal{I})}, \tag{5.13}$$

a fact that is crucial in the realization and analysis of adaptive schemes to be described next.

6. Adaptive wavelet schemes

6.1. The background

The design of adaptive solvers is perhaps one of the most prominent areas where modern mathematical concepts can greatly contribute to scientific large-scale computation. Adaptivity has therefore been the subject of numerous studies from different perspectives. The experiences gathered in the finite element context indicate the potential of a tremendous reduction of complexity, see e.g. [19,6]. However, very little is known about the *convergence* of adaptive schemes in a rigorous sense even for simple model problems. Of course, convergence means to relate the accomplished accuracy to the adaptively generated degrees of freedom and the accumulated computational work. Considering the possibly significant overhead caused by managing adaptive data structures it is obviously important to know whether a scheme converges at all and if so how fast. The availability of a posteriori local error estimators or indicators does not yet answer this question. In fact, one usually *assumes* the so called *saturation property* to conclude a fixed error reduction resulting from a mesh refinement which in effect is the same as *assuming* convergence rather than proving it [19]. To my knowledge the only exception is the analysis of the special case of Poisson's equation on planar polygonal domains in [72]. However, in this context there appear to be no error estimates for adaptive schemes that relate the achieved accuracy to the size of the adaptively generated systems.

Also the inherent adaptive potential of wavelets has often been praised and treated in numerous studies, see [5,10,11,16,73,82,92,102] for an incomplete list of references, see also [38] for first numerical implementations in a bivariate nonperiodic setting. But again, there are no convergence proofs. Inspired by the results in [10,72] it was shown first in [43] that a certain adaptive wavelet scheme converges for the scope of problems given in Section 4.1 under the assumption that $\mathscr{L}$ is symmetric positive definite, i.e., the symmetric bilinear form $a(v,w) := \langle v, \mathscr{L}w\rangle$ satisfies $a(v,v) \sim \|v\|_H^2$ or equivalently

$$c_1\|\mathbf{v}\|_{\ell_2(\mathscr{J})} \leqslant \|\mathbf{v}\| := (\mathbf{v}^{\mathrm{T}}\mathbf{A}\mathbf{v})^{1/2} \leqslant c_2\|\mathbf{v}\|_{\ell_2(\mathscr{J})} \tag{6.1}$$

for some positive constants c_1, c_2, which will be assumed throughout this section.

A conceptual breakthrough has been recently obtained for this setting in [32] in that rigorous *convergence rates* could be established which are optimal in a sense to be explained below. Before beginning with a brief summary of the results in [32], a few remarks concerning the interest in such theoretical results are in order. First, theoretical predictions from complexity theory are mostly pessimistic about the gain offered by adaptive schemes. Here it is important to understand under which *model assumptions* such a pessimistic statement holds. So it would be important to have positive results in a relevant setting.

Second, the results cover elliptic PDEs as well as elliptic *integral equations* even for operators of negative order such as single layer potentials. In this case even less than for PDEs is known about adaptive schemes for such problems in classical discretization settings.

Third, the attempts to prove such a convergence result have led to interesting wavelet specific algorithmic components that do not come up in a classical environment.

Finally, the key features described in Section 3 offer ways of extending the scope of applicability to *systems* of operator equations and *indefinite* problems.

6.2. Key questions

As pointed out in Section 4.3 best N-term approximation offers a natural bench mark. In fact, the best that could be achieved by an adaptive scheme is to produce errors that stay of the order of best N-term approximation in $\ell_2(\mathcal{J})$. Thus the first natural question is whether an adaptive scheme can match the rate of best N-term approximation asymptotically.

Best N-term approximation and adaptive schemes are *nonlinear* methods in the sense that the approximations are not taken from a *linear* space. Suppose for a moment that an adaptive scheme *does* match the rate of best N-term approximation. The second question remains: what is the potential gain over *linear methods*, i.e, methods for which the trial spaces are a-priorly prescribed? Such methods are, of course, much easier to realize in practice.

According to Section 4.3 the first question essentially concerns approximation in the sequence space $\ell_2(\mathcal{J})$. The second question will be seen to draw on *regularity theory* for solutions of elliptic problems in certain *nonclassical* scales of function spaces.

6.3. The basic paradigm

The practical realization of adaptive approximations to (4.2) in a finite element context is to refine step by step a given *mesh* according to a posteriori local error indicators. The point of view taken by wavelet schemes is somewhat different. Trial spaces are refined directly by incorporating *additional* basis functions whose selection depends on the previous step. Specifically, setting for any finite subset $\mathcal{I} \subset \mathcal{J}$

$$S_{\mathcal{I}} := \operatorname{span}\{\psi_I : I \in \mathcal{I}\}$$

and denoting by $u_{\mathcal{I}} = \boldsymbol{u}_{\mathcal{I}}^{\mathrm{T}} \boldsymbol{D}_{\mathcal{I}}^{-1} \Psi_{\mathcal{I}} \in S_{\mathcal{I}}$ always the Galerkin solution determined by (4.11), one starts with some small index set $\mathcal{I}_0$ (possibly the empty set) and proceeds as follows:

Given $\mathcal{I}_j$, $u_{\mathcal{I}_j}$ *and some* fixed $\theta \in (0,1)$, *find* $\mathcal{I}_{j+1} \supset \mathcal{I}_j$ *as* small as possible *such that the new error* $u - u_{\mathcal{I}_{j+1}}$ *in the energy norm is at most* θ *times the previous error. Obviously, iterating this step implies convergence of the resulting sequence of approximations in the energy norm.*

Successively growing index sets in this way, one hopes to track essentially the *most significant* coefficients in the true wavelet expansion $\boldsymbol{u}^{\mathrm{T}} \boldsymbol{D}^{-1} \Psi$ of the unknown solution u. In principle, the following observation has been used already earlier in the finite element context, see e.g. [19,72]. It was also the starting point in [43].

Remark 6.1. Defining as in (6.1) the *discrete energy norm* by $\|\boldsymbol{v}\|^2 := \boldsymbol{v}^{\mathrm{T}} \boldsymbol{A} \boldsymbol{v}$ with $\boldsymbol{A}$ from (4.7), observe that when $\hat{\mathcal{I}} \supset \mathcal{I}$ Galerkin orthogonality implies the equivalence of

$$\|\boldsymbol{u} - \boldsymbol{u}_{\hat{\mathcal{I}}}\| \leqslant \theta \|\boldsymbol{u} - \boldsymbol{u}_{\mathcal{I}}\| \tag{6.2}$$

and

$$\|\boldsymbol{u}_{\hat{\mathcal{I}}} - \boldsymbol{u}_{\mathcal{I}}\| \geqslant \beta \|\boldsymbol{u} - \boldsymbol{u}_{\mathcal{I}}\|, \tag{6.3}$$

where θ and β are related by

$$\theta := \sqrt{1 - \beta^2}. \tag{6.4}$$

Eq. (6.3) is often referred to as *saturation property* and is usually *assumed* rather than inferred. In the context of wavelet schemes such *assumptions* can be avoided.

6.4. Residual estimates

Our goal is to realize an estimate of type (6.3). Again this will rely crucially on (6.1) (and hence on (3.2)) which, in particular means that

$$c_1 c_2^{-1} \|\boldsymbol{A}\boldsymbol{v}\|_{\ell_2(\mathscr{I})} \leqslant \|\boldsymbol{v}\| \leqslant c_2 c_1^{-2} \|\boldsymbol{A}\boldsymbol{v}\|_{\ell_2(\mathscr{I})}. \tag{6.4}$$

In fact, for any $\hat{\mathscr{I}} \supset \mathscr{I}$ one has

$$\begin{aligned}\|\boldsymbol{u}_{\hat{\mathscr{I}}} - \boldsymbol{u}_{\mathscr{I}}\| &\geqslant c_1 c_2^{-2} \|\boldsymbol{A}(\boldsymbol{u}_{\hat{\mathscr{I}}} - \boldsymbol{u}_{\mathscr{I}})\|_{\ell_2(\mathscr{I})} \geqslant c_1 c_2^{-2} \|\boldsymbol{A}(\boldsymbol{u}_{\hat{\mathscr{I}}} - \boldsymbol{u}_{\mathscr{I}})|_{\hat{\mathscr{I}}}\|_{\ell_2(\mathscr{I})} \\ &= c_1 c_2^{-2} \|\boldsymbol{A}(\boldsymbol{u} - \boldsymbol{u}_{\mathscr{I}})|_{\hat{\mathscr{I}}}\|_{\ell_2(\mathscr{I})},\end{aligned}$$

since $(\boldsymbol{A}\boldsymbol{u})|_{\hat{\mathscr{I}}} = (\boldsymbol{A}\boldsymbol{u}_{\hat{\mathscr{I}}})|_{\hat{\mathscr{I}}}$. Thus in terms of the residual

$$\boldsymbol{r}_{\mathscr{I}} := \boldsymbol{A}(\boldsymbol{u} - \boldsymbol{u}_{\mathscr{I}}) = \boldsymbol{f} - \boldsymbol{A}\boldsymbol{u}_{\mathscr{I}},$$

the above estimate (6.4) says that

$$\|\boldsymbol{u}_{\hat{\mathscr{I}}} - \boldsymbol{u}_{\mathscr{I}}\| \geqslant c_1 c_2^{-2} \|\boldsymbol{r}_{\mathscr{I}}|_{\hat{\mathscr{I}}}\|_{\ell_2(\mathscr{I})}. \tag{6.5}$$

Key strategy: *If* $\hat{\mathscr{I}}$ *can be chosen such that*

$$\|\boldsymbol{r}_{\mathscr{I}}|_{\hat{\mathscr{I}}}\|_{\ell_2(\mathscr{I})} \geqslant a \|\boldsymbol{r}_{\mathscr{I}}\|_{\ell_2(\mathscr{I})} \tag{6.6}$$

holds for some fixed $a \in (0,1)$ *then again* (6.4) *combined with* (6.5) *yields a constant* $\beta := a c_1^3 c_2^{-2} \in (0,1)$ *such that* (6.3) *and hence, by Remark* 6.1, *also* (6.2) *holds.*

Thus, the reduction of the error has been reduced to catching the *bulk* of the *residual* $\boldsymbol{r}_{\mathscr{I}}$. This is a principal improvement since the residual involves only *known* quantities like the right-hand side $\boldsymbol{f}$ and the current solution $\boldsymbol{u}_{\mathscr{I}}$. However, in the present setting catching the bulk of the current residual requires knowing *all* coefficients of the infinite sequence $\boldsymbol{r}_{\mathscr{I}}$. Nevertheless, it will make things more transparent to neglect this latter issue for a moment when formulating the following core ingredient of the refinement strategy as the (idealized) routine:

GROW $(\mathscr{I}, \boldsymbol{u}_{\mathscr{I}}) \to (\hat{\mathscr{I}}, \boldsymbol{u}_{\hat{\mathscr{I}}})$
Given $(\mathscr{I}, \boldsymbol{u}_{\mathscr{I}})$ *find the* **smallest** $\hat{\mathscr{I}} \supset \mathscr{I}$ *such that for some fixed* $a \in (0,1)$ *one has* $\|\boldsymbol{r}_{\mathscr{I}}|_{\hat{\mathscr{I}}}\|_{\ell_2(\mathscr{I})} \geqslant a \|\boldsymbol{r}_{\mathscr{I}}\|_{\ell_2(\mathscr{I})}$.

In order to see how much the current index set is enlarged by **GROW** suppose that the error of best N-term approximation to $\boldsymbol{u}$ behaves like N^{-s}. Note that (6.6) is implied by finding a possibly small index set $\hat{\mathscr{I}}$ such that $\|\boldsymbol{r}_{\mathscr{I}}|_{\hat{\mathscr{I}}} - \boldsymbol{r}_{\mathscr{I}}\|_{\ell_2(\mathscr{I})} \leqslant \sqrt{1-a^2} \|\boldsymbol{r}_{\mathscr{I}}\|_{\ell_2(\mathscr{I})}$ which is again a task of best N-term approximation in $\ell_2(\mathscr{I})$, this time for the residual. It follows now from Proposition 4.3 that

$$\#(\hat{\mathscr{I}} \backslash \mathscr{I}) \sim \left(\frac{\|\boldsymbol{r}_{\mathscr{I}}\|_{\ell_\tau^w(\mathscr{I})}}{\|\boldsymbol{r}_{\mathscr{I}}\|_{\ell_2(\mathscr{I})}} \right)^{1/s}. \tag{6.7}$$

Since, by (6.4) $\|\boldsymbol{u} - \boldsymbol{u}_\Lambda\| \sim \|\boldsymbol{r}_\Lambda\|_{\ell_2(\mathscr{I})}$ one would retrieve the right growth $(\mathbf{error})^{-1/s}$ of the degrees of freedom *provided that* $\|\boldsymbol{r}_{\mathscr{I}}\|_{\ell_\tau^w(\mathscr{I})}$ *stays uniformly bounded.* Recall from (4.19) that for s and τ

related by (4.20) $\|\cdot\|_{\ell_\tau^w(\mathcal{J})}$ is a stronger norm than $\|\cdot\|_{\ell_2(\mathcal{J})}$. Nevertheless, it turns out that the uniform boundedness of $\|\boldsymbol{r}_\mathcal{J}\|_{\ell_\tau^w(\mathcal{J})}$ can indeed be enforced by a *clean up* step. This simply means that after several applications of **GROW** one has to discard all coefficients in the current approximation $\boldsymbol{u}_\mathcal{J}$ whose modulus is below a certain threshold. This threshold is chosen so that the current error is at most multiplied by a fixed uniform constant while $\|\boldsymbol{r}_\mathcal{J}\|_{\ell_\tau^w(\mathcal{J})}$ remains bounded. We summarize this clean up or *thresholding step* as follows:

THRESH $(\mathcal{J}, \boldsymbol{u}_\mathcal{J}) \to (\tilde{\mathcal{J}}, \boldsymbol{u}_{\tilde{\mathcal{J}}})$.
If $\|\boldsymbol{u} - \boldsymbol{u}_\mathcal{J}\|_{\ell_2(\mathcal{J})} \leqslant \varepsilon$ *find the* **smallest** $\tilde{\mathcal{J}} \subset \mathcal{J}$ *such that* $\|\boldsymbol{u}_\mathcal{J} - \boldsymbol{u}_\mathcal{J}|_{\tilde{\mathcal{J}}}\|_{\ell_2(\mathcal{J})} \leqslant 4\varepsilon$.

6.5. An optimal (idealized) algorithm

We next give a rough idealized version of the adaptive wavelet scheme from [32].

Algorithm

- $\mathcal{J}_0 = \emptyset$, $\boldsymbol{r}_{\mathcal{J}_0} = \boldsymbol{f}$, $\varepsilon_0 := \|\boldsymbol{f}\|_{\ell_2(\mathcal{J})}$;
- For $j = 0, 1, 2, \ldots$ determine $(\mathcal{J}_{j+1}, \boldsymbol{u}_{\mathcal{J}_{j+1}})$ from $(\mathcal{J}_j, \boldsymbol{u}_{\mathcal{J}_j})$ such that

$$\|\boldsymbol{u} - \boldsymbol{u}_{\mathcal{J}_{j+1}}\|_{\ell_2(\mathcal{J})} \leqslant \varepsilon_j/2 := \varepsilon_{j+1}$$

as follows:
Set $\mathcal{J}_{j,0} := \mathcal{J}_j$, $\boldsymbol{u}_{j,0} := \boldsymbol{u}_j$;
For $k = 1, 2, \ldots, K$ apply
GROW $(\mathcal{J}_{j,k-1}, \boldsymbol{u}_{\mathcal{J}_{j,k-1}}) \to (\mathcal{J}_{j,k}, \boldsymbol{u}_{\mathcal{J}_{j,k}})$;

Apply **THRESH** $(\mathcal{J}_{j,K}, \boldsymbol{u}_{\mathcal{J}_{j,K}}) \to (\mathcal{J}_{j+1}, \boldsymbol{u}_{\mathcal{J}_{j+1}})$.

The maximal number K of applications of **GROW** can be shown to be uniformly bounded depending only on the constants in (6.4).

6.6. Computational tasks

The core task in the above algorithm is the evaluation of the ℓ_2-norm of the dual wavelet coefficients of the residual. Since this is an infinite array one has to resort to suitable approximations. To indicate the actual concrete computational tasks required by a fully computable version of **ALGORITHM** we will assume that one has full information about the right-hand side $\boldsymbol{f}$ so that one can determine for any threshold $\eta > 0$ a finite array $\boldsymbol{f}_\eta$ such that $\|\boldsymbol{f}_\eta - \boldsymbol{f}\|_{\ell_2(\mathcal{J})} \leqslant \eta$. Nevertheless, one faces the following difficulties in evaluating residuals. The Galerkin solutions $\boldsymbol{u}_\mathcal{J}$ cannot be computed exactly but only approximately as a result of finitely many steps of an iterative scheme yielding $\bar{\boldsymbol{u}}_\mathcal{J}$. Neither is it possible to exactly evaluate the application of the infinite matrix $\boldsymbol{A}$ to a finitely supported vector which appears in the residual. Moreover, in order to keep the computational work proportional to the number of unknowns one cannot afford applying the finite matrices $\boldsymbol{A}_\mathcal{J}$ exactly as required in the iterative process since these matrices are not sparse in a strict sense. Therefore in both cases the fast *approximate* matrix/vector multiplication from (5.10) plays a crucial role. In particular, it yields an approximation $\boldsymbol{w}_\eta$ for $\boldsymbol{A}\bar{\boldsymbol{u}}_\mathcal{J}$. This suggests to split the true residual $\boldsymbol{r}_\mathcal{J}$ into an approximate residual $\bar{\boldsymbol{r}}_{\mathscr{I}}$ and a remainder to be controlled in the course of the calculations as follows:

$$\boldsymbol{r}_{\mathscr{I}} = \underbrace{\boldsymbol{f}_\eta - \boldsymbol{w}_\eta}_{\bar{\boldsymbol{r}}_{\mathscr{I}}} + \underbrace{\boldsymbol{f} - \boldsymbol{f}_\eta + \boldsymbol{A}(\bar{\boldsymbol{u}}_{\mathscr{I}} - \boldsymbol{u}_{\mathscr{I}}) + \boldsymbol{w}_\eta - \boldsymbol{A}\bar{\boldsymbol{u}}_{\mathscr{I}}}_{\text{error}}. \tag{6.8}$$

The routine **GROW** then works with the approximate residuals $\bar{\boldsymbol{r}}_{\mathscr{I}} := \boldsymbol{f}_\eta - \boldsymbol{w}_\eta$.

Furthermore, note that thresholding needed, e.g., when approximating the right-hand side as well as when determining the $\boldsymbol{v}_{[j]}$ in the fast matrix/vector multiplication (5.10) requires *sorting* the entries of arrays by size. On the other hand, carrying out the products in (5.10), one has to be able to efficiently switch back to the ordering induced by the index sets, a point to be addressed again below in Section 6.8.

A detailed description of all these algorithmic ingredients is given in [32] leading to a computable version **ALGORITHM**^{c} of the above idealized scheme. It is designed to converge without *any* a priori assumptions on the solution. Its *performance* is analysed for solutions whose best N-term approximation behaves like N^{-s} when $\boldsymbol{A} \in \mathscr{C}_s$. It makes heavy use of concepts from nonlinear approximation as indicated in Section 4.3. The main result reads as follows.

Theorem 6.2 (Cohen et al. [32]). *The computable version* **ALGORITHM**^{c} *always produces a solution with the desired accuracy after a finite number of steps.*

Moreover, assume that $\boldsymbol{A} \in \mathscr{C}_s$ *(e.g. when* $s < s^* := \min\{(\sigma - d/2)/d, 2(\tilde{m}+t)/d\}$ *and* $\boldsymbol{A}$ *satisfies* (5.3). *If the solution u to the operator equation* (4.2) *has the property that for some* $s < s^*$:

$$\sigma_N(u) := \inf_{d_\lambda, \lambda \in \mathscr{I}, \#\mathscr{I} \leqslant N} \left\| u - \sum_{I \in \mathscr{I}} d_I \psi_I \right\| \lesssim N^{-s}, \tag{6.9}$$

then **ALGORITHM**^{c} *generates a sequence* $u_{\mathscr{I}_j}$ *of Galerkin solutions* (4.11) *satisfying*

$$\|u - u_{\mathscr{I}_j}\| \lesssim (\#\mathscr{I}_j)^{-s}. \tag{6.6.3}$$

Furthermore, the number of arithmetic operations *needed to compute* $u_{\mathscr{I}_j}$ *stays* proportional *to* $\#\mathscr{I}_j$ *provided that the entries of* $\boldsymbol{A}$ *can be computed at unit cost. The number of operations for* sorting *stays bounded by a constant multiple of* $(\#\mathscr{I}_j)\log(\#\mathscr{I}_j)$.

The first question in Section 6.2 has now an affirmative answer in that the adaptive scheme realizes within a certain range (depending on $\boldsymbol{A}$ and Ψ) the best possible N-term rate at nearly minimal cost. We also know from Proposition 4.5 under which circumstances error decay rates like (6.9) hold, namely e.g. when $\|\cdot\| \sim \|\cdot\|_{H^t}$ and $u \in B^{sd+t}_\tau(L_\tau(\Omega))$ with τ and s related by (4.20).

This latter observation also leads to an answer to the second question in Section 6.2. Recall that the number of degrees of freedom required by a uniform mesh of mesh size h is $N \sim h^{-d}$. A rate $h^{sd} \sim N^{-s}$ in H^t can only be achieved by such uniform discretizations when the solution belongs to the space $H^{t+sd}(\Omega)$ which is *much smaller* than $B^{sd+t}_\tau(L_\tau(\Omega))$. In fact, by the Sobolev embedding Theorem, (4.20) means that $B^{sd+t}_\tau(L_\tau(\Omega))$ is for all $s > 0$ just embedded in $H^t(\Omega)$ but has no excess Sobolev regularity.

Thus when the solution of (4.2) has a *higher* Besov-regularity in the scale $B^{sd+t}_\tau(L_\tau(\Omega))$ than in the Sobolev scale $H^{sd+t}(\Omega)$ the above adaptive scheme produces an asymptotically *better* error decay in terms of the used unknowns than linear methods. In quantitative terms, even when the solution is pointwise smooth the adaptive scheme is expected to perform better when the norm $\|u\|_{H^{sd+t}}$ is much larger than $\|u\|_{B_\tau^{sd+t}(L_\tau)}$.

6.7. Besov regularity

Therefore the next natural question is, does it occur in the context of elliptic problems that the solution has deficient Sobolev regularity compared with the scale $B_\tau^{sd+t}(L_\tau(\Omega))$? The answer is *yes* as shown, e.g., in [39,40,42]. So there is a scope of problems where the above adaptive scheme would do asymptotically strictly better than linear methods. As an example, let us discuss a typical result in this direction which is concerned with Poisson's equation in a Lipschitz domain $\Omega \subset \mathbb{R}^d$,

$$-\Delta u = f \quad \text{in } \Omega, \quad u|_{\partial\Omega} = 0. \tag{6.10}$$

In this case, $\mathcal{L} = -\Delta$ is an isomorphism from $H_0^1(\Omega)$ onto $H^{-1}(\Omega)$, so that it is natural to consider the best N-term approximation in $H^1(\Omega)$. Based on the investigations in [42], the following theorem was established in [41].

Theorem 6.3. *Let Ω be a bounded Lipschitz domain in $\mathbb{R}^d$. Let u denote the solution of* (6.10) *with $f \in B_2^{\mu-1}(L_2(\Omega))$ for some $\mu \geqslant 1$. Then the following holds*:

$$u \in B_{\tau^*}^{\alpha}(L_{\tau^*}(\Omega)), \quad \frac{1}{\tau^*} = \left(\frac{\alpha-1}{d} + \frac{1}{2}\right), \quad 0 < \alpha < \min\left\{\frac{d}{2(d-1)}, \frac{(\mu+1)}{3}\right\} + 1.$$

It is well known that for domains with reentrant corners the Sobolev regularity of the solution may go down to $\frac{3}{2}$. In such a case Theorem 6.3 says that for data f with $\mu > \frac{1}{2}$ the Besov regularity of u is higher than its Sobolev regularity so that adaptive methods should provider better asymptotic accuracy. Specifically, for polygonal domains in $\mathbb{R}^2$ the Besov regularity may become *arbitrarily high* for correspondingly smooth right-hand side while the Sobolev regularity stays limited independently of f [40].

6.8. Implementation and numerical results

The theoretical analysis outlined above suggests new data structures which, in particular, take the need for *sorting* arrays of wavelet coefficients into account. *Key-based* data structures seem to fit these purposes best. The data are of two types, namely the *key*, for instance the index I and the *value*, i.e., the corresponding wavelet coefficient u_I. Thus every item is a pair (*key*, *value*) and the core structure can be viewed as a mapping from the set of keys to the set of admissible values. The data structures 'map' and 'multimap' from the *Standard Template Library* [105] matches the requirements. A first prototype implementation of an adaptive wavelet solver based on these STL libraries is developed in [7]. There one can find a detailed discussion of first numerical tests for one and two dimensional Poisson type problems. The performance of the scheme confirms for these examples the claimed optimality with a surprisingly tight relation between best N-term approximation and adapted Galerkin solutions.

Recall that in the theoretical analysis it is assumed that the entries of the stiffness matrices can be computed at unit cost. This is realistic for simple problems like constant coefficient differential operators on simple domains. In general the validity of this assumptions will depend on the concrete application and may very well pose serious difficulties. Of course, in principle, the same problem would arise for finite element dicretizations on highly nonuniform meshes when variable coefficients or isoparametric elements require sufficiently accurate quadrature at locations with large mesh cells. For wavelet discretizations adaptive schemes for the efficient computation of matrix and right-hand side entries have been recently developed in [13,9].

6.9. Expanding the range of applicability

Self-adjointness and definiteness of $\mathscr{L}$ is essential for the analysis of the above adaptive scheme. Although differential *and* integral equations are covered this is still a serious limitation. In [44,79] the ideas of [43] have been extended to *saddle point problems* arising, for instance, from the weak formulation of the Stokes problem. It is well known that for such problems stable discretizations have to satisfy the *Ladychenskaja–Babuška–Brezzi condition* (LBB condition). For uniformly refined discretizations wavelet concepts can be used to construct pairs of trial spaces for velocity and pressure that satisfy the LBB condition for any order of accuracy, see [55,88]. In contrast the treatment of adaptively refined index sets is more complicated [44,79,94]. Nevertheless, an adaptive wavelet scheme for saddle point problems based on an *Uzawa technique* is shown in [44,79] to converge in the spirit of [43], without establishing, however, convergence rates.

An alternative strategy is to consider *least-squares formulations* of operator equations. The possibility of using adaptive wavelet techniques for certain least-squares formulations of *first-order elliptic systems* has been indicated in [79]. In the finite element context least-squares formulations are currently a very active field of research, see e.g. [20]. An important issue in this context is to identify appropriate least-squares functionals which, in particular, do not pre-impose too strong regularity properties of the solution. The problem is that usually these functionals involve Sobolev norms of negative or fractional index which are hard to evaluate numerically. The techniques to overcome these difficulties are rather sophisticated and problem dependent.

It has recently been shown in [54] that wavelet concepts offer very promising perspectives in the following respects: (a) A large class of problems, again permitting mixtures of integral and differential operators, can be treated in a fairly unified fashion. (b) The norm equivalences (3.3) facilitate an efficient numerical evaluation of 'difficult' norms and lead to well-conditioned discrete problems. Specifically, norm equivalences of type (3.3) facilitate the approximate evaluation of Sobolev norms for negative or noninteger regularity indices. This fact has been already exploited in [43,32] in connection with adaptive schemes as well as in [12] for the purpose of stabilizing the numerical treatment of semidefinite problems. A variety of problems is covered by the setting in [54], e.g. elliptic boundary value problems, in particular, *first-order mixed formulations* of second-order elliptic boundary value problems or the following *transmission problem*. Let Ω be the *exterior* of some bounded domain in $\mathbb{R}^d$ with piecewise smooth boundary Γ_D and let Ω_0 denote some annular domain with inner boundary Γ_D and piecewise smooth outer boundary Γ. Consider

$$
\begin{aligned}
&-\nabla\cdot(\boldsymbol{a}\nabla u)=f \quad \text{in } \Omega_0,\\
&-\Delta u=0 \quad \text{in } \Omega_1:=\Omega\backslash\Omega_0,\\
&u|_{\Gamma_D}=0, \quad \partial_n u|_{\Gamma_N}=0
\end{aligned}
\tag{6.11}
$$

with transmission conditions $u^- = u^+$, $\partial_n u^- = \partial_n u^+$ on the interface Γ, where $+$, respectively — denote the limits from Ω_1, respectively, Ω_0. The corresponding weak formulation reads: find $U = (u,\sigma) \in H^1_{0,\Gamma_D}(\Omega_0) \times H^{-1/2}(\Gamma)$ such that

$$\begin{aligned} &\langle \boldsymbol{a}\nabla u, \nabla v\rangle_{L_2(\Omega_0)} + \langle \mathscr{W}u - (\tfrac{1}{2}\mathrm{id} - \mathscr{K}')\sigma, v\rangle_{L_2(\Gamma)} = \langle f, v\rangle_{L_2(\Omega_0)}, \quad v \in H^1_{0,\Gamma_D}(L_2(\Omega_0)),\\ &\langle (\tfrac{1}{2}\mathrm{id} - \mathscr{K})u, \delta\rangle_{L_2(\Gamma)} + \langle \mathscr{V}\sigma, \delta\rangle_{L_2(\Gamma)} = 0, \quad \delta \in H^{-1/2}(\Gamma), \end{aligned} \tag{6.12}$$

where the singular integral operators $\mathscr{K}, \mathscr{V}, \mathscr{W}$ have been defined in Section 4.1.

Another example is the *Stokes problem* with *inhomogeneous* Dirichlet boundary conditions

$$-\nu\Delta \boldsymbol{u} + \nabla p = \boldsymbol{f}, \quad \mathrm{div}\, \boldsymbol{u} = 0 \quad \text{in } \Omega, \quad \boldsymbol{u}|_\Gamma = \boldsymbol{g}, \tag{6.13}$$

where as usual $p \in L_{2,0}(\Omega) := \{q \in L_2(\Omega)\colon \int_\Omega q\,\mathrm{d}x = 0\}$. To describe the weak formulation note that for $\boldsymbol{v} \in \boldsymbol{H}^1(\Omega) := H^1(\Omega)^d$ the gradient is now a $d \times d$ matrix $\nabla\boldsymbol{v} := (\nabla v_1, \dots, \nabla v_d)$. Moreover, for any such $d \times d$ matrix valued functions $\underline{\boldsymbol{\theta}}, \underline{\boldsymbol{\eta}}$ whose columns are denoted by $\boldsymbol{\theta}_i$, $\boldsymbol{\eta}_i$ we define $\langle \underline{\boldsymbol{\theta}}, \underline{\boldsymbol{\eta}}\rangle := \sum_{i=1}^d \langle \boldsymbol{\theta}_i, \boldsymbol{\eta}_i\rangle = \sum_{i=1}^d \int_\Omega \boldsymbol{\theta}_i \cdot \boldsymbol{\eta}_i \,\mathrm{d}x$. Again the weak formulation requires finding $U = (\boldsymbol{u}, \boldsymbol{\lambda}, p) \in \boldsymbol{H}^1(\Omega) \times \boldsymbol{H}^{-1/2}(\Gamma) \times L_{2,0}(\Omega)$, such that

$$\begin{aligned} &\nu\langle \nabla\boldsymbol{v}, \nabla\boldsymbol{u}\rangle_{L_2(\Omega)} + \langle \boldsymbol{v}, \boldsymbol{\lambda}\rangle_{L_2(\Gamma)} + \langle \mathrm{div}\,\boldsymbol{v}, p\rangle_{L_2(\Omega)} = \langle \boldsymbol{f}, \boldsymbol{v}\rangle \quad \forall \boldsymbol{v} \in \boldsymbol{H}^1(\Omega),\\ &\langle \boldsymbol{u}, \boldsymbol{\mu}\rangle_{L_2(\Gamma)} = \langle \boldsymbol{g}, \boldsymbol{\mu}\rangle \quad \forall \boldsymbol{\mu} \in \boldsymbol{H}^{-1/2}(\Gamma),\\ &\langle \mathrm{div}\,\boldsymbol{u}, q\rangle_{L_2(\Omega)} = \boldsymbol{0} \quad \forall q \in L_{2,0}(\Omega). \end{aligned} \tag{6.14}$$

Similarly one could again consider first-order formulations of the Stokes problem [54,22].

These examples all have the common format of a (typically weakly defined) *system*

$$\mathscr{L}U = F \tag{6.15}$$

of operator equations where $\mathscr{L} = (\mathscr{L}_{i,l})_{i,l=1}^n$ defines a mapping from some product space $\mathscr{H} = H_{1,0} \times \cdots \times H_{n,0}$ into its dual $\mathscr{H}'$. In all the above cases the $H_{i,0}$ are closed subspaces of some Sobolev space $H_{i,0} = H_{i,0} \cap H^{s_i}(\Omega_i)$ of possibly negative or fractional order s_i with respect to some domain Ω_i that could be a bounded domain in Euclidean space or a boundary manifold. The point stressed in [54] is that once it is shown that $\mathscr{L}$ is actually an *isomorphism*, i.e.,

$$\|\mathscr{L}V\|_{\mathscr{H}'} \sim \|V\|_{\mathscr{H}}, \tag{6.16}$$

(recall (4.1)) then an appropriate equivalent least-squares formulation of (6.15) reads

$$\mathrm{LS}(U) := \sum_{i=1}^m \|\mathscr{L}_i U - f_i\|^2_{H'_{i,0}} \to \min. \tag{6.17}$$

In the above examples (6.16) holds indeed for $\mathscr{H} = H^1_{0,\Gamma_D}(\Omega_0) \times H^{-1/2}(\Gamma)$ and $\mathscr{H} := \boldsymbol{H}^1(\Omega) \times \boldsymbol{H}^{-1/2}(\Gamma) \times L_{2,0}(\Omega)$, see e.g. [54]. Now whenever norm equivalences of the form (3.3) for the individual component spaces are available the inner products inducing the dual norms $\|\cdot\|_{H'_{i,0}}$ can be replaced by equivalent inner products

$$\langle v, w\rangle_i := \sum_{I \in \mathscr{J}_i} (\boldsymbol{D}_i)_{I,I}^{-2} \langle v, \psi_I^i\rangle \langle \psi_I^i, w\rangle = \langle v, \Psi^i\rangle \boldsymbol{D}_i^{-2} \langle \Psi^i, w\rangle, \tag{6.18}$$

see also [12] for related stabilization strategies. Thus defining

$$Q(V,W) := \sum_{i=1}^n \langle \mathscr{L}_i V, \mathscr{L}_i W\rangle_i, \quad F(V) := \sum_{i=1}^m \langle \mathscr{L}_i V, f_i\rangle_i, \tag{6.19}$$

standard variational arguments say that U solves (6.17) (and hence (6.15)) if and only if

$$Q(U,V)=F(V), \quad V\in\mathscr{H}. \tag{6.20}$$

In wavelet coordinates with a proper scaling based again on the norm equivalences (3.2), (6.20) in turn is equivalent to

$$\boldsymbol{Q}\boldsymbol{U}=\boldsymbol{A}^*\boldsymbol{F}, \tag{6.21}$$

where $\boldsymbol{Q}=\boldsymbol{A}^*\boldsymbol{A}$, $\boldsymbol{A}:=(\boldsymbol{A}^{i,l})_{i,l=1}^m$, $\boldsymbol{A}^{i,l}:=\boldsymbol{D}_i^{-1}A_{i,l}(\Psi^i,\Psi^l)\boldsymbol{D}_l^{-1}$ with $A_{i,l}(w,v)=\langle w,\mathscr{L}_{i,l}v\rangle$. Moreover, system (6.21) is well-posed in $\ell_2(\mathscr{J})$, i.e.,

$$\operatorname{cond}_2(\boldsymbol{Q})\leqslant\frac{C_{\mathscr{L}}^2C_1^4}{c_{\mathscr{L}}^2c_1^4}, \tag{6.22}$$

see [54]. In all examples under consideration the matrices $\boldsymbol{A}_{i,l}$ can be shown to be compressible in the sense of (5.8). Hence the wavelet representation $\boldsymbol{A}$ of $\mathscr{L}$ is compressible. Moreover, it is shown in [54] that certain *finite sections* of $\boldsymbol{A}$ lead to uniformly well-conditioned finite-dimensional problems whose solutions give rise to asymptotically optimal error bounds.

Thus, once having established the equivalence of the original variational problem with an operator equation (6.15) satisfying (6.16), the wavelet framework allows one to transform this problem into a well posed symmetric positive-definite problem (6.21) on ℓ_2. Now the adaptive concepts from Section 6 can be applied again. Once it is shown that the accurate application of $\boldsymbol{Q}$ can be realized with comparable complexity as in the elliptic case one obtains the same optimal convergence results. The precise formulation of such strategies is in preparation.

Fictitious domains: Note that in the above example of the Stokes problem inhomogeneous essential boundary conditions have been incorporated in the variational formulation. One can show that (6.16) still holds when replacing the domain Ω in (6.14) by a larger simple domain $\square$ such as a cube while Γ is still the boundary of the physical domain Ω. This suggests combining the above wavelet least-squares formulation with a *fictitious domain* method. This is also a recurring theme in [54,84]. In particular, this permits employing simple and efficient wavelet bases on $\square$ and, e.g., for $d=2$ periodic wavelets on the boundary Γ. Such an approach is particularly promising when the boundary changes or when boundary values vary, for instance, when they serve as control parameters in optimal control problems. A detailed investigation of this direction is given in [84] where optimal control problems are transformed into format (6.15), (6.16) with boundary conditions incorporated in a variational formulation as in (6.14).

7. Nonlinear problems

Now consider problems of the form

$$\partial_t u+\mathscr{L}u+\mathscr{N}(u)=g \tag{7.1}$$

with suitable initial and boundary conditions where $\mathscr{N}(u)$ denotes a *nonlinear* operator of lower order while as before $\mathscr{L}$ is a dominant elliptic operator. First, adaptive wavelet schemes for such problems have been proposed in [92]. A somewhat different approach based on semigroup representations of the solution to (7.1) has been pursued in [16]. Using the compressibility of operators in wavelet coordinates, specifically in terms of the nonstandard representation from Section 5.3, as well as the potential near sparsity of wavelet coefficient sequences, a first key strategy is to evaluate quantities like $e^{-(t-t_0)\mathscr{L}}u_0$ within any desired tolerance of accuracy. Related truncation arguments have to be based, of course, on a proper choice of threshold parameters. It seems to remain open though how to relate these threshold parameters to the current number of significant coefficients in order to realize asymptotically optimal accuracy for a given norm. Thus, a rigorous asymptotic error analysis similar to the one outlined in Section 6 is yet to be provided.

The second issue concerns the treatment of the *nonlinear term*. For simplicity, let us even assume that $\mathscr{N}(u) = f(u)$ where f is a *smooth* but *nonlinear* function. Thus a Galerkin discretization of (7.1) requires evaluating quantities of the form $\langle f(u_{\mathscr{I}}), \psi_I\rangle$, $I \in \mathscr{I}'$, where $\mathscr{I}$, $\mathscr{I}'$ are possibly different index sets. In a finite element context the computation of such quantities reduces to local quadrature and poses no difficulty at the first glance. Of course, when dealing with highly nonuniform adaptive meshes the work required for this quadrature may no longer be always of order one when preserving the overall desired accuracy. This difficulty is very explicit when dealing with such nonlinear expressions of *multiscale expansions* $u_{\mathscr{I}} = \boldsymbol{d}_{\mathscr{I}}^{\mathrm{T}}\Psi_{\mathscr{I}}$. There are two strategies that may come to mind first: (i) Of course, it would be easy to evaluate single scale coefficients $\langle f(u_{\mathscr{I}}), \phi_{J,k}\rangle$, where $J = J(\mathscr{I}) = \max\{|I| + 1 \colon I \in \mathscr{I}\}$ is the highest scale occurring in $\mathscr{I}$. However, when $\mathscr{I}$ is very *lacunary*, i.e., $\#\mathscr{I} \ll \dim S(\Phi_J)$ a transformation of $u_{\mathscr{I}}$ into single-scale representation would completely waste the complexity reduction gained by the sparse approximation in $S_{\mathscr{I}}$. (ii) On the other hand, a naive application of quadrature to the quantities $\langle f(u_{\mathscr{I}}), \psi_I\rangle$ would also severely spoil complexity gains because some of the quadrature domains are comparable to the whole domain so that sufficient accuracy would require a computational expense of the order of the size of the problem. So there seems to be a serious bottleneck jeopardizing the advantages of wavelet concepts gained in other respects.

This problem has long been recognized and discussed in several papers, see e.g. [16,74,89,92]. A key idea is that the mixing of different scales caused by the nonlinear mapping stays limited. To make such statements quantitative, however, requires again truncation arguments based on a proper choice of threshold parameters and certain regularity assumptions. Both issues pose difficulties. The threshold parameters have to be dynamically adapted to the increasing number of degrees of freedom required by an overall increasing accuracy. As for the regularity assumptions, local Sobolev norms are often used which, by the discussion in Section 6.6 is not quite appropriate since highly lacunary index sets are only expected to pay for deficient Sobolev regularity.

A first analysis of the problem of evaluating nonlinear functionals of multiscale expansions along these lines was given in [65] leading to a concrete algorithm which could be shown to be optimal in a sense to be explained below. The first observation is that for any dual pair Ψ and $\tilde{\Psi}$ of wavelet bases the searched quantities $\langle f(u_{\mathscr{I}}), \psi_I\rangle$ are just the expansion coefficients of $f(u_{\mathscr{I}})$ with respect to the *dual basis* $f(u_{\mathscr{I}}) = \langle f(u_{\mathscr{I}}), \Psi\rangle\tilde{\Psi}$. The second point is to exploit again a norm equivalence of type (3.2) but this time for $\boldsymbol{D} = \mathrm{id}$ and $H = L_2(\Omega)$. In fact, whenever $\boldsymbol{f}^{\mathrm{T}}\tilde{\Psi}$ is an approximation to $f(u_{\mathscr{I}})$, (3.2) implies

$$\|\boldsymbol{f}^{\mathrm{T}}\tilde{\Psi} - f(u_{\mathscr{I}})\|_{L_2} \leqslant \varepsilon \Rightarrow \|\boldsymbol{f}^{\mathrm{T}} - \langle f(u_{\mathscr{I}}), \Psi\rangle\|_{\ell_2} \lesssim \varepsilon. \tag{7.2}$$

The idea is then, instead of computing the individual inner products $\langle f(u_{\mathscr{I}}), \psi_I\rangle$, to find a good approximation to the whole function $f(u_{\mathscr{I}})$ with respect to the *dual system* and use its wavelet coefficients as correspondingly good approximations to the searched quantities. This approximation, in turn, has to be efficient in the sense that the computational work needed for its realization should remain (asymptotically) proportional to the number of significant coefficients of $f(u_{\mathscr{I}})$ for a given accuracy.

The second important ingredient is the notion of *tree approximation*. In contrast to the treatment of linear elliptic systems it is important in the present context to impose a *tree structure* on the index sets $\mathscr{I}$. Here this means that when $I \in \mathscr{I}$ then $\operatorname{supp}\psi_{I'} \cap \operatorname{supp}\psi_I \neq \emptyset$ implies $I' \in \mathscr{I}$. This constraint on the index sets turns out not to be too severe. In fact, it can be shown that when $u \in B_q^s(L_\eta)$ with $1/\eta < s/d + 1/p$ then up to a uniform constant the error of best N-term approximation in L_p which behaves like N^{-s} can be realized by wavelet expansions with tree-like index sets $\mathscr{I}$, [33,65,79]. Thus, roughly speaking, as soon as a function belongs to a Besov space just strictly left of the Sobolev embedding line (3.12) significant wavelet coefficients start to gather in tree-like index sets whose cardinality, of course, can be much smaller than that of corresponding fully refined sets determined by the highest occurring frequency. The main result in [65] says that when u belongs to such a Besov space $B_q^s(L_\eta)$ and a best tree approximation $u_{\mathscr{I}} = \mathbf{d}_{\mathscr{I}}^{\mathrm{T}} \Psi_{\mathscr{I}}$ satisfies $\|u - u_{\mathscr{I}}\|_{L_2} \lesssim N^{-s}$, where $\mathscr{I}$ is a tree-like index set of cardinality N, then one can determine a somewhat expanded tree-like set $\hat{\mathscr{I}}$ still with $\#\hat{\mathscr{I}} \lesssim N$ and an approximation $A_{\hat{\mathscr{I}}}(f(u_{\mathscr{I}}))$ to $f(u_{\mathscr{I}})$ at the expense of $\mathcal{O}(N)$ operations such that

$$\|A_{\hat{\mathscr{I}}}(f(u_{\mathscr{I}})) - f(u_{\mathscr{I}})\|_{L_2} \lesssim N^{-s}. \tag{7.3}$$

The resulting algorithm consists of two parts. First, denoting by $\partial\mathscr{I}$ the set of *leaves* in the tree $\mathscr{I}$, i.e., the elements of $\partial\mathscr{I}$ have no children in $\mathscr{I}$, the approximations to $f(u_{\mathscr{I}})$ are given in the form

$$A_{\hat{\mathscr{I}}}(f(u_{\mathscr{I}})) = \sum_{j=j_0}^{J(\hat{\mathscr{I}})} \sum_{k \in \Gamma_j(\mathscr{I})} c_{j,k}\tilde{\phi}_{j,k},$$

where $\Gamma_j(\mathscr{I})$ contains those indices k for which $\operatorname{supp}\tilde{\phi}_{j,k}$ overlaps the support of a wavelet ψ_I, $|I| = j - 1$, $I \in \partial\hat{\mathscr{I}}$. The coefficients $c_{j,k}$ result from *quasi-interpolation* and best local polynomial reconstructions.

Once such approximations are obtained, in a second step *localized* versions of the multiscale transformations of type (2.15),(2.16) are applied to transform $A_{\hat{\mathscr{I}}}(f(u_{\mathscr{I}}))$ into the form

$$A_{\hat{\mathscr{I}}}(f(u_{\mathscr{I}})) = \sum_{I \in \hat{\mathscr{I}}} d_I \tilde{\psi}_I$$

at the expense of $\mathcal{O}(\#\mathscr{I})$ operations. The coefficients d_I are the searched approximations to the inner products $\langle f(u_{\mathscr{I}}), \psi_I \rangle$ whose accuracy is ensured by (7.3) and (7.2). Thus, in summary, the complexity of the problem remains proportional to the number of significant coefficients. The above concepts have been implemented. Tests of the resulting algorithm presented in [65] confirm and illustrate the theoretical analysis. For several extensions to estimates with respect to L_p norms see [65].

8. About the tools — some basic concepts

So far we have merely *assumed* that wavelet bases with certain properties are available and have derived some consequences of these properties. It is clear that the main features such as locality, norm equivalences and cancellation properties may not be easily realized for nontrivial domain geometries.

There are two basic strategies to cope with such domains. The first one is to employ a *fictitious domain* such as a cube and use periodic wavelets for an appropriately extended problem. Periodized wavelets immediately inherit all the nice properties of classical wavelets on $\mathbb{R}^d$ constructed with the aid of Fourier techniques. Essential boundary conditions can be treated in various ways. For problems with constant coefficients they can be corrected via an integral equation [5]. They can also be enforced approximately by appending them as a *penalty* term to the variational formulation, see [101] and the literature cited there. The problem then is to restore the condition of the resulting discrete systems. A further alternative is to append boundary conditions by Lagrange multipliers. This leads to a saddle point problem so that definiteness is lost. So one has to make sure that the discretizations of the Lagrange multipliers and those on the domain satisfy the *LBB condition.* Moreover, one has to come up with suitable preconditioners. Corresponding asymptotically optimal schemes are developed in [83,53] for the case that the discretizations on the boundary and the location of the boundary in the fictitious domain $\square$ are *independent* of the traces of the trial spaces on $\square$. This allows one to take best possible advantage of simple periodic wavelet bases for the bulk of degrees of freedom which stay fixed even when the boundary moves or the boundary values vary. This is, for instance, important in optimal control for boundary value problems where boundary values act as control parameters, see [84] for the development of wavelet methods for such problems. To avoid possible constraints imposed by the LBB condition one can also use least-squares formulations for incorporating boundary conditions as mentioned at the end of Section 6.9, see also [54]. Note that in all these variants for $d = 2$ the discretization of the Lagrange multipliers living on the boundary involves again *periodic* univariate wavelets. Hence in this case all ingredients are conveniently based on efficient periodic tools.

Of course, for $d > 2$ the boundary is, in general, no longer a torus so that at some point one faces the need for wavelets defined on closed (sphere-like) surfaces as well. In fact, browsing through the examples discussed so far, it would be desirable to have wavelet bases on bounded domains in $\mathbb{R}^d$ as well as on manifolds such as closed surfaces. In this regard, two major issues arise. First, one has to develop suitable criteria for the realization of norm equivalences and cancellation properties. Second, constructive concepts are needed that meet these criteria. We will briefly sketch some ideas and developments in this direction.

8.1. Fourier-free concepts for norm equivalences

Given complement bases Ψ_j it is typically not so difficult to show that they are *uniformly H-stable* in the sense that

$$\|\boldsymbol{d}_j\|_{\ell_2} \sim \|\boldsymbol{d}_j^{\mathrm{T}}\Psi_j\|_H \tag{8.1}$$

holds uniformly in j. Unless the W_j are orthogonal complements (which will typically not be the case) it is usually more difficult to ensure *stability over all levels* which is $\|\boldsymbol{d}\|_{\ell_2} \sim \|\boldsymbol{d}^{\mathrm{T}}\Psi\|_H$. (Here Ψ plays the role of the scaled basis $\boldsymbol{D}^{-1}\Psi$ in (3.2)). Since (8.1) means $\|(Q_{j+1} - Q_j)v\|_H^2 \sim \sum_{|I|=j} |d_I|^2$ it suffices to confirm that

$$\|v\|_H \sim \left(\sum_{j=0}^{\infty} \|(Q_j - Q_{j-1})v\|_H^2 \right)^{1/2}, \tag{8.2}$$

recall (2.7) in Section 2.4.

We adhere to this 'basis-free' formulation in order to stress that the *type* of splitting of the *spaces* S_j rather than the choice of the particular bases will be seen to matter, recall the discussion at the end of Section 4.2. Of course, when the Q_j are orthogonal projectors (8.2) even holds with '$\sim$' replaced by '$=$'. In general, orthogonal decompositions will not be available and then the validity of (8.2) is not clear.

Criteria for the validity of (8.2) in a general Hilbert space context have been derived in [48], see [47] for other function spaces. Roughly speaking, the spaces S_j must have some quantifiable *regularity* and *approximation properties* stated in terms of *inverse* and *direct estimates*.

As pointed out in Section 4.2 the case that H is the energy space related to (4.2) is of primary interest. Since in many classical examples the energy space agrees with a (closed subspace of a) Sobolev space the following specialization is important [48].

Theorem 8.1. *Assume that $S_j \subset H^s(\Omega)$ for $s < \gamma$. Let $\mathcal{Q}$ be any sequence of projectors onto $\mathcal{S}$ which are bounded in $H^s(\Omega)$ for $s < \gamma$ and satisfy the commutator property* (4.2). *Moreover assume that the* Jackson *or* direct *estimate*

$$\inf_{v_j \in V_j} \|v - v_j\|_{L_2(\Omega)} \lesssim 2^{-mj} \|v\|_{H^{m'}(\Omega)}, \quad v \in H^{m'}(\Omega), \tag{8.3}$$

as well as the Bernstein *or* inverse *estimate*

$$\|v_j\|_{H^s(\Omega)} \lesssim 2^{sj} \|v_j\|_{L_2(\Omega)}, \quad v_j \in V_j, \; s < \gamma' \tag{8.4}$$

hold for both $V_j = S_j$ and $V_j = \tilde{S}_j := \mathrm{range}\, Q_j'$ with $m' = m > \gamma' = \gamma$ and $m' = \tilde{m} > \gamma' = \tilde{\gamma} > 0$, respectively. Then one has

$$\|v\|_{H^s(\Omega)} \sim \left(\sum_{j=0}^{\infty} 2^{2sj} \|(Q_j - Q_{j-1})v\|^2_{L_2(\Omega)} \right)^{1/2}, \quad -\tilde{\gamma} < s < \gamma, \tag{8.5}$$

where for $s < 0$ it is understood that $H^s(\Omega) = (H^{-s}(\Omega))'$.

Note that standard interpolation arguments provide estimates of the type (8.3) and (8.4) for $0 < s < \gamma$. This implies that $\|(Q_j - Q_{j-1})v\|^2_{H^s(\Omega)} \sim 2^{2sj} \|(Q_j - Q_{j-1})v\|^2_{L_2(\Omega)}$ for $0 < s < \gamma$ so that (8.5) takes indeed the form (8.2) for $H = H^s(\Omega)$. Thus, such norm equivalences are typically obtained for a *whole range* of function spaces.

Discrete norms of the above type and their equivalence to norms like (3.10) (with $q = p = 2$) are familiar in the theory of function spaces as long as the Sobolev indices s are *positive*, see e.g. [47,71,77,96,97,114]. The essential difficulty is to include $s = 0$, i.e., $H^0(\Omega) = L_2(\Omega)$ or even *negative* indices. This expresses itself through conditions on *both* the primal multi-resolution $\mathcal{S}$ *and* on the dual multi-resolution $\tilde{\mathcal{S}}$. In particular, choosing $\mathcal{Q}$ fixes $\tilde{\mathcal{S}}$. We will encounter later cases where $\tilde{\mathcal{S}}$ is explicitly constructed. However, in other situations of interest this is not easily possible. It is therefore important to reduce the need for explicit a priori knowledge about $\mathcal{Q}$ and $\mathcal{Q}'$ as much as possible. The following result reflects such attempts.

Theorem 8.2 (Dahmen and Stevenson [66]). *Let $\mathscr{S}$ and $\tilde{\mathscr{S}}$ be two given multi-resolution sequences satisfying* $\dim S_j = \dim \tilde{S}_j$ *and*

$$\inf_{v_j \in S_j} \sup_{\tilde{v}_j \in \tilde{S}_j} \frac{|\langle v_j, \tilde{v}_j\rangle|}{\|v_j\|_{L_2(\Omega)}\|\tilde{v}_j\|_{L_2(\Omega)}} \gtrsim 1. \tag{8.6}$$

Then there exists a sequence $\mathscr{Q}$ of uniformly L_2-bounded projectors with ranges $\mathscr{S}$ such that $\mathrm{range}(\mathrm{id} - Q_j) = (\tilde{S}_j)^{\perp_{L_2}}$ *and likewise for the dual projectors $\mathscr{Q}'$ one has* $\mathrm{range}(\mathrm{id} - Q_j') = (S_j)^{\perp_{L_2}}$. *Moreover, $\mathscr{Q}$ satisfies the commutator property* (4.2).

Furthermore, if $\mathscr{S}$ and $\tilde{\mathscr{S}}$ satisfy (8.3) *and* (8.4) *with respective parameters* $m, \tilde{m} \in \mathbb{N}$ and $\gamma, \tilde{\gamma} > 0$ *then one has for any* $w_j \in \mathrm{range}(Q_j - Q_{j-1})$

$$\left\|\sum_{j=0}^{\infty} w_j\right\|_{H^s(\Omega)}^2 \lesssim \sum_{j=0}^{\infty} 2^{2sj}\|w_j\|_{L_2(\Omega)}^2, \quad s \in (-\tilde{m}, \gamma) \tag{8.7}$$

and

$$\sum_{j=0}^{\infty} 2^{2sj}\|(Q_j - Q_{j-1})v\|_{L_2(\Omega)}^2 \lesssim \|v\|_{H^s(\Omega)}^2, \quad s \in (-\tilde{\gamma}, m). \tag{8.8}$$

Thus for $s \in (-\tilde{\gamma}, \gamma)$, $v \to \{(Q_j - Q_{j-1})v\}_j$ is a bounded mapping from H^s onto $\ell_2(\mathscr{Q}) := \{\{w_j\}_j : w_j \in$ $\mathrm{range}(Q_j - Q_{j-1})$, $\|\{w_j\}_j\|_{\ell_{2,s}(\mathscr{Q})} := (\sum_{j=0}^{\infty} 2^{2sj}\|w_j\|_{L_2(\Omega)}^2)^{1/2} < \infty\}$, *with bounded inverse* $\{w_j\}_j \to \sum_{j=0}^{\infty} w_j$, *i.e., for $s \in (-\tilde{\gamma}, \gamma)$* (8.7) *and* (8.8) *hold with '$\lesssim$' replaced by '$\sim$'.*

We have assumed above that (as in all practical cases) the spaces S_j are finite dimensional and that $\dim S_j = \dim \tilde{S}_j$. In order to cover classical multi-resolution on $\mathbb{R}^d$ one can require in addition the validity of (8.6) with interchanged roles of S_j and $\tilde{S}_j$. Theorem 8.2 will be seen to be tailored to the construction of finite element-based wavelets [66].

If one is only interested in norm equivalences (8.5) for *positive* $s < \gamma$ estimates (8.4) and (8.3) are only required for $\mathscr{S}$. In this case the right-hand side of (8.5) can be shown to be equivalent to norms of type (3.10) [52,71,96].

Of course, typical applications do not concern *full* Sobolev spaces but closed subspaces H^s. To this end, recall that $H^s_{0,\Gamma_D}(\Omega)$ denotes the closure in the H^s-norm of smooth functions vanishing on Γ_D. Then for some $r \in \mathbb{N}$ one typically has $H^s := H^s(\Omega)$ for $0 \leqslant s < r$ and $H^s := H^r_{0,\Gamma_D}(\Omega) \cap H^s(\Omega)$ for $s \geqslant r$. Analogous norm equivalences can be established in those cases as well for a proper interpretation of H^{-s}, $s > 0$, see [64] for details.

In summary, once estimates of the type (8.5) have been obtained, norm equivalences of the form (3.2) follow for the complement basis functions satisfying (4.1) provided that the complement bases are uniformly stable for each level. This latter fact, in turn can be confirmed by criteria that will be explained next in Section 8.2.

8.2. Construction principles — stable completions

It remains to identify suitable complements W_j or equivalently appropriate projectors Q_j for a given multi-resolution sequence $\mathscr{S}$. We outline next a general concept that is again flexible enough to work in absence of Fourier techniques for realistic domain geometries. In a special setting this was presented first in [45] and in full generality in combination with stability concepts in [26]. The central ingredient is called there *stable completion*. In a slightly restricted form it was independently developed in [108,109] under the name of *lifting scheme* for *second generation wavelets*. The key is to identify the search for a complement basis with the completion of a rectangular matrix to a nonsingular square matrix. We will sketch the development in [26].

Not every complement W_j, or equivalently, not every stable completion $\boldsymbol{M}_{j,1}$ of $\boldsymbol{M}_{j,0}$ gives rise to a Riesz basis. Therefore we will proceed in two steps:

(I) Find *some initial* stable completion.
(II) Modify the initial stable completion so as to realize desirable properties such as the Riesz basis property or other features to be identified later.

The first point is that in practical situations some stable completion is often easy to find. In fact, returning to our example of piecewise linear finite elements on uniformly refined triangulations, Φ_j consists of the classical Courant hat functions. Identifying the indices $k \in \mathscr{I}_j$ with the vertices of the jth level triangulation an initial complement basis is given by the *hierarchical basis*

$$\check{\Psi}_j := \{\check{\psi}_{j,k} := \phi_{j+1,k}\colon k \in \mathscr{I}_{j+1} \backslash \mathscr{I}_j\} \tag{8.9}$$

[115]. The interpolation properties and the locality of the hat functions readily allows us to identify the entries of $\boldsymbol{M}_{j,1}$ and also of $\boldsymbol{G}_j$. Both $\boldsymbol{M}_j$ and $\boldsymbol{G}_j$ are uniformly sparse and the entries remain uniformly bounded which implies the validity of (2.13). Thus the hierarchical complement bases form a stable completion. They do *not* give rise to a Riesz basis in L_2 though.

The second point is that once an initial stable completion $\check{\boldsymbol{M}}_{j,1}$ $(\check{\boldsymbol{M}}_j := (\boldsymbol{M}_{j,0}, \check{\boldsymbol{M}}_{j,1}))$ along with the inverses $\check{\boldsymbol{G}}_j$ has been identified *all others* can be obtained as follows. Note that for any $(\#\Phi_j \times \#\Psi_j)$-matrix $\boldsymbol{L}_j$ and any invertible $(\#\Psi_j \times \#\Psi_j)$-matrix $\boldsymbol{K}_j$ one has

$$\mathrm{id} = \check{\boldsymbol{M}}_j \check{\boldsymbol{G}}_j = \check{\boldsymbol{M}}_j \begin{pmatrix} \mathrm{id} & \boldsymbol{L}_j \\ \boldsymbol{0} & \boldsymbol{K}_j \end{pmatrix} \begin{pmatrix} \mathrm{id} & -\boldsymbol{L}_j \boldsymbol{K}_j^{-1} \\ \boldsymbol{0} & \boldsymbol{K}_j^{-1} \end{pmatrix} \check{\boldsymbol{G}}_j =: \boldsymbol{M}_j \boldsymbol{G}_j. \tag{8.10}$$

which means

$$\boldsymbol{M}_{j,1} = \boldsymbol{M}_{j,0} \boldsymbol{L}_j + \check{\boldsymbol{M}}_{j,1} \boldsymbol{K}_j, \quad \boldsymbol{G}_{j,0} = \check{\boldsymbol{G}}_{j,0} - \boldsymbol{L}_j \boldsymbol{K}_j^{-1} \boldsymbol{G}_{j,1}, \quad \boldsymbol{G}_{j,1} = \boldsymbol{K}_j^{-1} \check{\boldsymbol{G}}_{j,1}. \tag{8.11}$$

The case $\boldsymbol{K}_j = \mathrm{id}$ corresponds to the lifting scheme [108,109]. Obviously, again by (2.13), the new completions $\boldsymbol{M}_{j,1}$ are stable if $\boldsymbol{L}_j$, $\boldsymbol{K}_j$, $\boldsymbol{K}_j^{-1}$ have uniformly bounded spectral norms. Moreover, sparseness is retained when the $\boldsymbol{L}_j$, $\boldsymbol{K}_j$ and $\boldsymbol{K}_j^{-1}$ are all sparse. We will show several applications of (8.11) later, see also [49].

Of course, any choice of nontrivial $\boldsymbol{L}_j, \boldsymbol{K}_j$ enlarges the supports of the wavelets and makes the multiscale transformations (2.15), (2.16) more expensive. However, additional cost can be kept small by factoring the new transformation [109]. Taking for simplicity $\boldsymbol{K}_j = \mathrm{id}$, for instance (2.15) takes, in view of (8.11), the form

$$\boldsymbol{c}_{j+1} = \boldsymbol{M}_{j,0}(\boldsymbol{c}_j + \boldsymbol{L}_j \boldsymbol{d}_j) + \check{\boldsymbol{M}}_{j,1} \boldsymbol{d}_j, \tag{8.12}$$

which is the *old* (supposively faster) transformation for *modified* coarse scale data $\boldsymbol{c}_j + \boldsymbol{L}_j \boldsymbol{d}_j$. Recall from Section 5.2 that this is crucial for the computational cost of the change of bases preconditioner.

As mentioned before a necessary condition for a multiscale basis Ψ to be a Riesz basis in L_2 is that it has a biorthogonal or dual basis $\tilde{\Psi}$ which also belongs to L_2. The above concept of changing stable completions can be used to construct biorthogonal bases.

Proposition 8.3 (Carnicer et al. [26]). *Suppose that in addition to the (stable) generator bases Φ_j one has (stable) dual generator bases $\tilde{\Phi}_j$, i.e., $\langle \Phi_j, \tilde{\Phi}_j \rangle = \mathrm{id}$, with refinement matrices $\tilde{\boldsymbol{M}}_{j,0}$. Furthermore, let as above $\check{\boldsymbol{M}}_{j,1}$ be some initial stable completion for $\boldsymbol{M}_{j,0}$. Then*

$$\boldsymbol{M}_{j,1} := (\mathrm{id} - \boldsymbol{M}_{j,0}\tilde{\boldsymbol{M}}^*_{j,0})\check{\boldsymbol{M}}_{j,1} \tag{8.13}$$

is also a stable completion and $\boldsymbol{G}_{j,0} = \tilde{\boldsymbol{M}}^*_{j,0}$, $\boldsymbol{G}_{j,1} = \check{\boldsymbol{G}}_{j,1} =: \tilde{\boldsymbol{M}}^*_{j,1}$. Moreover, the collections $\Psi := \Phi_{j_0} \cup_{j \geqslant j_0} \Psi_j$, $\tilde{\Psi} := \tilde{\Phi}_{j_0} \cup_{j \geqslant j_0} \tilde{\Psi}_j$ where $\Psi_j^{\mathrm{T}} := \Phi_{j+1}^{\mathrm{T}} \boldsymbol{M}_{j,1}$, $\tilde{\Psi}_j^{\mathrm{T}} := \tilde{\Phi}_{j+1}^{\mathrm{T}} \tilde{\boldsymbol{M}}_{j,1}$, are biorthogonal, i.e., $\langle \Psi, \tilde{\Psi} \rangle = \mathrm{id}$.

Thus, the input for this construction is some initial stable completion and a dual pair of single-scale bases which are usually much easier to construct than biorthogonal infinite collections.

Remark 8.4. Since for $|I| = j$ one has

$$|\langle v, \psi_I \rangle| = |\langle v, (Q_{j+1} - Q_j)\psi_I \rangle| = |\langle (Q'_{j+1} - Q'_j)v, \psi_I \rangle| \leqslant \|(Q'_{j+1} - Q'_j)v\|_{L_2(\operatorname{supp} \psi_I)},$$

using locality (2.3) we see that the order $\tilde{m}$ of the direct estimate (8.3) for the dual multi-resolution determines the order of the cancellation property in (3.1).

It is often not so easy to determine for a given Φ_j an explicit biorthogonal basis $\tilde{\Phi}_j$ satisfying (2.3). A remedy is offered by the following useful variant from [66].

Proposition 8.5. *Suppose that the multi-resolution sequences $\mathcal{S}$, $\tilde{\mathcal{S}}$ (with $\dim S_j = \dim \tilde{S}_j$) satisfy* (8.6) *and that $S_j = S(\Phi_j)$, $\tilde{S}_j = S(\tilde{\Phi}_j)$ where, however, $\Phi_j, \tilde{\Phi}_j$ are* not *biorthogonal yet. Let $\check{\Psi}_j^{\mathrm{T}} = \Phi_{j+1}^{\mathrm{T}} \check{\boldsymbol{M}}_{j,1}$ be an initial complement basis and assume that $\Theta_j \subset S_{j+1}$ is dual to $\tilde{\Phi}_j$, i.e., $\langle \Theta_j, \tilde{\Phi}_j \rangle = \mathrm{id}$. Then*

$$\Psi_j^{\mathrm{T}} := \check{\Psi}_j^{\mathrm{T}} - \Theta_j^{\mathrm{T}} \langle \tilde{\Phi}_j, \check{\Psi}_j \rangle \tag{8.14}$$

form stable bases of the complements $W_j := S_{j+1} \cap (\tilde{S}_j)^{\perp_{L_2}}$.

Note that this construction fits into the framework of Theorem 8.2. One way to derive it is to apply Proposition 8.3 to the dual pair $(\Theta_j, \tilde{\Phi}_j)$. Even when fixing the spaces S_j and $\tilde{S}_j$ beforehand it is in many cases easy to construct the Θ_j [66].

9. Construction of wavelets on bounded domains

We will briefly outline three types of constructions based on the above concepts that work for bounded domains as well as for closed surfaces.

9.1. Patchwise defined domains

The first two concepts apply to domains Ω which can be represented as a union of (essentially) disjoint smooth parametric images of the unit d-cube $\square := (0,1)^d$, i.e.,

$$\bar{\Omega} = \bigcup_{i=1}^{M} \bar{\Omega}_i, \quad \Omega_i \cap \Omega_l = \emptyset, \ i \neq l,$$

where $\Omega_i = \kappa_i(\square)$, $\kappa_i : \mathbb{R}^d \to \mathbb{R}^{d'}$, $d' \geqslant d$, $i = 1, \ldots, M$, and the κ_i are regular parametrizations. Of course, this covers a wide class of domains. For instance, every triangle can be decomposed into quadrilaterals. Moreover, typical free-form surfaces in *computer aided geometric design* are of this form where each κ_i is a polynomial or rational mapping. One can then proceed in three steps:

(i) Construct wavelet bases on $(0,1)$.
(ii) Employ tensor products to form bases on $\square$ and then on each patch Ω_i by means of parametric lifting.
(iii) From the local bases on Ω_i build global bases on Ω.

In order to retain the option of realizing high-order cancellation properties suggested by Theorem 5.1 all presently known approaches along the above receipe concern *biorthogonal wavelets*.

(i): Wavelets on $(0,1)$ have been studied in several papers [2,27,35,56,94]. The starting point is a multi-resolution setting for $\mathbb{R}$ generated by a suitable scaling function ϕ. The common strategy is then to construct generator bases $\Phi_j^{[0,1]}$ on $[0,1]$ consisting of three groups of basis functions. The first two are formed by the left and right *boundary functions* $\Phi_{j,\mathrm{L}}^{[0,1]}, \Phi_{j,\mathrm{R}}^{[0,1]}$ consisting of a fixed finite number of scaled versions of linear combinations of translates $2^{j/2}\phi(2^j \cdot -k)$ overlapping the left respectively right end point of the interval. The third group consists of *interior basis functions* $\Phi_{I,j}^{[0,1]} = \{2^{j/2}\phi(2^j \cdot -k)\colon k = \ell_1 + 1, \ldots, 2^j - \ell_2\}$ for fixed ℓ_1, ℓ_2, whose support is contained in $[0,1]$. The boundary functions are formed in such a way that $S(\Phi_j^{[0,1]})$ contains all polynomials up to some desired order m. Specifically, the construction of the $\Phi_j^{[0,1]}$ in [56] employs the biorthogonal multi-resolution from [34] based on B-splines. It is shown that for any order m of $\Phi_j^{[0,1]}$ and any $\tilde{m} \geqslant m$, $m + \tilde{m}$ even, the dual generators from [34] can be used to construct a biorthogonal generator basis $\tilde{\Phi}_j^{[0,1]}$ which has polynomial exactness $\tilde{m}$ on *all of* $[0,1]$. After constructing some initial stable completions Proposition 8.3 can be employed to construct biorthogonal wavelet bases $\Psi^{[0,1]}$, $\tilde{\Psi}^{[0,1]}$ satisfying (2.2) and having vanishing moments of order $\tilde{m}$ [56]. Moreover, the polynomial exactness combined with the locality (2.3) allows one to establish direct estimates of the form (8.3) for *both* the primal and dual multi-resolution sequences, while (8.4) is a consequence of stability. Hence Theorem 8.1 applies and ensures the validity of the norm equivalence (8.5) [56].

There are several ways of incorporating homogeneous boundary conditions. Since the boundary functions in $\Phi_{j,\mathrm{L}}^{[0,1]}$ and $\tilde{\Phi}_{j,\mathrm{L}}^{[0,1]}$ near 0 say agree locally with monomials, discarding the first l functions in both bases gives rise to a smaller pair of biorthogonal functions whose elements now have vanishing derivatives of order $< l$ at zero. In [64] an alternative concept is analyzed realizing *complementary boundary conditions*, see [30] for such earlier realizations based on spline systems.

This means that when the primal basis functions satisfy certain boundary conditions at an end point of the interval then the dual collection of basis functions has no boundary constraints at this end point but spans there locally all polynomials of order $\tilde{m}$ and vice versa. The importance of this construction will be explained later.

(ii): It is straightforward to construct now biorthogonal bases $\Psi^{\square}, \tilde{\Psi}^{\square}$ on $\square$ as tensor products of the constructions on $(0,1)$ (again with a desired type of boundary conditions on the faces of $\square$). Moreover, the collections $\Psi^{\Omega_i} := \Psi^{\square}\circ\kappa_i^{-1} := \{\psi^{\square}\circ\kappa_i^{-1}\colon \psi^{\square}\in\Psi^{\square}\}$ and $\tilde{\Psi}^{\Omega_i} := \tilde{\Psi}^{\square}\circ\kappa_i^{-1}$ are biorthogonal with respect to the *modified* inner products

$$(v,w)_i := \int_{\square} v(\kappa_i(x))\overline{w(\kappa_i(x))}\,\mathrm{d}x. \tag{9.1}$$

(iii): At least two essentially different concepts regarding (iii) have been proposed.

9.2. Globally continuous wavelets on Ω

The first one pursued for instance in [23,25,38,62,81,94] leads to *globally continuous* pairs of biorthogonal wavelet bases $\Psi^{\Omega}, \tilde{\Psi}^{\Omega}$ where, however, biorthogonality holds with respect to the inner product $(v,w) := \sum_{i=1}^{m}(v|_{\Omega_i}, w|_{\Omega_i})_i$ with $(\cdot,\cdot)_i$ defined by (9.1).

The first step in all constructions is to form globally continuous pairs of generator bases $\Phi_j^{\Omega}, \tilde{\Phi}_j^{\Omega}$. Due to the particular nature of the univariate boundary near basis functions in $\Phi_{j,K}^{[0,1]}, \tilde{\Phi}_{j,K}^{[0,1]}$, $K\in\{\mathrm{L},\mathrm{R}\}$, this is actually very easy since all but one generator function from each side vanish on the interface.

Given globally continuous generator bases on Ω the subsequent construction of wavelets differs in the above-mentioned papers. Gluing wavelets and scaling functions across patch boundaries subject to biorthogonality constraints boils down in [23,25] to studying certain systems of homogeneous linear equations whose structure depends on the combinatorial properties of the patch complex. A different strategy is used in [62]. First of all the univariate ingredients are symmetrized so that global constraints in connection with the parametric lifting through the mappings κ_i are avoided. Second, in view of Proposition 8.3, the gluing process is *completely confined* to generator basis functions which, as mentioned before, is easy. In addition, based on suitable stable completions for the individual patches some *global initial stable completion* $\check{\boldsymbol{M}}_{j,1}^{\Omega}$ is explicitly constructed so that (8.13) provides the desired stable completion $\boldsymbol{M}_{j,1}^{\Omega}$ which gives rise to biorthogonal wavelet bases. Of course, $\boldsymbol{M}_{j,1}^{\Omega}$ does not have to be assembled globally. In fact, expressing $\boldsymbol{M}_{j,1}^{\Omega}$ in the form (8.11) (with $\boldsymbol{K}_j = \mathrm{id}$) the entries of the corresponding matrix $\boldsymbol{L}_j$ can be identified as linear combinations of inner products of primal and dual generator functions [26].

Due to the locality of the generator bases Φ_j^{Ω}, $\tilde{\Phi}_j^{\Omega}$ it is not hard to confirm the validity of direct and inverse estimates of the form (8.3), (8.4) for the primal and dual multi-resolution sequences $\mathscr{S}$, $\tilde{\mathscr{S}}$. However, due to the nature of the modified inner products (9.1) to which biorthogonality refers, duality arguments apply only in a restricted range. As a consequence the norm equivalences (8.5) can be shown to hold only in the range $-\frac{1}{2} < s < \frac{3}{2}$. Thus Theorems 4.1 or 5.1 say that the single-layer potential is not covered with respect to optimal preconditioning. Nevertheless, these bases are appropriate for operator equations of order between zero and two. Applications to second-order elliptic boundary value problems on bivariate L-shaped domains and analogous three-dimensional domains are presented in [24,113] where the change-of-bases-preconditioner described in Section 5.2 is used.

Moreover, since gluing across patch interfaces fixes a global parametrization of Ω an extension of this concept to higher degrees of global regularity would only work for domains that are homeomorphic to some domain in Euclidean space.

9.3. The theory of Ciesielski and Figiel and new wavelet bases

To overcome the above limitations an alternative approach to the construction of bases on Ω is developed in [63] still for the above patchwise representations of domains. The main difference is that the construction of bases is now closely intertwined with the *characterization* of function spaces defined on Ω as Cartesian products of certain local spaces based on the *partition* of Ω into the patch complex $\{\Omega_i\}_{i=1}^M$ and *not* as usual on an atlas and charts. This characterization was first given in [29,30] for the purpose of constructing *one* good basis for various function spaces on compact C^∞-manifolds. [30] already contains the explicit construction of such bases in terms of spline systems. Since these bases however lack the desired locality the objective of [63] has been to adapt the concepts from [30] to applications in numerical analysis. In particular, this concerns isolating minimal constructive requirements on ingredients like extension operators and tying them with more recent technologies around the construction of suitable local wavelet bases.

These concepts work for a wide range of Sobolev and Besov spaces. For simplicity, we sketch some ideas here only for Sobolev spaces. The point is that $H^s(\Omega)$ is equivalent to the product of certain closed subspaces of the local spaces $H^s(\Omega_i)$. To be a bit more precise one assigns to each interface $\bar{\Omega}_i \cap \bar{\Omega}_l =: \Omega_{i,l}$ a *direction*. In other words, the patch complex is considered as an oriented graph whose edges are the interfaces. Now let $\Omega_i^\uparrow$ denote a slightly larger patch in Ω which contains Ω_i in such a way that the *inflow boundary*, i.e., the union of those faces of Ω_i whose direction points into Ω_i is also part of the boundary of $\Omega_i^\uparrow$. Hence $\Omega_i^\uparrow$ contains the (relative interior of) the *outflow boundary* of Ω_i in its interior. Thus one can consider *extensions* across the outflow boundary of functions on Ω_i to $\Omega_i^\uparrow$. The patch $\Omega_i^\downarrow$ is defined in complete analogy for reversed directions. Let $H^s(\Omega_i)^\uparrow := \{v \in H^s(\Omega_i)\colon \chi_{\Omega_i} v \in H^s(\Omega_i^\uparrow)\}$ and analogously $H^s(\Omega_i)^\downarrow$, where χ_{Ω_i} is the indicator function on Ω_i. Hence $H^s(\Omega_i)^\uparrow$ consists of those functions whose trivial extension by zero across the outflow boundary has the same Sobolev regularity on the larger domain $\Omega_i^\uparrow$. When $s \neq n + 1/2$, $n \in \mathbb{N}$, these spaces agree with $H^s_{0,\partial\Omega_i^\uparrow}(\Omega_i)$. In that sense the elements of $H^s(\Omega_i)^\uparrow$ satisfy certain homogeneous boundary conditions at the outflow boundary. These spaces are endowed with the norm $\|v\|_{H^s(\Omega_i)^\uparrow} := \|\chi_{\Omega_i} v\|_{H^s(\Omega_i^\uparrow)}$. Based on suitable bounded extension operators E_i from $H^s(\Omega_i)$ to $H^s(\Omega_i^\uparrow)$ whose adjoints E_i' are also continuous one can construct bounded projectors $P_i : H^s(\Omega) \to H^s(\Omega)$ such that

$$(P_i(H^s(\Omega))|_{\Omega_i} = H^s(\Omega_i)^\downarrow, \quad (P_i'(H^s(\Omega))|_{\Omega_i} = H^s(\Omega_i)^\uparrow. \tag{9.2}$$

A central result concerns the validity of the following topological isomorphisms:

$$H^s(\Omega) \simeq \prod_{i=1}^{M} H^s(\Omega_i)^{\downarrow}, \quad \|v\|_{H^s(\Omega)} \sim \left(\sum_{i=1}^{M} \|(P_i v)|_{\Omega_i}\|^2_{H^s(\Omega_i)^{\downarrow}}\right)^{1/2},$$

$$(H^s(\Omega))' \simeq \prod_{i=1}^{M} H^{-s}(\Omega_i)^{\uparrow}, \quad \|v\|_{H^s(\Omega)} \sim \left(\sum_{i=1}^{M} \|(P_i' v)|_{\Omega_i}\|^2_{H^{-s}(\Omega_i)^{\uparrow}}\right)^{1/2}, \tag{9.3}$$

where $H^{-s}(\Omega_i)^{\uparrow}$ is the dual of $H^s(\Omega_i)^{\downarrow}$.

Once (9.3) is given the idea is to construct local wavelet bases $\Psi^{\Omega_i} := \Psi^{\Box_{i,\downarrow}} \circ \kappa_i^{-1}$, $\tilde{\Psi}^{\Omega_i} := \tilde{\Psi}^{\Box_{i,\uparrow}} \circ \kappa_i^{-1}$, where $\Box_{i,\downarrow}$ indicates that the wavelets in $\Psi^{\Box_{i,\downarrow}}$ satisfy those boundary conditions induced by the inflow and outflow boundaries of Ω_i through the parametric pullback κ_i^{-1}. In particular, the elements in the dual basis $\tilde{\Psi}^{\Omega_i}$ satisfy corresponding *complementary boundary conditions* as indicated by the reversed arrow. Thus, one needs dual pairs of wavelet bases with complementary boundary conditions on the unit cube $\Box$. Employing spline systems, such bases were constructed first already in [29,30]. An alternative construction of local wavelet bases was given in [64]. The lifted bases Ψ^{Ω_i}, $\tilde{\Psi}^{\Omega_i}$ satisfy norm equivalences of the form

$$\|v\|_{H^s(\Omega_i)^{\downarrow}} \sim \|\{2^{2s|I|}(v, \tilde{\psi}_I^{\Omega_i})_i\}\|_{\ell_2}, \tag{9.4}$$

see [64].

Setting $g_i := |\partial\kappa_i(\kappa_i^{-1}(\cdot))|$ which by assumption is a smooth function on Ω_i, the global wavelet bases $\Psi^{\Omega}, \tilde{\Psi}^{\Omega}$ are now formed as follows. The indices I for $\psi_I^{\Omega_i}$ have the form $I = (I^i, i)$ where I^i is an index for the ith basis Ψ^{Ω_i}. Thus for any such I one sets

$$\psi_I^{\Omega} := P_i \chi_{\Omega_i} \psi_{I^i}^{\Omega_i}, \quad \tilde{\psi}_I^{\Omega} := P_i' \chi_{\Omega_i} g_i^{-1} \tilde{\psi}_{I^i}^{\Omega_i}. \tag{9.5}$$

The factor g_i^{-1} in the definition of $\tilde{\psi}_I^{\Omega}$ can be seen to ensure biorthogonality with respect to the canonical inner product $\langle \cdot, \cdot \rangle$ for $L_2(\Omega)$, i.e.,

$$\langle \Psi^{\Omega}, \tilde{\Psi}^{\Omega} \rangle = \mathrm{id}. \tag{9.6}$$

Moreover, combining the local norm equivalences (9.4) with relations (9.3) yields the desired relations

$$\|v\|_{H^s(\Omega)} \sim \|\langle v, \tilde{\Psi}^{\Omega} \rangle \boldsymbol{D}^s\|_{\ell_2}, \quad s \in (-\tilde{\gamma}, \gamma), \tag{9.7}$$

where $(\boldsymbol{D}^s)_{I,I'} = 2^{s|I|}\delta_{I,I'}$ and $H^s(\Omega) = (H^{-s}(\Omega))'$ for $s < 0$. Here the bounds $\tilde{\gamma}, \gamma$ depend on the regularity of the wavelets in $\Psi^{\Box_{i,\downarrow}}, \tilde{\Psi}^{\Box_{i,\uparrow}}$ as well as on the regularity of the manifold Ω, see [63] for details.

A few comments on this concept are in order:

- In contrast to the previously mentioned constructions, biorthogonality holds with respect to the canonical inner product so that duality arguments apply in the full range of validity in (9.7).
- In principle, any range for (9.7) permitted by the regularity of Ω can be realized.

- A *global parametrization* of Ω is never needed. In fact, the mappings κ_i should rather be viewed as equivalence classes of parametrizations whose elements are only identified up to regular reparametrizations. The regularity of Ω is only reflected by the ability of reparametrizing the mappings from neighboring patches in such a way that resulting locally combined reparametrization of a neighborhood of any patch Ω_i has the regularity of Ω. These reparametrizations only enter the concrete realization of the projections P_i, see [63].
- Relations (9.3) induce automatically a *domain decomposition* scheme also for *global operators* which is sketched next.

9.4. Domain decomposition

Abbreviating $\Pi_\downarrow := \prod_{i=1}^M H^t(\Gamma_i)^\downarrow$ and its dual $\Pi'_\uparrow := \prod_{i=1}^M H^{-t}(\Gamma_i)^\uparrow$, it is shown in [63] that for $t\in(-\tilde{\gamma},\gamma)$ the mapping $S(v_i)_{i=1}^N = \sum_{i=1}^N P_i\chi_{\Omega_i}v_i$ is a topological isomorphism from $\Pi_\downarrow$ to $H^t(\Omega)$. Thus

$$\mathscr{L}_\Pi := S'\mathscr{L}S : \Pi_\downarrow \to \Pi'_\uparrow, \quad \boldsymbol{f} := S'f\in\Pi'_\uparrow, \tag{9.8}$$

so that (4.2) is equivalent to

$$\mathscr{L}_\Pi\boldsymbol{u} = \boldsymbol{f}. \tag{9.9}$$

Of course, given $\boldsymbol{u}\in\Pi_\downarrow$ satisfying (9.9), $u = S\boldsymbol{u}$ solves (4.2). One can show that $\mathscr{L}_\Pi = (\mathscr{L}_{i,l})_{i,l=1}^M$ where $\mathscr{L}_{i,l}w := (P'_i\mathscr{L}P_l\chi_{\Omega_l}w)|_{\Omega_i}$. Moreover, setting $\langle\{v_i\},\{u_i\}\rangle_\Pi := \sum_{i=1}^M\langle v_i,u_i\rangle_{\Omega_i}$ where $\langle\cdot,\cdot\rangle_{\Omega_i}$ denotes the canonical inner product for $L_2(\Omega_i)$, one has by definition $\langle\mathscr{L}_\Pi\{u_i\},\{v_i\}\rangle_\Pi = \langle\mathscr{L}u,v\rangle_\Omega$ for $S\{u_i\} = u$. Thus the weak formulation $\langle\mathscr{L}_\Pi\boldsymbol{u},\boldsymbol{v}\rangle_\Pi = \langle\boldsymbol{f},\boldsymbol{v}\rangle_\Pi$, $\boldsymbol{v}\in\Pi_\downarrow$, of (9.9) takes the form

$$\sum_{l=1}^N\left\langle\sum_{i=1}^N\mathscr{L}_{l,i}u_i - f_l, v_l\right\rangle_{\Omega_l} = 0, \quad \boldsymbol{v} = \{v_i\}_{i=1}^N\in\Pi_\downarrow. \tag{9.10}$$

The simplest way to solve (9.10) iteratively is a Jacobi-type scheme yielding $\boldsymbol{u}^n$ by

$$\mathscr{L}_{i,i}u_i^{n+1} = f_i - \sum_{l\neq i}\mathscr{L}_{i,l}u_l^n, \quad i = 1,\ldots,N. \tag{9.11}$$

Each step requires the solution of an operator equation which, however, now lives only on *one* patch, or equivalently, on the parameter domain $\square$.

Let us briefly discuss now the convergence properties of the iteration (9.11) (or a corresponding relaxation version) first on the level of operator equations. To this end, note that the ellipticity (4.1) for $H = H^t(\Omega)$ combined with (9.3) yields $\|\mathscr{L}_\Pi\boldsymbol{v}\|_{\Pi'_\uparrow}\sim\|\boldsymbol{v}\|_{\Pi_\downarrow}$ for $\boldsymbol{v}\in\Pi_\downarrow$. Specifically, when $\mathscr{L}$ is self-adjoint and positive definite this means

$$\|\boldsymbol{v}\|_{\Pi_\downarrow}^2\sim\langle\mathscr{L}_\Pi\boldsymbol{v},\boldsymbol{v}\rangle_\Pi, \quad \boldsymbol{v}\in\Pi_\downarrow,$$

which, in particular, implies that

$$\|\mathscr{L}_{i,i}v\|_{H^{-t}(\Omega_i)^\uparrow}\sim\|v\|_{H^t(\Omega_i)^\downarrow}, \quad i = 1,\ldots,N. \tag{9.12}$$

Hence, the *local problems*

$$\mathscr{L}_{i,i}u_i = g_i, \quad i = 1,\ldots,N, \tag{9.13}$$

on $\square$ are elliptic. Thus, in view of (9.4), each of these local problems can be solved adaptively with optimal complexity for instance by the scheme from Section 6. Due to isomorphisms (9.3) and the properties of the wavelet schemes in combination with (9.12) one can invoke the theory of stable splittings and Schwarz schemes to conclude that this strategy gives rise to an asymptotically optimal solver, [77,97]. It should also be noted that this just as the adaptive concepts work for differential as well as integral operators.

Not only because integral operators are usually much less accessible by domain decomposition techniques this case deserves special attention for the following reason. Note that for $I=(i,I_i)$ and $I'=(l,I_l)$

$$\langle \mathscr{L}\psi_{I'}^{\Omega},\psi_{I}^{\Omega}\rangle_{\Omega}=\int_{\square}\int_{\square}L_{l,i}(x,y)\psi_{I_l}^{\square_{l,\downarrow}}(y)\psi_{I_i}^{\square_{i,\downarrow}}(x)\,\mathrm{d}x\,\mathrm{d}y, \tag{9.14}$$

where $L_{l,i}(x,y)=((P_i'\otimes P_l')K)(\kappa_i^{\uparrow}(x),\kappa_l^{\uparrow}(y))$ and $\kappa_i^{\uparrow}$ is the above mentioned extended parametrization of $\Omega_i^{\uparrow}$. More precisely, the adjoints P_i' only enter when ψ_I^{Ω} does not vanish at the outflow boundary of its support patch. Since due to the locality of the wavelets $\psi_{I_i}^{\Omega_i}$ in the local bases the P_i actually reduce to extensions across the outflow boundary $\Omega_i^{\uparrow}$ so that P_i' is a *restriction operator*. At any rate, computations reduce completely to computing with the wavelets on the unit cube $\square$ regardless of the appearance of the projections P_i' in (9.14). Hence the local problems can be processed in *parallel* after possible prior modifications of the kernels.

As pointed out above it is crucial that the biorthogonal pairs of wavelet bases $\Psi^{\square_{i,\downarrow}},\tilde{\Psi}^{\square_{i,\uparrow}}$ on $\square$ have complementary boundary conditions. However, therefore the elements of $\Psi^{\square_{i,\downarrow}}$ are only orthogonal to those polynomials which satisfy the boundary conditions corresponding to $\square_{i,\uparrow}$. Hence cancellation properties will only hold for functions with such boundary conditions. Due to the nature of the extension operators E_i in the above construction, the modified kernels $L_{l,i}(x,y)$ in (9.14) turn out to satisfied exactly the right type of boundary conditions. As a consequence of the resulting *optimal order patchwise cancellation properties* the entries of the stiffness matrices exhibit the desired decay properties of type (5.8) which was important for the adaptive concepts [63].

9.5. Finite element-based wavelets

There have been various attempts to merge finite element discretizations with wavelet concepts, see e.g. [49,90,91] and the discussion in Section 10.1 below. Especially the investigations in [106,107] aim at robust preconditioners based on wavelets with short masks which therefore give rise to efficient multiscale transformations. Here we briefly indicate the approach proposed in [66] which, in particular, led to Theorem 8.2. Recall that for the constructions in Sections 9.2, 9.3 the primal multi-resolution $\mathscr{S}$ was chosen first and the dual multi-resolution $\tilde{\mathscr{S}}$ was essentially determined through the construction of biorthogonal projectors Q_j, namely as the range of their adjoints. By contrast, in [66] both the primal and dual multi-resolution $\mathscr{S}$ and $\tilde{\mathscr{S}}$ are a priorily chosen to be spanned by Lagrange finite element bases on uniform subdivisions of some *arbitrary* initial triangulations of a two- or three-dimensional polygonal domain. In particular, the order of the dual sequence $\tilde{\mathscr{S}}$ reflects the desired order of cancellation properties. In this case the projectors Q_j are *not* explicitly given but only the primal wavelet basis Ψ is directly constructed according to (8.14) in Proposition 8.5. After some start up calculations on the *reference element* the construction is explicit and uses only information on the domain geometry, see [66] for details. In particular, the stability condition (8.6) has to be checked only on the reference element. This condition means that the bases for the spaces S_j and $\tilde{S}_j$ *can* be biorthogonalized. One can use Theorem 8.2 to establish norm equivalences of the form (8.7), (8.8) for $\gamma=\tilde{\gamma}=\frac{3}{2}$, when Ω is a domain in Euclidean space and $\gamma=\tilde{\gamma}=1$ when Ω is a piecewise affine Lipschitz manifold, without actually ever computing the dual bases. All biorthogonality relations refer to the standard inner product for $L_2(\Omega)$. The prize that has to be paid is that the dual basis $\tilde{\Psi}$ is not explicitly available and will in general not be local in the sense of (2.2). However, recall from (5.2) that only the transformation $\boldsymbol{T}_J$ from (2.14), (2.15) is needed which is efficient due to the locality of the primal basis Ψ. For more details on the efficient implementation of the multiscale transformation $\boldsymbol{T}_J$, using ideas similar to (8.12), see [66].

10. Applications in conventional discretizations

Except perhaps for Section 9.5 the above developments concern *pure* wavelet settings in the sense that little use can be made of standard discretizations. I would like to indicate next briefly some instances where wavelet concepts are *combined* with conventional discretization schemes without requiring sophisticated basis constructions. Specifically, the tools from Section 8.2 are applied.

10.1. Stabilizing hierarchical bases

The first example concerns standard elliptic second-order boundary value problems of the type

$$\mathscr{L}u := -\operatorname{div}(\varepsilon\nabla u)u + au = f \quad \text{in } \Omega\subset\mathbb{R}^d,$$

$$u=0 \quad \text{on } \partial\Omega,$$

which are to be solved by Galerkin schemes with respect to a multi-resolution sequence $\mathscr{S}$ consisting of nested standard piecewise linear finite element spaces on a hierarchy of uniformly refined triangulations of a polygonal domain Ω, recall Section 2.4. In principle, the *hierarchical basis preconditioner* from [115] is based on a change of bases as described in Section 5.2. The particular complement bases of the form (8.9) give rise to very efficient multiscale transformations in (5.2). However, as mentioned earlier, the hierarchical basis does not satisfy relations of the form (3.2) for the relevant function space $H_0^1(\Omega)$, say. Therefore, moderately growing condition numbers are encountered in the bivariate case $d=2$ while for $d\geqslant 3$ the condition numbers increase with decreasing mesh size at a larger unacceptable rate.

Instead stable bases for orthogonal complements between successive finite element spaces would satisfy (3.2) and thus would give rise to asymptotically optimal preconditioners, which has also been the starting point in [21]. The idea in [111] has therefore been to determine at least *approximate* orthogonal complements while preserving as much efficiency of the classical hierarchical basis preconditioner as possible. This can be viewed as choosing hierarchical complement bases as *initial* stable completions $\check{\boldsymbol{M}}_{j,1}$ which are then to be followed by a *change* of stable completions according to (8.10). To obtain corresponding matrices $\boldsymbol{L}_j$ note first that

$$\boldsymbol{M}_{j,1} := (\mathrm{id} - (\boldsymbol{M}_{j,0}\langle\Phi_j,\Phi_j\rangle^{-1}\langle\Phi_j,\Phi_{j+1}\rangle))\check{\boldsymbol{M}}_{j,1}$$

gives rise to bases $\Psi_j^{\mathrm{T}} = \Phi_{j+1}^{\mathrm{T}}\boldsymbol{M}_{j,1}$ that span *orthogonal* complements. In other words, the *ideal* $\boldsymbol{L}_j$ in (8.11) would have the form $-\langle\Phi_j,\Phi_j\rangle^{-1}\langle\Phi_j,\Phi_{j+1}\rangle\check{\boldsymbol{M}}_{j,1}$ with $\boldsymbol{K}_j=\mathrm{id}$. Of course, the inverse

$\langle \Phi_j, \Phi_j \rangle^{-1}$ of the mass matrix is dense and so is $\boldsymbol{M}_{j,1}$. However, to compute $\boldsymbol{M}_{j,1}\boldsymbol{v}$ for any coefficient vector $\boldsymbol{v}$, which is a typical step in the multiscale transformation (2.14) needed for the change-of-bases-preconditioner (see (5.2)), one can proceed as follows. First, compute $\check{\boldsymbol{M}}_{j,1}\boldsymbol{v} =: \hat{\boldsymbol{v}}$ and $\boldsymbol{w} := \langle \Phi_j, \Phi_{j+1} \rangle \hat{\boldsymbol{v}} = \boldsymbol{M}_{j,0} \langle \Phi_{j+1}, \Phi_{j+1} \rangle \hat{\boldsymbol{v}}$. Next, instead of computing $\langle \Phi_j, \Phi_j \rangle^{-1} \boldsymbol{w}$ exactly one performs only a few iterations on the well-conditioned system $\langle \Phi_j, \Phi_j \rangle \boldsymbol{y} = \boldsymbol{w}$, followed by $\hat{\boldsymbol{v}} - \boldsymbol{M}_{j,0}\boldsymbol{y}$, see [111,112].

10.2. Convection diffusion equations

The next example concerns *convection–diffusion-reaction equations* of the form

$$\mathscr{L}u := -\operatorname{div}(\varepsilon \nabla u) + \boldsymbol{\beta}^{\mathrm{T}} \nabla u + au = f \quad \text{in } \Omega \subset \mathbb{R}^d,$$
$$u = 0 \quad \text{on } \partial\Omega$$

and again the same finite element multi-resolution as in the previous section. Due to the additional first-order term $\mathscr{L}$ is no longer self-adjoint. Moreover, when ε is small compared to $\boldsymbol{\beta}$, i.e., when the *grid Peclet numbers* $P_h := \|\boldsymbol{\beta}\| h/2\|\varepsilon\|$ (h denoting the mesh size) are larger than one, it is well known that several severe obstructions arise. Firstly, the efficiency of standard multilevel schemes degrades or even ceases to work at all. Secondly, standard Galerkin discretizations become *unstable* and the solutions to the discrete systems exhibit unphysical oscillations. Therefore, one usually adds *stabilization terms* which, however, entails a loss of accuracy, in particular, in critical areas such as near layers. The concept proposed in [57,58] aims at combining an improved robustness of the solution process with preserving stability of the discretization through *adaptivity* without a priorily chosen stabilization.

The realization is based on an *adaptive full multi-grid scheme* with scale dependent grid transfer and smoothing operators. Setting $a(v,u) := \langle v, \mathscr{L}u \rangle$, the starting point is a two-level splitting of the fine grid space $S_{j+1} =: S_h$ into the coarse space $S_H := S_j$ and the *hierarchical* complement $W := S(\check{\Psi}_j)$, where $\check{\Psi}$ is given by (8.9). The stiffness matrix $\boldsymbol{A}$ relative to the basis $\Phi_H \cup \check{\Psi}$ has then a two by two block form $(\boldsymbol{A}_{i,l})_{i,l=0}^{1}$ where $\boldsymbol{A}_{0,0} = a(\Phi_H, \Phi_H)$, $\boldsymbol{A}_{0,1} = a(\Phi_H, \check{\Psi})$, $\boldsymbol{A}_{1,0} = a(\check{\Psi}, \Phi_H)$, $\boldsymbol{A}_{1,1} = a(\check{\Psi}, \check{\Psi})$. In principle, the test space could, of course, be chosen differently from the beginning, which means to start with a Petrov–Galerkin discretization in order to incorporate some stabilization. The next step is to apply a *change of stable completions* of the form (8.11) with $\check{\boldsymbol{K}} = \mathrm{id}$ and some $\check{\boldsymbol{L}}$ to achieve an L_2 or H^1-stabilization of the complement based, e.g., along the lines described in the previous section. Furthermore, a *dual change* of stable completions is performed on the splitting of the *test space* with some $\hat{\boldsymbol{L}}$ (and again for simplicity $\hat{\boldsymbol{K}} = \mathrm{id}$). This means that the roles of the matrices $\boldsymbol{M}_j$ and $\boldsymbol{G}_j$ from Section 8.2 are interchanged so that one obtains a *modified coarse grid basis*. Thus on the next coarser level one actually does use a Petrov–Galerkin scheme. This yields a modified block system of the form

$$\begin{pmatrix} \mathrm{id} & -\hat{\boldsymbol{L}} \\ \boldsymbol{0} & \mathrm{id} \end{pmatrix} \begin{pmatrix} \boldsymbol{A}_{0,0} & \boldsymbol{A}_{0,1} \\ \boldsymbol{A}_{1,0} & \boldsymbol{A}_{1,1} \end{pmatrix} \begin{pmatrix} \mathrm{id} & -\check{\boldsymbol{L}} \\ \boldsymbol{0} & \mathrm{id} \end{pmatrix} = \begin{pmatrix} \boldsymbol{A}_{0,0} - \boldsymbol{A}_{1,0} & \check{\boldsymbol{A}}_{0,1} - \hat{\boldsymbol{L}}\check{\boldsymbol{A}}_{1,1} \\ \boldsymbol{A}_{1,0} & \check{\boldsymbol{A}}_{1,1} \end{pmatrix}.$$

The matrices $\hat{\boldsymbol{L}}, \check{\boldsymbol{L}}$ are roughly chosen as follows. While $\check{\boldsymbol{L}}$ is to stabilize the complement bases (spanning the high-frequency components) for the trial spaces the matrix $\hat{\boldsymbol{L}}$ is used to *decouple* the block system, i.e.,

$$\hat{L}\check{A}_{1,1} \approx \check{A}_{0,1}. \tag{10.1}$$

Specifically, a SPAI technique is used in [58] to realize (10.1).

This approximate decoupling turns out to cause an automatic *up-wind* effect of increasing strength when descending to lower levels, see [57]. Specifically, the refinement matrix of the test bases changes according to the second relation in (8.11) which corresponds to an *adapted, scale-dependent* prolongation operator.

Exact decoupling in (10.1) would give rise via block elimination to an exact two-grid method which consists of a *coarse grid correction*

$$\mathring{A}_{0,0}e_0 = r_0^\mu = \hat{f}_0 - \mathring{A}_{0,0}u_H^\mu \rightsquigarrow u_H^{\mu+1} = u_H^\mu + e_0$$

of the coarse grid component and a *high-frequency correction* for $d^\mu = \hat{f}_1 - \mathring{A}_{1,0}u_H^\mu - \check{A}_{1,1}d^\mu$ of the form

$$d^{\mu+1} = d^\mu + \check{A}_{1,1}^{-1}(r_1^\mu - A_{1,0}e_0).$$

Approximate decoupling is shown in [58] to lead to a multi-grid scheme with a *smoother* of the form

$$S = \begin{pmatrix} I & 0 \\ -(M\mathring{A}_{0,0} + QA_{1,0}) & I - Q\check{A}_{1,1} \end{pmatrix}$$

which for

$$S = (I - G\mathring{A}), \quad G = \begin{pmatrix} 0 & 0 \\ M & Q \end{pmatrix}$$

corresponds to the basic iteration $u^{\mu+1} = (I - G\mathring{A})u^\mu + Gf = u^\mu + Gr^\mu$. Here the application of Q and M correspond to an approximate inversion of the *complement matrix* $\check{A}_{1,1}$, respectively the approximate inversion of the coarse grid system $\mathring{A}_{0,0}$ followed by the application of $-QA_{1,0}$.

Note that high-frequency components in the form of wavelet-type coefficients are approximated in each step. Their behavior can then serve as an indicator for local mesh refinements quite in the spirit of the adaptive wavelet scheme from Section 6. The details of such a strategy are given in [58] which leads to an adaptive full multi-grid method. In an oversimplified way it can be summarized as follows:

- *Add* artificial viscosity on the coarsest level and solve the perturbed problem.
- Refine the current discretization (locally) based on the magnitude of *wavelet coefficients*; use the current approximation as initial guess for the next level and reduce viscosity.
- Continue until original viscosity is restored and accuracy check is verified.

Thus regardless of the size of the local grid Peclet numbers on the final possibly highly nonuniform mesh one ends up with a Galerkin approximation. In [58] various bivariate examples for different flow fields such as constant directions, circulating flow, variable directions and combinations with reaction terms are tested. One can see that in all cases the refined mesh accurately reflects the behavior of the solution and provides oscillation free Galerkin approximations. The number of multi-grid cycles also stays bounded while, however, due to the SPAI technique, the work for each cycle may still vary significantly.

10.3. Conservation laws

The last example concerns problems of rather different type namely systems of hyperbolic conservation laws of the form

$$\frac{\partial u(t,\boldsymbol{x})}{\partial t}+\sum_{m=1}^{d}\frac{\partial f_m(u(t,\boldsymbol{x}))}{\partial x_m}=\mathbf{0},\quad 0<t<T,\ \boldsymbol{x}\in\Omega \tag{10.2}$$

with suitable initial and boundary data. A common way of discretizing such equations is to view (10.2) as an evolution of *cell averages* in the form

$$\bar{u}_k^{n+1}=\bar{u}_k^n-\lambda_k\sum_{m=1}^{d}(\bar{f}^n_{m,k+em}-\bar{f}^n_{m,k}),\quad \lambda_k:=\Delta t/|V_k|, \tag{10.3}$$

where here the V_k denote cells in some hexaedral mesh and

$$\bar{u}_k^n:=\frac{1}{|V_k|}\int_{V_k u}(t^n,\boldsymbol{x})\,\mathrm{d}\boldsymbol{x},\quad \bar{f}^n_{m,k}:=\frac{1}{\Delta t}\int_{t^n}^{t^{n+1}}\int_{\Gamma_{m,k}}\boldsymbol{f}(u(t,\boldsymbol{x}))^{\mathrm{T}}\boldsymbol{n}_{m,k}(\boldsymbol{x})\,\mathrm{d}\Gamma\,\mathrm{d}t.$$

Here $\Gamma_{m,k}$ denotes the cell interface with mth fixed coordinate. Approximating the exact flux balances $\bar{f}^n_{m,k+em}-\bar{f}^n_{m,k}$ by *numerical flux balances* $\bar{F}^n_{m,k+em}-\bar{F}^n_{m,k}$ depending typically on neighboring cell averages and collecting cell averages and numerical fluxes in arrays $\bar{\boldsymbol{v}}^n$ and $\bar{\boldsymbol{F}}_{m,\pm}$, one arrives at a discrete evolution scheme of the form

$$\bar{\boldsymbol{v}}_L^{n+1}=\bar{\boldsymbol{v}}_L^n-\Lambda_L\sum_{m=1}^{d}(\bar{\boldsymbol{F}}_{m,L,+}-\bar{\boldsymbol{F}}_{m,L,-})=:\bar{\boldsymbol{v}}_L^n-\Lambda_L\Delta\bar{\boldsymbol{F}}_L\equiv E_L\bar{\boldsymbol{v}}_L^n, \tag{10.4}$$

where the subscript L now indicates that cell averages and fluxes refer to the Lth level of a hierarchy of meshes and where Λ_L is the diagonal matrix with entries given in (10.3).

To realize high accuracy, e.g., through ENO reconstructions one usually faces expensive flux calculations typically based on approximate local Riemann solvers. Therefore A. Harten has proposed a *multiscale* methodology for accelerating the computation of *numerical fluxes* in higher-order *finite volume* discretization [78]. His approach was based on a rather general *discrete multi-resolution concept* closely related to the framework in [26] whose ingredients have to be adapted to essentially *any* given finite volume scheme, see also [3,4]. Specializations to bivariate applications and unstructured meshes are given in [1,17,104]. The key idea is to transform the array $\bar{\boldsymbol{v}}_L$ (suppressing for the moment the time index n) of cell averages on the finest level L into an array

$$\bar{\boldsymbol{v}}_L\to\bar{\boldsymbol{v}}_{MS}=(\bar{\boldsymbol{v}}_0,\bar{\boldsymbol{d}}_0,\ldots,\bar{\boldsymbol{d}}_{L-1}),$$

of cell averages $\bar{\boldsymbol{v}}_0$ on some coarsest level and *detail* coefficients $\bar{\boldsymbol{d}}_j$ on higher levels with the aid of a suitable multiscale transformation $\boldsymbol{T}_L^{-1}$ of the form (2.16). Here 'suitable' means (a) that $\boldsymbol{T}_L^{-1}$ should be efficient, i.e., the cost should remain proportional to the length of $\bar{\boldsymbol{v}}_L$ and (b) that whenever the function v behind the given cell averages is very smooth corresponding detail coefficients of the right length scale should be very small. A realization for uniform triangular meshes are developed in [36]. In [50] multiscale transformations are constructed that satisfy both requirements (a) and (b) for hierarchies of curvilinear boundary fitted quadrilateral meshes in $\mathbb{R}^2$. On account of the very nature of the cell averages as an inner product of a properly normalized characteristic function on a quadrilateral with the underlying function, the *Haar basis* for such meshes serves as a natural *initial stable completion*. Since Haar wavelets have only first order cancellation properties the requirement (b) will not be fulfilled to a satisfactory degree. Therefore a *change of stable completions* (8.11) is applied that realizes higher order vanishing moments of the corresponding new complement functions and therefore a correspondingly higher order of cancellation properties. One can choose always $\boldsymbol{K}_i=\mathrm{id}$ while the matrices $\boldsymbol{L}_j$ are determined completely locally, see [50].

Given such multiscale transformations a suitable time evolution law for detail coefficients could be derived in [50] which the following procedure can be based upon:

- Threshold the detail coefficients in the arrays $\bar{\boldsymbol{d}}_j^n$ (for suitable level dependent thresholds). From the resulting set $\mathscr{D}^n$ of *significant* coefficients predict a set $\tilde{\mathscr{D}}^{n+1}$ for the time level $n+1$ which consists of $\mathscr{D}^n$ padded by a security margin determined by the CFL condition. Here one uses the finite propagation speed for hyperbolic problems.
- Compute *flux balances* on the coarsest level taking however information from the finest level by adding suitable flux balances; propagate the cell averages on the coarsest level in time using these flux balances.
- Based on suitable evolution relations for detail coefficients derived in [50] propagate the detail coefficients to time level $n+1$ for $j=0,1,\ldots,L-1$. For coefficients corresponding to significant indices in $\tilde{\mathscr{D}}^{n+1}$ one has to perform expensive flux calculations. For coefficients not in $\tilde{\mathscr{D}}^{n+1}$ the flux values are obtained by cheap interpolation which permits the intended savings.

This scheme has been applied to various problems such as the reentry of a blunt body into the atmosphere where the conserved variables change by orders of magnitude near a detached bow shock, [50]. Again the size of wavelet-like coefficients is used to *adapt* the computation namely at that point to decide whether expensive flux calculations are needed or instead inexpensive interpolated values suffice. An extended analysis of the above procedure which, in particular, substantiates the choice of $\tilde{\mathscr{D}}^{n+1}$ is given in [37]. Of course, since eventually flux values are assigned to *every* cell on the finest level the complexity of the calculations can only be reduced by a fixed factor. Therefore, first attempts have been made meanwhile to turn this scheme into a *fully adaptive* one roughly speaking by not refining the mesh any further at locations where exact flux calculations are no longer needed, [76]. A central problem here is to ensure at any stage sufficiently accurate data for the flux calculations. This and the analysis of the ultimate complexity of such schemes are currently under investigation.

11. Concluding remarks

The key features of wavelet bases *cancellation properties* and *isomorphisms* between function and sequence spaces have been shown to play a pivotal role for the development of numerical multiscale methods for the solution of operator equations. Specifically, the combination of these isomorphisms with the mapping properties of the operators has been identified as the main driving mechanism which also determines the scope of problems for which wavelet concepts lead to asymptotically optimal schemes. Special emphasis has been put on *adaptivity* which offers perhaps the most promising perspectives for future developments. Therefore, this latter issue has been discussed in greater detail for a pure wavelet setting while I have tried to indicate more briefly a variety of further ideas and directions which branch off the core concepts. This includes recent attempts to apply wavelet concepts in traditional discretization settings, e.g., in connection with convection diffusion equations and conservation laws. All this reflects the potential strength of wavelets not only as a discretization but also as an *analysis tool* which helps bringing analysis and solution closer together. The prize one has to pay is that the wavelets themselves are usually much more sophisticated than standard discretization tools. Therefore the realization of the key features in a Fourier free context that can host realistic domain geometries has been the second guiding theme. In summary during the past ten years enormous progress has been made in the developments of sound theoretical concepts for wavelets in numerical analysis. Meanwhile, as I have indicated on several occasions in the course of the discussion many of these ideas have been implemented and tested although mostly for simple model problems. So it is fair to say that theory is still way ahead practical realizations. Moreover, most of the existing implementations are research codes that are far from exhausting their full potential. This has several reasons. First, as indicated by the adaptive scheme in Section 6 new data structures and evaluation schemes have to be developed that differ significantly from existing well established software tools. Second, for a wavelet method to work really well *all* its ingredients have to be coordinated well. Since little can be borrowed from standard software this is a time consuming process. It also leaves us with serious challanges with regard to the design of interfaces with real life applications. Finally, the sophistication of the tool offers many possibilities to do things wrong. For instance, working with domain adapted bases a proper tuning may be required to reduce the condition of the bases significantly. Therefore, exporting wavelet concepts into conventional discretization settings as indicated in Section 10 appears to be a very promising direction for future reasearch.

References

[1] R. Abgrall, Multi-resolution analysis on unstructured meshes — application to CFD, in: B.N. Chetverushkin (Ed.), Experimentation, Modeling and Combustion, Wiley, New York, 1997, pp. 147–156.

[2] L. Andersson, N. Hall, B. Jawerth, G. Peters, Wavelets on closed subsets of the real line, in: L.L. Schumaker, G. Webb (Eds.), Topics in the Theory and Applications of Wavelets, Academic Press, Boston, 1994, pp. 1–61.

[3] F. Arandiga, R. Donat, A. Harten, Multi-resolution based on weighted averages of the hat function, I — Linear reconstruction techniques, SIAM J. Numer. Anal. 36 (1998) 160–203.

[4] F. Arandiga, R. Donat, A. Harten, Multiresolution based on weighted averages of the hat function, II — Nonlinear reconstruction techniques, SIAM J. Sci. Comput. 20 (1999) 1053–1093.

[5] A. Averbuch, G. Beylkin, R. Coifman, M. Israeli, Multiscale inversion of elliptic operators, in: J. Zeevi, R. Coifman (Eds.), Signal and Image Representation in Combined Spaces, Academic Press, New York, 1995, pp. 1–16.

[6] R.E. Bank, A. Weiser, Some a posteriori error estimates for elliptic partial differential equations, Math. Comp. 44 (1985) 283–301.

[7] A. Barinka, T. Barsch, P. Charton, A. Cohen, S. Dahlke, W. Dahmen, K. Urban, Adaptive wavelet schemes for elliptic problems — Implementation and numerical experiments, IGPM Report # 173, RWTH, Aachen, June 1999.

[8] G. Berkooz, J. Elezgaray, P. Holmes, Wavelet analysis of the motion of coherent structures, in: Y. Meyer, S. Roques (Eds.), Progress in Wavelet Analysis and Applications, Editions Frontières, Dreux, 1993, pp. 471–476.
[9] S. Berrone, K. Urban, Adaptive wavelet Galerkin methods on distorted domains — setup of the algebraic system, IGPM Report # 178, RWTH Aachen, October 1999.
[10] S. Bertoluzza, A-posteriori error estimates for wavelet Galerkin methods, Appl. Math. Lett. 8 (1995) 1–6.
[11] S. Bertoluzza, An adaptive collocation method based on interpolating wavelets, in: W. Dahmen, A.J. Kurdila, P. Oswald (Eds.), Multiscale Wavelet Methods for PDEs, Academic Press, New York, 1997, pp. 109–135.
[12] S. Bertoluzza, Wavelet stabilization of the Lagrange multiplier method, Numer. Math., to appear.
[13] S. Bertoluzza, C. Canuto, K. Urban, On the adaptive computation of integrals of wavelets, Preprint No. 1129, Istituto di Analisi Numerica del C.N.R. Pavia, 1999. Appl. Numer. Math., to appear.
[14] G. Beylkin, On the representation of operators in bases of compactly supported wavelets, SIAM J. Numer. Anal. 29 (1992) 1716–1740.
[15] G. Beylkin, R.R. Coifman, V. Rokhlin, Fast wavelet transforms and numerical algorithms I, Comm. Pure Appl. Math. 44 (1991) 141–183.
[16] G. Beylkin, J.M. Keiser, An adaptive pseudo-wavelet approach for solving nonlinear partial differential equations, in: W. Dahmen, A.J. Kurdila, P. Oswald (Eds.), Multiscale Wavelet Methods for PDEs, Academic Press, New York, 1997, pp. 137–197.
[17] B.L. Bihari, A. Harten, Multiresolution schemes for the numerical solutions of 2-d conservation laws, I, SIAM J. Sci. Comput. 18 (1997) 315–354.
[18] H. Bockhorn, J. Fröhlich, K. Schneider, An adaptive two-dimensional wavelet-vaguelette algorithm for the computation of flame balls, Combust. Theory Modelling 3 (1999) 1–22.
[19] F. Bornemann, B. Erdmann, R. Kornhuber, A posteriori error estimates for elliptic problems in two and three space dimensions, SIAM J. Numer. Anal. 33 (1996) 1188–1204.
[20] J.H. Bramble, R.D. Lazarov, J.E. Pasciak, A least-squares approach based on a discrete minus one inner product for first order systems, Math. Comp. 66 (1997) 935–955.
[21] J.H. Bramble, J.E. Pasciak, J. Xu, Parallel multilevel preconditioners, Math. Comp. 55 (1990) 1–22.
[22] Z. Cai, T.A. Manteuffel, S.F. McCormick, First-order system least squares for the Stokes equations, with application to linear elasticity, SIAM J. Numer. Anal. 34 (1997) 1727–1741.
[23] A. Canuto, A. Tabacco, K. Urban, The wavelet element method, part II: realization and additional features, Preprint # 1052, Istituto del Analisi Numerica del C.N.R. Pavia, 1997, Appl. Comput. Harm. Anal., to appear.
[24] A. Canuto, A. Tabacco, K. Urban, Numerical solution of elliptic problems by the wavelet element method, in: H.G. Bock, et al., (Eds.), ENUMATH 97, World Scientific, Singapore, 1998, pp. 17–37.
[25] A. Canuto, A. Tabacco, K. Urban, The wavelet element method, part I: construction and analysis, Appl. Comput. Harm. Anal. 6 (1999) 1–52.
[26] J.M. Carnicer, W. Dahmen, J.M. Peña, Local decomposition of refinable spaces and wavelets, Appl. Comput. Harm. Anal. 3 (1996) 127–153.
[27] C.K. Chui, E. Quak, Wavelets on a bounded interval, in: D. Braess, L.L. Schumaker (Eds.), Numerical Methods of Approximation Theory, Birkhauser, Basel, 1992, pp. 1–24.
[28] Z. Ciesielski, On the isomorphisms of the spaces H_α and m, Bull. Acad. Pol. Sci. 4 (1960) (Math. Ser.) 217–222.
[29] Z. Ciesielski, Spline bases in classical function spaces on C^∞ manifolds, Part III, Constructive Theory of Functions 84, Sofia, 1984, pp. 214–223.
[30] Z. Ciesielski, T. Figiel, Spline bases in classical function spaces on compact C^∞ manifolds, part I & II, Studia Math. 76 (1983) 1–58,95–136.
[31] A. Cohen, Wavelets in numerical analysis, in: P.G. Ciarlet, J.L. Lions (Eds.), The Handbook of Numerical Analysis, Vol. VII, Elsevier, Amsterdam, 1999.
[32] A. Cohen, W. Dahmen, R. DeVore, Adaptive wavelet methods for elliptic operator equations — convergence rates, IGPM Report # 165, RWTH Aachen, September 1998, Math. Comp., to appear.
[33] A. Cohen, W. Dahmen, I. Daubechies, R. DeVore, Tree approximation and optimal encoding, IGPM Report, # 174, RWTH Aachen, September 1999.
[34] A. Cohen, I. Daubechies, J.-C. Feauveau, Biorthogonal bases of compactly supported wavelets, Comm. Pure Appl. Math. 45 (1992) 485–560.

[35] A. Cohen, I. Daubechies, P. Vial, Wavelets on the interval and fast wavelet transforms, Appl. Comput. Harm. Anal. 1 (1993) 54–81.

[36] A. Cohen, N. Dyn, S.M. Kaber, M. Postel, Multiresolution schemes on triangles for scalar conservation laws, LAN Report, University Paris VI, 1999.

[37] A. Cohen, S.M. Kaber, S. Müller, M. Postel, Accurate adaptive multiresolution scheme for scalar conservation laws, preprint, LAN Univeristy of Paris VI, 1999.

[38] A. Cohen, R. Masson, Wavelet adaptive methods for second order elliptic problems, boundary conditions and domain decomposition, preprint, 1997.

[39] S. Dahlke, Besov regularity for elliptic boundary value problems with variable coefficients, Manuscripta Math. 95 (1998) 59–77.

[40] S. Dahlke, Besov regularity for elliptic boundary value problems on polygonal domains, Appl. Math. Lett. 12 (1999) 31–36.

[41] S. Dahlke, W. Dahmen, R. DeVore, Nonlinear approximation and adaptive techniques for solving elliptic operator equations, in: W. Dahmen, A. Kurdila, P. Oswald (Eds.), Multiscale Wavelet Methods for PDEs, Academic Press, London, 1997, pp. 237–283.

[42] S. Dahlke, R. DeVore, Besov regularity for elliptic boundary value problems, Comm. Partial Differential Equations 22 (1997) 1–16.

[43] S. Dahlke, W. Dahmen, R. Hochmuth, R. Schneider, Stable multiscale bases and local error estimation for elliptic problems, Appl. Numer. Math. 23 (1997) 21–47.

[44] S. Dahlke, R. Hochmuth, K. Urban, Adaptive wavelet methods for saddle point problems, IGPM Report # 170, RWTH Aachen, 1999.

[45] W. Dahmen, Decomposition of refinable spaces and applications to operator equations, Numer. Algorithms 5 (1993) 229–245.

[46] W. Dahmen, Some remarks on multiscale transformations, stability and biorthogonality, in: P.J. Laurent, A. Le Méhauté, L.L. Schumaker (Eds.), Wavelets, Images and Surface Fitting, AK Peters, Wellesley, MA, 1994, pp. 157–188.

[47] W. Dahmen, Multiscale analysis, approximation, and interpolation spaces, in: C.K. Chui, L.L. Schumaker (Eds.), Approximation Theory VIII, Wavelets and Multilevel Approximation, World Scientific, Singapore, 1995, pp. 47–88.

[48] W. Dahmen, Stability of multiscale transformations, J. Fourier Anal. Appl. 2 (1996) 341–361.

[49] W. Dahmen, Wavelet and multiscale methods for operator equations, Acta Numer. 6 (1997) 55–228.

[50] W. Dahmen, B. Gottschlich-Müller, S. Müller, Multiresolution schemes for conservation laws, IGPM Report # 159, RWTH Aachen, April, 1998, Numer. Math., to appear.

[51] W. Dahmen, B. Kleemann, S. Prößdorf, R. Schneider, A multiscale method for the double layer potential equation on a polyhedron, in: H.P. Dikshit, C.A. Micchelli (Eds.), Advances in Computational Mathematics, World Scientific, Singapore, 1994, pp. 15–57.

[52] W. Dahmen, A. Kunoth, Multilevel preconditioning, Numer. Math. 63 (1992) 315–344.

[53] W. Dahmen, A. Kunoth, Appending boundary conditions by Lagrange multipliers: analysis of the LBB condition, IGPM Report # 164, RWTH Aachen, August 1998.

[54] W. Dahmen, A. Kunoth, R. Schneider, Wavelet least squares methods for boundary value problems, IGPM Report # 175, RWTH Aachen, September 1999.

[55] W. Dahmen, A. Kunoth, K. Urban, A wavelet Galerkin method for the Stokes problem, Computing 56 (1996) 259–302.

[56] W. Dahmen, A. Kunoth, K. Urban, Biorthogonal spline-wavelets on the interval — stability and moment conditions, Appl. Comput. Harm. Anal. 6 (1999) 132–196.

[57] W. Dahmen, S. Müller, T. Schlinkmann, Multigrid and multiscale decompositions, in: M. Griebel, O.P. Iliev, S.D. Margenov, P.S. Vassilevski (Eds.), Large-Scale Scientific Computations of Engineering and Environmental Problems, Notes on Numerical Fluid Mechanics, Vol. 62, Vieweg, Braunschweig/Wiesbaden, 1998, pp. 18–41.

[58] W. Dahmen, S. Müller, T. Schlinkmann, On a robust adaptive solver for convection-dominated problems, IGPM Report # 171, RWTH Aachen, April 1999.

[59] W. Dahmen, S. Prößdorf, R. Schneider, Wavelet approximation methods for pseudodifferential equations II: matrix compression and fast solution, Adv. Comput. Math. 1 (1993) 259–335.

[60] W. Dahmen, S. Prößdorf, R. Schneider, Wavelet approximation methods for pseudodifferential equations I: stability and convergence, Math. Z. 215 (1994) 583–620.

[61] W. Dahmen, S. Prössdorf, R. Schneider, Multiscale methods for pseudo-differential equations on smooth manifolds, in: C.K. Chui, L. Montefusco, L. Puccio (Eds.), Wavelets: Theory, Algorithms, and Applications, Academic Press, New York, 1994, pp. 385–424.

[62] W. Dahmen, R. Schneider, Composite Wavelet Bases for Operator Equations, Math. Comp. 68 (1999) 1533–1567.

[63] W. Dahmen, R. Schneider, Wavelets on manifolds I: construction and domain decomposition, SIAM J. Math. Anal. 31 (1999) 184–230.

[64] W. Dahmen, R. Schneider, Wavelets with complementary boundary conditions — function spaces on the cube, Results Math. 34 (1998) 255–293.

[65] W. Dahmen, R. Schneider, Y. Xu, Nonlinear functions of wavelet expansions — adaptive reconstruction and fast evaluation, Numer. Math. 86 (2000) 49–101.

[66] W. Dahmen, R. Stevenson, Element-by-element construction of wavelets — stability and moment conditions, SIAM J. Numer. Anal. 37 (1999) 319–352.

[67] I. Daubechies, Ten Lectures on Wavelets, CBMS-NSF Regional Conference Series in Applied Mathematics, Vol. 61, SIAM, Philadelphia, 1992.

[68] R. DeVore, Nonlinear approximation, Acta Numer. 7 (1998) 51–150.

[69] R. DeVore, B. Jawerth, V. Popov, Compression of wavelet decompositions, Amer. J. Math. 114 (1992) 737–785.

[70] R. DeVore, G.G. Lorentz, in: Constructive Approximation, Grundlehren der Mathematischen Wissenschaften, Vol. 303, Springer, New York, 1991.

[71] R. DeVore, V. Popov, Interpolation of Besov spaces, Trans. Amer. Math. Soc. 305 (1988) 397–414.

[72] W. Dörfler, A convergent adaptive algorithm for Poisson's equation, SIAM J. Numer. Anal. 33 (1996) 1106–1124.

[73] J. Fröhlich, K. Schneider, Numerical simulation of decaying turbulence in an adaptive wavelet basis, Appl. Comput. Harm. Anal. 3 (1996) 393–397.

[74] J. Fröhlich, K. Schneider, An adaptive wavelet-vaguelette algorithm for the solution of PDEs, J. Comput. Phys. 130 (1997) 174–190.

[75] G.N. Gatica, G.C. Hsiao, in: Boundary-field Equation Methods for a Class of Nonlinear Problems, Pitman Research Notes in Mathematics Series, Vol. 331, Wiley, New York, 1995.

[76] B. Gottschlich-Müller, S. Müller, Adaptive finite volume schemes for conservation laws based on local multiresolution techniques, in: M. Fey, R. Jeltsch (Eds.), Hyperbolic Problems, Theory, Numerics, Applications, Proceedings of the 7th International Conference on Hyperbolic Problems, February 9–13, 1998, Zürich, Birkhauser, Basel, 1999.

[77] M. Griebel, P. Oswald, Remarks on the abstract theory of additive and multiplicative Schwarz algorithms, Numer. Math. 70 (1995) 163–180.

[78] A. Harten, Multiresolution algorithms for the numerical solution of hyperbolic conservation laws, Comm. Pure Appl. Math. 48 (1995) 1305–1342.

[79] R. Hochmuth, Wavelet bases in numerical analysis and restricted nonlinear approximation, Habilitation Thesis, FU Berlin, 1999.

[80] S. Jaffard, Wavelet methods for fast resolution of elliptic equations, SIAM J. Numer. Anal. 29 (1992) 965–986.

[81] A. Jouini, P.G. Lemarié-Rieusset, Analyse multi-resolution bi-orthogonale sur l'intervalle et applications, Ann. Inst. H. Poincaré, Anal. Non Lineaire 10 (1993) 453–476.

[82] F. Koster, M. Griebel, N. Kevlahan, M. Farge, K. Schneider, Towards an adaptive Wavelet based 3D Navier–Stokes solver, in: E.H. Hirschel (Ed.), Notes on Numerical Fluid Mechanics, Vieweg-Verlag, Wiesbaden, 1988, pp. 339–364.

[83] A. Kunoth, Multilevel preconditioning — appending boundary conditions by Lagrange multipliers, Adv. Comput. Math. 4 (1995) 145–170.

[84] A. Kunoth, Wavelet Methods for Minimization Problems Involving Elliptic Partial Differential Equations, Habilitationsschrift, RWTH Aachen, 1999.

[85] C. Lage, Concept oriented design of numerical software, in: Boundary Elements: Implementation and Analysis of Advanced Algorithms, Proceedings of the 12th GAMM-Seminar, Kiel, Germany, January 19–21, 1996, Vieweg Notes Numer. Fluid Mech. 54 (1996) 159–170.

[86] C. Lage, C. Schwab, Wavelet Galerkin algorithms for boundary integral equations, SIAM J. Sci. Statist. Comput. 20 (1998) 2195–2222.

[87] S. Lazaar, J. Liandrat, Ph. Tchamitchian, Algorithme à base d'ondelettes pour la résolution numérique d'équations aux dérivées partielle à coefficients variables, C. R. Acad. Sci., Sér. I 319 (1994) 1101–1107.

[88] P.G. Lemarié-Rieusset, Analyses, multi-résolutions nonorthogonales, Commutation entre projecteurs et derivation et ondelettes vecteurs à divergence nulle, Rev. Mat. Iberoamericana 8 (1992) 221–236.

[89] J. Liandrat, Ph. Tchamitchian, On the fast approximation of some nonlinear operators in nonregular wavelet spaces, Adv. Comput. Math. 8 (1998) 179–192.

[90] R. Lorentz, P. Oswald, Multilevel finite element Riesz bases in Sobolev spaces, in: P. Bjorstad, M. Espedal, D. Keyes (Eds.), DD9 Proceedings, Domain Decomposition Press, Bergen, 1998, pp. 178–187.

[91] R. Lorentz, P. Oswald, Criteria for hierarchical bases in Sobolev spaces, Appl. Comput. Harm. Anal. 6 (1999) 219–251.

[92] Y. Maday, V. Perrier, J.C. Ravel, Adaptivité dynamique sur base d'ondelettes pour l'approximation d'équation aux dérivée partielles, C. R. Acad. Sci. Paris Sér. I Math. 312 (1991) 405–410.

[93] S. Mallat, Multiresolution approximations and wavelet orthonormal bases of $L_2(\mathbb{R})$, Trans. Amer. Math. Soc. 315 (1989) 69–87.

[94] R. Masson, Wavelet methods in numerical simulation for elliptic and saddle point problems, Ph.D. Thesis, University of Paris VI, January 1999.

[95] Y. Meyer, Ondelettes et opérateurs 1-3: Ondelettes, Hermann, Paris, 1990.

[96] P. Oswald, On discrete norm estimates related to multilevel preconditioners in the finite element method, in: K.G. Ivanov, P. Petrushev, B. Sendov (Eds.), Constructive Theory of Functions, Proceedings of International Conference Varna, 1991, Bulg. Acad. Sci., Sofia, 1992, pp. 203–214.

[97] P. Oswald, Multilevel Finite Element Approximations, Teubner Skripten zur Numerik, Teubner-Verlag, Stuttgart, 1994.

[98] T. von Petersdorff, C. Schwab, Wavelet approximation for first kind integral equations on polygons, Numer. Math. 74 (1996) 479–516.

[99] T. von Petersdorff, C. Schwab, Fully discrete multiscale Galerkin BEM, in: W. Dahmen, A. Kurdila, P. Oswald (Eds.), Multiscale Wavelet Methods for PDEs, Academic Press, San Diego, 1997, pp. 287–346.

[100] T. von Petersdorff, C. Schwab, R. Schneider, Multiwavelets for second kind integral equations, SIAM J. Numer. Anal. 34 (1997) 2212–2227.

[101] A. Rieder, A domain embedding method for Dirichlet problems in arbitrary space dimensions, Modélisation Mathématique et Analyse Numérique 32 (1998) 405–431.

[102] K. Schneider, M. Farge, Wavelet forcing for numerical simulation of two-dimensional turbulence, C. R. Acad. Sci. Paris Sér. II 325 (1997) 263–270.

[103] R. Schneider, Multiskalen- und Wavelet-Matrixkompression: Analysisbasierte Methoden zur effizienten Lösung großer vollbesetzter Gleichungssysteme, Habilitationsschrift, Technische Hochschule, Darmstadt, 1995, Advances in Numerical Mathematics, Teubner,1998.

[104] F. Schröder-Pander, T. Sonar, Preliminary investigations on multiresolution analysis on unstructured grids, Technical Report, Forschungsbericht IB, 223-95, A36, DLR Göttingen, 1995.

[105] Standard Template Library Programmer's Guide, SILICON GRAPHICS INC., 1996, "http://www.sgi.com/Technology/STL/".

[106] R.P. Stevenson, Experiments in 3D with a three-point hierarchical basis preconditioner, Appl. Numer. Math. 23 (1997) 159–176.

[107] R.P. Stevenson, Piecewise linear (pre-) wavelets on non-uniform meshes, in: W. Hackbusch, G. Wittum (Eds.), Multigrid Methods V, Proceedings of the Fifth European Multigrid Conference held in Stuttgart, Germany, October 1–4, 1996, Lecture Notes in Computational Science and Engineering, Vol. 3, Springer, Heidelberg, 1998.

[108] W. Sweldens, The lifting scheme: a custom-design construction of biorthogonal wavelets, Appl. Comput. Harm. Anal. 3 (1996) 186–200.

[109] W. Sweldens, The lifting scheme: a construction of second generation wavelets, SIAM J. Math. Anal. 29 (1998) 511–546.

[110] P. Tchamitchian, Wavelets, Functions, Operators, in: G. Erlebacher, M.Y. Hussaini, L. Jameson (Eds.), Wavelets: Theory and Applications, ICASE/LaRC Series in Computational Science and Engineering, Oxford University Press, Oxford, 1996, pp. 83–181.

[111] P.S. Vassilevski, J. Wang, Stabilizing the hierarchical basis by approximate wavelets, I: theory, Numer. Linear Algebra Appl. 4 (1997) 103–126.

[112] P.S. Vassilevski, J. Wang, Stabilizing the hierarchical basis by approximate wavelets, II Implementation and numerical results, SIAM J. Sci. Comput. 20 (1999) 490–514.

[113] J. Vorloeper, Multiskalenverfahren und Gebietszerlegungsmethoden, Diploma Thesis, RWTH Aachen, August 1999.

[114] J. Xu, Iterative methods by space decomposition and subspace correction, SIAM Rev. 34 (1992) 581–613.

[115] H. Yserentant, H. Yserentant, On the multilevel splitting of finite element spaces, Numer. Math. 49 (1986) 379–412.

ELSEVIER

Journal of Computational and Applied Mathematics 128 (2001) 187–204

JOURNAL OF
COMPUTATIONAL AND
APPLIED MATHEMATICS

www.elsevier.nl/locate/cam

Devising discontinuous Galerkin methods for non-linear hyperbolic conservation laws ☆

Bernardo Cockburn

School of Mathematics, University of Minnesota, 127 Vincent Hall, Minneapolis, MN, 55455, USA

Received 14 January 2000; received in revised form 14 March 2000

Abstract

In this paper, we give a simple introduction to the devising of discontinuous Galerkin (DG) methods for nonlinear hyperbolic conservation laws. These methods have recently made their way into the main stream of computational fluid dynamics and are quickly finding use in a wide variety of applications. The DG methods, which are extensions of finite volume methods, incorporate into a finite element framework the notions of *approximate Riemann solvers, numerical fluxes* and *slope limiters* coined during the remarkable development of the high-resolution finite difference and finite volume methods for nonlinear hyperbolic conservation laws. We start by stressing the fact that nonlinear hyperbolic conservation laws are usually obtained from well-posed problems by neglecting terms modeling nondominant features of the model which, nevertheless, are essential in crucial, small parts of the domain; as a consequence, the resulting problem becomes ill-posed. The main difficulty in devising numerical schemes for these conservation laws is thus how to re-introduce the neglected physical information in order to approximate the physically relevant solution, usually called the entropy solution. For the classical case of the entropy solution of the nonlinear hyperbolic scalar conservation law, we show how to carry out this process for two prototypical DG methods. The first DG method is the so-called shock-capturing DG method, which does not use slope limiters and is implicit; the second is the Runge–Kutta DG method, which is an explicit method that does not employ a shock-capturing term but uses a slope limiter *instead.* We then focus on the Runge–Kutta DG methods and show how to obtain a key stability property which holds independently of the accuracy of the scheme and of the nonlinearity of the conservation law; we also show some computational results. © 2001 Elsevier Science B.V. All rights reserved.

Keywords: Discontinuous Galerkin methods; Hyperbolic problems; Conservation laws

1. Introduction

The purpose of this paper is to give a brief introduction to the *devising* of DG methods for nonlinear hyperbolic conservation laws. These are methods that have recently moved into the main

E-mail address: cockburn@math.umn.edu (B. Cockburn).

☆ Supported in part by the National Science Foundation DMS-9807491 and by the Minnesota Supercomputing Institute.

0377-0427/01/$ - see front matter © 2001 Elsevier Science B.V. All rights reserved.
PII: S0377-0427(00)00512-4

stream of computational fluid dynamics and are being applied to problems of practical interest in which *convection* plays an important role like gas dynamics, aeroacoustics, turbomachinery, granular flows, semiconductor device simulation, magneto-hydrodynamics, and electro-magnetism, among many others. The distinctive feature of the DG methods that sets them apart from other finite element methods for hyperbolic problems is that DG methods enforce the nonlinear hyperbolic conservation law *locally*. This allows them to have a mass matrix that can be easily made to be the identity, while being highly accurate and nonlinearly stable. In this paper, we present a short introduction to the subject for the nonspecialist in the matter.

Implicit DG methods, like the shock-capturing DG (SCDG) method, and other methods like the streamline diffusion method and the characteristic Galerkin method, are studied in the monograph [16] on adaptive finite element methods for conservation laws. An introduction to the Runge–Kutta DG (RKDG) method, which is explicit, for hyperbolic conservation can be found in [2] and [1]. In this paper, we propose a *new* way of understanding the *heuristics* of the construction of the SCDG and RKDG methods; then we study a key stability property of the RKDG methods and briefly discuss their accuracy and convergence properties. For other finite element methods for nonlinear conservation laws, like evolution-Galerkin and semi-Lagrangian methods, and Petrov–Galerkin methods, see the monograph [17] which studies them in the context of convection–diffusion problems.

The organization of this paper is as follows. In Section 2, we consider a classical model problem of traffic flow and show how by dropping a second-order term from the equations a nonlinear hyperbolic conservation law is obtained which gives rise to ill-posed problems. In Section 3, we display the heuristics used by DG space discretizations to re-incorporate the information of the second-order term in order to approximate the physically relevant solution. It is based on the use of (i) approximate Riemann solvers and the associated numerical fluxes, which are nothing but suitably defined approximations to the traces of the real fluxes at the *borders* of the elements; and (ii) shock capturing terms (giving rise to SCDG methods) or generalized slope limiting procedures (giving rise to RKDG methods) which are different ways of incorporating the information of the dissipation effect of the second-order term at the *interior* of the elements. In Section 4, we focus on the RKDG method which is explicit, fully parallelizable and gives impressive computational results. We end in Section 5 with some concluding remarks.

2. The main difficulty: the loss of well-posedness

In this section, we consider a well-posed model of traffic flow proposed in [24] and illustrate how the typical modification of its conservation law, which results in a nonlinear hyperbolic conservation law, gives rise to an *ill-posed* problem. Thus, when the term modeling the driver's awareness of the conditions ahead is considered to be negligible, it is a wide spread practice to simply drop it from the nonlinear conservation law which now becomes hyperbolic. Although the neglected physical phenomenon can be correctly considered to be unimportant in most parts of the domain, *it is still crucial in small, key parts of the domain.* Indeed, the driver's awareness of the conditions ahead is essential near a strong variation of the density of cars which, as we all know, usually takes place in a small part of the highway. The *formal* modification of the equations is thus equivalent to the removal of essential physical information and this, not surprisingly, induces the loss of the *well-posedness* of the resulting problem.

2.1. A model of traffic flow

If ρ represents the density of cars in a highway and v represents the flow velocity, the fact that cars do not appear or vanish spontaneously in the middle of the highway can be written mathematically as follows: $\rho_t + (\rho v)_x = 0$.

A simple model for the flow velocity v is to take $\rho v = f(\rho) - \nu\rho_x$, where $f(\rho) = \rho V(\rho)$ is the so-called density flow. It is reasonable to assume that $\rho \mapsto V(\rho)$ is a decreasing mapping and that for a given density, say $\rho^\star$, the velocity V is equal to zero; this corresponds to the situation in which the cars are bumper to bumper. The simplest case is the following: $V(\rho) = v_{\max}(1 - \rho/\rho^\star)$, where $v_{\max}$ represents the maximum velocity, and it corresponds to a quadratic concave density flow, $f(\rho) = (\rho^\star v_{\max})(\rho/\rho^\star)(1 - \rho/\rho^\star)$. The term $\nu\rho_x$ models, see [24], our 'awareness of conditions ahead', since when we perceive a high density of cars ahead, we try to suitably decrease our speed to avoid a potentially dangerous situation; of course, for this to happen, the coefficient ν must be positive. With this choice of flow velocity, our conservation law becomes

$$\rho_t + (f(\rho))_x - \nu\rho_{xx} = 0, \tag{1}$$

which gives rise to mathematically *well-posed* initial-value problems.

2.2. The 'awareness of conditions ahead' term $\nu\rho_{xx}$

It is reasonable to expect that when the convection is dominant, that is, when the number $\rho^\star v_{\max}/\nu$ is big, the effects of the terms $\nu\rho_{xx}$ are negligible. However, it is clear that this cannot happen where ρ_x changes rapidly in an interval whose size is comparable to ν. The simplest way to illustrate this fact is to look for solutions of our conservation law (1) of the form $\rho(x,t) = \phi((x - ct)/\varepsilon)$; these are the so-called traveling wave solutions.

If we insert this expression for ρ in the conservation law and set $\varepsilon = \nu$, we obtain a simple equation for ϕ, namely, $-c\phi' + (f(\phi))' - \phi'' = 0$. If we now assume that $\lim_{z\to\infty^\pm} \phi(z) = \rho^\pm$, and that $\lim_{z\to\infty^\pm} \phi'(z) = 0$, we can integrate once the equation for ϕ *provided the speed of propagation* of the traveling wave is taken to be

$$c = \frac{f(\rho^+) - f(\rho^-)}{\rho^+ - \rho^-}. \tag{2}$$

In this case, we get that ϕ must satisfy the following simple first-order ordinary differential equation:

$$\phi' = f(\phi) - \mathscr{L}f(\phi), \tag{3}$$

where $\mathscr{L}f(\phi) = f(\rho^+) - c(\rho^+ - \phi)$ is nothing but the Lagrange interpolant of f at $\phi = \rho^\pm$. The equilibrium points $\phi = \rho^-$ and $\phi = \rho^+$ of this differential equation can be connected by a single orbit if and only if sign $(\rho^+ - \rho^-)\phi' > 0$. In other words, a traveling wave solution exists if and only if the graph of f on the interval (ρ^-, ρ^+) (resp. (ρ^+, ρ^-)) lies above (resp. below) the straight line joining the points $(\rho^\pm, f(\rho^\pm))$. If f is the quadratic concave function considered above, a traveling wave exists if and only if $\rho^- < \rho^+$. This corresponds, roughly speaking, to the case in which the density of cars ahead of us is higher than the density of cars behind.

If there is a traveling wave solution of the form $\rho(x,t) = \phi((x - ct)/\nu)$, it is easy to verify that $\nu\rho_{xx}$ is of order $1/\nu$ only for points (x,t) such that $|x - ct|$ less than a quantity of order ν. On the other hand, $\nu\rho_{xx}$ decays to zero like a quantity of order $\exp(-|\alpha||x-ct|/\nu)/\nu$ where α depends solely on $f'(\rho^{\pm})-c$ — for the case in which the order of contact of f and $\mathscr{L}f$ is one at the equilibrium points of Eq. (3). This indicates that, as ν goes to zero, $\nu\rho_{xx}(x,t)$ tends to zero wherever $x \neq ct$, and so the influence of the term $\nu\rho_{xx}$ is relevant only on a neighborhood of measure of order ν around the line $x=ct$. The additional fact that $\int_{-\infty}^{\infty} \nu\rho_{xx}(x,t)\,\mathrm{d}x=0$ implies that $\nu\rho_{xx}(x,t)$ does *not* tend to a Dirac delta as ν tends to zero; this renders its effect in the limit case quite subtle [23]. These are the facts that support the idea of dropping the term $\nu\rho_{xx}$ from the conservation law when ν is very small; however, to do that entails disastrous consequences as we show next.

2.3. The loss of well-posedness

It is very easy to see that when we let the diffusion coefficient to zero in the traveling wave solution, we obtain $\lim_{\nu\downarrow 0}\phi((x-ct)/\nu)=\rho^+$ if $x/t > c$ and $\lim_{\nu\downarrow 0}\phi((x-ct)/\nu)=\rho^-$ if $x/t < c$. Since this limit can be proven to be a *weak* solution of the following equation:

$$\rho_t + (f(\rho))_x = 0 \tag{4}$$

with initial data $\rho(x,0)=\rho^+$ if $x>0$, and $\rho(x,0)=\rho^-$ if $x<0$, this fact could be thought to be an indication that *formally* dropping the second-order term $\nu\rho_{xx}$, from Eq. (1) could be mathematically justified. Indeed, it is a well-known fact that piecewise-smooth *weak* solutions of the nonlinear hyperbolic conservation law (4) are strong solutions except at the discontinuity curves $(x(t),t)$ where the so-called jump condition is satisfied: $(\mathrm{d}/\mathrm{d}t)x(t)=[f(\rho)]/[\rho](x(t),t)$, where $[g]$ denotes the jump of the function g across the discontinuity. However, it is easy to construct infinitely many *weak* solutions for the same Cauchy problem.

To do that, let us fix ideas and set $f(\rho)=\rho(1-\rho)$, and $\rho^-=\frac{1}{4}$, $\rho^+=\frac{3}{4}$. Note that this gives $c=0$. A simple computation shows that the following functions are *weak* solutions of the same Cauchy problem, for *all* nonnegative values of the parameter δ:

$$\rho^{\delta}(x,t)=\begin{cases}\frac{3}{4} & \text{if } c_1 < x/t,\\ \frac{1}{4}-\delta & \text{if } c_2 < x/t < c_1,\\ \frac{3}{4}+\delta & \text{if } c_3 < x/t < c_2,\\ \frac{1}{4} & \text{if } x/t < c_3,\end{cases} \tag{5}$$

where $c_1=\delta$, $c_2=0$, $c_3=-\delta$. Note that the discontinuities $x/t=c_1$ and $x/t=c_3$ do satisfy the condition for the existence of traveling waves of the original conservation law (1). However, this is *not* true for the discontinuity $x/t=c_2$, except for $\delta=0$, of course; in other words, this discontinuity does not 'remember' anything about the physics contained in the modeling of the 'awareness of the conditions lying ahead'. Thus, because of the loss of this crucial information, *formally* dropping the second-order term $\nu\rho_{xx}$ from Eq. (1) results in the *loss of the well-posedness* of the Cauchy problem associated with the nonlinear hyperbolic conservation law (4), as claimed.

3. Devising discontinuous Galerkin methods: heuristics

3.1. The re-incorporation of the neglected term $\nu\rho_{xx}$

The unfortunate loss of well-posedness described for traffic flow is present, as a rule, in *all* nonlinear hyperbolic conservation laws. To re-incorporate the relevant physics into the numerical scheme constitutes the main difficulty in devising numerical schemes for nonlinear hyperbolic conservation laws. To show how to do this, we restrict ourselves to the simple framework of our model for traffic flow and, to render the presentation even simpler, we assume that the space domain is the interval (a,b) and that the boundary conditions are periodic. Since DG methods are extensions of finite-volume methods, we start by considering the celebrated *Godunov* finite-volume scheme; we pay special attention to the issue of re-incorporating the information of the term $\nu\rho_{xx}$ and show that this is achieved by using *Riemann solvers* and the corresponding *numerical fluxes*. This finite-volume scheme is then extended in two ways by using discontinuous Galerkin methods. In the first, the term $\nu\rho_{xx}$ is replaced in the weak formulation by a *shock-capturing* term; in the second, the term $\nu\rho_{xx}$ is removed from the weak formulation by using an operator splitting technique and then transformed into a *generalized slope limiter*. The first approach gives rise to the SCDG method, which is implicit, and the second to RKDG method, which is explicit.

3.2. The Godunov scheme

Let $\{x_{j+1/2}\}_{j=0}^{M}$ be a partition of $[a,b]$, and let us set $I_j = (x_{j-1/2}, x_{j+1/2})$ and $\Delta x_j = x_{j+1/2} - x_{j-1/2}$. Similarly, let $\{t^n\}_{n=0}^{N}$ be a partition of $[0,T]$, and let us set $J^n = (t^n, t^{n+1})$ and $\Delta t^n = t_{n+1} - t^n$. We want to find a *weak formulation* with which we will define the Godunov scheme. To do that we proceed as follows. Let us denote by ρ^ν the exact solution of the conservation law (1). Integrating Eq. (1) over the box $I_j \times J^n$, integrating by parts and formally *taking the limit* as ν goes to zero, we obtain

$$\Delta x_j(\rho_j^{n+1} - \rho_j^n) + \Delta t^n(\hat{f}_{j+1/2}^{n} - \hat{f}_{j-1/2}^{n}) = 0, \tag{6}$$

where $\rho_j^n = \lim_{\nu\to 0}(1/\Delta x_j)\int_{I_j}\rho^\nu(s,t^n)\,\mathrm{d}s$ and

$$\hat{f}_{j+1/2}^{n} = \lim_{\nu\to 0}\frac{1}{\Delta t^n}\int_{J^n}\{f(\rho^\nu(x_{j+1/2},\tau)) - \nu\rho_x^\nu(x_{j+1/2},\tau)\}\,\mathrm{d}\tau. \tag{7}$$

Note that the numerical flux $\hat{f}_{j+1/2}^{n}$ does contain information associated with the term $\nu\rho_{xx}$. To get a better feel for this quantity, let us consider the traveling wave solution $\rho^\nu = \phi((x-ct)/\nu)$ considered in the previous section. A simple computation gives, for $x_{j+1/2} = 0$,

$$\begin{aligned}
\hat{f}_{j+1/2}^{n} &= \lim_{\nu\to 0}\frac{1}{\Delta t^n}\int_{J^n}\{f(\phi(-c\tau/\nu)) - \phi'(-c\tau/\nu)\}\,\mathrm{d}\tau \\
&= \lim_{\nu\to 0}\frac{1}{\Delta t^n}\int_{J^n}\mathscr{L}(\phi(-c\tau/\nu))\,\mathrm{d}\tau \quad \text{by (3)}, \\
&= \begin{cases} f(\rho^+) & \text{if } c\leqslant 0, \\ f(\rho^-) & \text{if } c\geqslant 0, \end{cases}
\end{aligned}$$

which is nothing but *the average of the trace of* $f(\rho^0)$ on the edge $x_{j+1/2} \times J^n$, where ρ^0 is the entropy solution given by (5).

The Godunov scheme is now obtained as follows. Knowing the piecewise-constant approximation at time $t = t^n$, $\rho_h(x,t^n) = \rho_j^n$ for $x \in I_j$, compute another piecewise-constant approximation at time $t = t^{n+1}$ by using the weak formulation (6) where the *numerical flux* $\hat{f}$ given by (7) is evaluated by taking ρ^ν to be the solution of the following initial-value problem:

$$\rho_t^\nu + (f(\rho^\nu))_x - \nu\rho_{xx}^\nu = 0, \quad \text{in } (a,b) \times J^n, \quad \rho^\nu(t^n) = \rho_h(t^n) \quad \text{on } (a,b).$$

Of course, as written above, the computation of such a numerical flux does not look easy at all. Fortunately, it can be shown that, for Δt^n small enough, $\hat{f}_{j+1/2}^n = f(\rho(x_{j+1/2}, t^n))$ where ρ is the entropy solution of the following Riemann problem:

$$\rho_t + (f(\rho))_x = 0, \quad \text{in } R \times J^n, \qquad \rho(x,t^n) = \begin{cases} \rho_j^n & \text{if } x < x_{j+1/2}, \\ \rho_{j+1}^n & \text{if } x > x_{j+1/2} \end{cases}$$

and that

$$\hat{f}_{j+1/2}^n = \hat{f}^{\mathrm{G}}(\rho_{j-1/2}^n, \rho_{j+1/2}^n) \equiv \begin{cases} \min f(s), & \rho_{j-1/2}^n \leqslant s \leqslant \rho_{j+1/2}^n, \\ \max f(s), & \rho_{j-1/2}^n \geqslant s \geqslant \rho_{j+1/2}^n, \end{cases} \tag{8}$$

which generalizes the particular case of the traveling wave solution treated above; see [19].

Note that the main effort in devising the Godunov scheme (6), (8) has been invested in making sure that the influence of the term $\nu\rho_{xx}$ is well captured. This effort does pay off since it can be proven that the numerical solution obtained by use of the Godunov scheme converges to the entropy solution when the discretization parameters go to zero.

Fortunately, the Godunov scheme is not the only scheme with this property. Indeed, there are several schemes defined by the weak formulation (6) and numerical fluxes obtained by solving Riemann problems only *approximately* [14] which also converge to the entropy solution. Maybe the main two examples are the Engquist–Osher flux:

$$\hat{f}^{\mathrm{EO}}(a,b) = \int_0^b \min(f'(s),0)\,\mathrm{d}s + \int_0^a \max(f'(s),0)\,\mathrm{d}s + f(0)$$

and the Lax–Friedrichs flux:

$$\hat{f}^{\mathrm{LF}}(a,b) = \tfrac{1}{2}[f(a) + f(b) - C(b-a)], \quad C = \max_{\inf \rho_0 \leqslant s \leqslant \sup \rho_0} |f'(s)|,$$

which is particularly easy to compute. Moreover, the three methods above are such that, if $\|f'(\rho_0)\|_{L^\infty}(\Delta t^n/\Delta x_j) \leqslant 1$ then we have the *local* maximum principle

$$\rho_j^{n+1} \in [\![\rho_{j-1}^n, \rho_j^n, \rho_{j+1}^n]\!], \tag{9}$$

where $[\![a,b,c]\!] = [\min\{a,b,c\}, \max\{a,b,c\}]$ and the so-called total variation diminishing (TVD) property

$$|\rho_h^{n+1}|_{\mathrm{TV}} \leqslant |\rho_h^n|_{\mathrm{TV}}, \tag{10}$$

where $|\rho_h^m|_{\mathrm{TV}} = \sum_j |\rho_{j+1}^m - \rho_j^m|$. Unfortunately, these methods are at most first-order accurate [13].

The SCDG method was devised in an effort to obtain a high-order accurate method that converges to the entropy solution.

3.3. The SCDG method

In order to devise the SCDG method, we proceed as follows. First, we multiply Eq. (1) by a test function φ, integrate over the box $B_j^n = I_j \times J^n$ and then integrate by parts to obtain

$$\int_{\partial B_j^n} \{\rho^\nu n_t + (f(\rho^\nu) - \nu\rho_x^\nu)n_x\}\varphi \, \mathrm{d}\Gamma - \int_{B_j^n} \{\rho^\nu \varphi_t + (f(\rho^\nu) - \nu\rho_x^\nu)\varphi_x\} \, \mathrm{d}x \, \mathrm{d}t = 0.$$

To deal with the first term, we proceed as we did in the case of the Godunov scheme and replace it by

$$\int_{\partial B_j^n} \{\hat{\rho} n_t + \hat{f}(\rho)n_x\}\phi \, \mathrm{d}\Gamma,$$

where (n_x, n_t) denotes the outward unit normal to ∂B_j^n, $\hat{\rho}(x, t^m)$ is nothing but $\rho(x, t^m - 0))$, since this is precisely the Godunov flux for the identity function, and $\hat{f}(\rho)(x_{\ell+1/2}, t)$ is $\hat{f}^{\mathrm{G}}(\rho(x_{\ell+1/2} + 0, t)$, $\rho(x_{\ell+1/2} - 0, t))$.

The second term is simply replaced by

$$-\int_{B_j^n} (\rho\varphi_t + f(\rho)\varphi_x) \, \mathrm{d}x \, \mathrm{d}t + \int_{B_j^n} \hat{\nu}(\rho)\rho_x\varphi_x \, \mathrm{d}x \, \mathrm{d}t,$$

where the term containing the *artificial viscosity coefficient* $\hat{\nu}(\rho)$ is the so-called *shock-capturing term*. To obtain an idea of the form of the artificial viscosity coefficient, we note that

$$\begin{aligned}\int_{B_j^n} \nu\rho_x^\nu\varphi_x \, \mathrm{d}x \, \mathrm{d}t &= \int_{B_j^n} \left\{ \int_{-\infty}^x (\rho_t^\nu + (f(\rho^\nu))_x) \, \mathrm{d}X \right\} \varphi_x \, \mathrm{d}x \, \mathrm{d}t \\ &= \int_{B_j^n} \hat{\nu}(\rho^\nu)\rho_x^\nu\varphi_x \, \mathrm{d}x \, \mathrm{d}t,\end{aligned}$$

where

$$\hat{\nu}(\rho) = \frac{\int_{-\infty}^x (\rho_t + (f(\rho))_x) \, \mathrm{d}X}{\rho_x}.$$

This motivates the following (typical) choice:

$$\hat{\nu}(\rho)|_{I_j} = \delta_j \frac{\|\rho_t + (f(\rho))_x\|_{L^1(I_j)}}{|\rho_x| + \varepsilon},$$

where the auxiliary parameter δ_j is usually taken to be of the order of the diameter of I_j and the small positive number ε is fixed — its sole purpose is to avoid dividing by zero when $\rho_x = 0$.

We are now ready to define the SCDG method. The approximate solution ρ_h is taken to be a function whose restriction to the box B_j^n is in the polynomial space $\mathscr{P}(B_j^n)$ and is determined as follows: Given $\rho_h(\cdot, t^n - 0)$, we compute ρ_h in $\bigcup_j B_j^n$ as the only solution of

$$\begin{aligned}&\int_{\partial B_j^n} \{\hat{\rho}_h n_t + \hat{f}(\rho_h)n_x\}\varphi \, \mathrm{d}\Gamma - \int_{B_j^n} \{\rho_h\varphi_t + f(\rho_h)\varphi_x\} \, \mathrm{d}x \, \mathrm{d}t \\ &\quad + \int_{B_j^n} \hat{\nu}(\rho_h)(\rho_h)_x\varphi_x \, \mathrm{d}x \, \mathrm{d}t = 0 \quad \forall \varphi \in \mathscr{P}(B_j^n). \qquad (11)\end{aligned}$$

Note that the data is contained only in $\hat{\rho}_h(x,t^n) = \rho_h(x,t^n - 0)$. Note also that if the approximate solution is taken to be piecewise constant in space and time, the above weak formulation becomes

$$\int_{\partial B_j^n} \{\hat{\rho}_h n_t + \hat{f}(\rho_h) n_x\} \varphi \, d\Gamma = 0,$$

or,

$$\Delta x_j(\rho_j^n - \rho_j^{n-1}) + \Delta t^n(\hat{f}_{j+1/2}^n - \hat{f}_{j-1/2}^n) = 0, \tag{12}$$

which is nothing else but the implicit version of the Godunov scheme (6). Since we can use numerical fluxes other than the Godunov flux, this shows that the SCDG method is an extension of finite-volume schemes.

Finally, let us point out that the effort invested in re-introducing the information of the term $\nu\rho_{xx}$ into the SCDG method pays off; indeed, both a priori and a posteriori error estimates for this method have been obtained [5]. Also, because of the form of the artificial viscosity coefficient, the method is high-order accurate when the exact solution is smooth, as wanted.

On the other hand, no maximum principle or TVD property similar to (9) and (10), respectively, hold for this method which could be very convenient in practical computations. Another disadvantage of the SCDG method stems from it being implicit. Implicit methods are less popular than explicit methods when solving hyperbolic problems because (i) they contain more artificial viscosity, which results in a worse approximation of shocks, and (ii) they give rise to global systems of equations whose resolution becomes very inefficient when discontinuities are present. Indeed, when solving the implicit Godunov scheme with Newton's method in the simple case in which $f' \geqslant 0$, $f \in C^2$, and uniform grids, it can be proven that the convergence of Newton's method is ensured by the Kantorovich sufficient condition if

$$\|f'\|_{L^\infty} \frac{\Delta t^n}{\Delta x} \leqslant C(f, \rho_h),$$

where

$$c(f,\rho_h) = \left\{ \frac{1}{2} \left(\frac{\|f'\|_{L^\infty}}{\operatorname{disc}(\rho_h^{n-1}) \cdot \|f''\|_{L^\infty}} \right)^{1/2} + \frac{1}{4} \right\}^{1/2} - \frac{1}{2}$$

and $\operatorname{disc}(\rho_h^{n-1}) = \max|\rho_{j+1}^{n-1} - \rho_j^{n-1}|$. If ρ_h^{n-1} is smooth in the sense that $\operatorname{disc}(u_h^{n-1})$ is of order Δx, the above condition states that Δt^n must be of order $\Delta x^{3/4}$, which explains the fast convergence of the method in this case. However, if a discontinuity is present, then $\operatorname{disc}(\rho_h^{n-1})$ might be of order one (like in the case of strictly convex or concave nonlinearities), and then the above condition states that Δt^n must be of order Δx only. For the explicit schemes, a similar relation holds between Δt^n and Δx, but no iterative procedure has to be put in place and no Jacobian matrix has to be assembled and inverted at each time step. Even when these considerations are put aside, practical experience shows that the possibly larger Δt^n that implicit schemes can use is not significantly bigger than the Δt^n for their explicit counterparts.

The RKDG methods were devised [8–10,6,11] in an effort to obtain explicit, high-order accurate methods that converge to the entropy solution for which provable stability properties similar to (9) and (10) hold.

3.4. The RKDG method

To construct the RKDG method we first discretize the equations in space and then we discretize them in time. Thus, we multiply Eq. (1) by a test function φ, integrate over the interval $B_j^n = I_j$ and then integrate by parts to obtain

$$\{f(\rho^\nu) - \nu\rho_x^\nu)\}\varphi|_{x_{j-1/2}}^{x_{j+1/2}} + \int_{I_j} \{\rho_t^\nu\varphi - (f(\rho^\nu) - \nu\rho_x^\nu)\varphi_x\}\,\mathrm{d}x = 0.$$

Next, we pass to the limit in ν as we did for the SCDG method, to obtain

$$\hat{f}(\rho)\varphi|_{x_{j-1/2}}^{x_{j+1/2}} + \int_{I_j} \{\rho_t\varphi - (f(\rho) - \hat{\nu}(\rho)\rho_x)\varphi_x\}\,\mathrm{d}x = 0.$$

Now, we use *operator splitting* and approximate the solution of the above weak formulation. Suppose we know $\rho(\cdot, t_0)$; we compute $\rho(\cdot, t_1)$ as follows. First, we advance in time from t_0 to $t_{1/2}$ by solving the equation

$$\hat{f}(\rho)\varphi|_{x_{j-1/2}}^{x_{j+1/2}} + \int_{I_j} \{\rho_t\varphi - f(\rho)\varphi_x\}\,\mathrm{d}x = 0,$$

then, starting from $\rho(\cdot, t_{1/2})$, we advance in time from $t_{1/2}$ to t_1 by solving the equation

$$\int_{I_j} \{\rho_t\varphi - \hat{\nu}(\rho)\rho_x\varphi_x\}\,\mathrm{d}x = 0.$$

Note how what was the shock-capturing term in the SCDG method has been split off from the weak formulation and is now used in the second step to advance in time the approximate solution.

Let us emphasize that by assuming the function ρ to be very smooth and taking $\varphi = -(|\rho_x|)_x$, we can obtain that

(i) $|\rho(t_{1/2})|_{\mathrm{TV}} \leqslant |\rho(t_0)|_{\mathrm{TV}}$.

In a similar way, we can get that $\bar{\rho}(t_1)_j = \int_{I_j} \rho(t_1)\,\mathrm{d}x/\Delta x_j = \bar{\rho}(t_{1/2})_j$ by taking $\varphi = 1$, and that $|\rho(t_1)|_{\mathrm{TV}} \leqslant |\rho(t_{1/2})|_{\mathrm{TV}}$, by taking $\varphi = -(|\rho_x|)_x$.

Another point worth emphasizing is that for the RKDG method the artificial viscosity coefficient *does not* depend on the residual; instead, it depends on the *local smoothness* of the function ρ. For example, if ρ is locally $\mathscr{C}^1$, we take $\hat{\nu}(\rho) = 0$; this is motivated by the fact that $\lim_{\nu\downarrow 0} \hat{\nu}(\rho)\rho_x^\nu = 0$ if the entropy solution is $\mathscr{C}^1$.

Before discretizing in space the above equations, we transform the second step. To do that, we use a simple Euler forward time discretization

$$\int_{I_j} \left\{ \left(\frac{\rho(t_1) - \rho(t_{1/2})}{t_1 - t_{1/2}} \right) \varphi - \hat{\nu}(\rho(t_{1/2}))\rho_x(t_{1/2})\varphi_x \right\} \mathrm{d}x = 0,$$

which we rewrite as follows:

$$\int_{I_j} \rho(t_1)\varphi\,\mathrm{d}x = \int_{I_j} \{\rho(t_{1/2})\varphi - \mathrm{SL}\rho_x(t_{1/2})\varphi_x\}\,\mathrm{d}x, \tag{13}$$

where the *slope limiter coefficient* SL is nothing but $(t_1 - t_{1/2})\hat{\nu}(\rho(t_{1/2}))$. In other words, $\rho(t_1)$ is obtained directly from $\rho(t_{1/2})$ by a simple and *local* operator; form now on, we write

$\rho(t_1) = \Lambda\Pi(\rho(t_{1/2}))$ and call $\Lambda\Pi$ a generalized slope limiter. Note that the above properties of the mapping $\rho(t_{1/2}) \mapsto \rho(t_1)$ and that of the artificial viscosity $\hat{v}$ can be rewritten in terms of the limiter as follows:

(ii) $\overline{\Lambda\Pi}(\rho)_j = \bar{\rho}_j$,
(iii) $|\Lambda\Pi(\rho)|_{\mathrm{TV}} \leqslant |\rho|_{\mathrm{TV}}$,
(iv) $\Lambda\Pi(\rho) = \rho$ if ρ is *smooth*.

We are now ready to display the space discretization. The approximate solution $\rho_h(\cdot,t)$, $t_0 \leqslant t \leqslant t_1$, is taken to be a function whose restriction to the interval I_j is in the polynomial space $\mathscr{P}(I_j)$ and is determined as follows. Knowing $\rho_h(\cdot,t_0)$, we compute the auxiliary function $\rho_h(\cdot,t_{1/2})$ as the solution of

$$\hat{f}(\rho_h)\varphi|_{x_{j-1/2}}^{x_{j+1/2}} + \int_{I_j} \{(\rho_h)_t\varphi - f(\rho_h)\varphi_x\}\,\mathrm{d}x = 0 \quad \forall\varphi \in \mathscr{P}(I_j), \tag{14}$$

then, we set $\rho_h(\cdot,t_1) = \Lambda\Pi_h(\rho_h(\cdot,t_{1/2}))$ where $\Lambda\Pi_h$ is a discrete version of $\Lambda\Pi$.

Finally, we discretize in time by using a special Runge–Kutta discretization; see [21,12,22]. If we rewrite the first equation as $\mathrm{d}\rho_h/\mathrm{d}t = L_h(\rho_h)$, the RKDG method can be described as follows. Knowing the approximate solution t^n, ρ_h^n, we compute ρ_h^{n+1} as indicated below:

(1) set $\rho_h^{(0)} = \rho_h^n$;
(2) for $i = 1,\ldots,K$ compute the intermediate functions:

$$\rho_h^{(i)} = \Lambda\Pi_h\left(\sum_{l=0}^{i-1} \alpha_{il} w_h^l\right), \quad w_h^l = \rho_h^{(l)} + \frac{\beta_{il}}{\alpha_{il}}\Delta t^n L_h(\rho_h^{(l)});$$

(3) set $\rho_h^{n+1} = \rho_h^K$.

The fact that, given the polynomial spaces $\mathscr{P}(I_j)$, it is possible to find limiters $\Lambda\Pi_h$ and coefficients α_{il} and β_{il} such the nonlinear stability of the scheme is ensured without degradation of high-order accuracy is a nontrivial distinctive feature of the RKDG method which we consider in detail in the next section.

To end this section, let us stress once again that the term $v\rho_{xx}$ was incorporated into the RKDG method by using Riemann solvers and their corresponding numerical fluxes *and* by means of the limiter $\Lambda\Pi_h$ which incorporates the effect of $v\rho_{xx}$ in the interior of the elements.

4. The RKDG method

In this section, we take a closer look to the RKDG method; we keep our one dimensional framework for the sake of simplicity. We consider the RKDG method with the Engquist–Osher numerical flux and study its stability for a simple limiter. We then discuss the accuracy of the method, its convergence properties and, finally, show some typical numerical results displaying its performance.

4.1. Nonlinear stability: the TVDM property

We proceed in several steps; we follow [9]. First, we describe the operator L_h and find the properties of the function ρ_h for which $|\bar{w}_h|_{TV} \leqslant |\bar{\rho}_h|_{TV}$, where $w_h = \rho_h + \delta L_h(\rho_h)$. As a second step, we construct a limiter $\Lambda\Pi_h^k$ such that $\rho_h = \Lambda\Pi_h^k(\tilde{\rho}_h)$ satisfies those properties. Finally, we show that the RKDG method is TVDM provided the RK time discretization satisfies certain simple conditions.

Step 1: *The operator* L_h *and the function* $w_h = \rho_h + \delta L_h(\rho_h)$. In what follows, we consider approximations ρ_h such that for each time t and each interval I_j, $\rho_h(t)|_{I_j}$ is a polynomial of degree k. We take as the local basis function the suitably scaled Legendre polynomials, that is, for $x \in I_j$, we write $\rho_h(x,t) = \sum_{\ell=0}^{k} u_j^\ell(t)\varphi_j^\ell(x)$, where $\varphi_j^\ell(x) = P_\ell(2(x - x_j)/\Delta x_j)$ and P_ℓ is the ℓth Legendre polynomial. Since these polynomials are orthogonal, that is, since $\int_{-1}^{1} P_\ell(s)P_{\ell'}(s)\,ds = 2\delta_{\ell\ell'}/(2\ell+1)$, the mass matrix is *diagonal*. Indeed, the weak formulation (14) takes the following simple form: For each interval I_j and each $\ell = 0,\ldots,k$, we have

$$\frac{d}{dt}u_j^\ell(t) - \frac{2\ell+1}{\Delta x_j}\int_{I_j} f(\rho_h(x,t))\frac{d}{dx}\varphi_\ell(x)\,dx$$

$$+\frac{2\ell+1}{\Delta x_j}\{\hat{f}^{\mathrm{EO}}(\rho_h(x_{j+1/2}))(t) - (-1)^\ell \hat{f}^{\mathrm{EO}}(\rho_h(x_{j-1/2}))(t)\} = 0,$$

where we have used that $P_\ell(1)=1$ and $P_\ell(-1)=(-1)^\ell$. We rewrite the above equations as follows: $(d/dt)\rho_h = L_h(\rho_h)$; the function $L_h(\rho_h)$ is piecewise polynomial of degree k and is nothing but the approximation to $-f(u)_x$ provided by the DG-space discretization.

Next, we consider the stability properties of the mapping $\rho_h \mapsto w_h = \rho_h + \delta L_h(\rho_h)$. We have the following result, which is a discrete version of property (i) of the finite volume methods of Section 3.

Proposition 1 (The sign conditions). *If* $|\delta|((|f^+|_{\mathrm{Lip}}/\Delta x_{j+1})+(|f^-|_{\mathrm{Lip}}/\Delta x_j)) \leqslant \frac{1}{2}$, *then* $|\bar{w}_h|_{TV} \leqslant |\bar{\rho}_h|_{TV}$ *provided the following conditions are satisfied*:

$$\mathrm{sgn}(\bar{\rho}_j - \bar{\rho}_{j-1}) = \mathrm{sgn}(\rho_{j+1/2}^{n,-} - \rho_{j-1/2}^{n,-}),$$

$$\mathrm{sgn}(\bar{\rho}_{j+1} - \bar{\rho}_j) = \mathrm{sgn}(\rho_{j+1/2}^{n,+} - \rho_{j-1/2}^{n,+}).$$

Of course, the above sign conditions are not guaranteed to be satisfied at all; in order to enforce them, we shall use the limiter $\Lambda\Pi_h^k$.

Step 2: *Construction of a limiter* $\Lambda\Pi_h^k$. We construct our limiter $\Lambda\Pi_h^k$ in two steps. First, we consider the piecewise-linear case $k=1$ and set $\Lambda\Pi_h^1(\rho_h)|_{I_j} = \bar{\rho}_j + m(u_j^1, \bar{\rho}_{j+1} - \bar{\rho}_j, \bar{\rho}_j - \bar{\rho}_{j-1})\varphi_j^1(x)$, where the so-called *minmod* function is $m(a_1,\ldots,a_\nu) = s\min_{1\leqslant n\leqslant\nu}|a_n|$ if $s = \mathrm{sign}(a_1) = \cdots = \mathrm{sign}(a_\nu)$, and $m(a_1,\ldots,a_\nu) = 0$ otherwise. Note that $\Lambda\Pi_h^1(\rho_h)$ is always a piecewise-linear function (see Fig. 1).

It is not a coincidence that, for piecewise linear functions, the slope limiter $\Lambda\Pi_h^1$ can be defined by using a discrete version of Eq. (13), namely,

$$\int_{I_j} \Lambda\Pi_h^1(\rho)\varphi\,dx = \int_{I_j}\{\rho\varphi - \mathrm{SL}\;\rho_x\varphi_x\}\,dx \quad \forall \text{ linear functions } \varphi. \tag{15}$$

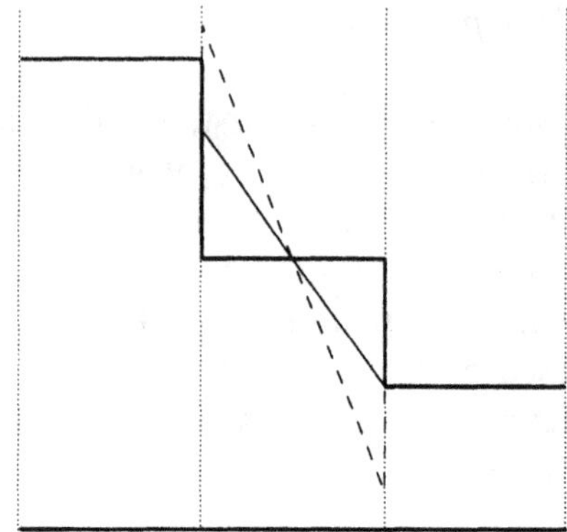

Fig. 1. The $\Lambda\Pi_h^1$ limiter: The local means of ρ_h (thick line), the linear function ρ_h in the element of the middle before limiting (dotted line) and the resulting function after limiting (solid line).

Indeed, a simple computation shows that the slope limiter coefficient is nothing but

$$\mathrm{SL} = \frac{\Delta x_j^2}{12}\left(1 - \frac{m(u_j^1, \bar{\rho}_{j+1} - \bar{\rho}_j, \bar{\rho}_j - \bar{\rho}_{j-1})}{u_j^1}\right),$$

which is a nonnegative number, as expected, and is a discrete measure of the local smoothness of the approximate solution ρ_h.

We now define $\Lambda\Pi_h^k(\rho_h)$, element by element, as follows:

(1) Compute

$$\tilde{\rho}_{j+1/2}^- = \bar{\rho}_j + m(\rho_{j+1/2}^- - \bar{\rho}_j, \bar{\rho}_{j+1} - \bar{\rho}_j, \bar{\rho}_j - \bar{\rho}_{j-1}),$$

$$\tilde{\rho}_{j-1/2}^+ = \bar{\rho}_j - m(\bar{\rho}_j \rho_{j-1/2}^+, -\bar{\rho}_{j+1} - \bar{\rho}_j, \bar{\rho}_j - \bar{\rho}_{j-1}).$$

(2) If $\tilde{\rho}_{j+1/2}^- = \rho_{j+1/2}^-$ and $\tilde{\rho}_{j-1/2}^+ = \rho_{j-1/2}^+$ set $\Lambda\Pi_h^k(\rho_h)|_{I_j} = \rho_h|_{I_j}$.
(3) If not, set $\Lambda\Pi_h^k(\rho_h)|_{I_j} = \Lambda\Pi_h^1(\rho_h)|_{I_j}$.

Note that this algorithm can be carried out in parallel.

Next, we put together the main properties of our limiter some of which are discrete versions of properties (ii)–(iv) of Section 3.

Proposition 2 (Properties of the limiter $\Lambda\Pi_h^k$). *Given any piecewise polynomial function η_h, the function $\rho_h := \Lambda\Pi_h^k(\eta_h)$ satisfies the sign conditions of Proposition* 1. *Moreover,*

(i) $\overline{\Lambda\Pi_h^k(\eta_h)}_j = \overline{\eta_{h_j}}$,
(ii) $|\Lambda\Pi_h^k(\eta_h)|_{\mathrm{TV}} \leqslant |\eta_h|_{\mathrm{TV}}$, *if η_h is piecewise linear,*
(iii) $\Lambda\Pi_h^k(\eta_h) = \eta_h$ *if η_h is linear.*

Step 3: *The TVDM property of the RKDG method.* Now, we collect the results obtained in the previous steps and obtain a remarkable stability property of the RKDG method similar to the TVD property (10) of the finite-volume methods of Section 3.

Table 1
Runge–Kutta discretization parameters

Order	α_{il}	β_{il}	$\max\{\beta_{il}/\alpha_{il}\}$
1	1	1	1
2	$\begin{pmatrix} 1 & \\ 1/2 & 1/2 \end{pmatrix}$	$\begin{pmatrix} 1 & \\ 0 & 1/2 \end{pmatrix}$	1
3	$\begin{pmatrix} 1 & & \\ 3/4 & 1/4 & \\ 1/3 & 0 & 2/3 \end{pmatrix}$	$\begin{pmatrix} 1 & & \\ 0 & 1/4 & \\ 0 & 0 & 2/3 \end{pmatrix}$	1

Theorem 3 (TVDM property of the RKDG method). *Assume that all the coefficients α_{il} are nonnegative and such that $\sum_{l=0}^{i-1}\alpha_{il}=1$, $i=1,\ldots,K+1$. Assume also that $\Delta t^n(|\beta^{il}/\alpha^{il}|)|((|f^+|_{\text{Lip}}/\Delta x_{j+1})+(|f^-|_{\text{Lip}}/\Delta x_j))\leqslant 1/2$. Then we have that*

$$|\bar{\rho}_h^n|_{\text{TV}}\leqslant|\rho_0|_{\text{TV}},\quad \forall n\geqslant 0.$$

This result states that the RKDG method produces an approximate solution whose element-by-element average is total variation bounded regardless of the degree of the polynomial, k, and the accuracy of the RK method used to march in time. Examples of RK methods satisfying the conditions of Theorem 3, which are the so-called TVD Runge–Kutta methods [21,12,22], are displayed in Table 1 below; for other higher-order TVD-RK methods, see [22,12].

It is interesting to point out that under the conditions of the above theorem, the local maximum principle (9) also holds for the local averages, $\bar{\rho}_h$. We include a proof [22,9] of the above theorem because it is extremely simple and because it shows how the different ingredients of the RKDG method come together.

Proof. Recall that the RKDG method computes ρ_h^{n+1} from ρ_h^n by setting $\rho_h^{n+1}=\rho_h^K$ and by computing recursively $\rho_h^{(i)}=\Lambda\Pi_h^k\{\sum_{l=0}^{i-1}\alpha_{il}w_h^{(il)}\}$, where $w_h^{(il)}=\rho_h^{(l)}+(\beta_{il}/\alpha_{il})\Delta t^n L_h(\rho_h^{(l)})$ for $i=1,\ldots,K$ and $\rho_h^0=\rho_h^n$. Hence,

$$\begin{aligned}
|\bar{\rho}_h^{(i)}|_{\text{TV}} &= \left|\sum_{l=0}^{i-1}\alpha_{il}\bar{w}_h^{(il)}\right|_{\text{TV}} \quad \text{by Proposition 2,}\\
&\leqslant \sum_{l=0}^{i-1}\alpha_{il}|\bar{w}_h^{(il)}|_{\text{TV}} \quad \text{since } \alpha_{il}\geqslant 0,\\
&\leqslant \sum_{l=0}^{i-1}\alpha_{il}|\bar{\rho}_h^{(l)}|_{\text{TV}} \quad \text{by Propositions 1 and 2,}\\
&\leqslant \max_{0\leqslant l\leqslant i-1}|\bar{\rho}_h^{(l)}|_{\text{TV}},
\end{aligned}$$

since $\sum_{l=0}^{i-1} \alpha_{il} = 1$. By induction, we get that $|\bar{\rho}_h^{n+1}|_{TV} \leqslant |\bar{\rho}_h^n|_{TV}$, and so $|\bar{\rho}_h^n|_{TV} \leqslant |\bar{\rho}_h^0|_{TV}$. Since ρ_h^0 is the L^2-projection of ρ_0, we have $|\bar{\rho}_h^0|_{TV} \leqslant |\rho_0|_{TV}$, and the result follows.

4.2. Accuracy, convergence properties and some numerical results

Let us begin our discussion about the accuracy of the method by considering the linear case f' constant. If we simply consider the DG space discretization and do not employ limiters, it can be proven that when polynomials of degree k are used, the method is of order $k+1$ in the L^2-norm, uniformly in time, provided that the exact solution is smooth enough. If not a TVD-RK time discretization of order $k+1$ is used, there is ample numerical evidence that indicates that the accuracy remains the same. When the complete RKDG method is used and the exact solution does not present any local critical points the accuracy is again of order $k+1$. However, in the presence of local critical points, the use of our limiter does degrade the order of accuracy. This happens because at those critical points, the approximation solution, which is a polynomial of degree k, is forcefully set equal to a constant by the limiter. Fortunately, this difficulty can be overcome by means of a slight *modification* of the limiter $\Lambda\Pi_h^k$- the resulting RKDG scheme can be proven to be, not a TVDM scheme, but a total variation bounded in the means (TVBM) scheme; see [20,9]. Moreover, the observed order of accuracy is now $k+1$. This has also been observed in all the main prototypical nonlinear cases; see [9].

Now, let us address the issue of convergence to the entropy solution. From Theorem 3, it is easy to see that the RKDG method (with the $\Lambda\Pi_h^k$ limiter or its TVBM modification) generates a subsequence that converges to a weak solution. However, unlike the case of the SCDG method, it has not been proven that such a weak solution is the entropy solution. Since all the numerical experiments do indicate that this is actually the case, it is reasonable to expect that such a proof could be obtained. This would require, however, new techniques; see [15,3,18].

Let us end this section by showing two numerical examples, both on the domain (0,1). The first deals with the simple, but numerically difficult to approximate, linear case $f(\rho) = \rho$ with initial condition $\rho_0(x) = 1$ if $0.4 < x < 0.6$ and $\rho_0(x) = 0$ otherwise. Our purpose is to illustrate how the amount of numerical dissipation varies with the polynomial degree k of the RKDG method; we use our $\Lambda\Pi_h^k$ limiter. The results, for $T = 100$, are displayed in Fig. 2 where we see that the numerical dissipation diminishes as the polynomial degree increases. The second example deals with the classical inviscid Burgers equation $f(\rho) = \rho^2/2$ with initial data $\rho_0(x) = 0.25 + 0.5\sin(\pi(2x-1))$. Our purpose is to display the performance of the RKDG method around the shock. We use a TVBM modification of our limiter which leaves untouched local critical points whose second-order derivative is, in absolute value, smaller than $M = 20$. In Fig. 3 we display how a shock passes through a single element. Note the excellent approximation to the exact solution only *one* element away form the exact shock for both $k = 1$ and 2.

5. Concluding remarks

The RKDG method has been extended to multidimensional nonlinear hyperbolic systems and has proven to be a highly parallelizable, stable and accurate method. It has also been extended to convection–diffusion equations, the Hamilton–Jacobi equations, nonlinear possibly degenerate

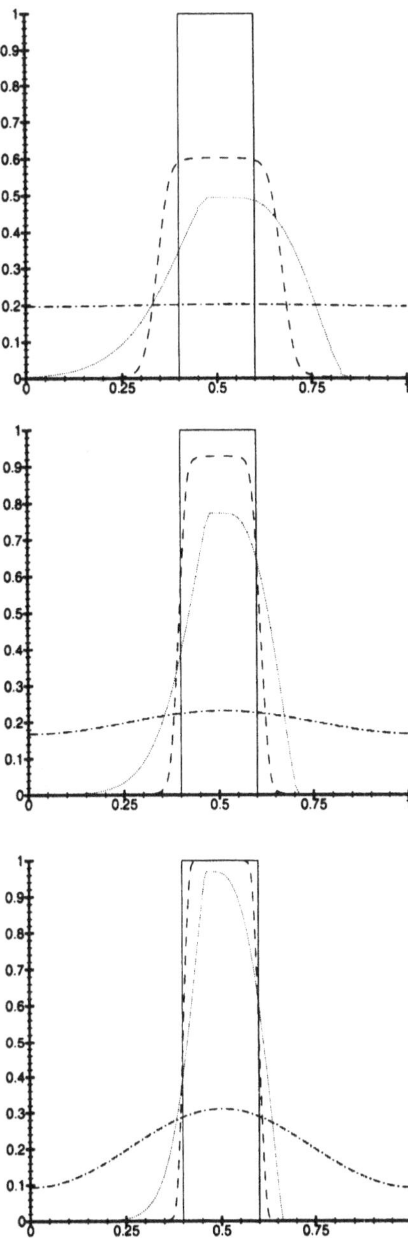

Fig. 2. Comparison of the exact and the approximate solutions for the linear case $f(\rho)=\rho$. Top: $\Delta x=\frac{1}{40}$, middle: $\Delta x=1.80$, bottom: $\Delta x=1/160$. Exact solution (solid line), piecewise-constant elements (dash/dotted line), piecewise-linear elements (dotted line) and piecewise-quadratic elements (dashed line).

second-order parabolic equations and even to elliptic problems. The reader is referred to [7] for an overview of the historical evolution of DG methods, for an account of the state of the art, and for a short description of the main challenges of their future development.

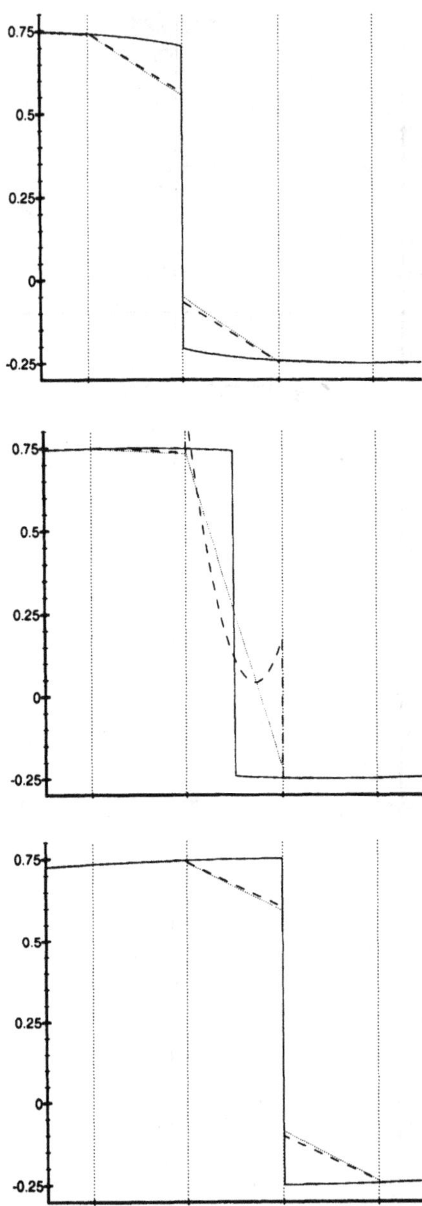

Fig. 3. Comparison of the exact and the approximate solutions for the Burgers equation $f(\rho) = \rho^2/2$, $\Delta x = 1/40$ as the shock passes through one element. Exact solution (solid line), piecewise linear elements (dotted line) and piecewise quadratic elements (dashed line). Top: $T = 0.40$, middle: $T = 0.45$, and bottom: $T = 0.50$.

Let us give an idea of the computational results this method produces and let us consider the classical double Mach reflection test problem. In Fig. 4, we display the results obtained by using P^2 polynomials on squares. Note the rich structure around the contact discontinuities that can be

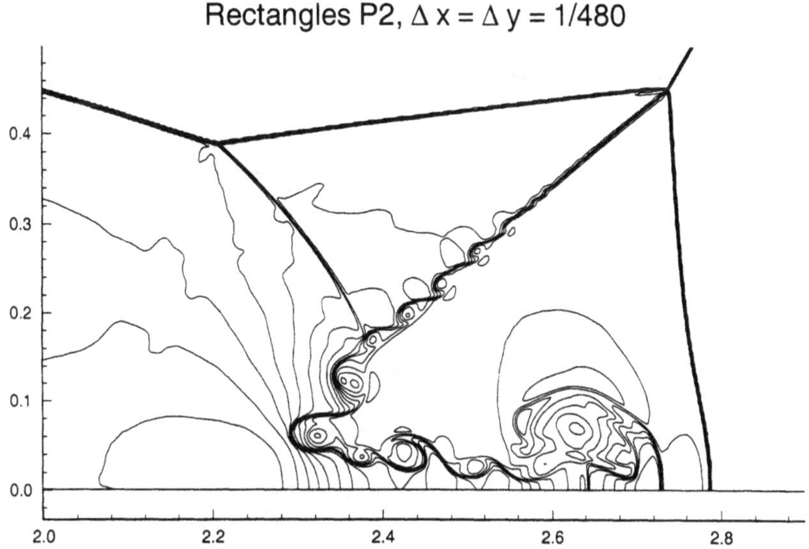

Fig. 4. Approximate density obtained by the RKDG method.

captured because of the small amount of artifical dissipation (recall Fig. 2) and how sharply the strong shocks have been approximated (recall Fig. 3).

In this paper we have tried to give a flavor of the techniques used to devise DG methods for nonlinear hyperbolic conservation laws in the framework of a one-dimensional scalar conservation law. Our main point is that the non-linear scalar conservation law obtained by dropping the term $\nu\rho_{xx}$ from the equation defines an ill-posed problem and that the main effort to devise DG methods for the hyperbolic conservation law is to re-incorporate into the numerical scheme the information provided by the neglected term. We have seen that this can be accomplished by means of (i) approximate Riemann solvers and their corresponding numerical fluxes and by means of (ii) shock-capturing terms and generalized slope limiters. It is our hope that these ideas could be of use when devising DG methods for other nonlinear hyperbolic systems and even for more challenging ill-posed problems like, for example, the model of phase transition in solids; see [4] and the references therein.

Acknowledgements

The author would like to thank Endre Süli for the kind invitation to incorporate this paper into this special volume of the Journal of Computational and Applied Mathematics. The author would also like to thank Franco Brezzi, Pierre A. Gremaud, Donatella Marini, K.W. Morton, Petr Plecháč, Chi-Wang Shu, and Endre Süli for the many conversations this paper is the product of.

References

[1] B. Cockburn, An introduction to the discontinuous Galerkin method for convection-dominated problems, in: A. Quarteroni (Ed.), Advanced Numerical Approximation of Nonlinear Hyperbolic Equations, Lecture Notes in Mathematics, Vol. 1697; Subseries Fondazione C.I.M.E., Firenze, Springer, Berlin, 1998, pp. 151–268.

[2] B. Cockburn, Discontinuous Galerkin methods for convection-dominated problems, in: T. Barth, H. Deconink (Eds.), High-Order Methods for Computational Physics, Lecture Notes in Computational Science and Engineering, Vol. 9, Springer, Berlin, 1999, pp. 69–224.

[3] B. Cockburn, F. Coquel, P. LeFloch, An error estimate for finite volume methods for multidimensional conservation laws, Math. Comp. 63 (1994) 77–103.

[4] B. Cockburn, H. Gau, A model numerical scheme for the propagation of phase transitions in solids, SIAM J. Sci. Comput. 17 (1996) 1092–1121.

[5] B. Cockburn, P.A. Gremaud, Error estimates for finite element methods for nonlinear conservation laws, SIAM J. Numer. Anal. 33 (1996) 522–554.

[6] B. Cockburn, S. Hou, C.W. Shu, TVB Runge–Kutta local projection discontinuous Galerkin finite element method for conservation laws IV: the multidimensional case, Math. Comp. 54 (1990) 545–581.

[7] B. Cockburn, G.E. Karniadakis, C.-W. Shu (Eds.), First International Symposium on Discontinuous Galerkin Methods, Lecture Notes in Computational Science and Engineering, Vol. 11, Springer, Berlin, February 2000.

[8] B. Cockburn, S.Y. Lin, C.W. Shu, TVB Runge–Kutta local projection discontinuous Galerkin finite element method for conservation laws III: one dimensional systems, J. Comput. Phys. 84 (1989) 90–113.

[9] B. Cockburn, C.W. Shu, TVB Runge–Kutta local projection discontinuous Galerkin finite element method for scalar conservation laws II: general framework, Math. Comp. 52 (1989) 411–435.

[10] B. Cockburn, C.W. Shu, The Runge–Kutta local projection P^1-discontinuous Galerkin method for scalar conservation laws, RAIRO Modél. Math. Anal. Numér. 25 (1991) 337–361 .

[11] B. Cockburn, C.W. Shu, The Runge–Kutta discontinuous Galerkin finite element method for conservation laws V: Multidimensional systems, J. Comput. Phys. 141 (1998) 199–224.

[12] S. Gottlieb, C.-W. Shu, Total variation diminishing Runge–Kutta schemes, Math. Comp. 67 (1998) 73–85.

[13] A. Harten, J.M. Hyman, P. Lax, On finite difference approximations and entropy conditions for shocks, Comm. Pure Appl. Math. 29 (1976) 297–322.

[14] A. Harten, P. van Leer, On upstream differencing and godunov-type schemes for hyperbolic conservation laws, SIAM Rev. 25 (1983) 35–61.

[15] G. Jiang, C.-W. Shu, On cell entropy inequality for discontinuous Galerkin methods, Math. Comp. 62 (1994) 531–538.

[16] C. Johnson, Adaptive finite element methods for conservation laws, in: A. Quarteroni (Ed.), Advanced Numerical Approximation of Nonlinear Hyperbolic Equations, Lecture Notes in Mathematics, Vol. 1697; Subseries Fondazione C.I.M.E., Firenze, Springer, Berlin, 1998, pp. 269–323.

[17] K.W. Morton, Numerical Solution of Convection-Diffusion Problems, Chapman & Hall, London, 1996.

[18] S. Nöelle, A note on entropy inequalities and error estimates for higher-order accurate finite volume schemes on irregular families of grids, Math. Comp. 65 (1996) 1155–1163.

[19] S. Osher, Riemann solvers, the entropy condition and difference approximations, SIAM J. Numer. Anal. 21 (1984) 217–235.

[20] C.W. Shu, TVB boundary treatment for numerical solutions of conservation laws, Math. Comp. 49 (1987) 123–134.

[21] C.-W. Shu, Essentially non-oscillatory and weighted essentially non-oscillatory schemes for hyperbolic conservation laws, in: A. Quarteroni (Ed.), Advanced Numerical Approximation of Nonlinear Hyperbolic Equations, Lecture Notes in Mathematics, Vol. 1697, Subseries Fondazione C.I.M.E., Firenze, Springer, Berlin, 1998, pp. 325–432.

[22] C.-W. Shu, S. Osher, Efficient implementation of essentially non-oscillatory shock-capturing schemes, J. Comput. Phys. 77 (1988) 439–471.

[23] E. Tadmor, Approximate solutions of nonlinear conservation laws, in: A. Quarteroni (Ed.), Advanced Numerical Approximation of Nonlinear Hyperbolic Equations, Lecture Notes in Mathematics, Vol. 1697; Subseries Fondazione C.I.M.E., Firenze, Springer, Berlin, 1998, pp. 1–149.

[24] G.B. Witham, Linear and Nonlinear Waves, Wiley, New York, 1974.

ELSEVIER Journal of Computational and Applied Mathematics 128 (2001) 205–233

JOURNAL OF COMPUTATIONAL AND APPLIED MATHEMATICS

www.elsevier.nl/locate/cam

Adaptive Galerkin finite element methods for partial differential equations

R. Rannacher

Institute of Applied Mathematics, University of Heidelberg, INF 293/294, D-69120 Heidelberg, Germany

Received 10 November 1999; received in revised form 28 February 2000

Abstract

We present a general method for error control and mesh adaptivity in Galerkin finite element discretizations of partial differential equations. Our approach is based on the variational framework of projection methods and uses concepts from optimal control and model reduction. By employing global duality arguments and Galerkin orthogonality, we derive a posteriori error estimates for quantities of physical interest. These residual-based estimates contain the dual solution and provide the basis of a feed-back process for successive mesh adaptation. This approach is developed within an abstract setting and illustrated by examples for its application to different types of differential equations including also an optimal control problem. © 2001 Elsevier Science B.V. All rights reserved.

Keywords: Finite elements; Partial differential equations; Error control; Mesh adaptation; Model reduction; Optimal control; Wave propagation

1. Introduction

Solving complex systems of partial differential equations by discretization methods may be considered in the context of *model reduction*; a conceptually infinite-dimensional model is approximated by a finite-dimensional one. As an example, we may think of a finite volume or a Galerkin finite element method applied to compute the drag coefficient of a body immersed in a viscous fluid where the governing *continuous* model is given by the classical Navier–Stokes equations. Here, the quality of the approximation depends on the proper choice of the discretization parameters (the mesh width, the polynomial degree of the trial functions, the size of certain stabilization parameters, etc.). As the result of the computation, we obtain an approximation to the desired output quantity of the simulation and besides that certain accuracy indicators like cell-truncation errors or cell-residuals. Controlling the error in such an approximation of a continuous model of a physical system requires to determine

E-mail address: rannacher@iwr.uni-heidelberg.de (R. Rannacher).

PII: S0377-0427(00)00513-6

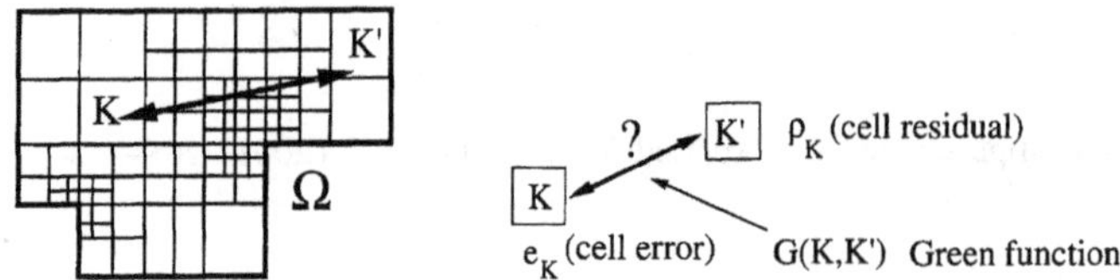

Fig. 1. Scheme of error propagation.

the influence factors for the *local* error indicators on the target functional. Such a sensitivity analysis with respect to local perturbations of the model is common in optimal control theory and introduces the notion of a *dual* (or *adjoint*) problem. It is used to describe the following two features of the approximation process:

(i) *Global error transport*: The local error e_K at some mesh cell K may be strongly affected by the residuals $\rho_{K'}$ at distant cells K' (so-called "pollution effect").

(ii) *Interaction of error components*: The error in one component of the solution may depend on the different components of the cell-residuals in a complicated way.

An effective method for error estimation should take all these dependencies into account. The effect of the cell residuals ρ_K on the local error components $e_{K'}$, at another cell K', is governed by the global Green tensor of the continuous problem. Capturing this dependence by *numerical* evaluation is the general philosophy underlying our approach to error control (Fig. 1).

The mechanisms of error propagation can be rather different depending on the characteristics of the differential operator:

- Diffusion terms cause slow isotropic error decay, but global error pollution may occur from local irregularities.
- Advection terms propagate local errors in the transport direction, but errors decay exponentially in the crosswind direction.
- Reaction terms cause isotropic exponential error decay, but "stiff" behavior may occur in the coupling of error components.

For models in which all these mechanisms are present it is mostly impossible to determine the complex error interaction by analytical means, but rather has to be aided by computation. This automatically leads to a feed-back process in which error estimation and mesh adaptation goes hand-in-hand leading to economical discretizations for computing the quantities of interest. Such an approach seems indispensable for the numerical simulation of large-scale problems. It is particularly designed for achieving high solution accuracy at minimum computational costs.

Traditionally, a posteriori error estimation in Galerkin finite element methods is done with respect to the natural *energy norm* induced by the underlying differential operator; for references see the survey articles by Ainsworth and Oden [1] and Verfürth [22]. This approach seems rather generic as it is directly based on the variational formulation of the problem and allows to exploit its coercivity properties. However, in most applications the error in the energy norm does not provide a useful bound on the error in the quantities of real physical interest. A more versatile method for a posteriori error estimation with respect to more relevant error measures (L^2 norm, point values, line averages, etc.) is obtained by using duality arguments as common from a priori error analysis of finite element methods. This approach has first been systematically developed by Johnson and his co-workers [16,12] and was then extended by the author and his group to a practical feedback method for mesh optimization ([11,10]; see also [19] for a survey of this method). Below, we will describe this general approach to error control and adaptivity first within an abstract setting and then illustrate it by simple examples of its use for different types of differential equations including also an optimal control problem. More involved applications have been considered by Becker [4,5] (viscous incompressible flows), Becker et al. [6,7,3] (chemically reactive flows), Kanschat [18] (radiative transfer), Rannacher and Suttmeier [20] (elasto-plasticity), Bangerth and Rannacher [3] (acoustic waves), and Becker et al. [9] (optimal control).

2. A general paradigm for a posteriori error estimation

We present our approach to residual-based adaptivity in an abstract variational setting following the general paradigm described by Johnson [16] and Eriksson et al. [12]. For a detailed discussion of various aspects of this method, we refer to [11,10].

Let V be a Hilbert space with inner product $(\cdot,\cdot)$ and norm $\|\cdot\|$. Further, let $A(\cdot;\cdot)$ be a semi-linear form with derivatives $A'(\cdot;\cdot,\cdot)$ and $A''(\cdot;\cdot,\cdot,\cdot)$ defined on V. We seek a solution $u \in V$ to the variational equation

$$A(u;\varphi)=0 \quad \forall \varphi \in V. \tag{1}$$

This problem is approximated by a Galerkin method using a sequence of finite-dimensional subspaces $V_h \subset V$, where $h \in \mathbb{R}_+$ is a discretization parameter.

The discrete problems seek $u_h \in V_h$ satisfying

$$A(u_h;\varphi_h)=0 \quad \forall \varphi_h \in V_h. \tag{2}$$

The key feature of this approximation is the "Galerkin orthogonality" which in this nonlinear case is expressed as

$$A(u;\varphi_h)-A(u_h;\varphi_h)=0, \quad \varphi_h \in V_h. \tag{3}$$

Suppose that we want to bound the error $E(u_h):=J(u)-J(u_h)$ with respect to some output functional $J(\cdot)$ defined on V with derivatives $J'(\cdot;\cdot)$ and $J''(\cdot;\cdot,\cdot)$. For this situation, we have the following general result.

Proposition 1. *For the Galerkin scheme* (2) *there holds in first-order approximation the a posteriori error estimate*

$$|E(u_h)| \simeq \eta(u_h) := \inf_{\varphi_h \in V_h} |A(u_h; z-\varphi_h)|, \tag{4}$$

where $z \in V$ is the solution of the linearized dual problem

$$A'(u_h;\varphi,z)=J'(u_h;\varphi) \quad \forall \varphi \in V. \tag{5}$$

The a posteriori error estimate (4) *becomes exact if the form $A(\cdot;\cdot)$ and the functional $J(\cdot)$ are (affine-) linear.*

Proof. We set $e := u - u_h$. By elementary calculus, there holds

$$A(u;\cdot) = A(u_h;\cdot) + A'(u_h;e,\cdot) - \int_0^1 A''(u_h + se; e, e, \cdot)(s-1)\,\mathrm{d}s,$$

$$J(u) = J(u_h) + J'(u_h;e) - \int_0^1 J''(u_h + se; e, e)(s-1)\,\mathrm{d}s.$$

Hence, setting $\varphi = e$ in the dual problem (5), it follows that

$$\begin{aligned} E(u_h) &= J'(u_h;e) - \int_0^1 J''(u_h + se; e, e)(s-1)\,\mathrm{d}s \\ &= A'(u_h;e,z) - \int_0^1 J''(u_h + se; e, e)(s-1)\,\mathrm{d}s \\ &= A(u;z) - A(u_h;z) + r(e,e) \end{aligned} \tag{6}$$

with the remainder term

$$r(e,e) := \int_0^1 \{A''(u_h + se; e, e, z) - J''(u_h + se; e, e)\}(s-1)\,\mathrm{d}s.$$

Hence, using the Galerkin orthogonality (3) and that u solves (1), we find

$$E(u_h) = -A(u_h; z - \varphi_h) + \mathrm{O}(\|e\|^2), \tag{7}$$

provided that the second derivatives of $A(\cdot;\cdot)$ and $J(\cdot)$ are bounded and that also the dual solution z is bounded uniformly with respect to h. Clearly, the $\mathrm{O}(\|e\|)^2$-term is not present if $A(\cdot;\cdot)$ and $J(\cdot)$ are (affine-) linear. □

The a posteriori error estimate (4) holds in general only approximately due to the use of a linearized duality argument. Controlling the effect of this perturbation may be a delicate task and depends strongly on the particular problem considered. Our experiences with different types of problems (including the Navier–Stokes equations and problems in elasto-plasticity) indicate that this problem is less critical as long as the continuous solution is stable. The crucial problem is the numerical computation of the linearized dual solution z by solving a discretized dual problem

$$A'(u_h; \varphi_h, z_h) = J'(u_h; \varphi_h) \quad \forall \varphi_h \in V_h. \tag{8}$$

This results in practically useful error estimators as we will see below.

Next, we consider the special situation that the semi-linear form $A(\cdot;\cdot)$ is given as differential of some "energy" functional $L(\cdot)$ on V, i.e., $A(\cdot;\cdot) = L'(\cdot;\cdot)$. Then, the analogue of problem (1),

$$L'(u;\varphi) = 0 \quad \forall \varphi \in V, \tag{9}$$

determines stationary points of $L(\cdot)$. Its Galerkin approximation reads

$$L'(u_h;\varphi_h) = 0 \quad \forall \varphi_h \in V_h. \tag{10}$$

In this case the general a posteriori error estimate (4) takes the form

$$|E(u_h)| \simeq \inf_{\varphi_h \in V_h} |L'(u_h; z - \varphi_h)|. \tag{11}$$

Now, we restrict the situation even more by taking the generating functional $L(\cdot)$ also for error control, i.e., we consider the error

$$E(u_h) := L(u) - L(u_h).$$

We want to drive the corresponding analogue of the estimate (11).

Proposition 2. *Suppose that the variational equation is derived from an "energy" functional $L(\cdot)$ and that the same functional is used for error control in the Galerkin method. In this case, the corresponding linearized dual problem has the solution $z = -\frac{1}{2}e$ and the a posteriori error estimate (11) becomes*

$$|E(u_h)| \simeq \eta(u_h) := \inf_{\varphi_h \in V_h} \tfrac{1}{2}|L'(u_h; u - \varphi_h)|. \tag{12}$$

This error bound is exact if the functional $L(\cdot)$ is quadratic.

Proof. In virtue of the particular relations $J(\cdot) = L(\cdot)$ and $A(\cdot;\cdot) = L'(\cdot;\cdot)$, we can refine the argument used in the proof of Proposition 1. First, integrating by parts and observing $L'(u;\cdot) \equiv 0$, we can write

$$\begin{aligned} E(u_h) &= \int_0^1 L'(u_h + se; e)\,\mathrm{d}s = L'(u; e) - \int_0^1 L''(u_h + se; e, e)\, s\,\mathrm{d}s \\ &= -\frac{1}{2}L''(u_h; e, e) + \frac{1}{2}\int_0^1 L'''(u_h + se; e, e, e)\,(s^2 - 1)\,\mathrm{d}s, \end{aligned}$$

with the third derivative $L'''(\cdot;\cdot,\cdot,\cdot)$ of $L(\cdot)$. This suggests to work with the linearized dual problem

$$L''(u_h; \varphi, z) = -\tfrac{1}{2}L''(u_h; \varphi, e) \quad \forall \varphi \in V, \tag{13}$$

which has the solution $z = -\frac{1}{2}e$. Further, noting again that $L'(u;\cdot) \equiv 0$, there holds

$$0 = L'(u_h; z) + L''(u_h; e; z) + \int_0^1 L'''(u_h + se; e, e, z)\,(s - 1)\,\mathrm{d}s.$$

From the foregoing relations, we conclude that

$$\begin{aligned} E(u_h) &= L''(u_h; e, z) + \frac{1}{2}\int_0^1 L'''(u_h + se; e, e, e)\,(s^2 - 1)\,\mathrm{d}s \\ &= -L'(u_h; z) - \int_0^1 L'''(u_h + se; e, e, z)\left\{(s - 1) + \frac{1}{2}(s^2 - 1)\right\}\,\mathrm{d}s. \end{aligned}$$

Next, employing Galerkin orthogonality, there holds

$$L'(u_h; z) = L'(u_h; z - \varphi_h), \quad \varphi_h \in V_h. \tag{14}$$

Hence, recalling that $z = -\frac{1}{2}e$, we conclude

$$|E(u_h)| = \inf_{\varphi_h \in V_h} \tfrac{1}{2}|L'(u_h; u - \varphi_h)| + \mathrm{O}(\|e\|^3), \tag{15}$$

where the $\mathrm{O}(\|e\|^3)$ term vanishes if the functional $L(\cdot)$ is quadratic. This proves the assertion. □

Again, in the a posteriori error estimate (12) the quantity $u - \varphi_h$ has to be approximated as described above by using the computed solution $u_h \in V_h$. We emphasize that in this particular case the evaluation of the a posteriori error estimate with respect to the "energy" functional $L(\cdot)$ does not require the explicit solution of a dual problem. We will illustrate this general reasoning for some model situations below.

3. Evaluation of the a posteriori error estimates

The goal is to evaluate the right-hand side of (4) or (12) numerically, in order to get a criterion for the local adjustment of the discretization. For the further discussion, we need to become more specific about the setting of the variational problem. For example, let the variational problem (1) originate from a quasi-linear elliptic partial differential equation of the form

$$A(u) := -\nabla \cdot a(\nabla u) = f, \tag{16}$$

on a bounded domain $\Omega \subset \mathbb{R}^d$, with some $f \in L^2(\Omega)$ and homogeneous Dirichlet boundary conditions, $u|_{\partial\Omega} = 0$. Let $(\cdot,\cdot)_D$ denote the L^2 scalar product on some domain D and $\|\cdot\|_D$ the corresponding norm; in the case $D=\Omega$, we will usually omit the subscript D. The function $a : \mathbb{R}^d \to \mathbb{R}^d$ is assumed to be sufficiently regular, such that the corresponding "energy" form

$$A(u;\varphi) := (a(\nabla u), \nabla\varphi) - (f,\varphi)$$

and its derivative

$$A'(u;\varphi,\psi) := \sum_{i=1}^{d} (a'(\nabla u)\nabla\varphi, \nabla\psi),$$

are well defined on the Sobolev space $V := H_0^1(\Omega)$. The following discussion assumes (16) to be a scalar equation, but everything carries directly over to systems.

We consider the approximation of (16) by standard low-order conforming (linear or bilinear) finite elements defined on quasi-regular meshes $\mathbb{T}_h = \{K\}$ consisting of non-degenerate cells K (triangles or rectangles in two and tetrahedra or hexahedra in three dimensions) as described in the standard finite element literature; see, e.g., [15]. The local mesh width is denoted by $h_K = \operatorname{diam}(K)$. We also use the notation $h = h(x)$ for a global mesh-size function defined by $h_{|K} \equiv h_K$. For the ease of local mesh refinement and coarsening, we allow cells with "hanging nodes" as shown in Fig. 2.

In this setting, the error representation (11) can be developed into a more concrete form. By cellwise integration by parts, we obtain

$$\begin{aligned} |E(u_h)| &\simeq |A(u_h; z - \varphi_h)| \\ &\leqslant \sum_{K\in\mathbb{T}_h} |(f - A(u_h), z - \varphi_h)_K - \tfrac{1}{2}(n\cdot[a(\nabla u_h)], z - \varphi_h)_{\partial K}|, \end{aligned}$$

with the corresponding dual solution $z \in V$ and an arbitrary $\varphi_h \in V_h$. Here, $[a(\nabla u_h)]$ denotes the jump of the (generally discontinuous) flux $a(\nabla u_h)$ across the cell boundaries ∂K, with the convention $[a(\nabla u_h)] := a(\nabla u_h)$ along $\partial\Omega$.

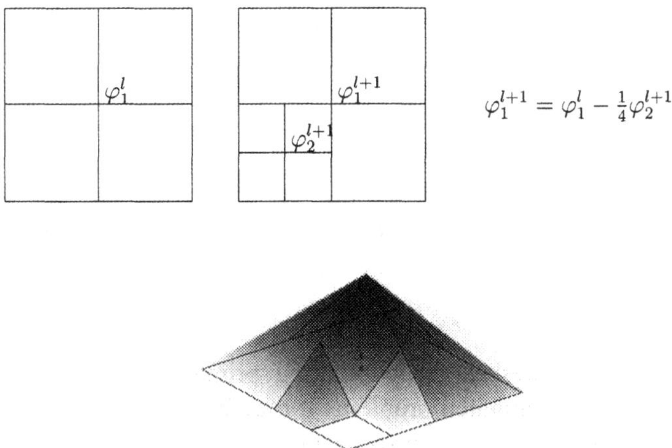

Fig. 2. Q_1 nodal basis function on a 2-d cell patch with hanging nodes.

Notice that in the nonlinear case, this error estimate only holds in first-order approximation; see Proposition 1. For later use in mesh adaptation algorithms, we write it in the form

$$|E(u_h)| \leqslant \eta(u_h) := \sum_{K\in\mathbb{T}_h} \eta_K(u_h), \tag{17}$$

with the cellwise *error indicators*

$$\eta_K(u_h) := |(f - A(u_h), z - \varphi_h)_K - \tfrac{1}{2}(n\cdot[a(\nabla u_h)],\ z - \varphi_h)_{\partial K}|.$$

The interpretation of this is as follows. On each cell, we have an "equation residual" $f - A(u_h)$ and a "flux residual" $n\cdot[a(\nabla u_h)]$, the latter one expressing smoothness of the discrete solution. Both residuals can easily be evaluated. They are multiplied by the weighting function $z - \varphi_h$ which provides quantitative information about the impact of these cell-residuals on the error $E(u_h)$ in the target quantity. In this sense $z - \varphi_h$ may be viewed as *sensitivity factors* like in optimal control problems. We recall the local approximation properties of finite elements, in the present case of linear or d-linear shape functions (for references see, e.g., [15]),

$$\|v - I_h v\|_K + h_K^{1/2}\|v - I_h v\|_{\partial K} \leqslant c_I h_K^2 \|\nabla^2 v\|_K, \tag{18}$$

where I_h denotes the natural nodal interpolation operator, and c_I is an *interpolation constant* usually of size $c_I \sim 0.1$–1. Hence, taking $\varphi_h = I_h z$ in (17), there holds

$$\eta_K(u_h) \leqslant \rho_K(u_h)\omega_K(z), \tag{19}$$

with the cell residuals

$$\rho_K(u_h) := \|f - A(u_h)\|_K + \tfrac{1}{2}h_K^{-1/2}\|n\cdot[a(\nabla u_h)]\|_{\partial K},$$

and the weights

$$\omega_K(z) := c_I h_K^2 \|\nabla^2 z\|_K.$$

Accordingly, the influence factors have the behavior $h_K^2\|\nabla^2 z\|_K$ which is characteristic for the finite element approximation being a *projection method.* We note that for a *finite difference discretization* which lacks the Galerkin orthogonality property the corresponding influence factors would behave like $\|z\|_K$.

4. Algorithmic aspects of mesh adaptation

For evaluating the a posteriori error estimator $\eta(u_h)$, one may solve the linearized dual problem numerically by the same method as used in computing u_h yielding an approximation $z_h \in V_h$,

$$A'(u_h;\varphi_h,z_h)=J'(u_h;\varphi_h) \quad \forall \varphi_h \in V_h. \tag{20}$$

However, the use of the same meshes for computing primal and dual solution is by no means obligatory. In fact, for transport-dominated problems it may be advisable to compute the dual solution on a different mesh or with higher-order approximation; see [14,2,3] for examples.

There are various possibilities for evaluating the cell-error indicators $\eta_K(u_h)$ and to use them for local mesh refinement and as stopping criterion for the adaptation process.

4.1. Strategies for estimator evaluation

- Approximation by second-order difference quotients of the discrete dual solution $z_h \in V_h$, e.g.,

$$\omega_K(z) \leqslant c_I h_K^2 \|\nabla^2 z\|_K \approx c_I h_K^{2+d/2} |\nabla_h^2 z_h(x_K)|, \tag{21}$$

  x_K being the center point of K, where $\nabla^2 z$ is the tensor of second derivatives of z and $\nabla_h^2 z_h$ a suitable difference approximation.
- Computation of a discrete dual solution $\tilde{z}_{h'} \in V_{h'}$ in a richer space $V_{h'} \supset V_h$ (e.g., on a finer mesh or using higher-order finite elements) and setting, e.g.,

$$\omega_K(z) \approx \|\tilde{z}_{h'} - I_h \tilde{z}_{h'}\|_K, \tag{22}$$

  where $I_h \tilde{z}_{h'} \in V_h$ is the generic nodal interpolation.
- Interpolation of the discrete dual solution $z_h \in V_h$ by higher-order polynomials on certain cell-patches, e.g., bi-quadratic interpolation $I_h^{(2)} z_h$:

$$\omega_K(z) \approx \|I_h^{(2)} z_h - z_h\|_K. \tag{23}$$

  The second option is quite expensive and rarely used. Notice that the third option does not involve an interpolation constant which needs to be specified. Our experience is that the use of bi-quadratic interpolation on patches of four quadrilaterals is more accurate than using the finite difference approximation (21). One may try to further improve the quality of the error estimate by solving local (patchwise) defect equations, either Dirichlet problems (à la Babuška and Miller) or Neumann problems (à la Bank and Weiser); for details see [11]. General references for these approaches are Ainsworth and Oden [1] and Verfürth [22].

4.2. Strategies for mesh adaptation

The mesh design strategies are oriented towards a prescribed tolerance TOL for the error quantity $E(u_h)=J(u)-J(u_h)$ and the number of mesh cells N which measures the complexity of the reduced computational model. Usually the admissible complexity is constrained by some maximum value $N_{\max}$.

- *Error balancing strategy*: Cycle through the mesh and seek to equilibrate the local error indicators,

$$\eta_K(u_h) \approx \frac{\text{TOL}}{N}. \tag{24}$$

  This process requires iteration with respect to the number of mesh cells N and eventually leads to $\eta(u_h) \approx \text{TOL}$.
- *Fixed fraction strategy*: Order cells according to the size of $\eta_K(u_h)$ and refine a certain percentage (say 20%) of cells with largest $\eta_K(u_h)$ (or those which make up 20% of the estimator value $\eta(u_h)$) and coarsen those cells with smallest $\eta_K(u_h)$. By this strategy, we may achieve a prescribed rate of increase of N (or keep it constant in solving nonstationary problems).
- *Mesh optimization strategy*: Use the error representation

$$\eta(u_h) \approx \sum_{K \in \mathbb{T}_h} \eta_K(u_h) \tag{25}$$

  directly for generating a formula for an optimal mesh-size function $h_{\text{opt}}(x)$.

We want to discuss the "mesh optimization strategy" in more detail. As a by-product, we will also obtain the justification of the indicator equilibration strategy. Let $N_{\max}$ and TOL be prescribed. We assume that for $h \to 0$, the normalized cell residuals approach certain mesh-independent limits,

$$h_K^{-d/2} \rho_K(u_h) \approx \Psi_u(x_K), \tag{26}$$

which involve the second derivatives of the solution and the data. This property can rigorously be proven on uniformly refined meshes by exploiting super-convergence effects, but they still need theoretical justification on locally refined meshes as constructed by the strategies described above. For the weights, we know from (21) that

$$h_K^{-d/2-2} \omega_K(z) \leqslant c_I |\nabla^2 z(x_K)| =: \Psi_z(x_K). \tag{27}$$

This suggest to use the relation

$$\eta(u_h) \approx E(h) := \int_\Omega h(x)^2 \Psi(x)\, dx, \tag{28}$$

with the mesh-independent function $\Psi(x) = \Psi_u(x)\Psi_z(x)$.

Proposition 3. *Suppose that in the limit* TOL $\to 0$ *or* $N_{\max} \to 0$ *the error estimator takes on the form* (28), *and let*

$$W := \int_\Omega \Psi(x)^{d/(2+d)}\, dx < \infty. \tag{29}$$

(I) *The optimization problem* $E(h) \to \min!$, $N \leqslant N_{\max}$, *has the solution*

$$h_{\text{opt}}(x) \approx \left(\frac{W}{N_{\max}}\right)^{1/d} \Psi(x)^{-1/(2+d)}. \tag{30}$$

(II) *The optimization problem* $N \to \min!$, $E(h) \leqslant \mathrm{TOL}$, *has the solution*

$$h_{\mathrm{opt}}(x) \approx \left(\frac{\mathrm{TOL}}{W}\right)^{1/d} \Psi(x)^{-1/(2+d)}. \tag{31}$$

Proof. First, we note the crucial relation

$$N(h) = \sum_{K \in \mathbb{T}_h} h_K^d h_K^{-d} \approx \int_\Omega h(x)^{-d}\, \mathrm{d}x. \tag{32}$$

Then with the Lagrangian functional

$$L(h, \lambda) := E(h) + \lambda\{N(h) - N_{\max}\},$$

the first-order optimality condition is

$$\frac{\mathrm{d}}{\mathrm{d}t} L(h + t\varphi, \lambda + t\mu)_{|t=0} = 0,$$

for all admissible variations φ and μ. This implies that

$$2h(x)\Psi(x) - \mathrm{d}\lambda h(x)^{-d-1} = 0, \quad \int_\Omega h(x)^{-d}\, \mathrm{d}x - N_{\max} = 0.$$

Consequently,

$$h(x) = \left(\frac{2}{\mathrm{d}\lambda}\Psi(x)\right)^{-1/(2+d)}, \quad \left(\frac{2}{\mathrm{d}\lambda}\right)^{d/(2+d)} \int_\Omega \Psi(x)^{d/(2+d)}\, \mathrm{d}x = N_{\max}.$$

From this, we conclude the desired relations

$$\lambda \equiv \frac{2}{d} \frac{\Psi(x)}{h(x)^{2+d}}, \quad h_{\mathrm{opt}}(x) = \left(\frac{W}{N_{\max}}\right)^{1/d} \Psi(x)^{-1/(2+d)}.$$

In an analogous way, we can also treat the optimization problem (II). □

We note that even for rather "irregular" functionals $J(\cdot)$ the quantity W is bounded. For example, the evaluation of derivative point values $J(u) = \partial_i u(a)$ for smooth u in two dimensions leads to $\Psi(x) \approx |x - a|^{-3}$ and, consequently,

$$W \approx \int_\Omega |x - a|^{-3/2}\, \mathrm{d}x < \infty.$$

The explicit formulas for $h_{\mathrm{opt}}(x)$ have to be used with care in designing a mesh. Their derivation implicitly assumes that they actually correspond to *scalar* mesh-size functions of isotropic meshes such that $h_{\mathrm{opt}|K} \approx h_K$. However, this condition is not incorporated into the formulation of the mesh-optimization problems (I) and (II). Anisotropic meshes containing stretched cells require a more involved concept of mesh description and optimization. This is subject of current research.

5. A nested solution approach

For solving a nonlinear problem like (16) by the adaptive Galerkin finite element method (2), we may employ the following scheme. Let a desired error tolerance TOL or a maximum mesh

complexity $N_{\max}$ be given. Starting from a coarse initial mesh $\mathbb{T}_0$, a hierarchy of successively refined meshes $\mathbb{T}_i$, $i \geqslant 1$, and corresponding finite element spaces V_i is generated as follows:
(0) *Initialization* $i=0$: Compute an initial approximation $u_0 \in V_0$.
(i) *Defect correction iteration*: For $i \geqslant 1$, start with $u_i^{(0)} := u_{i-1} \in V_i$.
(ii) *Iteration step*: For $j \geqslant 0$ evaluate the defect

$$(d_i^{(j)}, \varphi) := F(\varphi) - A(u_i^{(j)}; \varphi), \quad \varphi \in V_i. \tag{33}$$

Choose a suitable approximation $\tilde{A}'(u_i^{(j)}; \cdot, \cdot)$ to the derivative $A'(u_i^{(j)}; \cdot, \cdot)$ (with good stability and solubility properties) and compute a correction $v_i^{(j)} \in V_i$ from the linear equation

$$\tilde{A}'(u_i^{(j)}; v_i^{(j)}, \varphi) = (d_i^{(j)}, \varphi) \quad \forall \varphi \in V_i.$$

For this, Krylov-space or multi-grid methods are employed using the hierarchy of meshes $\{\mathbb{T}_i, \ldots, \mathbb{T}_0\}$. Then, update $u_i^{(j+1)} = u_i^{(j)} + \lambda_i v_i^{(j)}$, with some relaxation parameter $\lambda_i \in (0,1]$, set $j := j+1$ and go back to (2). This process is repeated until a limit $\tilde{u}_i \in V_i$, is reached with a certain prescribed accuracy.

(iii) *Error estimation*: Accept $\tilde{u}_i = u_i$ as the solution on mesh $\mathbb{T}_i$ and solve the discrete linearized dual problem

$$z_i \in V_i: \quad A'(u_i; \varphi, z_i) = J'(u_i; \varphi) \quad \forall \varphi \in V_i,$$

and evaluate the a posteriori error estimate

$$|E(u_i)| \approx \eta(u_i). \tag{34}$$

For controlling the reliability of this bound, i.e., the accuracy in the determination of the dual solution z, one may check whether $\|z_i - z_{i-1}\|$ is sufficiently small; if this is not the case, additional global mesh refinement is advisable. If $\eta(u_i) \leqslant \mathrm{TOL}$ or $N_i \geqslant N_{\max}$, then stop. Otherwise, cell-wise mesh adaptation yields the new mesh $\mathbb{T}_{i+1}$. Then, set $i := i+1$ and go back to (i).

We note that the evaluation of the a posteriori error estimate (34) involves only the solution of *linearized* problems. Hence, the whole error estimation may amount only to a relatively small fraction of the total cost for the solution process. This has to be compared to the usually much higher cost when working on non-optimized meshes.

In using the a posteriori error estimate (34), it is assumed that the exact discrete solution $u_i \in V_i$ on mesh $\mathbb{T}_i$ is available. This asks for estimation of the unavoidable iteration error $\tilde{u}_i - u_i$ and its effect on the accuracy of the estimator for the discretization error. This can be achieved in the case of a Galerkin finite-element multigrid iteration by exploiting the projection properties of the combined scheme; for details see [8].

6. Applications to model problems

6.1. An elliptic model problem

We begin with the Poisson diffusion problem

$$-\Delta u = f \text{ in } \quad \Omega, \quad u = 0 \quad \text{on } \partial\Omega, \tag{35}$$

posed on a polygonal domain $\Omega \subset \mathbb{R}^2$. In this case the "energy" form is defined by $A(\cdot;\cdot) := (\nabla\cdot, \nabla\cdot) - (f, \cdot)$ and the discrete problems read

$$(\nabla u_h, \nabla \varphi_h) = (f, \varphi_h) \quad \forall \varphi_h \in V_h. \tag{36}$$

Here, $V_h \subset V := H_0^1(\Omega)$ are the finite element subspaces as defined above. Now, let $J(\cdot)$ be an arbitrary *linear* error functional defined on V and $z \in V$ the solution of the corresponding dual problem

$$(\nabla \varphi, \nabla z) = J(\varphi) \quad \forall \varphi \in V. \tag{37}$$

From the general a posteriori error estimate (17), we infer the following result.

Proposition 4. *For the approximation of the Poisson problem* (35) *by the finite element scheme* (36), *there holds the a posteriori error estimate*

$$|J(e)| \leqslant \eta(u_h) := \sum_{K \in \mathbb{T}_h} \left\{ \sum_{i=1,2} \rho_K^{(i)}(u_h) \omega_K^{(i)}(z) \right\}, \tag{38}$$

where the cellwise residuals and weights are defined by

$$\rho_K^{(1)}(u_h) := \|f + \Delta u_h\|_K, \quad \omega_K^{(1)}(z) := \|z - I_h z\|_K,$$
$$\rho_K^{(2)}(u_h) := \tfrac{1}{2} h_K^{-1/2} \|n \cdot [\nabla u_h]\|_{\partial K}, \quad \omega_K^{(2)}(z) := h_K^{1/2} \|z - I_h z\|_{\partial K},$$

with some nodal interpolation $I_h z \in V_h$ of z.

Example 1. We want to use Proposition 4 to derive an a posteriori bound for the L^2-error. Using the functional $J(\varphi) := \|e\|^{-1}(e, \varphi)$ in the dual problem, we obtain the estimate

$$J(e) = \|e\| \leqslant \sum_{K \in \mathbb{T}_h} \left\{ \sum_{i=1,2} \rho_K^{(i)}(u_h) \omega_K^{(i)}(z) \right\}. \tag{39}$$

In view of the interpolation property (18) the weights my be estimated by

$$\omega_K^{(i)}(z) \leqslant c_I h_K^2 \|\nabla^2 z\|_K^2.$$

This results in the error bound

$$\|e\| \leqslant c_I \sum_{K \in \mathbb{T}_h} h_K^2 \left\{ \sum_{i=1,2} \rho_K^{(i)}(u_h) \right\} \|\nabla^2 z\|_K$$
$$\leqslant c_I \left(\sum_{K \in \mathbb{T}_h} h_K^4 \left\{ \sum_{i=1,2} \rho_K^{(i)}(u_h) \right\}^2 \right)^{1/2} \|\nabla^2 z\|.$$

In view of the a priori bound $\|\nabla^2 z\| \leqslant c_S$ ($c_S = 1$ if Ω is convex), this implies the a posteriori error estimate

$$\|e\| \leqslant \eta_{L^2}(u_h) := c_I c_S \left(\sum_{K \in \mathbb{T}_h} h_K^4 \left\{ \sum_{i=1,2} \rho_K^{(i)}(u_h) \right\}^2 \right)^{1/2}, \tag{40}$$

which is well-known from the literature; see e.g., [12] or [22].

Example 2. We now consider a highly localized error functional. As concrete example, we choose the square domain $\Omega = (-1,1)^2$ and the error functional

$$J(u) = \partial_1 u(a)$$

at the mid point $a = (0,0)$. In this case, the dual solution does not exist in the sense of $H_0^1(\Omega)$, such that for practical use, we have to regularize the functional $J(u) = \partial_1 u(a)$, for example like

$$J_\varepsilon(u) := |B_\varepsilon|^{-1} \int_{B_\varepsilon} \partial_1 u \, \mathrm{d}x = \partial_1 u(a) + \mathrm{O}(\varepsilon),$$

where $B_\varepsilon = \{x \in \Omega \mid |x-a| < \varepsilon\}$, and $\varepsilon := \mathrm{TOL}$ is a suitable error tolerance. The corresponding dual solution z behaves like

$$|\nabla^2 z(x)| \approx d(x)^{-3}, \quad d(x) := |x-a| + \varepsilon.$$

From the general a posteriori error estimate (38), we obtain for the present case

$$|\partial_1 e(a)| \approx \eta(u_h) := c_I \sum_{K \in \mathbb{T}_h} \frac{h_K^3}{d_K^3} \left\{ \sum_{i=1,2} \rho_K^{(i)}(u_h) \right\}. \tag{41}$$

Now, we want to apply Proposition 3 for determining a priori an optimal meshsize distribution. To this end, let us again assume that the residuals behave like $h_K^{-1}\{\rho_K^{(1)}(u_h) + \rho_K^{(2)}(u_h)\} \approx \Psi_u(x_K)$, with a mesh-independent function Ψ_u, such that

$$\eta(u_h) \approx \int_\Omega h(x)^2 \Psi(x) \mathrm{d}x, \quad \Psi(x) := \frac{\Psi_u(x)}{d(x)^3}.$$

Hence, from Proposition 3, we obtain

$$h_{\mathrm{opt}}(x) \approx \left(\frac{W}{N_{\max}} \right)^{1/2} \Psi(x)^{-1/4} \approx N_{\max} d(x)^{-3/4},$$

with $W := \int_\Omega \Psi_u(x)^{1/2} \mathrm{d}x < \infty$. This implies the relation

$$N_{\mathrm{opt}} = \sum_{K \in \mathbb{T}_h} h_K^2 h_K^{-2} = \left(\frac{N_{\mathrm{opt}}}{\mathrm{TOL}} \right)^{1/2} \sum_{K \in \mathbb{T}_h} h_K^2 d_K^{-3/2} \approx \left(\frac{N_{\mathrm{opt}}}{\mathrm{TOL}} \right)^{1/2}.$$

and, consequently, $N_{\mathrm{opt}} \approx \mathrm{TOL}^{-1}$. This is better than what could be achieved on uniformly refined meshes. In fact, mesh refinement on the basis of global energy-error control results in almost uniform refinement, i.e., $N_{\mathrm{opt}} \approx \mathrm{TOL}^{-2}$. This predicted asymptotic behavior is well confirmed by the results of our computational test. Fig. 3 shows the balanced mesh for $\mathrm{TOL} = 4^{-4}$ and the approximation to the dual solution $z_\varepsilon, \varepsilon = \mathrm{TOL}$, computed on this mesh. The corresponding errors for a sequence of tolerances are listed in Table 1. For the problem considered, the weighted a posteriori estimate (38) for the point error is obviously asymptotically optimal and the predicted dependence $N_{\mathrm{opt}} \approx \mathrm{TOL}^{-1}$ is confirmed; for more details, we refer to [11].

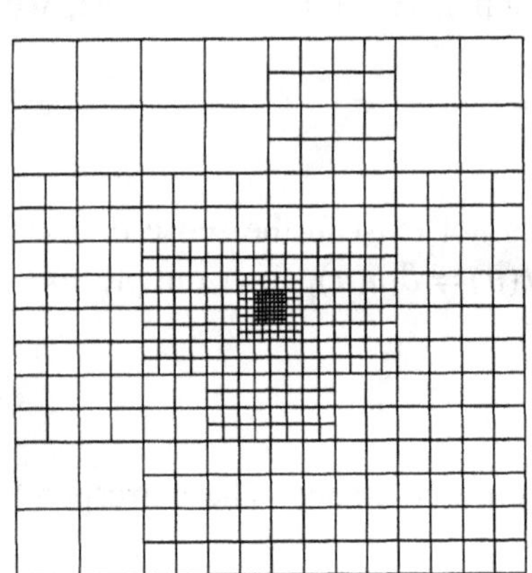
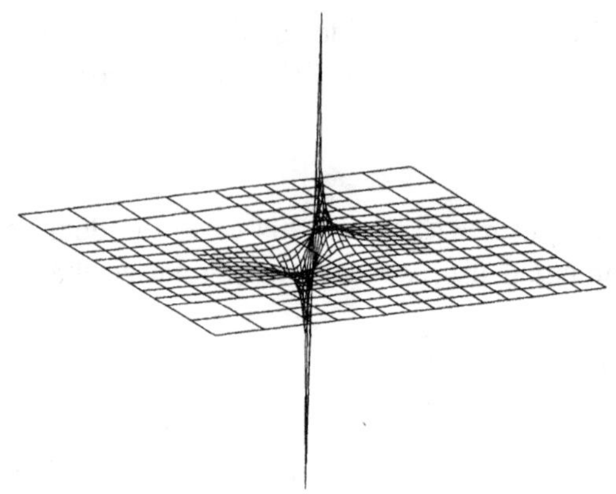

Fig. 3. Refined mesh and approximate dual solution for computing $\partial_1 u(0)$ in Example 2, using the a posteriori error estimator $\eta(u_h)$, with $\mathrm{TOL} = 4^{-4}$.

Table 1
Results for computing $\partial_1 u(0)$ using the a posteriori error estimator $\eta(u_h)$ for several levels of refinement L; the "effectivity index" is defined by $I_{\mathrm{eff}} := |\partial_1 e(0)|/\eta(u_h)$

TOL	N	L	$\|\partial_1 e(0)\|$	$\eta(u_h)$	I_{eff}
4^{-2}	148	6	7.51e − 1	5.92e − 2	12.69
4^{-3}	940	9	4.10e − 1	1.42e − 2	28.87
4^{-4}	4912	12	4.14e − 3	3.50e − 3	1.18
4^{-5}	20 980	15	2.27e − 4	9.25e − 4	0.24
4^{-6}	86 740	17	5.82e − 5	2.38e − 4	0.24

Example 3. The third example is meant as an illustrative exercise. For problem (35) on a smoothly bounded domain $\Omega \subset \mathbb{R}^2$, we consider the functional

$$J(u) := \int_{\partial\Omega} \partial_n u \,\mathrm{d}s \,\left(= \int_\Omega f \,\mathrm{d}x \right),$$

and pose the question: *What is an optimal mesh-size distribution for computing* $J(u)$? The answer is based on the observation that the corresponding dual problem

$$(\nabla\varphi, \nabla z) = \int_{\partial\Omega} \partial_n \varphi \,\mathrm{d}s \quad \forall \varphi \in V \cap W^{2,1}(\Omega),$$

has a measure solution with density of the form $z \equiv 1$ in Ω, $z = 0$ on $\partial\Omega$. In order to avoid the use of measures, we consider the regularized functional

$$J_\varepsilon(\varphi) = |S_\varepsilon|^{-1} \int_{S_\varepsilon} \partial_n \varphi \,\mathrm{d}x = \int_{\partial\Omega} \partial_n \varphi \,\mathrm{d}s + \mathrm{O}(\varepsilon),$$

where $\varepsilon = \mathrm{TOL}$, $S_\varepsilon := \{x \in \Omega,\ \mathrm{dist}\{x, \partial\Omega\} < \varepsilon\}$, and $\partial_n \varphi$ the generic continuation of $\partial_n \varphi$ to S_ε. The corresponding regularized dual solution is

$$z_\varepsilon = 1 \quad \text{in } \Omega \setminus S_\varepsilon, \qquad z_\varepsilon(x) = \varepsilon^{-1} \mathrm{dist}\{x, \partial\Omega\} \quad \text{on } S_\varepsilon.$$

This implies that

$$J_\varepsilon(e) \leqslant c_I \sum_{K\in\mathbb{T}_h,\, K\cap S_\varepsilon\neq\emptyset} h_K^2 \left\{ \sum_{i=1,2} \rho_K^{(i)}(u_h) \right\} \|\nabla^2 z_\varepsilon\|_K,$$

i.e., there is no contribution to the error from the interior of Ω. Hence, independent of the form of the forcing f, the optimal strategy is to refine the elements adjacent to the boundary and to leave the others unchanged, assuming of course, that the force term f is integrated exactly.

Example 4. Finally, we apply the abstract theory developed above to derive an "energy-norm" error estimate. This is intended to prepare for the application to error estimation in approximating optimal control problems.

The solution $u \in V$ of (35) minimizes the quadratic "energy" functional

$$L(u) := \tfrac{1}{2}\|\nabla u\|^2 - (f,u),$$

on the function space $V = H_0^1(\Omega)$, i.e.,

$$L'(u,\varphi) = (\nabla u; \nabla\varphi) - (f,\varphi) = 0, \quad \varphi \in V. \tag{42}$$

Further, we note that

$$\begin{aligned} L(u) - L(u_h) &= \tfrac{1}{2}\|\nabla u\|^2 - (f,u) - \tfrac{1}{2}\|\nabla u_h\|^2 + (f,u_h) \\ &= -\tfrac{1}{2}\|\nabla u\|^2 - \tfrac{1}{2}\|\nabla u_h\|^2 + (\nabla u, \nabla u_h) \\ &= -\tfrac{1}{2}\|\nabla e\|^2. \end{aligned}$$

Hence, energy-error control means control of the error with respect to the "energy" functional $L(\cdot)$. Applying Proposition 2 to this situation, we obtain the a posteriori error estimate

$$|L(u) - L(u_h)| \leqslant \inf_{\varphi_h\in V_h} \tfrac{1}{2}|L'(u_h; u-\varphi_h)| \tag{43}$$

$$\leqslant \sum_{K\in\mathbb{T}_h} \left\{ \sum_{i=1,2} \rho_K^{(i)}(u_h)\omega_K^{(i)}(u) \right\}. \tag{44}$$

which is exact since the functional $L(\cdot)$ is quadratic. Then, using the local approximation estimate (see, e.g., [22])

$$\inf_{\varphi_h\in V_h} \left(\sum_{K\in\mathbb{T}_h} \{ h_K^{-2}\|u-\varphi_h\|_K^2 + h_K^{-1}\|u-\varphi_h\|_{\partial K}^2 \} \right)^{1/2} \leqslant c_I \|\nabla e\|, \tag{45}$$

we conclude that

$$|L(u) - L(u_h)| \leqslant c_I \left(\sum_{K\in\mathbb{T}_h} \left\{ \sum_{i=1,2} \rho_K^{(i)}(u_h) \right\}^2 \right)^{1/2} \|\nabla e\|. \tag{46}$$

This implies the standard energy-norm a posteriori error estimate (see, e.g., [22] or [1])

$$\|\nabla e\| \leqslant 2c_I \left(\sum_{K\in\mathbb{T}_h} \left\{ \sum_{i=1,2} \rho_K^{(i)}(u_h) \right\}^2 \right)^{1/2}. \tag{47}$$

6.2. A nonlinear test case: Hencky model in elasto-plasticity

The results in this section are taken from Suttmeier [21] and Rannacher and Suttmeier [20]. The fundamental problem in the static deformation theory of linear-elastic perfect-plastic material (so-called *Hencky* model) reads

$$\begin{aligned} &\nabla\cdot\sigma = -f, \qquad \varepsilon(u) = A:\sigma + \lambda \quad \text{in } \Omega,\\ &\lambda:(\tau-\sigma)\leqslant 0 \quad \forall\tau \text{ with } F(\tau)\leqslant 0,\\ &u=0 \quad \text{on } \Gamma_D, \qquad \sigma\cdot n = g \quad \text{on } \Gamma_N, \end{aligned} \tag{48}$$

where σ and u are the stress tensor and displacement vector, respectively, while λ denotes the plastic growth. This system describes the deformation of an elasto-plastic body occupying a bounded domain $\Omega\subset\mathbb{R}^d$ ($d=2$ or 3) which is fixed along a part Γ_D of its boundary $\partial\Omega$, under the action of a body force with density f and a surface traction g along $\Gamma_N=\partial\Omega\setminus\Gamma_D$. The displacement u is supposed to be small in order to neglect geometric nonlinear effects, so that the strain tensor can be written as $\varepsilon(u)=\frac{1}{2}(\nabla u+\nabla u^{\mathrm{T}})$. The material tensor A is assumed to be symmetric and positive definite. We assume a linear-elastic isotropic material law $\sigma=2\mu\varepsilon^{\mathrm{D}}(u)+\kappa\nabla\cdot uI$, with material-dependent constants $\mu>0$ and $\kappa>0$, while the plastic behavior follows the von Mises flow rule $F(\sigma):=|\sigma^{\mathrm{D}}|-\sigma_0\leqslant 0$, with some $\sigma_0>0$. Here, ε^{D} and σ^{D} denote the deviatoric parts of ε and σ, respectively.

The *primal* variational formulation of problem (48) seeks a displacement $u\in V:=\{u\in H^1(\Omega)^d,\ u_{|\Gamma_D}=0\}$, satisfying

$$A(u;\varphi)=0 \quad \forall\varphi\in V, \tag{49}$$

with the semi-linear form

$$A(u;\varphi):=(\Pi(2\mu\varepsilon^{\mathrm{D}}(u)),\varepsilon(\varphi))+(\kappa\nabla\cdot u,\nabla\cdot\varphi)-(f,\varphi)-(g,\varphi)_{\Gamma_N},$$

and the projection

$$\Pi(2\mu\varepsilon^{\mathrm{D}}(u)):=\begin{cases} 2\mu\varepsilon^{\mathrm{D}}(u) & \text{if } |2\mu\varepsilon^{\mathrm{D}}(u)|\leqslant\sigma_0,\\ \dfrac{\sigma_0}{|\varepsilon^{\mathrm{D}}(u)|}\varepsilon^{\mathrm{D}}(u) & \text{if } |2\mu\varepsilon^{\mathrm{D}}(u)|>\sigma_0. \end{cases}$$

The finite element approximation of problem (49) reads

$$A(u_h;\varphi_h)=0 \quad \forall\varphi_h\in V_h, \tag{50}$$

where V_h is the finite element space of bilinear shape functions as descibed above. Having computed the displacement u_h, we obtain a corresponding stress by $\sigma_h:=\Pi(2\mu\varepsilon^{\mathrm{D}}(u_h))+\kappa\nabla\cdot u_hI$. Details of the solution process can be found in [21,20]. Given an error functional $J(\cdot)$, we may apply the general a posteriori error estimate (17) to the present situation.

Proposition 5. *For the approximation of the Hencky model* (48) *by the finite element scheme* (50), *there holds the a posteriori error estimate*

$$|J(e)| \leqslant \sum_{K \in \mathbb{T}_h} \left\{ \sum_{i=1,2} \rho_K^{(i)}(u_h)\omega_K^{(i)}(z) \right\}, \tag{51}$$

where the cellwise residuals and weights are defined by

$$\rho_K^{(1)}(u_h) := \|f - \nabla \cdot C(\varepsilon(u_h))\|_K, \qquad \omega_K^{(1)}(z) := \|z - I_h z\|_K,$$

$$\rho_K^{(2)}(u_h) := \tfrac{1}{2} h_K^{-1/2} \|n \cdot [C(\varepsilon(u_h))]\|_{\partial K}, \qquad \omega_K^{(2)}(z) := h_K^{-1/2} \|z - I_h z\|_{\partial K},$$

with $C(\varepsilon) := \Pi(2\mu\varepsilon^{\mathrm{D}}) + \kappa\, \mathrm{tr}(\varepsilon)$ *and some nodal interpolation* $I_h z \in V_h$ *of* z.

We compare the *weighted* error estimate (51) with two heuristic ways of estimating the stress error $e_\sigma := \sigma - \sigma_h$:

(1) The heuristic ZZ-error indicator of Zienkiewicz and Zhu (see [1]) uses the idea of higher-order stress recovery by local averaging,

$$\|e_\sigma\| \approx \eta_{ZZ}(u_h) := \left(\sum_{K \in \mathbb{T}_h} \|M_h \sigma_h - \sigma_h\|_K^2 \right)^{1/2}, \tag{52}$$

where $M_h \sigma_h$ is a local (super-convergent) approximation of σ.

(2) The heuristic energy error estimator of Johnson and Hansbo [17] is based on a decomposed of the domain Ω into "discrete" plastic elastic zones, $\Omega = \Omega_h^p \cup \Omega_h^e$. Accordingly the error estimator has the form

$$\|e_\sigma\| \approx \eta_E(u_h) := c_I \left(\sum_{K \in \mathbb{T}_h} \eta_K^2 \right)^{1/2}, \tag{53}$$

with the local error indicators defined by

$$\eta_K(u_h)^2 := \begin{cases} h_K^2 \{\rho_K^{(1)}(u_h) + \rho_K^{(2)}(u_h)\}^2 & \text{if } K \subset \Omega_h^e, \\ \{\rho_K^{(1)}(u_h) + \rho_K^{(2)}(u_h)\} \|M_h \sigma_h - \sigma_h\|_K & \text{if } K \subset \Omega_h^p. \end{cases}$$

6.2.1. Numerical test

A geometrically two-dimensional square disc with a hole is subjected to a constant boundary traction acting upon two opposite sides. We use the two-dimensional plain-strain approximation, i.e., the components of $\varepsilon(u)$ in z-direction are assumed to be zero. In virtue of symmetry the consideration can be restricted to a quarter of the domain shown in Fig. 4; for the precise parameters in this model see [20]. Among the quantities to be computed is the component σ_{22} of the stress tensor at point a_2. The result on a very fine adapted mesh with about 200,000 cells is taken as reference solution u_{ref}.

The result of this benchmark computation is summarized in Fig. 5. We see that the weighted a posteriori error estimate leads to more economical meshes, particularly if high accuracy is required. For more details as well as for further results also for the time-dependent. Prandtl–Reuss model in perfect plasticity, we refer to [20].

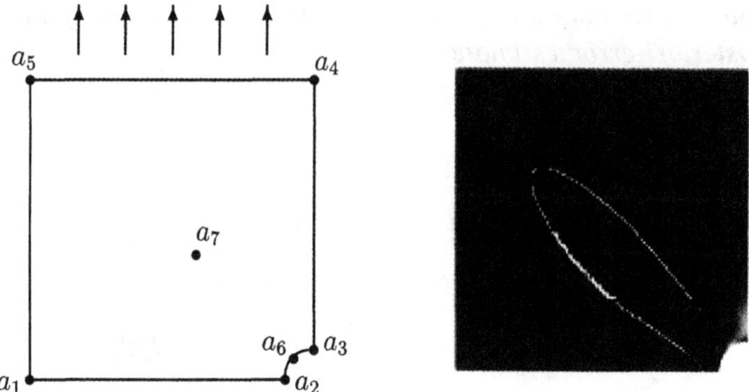

Fig. 4. Geometry of the benchmark problem and plot of $|\sigma^D|$ (plastic region black, transition zone white) computed on a mesh with $N \approx 10\,000$ cells.

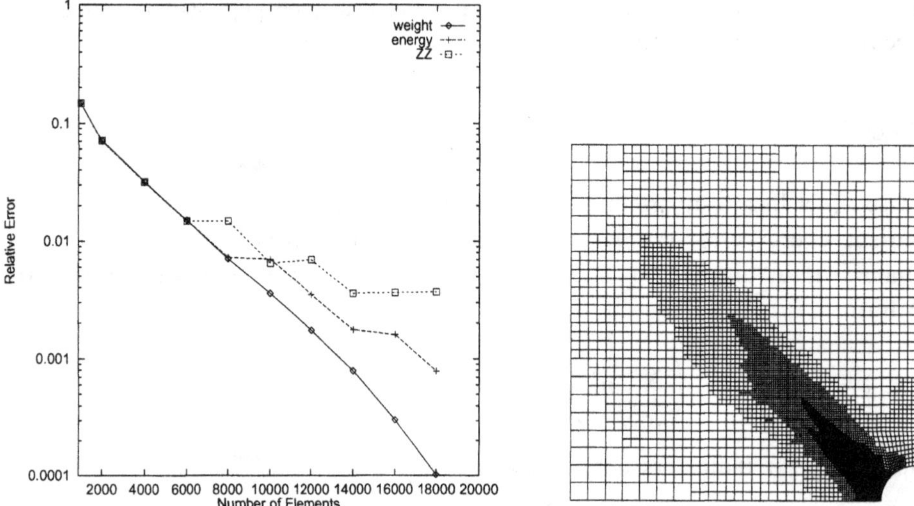

Fig. 5. Relative error for computation of σ_{22} at point a_2 using different estimators and optimized grid with $N \approx 10\,000$ cells.

6.3. A parabolic model problem

We consider the heat-conduction problem

$$\begin{aligned} &\partial_t u - \nabla \cdot (a\nabla u) = f \quad \text{in } Q_T, \\ &u_{|t=0} = u^0 \quad \text{on } \Omega, \\ &u_{|\partial\Omega} = 0 \quad \text{on } I, \end{aligned} \tag{54}$$

on a space–time region $Q_T := \Omega \times I$, where $\Omega \subset \mathbb{R}^d$, $d \geqslant 1$, and $I = [0, T]$; the coefficient a may vary in space. This model is used to describe diffusive transport of energy or certain species concentrations. The results of this section are taken from Hartmann [13]; see also [19].

The discretization of problem (54) is by a Galerkin method in space–time. We split the time interval $[0,T]$ into subintervals $I_n = (t_{n-1}, t_n]$ according to

$$0 = t_0 < \cdots < t_n < \cdots < t_N = T, \qquad k_n := t_n - t_{n-1}.$$

At each time level t_n, let $\mathbb{T}_h^n$ be a regular finite element mesh as defined above with local mesh width $h_K = \operatorname{diam}(K)$, and let $V_h^n \subset H_0^1(\Omega)$ be the corresponding finite element subspace with d-linear shape functions. Extending the spatial mesh to the corresponding space–time slab $\Omega \times I_n$, we obtain a global space–time mesh consisting of $(d+1)$-dimensional cubes $Q_K^n := K \times I_n$. On this mesh, we define the global finite element space

$$V_h^k := \{v \in W, v(\cdot,t)_{|Q_K^n} \in \tilde{Q}_1(K), v(x,\cdot)_{|Q_K^n} \in P_r(I_n), \forall Q_K^n\},$$

where $W := L^2((0,T); H_0^1(\Omega))$ and $r \geqslant 0$. For functions from this space and their time-continuous analogues, we use the notation

$$v^{n+} := \lim_{t \to t_n + 0} v(t), \quad v^{n-} := \lim_{t \to t_n - 0} v(t), \quad [v]^n := v^{n+} - v^{n-}.$$

The discretization of problem (54) is based on a variational formulation which allows the use of piecewise discontinuous functions in time. This method, termed "dG(r) method" (*discontinuous* Galerkin method in time), determines approximations $u_h \in V_h^k$ by requiring

$$A(u_h, \varphi_h) = 0 \quad \forall \varphi_h \in V_h^k, \tag{55}$$

with the semi-linear form

$$A(u,\varphi) := \sum_{n=1}^{N} \int_{I_n} \{(\partial_t u, \varphi) + (a\nabla u, \nabla \varphi) - (f, \varphi)\}\,dt + \sum_{n=1}^{N} ([u]_{n-1}, \varphi_{n-1}^{+}),$$

where $u_0^- := u_0$. We note that the continuous solution u also satisfies equation (55) which again implies Galerkin orthogonality for the error $e := u - u_h$ with respect to the bilinear form $A(\cdot,\cdot)$. Since the test functions $\varphi_h \in V_h^k$ may be discontinuous at times t_n, the global system (55) decouples and can be written in form of a time-stepping scheme,

$$\int_{I_n} \{(\partial_t u_h, \varphi_h) + (a\nabla u_h, \nabla \varphi_h)\}\,dt + ([u_h]^{n-1}, \varphi_h^{(n-1)+}) = \int_{I_n} (f, \varphi_h)\,dt,$$

for all $\varphi_h \in V_h^n$, $n = 1, \ldots, N$. In the following, we consider only the lowest-order case $r = 0$, the "dG(0) method" which is equivalent to the backward Euler scheme. We concentrate on the control of the spatial L^2-error $\|e^{N-}\|$ at the end time $T = t_N$. To this end, we use a duality argument in space–time,

$$\begin{aligned} &\partial_t z - a\Delta z = 0 \quad \text{in } \Omega \times I, \\ &z_{|t=T} = \|e^{N-}\|^{-1} e^{N-} \quad \text{in } \Omega, \quad z_{|\partial\Omega} = 0 \quad \text{on } I, \end{aligned} \tag{56}$$

which can also be written in variational form as

$$A(\varphi, z) = J(\varphi) := \|e^{N-}\|^{-1}(\varphi^{N-}, e^{N-}) \quad \forall \varphi \in W. \tag{57}$$

Then, from Proposition 1, we obtain the estimate

$$\|e^{N-}\| \leqslant \sum_{n=1}^{N} \sum_{K \in \mathbb{T}_h^n} |(R(u_h), z - \varphi_h)_{K \times I_n} - \frac{1}{2}(n \cdot [a\nabla u_h], z - \varphi_h)_{\partial K \times I_n} - ([u_h]^{n-1}, (z - \varphi_h)^{(n-1)+})_K|,$$

with the residual $R(u_h) := f + \nabla\cdot(a\nabla u_h) - \partial_t u_h$ and a suitable approximation $\varphi_h \in V_h^k$ to the dual solution z. From this, we infer the following result.

Proposition 6. *For the $dG(0)$ finite element method applied to the heat conduction equation* (54), *there holds the a posteriori error estimate*

$$\|e^{N+}\| \leqslant \sum_{n=1}^{N} \sum_{K\in\mathbb{T}_h^n} \left\{ \sum_{i=1,2,3} \rho_K^{n,i}(u_h)\omega_K^{n,i}(z) \right\}, \tag{58}$$

where the cellwise residuals and weights are defined by

$$\begin{aligned}
\rho_K^{n,1}(u_h) &:= \|R(u_h)\|_{K\times I_n}, & \omega_K^{n,1}(z) &:= \|z - I_h^k z\|_{K\times I_n},\\
\rho_K^{n,2}(u_h) &:= \tfrac{1}{2} h_K^{-1/2}\|n\cdot[a\nabla u_h]\|_{\partial K\times I_n}, & \omega_K^{n,2}(z) &:= h_K^{1/2}\|z - I_h^k z\|_{\partial K\times I_n},\\
\rho_K^{n,3}(u_h) &:= k_n^{-1/2}\|[u_h]_{n-1}\|_K, & \omega_K^{n,3}(z) &:= k_n^{1/2}\|(z - I_h^k z)^{(n-1)+}\|_K.
\end{aligned}$$

with some nodal interpolation $I_h^k z \in W_h$ of z.

The weights are evaluated numerically as described in Section 3 above.

6.3.1. Numerical test

The performance of the error estimator (58) is illustrated by a simple test in two space dimensions where the (known) exact solution represents a smooth rotating bump with a suitably adjusted force f on the unit square; for details see [13]. Fig. 6 shows a sequence of adapted meshes at successive times obtained by controlling the spatial L^2 error at the end time $t_N = 0.5$. We clearly see the effect of the weights in the error estimator which suppress the influence of the residuals during the initial period.

6.4. A hyperbolic model problem

We consider the acoustic wave equation

$$\begin{aligned}
&\partial_t^2 u - \nabla\cdot\{a\nabla u\} = 0 && \text{in } Q_T,\\
&u_{|t=0} = u^0,\ \partial_t u_{|t=0} = v^0 && \text{on } \Omega,\\
&n\cdot a\nabla u_{|\partial\Omega} = 0 && \text{on } I,
\end{aligned} \tag{59}$$

on a space–time region $Q_T := \Omega\times I$, where $\Omega\subset\mathbb{R}^d$, $d\geqslant 1$, and $I = [0.T]$; the elastic coefficient a may vary in space. This equation frequently occurs in the simulation of acoustic waves in gaseous or fluid media, seismics, electro-dynamics and many other applications. The material of this section is taken from Bangerth [2] and Bangerth and Rannacher [3].

We approximate problem (59) by a "velocity-displacement" formulation which is obtained by introducing a new velocity variable $v := \partial_t u$. Then, the pair $w = \{u, v\}$ satisfies the linear variational equation

$$A(w,\tau) = 0 \quad \forall\tau = \{\varphi,\psi\}\in T, \tag{60}$$

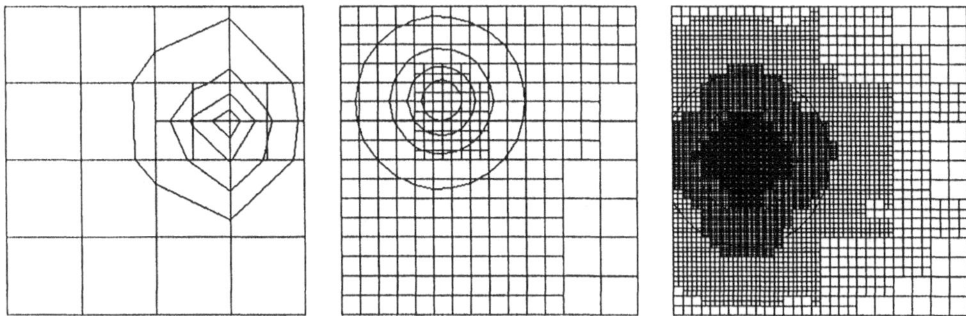

Fig. 6. Sequence of refined meshes at time $t_n = 0.125,\ 0.421875,\ 0.5$, for controlling the end-time error $\|e^{N-}\|$.

with the bilinear form

$$A(\boldsymbol{w},\boldsymbol{\tau}) := (\partial_t u, \varphi)_{Q_T} - (v,\varphi)_{Q_T} + (\partial_t v, \psi)_{Q_T} + (a\nabla u, \nabla\psi)_{Q_T},$$

and a suitable space T of test functions $\boldsymbol{\tau}$ for which $A(\boldsymbol{w},\boldsymbol{\tau})$ is defined. This formulation is the basis for a Galerkin discretization in space–time similar to the one described in the preceding section for the heat-conduction equation. We decompose the time interval $I = [0, T]$ again into subintervals $I_n = (t_{n-1}, t_n]$ with length $k_n = t_n - t_{n-1}$. On each time slab $Q^n := \Omega \times I_n$, we use meshes $\mathbb{T}_h^n$ consisting of $(d+1)$-dimensional cubes $Q_K^n = K \times I_n$ with spatial width h_K; these meshes may vary between the time levels in order to allow for grid refinement moving with the wave field. The discrete "trial spaces" $W_h = V_h \times V_h$ in space–time domain consist of functions which are $(d+1)$-linear on each space–time cell Q_K^n and globally *continuous* on Q_T. This prescription requires the use of "hanging nodes" if the spatial mesh changes across a time level t_n. The corresponding discrete "test spaces" $T_h \subset T$ consist of functions which are constant in time on each cell Q_K^n, while they are d-linear in space and globally continuous on Ω. We further assume that the test space T is chosen large enough to contain the elements from $\boldsymbol{w} + W_h$. The Galerkin approximation of problem (59) seeks pairs $\boldsymbol{w}_h = \{u_h, v_h\} \in W_h$ satisfying

$$A(\boldsymbol{w}_h, \boldsymbol{\tau}_h) = 0 \quad \forall \boldsymbol{\tau}_h = \{\varphi_h, \psi_h\} \in \mathbb{T}_h. \tag{61}$$

For more details, we refer to [2,3]. Since the continuous solution $\boldsymbol{w}$ also satisfies (61), we have again Galerkin orthogonality for the error $\boldsymbol{e} := \{e_u, e_v\}$. This time-discretization scheme is termed "$cG(1)$-method" (*continuous* Galerkin method) in contrast to the dG-method used in the preceding section. We note that from this scheme, we can recover the standard Crank–Nicolson scheme in time combined with a spatial finite element method:

$$\begin{aligned} &(u^n - u^{n-1}, \varphi) - \tfrac{1}{2}k_n(v^n + v^{n-1}, \varphi) = 0, \\ &(v^n - v^{n-1}, \psi) + \tfrac{1}{2}k_n(a\nabla(u^n + u^{n-1}), \nabla\psi) = 0. \end{aligned} \tag{62}$$

The system (62) splits into two equations, a discrete Helmholtz equation and a discrete L^2-projection. We choose this time-stepping scheme because it is of second order and energy conserving, i.e.,

$$\|v^n\|^2 + \|\sqrt{a}\nabla u^n\|^2 = \|v^{n-1}\|^2 + \|\sqrt{a}\nabla u^{n-1}\|^2.$$

This conservation property carries over to the spatially discretized equations provided that the meshes do not change between time levels. In case of mesh coarsening a loss of energy may occur which has to be "seen" by a useful a posteriori error estimator; we refer to [3] for a discussion of this issue.

We want to control the error $\boldsymbol{e} = \{e_u, e_v\}$ by a functional of the form

$$J(\boldsymbol{e}) := (j, e_u)_{Q_T},$$

with some density function $j(x,t)$. To this end, we use again a duality argument in space–time written in variational form like

$$A(\boldsymbol{\tau}, \boldsymbol{z}) = J(\boldsymbol{\tau}) \quad \forall \boldsymbol{\tau} \in T, \tag{63}$$

where the dual solution is of the form $\boldsymbol{z} = \{-\partial_t z, z\}$. This means that z satisfies the backward-in-time wave equation

$$\begin{aligned} &\partial_t^2 z - \nabla \cdot \{a \nabla z\} = j \quad \text{in } Q_T, \\ &z_{|t=T} = 0, \quad -\partial_t z_{|t=T} = 0 \quad \text{on } \Omega, \\ &n \cdot a \nabla z_{|\partial\Omega} = 0 \quad \text{on } I \times \partial\Omega. \end{aligned} \tag{64}$$

Then, from Proposition 1, we have the abstract result

$$(j \cdot u)_{Q_T} \leqslant |A(\boldsymbol{w}_h, \boldsymbol{z} - I_h \boldsymbol{z})|, \tag{65}$$

with the natural nodal interpolation I_h in the space W_h. Recalling the definition of the bilinear form $A(\cdot,\cdot)$, we obtain

$$|(j,u)_{Q_T}| \leqslant \sum_{n=1}^{N} \sum_{K \in \mathbb{T}_h^n} |(r_1, \partial_t z - I_h \partial_t z)_{K \times I_n} - (r_2, z - I_h z)_{K \times I_n} - \tfrac{1}{2}(n \cdot [a \nabla u_h], z - I_h z)_{\partial K \times I_n}|,$$

where $r_1 = \partial_t u_h - v_h$ and $r_2 = \partial_t v_h - \nabla \cdot a \nabla u_h$ denote the cell residuals of the two equations and $n \cdot r_{\partial K} = n \cdot [a \nabla u_h]$ is the jump of the co-normal derivative across cell boundaries. From this, we infer the following result.

Proposition 7. *For the cG(1) finite element method applied to the acoustic wave equation* (59), *there holds the a posteriori error estimate*

$$|j(e_u^N)| \leqslant \sum_{n=1}^{N} \sum_{K \in \mathbb{T}_h^n} \left\{ \sum_{i=1,2,3} \rho_K^{n,i}(\boldsymbol{w}_h) \omega_K^{n,i}(\boldsymbol{z}) \right\}, \tag{66}$$

where the cellwise residuals and weights are defined by

$$\begin{aligned} &\rho_K^{n,1}(\boldsymbol{w}_h) := \|r_1(\boldsymbol{w}_h)\|_{K \times I_n}, && \omega_K^{n,1}(\boldsymbol{z}) := \|\partial_t z - I_h \partial_t z\|_{K \times I_n}, \\ &\rho_K^{n,2}(\boldsymbol{w}_h) := \|r_2(\boldsymbol{w}_h)\|_{K \times I_n}, && \omega_K^{n,2}(\boldsymbol{z}) := \|z - I_h z\|_{K \times I_n}, \\ &\rho_K^{n,3}(\boldsymbol{w}) := \tfrac{1}{2} h_K^{-1/2} \|n \cdot [a \nabla u_h]\|_{\partial K \times I_n}, && \omega_K^{n,3}(\boldsymbol{z}) := h_K^{1/2} \|z - I_h z\|_{\partial K \times I_n}, \end{aligned}$$

with some nodal interpolations $\{I_h \partial_t z, I_h z\} \in V_h \times V_h$ *of* $\{\partial_t z, z\}$.

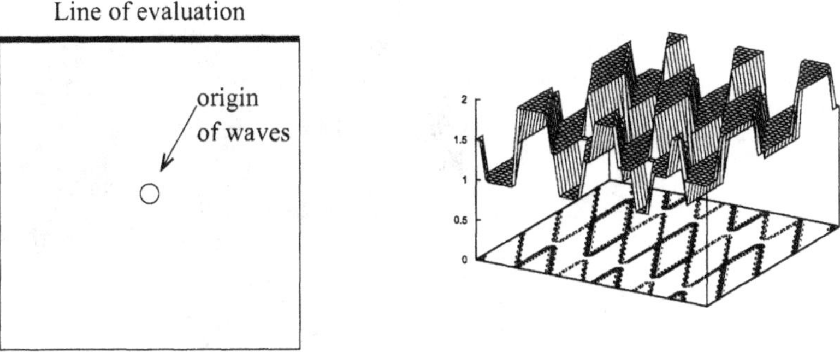

Fig. 7. Layout of the domain (left) and structure of the coefficient $a(x)$ (right).

We will compare the error estimator (66) with a simple heuristic "energy" error indicator which measures the spatial smoothness of the computed solution u_h:

$$\eta_E(w_h) := \left(\sum_{n=1}^{N} \sum_{K \in \mathbb{T}_h^n} \rho_K^{n,3}(w_h)^2 \right)^{1/2}. \tag{67}$$

6.4.1. Numerical test

The error estimator (66) is illustrated by a simple test: the propagation of an outward traveling wave on $\Omega = (-1, 1)^2$ with a strongly distorted coefficient. Layout of the domain and structure of the coefficient are shown in Fig. 7. Boundary and initial conditions were chosen as follows:

$$n \cdot a\nabla u = 0 \quad \text{on } y = 1, \quad u = 0 \quad \text{on } \partial\Omega \setminus \{y = 1\},$$
$$u_0 = 0, \quad v_0 = \theta(s - r)\exp(-|x|^2/s^2)(1 - |x|^2/s^2),$$

with s=0.02 and $\theta(\cdot)$ the jump function. The region of origin of the wave field is significantly smaller than shown in Fig. 7. Notice that the lowest frequency in this initial wave field has wavelength λ=4s; hence taking the common minimum 10 grid points per wavelength would yield 62,500 cells already for the largest wavelength. Uniform grids obviously quickly get to their limits in such cases.

If we consider this example as a model of propagation of seismic waves in a faulted region of rock, then we would be interested in recording seismograms at the surface, which we here choose as the top line Γ of the domain. A corresponding functional output is

$$J(w) = \int_0^T \int_\Gamma u(x,t)\omega(\xi,t)\,\mathrm{d}\xi\,\mathrm{d}t,$$

with a weight factor $\omega(\xi,t) = \sin(3\pi\xi)\sin(5\pi t/T)$, and end-time $T = 2$. The frequency of oscillation of this weight is chosen to match the frequencies in the wave field to obtain good resolution of changes. In Fig. 8, we show the grids resulting from refinement by the dual error estimator (66) compared with the energy error indicator (67). The first one resolves the wave field well, including reflections from discontinuities in the coefficient. The second additionally takes into account, that the lower parts of the domain lie outside the domain of influence of the target functional if we truncate the time domain at $T = 2$; this domain of influence constricts to the top as we approach the final time, as is reflected by the produced grids. The computational meshes obtained in this way are

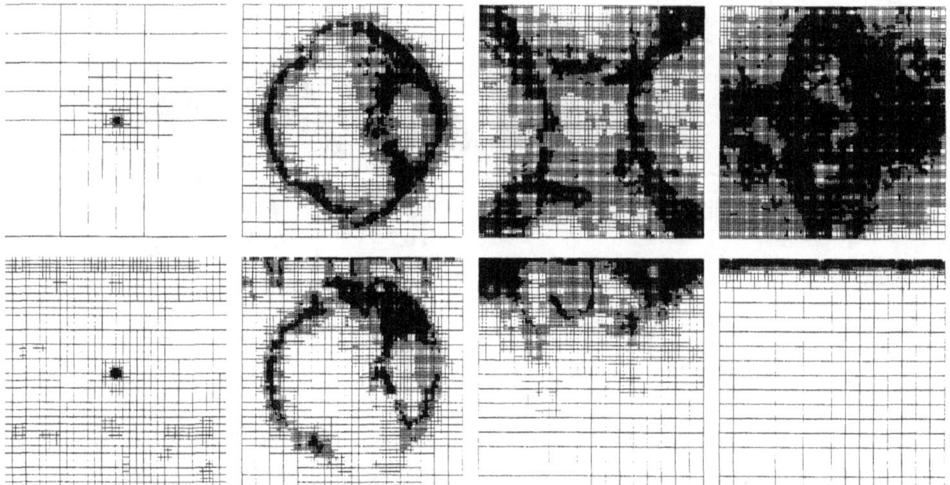

Fig. 8. Grids produced from refinement by the energy error indicator (67) (top row) and by the dual estimator (66) (bottom row) at times $t = 0, \frac{2}{3}, \frac{4}{3}, 2$.

obviously much more economical, without degrading the accuracy-in approximating the quantity of interest; for more examples see [3].

6.5. An optimal control model problem

As the last application of the general framework laid out in Section 2, we consider the finite element approximation of an optimal control problem. The state equations are

$$-\Delta u + u = f \quad \text{on } \Omega,$$
$$\partial_n u = q \quad \text{on } \Gamma_C, \quad \partial_n u = 0 \quad \text{on } \partial\Omega \setminus \Gamma_C, \tag{68}$$

defined on a bounded domain $\Omega \subset \mathbb{R}^2$ with boundary $\partial\Omega$. The control q acts on the boundary component Γ_C, while the observations $u_{|\Gamma_O}$ are taken on a component Γ_O; see Fig. 9. The cost functional is defined by

$$J(u,q) = \tfrac{1}{2}\|u - u_O\|^2_{\Gamma_O} + \tfrac{1}{2}\alpha\|q\|^2_{\Gamma_C}. \tag{69}$$

with a prescribed function u_O and a regularization parameter $\alpha > 0$.

We want to apply the general formalism of Section 2 to the Galerkin finite element approximation of this problem. This may be considered within the context of “model reduction” in optimal control theory. First, we have to prepare the corresponding functional analytic setting. The functional of interest is the Lagrangian functional of the optimal control problem,

$$L(v) = J(u,q) + (\nabla u, \nabla\lambda) + (u - f, \lambda) - (q,\lambda)_{\Gamma_C},$$

defined for triples $v = \{u, \lambda, q\}$ in the Hilbert space $W := V \times V \times Q$, where $V := H^1(\Omega)$ and $Q := L^2(\Gamma_C)$. The equation for stationary points $v = \{u, \lambda, q\} \in W$ of $L(\cdot)$ are (Euler–Lagrange equations)

$$L'(v; \varphi) = 0 \quad \forall \varphi \in W, \tag{70}$$

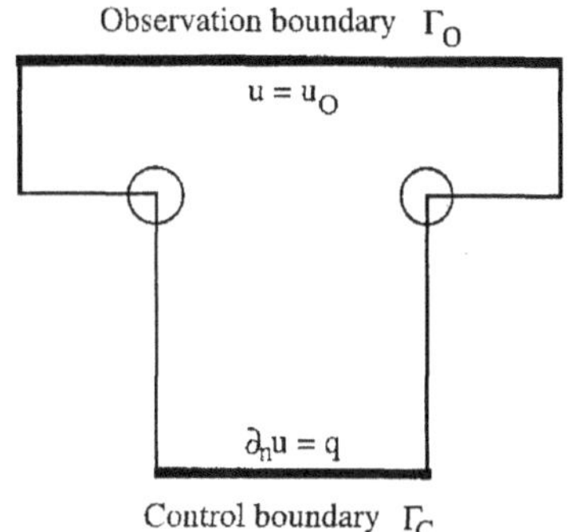

Fig. 9. Configuration of the boundary control model problem.

or written in explicit saddle-point form,

$$(\psi, u - u_O)_{\Gamma_O} + (\nabla\psi, \nabla\lambda) + (\psi, \lambda) = 0 \quad \forall \psi \in V, \tag{71}$$

$$(\nabla u, \nabla\pi) + (u - f, \pi) - (q, \pi)_{\Gamma_C} = 0 \quad \forall \pi \in V, \tag{72}$$

$$(\lambda - \alpha q, \chi)_{\Gamma_C} = 0 \quad \forall \chi \in Q. \tag{73}$$

The corresponding discrete approximations $\boldsymbol{v}_h = \{u_h, \lambda_h, q_h\}$ are determined in the finite element space $W_h = V_h \times V_h \times Q_h \subset V$ by

$$(\psi_h, u_h - u_O)_{\Gamma_O} + (\nabla\psi_h, \nabla\lambda_h) + (\psi_h, \lambda_h) = 0 \quad \forall \psi_h \in V_h, \tag{74}$$

$$(\nabla u_h, \nabla\pi_h) + (u_h - f, \pi_h) - (q_h, \pi_h)_{\Gamma_C} = 0 \quad \forall \pi_h \in V_h, \tag{75}$$

$$(\lambda_h - \alpha q_h, \chi_h)_{\Gamma_C} = 0 \quad \forall \chi_h \in Q_h. \tag{76}$$

Here, the trial spaces V_h for the state and co-state variables are as defined above in Section 3 (linear or bilinear shape functions), and the spaces Q_h for the controls consist of traces of Γ_C of V_h-functions, for simplicity.

Following the formalism of Section 2, we seek to estimate the error $\boldsymbol{e} = \{e_u, e_\lambda, e_q\}$ with respect to the Lagrangian functional $L(\cdot)$. Proposition 2 yields the following a posteriori estimate for the error $E(\boldsymbol{v}_h) := L(\boldsymbol{v}) - L(\boldsymbol{v}_h)$:

$$|E(\boldsymbol{v}_h)| \leqslant \eta(\boldsymbol{v}_h) := \inf_{\varphi_h \in V_h} \tfrac{1}{2}|L'(\boldsymbol{v}_h; \boldsymbol{v} - \boldsymbol{\varphi}_h)|. \tag{77}$$

Since $\{u, \lambda, q\}$ and $\{u_h, \lambda_h, q_h\}$ satisfy (72) and (75), respectively, there holds

$$\begin{aligned} L(\boldsymbol{v}) - L(\boldsymbol{v}_h) &= J(u, q) + (\nabla u, \nabla\lambda) + (u - f, \lambda) - (q, \lambda)_{\Gamma_C} \\ &\quad - J(u_h, q_h) - (\nabla u_h, \nabla\lambda_h) - (u_h - f, \lambda_h) + (q_h, \lambda_h)_{\Gamma_C} \\ &= J(u, q) - J(u_h, q_h). \end{aligned}$$

Hence, error control with respect to the Lagrangian functional $L(\cdot)$ and the cost functional $J(\cdot)$ are equivalent. Now, evaluation of the abstract error bound (77) employs again splitting the integrals into the contributions by the single cells, cellwise integration by parts and Hölder's inequality. In this way, we obtain the following result; for the detailed argument see [9].

Proposition 8. *For the finite element discretization of the system (71)–(73), there holds (in first-order approximation) the a posteriori error estimate*

$$|J(u,q)-J(u_h,q_h)| \leqslant \sum_{\Gamma\subset\partial\Omega}\{\rho_\Gamma^{(\lambda)}\omega_\Gamma^{(u)}+\rho_\Gamma^{(u)}\omega_\Gamma^{(\lambda)}\}+\sum_{\Gamma\subset\Gamma_C}\rho_\Gamma^{(q)}\omega_\Gamma^{(q)}$$
$$+\sum_{K\in\mathbb{T}_h}\{\rho_K^{(u)}\omega_K^{(\lambda)}+\rho_{\partial K}^{(u)}\omega_{\partial K}^{(\lambda)}+\rho_K^{(\lambda)}\omega_K^{(u)}+\rho_{\partial K}^{(\lambda)}\omega_{\partial K}^{(u)}\}, \tag{78}$$

where the cellwise residuals and weights are defined by

$$\rho_\Gamma^{(\lambda)}=\begin{cases} h_\Gamma^{-1/2}\|u_h-u_O+\partial_n\lambda_h\|_\Gamma & \text{if } \Gamma\subset\Gamma_O,\\ h_\Gamma^{-1/2}\|\partial_n,\lambda_h\|_\Gamma & \text{if } \Gamma\subset\partial\Omega\setminus\Gamma_O,\end{cases}\qquad \omega_\Gamma^{(u)}=h_\Gamma^{1/2}\|u-I_hu\|_\Gamma,$$

$$\rho_\Gamma^{(u)}=\begin{cases} h_\Gamma^{-1/2}\|\partial_nu_h-q_h\|_\Gamma & \text{if } \Gamma\subset\Gamma_C,\\ h_\Gamma^{-1/2}\|\partial_nu_h\|_\Gamma & \text{if } \Gamma\subset\partial\Omega\setminus\Gamma_C,\end{cases}\qquad \omega_\Gamma^{(\lambda)}=h_\Gamma^{1/2}\|\lambda-I_h\lambda\|_\Gamma,$$

$$\rho_\Gamma^{(q)}=h_\Gamma^{-1/2}\|\lambda_h-\alpha q_h\|_\Gamma,\quad \omega_\Gamma^{(q)}=h_\Gamma^{1/2}\|q-I_hq\|_\Gamma,$$

$$\rho_K^{(u)}=\|\Delta u_h-u_h+f\|_K,\quad \omega_K^{(\lambda)}=\|\lambda-I_h\lambda\|_K,$$

$$\rho_{\partial K}^{(u)}=\tfrac12 h_K^{-1/2}\|n\cdot[\nabla u_h]\|_{\partial K},\quad \omega_{\partial K}^{(\lambda)}=h_K^{1/2}\|\lambda-I_h\lambda\|_{\partial K},$$

$$\rho_K^{(\lambda)}=\|\Delta\lambda_h-\lambda_h\|_K,\quad \omega_K^{(u)}=\|u-I_hu\|_K,$$

$$\rho_{\partial K}^{(\lambda)}=\tfrac12 h_K^{-1/2}\|n\cdot[\nabla\lambda_h]\|_{\partial K},\quad \omega_{\partial K}^{(u)}=h_K^{1/2}\|u-I_hu\|_{\partial K},$$

with some nodal interpolations $\{I_hu,I_h\lambda,I_hq\}\in V_h\times V_h\times Q_h$ *of* $\{u,\lambda,q\}$.

We will compare the performance of the weighted error estimator (78) with a more traditional error indicator. Control of the error in the "energy norm" of the state equation alone leads to the a posteriori error indicator

$$\eta_E(u_h):=c_I\left(\sum_{K\in\mathbb{T}_h}h_K^2\{\rho_K^{(u)2}+\rho_{\partial K}^{(u)2}\}+\sum_{\Gamma\subset\partial\Omega}h_\Gamma^2\rho_\Gamma^{(u)2}\right)^{1/2}, \tag{79}$$

with the cell residuals $\rho_{\partial K}^{(u)}$ and $\rho_\Gamma^{(u)}$ as defined above. This ad-hoc criterion aims at satisfying the state equation uniformly with good accuracy. However, this concept seems questionable since it does not take into account the sensitivity of the cost functional with respect to the local perturbations introduced by discretization. Capturing these dependencies is the particular feature of our approach.

6.5.1. Numerical test

We consider the configuration as shown in Fig. 9 with a T-shaped domain Ω of width one. The control acts along the lower boundary Γ_C, whereas the observations are taken along the (longer) upper boundary Γ_O. The cost functional is chosen as in (69) with $u_O\equiv 1$ and $\alpha=1$, i.e., the stabilization term constitutes a part of the cost functional.

N	$E(v_h)$	I_{eff}
320	$1.0e-3$	1.1
1376	$3.5e-4$	0.7
4616	$3.2e-5$	0.7
11816	$1.6e-5$	1.0
23624	$6.4e-6$	0.8
48716	$2.8e-6$	0.7

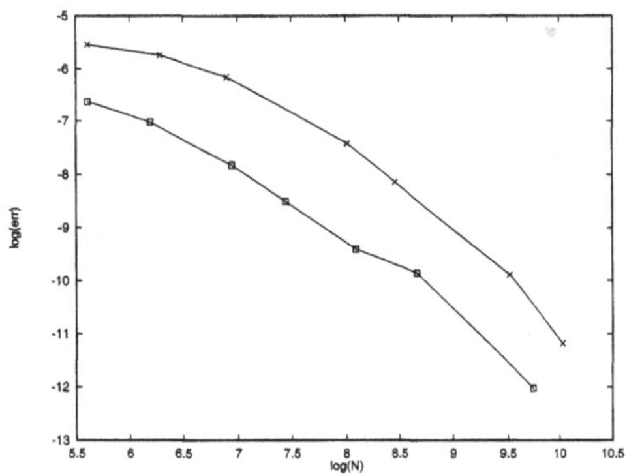

Fig. 10. Effectivity of the weighted error estimator (left), and comparison of the efficiency of the meshes generated by the two estimators, "x" error values by the energy estimator, "□" error values by the weighted estimator (log–log scale).

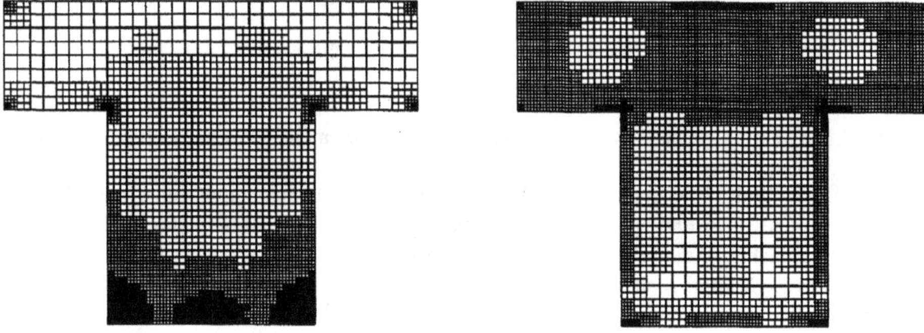

Fig. 11. Comparison between meshes obtained by the energy-error estimator (left) and the weighted error estimator (right); $N \sim 5000$ cells in both cases.

Fig. 10 shows the quality of the weighted error estimator (78) for quantitative error control. The *effectivity index* is again defined as $I_{\text{eff}} := |J(u,q) - J(u_h,q_h)|/\eta(u_h,q_h)$, whereas $\eta(u_h,q_h)$ is the value of the estimator. The reference value is obtained on a mesh with more than 200 000 elements. We compare the weighted error estimator with a simple ad-hoc strategy based on the energy-error estimator (79) applied only to the state equation. Fig. 11 shows meshes generated by the two estimators.

The difference in the meshes can be explained as follows. Obviously, the energy-error estimator observes the irregularities introduced on the control boundary by the jump in the nonhomogeneous Neumann condition, but it tends to over-refine in this region and to under-refine at the observation boundary. The weighted error estimator observes the needs of the optimization process by distributing the cells more evenly.

7. Conclusion and outlook

We have presented a universal method for error control and mesh adaptivity in Galerkin finite element discretization based on global duality arguments. This approach has been illustrated for simple examples of linear and nonlinear differential equations including also an optimal control problem. More challanging applications involving multi-physical effects like, for example, low-Mach number flows with chemical reactions and multidimensional energy transfer by radiation with recombination are presently studied. Of course, despite the generality of the method, there are several technical questions to be addressed in the future. The main problem is the accurate but cost-efficient determination of the dual solution especially in the presence of oscillatory solutions (wave propagation). Another question is the reliable control of the linearization in the neighborhood of a bifurcation point. Finally, the cost-efficient application of our method for truly nonstationary problems in two and particularly in three dimensions is still a largely unsolved problem.

Acknowledgements

The author acknowledges the support by the German Research Association (DFG) through the SFB 359 "Reactive Flow, Diffusion and Transport", University of Heidelberg.

References

[1] M. Ainsworth, J.T. Oden, A posteriori error estimation in finite element analysis, Comput. Methods Appl. Mech. Eng. 142 (1997) 1–88.

[2] W. Bangerth, Adaptive Finite-Elemente-Methoden zur Lösung der Wellengleichung mit Anwendungen in der Physik der Sonne, Diploma Thesis, Institut für Angewandte Mathematik, Universität Heidelberg, 1998.

[3] W. Bangerth, R. Rannacher, Finite element approximation of the acoustic wave equation: error control and mesh adaptation, East–West J. Numer. Math. 7 (1999) 263–282.

[4] R. Becker, An adaptive finite element method for the incompressible Navier–Stokes equations on time-dependent domains, Doctor Thesis, preprint 95-44, SFB 359, Universität Heidelberg, 1995.

[5] R. Becker, Weighted error estimators for finite element approximations of the incompressible Navier–Stokes equations, Technical Report 98-20, SFB 359, Universität Heidelberg, 1998.

[6] R. Becker, M. Braack, R. Rannacher, Numerical simulation of laminar flames at low Mach number with adaptive finite elements, Combust. Theory Modelling 3 (1999) 503–534.

[7] R. Becker, M. Braack, R. Rannacher, C. Waguet, Fast and reliable solution of the Navier–Stokes equations including chemistry, Comput. Visualization Sci. 2 (1999) 107–122.

[8] R. Becker, C. Johnson, R. Rannacher, Adaptive error control for multigrid finite element methods, Computing 55 (1995) 271–288.

[9] R. Becker, H. Kapp, R. Rannacher, Adaptive finite element methods for optimal control of partial differential equations: basic concepts, preprint 98-55, SFB 359, Universität Heidelberg, 1998, SIAM J. Control Optim., to appear.

[10] R. Becker, R. Rannacher, Weighted a posteriori error control in FE methods, lecture at ENUMATH 95, Paris, August 1995, in: H.G. Bock et al. (Eds.), Proceedings of ENUMATH 97, World Scient. Publ., Singapore, 1998.

[11] R. Becker, R. Rannacher, A feed-back approach to error control in finite element methods: basic analysis and examples, East–West J. Numer. Math. 4 (1996) 237–264.

[12] K. Eriksson, D. Estep, P. Hansbo, C. Johnson, Introduction to adaptive methods for differential equations, in: A. Iserles (Ed.), Acta Numerica 1995, Cambridge University Press, Cambridge, 1995, pp. 105–158.

[13] R. Hartmann, A posteriori Fehlerschätzung und adaptive Schrittweiten- und Ortsgittersteuerung bei Galerkin-Verfahren für die Wärmeleitungsgleichung, Diploma Thesis, Institut für Angewandte Mathematik, Universität Heidelberg, 1998.

[14] P. Houston, R. Rannacher, E. Süli, A posteriori error analysis for stabilised finite element approximation of transport problems, Report No. 99/04, Oxford University Computing Laboratory, Oxford, 1999, Comput. Methods Appl. Mech. Eng., to appear.

[15] C. Johnson, Numerical Solution of Partial Differential Equations by the Finite Element Method, Cambridge University Press, Cambridge-Lund, 1987.

[16] C. Johnson, A new paradigm for adaptive finite element methods, in: J. Whiteman (Ed.), Proceedings of MAFELAP 93, Wiley, New York, 1993.

[17] C. Johnson, P. Hansbo, Adaptive finite elemente methods in computational mechanics, Comput. Meth. Appl. Mech. Eng. 101 (1992) 143–181.

[18] G. Kanschat, Efficient and reliable solution of multi-dimensional radiative transfer problems, Proceedings of Multiscale Phenomena and Their Simulation, World Scient. Publ., Singapore, 1997.

[19] R. Rannacher, Error control in finite element computations, in: H. Bulgak, C. Zenger (Eds.), Proceedings of Summer School Error Control and Adaptivity in Scientific Computing, Kluwer Academic Publishers, Dordrecht, 1998, pp. 247–278.

[20] R. Rannacher, F.-T. Suttmeier, A posteriori error estimation and mesh adaptation for finite element models in elasto-plasticity, Comput. Methods Appl. Mech. Eng. 176 (1999) 333–361.

[21] F.-T. Suttmeier, Adaptive finite element approximation of problems in elasto-plasticity theory, Doctor Thesis, Preprint 97-11, SFB 359, Universität Heidelberg, 1997.

[22] R. Verfürth, A Review of A Posteriori Error Estimation and Adaptive Mesh-Refinement Techniques, Wiley/Teubner, New York, Stuttgart, 1996.

ELSEVIER

Journal of Computational and Applied Mathematics 128 (2001) 235–260

JOURNAL OF COMPUTATIONAL AND APPLIED MATHEMATICS

www.elsevier.nl/locate/cam

The p and hp finite element method for problems on thin domains

Manil Suri *,[1]

Department of Mathematics and Statistics, University of Maryland Baltimore County, Baltimore, MD 21250, USA

Received 27 August 1999; received in revised form 7 October 1999

Abstract

The p and hp versions of the finite element method allow the user to change the polynomial degree to increase accuracy. We survey these methods and show how this flexibility can be exploited to counter four difficulties that occur in the approximation of problems over *thin* domains, such as plates, beams and shells. These difficulties are: (1) control of modeling error, (2) approximation of corner singularities, (3) resolution of boundary layers, and (4) control of locking. Our guidelines enable the efficient resolution of these difficulties when a p/hp code is available. © 2001 Elsevier Science B.V. All rights reserved.

MSC: 65N30

Keywords: p version; hp version; Boundary layers; Hierarchical modeling; Plates; Singularities; Locking

1. Introduction

The classical 1956 reference [42] contains one of the first published systematic description of finite elements. The elements described in this paper (such as the Turner rectangle, the Timoshenko beam element and the linear triangle (first proposed by Courant in 1943)) use linear (or bilinear) piecewise polynomials to approximate the solution, and depend on mesh refinement for increased accuracy. This philosophy, of using low-order polynomials over successively finer meshes, has been the predominant one considered by researchers for many years, and was the one under which the development of the finite element method proceeded (with much success) through the 1970s.

* Fax: 1-410-455-1066.

E-mail address: suri@math.umbc.edu (M. Suri).

[1] Supported in part by the Air Force Office of Scientific Research. Air Force Systems Command, USAF, under grant F49620-98-1-0161 and the National Science Foundation under grant DMS-9706594.

PII: S0377-0427(00)00514-8

Experiments by Szabo and his group conducted in the mid-1970s [41] indicated that an alternative strategy might hold great promise as well. Their idea was to keep the mesh fixed, but increase the polynomial degree for accuracy. They called this the p version, to distinguish it from the classical method, which was labelled the h version. Computations on the elasticity problem indicated that this new philosophy was always competitive with, and often out-performed, the traditional h version. The first theoretical paper on the p version was published by Babuška et al. in 1981 [11], and showed, among other results, that the convergence rate was *double* that the h version for domains with corners, and *exponential* for smooth solutions.

The collaboration between Babuška and Szabo also led to the development of the so-called hp version, which combines both strategies. It was shown that with proper mesh selection/refinement coupled with increase of polynomial degrees, an exponential convergence rate could be achieved even for unsmooth solutions, i.e. in the presence of corner singularities. The first mathematical paper on the hp version was by Babuška and Dorr, and appeared in 1981 [2].

Since the advent of these first results, the p and hp versions have, over the past two decades, come into their own as viable complements to the h version. Various commercial codes, such as STRESS CHECK, POLYFEM, PHLEX, APPLIED STRUCTURE, MSC-NASTRAN (among others) have either been developed or modified to include p/hp capability. While no single strategy (such as h, p or hp refinement) can be expected to be optimal for all problems, having both h and p capability allows a level of flexibility that can often be exploited to significantly increase accuracy and efficiency. In particular, the high rates of convergence afforded by p/hp techniques when corner (r^α) singularities are present in the solution puts these methods ahead of traditional h-refinement for some important classes of problems.

In this paper, we bring out some of these advantages of p/hp methods. Rather than give a general survey of such methods (for which the reader is referred, e.g., to [9], and to the books [40,31]), we list some important properties in Section 2, and then concentrate mainly on one class of problems — that of linear elasticity posed on *thin* domains such as beams, plates and shells. This class of problems is one that comes up very frequently in engineering structural analysis. In Sections 3–6, we discuss four areas where p/hp capability leads to advantages in approximation: (1) control of *modeling error*, (2) good approximation of *singularities* occurring at the corners of the domain, (3) accurate resolution of *boundary layers*, and (4) control of *locking* phenomena. In the course of our discussion, we also consider related problems where p/hp versions have advantages, such as singularly perturbed second order elliptic PDEs, which can result in stronger boundary layers than the ones found in plate models.

2. *h*, *p* and *hp* finite element spaces

Suppose we are given a problem in variational form: Find $u \in V$ such that

$$B(u,v)=F(v), \quad v \in V. \tag{2.1}$$

Here F is a bounded linear functional on the (infinite dimensional) Hilbert space V, and $B(\cdot,\cdot)$ is a bounded, coercive, symmetric, bilinear form on $V \times V$. Then given a sequence of (finite element) subspaces $\{V_N\} \subset V$, we can define the finite element approximations $u_N \in V_N$ satisfying

$$B(u_N,v)=F(v), \quad v \in V_N. \tag{2.2}$$

It is easily shown that

$$\|u-u_N\|_{\mathrm{E}} \leqslant \inf_{v \in V_N} \|u-v\|_{\mathrm{E}}, \tag{2.3}$$

where $\|\cdot\|_{\mathrm{E}}$ is the *energy norm*

$$\|u\|_{\mathrm{E}} = (B(u,u))^{1/2}. \tag{2.4}$$

The coercivity and boundedness of B then gives

$$\|u-u_N\|_V \leqslant C \inf_{v \in V_N} \|u-v\|_V \tag{2.5}$$

but, as we shall see, the constant C in (2.5) can be large for some problems.

In all our examples, V will satisfy $H_0^1(\Omega) \subset V \subset H^1(\Omega)$ [2] where $\Omega \subset \mathbb{R}^t$, $t = 1,2,3$ is a bounded domain with piecewise analytic boundary. The finite element spaces V_N will then consist of continuous piecewise polynomials defined on some mesh $\mathcal{T}_N$ on Ω. We describe these in more detail for the case $\Omega \subset \mathbb{R}^2$.

Assume each $\mathcal{T}_N$ is a regular [13] mesh consisting of straight-sided triangles and parallelograms $\{\tau_i^N\}$, $i = 1,2,\ldots,I(N)$ (more general curvilinear elements could also be considered). For any element S, we define for $p \geqslant 0$ integer, $\mathcal{P}_p(S)$ ($\mathcal{Q}_p(S)$) to be the set of all polynomials of total degree (degree in each variable) $\leqslant p$. We also denote $\mathcal{Q}'_p(S) = \mathrm{span}\{\mathcal{P}_p(S), x^p y, xy^p\}$. Let $\boldsymbol{p}_N$ be a degree vector associating degree p_i^N to element τ_i^N (if p_i^N is independent of i, we write $\boldsymbol{p}_N = p_N$). Then the local polynomial spaces are denoted $\mathcal{R}_N(\tau_i^N)$ where $\mathcal{R}_N = \mathcal{P}_{p_i^N}$ if τ_i^N is a triangle and $\mathcal{R}_N = \mathcal{Q}_{p_i^N}$ or $\mathcal{Q}'_{p_i^N}$ if τ_i^N is a parallelogram. We then set

$$V_N = \{v \in V, v|_{\tau_i^N} \in \mathcal{R}_N(\tau_i^N)\}$$

($V_N \subset C^{(0)}(\Omega)$ for our examples).

Let us denote $h_i^N = \mathrm{diam}(\tau_i^N)$, $h_N = \max_i h_i^N$, $\underline{h}_N = \min_i h_i^N$. The sequence $\{\mathcal{T}_N\}$ is called *quasi-uniform* provided there exists α independent of N such that

$$\frac{h_N}{\underline{h}_N} \leqslant \alpha.$$

As we shall see, spaces on quasiuniform meshes do not have the best properties where approximation of corner singularities is concerned. Rather, such singularities must be treated by nonquasiuniform mesh refinement, where the mesh becomes finer as one approaches the point of singularity. The type of mesh used in the hp version is *geometric*, and is defined below for the case of refinement towards the origin $\mathbf{O}$. (Fig. 1 gives a more intuitive idea than the definition, which is technical.)

Let $0 < q < 1$ be a number called the *geometric ratio*. Let $n(=n_N)$ be the number of layers around the origin $\mathbf{O}$. We denote the elements of $\mathcal{T}_N$ by $\tau_{n,j,k}^N$ where $j = 1,\ldots,\mu(k)$, $\mu(k) \leqslant \mu_0$ and $k = 1,2,\ldots,n+1$. Let $h_{n,j,k} = \mathrm{diam}(\tau_{n,j,k}^N)$ and $d_{n,j,k} = \mathrm{dist}(\tau_{n,j,k}^N, \mathbf{O})$. Then

(A) if $\mathbf{O} \notin \bar{\tau}_{n,j,k}^N$, for $j = 1,\ldots,\mu(k)$, $k = 2,\ldots,n+1$,

$$C_1 q^{n+2-k} \leqslant d_{n,j,k} \leqslant C_2 q^{n+1-k},$$

$$\kappa_1 d_{n,j,k} \leqslant h_{n,j,k} \leqslant \kappa_2 d_{n,j,k}.$$

[2] We use standard Sobolev space notation: $H^k(\omega)$ is the space of functions with k derivatives that are square integrable over ω. $H_0^1(\omega)$ is the subset of functions in $H^1(\omega)$ with vanishing trace on $\partial\omega$. $\|\cdot\|_{H^k(\omega)}$ is the norm of $H^k(\omega)$.

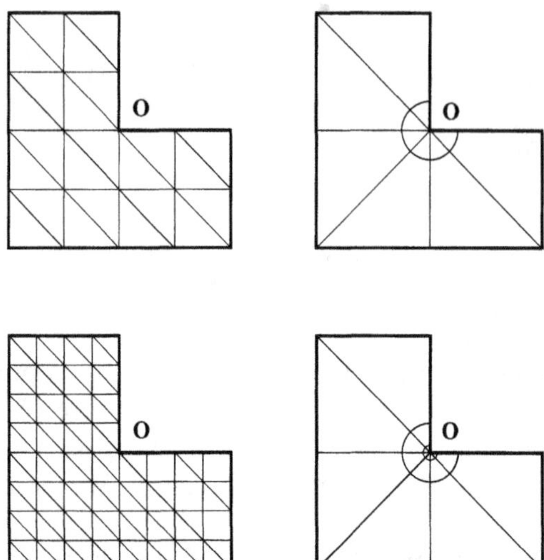

Fig. 1. Types of refinement: (a) quasiuniform, (b) geometric.

(B) if $\mathbf{O} \in \bar{\tau}^N_{n,j,k}$, then $k = 1$ and for $j = 1, \ldots, \mu(1)$,

$$\kappa_3 q^n \leqslant h_{n,j,k} \leqslant \kappa_4 q^n.$$

The constants C_i and κ_i are independent of N. See [21], where it is empirically shown that $q \approx 0.15$ is the best choice in terms of approximability. (In one dimension, this is theoretically established in [20].)

We can now talk about the following *extension procedures*, i.e. strategies for increasing the dimension of the space V_N to get a more accurate estimate in (2.3).

(i) *h version*: The most basic extension procedure consists of using quasiuniform meshes $\{\mathcal{T}_N\}$ with successively smaller h_N, with uniform $p_N = p$, kept fixed at $p = 1$ or 2 usually. Nonquasiuniform meshes (such as *radical* meshes, see [9]) could also be used.

(ii) *p version*: Here, $\mathcal{T}_N$ is the same for all N. Accuracy is achieved by increasing $\boldsymbol{p}_N$. The mesh could be uniform, but if a properly refined mesh (Fig. 1(b)) is used, the p version can often be made to yield the superior performance of the hp version (which is harder to implement).

(iii) *hp version*: Any $\{V_N\}$ for which both $\mathcal{T}_N$ and $\boldsymbol{p}_N$ are changed will give rise to an hp version. For instance, in Section 5, we consider an hp version designed to resolve boundary layers, where as $\boldsymbol{p}_N$ is increased, the number of elements remains the same, but the *size* of elements changes. For treating corner singularities, the geometric meshes with n layers described above are used, with p_N chosen uniformly to be $p_N = \kappa n_N$ (κ fixed), so that both p_N and n_N increase simultaneously. Even more efficient is the case that p_N is chosen to be the same over all elements in layer k, i.e. over all $\tau^N_{n,j,k}$ for $j = 1, \ldots, \mu(k)$, and increasing linearly with the index $k = 1, 2, \ldots, n+1$ (see [21] for more details).

Let us mention that there are several differences between implementation of p/hp methods and the traditional h methods (for instance, the basis functions used in the p version are often not of nodal type). We do not discuss these aspects here but instead refer the reader to [40].

To conclude this section, we present the following theorem, which gives the basic estimate for the infimum in (2.5) when the norm $\|\cdot\|_V = \|\cdot\|_{H^1(\Omega)}$ and when the h, p or hp version over quasiuniform meshes is used.

Theorem 2.1 (Babuška and Suri [7]). *Let the spaces V_N consist of piecewise polynomials of degree p_N over a quasiuniform family of meshes $\{\mathcal{T}_N\}$ on $\Omega \subset \mathbb{R}^t$, $t = 1,2,3$. Then for $u \in H^k(\Omega)$, $k \geqslant 1$,*

$$\inf_{v \in V_N} \|u - v\|_{H^1(\Omega)} \leqslant C h_N^{\min(k-1,\, p_N)} p_N^{-(k-1)} \|u\|_{H^k(\Omega)} \tag{2.6}$$

with C a constant independent of u and N. (For a nonuniform distribution $\boldsymbol{p}_N$, we replace p_N by $\min(\boldsymbol{p}_N)$ in (2.6)).

Theorem 2.1 combined with (2.5) immediately shows that the following rates of convergence hold:

$$h \text{ version}: \quad \|u - u_N\|_V = \mathrm{O}(h_N^{\min(k-1,\, p_N)}), \tag{2.7}$$

$$p \text{ version}: \quad \|u - u_N\|_V = \mathrm{O}(p_N^{-(k-1)}). \tag{2.8}$$

Since the number of degrees of freedom $N \sim \mathrm{O}(h_N^{-t} p_N^t)$, we see that asymptotically, the rate of convergence (in terms of N) of the p version is never lower than that of the h version. In fact, for smooth solutions, it is often much better, as seen from (2.8) ("spectral" convergence). As we shall see in Section 4, the rate can be better even when the solution is not smooth.

3. Control of modeling error

Structural analysis over a three-dimensional domain (thin or otherwise) involves solving the equations of three-dimensional elasticity on the domain. When three-dimensional finite elements are used, the number of degrees of freedom N grows quite rapidly. For instance, while $N \sim \mathrm{O}(h^{-2})$ or $\mathrm{O}(p^2)$ in two dimensions (for the h and p versions, respectively), we have $N \sim \mathrm{O}(h^{-3})$ or $\mathrm{O}(p^3)$ in three dimensions. In the case that one dimension of the domain is thin, the three-dimensional model is often replaced by a two-dimensional model, formulated generally on the mid-plane of the domain. Discretization of this two-dimensional model then requires fewer degrees of freedom.

Let us present an illustrative example, that of a thin plate. Let ω be a bounded domain in $\mathbb{R}^2$ with piecewise smooth boundary, which represents the midplane of the plate, which we assume to be of thickness d ($d \ll \mathrm{diam}(\omega)$). Then, we represent the three-dimensional plate as

$$\Omega = \{x = (x_1, x_2, x_3) \in \mathbb{R}^3 \mid (x_1, x_2) \in \omega, \ |x_3| < d/2\}.$$

The lateral surface S and top and bottom surfaces $R_\pm$ are given by (see Fig. 2)

$$S = \{x \in \mathbb{R}^3 \mid (x_1, x_2) \in \partial\omega, \ |x_3| < d/2\},$$

$$R_\pm = \{x \in \mathbb{R}^3 \mid (x_1, x_2) \in \omega, \ x_3 = \pm d/2\}.$$

Let us denote the displacement $\boldsymbol{u} = (u_1, u_2, u_3)$. Then we consider the problem of finding $\boldsymbol{u} \in \boldsymbol{H}_D^1(\Omega) = \{\boldsymbol{u} \in (H^1(\Omega))^3 \mid u_3 = 0 \text{ on } S\}$ which satisfies

$$B(\boldsymbol{u}, \boldsymbol{v}) = F(\boldsymbol{v}) \ \forall \boldsymbol{v} \in \boldsymbol{H}_D^1(\Omega), \tag{3.1}$$

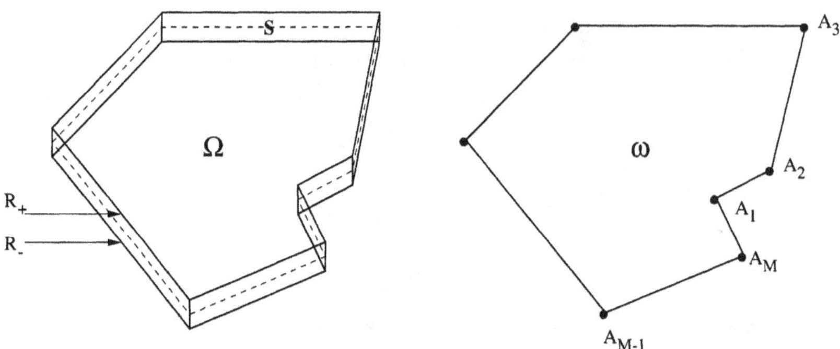

Fig. 2. The three-dimensional plate Ω and two-dimensional midplane ω.

where, with $t = 3$,

$$B(\boldsymbol{u},\boldsymbol{v}) = \frac{E}{1+\nu}\int_\Omega \left[\sum_{i,j=1}^{t} \varepsilon_{ij}(\boldsymbol{u})\varepsilon_{ij}(\boldsymbol{v}) + \frac{\nu}{1-2\nu}(\operatorname{div}\boldsymbol{u})(\operatorname{div}\boldsymbol{v})\right] \mathrm{d}x \tag{3.2}$$

and

$$F(\boldsymbol{v}) = \frac{1}{2}\int\int_\omega g(x_1,x_2)(v_3(x_1,x_2,d/2) + v_3(x_1,x_2,-d/2))\,\mathrm{d}x_1,\,\mathrm{d}x_2. \tag{3.3}$$

In the above, $E > 0$ is the Young's modulus of elasticity, $0 \leqslant \nu < \frac{1}{2}$ is the Poisson ratio and

$$\varepsilon_{ij}(\boldsymbol{u}) = \frac{1}{2}\left[\frac{\partial u_i}{\partial x_j} + \frac{\partial u_j}{\partial x_i}\right].$$

Eqs. (3.1)–(3.3) constitute the plate problem with *soft simple support* — we could use different constrained spaces to describe other physical problems (such as the *clamped* or *built-in* boundary condition which is obtained by using $\boldsymbol{u} = 0$ instead of $u_3 = 0$ on S).

Two-dimensional models are derived from (3.1)–(3.3) by making assumptions about the behavior of $\boldsymbol{u}$ with respect to the x_3 variable, substituting these into (3.1)–(3.3), and integrating in the x_3 variable to give a problem formulated on ω alone. The most basic classical plate model is the Kirchhoff–Love model, where we assume that

$$u_3(x_1,x_2,x_3) = w(x_1,x_2), \tag{3.4}$$

$$u_i(x_1,x_2,x_3) = -\frac{\partial w}{\partial x_i}(x_1,x_2)\left(\frac{x_3}{d/2}\right), \quad i = 1,2. \tag{3.5}$$

Substituting (3.4) and (3.5) into (3.1)–(3.3) and integrating in x_3, we get an equation for $w(x_1,x_2)$ alone, which is the weak form of the biharmonic problem.

One disadvantage of the KL model is that the corresponding variational form involves second derivatives of the FE functions, and hence requires $C^{(1)}$ continuous elements. The so-called Reissner–Mindlin (RM) model avoids this problem by replacing (3.5) by the expression

$$u_i(x_1,x_2,x_3) = \phi_i(x_1,x_2)\left(\frac{x_3}{d/2}\right), \quad i = 1,2. \tag{3.6}$$

Hence, we now have two more unknowns $(\phi_1,\phi_2)=\boldsymbol{\phi}$. This leads to the following variational form, after some modification of the elastic constants (see [5]): Find $U=(\boldsymbol{\phi},w)\in V=H^1(\Omega)\times H^1(\Omega)\times H^1_D(\Omega)$ satisfying for all $W=(\boldsymbol{\theta},\xi)\in V$,

$$B(U,W)=a(\boldsymbol{\phi},\boldsymbol{\theta})+\gamma\mu d^{-2}(\nabla w+\boldsymbol{\phi},\nabla\xi+\boldsymbol{\theta})=\int\int_\omega g\xi\,\mathrm{d}x_1\,\mathrm{d}x_2, \tag{3.7}$$

where

$$a(\boldsymbol{\phi},\boldsymbol{\theta})=D\int\int_\omega\left\{(1-v)\sum_{i,j=1}^{2}\varepsilon_{ij}(\boldsymbol{\phi})\varepsilon_{ij}(\boldsymbol{\theta})+v(\operatorname{div}\boldsymbol{\phi})(\operatorname{div}\boldsymbol{\theta})\right\}\mathrm{d}x_1\,\mathrm{d}x_2 \tag{3.8}$$

and where $D=E/12(1-v^2)$, $\mu=E/2(1+v)$ and γ is the shear correction factor.

It is seen from (3.7) that the RM model essentially enforces the Kirchhoff constraint,

$$KU=\nabla w+\boldsymbol{\phi}=0 \tag{3.9}$$

in a penalized form (while the KL model enforces it *exactly*).

The replacement of the three-dimensional elasticity problem by a two-dimensional model such as RM leads to a *modeling error* between the actual three-dimensional solution and the solution obtained e.g. by (3.4)–(3.6). This modeling error is unaffected by any subsequent discretization of the two-dimensional model. The only way to decrease it is to use a more accurate model. To derive such more accurate models, we expand $\boldsymbol{u}(x_1,x_2,x_3)$ in terms of higher-order polynomials of x_3.

More precisely, let $n=(n_1,n_2,n_3)$ by a triple of integers $\geqslant 0$. We then make the ansatz (for $i=1,2$)

$$u_3(x_1,x_2,x_3)=\sum_{j=0}^{n_3}w_j(x_1,x_2)L_j\left(\frac{x_3}{d/2}\right), \tag{3.10}$$

$$u_i(x_1,x_2,x_3)=\sum_{j=0}^{n_i}\phi_{ij}(x_1,x_2)L_j\left(\frac{x_3}{d/2}\right), \tag{3.11}$$

which when substituted into the three-dimensional equations gives us the so-called $n=(n_1,n_2,n_3)$ model upon integration. (Here $\{L_j\}$ are the Legendre polynomials.) It is possible to show that (under some assumptions)

(1) For $n_1\geqslant 1$, $n_2\geqslant 1$, $n_3\geqslant 2$, models $n=(n_1,n_2,n_3)$ converge, as $d\to 0$, to the same limiting solution as that obtained when $d\to 0$ in the three-dimensional elasticity equations. (For $n=(1,1,0)$, this is true with modified elastic constants.)
(2) For fixed d, as $\min(n_1,n_2,n_3)\to\infty$, the solutions of the models converge to the exact three-dimensional solution.

There are various factors such as boundary conditions, boundary layers, corner singularities, etc., which affect the convergence of a given model as $d\to 0$ [6]. For instance, in [17] it is shown that the RM model may not lead to good approximations of the boundary layer, while higher-order models significantly improve the convergence as $d\to 0$.

Let us denote for $n=(n_1,n_2,n_3)$

$${}^nH^1_D(\Omega)=\{\boldsymbol{u}=(u_1,u_2,u_3)\in H^1_D(\Omega)\text{ satisfying (3.10) and (3.11)}\}.$$

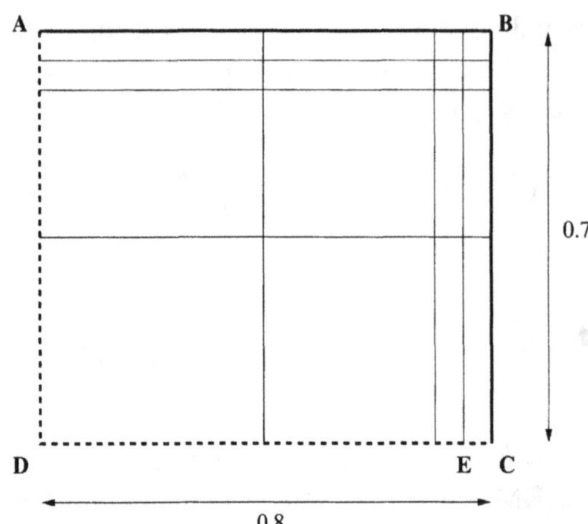

Fig. 3. Mid-plane of the quarter plate.

Then the solution of the (n_1, n_2, n_3) model can be expressed as the unique solution $\boldsymbol{u}^n$ of

$$B(\boldsymbol{u}^n, v) = F(\boldsymbol{v}) \quad \forall \boldsymbol{v} \in {}^n H_D^1(\Omega), \tag{3.12}$$

where B, F are as in (3.2) and (3.3). Eq. (3.12) represents a hierarchy of models of increasing accuracy.

Unfortunately, most finite element codes for plates (and shells) only discretize a *fixed* model of the above hierarchy (often the RM model, which is essentially the $(1,1,0)$ model — though sometimes the $(1,1,2)$ model is also used). This is generally the only option available in the h version. If, however, the code has p version capability, then it is possible to implement *hierarchical modeling* corresponding to (3.12) quite easily, provided the user can pick different polynomial degrees in different directions.

Assume that the midsurface ω is partitioned into elements $\{\tau_j\}$ which are either quadrilaterals or triangles. Then the domain Ω is partitioned into three-dimensional elements $\{T_j\} = \{\tau_j \times (-d/2, d/2)\}$. Let $R_j(p,q)$ be the corresponding polynomial space on T_j, where p is the degree in x_1, x_2 and q is the degree chosen in x_3. This space could be based on the choice $\mathscr{P}_p$, $\mathscr{Q}_p$ or $\mathscr{Q}'_p$ over τ_j in the (x_1, x_2) variables, and a (usually lower-degree) choice of q. Then, the corresponding FE solution with elements $R_j(p,q)$ of the three-dimensional plate problem will be exactly the same as the FE approximation (using elements of degree p) of the two-dimensional plate model (3.12) with $n_1 = n_2 = n_3 = q$. Hence, one can implement different plate models in the p/hp versions just by taking $q = 1, 2, 3, \ldots,$ in the discretization of the three-dimensional elasticity problem (this hierarchic modeling is available, e.g., in the code STRESS CHECK). The modeling error may now be controlled.

Let us present a simple example using the code STRESS CHECK. Consider the quarter rectangular plate shown in Fig. 3, which has soft simple support conditions on AB, BC and symmetric conditions on CD, DA. The plate is subjected to a uniform outward transverse force of 1. Elastic constants are $E = 3.0 \times 10^7$ and $v = 0.3$, with shear correction factor of unity. The plate has thickness $=0.1$, which puts it in the "moderately thin" class — the mesh is designed so as to approximate the boundary layer well (see Section 5). We use the $n = (n, n, n)$ plate model with $n = 1, 2, 3$ and 6.

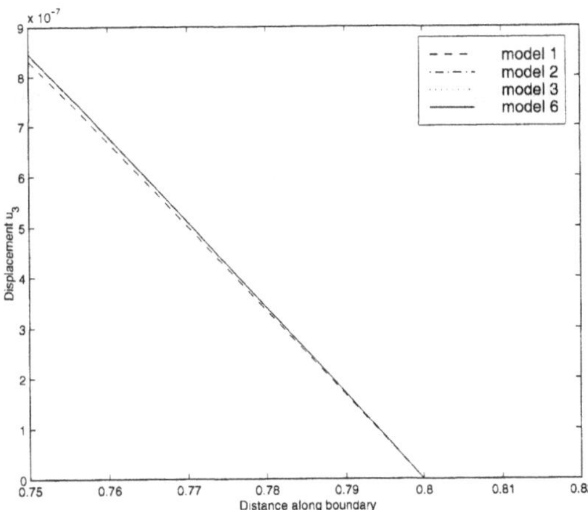

Fig. 4. Transverse displacement u_3 along EC for different models.

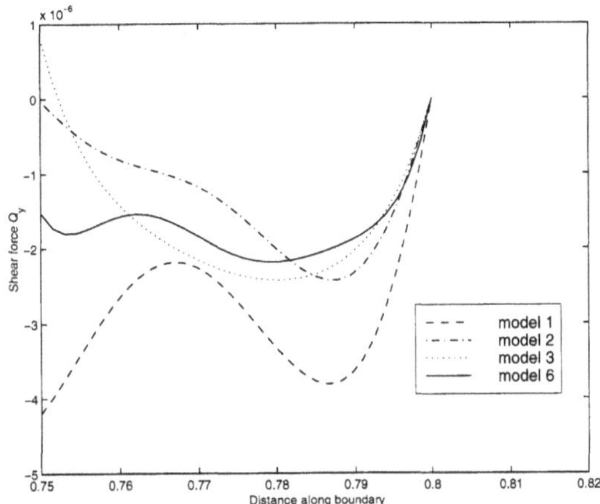

Fig. 5. Shear force distribution Q_y along EC for different models.

Figs. 4 and 5 show the extraction of two quantities — the transverse displacement u_3 and the shear force Q_y, along the lower edge of the element situated at the corner C. Degree $p=8$ is used for the finite element approximations. It is observed that while there is no appreciable difference for u_3 with the model order, the quantity Q_y does vary appreciably. The reason for this is that u_3 has a very weak boundary layer, while Q_y has a stronger layer — and this is dependent on the model order (see [17] for more detailed experiments).

The above ideas also apply to other thin domains, such as composite plates and shells [10]. For these more complicated domains, the differences observed between different model orders is usually even more striking.

4. Approximation of singularities

The previous section showed how the three-dimensional elasticity equations over the plate Ω could be reduced to a hierarchy of elliptic problems (such as (3.7)) over the two-dimensional domain ω (Fig. 2(b)). It is well known that linear elliptic problems will (in general) have singularities at the corners A_j, $j = 1,\ldots,M$ of the domain ω, and also at points on $\partial\omega$ where the type of boundary condition changes, e.g., from Dirichlet to Neumann. For the three-dimensional problem over Ω, the singularities will occur not only at vertices of Ω but also along its edges. One of the main advantages of p/hp methods is that with proper mesh design, such singularities can be approximated very well.

4.1. A decomposition result

An examination of Eq. (3.7) shows that plate models such as the RM model are singularly perturbed in the variable ϕ. This results in boundary layers when d is small. The interaction of corner singularities and boundary layers is a complicated phenomenon, for which decomposition and regularity results may be found, e.g., in [25,16]. We will postpone the consideration of boundary layer approximation to Section 5. Here, to simplify the exposition, we assume that the thickness d is a fixed positive constant and do not worry about our approximation results being uniform in d.

Accordingly, given a linear elliptic problem on ω (which is one of our (n_1, n_2, n_3) models), we may use the results of Kondratiev (see [26] and also [19,15]) to decompose the solution in the neighborhood of any vertex A_j. If U is the solution and (r,θ) are the polar coordinates with origin at A_j, then we may write

$$U = \sum_{\ell=1}^{L}\sum_{s=0}^{S}\sum_{t=0}^{T} c_{\ell st}\psi_{\ell st}(\theta) r^{\alpha_\ell + t}\ln^s r + U_0$$

$$= \sum_{\ell=1}^{L} U_\ell + U_0 \tag{4.1}$$

in the neighborhood of A_j. Here $\psi_{\ell st}$ is a vector with the same number of components as U, and is analytic in θ. The exponents α_ℓ can be complex, and the coefficients $c_{\ell st}$ will depend on d. By taking L, S, T large enough, we can make U_0 as smooth as the data will allow. In fact, taking the correct number of terms near each A_j, we may write

$$U = \sum_{j=1}^{M}\sum_{\ell} U_\ell^{(j)}\chi_j + U_0, \tag{4.2}$$

where χ_j is a smooth cut-off function in the neighborhood of A_j and U_0 satisfies the appropriate shift theorem (e.g., in the case of (3.7), we would have

$$\|U_0\|_{[H^{k+2}(\omega)]^3} \leqslant C(d)\|g\|_{H^k(\omega)},$$

i.e., the same theorem as for a smooth boundary). Eqs. (4.1) and (4.2) show that when the finite element method is used, the rate of best approximation (i.e., the right side of (2.3)) will be determined by the worst singularity $\min_{\ell,j}|\alpha_\ell^{(j)}|$ (and the corresponding S in decomposition (4.1)), since the other terms (including U_0) are all smoother and hence result in better approximation rates.

4.2. hp convergence: quasiuniform meshes

We now present a theorem for the approximation of one of the canonical singularities present in (4.2), which we write as

$$u = r^{\alpha} \ln^{s} r \psi(\theta), \tag{4.3}$$

where ψ is a vector and for α complex, we understand r^{α} to denote $\operatorname{Re} r^{\alpha}$.

First, we note that for $\omega \subset \mathbb{R}^2$, $u \in H^{1+\hat{\alpha}-\varepsilon}(\omega)$ for any $\varepsilon > 0$, where $\hat{\alpha} = \operatorname{Re} \alpha$. We could apply Theorem 2.1 to get an approximation rate of $O(h_N^{\min(\hat{\alpha}-\varepsilon, p_N)} p_N^{-\hat{\alpha}+\varepsilon})$, but as shown in [7], this result can be improved to give the following theorem.

Theorem 4.1. *Let u be given by* (4.3). *Let quasiuniform meshes be used. Then*

$$\inf_{v \in V_N} \|u - v\|_{H^1(\omega)} \leqslant Ck(h_N, p_N, s) \min \left\{ h_N^{\hat{\alpha}}, \frac{h^{\min(\hat{\alpha}, p_N - \hat{\alpha})}}{p_N^{2\hat{\alpha}}} \right\}, \tag{4.4}$$

where $k(h_N, p_N, s) = \max(|\ln^s h_N|, |\ln^s p_N|)$ *and C is a constant independent of N.*

Theorem 4.1 shows that one gets the following convergence rates:

$$h \text{ version: } O(h_N^{\hat{\alpha}}), \tag{4.5}$$

$$p \text{ version: } O(p_N^{-2\hat{\alpha}}). \tag{4.6}$$

Since $N = O(h_N^{-2} p_N^2)$, we see that the p version gives *twice* the convergence rate of the h version. This happens whenever the point of singularity $r = 0$ coincides with the node of an element. If the singularity point lies in the interior of an element, then the doubling effect does not take place.

Applying Theorem 4.1 to each component of (4.2) and assuming that U_0 is smooth enough to be better approximated, we see that (4.4)–(4.6) will also hold for the error $\|U - U_N\|_V$ where U_N is the FEM approximation to U.

4.3. hp convergence: nonquasiuniform meshes

For the h version with polynomials of fixed degree p, it is well known that the optimal convergence rate for smooth functions is $O(h_N^p)$. Hence, the rate (4.5) can be quite far from optimal if $\hat{\alpha}$ is small (for example, $\hat{\alpha} < 1$ in nonconvex domains for the Poisson equation). It turns out that the use of nonquasiuniform meshes, which are increasingly refined towards the points of singularity, can increase the convergence rate. The optimal meshes in this regard are the so-called *radical* meshes, which result in $O(h_N^p)$ convergence for function (4.3) when the mesh is designed taking into account the strength of the singularity $\hat{\alpha}$ and the degree of the polynomial p. We refer to [20] where a full analysis of such meshes is given in one dimension and to [4] where the case $p = 1$ is analyzed in two dimensions.

Here, we present a result for the hp version, which shows that with the *geometric* meshes described in Section 2, one obtains *exponential* convergence in terms of N, the number of degrees of freedom (note that $O(h_N^p)$ convergence is only *algebraic* in N). The spaces V_N are chosen as follows. In the vicinity of each singularity point, a geometric mesh with n_N layers (see Section 2) is constructed,

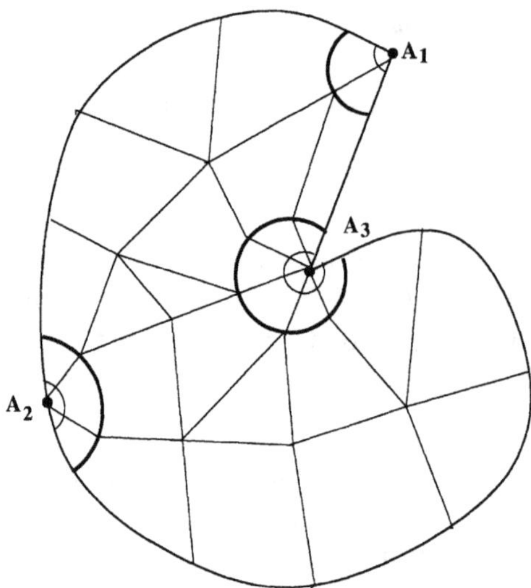

Fig. 6. hp mesh with $n_N = 2$ layers. A_2 is a singularity point where the type of boundary condition changes.

such that these meshes conform with a fixed mesh over the remainder of the domain (see Fig. 6). The (uniform) degree p_N is chosen to be κn_N over all elements on the domain. As N increases, n_N is increased (and hence so is p_N). The increase in the number of layers causes the singularities to be better approximated. The smooth parts are well-approximated by the increase in p_N.

Theorem 4.2. *Let U be the solution of the RM model* (3.7), (3.8). *Let V_N be the hp spaces described above. Then*

$$\inf_{W \in V_N} \|U - W\|_V \leqslant C e^{-\gamma \sqrt[3]{N}} \tag{4.7}$$

with C a constant independent of N.

The bound in estimate (4.7) in one dimension becomes $Ce^{-\gamma\sqrt{N}}$ [20], while in three dimensions it is $Ce^{-\gamma\sqrt[5]{N}}$ [3]. The design of geometric meshes in three dimensions is described further in [3]. The constant C in Eq. (4.7) will depend on d.

4.4. Numerical example

We illustrate some of the results of the past two theorems, by considering the bending of the L-shaped plate shown in Fig. 1. The longest side of the plate is taken to be 0.8, with $d = 0.1$, $E = 3.0 \times 10^7$, $v = 0.3$, $\gamma = 1$. A uniform $g = 1$ is applied, and the plate is *clamped* ($\boldsymbol{u} = 0$) along the entire boundary.

Fig. 7 shows the percentage relative energy norm error when various h and p versions are used. (The true energy was estimated by using a much larger number of degrees of freedom.) It is observed that with the h version on meshes of the type shown in Fig. 1(a), the error decreases algebraically. For the p version on a uniform mesh of only 6 triangles (i.e., no geometric refinement layers

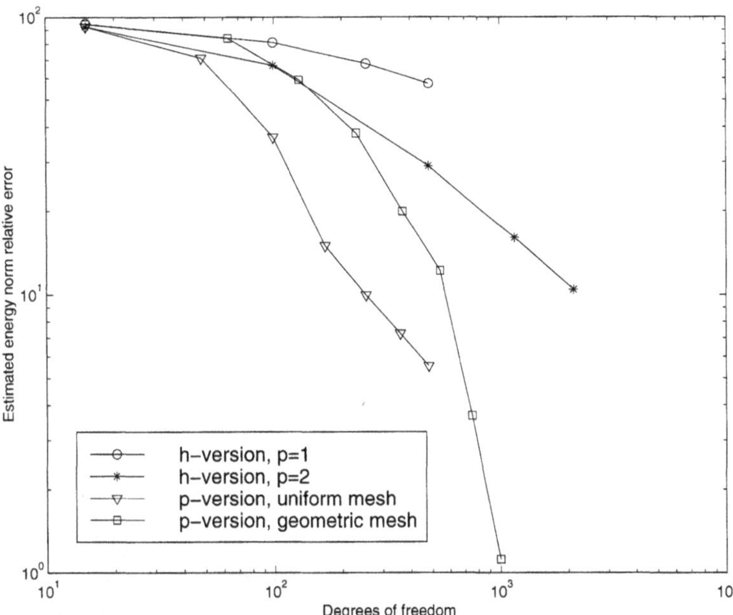

Fig. 7. Error curves for h and p versions.

in Fig. 1(b)), the error follows an 'S' curve, with the middle portion representing an exponential decrease, and the flat portion representing twice the algebraic h version rate. When one layer of mesh refinement is introduced (as in the top of Fig. 1(b)), it is observed the exponential rate continues through $p = 8$. Further mesh refinement would only be useful when this graph flattens out as well — essentially, in the range of parameters shown, the p version is displaying the same exponential rate of convergence expected from the hp version. For more detailed experiments, we refer, e.g., to [7].

5. Resolution of boundary layers

If one lets $d \to 0$ in Eqs. (3.1)–(3.3), then one gets a system of *singularly perturbed* equations [16]. This will also be true for the (n_1, n_2, n_3) model given by (3.12). For instance, it is clear from (3.7) and (3.8) that the RM model is singularly perturbed in ϕ (but not in w), since multiplying (3.7) through by d^2, the order of derivatives in ϕ decreases to 0 as $d \to 0$. This implies that for most boundary conditions ϕ will have a *boundary layer* at $\partial\omega$ for small d, i.e., components in the solution of the form (Fig. 8)

$$\phi_b(s,t) = C(d) f(t) e^{-ks/d}. \tag{5.1}$$

Here (s,t) are local coordinates in the normal and tangential directions at $\partial\omega$, k is a constant and f is a smooth function. The amplitude $C(d)$ of the boundary layer will depend upon the type of boundary conditions — for instance, the largest amplitude $C(d) = O(d)$ occurs for the case of the soft simple support and free plates, while for clamped (built-in) conditions, it is only $O(d^2)$ [1].

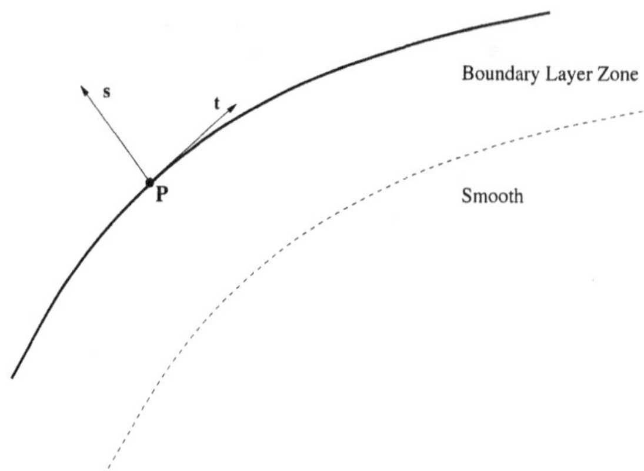

Fig. 8. The boundary layer.

The amplitude will have the same *order* (though not the same value [17]) for all (n_1, n_2, n_3) plate models. The only model free of boundary layers is the KL model. For a discussion of the relation of boundary layers in the three-dimensional case to those in (n_1, n_2, n_3) models, we refer to [17].

Our concern in this section is the *approximation* of components such as (5.1) by the FEM. We note that the boundary layer effect is essentially a one-dimensional effect, so that if we can efficiently approximate the functions ϕ_{b} in the normal direction, then we can expect to have an overall good approximation. Our first step, therefore, is to consider the approximation of the one-dimensional boundary layer function

$$\bar{u}_d(x) = \mathrm{e}^{-x/d} \tag{5.2}$$

by the space V_N of piecewise polynomials of degree $\boldsymbol{p}_N = (p_1^N, p_2^N, \ldots, p_I^N)$ on the mesh $\mathcal{T}_N = \{0 = x_0 < x_1, \ldots, < x_I = 1\}$.

Let us consider the approximation of (5.2) in the context of the one-dimensional singularly perturbed problem

$$-d^2 u_d''(x) + u_d(x) = f(x), \quad x \in I = (-1, 1), \tag{5.3}$$

$$u_d(\pm 1) = \alpha^{\pm}. \tag{5.4}$$

It is well known (see, e.g., [33]) that the solution of (5.3) and (5.4) can be decomposed as

$$u_d(x) = u_d^{\mathrm{s}} + A_d \bar{u}_d(1 + x) + B_d \bar{u}_d(1 - x), \tag{5.5}$$

i.e., it contains terms of form (5.2), plus a smooth (in d) component u_d^{s}. Writing (5.3) and (5.4) in variational form, we have that $u_d \in H_0^1(I)$ satisfies $\forall v \in H_0^1(I)$,

$$B_d(u_d, v) := \int_I \{d^2 u_d' v' + u_d v\} \,\mathrm{d}x = \int_I f v \,\mathrm{d}x. \tag{5.6}$$

We consider the approximation of $\bar{u}_d$ in the energy norm for (5.6), i.e., in the norm

$$\|v\|_d = (B_d(v, v))^{1/2} \approx d\|v'\|_{L_2(I)} + \|v\|_{L_2(I)}.$$

For the h version on a quasiuniform mesh, we may apply (2.6), together with an L_2 duality result, to obtain (for $k \geqslant p_N + 1$)

$$\inf_{v \in V_N} \|\bar{u}_d - v\|_d \leqslant C h_N^{k-1} d^{-k+3/2}, \tag{5.7}$$

where we have used (5.2) to estimate $\|\bar{u}_d\|_{H^k(I)}$. We see that for fixed d, we get the optimal $O(h_N^{k-1})$, but as $d \to 0$, this rate deteriorates. The best possible *uniform* rate for $0 < d \leqslant 1$ is $O(h_N^{1/2})$, obtained by taking $k = \frac{3}{2}$. This is the rate observed in practice for d small.

For the p version, it is shown in [33] that the following theorem holds. Here, $\tilde{p} := p + \frac{1}{2}$.

Theorem 5.1. (A) *Let* $r = (e/2\tilde{p}_N d) < 1$. *Then*

$$\inf_{v \in V_N} \|\bar{u}_d - v\|_d \leqslant C d^{1/2} r^{\tilde{p}_N} (1 - r^2)^{-1/2}, \tag{5.8}$$

where C *is a constant independent of* p_N *and* d.

(B) *We have*

$$\inf_{v \in V_N} \|\bar{u}_d - v\|_d \leqslant C p_N^{-1} \sqrt{\ln p_N}, \tag{5.9}$$

uniformly for $0 < d \leqslant 1$.

Theorem 5.1 shows that while rate (5.8) is exponential, it only holds for p_N very large (not generally seen in practice). The best uniform rate is given by (5.9), and is *twice* that for the h version (modulo the log factor).

Estimates (5.7) and (5.9) show that both the h and p versions give disappointing convergence rates uniform in d. One remedy for this is to use nonquasiuniform meshes, by which the $O(h_N^{k-1})$ rate of the h version can be recovered. One such mesh is given in [33], by $\mathcal{T}_N = \{-1, x_1, \ldots, x_{m-1}, 1\}$, where, for m even, and $p_N \equiv p$,

$$x_{m/2 \pm i} = \mp d\tilde{p} \ln(1 - 2ic/m), \quad i = 0, \ldots, m/2 \tag{5.10}$$

with $c = 1 - \exp(1 - 1/(d\tilde{p}))$. See, e.g., [35,43] for other examples.

Another solution, possible when both h and p capability is available, is to insert a single element of size $O(\tilde{p}d)$ at the boundary layer, and let $p \to \infty$. This gives exponential convergence. More precisely, the following theorem is established in [33].

Theorem 5.2. *Let* $(\mathcal{T}_N, \boldsymbol{p}_N)$ *be such that for* $p \geqslant 1$,

$$\begin{aligned} \boldsymbol{p}_N &= \{p, 1\}, \quad \mathcal{T}_N = \{-1, -1 + \kappa\tilde{p}d, 1\} \quad \textit{if } \kappa\tilde{p}d < 2, \\ \boldsymbol{p}_N &= \{p\}, \quad \mathcal{T}_N = \{-1, 1\} \quad \textit{if } \kappa\tilde{p}d \geqslant 2, \end{aligned} \tag{5.11}$$

where $0 < \kappa_0 \leqslant \kappa < 4/e$ *is a constant independent of* p *and* d. *Then with* $\bar{u}_d = \bar{u}_d(x+1)$, *there exists* $0 < \alpha < 1$ *such that*

$$\inf_{v \in V_N} \|\bar{u}_d - v\|_d \leqslant C d^{1/2} \alpha^{\tilde{p}}$$

with C, α *independent of* p *and* d.

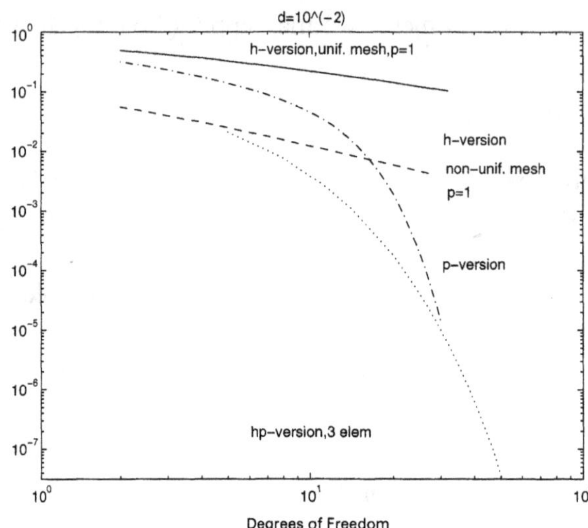

Fig. 9. Comparison of various methods, $d = 10^{-2}$.

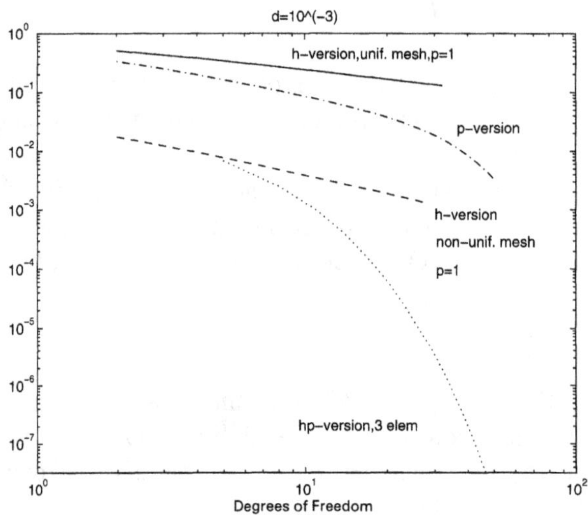

Fig. 10. Comparison of various methods, $d = 10^{-3}$.

We see from the above construction that what we have is an hp method, since the mesh changes as $p \to \infty$. (Experiments indicate that remeshing can be avoided by fixing $\mathcal{T}_N = \{-1, -1+\kappa\tilde{p}_{\max}d, 1\}$ in many cases.)

In Figs. 9 and 10, we compare four different methods discussed above, including the h version given by (5.10) (with $p = 1$), and the hp version described in Theorem 5.2. In each case, we treat problem (5.3), (5.4), with $f(x) = 1$, $\alpha^{\pm} = 0$. Since the solution has boundary layers at both end points of I, we must now refine near both end points, so that, e.g., instead of (5.11), we get the three-element mesh $\{-1, -1+\kappa\tilde{p}d, 1-\kappa\tilde{p}d, 1\}$. The superiority of the hp version is clearly noticed for d small.

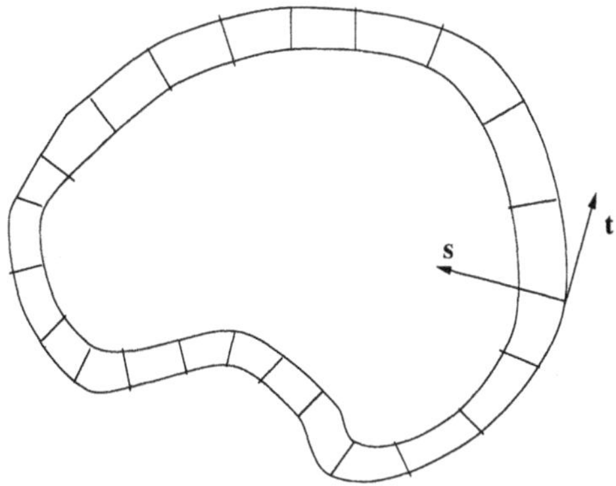

Fig. 11. Boundary-fitted mesh.

The above effect may also be observed for two-dimensional problems. Consider the singularly perturbed problem

$$-d^2\Delta u + u = f \quad \text{in } \omega, \tag{5.12}$$

$$u = 0 \quad \text{on } \partial\omega. \tag{5.13}$$

If ω is a smooth domain, then the solution will again have a component of the form (5.1), with $C(d) = O(1)$. If we now construct a mesh in boundary-fitted coordinates (s,t), such that the (tensor product) basis functions are polynomials in s multiplied by polynomials in t, then the problem of approximating (5.1) reduces once more to a one-dimensional approximation. Hence, for example, we should take the mesh to be such that in the s direction, the first layer of elements is of $O(\kappa\tilde{p}d)$ (see Fig. 11). Then one can again establish exponential hp convergence for the solution of (5.12) and (5.13) uniform in d [44,45,27]. This may also be done when the meshes and basis functions are suitably fitted in (x,y) rather than (s,t) coordinates, as long as the $O(\kappa\tilde{p}d)$ layer character is maintained.

In Figs. 13 and 14, we show experiments performed using STRESS CHECK for (5.12) and (5.13) on the unit circle, with $f = 1$, for which the exact solution can be expressed in terms of a modified Bessel function (Eq. (7.2) of [45]) and has a boundary layer at $\partial\omega$. We take $p_{\max} = 8$ and perform the p version on the two meshes shown in Fig. 12. The first approximates the one from Theorem 5.2, with a layer of elements of size $p_{\max}d$, while the second is a more uniform mesh, fixed for all d. The advantage of the first mesh is clearly demonstrated in Figs. 13 and 14.

The case of a nonsmooth domain is more subtle. For instance, if ω is a polygon, then the singularities for (5.12) and (5.13) now behave essentially like $(r/d)^{\alpha}$ [24]. Consequently, the mesh refinement for the singularities must now be carried out in an $O(d)$ region of the corners, as shown in Fig. 15. In [47], *spectral* convergence of $O(p^{-k})$, k arbitrary, has been established for such meshes as $p \to \infty$.

For the case of the RM plate, as noted before, the strongest boundary layer (in the case of the free or the soft simply supported plate) is only $O(d)$. Consequently, the refinement of the singularities (i.e., usual $O(1)$ hp refinement) is usually sufficient in practice when the error in the energy norm is of interest. However, if quantities involving the s-derivative of (5.1) are calculated (such as

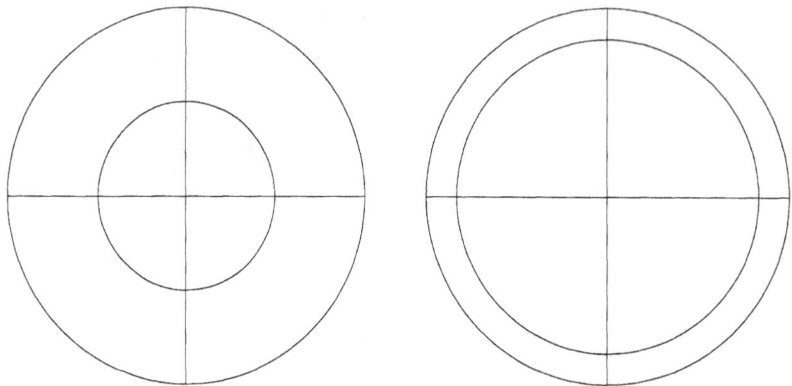

Fig. 12. The two meshes.

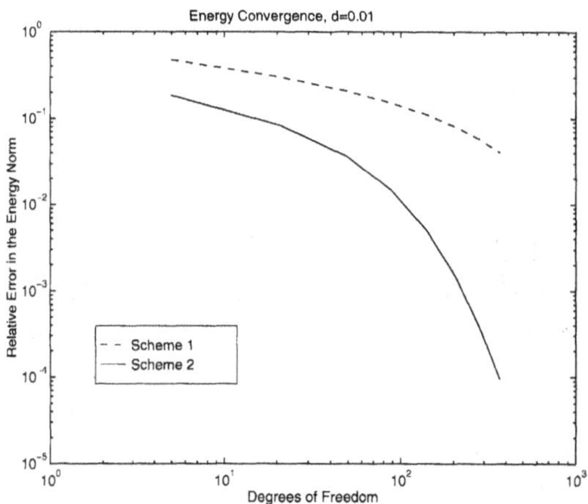

Fig. 13. Comparison of p version, $d = 10^{-2}$.

moments), then these will again contain an O(1) boundary layer, and refinements such as that in Fig. 15 will again be necessary (recall the example in Fig. 3, see, e.g., [34,46]).

For shell problems, the solutions have a rich array of boundary layers, which are often stronger than those encountered in plates. Ideas similar to the ones discussed above may be applied to such problems as well. We refer to [23,29,18,22] for some results.

6. The problem of locking

There is one more problem that occurs in the numerical approximation of thin domains, that of *numerical locking*. Consider once again the RM model (3.7). As $d \to 0$, we see that for the energy norm $\|U\|_{E,d} = (B(U,U))^{1/2}$ to remain finite, we must have the Kirchhoff constraint (3.9) to be satisfied in the limit. This just expresses the fact that the RM model converges to the KL model in energy norm.

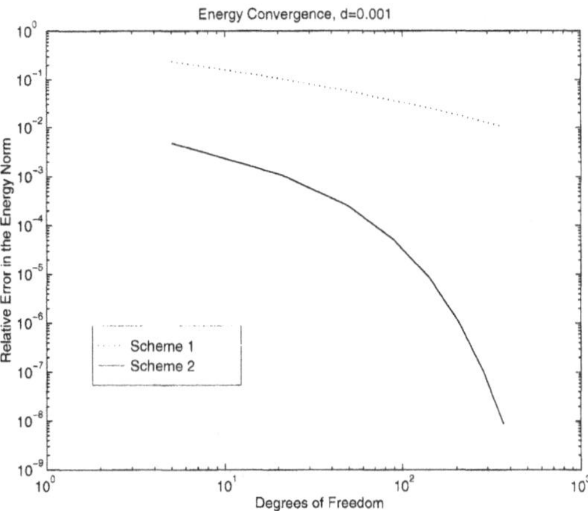

Fig. 14. Comparison of p version, $d = 10^{-3}$.

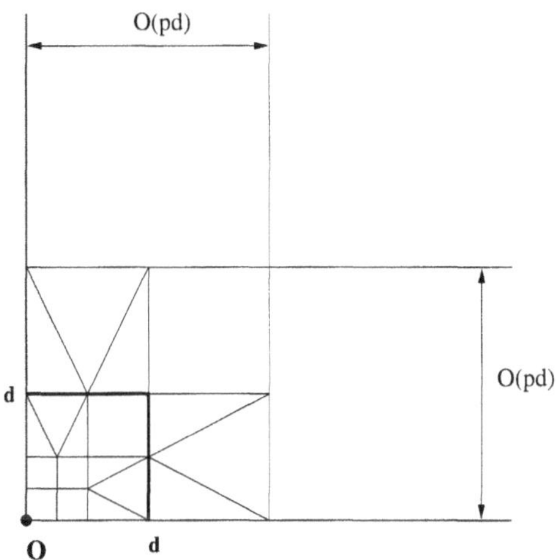

Fig. 15. The mesh for polygons.

If we now consider the finite element approximation of (3.7), then the finite element solution $U_N = (\boldsymbol{\phi}_N, w_N)$ must *also* satisfy (3.9) in the limit $d \to 0$ to keep the energy norm $\|U_N\|_{\mathrm{E},d}$ finite, i.e.,

$$KU_N = \nabla w_N + \boldsymbol{\phi}_N = 0. \tag{6.1}$$

As a result, only those functions $U_N \in V_N$ satisfying (6.1) will be relevant in approximating U in the limit. The quasioptimality estimate (2.3) must now be replaced by [8]

$$\|U - U_N\|_{\mathrm{E}} \leqslant \inf_{\substack{Y \in V_N \\ KY=0}} \|U - Y\|_{\mathrm{E}}, \tag{6.2}$$

which is the best estimate that holds for $\|\cdot\|_{\mathrm{E}} = \|\cdot\|_{\mathrm{E},0}$.

Defining $S^K = \{Y \in S,\ KY = 0\}$ for any space $S \subset V$, we see that the infimum in (6.2) is taken not over V_N but over a proper subset V_N^K. *Locking* ("shear" locking) is said to occur when this causes a deterioration in the resulting approximation.

One way of understanding this deterioration is to note that (6.1) essentially forces W_N to be a $C^{(1)}(\omega)$ function (since ϕ_N must lie in $H^1(\omega)$, i.e., W_N must be in $H^2(\omega)$). This, of course, is consistent with the fact that in the $d=0$ limit, we are approximating the fourth-order KL (biharmonic) problem, for which, as noted earlier, $C^{(1)}$ functions are needed for W_N. Suppose, for instance, we choose $V_N = S_N \times Z_N$ where S_N, Z_N both contain continuous piecewise linears. Then $(\psi_N, z_N) \in V_N^K$ only if $z_N \in C^{(1)}(\omega)$, so that z_N will be a linear function over *all* of ω. If, for instance, we have built-in boundary conditions, then $z_N = 0$ will be the only function satisfying this requirement in Z_N, and the finite element solution (ϕ_N, w_N) will be such that w_N is forced to converge to 0 as $d \to 0$. This is an example of *complete* locking, where there is *no* uniform rate of convergence. For other choices of spaces, locking may manifest itself as a degradation (but perhaps not a complete annihilation) of the uniform convergence rate.

A formal definition of locking may be found in [8]. The definition involves several factors. First, we are given a sequence of problems $\{P_d\}$ dependent on a parameter d in some set $S = (0,1]$ (say). For thin domains, d is, of course, the thickness. For linear elasticity, we could take $d = 1 - 2\nu$, with $\nu=$ Poisson's ratio. Next, we assume that the solution u_d to P_d lies in the solution space H_d — this characterizes the smoothness of the exact solutions $\{u_d\}$. Also, we are given a sequence of finite element spaces $\{V_N\}$ that comprise an extension procedure $\mathscr{F}$. Finally, we are interested in the errors $E_d(u_d - u_{d,N})$ where $u_{d,N} \in V_N$ is the FE solution and E_d is some given error functional (e.g., the energy norm, the $H^1(\Omega)$ norm, etc.).

We now define a function $L(d,N)$ called the *locking ratio* by

$$L(d,N) = \frac{\sup_{u_d \in H_d^B} E_d(u_d - u_{d,N})}{\inf_{d \in S_\alpha} \sup_{u_d \in H_d^B} E_d(u_d - u_{d,N})} \tag{6.3}$$

where $S_\alpha = S \cap [\alpha, \infty)$ for some $\alpha > 0$ such that $S_\alpha \neq \phi$. Here $H_d^B = \{u \in H_d,\ \|u\|_{H_d} \leqslant B\}$.

What $L(d,N)$ does is to compare the performance of the method at thickness $= d$ to the best possible performance for values of $d \geqslant \alpha$ for which locking does not occur. If S is restricted to a discrete set, it gives a *computable* measure of the amount of locking [37].

An alternative definition of locking can be based on the asymptotic rate of convergence [8]:

Definition 6.1. The extension procedure $\mathscr{F}$ is free from locking for the family of problems $\{P_d\}$, $d \in S = (0,1]$ with respect to the solution sets H_d and error measures E_d, if and only if

$$\limsup_{N \to \infty} \left(\sup_{d \in (0,1]} L(v,N) \right) = C < \infty.$$

$\mathscr{F}$ shows locking of order $f(N)$ if and only if

$$0 < \limsup_{N \to \infty} \left(\sup_{v} L(v,N) \frac{1}{f(N)} \right) = C < \infty$$

where $f(N) \to \infty$ as $N \to \infty$. It shows locking of at least (respectively, at most) order $f(N)$ if $C > 0$ (respectively, $C < \infty$).

Let us make a second definition, that of robustness:

Definition 6.2. The extension procedure $\mathscr{F}$ is robust for $\{P_d\}$, $d \in S=(0,1]$ with respect to solution sets H_d and error measures E_d if and only if

$$\lim_{N\to\infty} \sup_{d} \sup_{u_d \in H_d^B} E_d(u_d - u_{d,N}) = 0.$$

It is robust with uniform order $g(N)$ if and only if

$$\sup_{d} \sup_{u_d \in H_d^B} E_d(u_d - u_{d,N}) \leqslant g(N),$$

where $g(N) \to 0$ as $N \to \infty$.

We now discuss locking and robustness for the plate problem in the context of the above definitions. In order to concentrate solely on the issue of locking, we consider *periodic* boundary conditions, for which no boundary layers exist, and the solution is smooth. For $\omega = (-1,1)^2$, these conditions are given by

$$U_d(x_1,1) = U_d(x_1,-1), \quad U_d(1,x_2) = U_d(-1,x_2), \quad |x_1|, |x_2| \leqslant 1 \tag{6.4}$$

so that $V = [H^1_{\mathrm{per}}(\omega)]^3$ where $H^k_{\mathrm{per}}(\omega)$ denotes the set of functions on ω whose periodic extensions to $\mathbb{R}^2$ lie in $H^k_{\mathrm{loc}}(\mathbb{R}^2)$.

Let us consider Definitions 6.1 and 6.2 for the case that the error measure $E_d \equiv \|\cdot\|_{\mathrm{E},d}$ (the energy norm corresponding to (3.7)). For our solution sets, we take

$$H_d = \{U = (\phi, w),\ \phi \in H^{k+1}_{\mathrm{per}}(\omega), C_d U = 0\}, \tag{6.5}$$

where

$$C_d U = \frac{d^2}{12(1-v)}\{(1-v)\Delta\phi + (1+v)\nabla\nabla\cdot\phi\} + \frac{\gamma}{2(1+v)}\{\nabla w + \phi\}.$$

The constraint in (6.5) says that $U \in H_d$ satisfies the first equation in the usual strong form of (3.7).

We then have the following two theorems [39].

Theorem 6.3. *Let $\mathscr{F}$ be an h version extension procedure on a uniform mesh consisting either of triangles or rectangles, for problem* (3.7), (6.4). *Let polynomials of degree $p = p_N$ and $q = q_N$ be used for the rotations ϕ_d and displacements w_d respectively. Then with solution sets H_d and error measures $\|\cdot\|_{\mathrm{E},d}$, $\mathscr{F}$ is robust with uniform order $N^{-\ell}$ when $k \geqslant 2q+1$ and shows locking of order N^r when $k \geqslant p+1$, as summarized in Table* 1.

Theorem 6.4. *Let $\mathscr{F}$ be the p version on a mesh of triangles and parallelograms, with $p_N \geqslant 1$, $q_N \geqslant p_N$. Then with solution sets $H_d, k \geqslant 1, \mathscr{F}$ is free of locking in the energy norm and is robust with uniform order $N^{-(k-1)/2}$ as $p_N \to \infty$.*

Remark 6.5. The hp version will also be free of locking, provided (uniform) triangular meshes with $p_N \geqslant 5$ and $q_N \geqslant p_N + 1$ are used (Theorem 5.2 of [39]).

Table 1
Locking and robustness for the h version with uniform meshes

Type of element	Degree p	Degree q	Order of locking, r $f(N)=\mathcal{O}(N^r)$	Robustness order, l $g(N)=\mathcal{O}(N^{-l})$
Triangle ($\mathscr{P}_p$)	1	$q\geqslant 1$	$r=\frac{1}{2}$	$l=0$
	$2\leqslant p\leqslant 4$	$q=p$	$r=1$	$l=(p-2)/2$
	$p\geqslant 5$		$r=\frac{1}{2}$	$l=(p-1)/2$
	$2\leqslant p\leqslant 3$	$l\geqslant p+1$	$r=\frac{1}{2}$	$l=(p-1)/2$
	$p\geqslant 4$	$l\geqslant p+1$	$r=0$	$l=p/2$
Product ($\mathscr{Q}_p$)	1	$l\geqslant 1$	$r=\frac{1}{2}$	$l=0$
	$p\geqslant 2$	$q\geqslant p$	$r=\frac{1}{2}$	$l=(p-1)/2$
Trunk ($\mathscr{Q}'_p$)	1	$q\geqslant 1$	$r=\frac{1}{2}$	$l=0$
	2	$q=2,3$	$r=1$	$l=0$
		$q\geqslant 4$	$r=\frac{1}{2}$	$l=1/2$
	$p\geqslant 3$	$q=p$	$r=\frac{3}{2}$	$l=(p-3)/2$
		$q\geqslant p+1$	$r=1$	$l=(p-2)/2$

Let us briefly explain the idea behind Theorems 6.3 and 6.4. Suppose $F_0(N)$ represents the *optimal* rate of convergence for $\mathscr{F}$ in the absence of locking. For example, $F_0(N)$ would behave like the $\|\cdot\|_{E,d}$ error when d is set equal to 1. Let us also define

$$g(N)=\sup_{w\in H^{k+2,B}_{\rm per}}\ \inf_{z\in W_N}\|w-z\|_{H^2(\omega)} \tag{6.6}$$

where for $V_N=S_N$, $W_N=\{w\in Z_N, \operatorname{grad} w\in S_N\}$. Then it can be shown quite easily [39] that $\mathscr{F}$ is robust with uniform order $\max\{F_0(N),g(N)\}$ and shows locking of order $f(N)$ if and only if

$$C_1F_0(N)f(N)\leqslant g(N)\leqslant C_2F_0(N)f(N).$$

(The characterization (6.6) follows easily if we remember that $\boldsymbol{\phi}_d=\nabla w_d$, $\boldsymbol{\phi}_{d,N}=\nabla w_{d,N}$ in the limit $d\to 0$.)

Now the question of locking becomes one in pure approximation theory. For instance, it is known that to get optimal h-approximation over triangles using $C^{(1)}$ elements, one must use at least degree 5 piecewise polynomials, which shows why in Table 1, the locking disappears only when $q\geqslant 5$ is used for Z_N. In the p version, on the other hand, since we have p_N, $q_N\to\infty$, the condition $q_N\geqslant 5$ will not cause any problem asymptotically. Hence, in the asymptotic sense, there will be no locking, though as seen from Fig. 17 ahead, there *is* a shift in the error curves (i.e., there will be an increase in the locking ratios).

In Figs. 16 and 17, we illustrate the above rates of convergence for problem (3.7), (6.4) on $\omega=(-1,1)^2$. The Poisson ratio $v=0.3, E=1$, and the load is $g(x,y)=\cos(\pi x/2)\cos(\pi y/2)$. A uniform rectangular mesh is used with $p=q$ product spaces, and the errors presented are those estimated by STRESS CHECK.

Fig. 16 clearly shows the locking of $O(N^{1/2})$. Fig. 17 shows the parallel error curves for the p version, indicating an absence of asymptotic locking (but also indicating locking ratios $L(d,N)>1$).

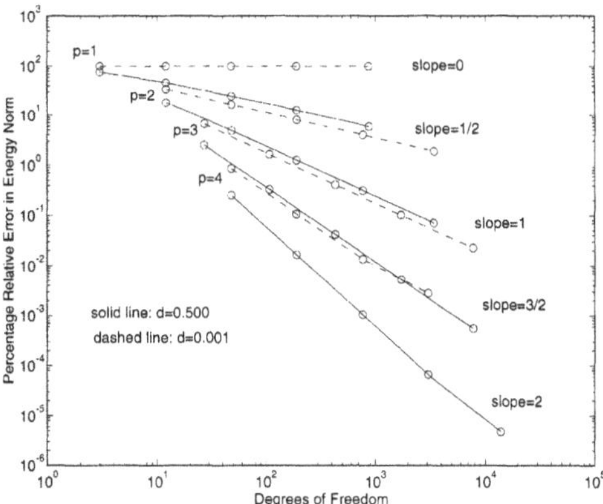

Fig. 16. The h version for the RM plate.

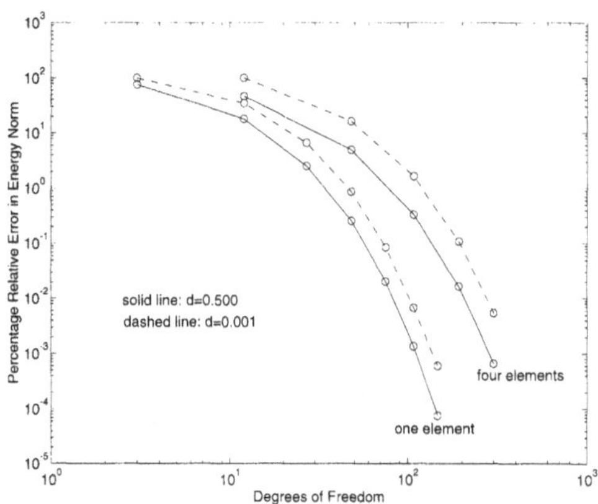

Fig. 17. The p version for the RM plate.

Let us remark now on some extension of these results:

1. *Non-periodic boundary conditions*: Similar results were observed computationally for clamped (built-in) plates. The locking theorems above will hold once more, provided the solution is smooth. But in general, it is not, and boundary layer effects must also be considered [32,5].

2. *Higher-order plate models*: In [39], it is shown that locking effects (and results) are identical to the RM case, since no additional constraints arise as $d \to 0$.

3. *Locking in other error measures*: Locking is very dependent on the error measure under consideration. If, for instance, we take the moments M_x, M_y, M_{xy} at a point, then the locking effects are similar to those for the energy norm. See Fig. 18, where the moment M_x at the point (0.3, 0.8) is considered. If, on the other hand, we consider the shear stresses Q_x, Q_y, then the locking is significantly worse, since these involve an extra power d^{-1}. It is seen from Fig. 19 that little convergence is obtained for the h version, though the p version is still robust.

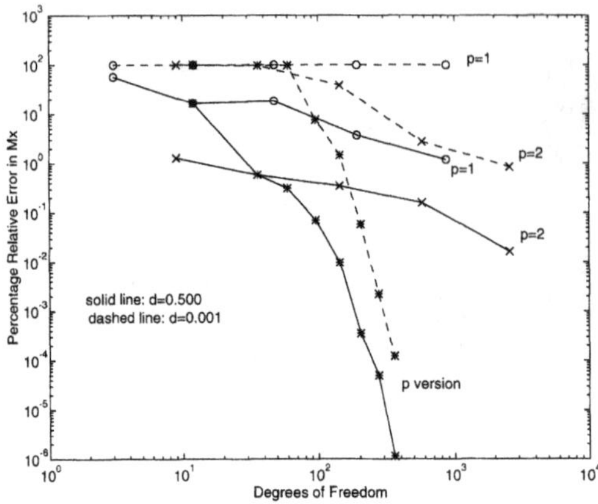

Fig. 18. Locking for the moment M_x at (0.3, 0.8).

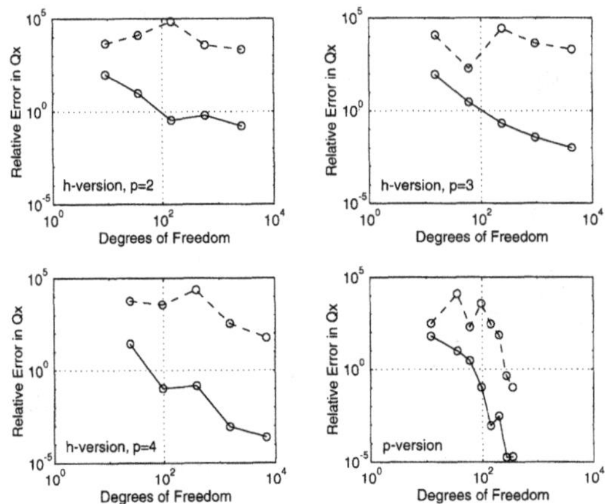

Fig. 19. Locking for the shear stress Q_x. Solid line: $d = 0.5$. Dashed line: $d = 0.001$.

4. *Mixed/reduced constraint methods*: For standard FEMs, the Kirchhoff constraint (6.1) must be satisfied *exactly* in the limit as $d \to 0$. In reduced constraint methods, we replace this constraint by

$$R_N K U_N = R_N(\nabla w_N + \boldsymbol{\phi}_N) = 0 \tag{6.7}$$

by defining a new bilinear form

$$B_N(U,V) = a(\boldsymbol{\phi},\boldsymbol{\theta}) + \gamma\mu d^{-2}(R_N KU, R_N KV)$$

and solving problem (3.7) with B_N instead of B. Here, R_N is a projection into a suitable space of polynomials, that is generally defined piecewise over each element, and which is designed so that (6.7) is easier to satisfy than (6.1). Various choices of projections are discussed, e.g., in [30]. The analysis of such methods can then proceed either by analyzing separately an *approximation* and a *consistency* error [30], or by re-writing the method as a mixed method with the shear stress

$$\boldsymbol{q}_d = d^{-2}(\nabla w_d + \boldsymbol{\phi}_d)$$

as a new unknown (see, e.g., [36]).

In [36], we have analyzed an hp method based on the MITC reduction operators R_N [12]. This method has the advantage of being locking free both in h and in p. Moreover, uniform estimates in the shear stress can now be obtained. Some computational results using these elements may be found in [14].

5. *Some remarks on shells*: Locking effects in shells can be more serious than those in plates. In addition to shear locking, we may also observe *membrane* locking, which is a harder phenomenon to deal with. Moreover, there is strong interaction between locking and boundary layers. We refer to [28,23,18,38] where some hp methods (both standard and mixed) have been discussed to overcome these effects.

References

[1] D.N. Arnold, R. Falk, Asymptotic analysis of the boundary layer for the Reissner-Mindlin plate model, SIAM J. Math. Anal. 27 (1996) 486–514.

[2] I. Babuška, M.R. Dorr, Error estimates for the combined h and p version of the finite element method, Numer. Math. 37 (1981) 252–277.

[3] I. Babuška, B. Guo, Approximation properties of the h–p version of the finite element method, Comput. Methods Appl. Mech. Eng. 133 (1996) 319–346.

[4] I. Babuška, R.B. Kellogg, J. Pitkaranta, Direct and inverse estimates for finite elements with mesh refinement, Numer. Math. 33 (1979) 447–471.

[5] I. Babuška, L. Li, The h–p version of the finite element method in the plate modeling problem, Comm. Appl. Numer. Methods 8 (1992) 17–26.

[6] I. Babuška, L. Li, The problem of plate modeling, theoretical and computational results, Comput. Math. Appl. Mech. Eng. 100 (1992) 249–273.

[7] I. Babuška, M. Suri, The h–p version of the finite element method with quasiuniform meshes, RAIRO, Math. Mod. Numer. Anal. 21 (1987) 199–238.

[8] I. Babuška, M. Suri, On locking and robustness in the finite element method, SIAM J. Numer. Anal. 29 (1992) 1261–1293.

[9] I. Babuška, M. Suri, The p and hp versions of the finite element method, basic principles and properties, SIAM Rev. 36 (1994) 578–632.

[10] I. Babuška, B.A. Szabo, R. Actis, Hierarchic models for laminated composites, Internat. J. Numer. Methods Eng. 33 (1992) 503–535.

[11] I. Babuška, B.A. Szabo, I.N. Katz, The p-version of the finite element method, SIAM J. Numer. Anal. 18 (1981) 515–545.

[12] F. Brezzi, M. Fortin, R. Stenberg, Error analysis of mixed-interpolated elements for Reissner-Mindlin plates, Math. Models Methods Appl. Sci. 1 (1991) 125–151.

[13] P.G. Ciarlet, The Finite Element Method for Elliptic Problems, North-Holland, Amsterdam, 1978.

[14] L.D. Croce, T. Scapolla, Some applications of hierarchic high-order MITC finite elements for Reissner-Mindlin plates, in: P. Neittaanmaki, M. Krizek, R. Stenberg (Eds.), Finite Element Methods. Fifty Years of the Courant Element, Marcel Dekker, New York, 1994, pp. 183–190.

[15] M. Dauge, in: Elliptic Boundary Value Problems on Corner Domains, Lecture Notes in Mathematics, Vol. 1341, Springer, New York, 1988.

[16] M. Dauge, I. Gruais, Asymptotics of arbitrary order for a thin elastic clamped plate, Asymptotic Anal. Part I: 13 (1996) 167–197, Part II: 16 (1998) 99–124.

[17] M. Dauge, Z. Yosibash, Boundary layer realization in thin elastic 3-d domains and 2-d hierarchic plate models, Internat. J. Solids Struct. 37 (2000) 2443–2471.

[18] K. Gerdes, A. Matache, C. Schwab, Analysis of membrane locking in hp FEM for a cylindrical shell, Z. Angew. Math. Mech. 78 (1998) 663–686.

[19] P. Grisvard, Elliptic Problems in Non-smooth Domains, Pitman, Boston, 1985.

[20] W. Gui, I. Babuška, The h, p and h–p versions of the finite element method in one dimension (Parts 1–3), Numer. Math. 40 (1986) 577–683.

[21] B. Guo, I. Babuška, The h–p version of the finite element method I and II, Comput. Mech. 1 (1986) 21–41 and 203–226.

[22] H. Hakula, Licentiate Thesis, Helsinki University of Technology, 1997.

[23] H. Hakula, Y. Leino, J. Pitkaranta, Scale resolution, locking, and high-order finite element modeling of shells, Comput. Methods Appl. Mech. Eng. 133 (1996) 157–182.

[24] H. Han, R.B. Kellogg, Differentiability properties of solutions of the equation $-\varepsilon^2\Delta u + ru = f(x,y)$ in a square, SIAM J. Math. Anal. 21 (1990) 394–408.

[25] R.B. Kellogg, Boundary layers and corner singularities for a self-adjoint problem, in: M. Costabel, M. Dauge, S. Nicaise (Eds.), Lecture Notes in Pure and Applied Mathematics, Vol. 167, Marcel Dekker, New York, 1995, pp. 121–149.

[26] V.A. Kondrat'ev, Boundary value problems for elliptic equations in domains with conic or corner points, Trans. Moscow Math. Soc. 16 (1967) 227–313.

[27] J.M. Melenk, C. Schwab, hp FEM for reaction-diffusion equations I: robust exponential convergence, SIAM J. Numer. Anal. 35 (1998) 1520–1557.

[28] J. Pitkaranta, The problem of membrane locking in finite element analysis of cylindrical shells, Numer. Math. 61 (1992) 523–542.

[29] J. Pitkaranta, Y. Leino, O. Ovaskainen, J. Piila, Shell deformation states and the finite element method: A benchmark study of cylindrical shells, Comput. Methods Appl. Mech. Eng. 128 (1995) 81–121.

[30] J. Pitkaranta, M. Suri, Design principles and error analysis for reduced-shear plate-bending finite elements, Numer. Math. 75 (1996) 223–266.

[31] C. Schwab, p- and hp- Finite Element Methods, Oxford Science Publications, Oxford, 1998.

[32] C. Schwab, M. Suri, Locking and boundary layer effects in the finite element approximation of the Reissner–Mindlin plate model, Proc. Sympos. Appl. Math. 48 (1994) 367–371.

[33] C. Schwab, M. Suri, The p and hp version of the finite element method for problems with boundary layers, Math. Comp. 65 (1996) 1403–1429.

[34] C. Schwab, M. Suri, C. Xenophontos, The hp finite element method for problems in mechanics with boundary layers, Comput. Methods Appl. Mech. Eng. 157 (1998) 311–333.

[35] G.I. Shishkin, Grid approximation of singularly perturbed parabolic equations with internal layers, Soviet J. Numer. Anal. Math. Modelling 3 (1988) 393–407.

[36] R. Stenberg, M. Suri, An hp error analysis of MITC plate elements, SIAM J. Numer. Anal. 34 (1997) 544–568.

[37] M. Suri, Analytical and computational assessment of locking in the hp finite element method, Comput. Methods Appl. Mech. Eng. 133 (1996) 347–371.

[38] M. Suri, A reduced constraint hp finite element method for shell problems, Math. Comp. 66 (1997) 15–29.

[39] M. Suri, I. Babuška, C. Schwab, Locking effects in the finite element approximation of plate models, Math. Comp. 64 (1995) 461–482.

[40] B.A. Szabo, I. Babuška, Finite Element Analysis, Wiley, New York, 1991.

[41] B.A. Szabo, A.K. Mehta, p-convergent finite element approximations in fracture mechanics, Internat. J. Numer. Methods Eng. 12 (1978) 551–560.

[42] M.J. Turner, R.W. Clough, H.C. Martin, L.J. Topp, Stiffness and deflection analysis of complex structures, J. Aeronaut. Sci. 23 (1956) 805–823.

[43] R. Vulanović, D. Herceg, N. Petrović, On the extrapolation for a singularly perturbed boundary value problem, Computing 36 (1986) 69–79.

[44] C. Xenophontos, The hp version of the finite element method for singularly perturbed boundary value problems, Doctoral Dissertation, University of Maryland, Baltimore County, May 1996.

[45] C. Xenophontos, The hp finite element method for singularly perturbed problems in smooth domains, Math. Models Methods Appl. Sci. 8 (1998) 299–326.

[46] C. Xenophontos, Finite element computations for the Reissner–Mindlin plate model, Comm. Numer. Methods Eng. 14 (1998) 1119–1131.

[47] C. Xenophontos, The hp finite element method for singularly perturbed problems in nonsmooth domains, Numer. Methods PDEs 27 (1999) 63–89.

ELSEVIER

Journal of Computational and Applied Mathematics 128 (2001) 261–279

JOURNAL OF COMPUTATIONAL AND APPLIED MATHEMATICS

www.elsevier.nl/locate/cam

Efficient preconditioning of the linearized Navier–Stokes equations for incompressible flow

David Silvester[a,*], Howard Elman[b], David Kay[c], Andrew Wathen[d]

[a]*Department of Mathematics, UMIST, Manchester M60 1QD, UK*

[b]*Computer Science Department and Institute of Advanced Computer Studies, University of Maryland, College Park, MD 20742, USA*

[c]*School of Mathematical Sciences, University of Sussex, Brighton BN1 9QH, UK*

[d]*Oxford University Computing Laboratory, Parks Road, Oxford OX1 3QD, UK*

Received 14 October 1999; received in revised form 25 January 2000

Abstract

We outline a new class of robust and efficient methods for solving subproblems that arise in the linearization and operator splitting of Navier–Stokes equations. We describe a very general strategy for preconditioning that has two basic building blocks; a multigrid V-cycle for the scalar convection–diffusion operator, and a multigrid V-cycle for a pressure Poisson operator. We present numerical experiments illustrating that a simple implementation of our approach leads to an effective and robust solver strategy in that the convergence rate is independent of the grid, robust with respect to the time-step, and only deteriorates very slowly as the Reynolds number is increased. © 2001 Elsevier Science B.V. All rights reserved.

Keywords: Navier–Stokes equations; Incompressible flow; Preconditioning; Multigrid iteration

1. Introduction

The underlying goal here is to compute solutions of incompressible flow problems modelled by the Navier–Stokes equations in a flow domain $\Omega \subset \mathbb{R}^d$ ($d = 2$ or 3) with a piecewise smooth boundary $\partial\Omega$:

$$\frac{\partial \boldsymbol{u}}{\partial t} + \boldsymbol{u} \cdot \nabla \boldsymbol{u} - \nu \nabla^2 \boldsymbol{u} + \nabla p = 0 \quad \text{in } \mathscr{W} \equiv \Omega \times (0, T), \tag{1.1}$$

$$\nabla \cdot \boldsymbol{u} = 0 \quad \text{in } \mathscr{W}, \tag{1.2}$$

* Corresponding author. Tel.: +44-161-200-3656; fax: +44-161-200-3669.
E-mail address: na.silvester@na-net.ornl.gov (D. Silvester).

PII: S0377-0427(00)00515-X

together with boundary and initial conditions of the form

$$u(x,t) = g(x,t) \quad \text{on } \bar{\mathscr{W}} \equiv \partial\Omega \times [0,T]; \tag{1.3}$$

$$u(x,0) = u_0(x) \quad \text{in } \Omega. \tag{1.4}$$

We use standard notation: u is the fluid velocity, p is the pressure, $v > 0$ is a specified viscosity parameter (in a nondimensional setting it is the inverse of the Reynolds number), and $T > 0$ is some final time. The initial velocity field u_0 is typically assumed to satisfy the incompressibility constraint, that is, $\nabla \cdot u_0 = 0$. The boundary velocity field satisfies $\int_{\partial\Omega} g \cdot n \, ds = 0$ for all time t, where n is the unit vector normal to $\partial\Omega$.

If the aim is to simply compute steady-state solutions of (1.1)–(1.2) then time accuracy is not an issue. In other cases however, having an accurate solution at each time-step is important and the requirements of the time discretisation will be more demanding; specifically, an accurate and unconditionally stable time-discretisation method is necessary to adaptively change the time-step to reflect the dynamics of the underlying flow. We will not attempt to describe the many possibilities — the recent monographs of Gresho and Sani [14] and Turek [29] are worth consulting in this respect — but will restrict attention here to the simplest unconditionally stable approach using a one-stage finite difference discretisation, as given below.

Algorithm 1. *Given* u^0, $\theta \in [1/2, 1]$, *find* u^1, $u^2, \ldots, u^n$ *via*

$$\begin{aligned} \frac{(u^{n+1} - u^n)}{\Delta t} + u^* \cdot \nabla\, u^{n+\theta} - v\nabla^2 u^{n+\theta} + \nabla p^{n+\theta} &= 0, \\ \nabla \cdot u^{n+\theta} = 0 \quad &\text{in } \Omega, \\ u^{n+\theta} = g^{n+\theta} \quad &\text{on } \partial\Omega. \end{aligned} \tag{1.5}$$

Here $u^{n+\theta} = \theta u^{n+1} + (1-\theta)u^n$ and $p^{n+\theta} = \theta p^{n+1} + (1-\theta)p^n$. Note that p^0 is required if $\theta \neq 1$ so the Algorithm 1 is not self-starting in general. In this case an approximation to p^0 must be computed explicitly by manipulation of the continuum problem, or alternatively it must be approximated by taking one (very small) step of a self-starting algorithm (e.g., with $\theta = 1$ above).

Algorithm 1 contains the well-known nonlinear schemes of backward Euler and Crank–Nicolson. These methods are given by ($u^{n+\theta} = u^{n+1}$, $u^* = u^{n+1}$), ($u^{n+\theta} = u^{n+1/2}$, $u^* = u^{n+1/2}$), and are first and second-order accurate, respectively. In either case, a nonlinear problem must be solved at every time-level. A well-known linearization strategy is to set $u^* = u^n$ above. This does not affect the stability properties of the time discretisation, but it does reduce the Crank–Nicolson accuracy to first order as $\Delta t \to 0$ (the first order accuracy of backward Euler is unchanged). To retain second-order accuracy in a linear scheme the Simo–Armero scheme [24] given by setting $u^{n+\theta} = u^{n+1/2}$ with $u^* = (3u^n - u^{n-1})/2$ in Algorithm 1 is recommended, see [26] for further details.

Using linearized backward Euler or the Simo-Armero scheme, a frozen-coefficient Navier–Stokes problem (or generalised *Oseen* problem) arises at each discrete time step: given a divergence-free vector field $w(x)$ (which we will refer to as the "wind"), the aim is to compute $u(x)$ and $p(x)$ such that

$$\frac{1}{\Delta t} u + w \cdot \nabla u - v\nabla^2 u + \nabla p = f \quad \text{in } \Omega \tag{1.6}$$

$$\nabla \cdot \boldsymbol{u} = 0 \quad \text{in } \Omega, \tag{1.7}$$

$$\boldsymbol{u} = \boldsymbol{g} \quad \text{on } \partial\Omega. \tag{1.8}$$

Notice that since (1.6)–(1.8) represents a linear elliptic PDE problem, the existence and uniqueness of a solution $(\boldsymbol{u}, p)$ can be established under very general assumptions. The development of efficient methods for solving discrete analogues of (1.6)–(1.8) is the focal point of this work.

An outline is as follows. The spatial discretisation of the generalised Oseen problem is discussed in Section 2. Some standard Krylov iteration methods that are applicable to the (nonsymmetric-) systems that arise after discretisation are briefly reviewed in Section 3. Our general preconditioning approach is then developed in Section 4. This approach builds on our research effort over the last decade on developing effective preconditioners for limiting cases of the Oseen problem (1.6)–(1.8): specifically steady-state Stokes problems ($\Delta t \to \infty$, $\boldsymbol{w} \to \boldsymbol{0}$) [22]; generalised Stokes problems ($\boldsymbol{w} \to \boldsymbol{0}$), see Silvester and Wathen [23]; and steady Oseen problems ($\Delta t \to \infty$) [5,6,16]. Some computational experiments that demonstrate the power of our solution methodology are presented in Section 5. Implementation of "pure" multigrid methods seems to be relatively complicated, and performance seems to be (discretisation-) method dependent by comparison. The derivation of analytic bounds on convergence rates for the general preconditioner is an ongoing project which will be treated in a forthcoming paper [7]; in the final section we give a flavour of the analysis by quoting results that we have established in two special cases; potential flow ($\boldsymbol{w} = \boldsymbol{0}$ and $\nu = 0$) and generalised Stokes flow ($\boldsymbol{w} = \boldsymbol{0}$). These cases typically arise using time-stepping methods for (1.1)–(1.2) based on *operator splitting* — showing the inherent generality of the preconditioning approach.

2. Spatial discretisation

Given that we would like to solve our model problem (1.6)–(1.8) over irregular geometries, the spatial discretisation will be done using finite element approximation (this also gives us more flexibility in terms of adaptive refinement via a posteriori error control, see, e.g., [17]). We note that the algorithm methodology discussed in the paper applies essentially verbatim to finite difference and finite volume discretisations. In the remainder of this section we briefly review the error analysis associated with mixed finite element approximation of (1.6)–(1.8). For full details see [13].

The weak formulation of (1.6)–(1.8) is defined in terms of the Sobolev spaces $H_0^1(\Omega)$ (the completion of $C_0^\infty(\Omega)$ in the norm $\|\cdot\|_1$) and $L_0^2(\Omega)$ (the set of functions in $L^2(\Omega)$ with zero mean value on Ω). Defining a velocity space $\boldsymbol{X} \equiv (H_0^1(\Omega))^d$ and a pressure space $M \equiv L_0^2(\Omega)$, it is easy to see that the solution $(\boldsymbol{u}, p)$ of (1.6)–(1.8) satisfies

$$\frac{1}{\Delta t}(\boldsymbol{u}, \boldsymbol{v}) + (\boldsymbol{w} \cdot \nabla \boldsymbol{u}, \boldsymbol{v}) + \nu(\nabla \boldsymbol{u}, \nabla \boldsymbol{v}) - (p, \nabla \cdot \boldsymbol{v}) = (\boldsymbol{f}, \boldsymbol{v}) \quad \forall \boldsymbol{v} \in \boldsymbol{X}, \tag{2.1}$$

$$(\nabla \cdot \boldsymbol{u}, q) = 0 \quad \forall q \in M, \tag{2.2}$$

where $(\cdot,\cdot)$ denotes the usual vector or scalar $L^2(\Omega)$ inner product. Since Ω is bounded and connected there exists a constant κ satisfying the continuous *inf–sup* condition:

$$\sup_{\boldsymbol{w} \in X} \frac{(p, \nabla \cdot \boldsymbol{w})}{\|\boldsymbol{w}\|_1} \geqslant \kappa \|p\| \quad \forall p \in M. \tag{2.3}$$

Furthermore, since $\boldsymbol{w}$ is divergence-free, the bilinear form $c(\cdot,\cdot)$ given by

$$c(\boldsymbol{u},\boldsymbol{v}) = \frac{1}{\Delta t}(\boldsymbol{u},\boldsymbol{v}) + (\boldsymbol{w}\cdot\nabla\boldsymbol{u},\boldsymbol{v}) + \nu(\nabla\boldsymbol{u},\nabla\boldsymbol{v}) \tag{2.4}$$

is coercive and bounded over $\boldsymbol{X}$;

$$c(\boldsymbol{v},\boldsymbol{v}) \geqslant \nu\|\nabla\boldsymbol{v}\|^2 \quad \forall \boldsymbol{v}\in\boldsymbol{X}, \tag{2.5}$$

$$|c(\boldsymbol{u},\boldsymbol{v})| \leqslant C_w\|\nabla\boldsymbol{u}\|\|\nabla\boldsymbol{v}\| \quad \forall \boldsymbol{u}\in\boldsymbol{X}, \forall\, \boldsymbol{v}\in\boldsymbol{X}. \tag{2.6}$$

Existence and uniqueness of a solution to (2.1)–(2.2) then follows from a generalisation of the usual Lax–Milgram lemma [13].

To generate a discrete system we take finite dimensional subspaces $\boldsymbol{X}_h\subset\boldsymbol{X}$ and $M_h\subset L^2(\Omega)$, where h is a representative mesh parameter, and enforce (2.1)–(2.2) over the discrete subspaces (again specifying that functions in M_h have zero mean to ensure uniqueness). Specifically, we look for a function $\boldsymbol{u}_h$ satisfying the boundary condition (1.8), and a function $p_h\in M_h$ such that

$$\frac{1}{\Delta t}(\boldsymbol{u}_h,\boldsymbol{v}) + (\boldsymbol{w}_h\cdot\nabla\boldsymbol{u}_h,\boldsymbol{v}) + \nu(\nabla\boldsymbol{u}_h,\nabla\boldsymbol{v}) - (p_h,\nabla\cdot\boldsymbol{v}) = (\boldsymbol{f},\boldsymbol{v}) \quad \forall\boldsymbol{v}\in\boldsymbol{X}_h, \tag{2.7}$$

$$(\nabla\cdot\boldsymbol{u}_h, q) = 0 \quad \forall q\in M_h, \tag{2.8}$$

where $\boldsymbol{w}_h$ represents the interpolant of $\boldsymbol{w}$ in X_h. Notice that this approximation means that the discrete wind is not actually pointwise divergence-free. From the linear algebra perspective the point is that in the case of the enclosed flow boundary condition (1.3), the discrete convection matrix corresponding to the term $(\boldsymbol{w}_h\cdot\nabla\boldsymbol{u}_h,\boldsymbol{v})$ is *skew symmetric.*

The well-posedness of (2.7)–(2.8) is not automatic since we do not have an internal approximation. A sufficient condition for the existence and uniqueness of a solution to (2.7)–(2.8) is that the following discrete inf–sup condition is satisfied: there exists a constant γ independent of h such that

$$\sup_{\boldsymbol{v}\in\boldsymbol{X}_h}\frac{(q,\nabla\cdot\boldsymbol{v})}{\|\nabla\boldsymbol{v}\|} \geqslant \gamma\|q\| \quad \forall q\in M_h. \tag{2.9}$$

Note that the semi-norm $\|\nabla\boldsymbol{v}\|$ in (2.9) is equivalent to the norm $\|\boldsymbol{v}\|_1$ for functions $\boldsymbol{v}\in\boldsymbol{X}$. The inf–sup condition also guarantees optimal approximation in the sense of the error estimate [13]

$$\|\nabla(\boldsymbol{u}-\boldsymbol{u}_h)\| + \|p-p_h\| \leqslant C\left(\inf_{\boldsymbol{v}\in\boldsymbol{X}_h}\|\nabla(\boldsymbol{u}-\boldsymbol{v})\| + \inf_{q\in M_h}\|p-q\|\right). \tag{2.10}$$

Note that the constant C is inversely proportional to the inf–sup constant γ in (2.9).

Since we want to use linear algebra tools it is convenient to express the discrete problem (2.7)–(2.8) as a matrix problem. To do this we introduce discrete operators $\mathscr{F}:\boldsymbol{X}_h\mapsto\boldsymbol{X}_h$ and $\mathscr{B}:\boldsymbol{X}_h\mapsto M_h$ defined via

$$(\mathscr{F}\boldsymbol{v}_h,\boldsymbol{z}_h) = \frac{1}{\Delta t}(\boldsymbol{v}_h,\boldsymbol{z}_h) + (\boldsymbol{w}_h\cdot\nabla\boldsymbol{v}_h,\boldsymbol{z}_h) + \nu(\nabla\boldsymbol{v}_h,\nabla\boldsymbol{z}_h) \quad \forall\boldsymbol{v}_h,\boldsymbol{z}_h\in\boldsymbol{X}_h, \tag{2.11}$$

$$(\mathscr{B}\boldsymbol{v}_h,q_h) = (\boldsymbol{v}_h,\mathscr{B}^*q_h) = -(\nabla\cdot\boldsymbol{v}_h,q_h) \quad \forall\boldsymbol{v}_h\in\boldsymbol{X}_h, \forall q_h\in M_h, \tag{2.12}$$

so that $\mathscr{B}^*$ is the adjoint of $\mathscr{B}$. With these definitions the discrete problem (2.7)–(2.8) can be rewritten as a matrix system: find $\boldsymbol{u}_h$ satisfying the boundary condition (1.8) such that

$$\begin{pmatrix}\mathscr{F} & \mathscr{B}^*\\ \mathscr{B} & 0\end{pmatrix}\begin{pmatrix}\boldsymbol{u}_h\\ p_h\end{pmatrix} = \begin{pmatrix}\boldsymbol{f}\\ 0\end{pmatrix}. \tag{2.13}$$

Furthermore, introducing $\mathscr{A} : \boldsymbol{X}_h \mapsto \boldsymbol{X}_h$, satisfying

$$(\mathscr{A}\boldsymbol{v}_h, \boldsymbol{z}_h) = (\nabla \boldsymbol{v}_h, \nabla \boldsymbol{z}_h) \quad \forall \boldsymbol{v}_h, \boldsymbol{z}_h \in \boldsymbol{X}_h, \tag{2.14}$$

the inf–sup inequality (2.9) simplifies to

$$\gamma \| q_h \| \leqslant \sup_{\boldsymbol{v}_h \in X_h} \frac{(\mathscr{B}\boldsymbol{v}_h, q_h)}{(\mathscr{A}\boldsymbol{v}_h, \boldsymbol{v}_h)^{1/2}} \quad \forall q_h \in M_h. \tag{2.15}$$

It is instructive to express (2.13) and (2.15) in terms of the actual finite element matrices that arise in practice. To this end, let us explicitly introduce the finite element basis sets, say,

$$\boldsymbol{X}_h = \mathrm{span}\{\boldsymbol{\phi}_i\}_{i=1}^n, \quad M_h = \mathrm{span}\{\psi_j\}_{j=1}^m; \tag{2.16}$$

and associate the functions $\boldsymbol{u}_h$, p_h, with the vectors $u \in \mathbb{R}^n$, $p \in \mathbb{R}^m$ of generalised coefficients, $p_h = \sum_{j=1}^m p_j \psi_j$, etc. Defining the $n \times n$ "convection", "diffusion" and "mass" matrices $N_{ij} = (\boldsymbol{w}_h \cdot \nabla \phi_i, \phi_j)$, $A_{ij} = (\nabla \phi_i, \nabla \phi_j)$ and $G_{ij} = (\phi_i, \phi_j)$, and also the $m \times n$ "divergence matrix" $B_{ij} = -(\nabla . \phi_j, \psi_i)$, gives the finite element version of (2.13):

$$\begin{pmatrix} \frac{1}{\Delta t} G + N + \nu A & B^{\mathrm{t}} \\ B & 0 \end{pmatrix} \begin{pmatrix} u \\ p \end{pmatrix} = \begin{pmatrix} f \\ g \end{pmatrix}, \tag{2.17}$$

where the RHS term g arises from enforcement of the (non-homogeneous) boundary condition on the function $\boldsymbol{u}_h$; see [14, pp. 440–448] for details.

Moreover, introducing the $m \times m$ pressure "mass" matrix $Q_{ij} = (\psi_i, \psi_j)$; leads to the finite element version of (2.9): for all $p \in \mathbb{R}^m$,

$$\gamma (p^{\mathrm{t}} Q p)^{1/2} \leqslant \max_u \frac{p^{\mathrm{t}} B u}{(u^{\mathrm{t}} A u)^{1/2}} \tag{2.18}$$

$$= \max_{w = A^{1/2} u} \frac{p^{\mathrm{t}} B A^{-1/2} w}{(w^{\mathrm{t}} w)^{1/2}} \tag{2.19}$$

$$= (p^{\mathrm{t}} B A^{-1} B^{\mathrm{t}} p)^{1/2}, \tag{2.20}$$

since the maximum is attained when $w = A^{-1/2} B^{\mathrm{t}} p$. Thus, we have a characterization of the inf–sup constant:

$$\gamma^2 = \min_{p \neq 0} \frac{p^{\mathrm{t}} B A^{-1} B^{\mathrm{t}} p}{p^{\mathrm{t}} Q p}. \tag{2.21}$$

In simple terms it is precisely the square root of the smallest eigenvalue of the preconditioned Schur complement $Q^{-1} B A^{-1} B^{\mathrm{t}}$. We also have that

$$(q, \nabla \cdot \boldsymbol{v}) \leqslant \|q\| \, \|\nabla \cdot \boldsymbol{v}\| \leqslant \sqrt{d} \|q\| \, \|\nabla \boldsymbol{v}\| \tag{2.22}$$

where $\Omega \subset \mathbb{R}^d$, and so there also exists a constant $\Gamma \leqslant \sqrt{d}$ satisfying

$$\Gamma^2 = \max_{p \neq 0} \frac{p^{\mathrm{t}} B A^{-1} B^{\mathrm{t}} p}{p^{\mathrm{t}} Q p}. \tag{2.23}$$

Note that the tight bound $\Gamma \leqslant 1$ was recently established (valid in the case of a conforming approximation space, $\boldsymbol{X}_h \subset \boldsymbol{X}$) by Stoyan [28].

In practice, the inf–sup condition (2.9) is extremely restrictive. Problems arise if the pressure space M_h is too rich compared to the velocity space X_h. Although many stable methods have been developed (see [14] for a complete list of possibilities), many natural low order conforming finite element methods like Q_1–P_0 (trilinear/bilinear velocity with constant pressure) are unstable in the sense that pressure vectors $p \in M_h$ can be constructed for which the inf–sup constant tends to zero under uniform grid refinement. This type of instability can be difficult to detect in practice since the associated discrete systems (2.17) are all nonsingular — so that every discrete problem is uniquely solvable — however they become ill-conditioned as $h \to 0$.

Another issue, which needs to be addressed when applying multigrid solution techniques to convection–diffusion problems of the form

$$c(\boldsymbol{u}_h, \boldsymbol{v}) = (\boldsymbol{f}_h, \boldsymbol{v}) \quad \forall \boldsymbol{v} \in \boldsymbol{X}_h, \tag{2.24}$$

(with $c(\cdot,\cdot)$ given by (2.4)), is that standard approximation methods may produce an unstable, possibly oscillating, solution if the mesh is too coarse in critical regions. In such cases, to give additional stability on coarse meshes used in the multigrid process, the discrete problem (2.24) needs to be stabilised. For example, using a *streamline-diffusion* method, we replace (2.24) by the regularised problem

$$c(\boldsymbol{u}_h, \boldsymbol{v}) + \delta(\boldsymbol{w}_h \cdot \nabla \boldsymbol{u}_h, \boldsymbol{w}_h \cdot \nabla \boldsymbol{v}) = (\boldsymbol{f}_h, \boldsymbol{v}) \quad \forall \boldsymbol{v} \in \boldsymbol{X}_h, \tag{2.25}$$

where δ is a locally defined stabilisation parameter, see [15] for further details.

The formulation (2.25) clearly has better stability properties than (2.24) since there is additional coercivity in the local flow direction. The local mesh Péclet number $P_T^e = ||\boldsymbol{w}_h||_{\infty,T} h_T/\nu$ determines the streamline-diffusion coefficient δ_T in a given element T via the "optimal" formula [10];

$$\delta_T = \begin{cases} \frac{1}{2} h_T \left(1 - \frac{1}{P_T^e}\right) & \text{if } P_T^e > 1, \\ 0 & \text{if } P_T^e \leqslant 1, \end{cases} \tag{2.26}$$

where h_T is a measure of the element length in the direction of the wind.

3. Krylov subspace solvers

Let $\mathscr{L}x = f$ denote a generic linear system of equations. Krylov subspace solution methods start with a guess $x^{(0)}$ for the solution, with residual $r^{(0)} = f - \mathscr{L}x^{(0)}$, and construct a sequence of approximate solutions of the form

$$x^{(k)} = x^{(0)} + p^{(k)} \tag{3.1}$$

where $p^{(k)}$ is in the k-dimensional *Krylov space*

$$\mathscr{K}_k(r^{(0)}, \mathscr{L}) = \operatorname{span}\{r^{(0)}, \mathscr{L}r^{(0)}, \ldots, \mathscr{L}^{k-1}r^{(0)}\}.$$

In this section, we give a brief overview of properties of Krylov subspace methods for solving the systems arising from the discretizations discussed in the previous section.

Problem (2.17) is nonsymmetric so that algorithms applicable to such problems are of primary concern, but the small Reynolds number limit leads to a symmetric indefinite Stokes problem, and we first briefly discuss this case. It is well-known that for symmetric indefinite problems, the MINRES algorithm [20] generates iterates of the form (3.1) for which the residual $r^{(k)}$ has minimal Euclidean norm. It follows that the residuals satisfy

$$\frac{\|r^{(k)}\|_2}{\|r^{(0)}\|_2} \leqslant \min_{\phi_k(0)=1} \max_{\lambda\in\sigma(\mathscr{L})} |\phi_k(\lambda)|,$$

where the minimum is taken over polynomials ϕ_k of degree k satisfying $\phi_k(0)=1$. This result leads to the following bound on the relative residual norm [18].

Theorem 3.1. *If the eigenvalues of $\mathscr{L}$ are contained in two intervals $[-a,-b]\cup[c,d]$ with $a-b=d-c>0$, then the residuals generated by MINRES satisfy*

$$\frac{\|r^{(k)}\|_2}{\|r^{(0)}\|_2} \leqslant 2\left(\frac{1-\sqrt{\beta}}{1+\sqrt{\beta}}\right)^{k/2},$$

where $\beta=(bc)/(ad)$.

We apply this result to the Stokes equations in the final section. We also point out that tighter bounds can be established when $a-b\neq d-c$ and b,d have some asymptotic behaviour, [32,33]. Each step of the computation entails only a matrix–vector product together with a small number, independent of the iteration count, of vector operations (scalar-vector products and inner products), so that the cost per step of the MINRES iteration is low.

For nonsymmetric problems, there is no Krylov subspace solver that is optimal with respect to some error norm for which the cost per step is independent of the iteration count [8,9]. The *generalized minimal residual algorithm* (GMRES) [21] is the most efficient "optimal" solver, producing the unique iterate of the form (3.1) for which the Euclidean norm of the residual is smallest. Step k requires one matrix–vector product together with a set of k vector operations, making its cost, in terms of both operation counts and storage, proportional to kN where N is the problem dimension. We summarize the main convergence properties of GMRES below. See [4,21] for proofs.

Theorem 3.2. *Let $x^{(k)}$ denote the iterate generated after k steps of GMRES, with residual $r^{(k)}=f-\mathscr{L}x^{(k)}$.*

(i) *The residual norms satisfy $\|r^{(k)}\|_2=\min_{\phi_k(0)=1}\|\phi_k(\mathscr{L})r^{(0)}\|_2$.*

(ii) *If $\mathscr{L}=X\Lambda X^{-1}$ is diagonalizable, where Λ is the diagonal matrix of eigenvalues of $\mathscr{L}$, then*

$$\|r^{(k)}\|_2 \leqslant \|X\|_2\,\|X^{-1}\|_2 \min_{\phi_k(0)=1}\max_{\lambda_j}|\phi_k(\lambda_j)|\,\|r^{(0)}\|_2.$$

Assertions (i) and (ii) follow from the optimality of GMRES with respect to the residual norm. Assertion (i) guarantees that GMRES will solve any nonsingular problem provided that the dimensions of the Krylov space is large enough. This differentiates GMRES from most other nonsymmetric Krylov subspace methods.

The GMRES iterate is computed as in (3.1) with $p^{(k)}$ of the form $p^{(k)}=V_k y^{(k)}$, where V_k is a matrix whose columns form an orthogonal basis for $\mathscr{K}_k$. The construction of the orthogonal basis is what makes the cost per step high, but once such a basis is available, the iterate with smallest residual norm can be computed cheaply. See [21] for details. Nonoptimal methods compromise on these points, reducing the cost per step by avoiding the construction of an orthogonal basis, but thereby making the construction of an optimal iterate too expensive. Numerous methods of this type have been proposed, for example, BiCGSTAB [30], BiCGSTAB (ℓ) [25], CGS [27], QMR [11].

The results of Theorems 3.1–3.2 indicate that if the eigenvalues of $\mathscr{L}$ are tightly clustered, then convergence will be rapid. In particular, for MINRES, it is desirable for the sizes of the two intervals (one on each side of the origin) to be as small as possible, and well separated from the origin. For GMRES, Theorem 3.2(ii) suggests that convergence will be fast if the eigenvalues can be enclosed in a region in the complex plane that is small. The spectra of the discrete problems of Section 2 are not well-behaved in this sense, and convergence must be enhanced by preconditioning. That is, we use an operator $\mathscr{P} \approx \mathscr{L}$ and solve an equivalent system such as $\mathscr{P}^{-1}\mathscr{L}x = \mathscr{P}^{-1}b$, with a more favorable distribution of eigenvalues, by Krylov subspace iteration.

We conclude this section with a few general observations concerning preconditioning for both symmetric indefinite and nonsymmetric problems. Sections 4 and 6 discuss and analyze some specific strategies suitable for (2.17). First, we note that preconditioning increases the cost per step, since the matrix–vector product now requires a preconditioning operation, i.e., application of the action of $\mathscr{P}^{-1}$ to a vector. Thus, for the preconditioner to be effective, the improved convergence speed must be enough to compensate for the extra cost.

The MINRES algorithm can be combined with preconditioning by a symmetric positive-definite operator $\mathscr{P}$. Formally, MINRES is then applied to the symmetric matrix $\hat{\mathscr{L}} = \mathscr{S}^{-1}\mathscr{L}\mathscr{S}^{-T}$, where $\mathscr{P} = \mathscr{S}\mathscr{S}^{t}$. The error bound analogous to that of Theorem 3.1 is

$$\frac{\|r^{(k)}\|_{\mathscr{P}^{-1}}}{\|r^{(0)}\|_{\mathscr{P}^{-1}}} \leqslant 2\left(\frac{1-\sqrt{\beta}}{1+\sqrt{\beta}}\right)^{k/2}, \tag{3.2}$$

where the intervals defining β now come from the eigenvalues of the preconditioned operator $\hat{\mathscr{L}}$. Thus, we seek a preconditioner for which the computation of the action of $\mathscr{P}^{-1}$ is inexpensive, and for which the eigenvalues of $\hat{\mathscr{L}}$ are tightly clustered, leading to smaller β. Note also that the norm in (3.2) is now different; for further details see [23]. It is also possible to apply the QMR algorithm to symmetric indefinite problems (with comparable complexity to that of MINRES). In this case a symmetric indefinite preconditioner can be used [12].

For nonsymmetric problems, there is some flexibility in how the preconditioned problem may be formulated, with three possible different "orientations":

Left orientation	$[\mathscr{P}^{-1}\mathscr{L}][x] = [\mathscr{P}^{-1}f]$;
Two-sided orientation	$[\mathscr{P}_1^{-1}\mathscr{L}\mathscr{P}_2^{-1}][\mathscr{P}_2 x] = [\mathscr{P}_2^{-1}f]$;
Right orientation	$[\mathscr{L}\mathscr{P}^{-1}][\mathscr{P}x] = [f]$.

The two-sided orientation depends on having an explicit representation of the preconditioner in factored form $\mathscr{P} = \mathscr{P}_1\mathscr{P}_2$. In our experience, there is little difference in the effectiveness of these choices. We tend to prefer the "right" variant, especially for use with GMRES, since the norm being minimized (the Euclidian norm of the residual) is then independent of the choice of the preconditioner.

4. Preconditioning strategy

Our starting point is the discrete system $\mathscr{L}x = f$ associated with (2.17), which we write in the form

$$\begin{pmatrix} F & B^t \\ B & 0 \end{pmatrix} \begin{pmatrix} u \\ p \end{pmatrix} = \begin{pmatrix} f \\ g \end{pmatrix} \tag{4.1}$$

so that $F = (1/\Delta t)G + N + vA \in \mathbb{R}^{n\times n}$, with $B \in \mathbb{R}^{m\times n}$. Our preconditioning strategy is based on the assumption that a fast solver (typically based on multigrid) is available for the convection–diffusion system $Fu = f$. This leads us to consider a block triangular preconditioning

$$\mathscr{P}^{-1} = \begin{pmatrix} F^{-1} & R \\ 0 & -S^{-1} \end{pmatrix}, \tag{4.2}$$

with matrix operators $R \in \mathbb{R}^{n\times m}$ and $S \in \mathbb{R}^{m\times m}$ chosen to provide clustering of the eigenvalues $\sigma(\mathscr{L}\mathscr{P}^{-1})$ of the right preconditioned system

$$\mathscr{L}\mathscr{P}^{-1} = \begin{pmatrix} I_n & FR - B^tS^{-1} \\ BF^{-1} & BR \end{pmatrix}. \tag{4.3}$$

The specific choice of R and S in (4.2) satisfying

$$FR - B^tS^{-1} = O, \quad BR = I_m,$$

that is, $R = F^{-1}B^tS^{-1}$ with $S = BF^{-1}B^t$, is the optimal choice, see [19]. For this choice, it follows from (4.3) that $\sigma(\mathscr{L}\mathscr{P}^{-1}) = \{1\}$, and preconditioned GMRES converges to the solution of (4.1) in at most two iterations.

Implementation of a right preconditioner for GMRES requires the solution of a system of the form $\mathscr{P}y = r$ at every step. (QMR also requires the solution of a system with $\mathscr{P}^t$.) With the optimal choice of R and S we need to compute the vector $\binom{v}{q}$ satisfying

$$\begin{pmatrix} v \\ q \end{pmatrix} = \begin{pmatrix} F^{-1} & F^{-1}B^tS^{-1} \\ 0 & -S^{-1} \end{pmatrix} \begin{pmatrix} r \\ s \end{pmatrix}, \tag{4.4}$$

for given vectors $r \in \mathbb{R}^n$, and $s \in \mathbb{R}^m$. Rewriting (4.4) shows that the optimal preconditioner is defined by a two-stage process:

$$\begin{aligned} &\text{Solve for } q\text{: } Sq = -s; \\ &\text{Solve for } v\text{: } Fv = r - B^tq. \end{aligned} \tag{4.5}$$

To get a practical method, we modify the preconditioning process (4.5) by replacing the matrix operators $S = BF^{-1}B^t$ and F, by approximations S_* and F_* respectively, designed so that the preconditioned Oseen operator has a tightly clustered spectrum. We are particularly interested in operators S_* and F_* derived from multigrid computations such that $\sigma(SS_*^{-1}) \in \omega_S$ and $\sigma(FF_*^{-1}) \in \omega_F$ where ω_S and ω_F represent small convex sets in the right half of the complex plane; ideally, these sets would be independent of the problem parameters v, h, and Δt.

The construction of the operator $F_* \approx F$ is relatively straightforward, see Section 5. The more difficult issue is the construction of a simple multigrid approximation to the Schur complement

$BF^{-1}B^t$, see, e.g., [29, p. 56]. The approach presented here was developed by Kay and Loghin [16] and represents an improved version of ideas in [6,5].

To motivate the derivation, suppose for the moment that we have an unbounded domain, and that differential operators arising in (1.6)–(1.8) commute:

$$\nabla(1/\Delta t + \mathbf{w}\cdot\nabla - \nu\nabla^2)_p \equiv (1/\Delta t + \mathbf{w}\cdot\nabla - \nu\nabla^2)_{\mathbf{u}}\nabla \tag{4.6}$$

where for any operator Θ, $\Theta_{\mathbf{u}}$ represents the vector analogue of the scalar operator Θ_p. If we further assume that a C^0 pressure approximation is used (so that $M_h \subset H^1(\Omega)$) then we can construct a discrete pressure convection–diffusion operator $\mathscr{F}_p : M_h \mapsto M_h$ such that

$$(\mathscr{F}_p q_h, r_h) = \frac{1}{\Delta t}(q_h, r_h) + (\mathbf{w}_h \cdot \nabla q_h, r_h) + \nu(\nabla q_h, \nabla r_h) \quad \forall q_h, r_h \in M_h. \tag{4.7}$$

Introducing the L_2-projection operators $\mathscr{G} : \mathbf{X}_h \mapsto \mathbf{X}_h$ and $\mathscr{Q} : M_h \mapsto M_h$

$$(\mathscr{G}\mathbf{v}_h, \mathbf{z}_h) = (\mathbf{v}_h, \mathbf{z}_h) \quad \forall \mathbf{v}_h,\ \mathbf{z}_h \in \mathbf{X}_h,$$
$$(\mathscr{Q}q_h, r_h) = (q_h, r_h) \quad \forall q_h,\ r_h \in M_h,$$

then gives the discrete analogue of (4.6)

$$(\mathscr{G}^{-1}\mathscr{B}^*)(\mathscr{Q}^{-1}\mathscr{F}_p) \equiv (\mathscr{G}^{-1}\mathscr{F})(\mathscr{G}^{-1}\mathscr{B}^*). \tag{4.8}$$

A simple rearrangement of (4.8) gives

$$(\mathscr{G}^{-1}\mathscr{F})^{-1}(\mathscr{G}^{-1}\mathscr{B}^*) \equiv (\mathscr{G}^{-1}\mathscr{B}^*)(\mathscr{Q}^{-1}\mathscr{F}_p)^{-1},$$
$$\mathscr{F}^{-1}\mathscr{B}^* \equiv \mathscr{G}^{-1}\mathscr{B}^*\mathscr{F}_p^{-1}\mathscr{Q}.$$

Hence, assuming that (4.8) is valid, we have an alternative expression for the Schur complement operator $\mathscr{B}\mathscr{F}^{-1}\mathscr{B}^* : M_h \mapsto M_h$, namely

$$\mathscr{B}\mathscr{F}^{-1}\mathscr{B}^* \equiv \mathscr{B}\mathscr{G}^{-1}\mathscr{B}^*\mathscr{F}_p^{-1}\mathscr{Q}. \tag{4.9}$$

For equivalence (4.9) to hold, it is necessary for the spaces $\mathbf{X}_h$ and M_h to be defined with periodic boundary conditions. In the case of an enclosed flow boundary condition like (1.8), the discrete operator $\mathscr{F}_p$ inherits natural boundary conditions (associated with the space M), and in this case (4.9) gives us a starting point for approximating the Schur complement matrix $S = BF^{-1}B^t$. Using basis (2.16), we have the approximation

$$BG^{-1}B^tF_p^{-1}Q = P_S \approx S. \tag{4.10}$$

The goal now is to design an efficient implementation of a preconditioner based on (4.10). This requires that fast solvers for the underlying operators $\mathscr{Q}$ and $\mathscr{B}\mathscr{G}^{-1}\mathscr{B}^*$ are available: we seek operators Q_* and H_* such that there exist constants θ, Θ, λ, Λ independent of h, satisfying

$$\theta^2 \leqslant \frac{p^tQp}{p^tQ_*p} \leqslant \Theta^2 \quad \forall p \in \mathbb{R}^m, \tag{4.11}$$

and

$$\lambda^2 \leqslant \frac{p^tBG^{-1}B^tp}{p^tH_*p} \leqslant \Lambda^2 \quad \forall p \in \mathbb{R}^m, \tag{4.12}$$

respectively. The practical version of the preconditioner is then defined by replacing the action of S^{-1} in the first step of (4.5) by the so called F_p approximation:

$$S_*^{-1} = Q_*^{-1} F_p H_*^{-1}. \tag{4.13}$$

Satisfying (4.11) is straightforward; the simple pressure scaling $Q_* = \mathrm{diag}\,(Q)$ does the trick [31]. The upshot is that the action of Q^{-1} in (4.10) can be approximated very accurately using a fixed (small) number of steps of diagonally scaled conjugate gradient iteration applied to the operator Q.

Relation (4.12) can also be satisfied using a multigrid approach. The crucial point is that the use of a C^0 pressure approximation space is associated with an alternative inf–sup condition, see, e.g., [1]: for a stable mixed approximation there exists a constant β independent of h, such that

$$\sup_{\mathbf{v}\in X_h} \frac{(\mathbf{v}, \nabla q)}{||\mathbf{v}||} \geqslant \beta ||\nabla q|| \quad \forall q \in M_h. \tag{4.14}$$

Thus, introducing the pressure Laplacian operator $\mathscr{A}_p : M_h \mapsto M_h$ such that

$$(\mathscr{A}_p q_h, r_h) = (\nabla q_h, \nabla r_h) \quad \forall q_h, r_h \in M_h,$$

we have that (4.14) is equivalent to

$$\beta(\mathscr{A}_p q_h, q_h)^{1/2} \leqslant \sup_{\mathbf{v}_h \in X_h} \frac{(\mathbf{v}_h, \mathscr{B}^* q_h)}{||\mathbf{v}_h||} \quad \forall q_h \in M_h.$$

Applying the same arguments used to get (2.21) and (2.23), we have a natural characterization in terms of the matrices associated with the finite element basis (2.16):

$$\beta^2 \leqslant \frac{p^t B G^{-1} B^t p}{p^t A_p p} \leqslant 1 \quad \forall p \in \mathbb{R}^m. \tag{4.15}$$

In simple terms, for a stable mixed discretisation, the operator $\mathscr{B}\mathscr{G}^{-1}\mathscr{B}^*$ is spectrally equivalent to the Poisson operator $\mathscr{A}_p$ defined on the pressure space M_h (with inherited Neumann boundary conditions) [14, p. 563]. We note in passing that an equivalence of the form (4.15) can also hold in cases when a discontinuous pressure approximation is used (with an appropriately defined matrix operator A_p). For example, in the case of well known MAC discretization on a square grid, we have $A_p = h^{-2} B B^t$ where A_p is the standard five-point Laplacian defined at cell centres.

The result (4.15) opens up the possibility of using a multigrid preconditioner. In particular, a single multigrid V-cycle with point Jacobi or (symmetric) Gauss–Seidel smoothing defines an approximation H_*, with spectral bounds

$$\kappa^2 \leqslant \frac{p^t A_p p}{p^t H_* p} \leqslant 1 \quad \forall p \in \mathbb{R}^m. \tag{4.16}$$

The combination of (4.15) and (4.16) shows that a simple multigrid cycle can be used as an approximation to $\mathscr{B}\mathscr{G}^{-1}\mathscr{B}^*$ in the sense that (4.12) holds with constants $\lambda = \beta\kappa$ and $\Lambda = 1$.

To end this section we would like to emphasize the simplicity of the practical implementation of the preconditioner associated with (4.5). The computation of q in the first stage entails an approximation of the action of P_S^{-1} defined by (4.10). This is done in three steps; the first is the approximation to the action of the inverse of $BG^{-1}B^t$ using a multigrid iteration applied to a system with coefficient matrix A_p (typically representing a Poisson operator with Neumann boundary conditions), the second step is a matrix–vector product involving the discrete convection–diffusion operator F_p, and the third step is essentially a scaling step corresponding to the solution of a system with coefficient matrix given by the pressure mass matrix Q. For the second stage of (4.5), the computation of v is approximated by a multigrid iteration for the convection–diffusion equation. Clearly, the overall cost of the preconditioner is determined by the cost of a convection–diffusion solve on the velocity space and of a Poisson solve on the pressure space; with multigrid used for each of these, the complexity is proportional to the problem size.

5. Computational results

We use P_2–P_1 mixed finite element approximation (see, e.g., [14, p. 462]), that is, we choose spaces

$$X_h = \{v \in H_0^1(\Omega) : v|_T \in \mathbb{P}^2(T)\ \forall T \in \mathcal{T}_h\},$$
$$M_h = \{q \in H^1(\Omega) : q|_T \in \mathbb{P}^1(T)\ \forall T \in \mathcal{T}_h\},$$

where T is a triangle in the mesh $\mathcal{T}_h$. (This mixed method is shown to be inf–sup stable in [1].) We restrict attention to uniformly refined meshes in this work, analogous results for adaptively refined meshes are given in [16].

We present results for three standard test flow problems below. The time discretization is backward Euler, and the linearization strategy is given by the choice $\boldsymbol{u}^* = \boldsymbol{u}^n$ in Algorithm 1. In all cases we run the time integrator for 15 time-steps, unless the stopping criterion $\|\boldsymbol{u}^{n+1} - \boldsymbol{u}^n\|_2 < 10^{-6}$ is satisfied. We solve the linear system that arises at each discrete time interval using GMRES with the preconditioner $\mathcal{P}$ that is defined below. The GMRES starting vector for the nth time-step is always taken to be the previous time-step solution $(\boldsymbol{u}^{n-1}, p^{n-1})$. GMRES iterations are performed until the relative residual is reduced by 10^{-6}.

We will denote the action of a single multigrid V-cycle using a point Gauss–Seidel smoother for the discrete velocity operator $\mathcal{F}$ in (2.11), by F_*^{-1}; where we perform one smoothing sweep before a fine to coarse grid transfer of the residual, and one smoothing sweep after a coarse to fine grid transfer of the correction. For details see e.g., [34]. Similarly we let H_*^{-1} denote the action of a single multigrid V-cycle using damped Jacobi as a smoother (with damping parameter 0.8) for the pressure Laplacian operator $\mathcal{A}_p$ in (4.7) (again with a single sweep of pre- and post-smoothing). We comment that although the use of multigrid as a solver for a Laplacian operator is very robust, using a simple multigrid cycle with point smoothing does not generally lead to an efficient solver for the convection–diffusion operator $\mathcal{F}$ when convection dominates (although the same strategy can still be an effective preconditioner [34]). If we let Q_*^{-1} denote two diagonally scaled conjugate gradient iterations applied to the discrete pressure identity, then our inverse preconditioner is of the form:

$$\mathcal{P}_*^{-1} = \begin{pmatrix} F_*^{-1} & 0 \\ 0 & I \end{pmatrix} \begin{pmatrix} I & B^T \\ 0 & -I \end{pmatrix} \begin{pmatrix} I & 0 \\ 0 & Q_*^{-1} F_p H_*^{-1} \end{pmatrix}.$$

Within the multigrid process we construct prolongation operators using interpolation that is consistent with the order of the velocity/pressure approximation spaces. Furthermore, the restriction operator is the usual transpose of the prolongation, and on the coarsest level ($h = 1/2$ below) we perform an exact solve. Finally, we emphasize that if the local mesh Péclet number is greater than

Table 1
$N^h_{\Delta t}$ for Stokes driven cavity flow

Δt	$h=1/4$	$h=1/8$	$h=1/16$
0.001	9	10	12
0.1	13	14	14
1	14	15	15
10	14	15	15
1000	14	15	15

unity on any grid, then streamline diffusion is included in the discrete system that is solved (as well as the discrete convection–diffusion problems defining the operator F_*^{-1}, see (2.26)).

To show the robustness of our solver we report below the maximum number of GMRES iterations required for the tolerance to be satisfied on a given mesh (with a given Δt) over all time iterations; this maximum iteration count is denoted by $N^h_{\Delta t}$.

5.1. Stokes: driven cavity flow

We firstly consider the (symmetric-) generalized Stokes problem, associated with a standard driven cavity flow problem defined on a unit domain $\Omega=(0,1)\times(0,1)$. The associated boundary condition is given by

$$\boldsymbol{u}(\partial\Omega,t)=\begin{cases}(1,0) & y=1,\\ \mathbf{0} & \text{otherwise},\end{cases}$$

and we "spin-up" to the steady state from the initial condition $\boldsymbol{u}(\boldsymbol{x},0)=\mathbf{0}$.

The performance of our preconditioned method is summarized in Table 1. These iteration counts are consistent with our expectation that the rate of convergence is independent of the degree of mesh refinement, and the size of the time-step. We note that in the limit $\Delta t\to\infty$, the system reduces to a stationary Stokes system in which case we have tight analytic bounds showing the effectiveness of the same preconditioning strategy in a MINRES context, see Section 6.

5.2. Navier–Stokes: driven cavity flow

We also consider the Navier–Stokes problem associated with the domain, boundary and initial conditions given above. These results are given in Table 2.

The obvious point to note here is that, as in the Stokes case, the performance is not affected by mesh refinement. (The trend is clearly evident even though the meshes are relatively coarse.) In contrast to the results in the Stokes case it can be seen that as Δt gets larger in Table 2, the iteration counts tend to an asymptotic maximum value. Moreover this maximum value becomes somewhat larger as ν is decreased. This behaviour is consistent with our expectations — steady-state iteration counts that are presented in [16] can be seen to slowly increase as the Reynolds number is increased. A complete theoretical explanation is not yet available, but see [7].

Table 2
$N_{\Delta t}^{h}$ for Navier–Stokes driven cavity flow

	$h=1/4$	$h=1/8$	$h=1/16$
$\nu=1/50$			
$\Delta t=0.1$	14	15	14
$\Delta t=1$	14	15	15
$\Delta t=10$	17	18	18
$\nu=1/100$			
$\Delta t=0.1$	14	15	14
$\Delta t=1$	14	16	16
$\Delta t=10$	19	21	21
$\nu=1/200$			
$\Delta t=0.1$	14	15	14
$\Delta t=1$	15	18	18
$\Delta t=10$	23	24	24

5.3. Navier–Stokes: backward facing step

We finally consider a Navier–Stokes problem on an L-shaped domain. We start with the coarse (level 0) mesh in Fig. 1, and generate subsequent meshes (i.e., levels 1–3) by successive uniform refinement. The total number of degrees of freedom on the respective levels 1, 2 and 3 are 309, 1092 and 4089, respectively. We again start from a “no flow” initial condition, and impose the following enclosed flow boundary condition:

$$\boldsymbol{u}(\partial\Omega,t)=\begin{cases}(2y-y^2,0) & x=-6,\\ (\frac{8}{27}(y+1)(2-y),0) & x=16,\\ \boldsymbol{0} & \text{otherwise.}\end{cases}$$

An important point here is that the initial condition is not divergence free in this case; i.e. $\nabla\cdot\boldsymbol{u}^0\neq 0$. This means that the first time-step is artificial (see [14]) — for small Δt it gives a projection to the discretely divergence-free space so that $\boldsymbol{u}^1$ looks like a potential flow field. The fact that we

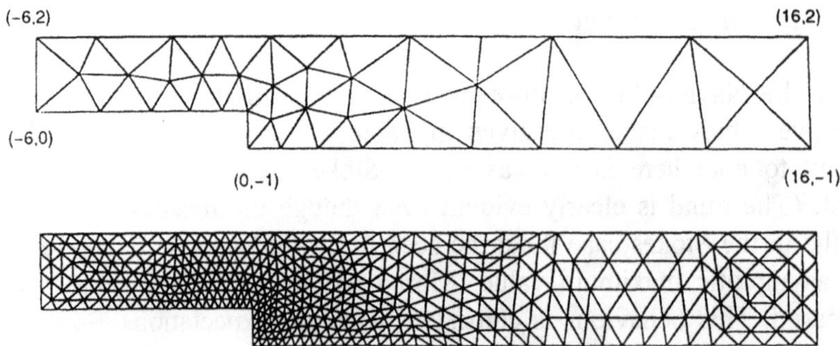

Fig. 1. Coarsest and finest grid triangulations for the backward facing step.

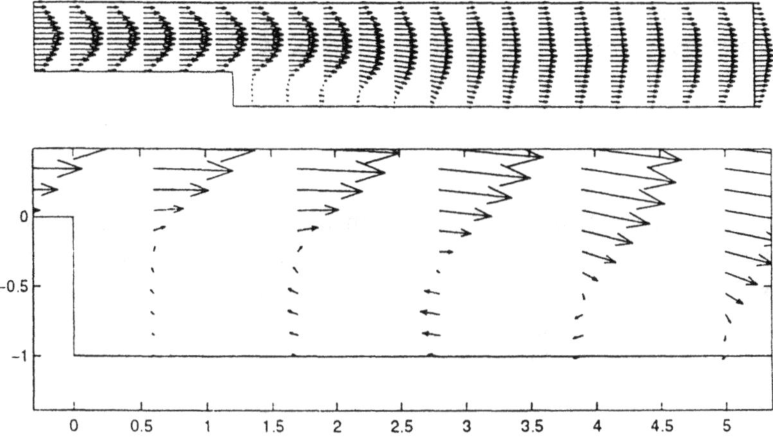

Fig. 2. Velocity solution for $\nu = 1/200$.

Table 3
$N^h_{\Delta t}$ for Navier–Stokes flow over a backward facing step

$\nu = 1/50$	Level 1	Level 2	Level 3
$\Delta t = 0.1$	26	32	33
$\Delta t = 1$	26	32	33
$\Delta t = 10$	26	43	40
$\nu = 1/100$	Level 1	Level 2	Level 3
$\Delta t = 0.1$	26	32	33
$\Delta t = 1$	26	32	33
$\Delta t = 10$	26	51	47
$\nu = 1/200$	Level 1	Level 2	Level 3
$\Delta t = 0.1$	26	32	33
$\Delta t = 1$	26	32	33
$\Delta t = 10$	33	64	59

have a prescribed outflow boundary condition is also emphasized here. Fig. 2 illustrates the computed steady flow (interpolated from the finest mesh) in the case $\nu = 1/200$, and shows that the downstream evolution from the inflow to the outflow profile is physically realistic.

The maximum iteration counts are given in Table 3. These results have the same general character as those in Table 2, although the iteration counts for a given ν and Δt are increased by a factor of about two. We attribute this difference to the fact that the longer flow domain means that the local mesh Péclet number is relatively large in this case. We remark that for the largest time-step there is a reduction in the iteration count when going from the second to the third level of refinement. Indeed the *average* GMRES iteration counts in the case $\nu = 1/200$, $\Delta t = 10$ are 27.3, 52.3 and 50.1, respectively. This phenomenon of increased mesh refinement being correlated with faster convergence is also evident in the steady-state results that are presented in [16].

6. Analytic results

For problems where the coefficient matrix is symmetric, specifically whenever $N = 0$ in (2.17), there is a well-established convergence analysis associated with preconditioners based on the Schur complement approximation (4.10). We outline this theory in this final section.

As discussed in Section 3, MINRES is the optimal Krylov solver in the case of a symmetric coefficient matrix $\mathscr{L}$, but it can only be used in conjuction with a symmetric positive definite preconditioning operator $\mathscr{P}$. For this reason, in place of the block triangular preconditioner (4.2), we introduce the simpler block diagonal variant

$$\mathscr{P}^{-1} = \begin{pmatrix} F_*^{-1} & 0 \\ 0 & S_*^{-1} \end{pmatrix}, \tag{6.1}$$

and insist that the block-diagonal entries F_* and S_* are themselves symmetric. The convergence analysis is based on the following result, which is established by Silvester and Wathen [22].

Theorem 6.1. *Assume that the blocks F_* and S_* in* (6.1) *satisfy*

$$\lambda_F \leqslant \frac{u^t F u}{u^t F_* u} \leqslant \Lambda_F \qquad \forall u \in \boldsymbol{X}_h, \tag{6.2}$$

$$\lambda_S \leqslant \frac{p^t B F^{-1} B^t p}{p^t S_* p} \leqslant 1 \qquad \forall p \in M_h, \tag{6.3}$$

then the eigenvalues of the preconditioned problem,

$$\begin{pmatrix} F & B^t \\ B & 0 \end{pmatrix} \begin{pmatrix} \boldsymbol{u} \\ p \end{pmatrix} = \lambda \begin{pmatrix} F_* & 0 \\ 0 & S_* \end{pmatrix} \begin{pmatrix} \boldsymbol{u} \\ p \end{pmatrix}, \tag{6.4}$$

lie in the union of intervals

$$E \equiv \left[\frac{1}{2}(\lambda_F - \sqrt{\lambda_F^2 + 4\Lambda_F}), \frac{1}{2}(\lambda_F - \sqrt{\lambda_F^2 + 4\lambda_S\lambda_F})\right] \cup \left[\lambda_F, \frac{1}{2}(\Lambda_F + \sqrt{\Lambda_F^2 + 4\Lambda_F})\right]. \tag{6.5}$$

We now consider two special cases; corresponding to potential flow and generalized Stokes flow, respectively.

6.1. Potential flow

In the simplest case of potential flow, $\nu = 0$ and $N = 0$ in (2.17) thus in (6.4) we have that $F = (1/\Delta t)G$, and the Schur complement matrix is $S = \Delta t B G^{-1} B^t$. Since F is simply a (scaled) velocity mass matrix, the choice of $F_* \equiv (1/\Delta t)\mathrm{diag}(G)$ ensures that (6.2) holds with λ_F and Λ_F independent of h. For the Schur complement, we consider a preconditioner corresponding to (4.10) with $F_p = (1/\Delta t)Q$, and with $BG^{-1}B^t$ replaced by the spectrally equivalent operator A_p, that is we take

$$P_S = BG^{-1}B^tF_p^{-1}Q \simeq A_pF_p^{-1}Q = \Delta t\, A_p. \tag{6.6}$$

Bound (4.16) suggests that a practical choice for the preconditioner in (6.3) is $S_* = \Delta t\, H_*$ corresponding to a (symmetric) multigrid approximation to the inverse of the pressure Poisson operator A_p. (With this choice of S_* bounds (4.15) and (4.16) show that (6.3) holds with $\lambda_S = \beta^2\kappa^2$.) Combining Theorems 3.1 and 6.1 then leads to the following result.

Theorem 6.2. *In the case of a potential flow problem, MINRES iteration with a velocity scaling together with a simple multigrid preconditioning for the pressure Poisson operator, converges to a fixed tolerance in a number of iterations that is independent of the mesh size h, and the time step Δt.*

6.2. The generalised Stokes equations

We now consider eigenvalue bounds in the case $N = 0$ in (2.17) so that $F = (1/\Delta t)G + \nu A$ in (4.1). Since F is essentially a scaled vector-Laplacian plus an identity operator, it is well-known that multigrid can be used to generate an approximation F_* satisfying (6.2). For the Schur complement, we consider a preconditioner corresponding to (4.10) with $F_p = (1/\Delta t)Q + \nu A_p$, that is we take

$$\begin{aligned} P_S^{-1} &= (BG^{-1}B^t)^{-1}F_pQ^{-1} \\ &\simeq A_p^{-1}F_pQ^{-1} \\ &\equiv (1/\Delta t)A_p^{-1} + \nu Q^{-1}. \end{aligned} \tag{6.7}$$

The optimality of this combination is well established [3]. Using (6.7) we have that the Rayleigh quotient in (6.3) satisfies

$$\frac{p^tBF^{-1}B^tp}{p^tP_sp} = \frac{p^tB((1/\Delta t)G + \nu A)^{-1}B^tp}{p^t((1/\Delta t)A_p^{-1} + \nu Q^{-1})^{-1}p}. \tag{6.8}$$

This shows the importance of the inf–sup condition (2.21) in the limiting case of steady flow — for large Δt the quotient (6.8) reduces to the quotient in (2.21), (2.23), and it follows that (6.3) is satisfied with $\lambda_S = \gamma^2$ in the steady-state limit $\Delta t \to \infty$. Recent work by Bramble and Pasciak [2] has formally established that for finite Δt, quotient (6.8) is bounded both above and below by constants independent of h and Δt, although careful consideration is required in the separate cases $\nu\Delta t < h^2$ and $\nu\Delta t \geqslant h^2$.

Our analysis in Section 4 suggests that a practical version of the generalized Stokes preconditioner is given by (6.1) with:

$$S_* = \frac{1}{\Delta t}H_*^{-1} + \nu Q_*^{-1}. \tag{6.9}$$

The point here is that P_S is spectrally equivalent to S_* so that (6.3) is satisfied for the choice (6.9), in which case Theorem 6.1 implies that the intervals defining E in (6.5) are independent of h and Δt. This fact can be combined with Theorem 3.1 to establish the following convergence result (corroborated by the iteration counts presented in Section 5.1).

Theorem 6.3. *In the case of a generalized Stokes problem, preconditioned MINRES iteration with a simple multigrid cycle approximating a Helmholtz operator for each velocity component and a Poisson operator for the pressure, converges to a fixed tolerance in a number of iterations that is independent of the mesh size h, and the time step* Δt.

7. Conclusion

The Navier–Stokes solution algorithm that is outlined in this work is rapidly converging, and the resulting iteration counts are remarkably insensitive to the mesh size and time-step. The algorithm is also robust with respect to variations in the Reynolds number of the underlying flow. An attractive feature of our approach is that it can be implemented using simple building blocks for solving the two subsidiary problems that arise, namely, a pressure Poisson problem and a scalar convection–diffusion problem. If multigrid is used for each of these then the overall complexity is proportional to the number of degrees of freedom.

Acknowledgements

This is ongoing research and is supported by the EPSRC via grants GR/K91262 (DS), GR/M59044 (AW) and GR/L05617 (DK) and by the US National Science Foundation via grant DMS9972490.

References

[1] M. Bercovier, O.A. Pironneau, Error estimates for finite element method solution of the Stokes problem in the primitive variables, Numer. Math. 33 (1977) 211–224.

[2] J. Bramble, J. Pasciak, Iterative techniques for time dependent Stokes problems, Math. Appl. 33 (1997) 13–30.

[3] J. Cahouet, J.P. Chabard, Some fast 3D finite element solvers for the generalised Stokes problem, Internat. J. Numer. Methods Fluids 8 (1988) 869–895.

[4] S.C. Eisenstat, H.C. Elman, M.H. Schultz, Variational iterative methods for nonsymmetric systems of linear equations, SIAM J. Numer. Anal. 20 (1983) 345–357.

[5] H. Elman, Preconditioning for the steady state Navier–Stokes equations with low viscosity, SIAM J. Sci. Comput. 20 (1999) 1299–1316.

[6] H. Elman, D. Silvester, Fast nonsymmetric iterations and preconditioning for Navier–Stokes equations, SIAM J. Sci. Comput. 17 (1996) 33–46.

[7] H.C. Elman, D.J Silvester, A.J. Wathen, Performance and analysis of saddle point preconditioners for the discrete steady-state Navier–Stokes equations, Technical Report UMIACS-TR-2000-54 (2000).

[8] V. Faber, T.A. Manteuffel, Necessary and sufficient conditions for the existence of a conjugate gradient method, SIAM J. Numer. Anal. 21 (1984) 352–362.

[9] V. Faber, T.A. Manteuffel, Orthogonal error methods, SIAM J. Numer. Anal. 24 (1987) 170–187.

[10] B. Fischer, A. Ramage, D. Silvester, A. Wathen, On parameter choice and iterative convergence for stabilised discretisations of advection-diffusion problems, Comput. Methods Appl. Mech. Eng. 179 (1999) 185–202.

[11] R. Freund, N.M. Nachtigal, QMR: a quasi-minimal residual method for non-Hermitian linear systems, Numer. Math. 60 (1991) 315–339.

[12] R. Freund, N.M. Nachtigal, A new Krylov-subspace method for symmetric indefinite linear systems, in: Ames, W.F. (Ed.), Proceedings of the 14th IMACS World Congress on Computational and Applied Mathematics, IMACS, 1994, pp. 1253–1256.

[13] V. Girault, P.A. Raviart, Finite Element Methods for Navier–Stokes Equations—Theory and Algorithms, Springer, Berlin, 1986.

[14] P.M. Gresho, R. Sani, Incompressible Flow and the Finite Element Method, Wiley, Chichester, 1998.

[15] C. Johnson, The streamline diffusion finite element method for compressible and incompressible flow, in: D.F. Griffiths, G.A. Watson (Eds.), Numerical Analysis 1989, Longman Scientific, 1989, pp. 155–181.

[16] D. Kay, D. Loghin, A Green's function preconditioner for the steady state Navier–Stokes equations, Oxford University Computing Laboratory Report 99/06, 1999, SIAM J. Sci. Comput., submitted for publication.

[17] D. Kay, D. Silvester, A posteriori error estimation for stabilised mixed approximations of the Stokes equations, SIAM J. Sci. Comput. 21 (2000) 1321–1336.

[18] V.I. Lebedev, Iterative methods for solving operator equations with a spectrum contained in several intervals, USSR Comput. Math. and Math. Phys. 9 (1969) 17–24.

[19] M.F. Murphy, G.H. Golub, A.J. Wathen, A note on preconditioning for indefinite linear systems, SIAM J. Sci. Comput. 21 (2000) 1969–1972.

[20] C.C. Paige, M.A. Saunders, Solution of sparse indefinite systems of linear equations, SIAM. J. Numer. Anal. 12 (1975) 617–629.

[21] Y. Saad, M.H. Schultz, GMRES: a generalized minimal residual algorithm for solving nonsymmetric linear systems, SIAM J. Sci. Statist. Comput. 7 (1986) 856–869.

[22] D. Silvester, A. Wathen, Fast iterative solution of stabilised Stokes systems part II: using general block preconditioners, SIAM J. Numer. Anal. 31 (1994) 1352–1367.

[23] D. Silvester, A. Wathen, Fast & robust solvers for time-discretised incompressible Navier–Stokes equations, in: D.F. Griffiths, G.A. Watson (Eds.), Numerical Analysis 1995, Longman Scientific, 1996, pp. 230–240.

[24] J. Simo, F. Armero, Unconditional stability and long-term behavior of transient algorithms for the incompressible Navier–Stokes and Euler equations, Comput. Methods Appl. Mech. Eng. 111 (1994) 111–154.

[25] G.L.G. Sleijpen, D.R. Fokkema, BiCGSTAB(ℓ) for linear equations involving unsymmetric matrices with complex spectrum, Electron. Trans. Numer. Anal. 1 (1993) 11–32.

[26] A. Smith, D. Silvester, Implicit algorithms and their linearisation for the transient Navier–Stokes equations, IMA J. Numer. Anal. 17 (1997) 527–543.

[27] P. Sonneveld, CGS, a fast Lanczos-type solver for nonsymmetric linear systems, SIAM J. Sci. Statist. Comput. 10 (1989) 36–52.

[28] G. Stoyan, Towards discrete Velte decompositions and narrow bounds for inf-sup constants, Comput. Math. Appl. 38 (1999) 243–261.

[29] S. Turek, Efficient Solvers for Incompressible Flow Problems, Springer, Berlin, 1999.

[30] H.A. Van der Vorst, BI-CGSTAB: a fast and smoothly converging variant of BI-CG for the solution of nonsymmetric linear systems, SIAM J. Sci. Statist. Comput. 10 (1992) 631–644.

[31] A.J. Wathen, Realistic eigenvalue bounds for the Galerkin mass matrix, IMA J. Numer. Anal. 7 (1987) 449–457.

[32] A. Wathen, B. Fischer, D. Silvester, The convergence rate of the minimal residual method for the Stokes problem, Numer. Math. 71 (1995) 121–134.

[33] A. Wathen, B. Fischer, D. Silvester, The convergence of iterative solution methods for symmetric and indefinite linear systems, in: D.F. Griffiths, D.J. Higham, G.A. Watson (Eds.), Numerical Analysis 1997, Longman Scientific, 1997, pp. 230–243.

[34] P. Wesseling, An Introduction to Multigrid, Wiley, New York, 1991.

ELSEVIER Journal of Computational and Applied Mathematics 128 (2001) 281–309

JOURNAL OF COMPUTATIONAL AND APPLIED MATHEMATICS

www.elsevier.nl/locate/cam

A review of algebraic multigrid

K. Stüben

German National Research Center for Information Technology (GMD), Institute for Algorithms and Scientific Computing (SCAI), Schloss Birlinghoven, D-53754 St. Augustin, Germany

Received 3 September 1999; received in revised form 9 November 1999

Abstract

Since the early 1990s, there has been a strongly increasing demand for more efficient methods to solve large sparse, *unstructured* linear systems of equations. For practically relevant problem sizes, classical *one-level* methods had already reached their limits and new *hierarchical* algorithms had to be developed in order to allow an efficient solution of even larger problems. This paper gives a review of the first hierarchical and *purely matrix-based* approach, *algebraic multigrid* (AMG). AMG can directly be applied, for instance, to efficiently solve various types of elliptic partial differential equations discretized on unstructured meshes, both in 2D and 3D. Since AMG does not make use of any geometric information, it is a "plug-in" solver which can even be applied to problems without any geometric background, provided that the underlying matrix has certain properties.

Keywords: Algebraic multigrid

1. Introduction

The efficient numerical solution of large systems of discretized elliptic partial differential equations (PDEs) requires hierarchical algorithms which ensure a rapid reduction of both short- and long-range error components. A break-through, and certainly one of the most important advances during the last three decades, was due to the multigrid principle. Any corresponding method operates on a hierarchy of grids, defined a priori by coarsening the given discretization grid in a geometrically natural way ("geometric" multigrid). Clearly, this is straightforward for logically regular grids. However, the definition of a natural hierarchy may become very complicated for highly complex, unstructured meshes, if possible at all.

The first attempt to automate the coarsening process took place in the early 1980s [10,12,13], at the time when the so-called *Galerkin-principle* and *operator-dependent interpolation* were combined in geometric multigrid to increase its robustness (aiming at the efficient solution of diffusion

E-mail address: stueben@gmd.de (K. Stüben).

0377-0427/01/$ - see front matter
PII: S0377-0427(00)00516-1

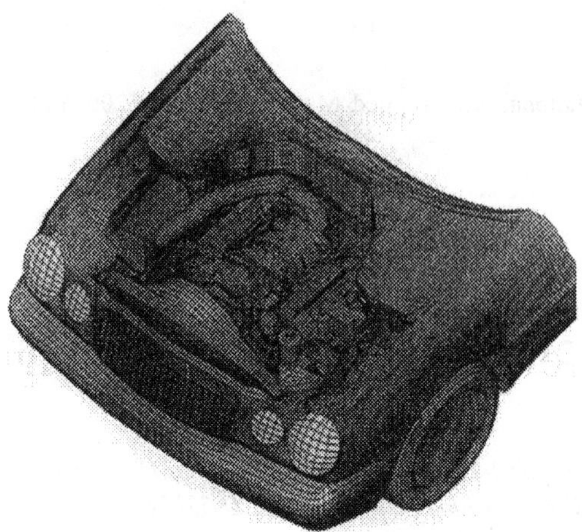

Fig. 1. Mesh for computing the underhood flow of a Mercedes-Benz E-Class.

problems with jumping coefficients [1,20]). This attempt was motivated by the observation that reasonable operator-dependent interpolation and the Galerkin operator can often be derived directly from the underlying matrices, without any reference to the grids. The result was a multigrid-like approach which did not merely allow an automatic coarsening process, but could be directly applied to (line sparse) algebraic equations of certain types, without any pre-defined hierarchy ("algebraic" multigrid,[1] AMG).

The first fairly general AMG program was described and investigated in [47,48,50]. Since this code was made publically available in the mid-1980s (AMG1R5), there had been no substantial further research and development in AMG for many years. However, since the early 1990s, and even more since the mid-1990s, there was a strong increase of interest in algebraically oriented multilevel methods. One reason for this was certainly the increasing geometrical complexity of applications which, technically, limited the immediate use of geometric multigrid. Another reason was the steadily increasing demand for efficient "plug-in" solvers. In particular, in commercial codes, this demand was driven by increasing problem sizes which made clear the limits of the classical one-level solvers still used in most packages.

For instance, CFD applications in the car industry involve very complicated flow regions. Flows through heating and cooling systems, complete vehicle underhood flows, or flows within passenger compartments are computed on a regular basis. Large complex meshes, normally unstructured, are used to model such situations. Requirements on the achievable accuracy are ever increasing, leading to finer and finer meshes. Locally refined grid patches are introduced to increase the accuracy with as few additional mesh points as possible. Fig. 1 shows an example.

In the recent past, several ways to realize concrete AMG algorithms have been investigated and there is still an ongoing rapid development of new AMG and AMG-like approaches and variants. Consequently, there is no unique and best approach yet. Whenever we talk about AMG in the

[1] We should actually use the term multi*level* rather than multi*grid*. It is just for historical reasons that we use the term multi*grid*.

context of concrete numerical results, we actually refer to the code RAMG05 [2] (described in detail in [51]), which is a successor of the original code AMG1R5 mentioned above. However, RAMG05 is completely new and, in particular, incorporates more efficient and more robust interpolation and coarsening strategies.

This paper gives a survey of the classical AMG idea [48], certain improvements and extensions thereof, and various new approaches. The focus in Sections 2–6 is on fundamental ideas and aspects, targeting classes of problems for which AMG is best-developed, namely, *symmetric positive-definite* (s.p.d.) problems of the type as they arise, for instance, from the discretization of *scalar* elliptic PDEs of second order. We want to point out, however, that the potential range of applicability is much larger. In particular, AMG has successfully been applied to various nonsymmetric (e.g. convection–diffusion) and certain indefinite problems. Moreover, important progress has been achieved in the numerical treatment of *systems* of PDEs (mainly Navier–Stokes and structural mechanics applications). However, major research is still ongoing and much remains to be done to obtain an efficiency and robustness comparable to the case of scalar applications. In particular in Section 7, we will set pointers to the relevant literature where one can find further information or more recent AMG approaches. Although we try to cover the most important references, the list is certainly not complete in this rapidly developing field of research.

2. Algebraic versus geometric multigrid

Throughout this paper, we assume the reader to have some basic knowledge of geometric multigrid. In particular, he should be familar with the fundamental principles (smoothing and coarse-grid correction) and with the recursive definition of multigrid cycles. This is because, for simplicity, we limit our main considerations to just two levels. Accordingly, whenever we talk about the efficiency of a particular approach, we implicitly always assume the underlying two-level method to be recursively extended to a real multi-level method, involving only a small number of variables (20–40, say) on the coarsest level. Regarding more detailed information on geometric multigrid, we refer to [61] and the extensive list of references given therein.

A two-level AMG cycle to solve (sparse) s.p.d. systems of equations

$$A_h u^h = f^h \quad \text{or} \quad \sum_{j\in\Omega^h} a_{ij}^h u_j^h = f_i^h \quad (i \in \Omega^h) \tag{1}$$

is formally defined in the same way as a Galerkin-based geometric two-grid cycle. The only difference is that, in the context of AMG, Ω^h is just an *index set* while it corresponds to a *grid* in geometric multigrid. Accordingly, a coarser level, Ω^H, just corresponds to a (much smaller) index set.

If we know how to map H-vectors into h-vectors by some (full rank) "interpolation" operator I_H^h, the (s.p.d.) coarse-level operator A_H is defined via

$$A_H := I_h^H A_h I_H^h \quad \text{with } I_h^H = (I_H^h)^{\mathrm{T}}.$$

One two-level correction step then runs as usual, that is

$$u_{\text{new}}^h = u_{\text{old}}^h + I_H^h e^H, \tag{2}$$

[2] The development of RAMG05 has partly been funded by Computational Dynamics Ltd., London.

where the correction e^H is the exact solution of

$$A_H e^H = r^H \quad \text{or} \quad \sum_{j\in\Omega^H} a_{ij}^H e_j^H = r_i^H \quad (i \in \Omega^H)$$

with $r^H = I_h^H(r_{\text{old}}^h)$ and $r_{\text{old}}^h = f^h - A_h u_{\text{old}}^h$. (Note that we normally use the letter u for *solution* quantities and the letter e for *correction* or *error* quantities.) For the corresponding errors $e^h = u_\star^h - u^h$ ($u_\star^h$ denotes the exact solution of (1)), this means

$$e_{\text{new}}^h = K_{h,H} e_{\text{old}}^h \quad \text{with } K_{h,H} := I_h - I_H^h A_H^{-1} I_h^H A_h, \tag{3}$$

being the *coarse-grid correction operator* (I_h denotes the identity).

We finally recall that – given any relaxation operator, S_h, for smoothing – the convergence of Galerkin-based approaches can most easily be investigated w.r.t. the *energy norm*, $\|e^h\|_{A_h} = (A_h e^h, e^h)^{1/2}$. Assuming v relaxation steps to be performed for (pre-) smoothing, the following well-known variational principle holds (see, for instance [51]),

$$\|K_{h,H} S_h^v e^h\|_{A_h} = \min_{e^H} \|S_h^v e^h - I_H^h e^H\|_{A_h}. \tag{4}$$

As a trivial consequence, convergence of two-level cycles and, if recursively extended to any number of levels, the convergence of complete V-cycles is always ensured as soon as the relaxation method converges. This is true for any sequence of coarser levels and interpolation operators. More importantly, (4) indicates that the *speed* of convergence strongly depends on the efficient interplay between *relaxation* and *interpolation*. Based on (4), we want to outline the basic conceptual difference between geometric and algebraic multigrid.

2.1. Geometric multigrid

In geometric multigrid, fixed coarsening strategies are employed and interpolation is usually defined geometrically, typically by linear interpolation. Depending on the given problem, this necessarily imposes strong requirements on the smoothing properties of S_h (in order for the right-hand side in (4) to become small), namely, that the error after relaxation varies in a geometrically smooth way from the fine-level grid points to the neighboring coarse-level ones. In other words, the error after relaxation *has to be geometrically smooth*, relative to the coarse grid.

As an illustration, let us assume the coarser levels to be defined by standard geometric $h \to 2h$ coarsening in each spatial direction. It is well known that pointwise relaxation geometrically smooths the error in each direction only if the given problem is essentially isotropic. In case of anisotropic problems, however, smoothing is only "in the direction of strong couplings". In such cases, more complex smoothers, such as alternating line-relaxation or ILU-type schemes, are required in order to still achieve a good interplay between smoothing and interpolation and, thus, fast multigrid convergence.

While the construction of "robust smoothers" is not difficult in 2D model situations, for 3D applications on complex meshes their realization tends to become rather cumbersome. For instance, the robust 3D analog of alternating line relaxation is alternating *plane* relaxation (e.g., realized by 2D multigrid within each plane) which, in complex geometric situations, becomes very complicated to implement, if possible at all. ILU smoothers, on the other hand, loose much of their smoothing property in general 3D situations. The only way to loosen the requirements on the smoothing properties of the relaxation and still maintain an efficient interplay relaxation and interpolation is to use more sophisticated coarsening techniques. In geometric multigrid, steps in this direction have been done by, for example, employing more than one coarser grid on each multigrid level ("multiple semi-coarsening" [35,59,21,34]).

2.2. Algebraic multigrid

While geometric multigrid essentially relies on the availability of robust smoothers, AMG takes the opposite point of view. It assumes a simple relaxation process to be given (typically plain Gauss–Seidel relaxation) and then attempts to construct a suitable operator-dependent interpolation I_H^h (including the coarser level itself). According to (4), this construction has to be such that error of the form $S_h^\nu e^h$ is sufficiently well represented in the *range of interpolation*, $\mathscr{R}(I_H^h)$. The better this is satisfied, the faster the convergence can be. Note that it is *not* important here whether relaxation smooths the error in any geometric sense. What *is* important, though, is that the error after relaxation can be characterized algebraically to a degree which makes it possible to automatically construct coarser levels and define interpolations which are *locally* adapted to the properties of the given relaxation. This local adaptation is the main reason for AMG's flexibility in adjusting itself to the problem at hand and its robustness in solving large classes of problems *despite using very simple point-wise smoothers.*

3. The classical AMG approach

In classical AMG, we regard the coarse-level variables as a subset of the fine-level ones. That is, we assume the set of fine-level variables to be split into two disjoint subsets, $\Omega^h = C^h \cup F^h$, with C^h representing those variables which are to be contained in the coarse level (*C-variables*) and F^h being the complementary set (*F-variables*). Given such a C/F-splitting, we define $\Omega^H = C^h$ and consider (full rank) interpolations $e^h = I_H^h e^H$ of the form

$$e_i^h = (I_H^h e^H)_i = \begin{cases} e_i^H & \text{if } i \in C^h, \\ \sum_{k \in P_i^h} w_{ik}^h e_k^H & \text{if } i \in F^h, \end{cases} \tag{5}$$

where $P_i^h \subset C^h$ is called the set of *interpolatory variables*. (For reasons of sparsity of A_H, P_i^h should be a reasonably small subset of C-variables "near" i.) Clearly, $\mathscr{R}(I_H^h)$ strongly depends on both the concrete selection of the C-variables *and* the definition of the interpolation. In a given situation, one can easily imagine "bad" C/F-splittings which just do not allow any interpolation which is suitable in the sense that was outlined in the previous section. That is, the construction of concrete C/F-splittings and the definition of interpolation are closely related processes.

Concrete algorithms used in practice are largely heuristically motivated. In Section 3.1, we mainly summarize the basic ideas as described in [48] and some modifications introduced in [51]. In Section 3.2, we take a closer look at some theoretical and practical aspects in case that A_h contains only nonpositive off-diagonal entries (M-matrix). To simplify notation, we usually omit the index h in the following, for instance, we write S, A, K and e instead of S_h, A_h, $K_{h,H}$ and e^h. Moreover, instead of (5), we simply write

$$e_i = \sum_{k \in P_i} w_{ik} e_k \quad (i \in F). \tag{6}$$

3.1. The basic ideas

Classical AMG uses plain Gauss–Seidel relaxation for smoothing. From some heuristic arguments, one can see that the error e, obtained after a few relaxation steps, is characterized by the fact that the (scaled) residual is, on the average for each i, much smaller than the error itself, $|r_i| \ll a_{ii}|e_i|$. This implies that e_i can *locally* be well approximated by

$$e_i \approx -\left(\sum_{j\in N_i} a_{ij}e_j\right)\bigg/ a_{ii} \quad (i\in\Omega), \tag{7}$$

where $N_i = \{j\in\Omega\colon j\neq i,\ a_{ij}\neq 0\}$ denotes the *neighborhood* of any $i\in\Omega$. Such an error is called *algebraically smooth.* According to the remarks at the end of Section 2, it is this kind of error which has to be well represented in $\mathscr{R}(I_H^h)$. That is, the general goal is to construct C/F-splittings and define sets of interpolatory variables $P_i\subset C$ $(i\in F)$ along with corresponding weights w_{ik} such that (6) yields a reasonable approximation for each algebraically smooth vector e.

Obviously, a very "accurate" interpolation in this sense is obtained by directly using (7), that is, by choosing $P_i = N_i$ and $w_{ik} = -a_{ik}/a_{ii}$. However, this would require selecting a C/F-splitting so that, for each $i\in F$, *all of its neighbors* are contained in C. Although any such selection can even be seen to yield a *direct* solver, this approach is of no real practical relevance since, in terms of computational work and memory requirement, the resulting method will generally be extremely inefficient if recursively extended to a hierarchy of levels [51].

In practice, we want to achieve rapid convergence with as small sets of interpolatory variables P_i as possible (in order to allow for a rapid coarsening and to obtain reasonably sparse Galerkin operators). Various approaches have been tested in practice which cannot be described in detail here. In the following, we just give an outline of some typical approaches.

3.1.1. Direct interpolation

We talk about *direct interpolation* if $P_i\subseteq N_i$. Such an interpolation can immediately be derived from (7) if we know how to approximate the "noninterpolatory part" (i.e. that part of the sum in (7) which refers to $j\in N_i\setminus P_i$) for an algebraically smooth error. This approximation is the most critical step in defining interpolation.

For M-matrices A, for instance, such an approximation can be obtained by observing that an algebraically smooth error varies slowly in the direction of *strong* (large) couplings. In particular, the more strong couplings of any variable i are contained in P_i, the better an algebraically smooth error satisfies

$$\frac{1}{\sum_{k\in P_i} a_{ik}}\sum_{k\in P_i} a_{ik}e_k \approx \frac{1}{\sum_{j\in N_i} a_{ij}}\sum_{j\in N_i} a_{ij}e_j \quad (i\in\Omega).$$

Inserting this into (7), we obtain an interpolation (6) with positive weights

$$w_{ik} = -\alpha_i a_{ik}/a_{ii} \quad \text{where } \alpha_i = \frac{\sum_{j\in N_i} a_{ij}}{\sum_{\ell\in P_i} a_{i\ell}}. \tag{8}$$

Practically, this means that we have to construct C/F-splittings so that each $i\in F$ has a sufficiently large number of *strongly* coupled C-neighbors which are then taken as the set of interpolatory variables P_i. (See Section 3.2 regarding some important additional aspects.)

In the case of (scalar) elliptic PDEs, the largest off-diagonal entries are usually negative. If there are also *positive* off-diagonal entries, a similar process as before can be applied as long as such entries are relatively small: Small positive couplings can simply be ignored by just considering them as *weak*. However, the situation becomes less obvious, if A contains *large* positive off-diagonal entries. In many such cases, an algebraically smooth error can be assumed to *oscillate* along such couplings (e.g. in case of *weakly diagonally dominant* s.p.d. matrices A [26,51]). This can be used to generalize the above approach by, for instance, a suitable separation of positive and negative couplings, leading to interpolation formulas containing both positive and negative weights. A corresponding approach has been proposed in [51] which can formally be applied to arbitrary s.p.d. matrices. However, the resulting interpolation is heuristically justified only as long as, for any i, those error components e_k which correspond to large positive couplings $a_{ik} > 0$, change slowly *among each other* (unless a_{ik} is very small in which case its influence can be ignored).

In practice, these simple algebraic approaches to construct an interpolation cover a large class of applications. However, there is no best way yet to automatically construct an interpolation which is good for *arbitrary* s.p.d. matrices, at least not by merely considering the size and sign of coefficients. For instance, in case of particular higher-order finite-element discretizations or discretizations by bilinear elements on quadrilateral meshes with large aspect ratios, the resulting matrices typically contain significant positive off-diagonal entries and are far from being weakly diagonally dominant. In such cases, the performance of AMG may be only suboptimal. If this happens, it often helps to exploit some information about the origin of these positive connections rather than to rely only on information directly contained in the matrix. For instance, one could try to structurally simplify the given matrix before applying AMG (see, e.g. [41]). Alternative approaches, defining the coarsening and interpolation are outlined in Section 7.

3.1.2. More complex interpolations

There are several ways to improve the direct interpolation of the previous section. To outline some possibilities, let us assume a C/F-splitting and, for each $i \in F$, a set of (strongly coupled) interpolatory variables $P_i \subseteq N_i \cap C$ to be given. Rather than to immediately approximate the noninterpolatory part of the ith equation (7) as done above, one may first (approximately) eliminate all strongly coupled $e_j \, (j \notin P_i)$ by means of the corresponding jth equation. The ideas outlined in the previous section can then analogously be applied to the resulting extended equation for e_i, leading to an interpolation with an increased set of interpolatory variables. The corresponding interpolation (called *standard interpolation* in [51]) is, in general, considerably more robust in practice. Alternatively, one may obtain an improved interpolation by applying one F-relaxation step (for more details, see "Jacobi-interpolation" in Section 5) to either the direct or the standard interpolation.

In both approaches, compared to the direct interpolation, the "radius" of interpolation increases which, in turn, will reduce the sparsity of the resulting Galerkin operator. However, interpolation weights corresponding to variables "far away" from variable i are typically much smaller than the largest ones. Before computing the Galerkin operator, one should therefore truncate the interpolation operator by ignoring all small entries (and re-scale the remaining weights so that the total sum remains unchanged). Note that, because of the variational principle, the truncation of interpolation is a "safe process"; in the worst case, overall convergence may slow down, but no divergence can occur. On the other hand, a truncation of the *Galerkin operators* can be dangerous since this destroys the validity of the variational principle and, if not applied with great care, may even cause strong divergence in practice.

Apart from other minor differences, the original AMG interpolation proposed in [48] (and realized in the code AMG1R5) can be regarded as a compromise between the direct interpolation and the standard interpolation described before. There, an attempt was made to replace all strongly coupled $e_j\,(j \notin P_i)$ in (7) by averages *involving only variables in* P_i. However, for this to be reasonable, based on certain criteria, new C-variables had to be added to a given splitting a posteriori ("final C-point choice" in [48]). Although this approach worked quite well in those cases treated in [48], typically too many additional C-variables are required in geometrically complex 3D situations, causing unacceptably high fill-in towards coarser levels (see Section 4.2 for examples). In practice, the standard interpolation outlined above (in combination with a reasonable truncation) has turned out to be more robust and often considerably more efficient.

The above improvements of interpolation generally lead to faster convergence but also increase the computational work per cycle and the required memory to some extent. Whether or not this finally pays, depends on the given application. If memory is an issue (as it is often in commercial environments), one may, instead, wish to *simplify* interpolation at the expense of a reduced convergence speed. One way to achieve this is to generally allow interpolation from variables which are *not* in the direct neighborhood. Such "long-range" interpolation [48] generally allows a much faster coarsening and drastically increases the sparsity on coarser levels. For details of a simple approach which has been tested in practice, see [51] ("aggressive coarsening" and "multi-pass interpolation").

3.2. The M-matrix case

In practice, it turns out that AMG V-cycle convergence is, to a large extent, independent of the problem size. Unfortunately, this cannot be proved based merely on algebraic arguments. Nevertheless, some important aspects can be investigated theoretically, in particular, matrix-independent two-level convergence can be proved for various classes of matrices if interpolation is defined properly. We consider here the class of M-matrices. Generalizations to other classes and the corresponding proofs can be found in [51].

3.2.1. Two-level considerations

The following theorem shows that direct interpolation based on (8) ensures matrix-independent two-level convergence if, for each $i \in F$, the connectivity represented in P_i is a *fixed fraction* of the total connectivity.

Theorem 1. *Let A be a symmetric, weakly diagonally dominant M-matrix. With fixed $0 < \tau \leqslant 1$ select a C/F-splitting so that, for each $i \in F$, there is a set $P_i \subseteq C \cap N_i$ satisfying*

$$\sum_{k \in P_i} |a_{ik}| \geqslant \tau \sum_{j \in N_i} |a_{ij}| \tag{9}$$

and define interpolation according to (8). *Then the two-level method, using one Gauss–Seidel relaxation step for* (*post-*) *smoothing, converges at a rate which depends only on τ but not on the*

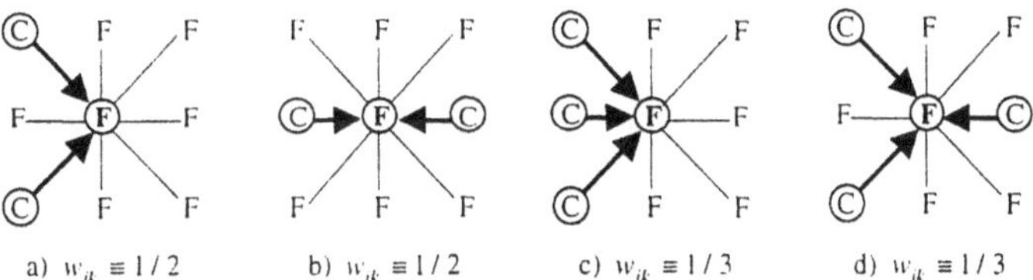

Fig. 2. Different C/F-arrangements and corresponding interpolation formulas.

given matrix,

$$\|SK\|_A \leqslant \sqrt{1-\tau/4}.$$

The above theorem confirms that it is the *strong* couplings which are important to interpolate from, while the use of *weak* couplings would increase the computational work but hardly affect the convergence. The more strong connections are used for interpolation, the better the convergence can be. Note that this implicitly means that coarsening will be "in the direction of smoothness" which is the main reason for the fact that AMG's convergence does not sensitively depend on anisotropies. Moreover, AMG's interpolation can be regarded as an algebraic generalization of the operator-dependent interpolation introduced in [1,20], which explains why the performance of AMG does not sensitively depend on large, discontinuous variations in the coefficients of the given system of equations. For an illustration, see Section 4.1.

From a practical point of view, the above convergence estimate is a worst-case estimate, at least if the given problem has some kind of geometric background (which it typically does). The reason is that the algebraic requirement (9) does not take the location of the interpolatory C-points, relative to the F-points, into account. For an illustration, consider the 9-point discretization of the Poisson operator

$$\frac{1}{3h^2}\begin{bmatrix} -1 & -1 & -1 \\ -1 & 8 & -1 \\ -1 & -1 & -1 \end{bmatrix}_h . \tag{10}$$

From geometric multigrid, we know that linear interpolation yields fast convergence. The algebraic interpolation, however, cannot distinguish between geometrically "good" and "bad" C/F-splittings. For instance, in Fig. 2a and b we use the same total weight for interpolation but the second arrangement will clearly result in much better convergence. Similarly, the arrangement in Fig. 2d, although it does not give exactly second order, will be much better than the one in Fig. 2c.

This illustrates that the concrete arrangement of a C/F-splitting will have a substantial influence on the quality of interpolation, and, through this, on the final convergence. In order to *strictly ensure* an optimal interpolation, we would have to exploit the geometric location of (strongly coupled) points among each other. In practice, however, it turns out to be sufficient to base the construction of a C/F-splitting on the following additional objective. As a rule, one should arrange the C/F-splitting so that the set of C-variables builds (approximately) a *maximal* set with the property that the C-variables are not strongly coupled among each other ("maximally independent set") and that the F-variables are "surrounded" by their interpolatory C-variables. This can be ensured to a sufficient extent by merely exploiting the connectivity information contained in the matrix (for an algorithm, see [48,51]). Note that strong connectivity does not necessarily have to be via direct couplings.

Observing this objective will, in practice, substantially enhance convergence even if only small sets of interpolatory variables are used.

3.2.2. Extension to multi-level cycles

Unfortunately, the assumptions on interpolation in Theorem 1 are sufficient for uniform two-level, but not for uniform V-cycle convergence. Although, by choosing $\tau \geqslant \frac{1}{2}$, one can ensure that all recursively defined Galerkin operators remain weakly diagonally dominant M-matrices and, hence, the formal extension to complete V-cycles is straightforward, A-independent V-cycle convergence cannot be proved. The reason is the limited accuracy of purely algebraically defined interpolation as discussed in the previous section. We will return to this problem in Section 6 where we consider a worst-case scenario in the context of "aggregation-type" AMG approaches.

In practice, however, one can observe V-cycle convergence which is, to a large extent, independent of the problem size if we take the additional objective of the previous section into account. Furthermore, it turns out that it is not important to force the coarse-level matrices to exactly remain M-matrices. On the contrary, such a requirement puts too many restrictions on the coarsening process, in particular on lower levels, where the size of the Galerkin operators then may grow substantially.

In this context, we want to emphasize again that, for an efficient overall performance, convergence speed is only one aspect. An equally important aspect is the complexity (sparsity) of the coarser level matrices produced by AMG (which directly influences both the run time and the overall memory requirement). Only if both the convergence speed and the *operator complexity*

$$c_A = \sum_{\ell} |m_\ell| / |m_1|, \tag{11}$$

(m_ℓ denotes the number of nonzero entries contained in the matrix on level ℓ) are bounded independent of the size of A, do we have an asymptotically optimal performance. The typical AMG performance in case of some complex problems is given in Section 4.2.

3.3. AMG as a pre-conditioner

In order to increase the robustness of geometric multigrid approaches, it has become very popular during the last years, to use multigrid not as a stand-alone solver but rather combine it with acceleration methods such as conjugate gradient. BI-CGSTAB or GMRES. This development was driven by the observation that it is often not only simpler but also more efficient to use accelerated multigrid approaches rather than to try to optimize the interplay between the various multigrid components in order to improve the convergence of stand-alone multigrid cycles.

This has turned out to be similar for AMG which, originally, was designed to be used stand-alone. Practical experience has clearly shown that AMG is also a very good pre-conditioner, much better than standard (one-level) ILU-type pre-conditioners, say. Heuristically, the major reason is due to the fact that AMG, in contrast to any one-level pre-conditioner, aims at the efficient reduction of *all* error components, short range as well as long range. However, although AMG tries to capture all relevant influences by proper coarsening and interpolation, its interpolation will hardly ever be optimal. It may well happen that error reduction is significantly less efficient for some very specific error components. This may cause a few eigenvalues of the AMG iteration matrix to be considerably closer to 1 than all the rest. If this happens, AMG's convergence factor is limited by the slow convergence of just a few exceptional error components while the majority of the error components is reduced very quickly. Acceleration by, for instance, conjugate gradient typically eliminates these particular frequencies very efficiently. The alternative, namely, to try to prevent such situations by putting more effort into the construction of interpolation, will generally be much more expensive. Even then, there is no final guarantee that such situations can be avoided. (We note that this even happens with "robust" geometric multigrid methods.)

4. Applications and performance

The flexibility of AMG and its simplicity of use, of course, have a price: A *setup phase*, in which the given problem is analyzed, the coarse levels are constructed and all operators are assembled, has to be concluded before the actual *solution phase* can start. This extra overhead is one reason for the fact that AMG is usually less efficient than geometric multigrid approaches (if applied to problems for which geometric multigrid *can* be applied efficiently). Another reason is that AMG's components can, generally, not be expected to be "optimal". They are always constructed on the basis of compromises between numerical work and overall efficiency. Nevertheless, if applied to standard elliptic test problems, the computational cost of AMG's solution phase (ignoring the setup cost) is typically comparable to the solution cost of a *robust* geometric multigrid solver [47].

However, AMG should not be regarded as a competitor of geometric multigrid. The strengths of AMG are its robustness, its applicability in complex geometric situations and its applicability to even solve certain problems which are out of the reach of geometric multigrid, in particular, problems with no geometric or continuous background at all. In such cases, AMG should be regarded as an efficient alternative to standard numerical methods such as conjugate gradient accelerated by typical (one-level) pre-conditioners. We will show some concrete performance comparisons in Section 4.2. Before, however, we want to illustrate the flexibility of AMG in adjusting its coarsening process locally to the smoothing properties of relaxation by means of a simple but characteristic model equation.

4.1. A model problem for illustration

We consider the model equation

$$-(au_x)_x-(bu_y)_y+cu_{xy}=f(x,y), \tag{12}$$

defined on the unit square with Dirichlet boundary conditions. We assume $a=b=1$ everywhere except in the upper left quarter of the unit square (where $b=10^3$) and in the lower right quarter (where $a=10^3$). The coefficient c is zero except for the upper right quarter where we set $c=2$. The resulting discrete system is isotropic in the lower left quarter of the unit square but strongly anisotropic in the remaining quarters. Fig. 3a shows what a "smooth" error looks like on the finest level after having applied a few Gauss–Seidel point relaxation steps to the homogeneous problem,

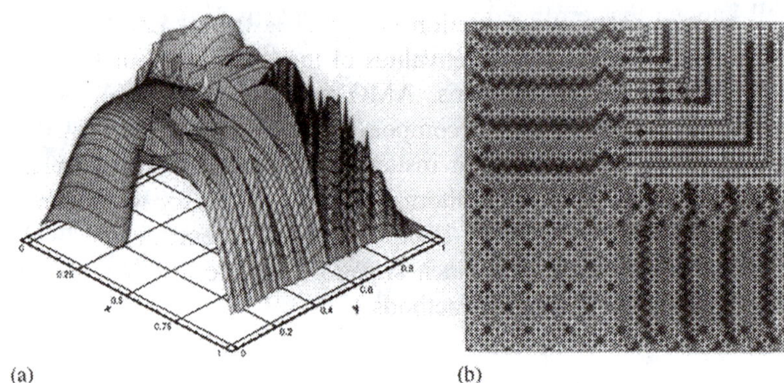

Fig. 3. (a) "Smooth" error in case of problem (12). (b) The finest and three consecutive levels created by the standard AMG coarsening algorithm.

starting with a random function. The different anisotropies as well as the discontinuities across the interface lines are clearly reflected in the picture.

It is heuristically clear that such an error can only be effectively reduced by means of a coarser grid if that grid is obtained by essentially coarsening in the directions in which the error really changes smoothly in the geometric sense and if interpolation treats the discontinuities correctly. Indeed, see Section 3.2, this is exactly what AMG does. First, the operator-based interpolation ensures the correct treatment of the discontinuities. Second, AMG coarsening is in the direction of strong connectivity, that is, in the direction of smoothness.

Fig. 3b depicts the finest and three consecutive grids created by using standard AMG coarsening and interpolation. The smallest dots mark grid points which are contained *only* on the finest grid, the squares mark those points which are also contained on the coarser levels (the bigger the square, the longer the corresponding grid point stays in the coarsening process). The picture shows that coarsening is uniform in the lower left quarter where the problem is isotropic. In the other quarters, AMG adjusts itself to the different anisotropies by locally coarsening in the proper direction. For instance, in the lower right quarter, coarsening is in the x-direction only. Since AMG takes only *strong* connections in coarsening into account and since all connections in the y-direction are *weak*, the individual lines are coarsened *independently of each other*. Consequently, the coarsening of neighboring x-lines is not "synchronized"; it is actually a matter of "coincidence" where coarsening starts within each line. This has to be observed in interpreting the coarsening pattern in the upper right quarter: within each diagonal line, coarsening is essentially in the direction of this line.

4.2. Complex applications

For a demonstration of AMG's efficiency, we consider some complex problems of the type typically solved in two commercial codes designed for *oil reservoir simulation* and for *computational fluid dynamics*, respectively. In both codes, the numerical kernel requires the fast solution of scalar elliptic equations. While, in oil reservoir simulation, geometries are typically fairly simple but the underlying problems have strongly anisotropic and discontinuous coefficients (jumps by several orders of magnitude in a nearly random way), in computational fluid dynamics these problems are

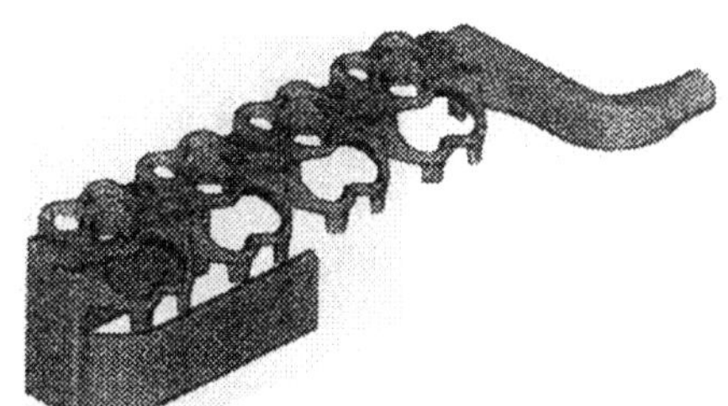

Fig. 4. Cooling jacket of a four-cylinder engine.

Poisson-like but defined on very complex, unstructured grids. For more details on these codes, see [51].

The following test cases are considered: [3]

(1) *Reservoir*. The underlying reservoir corresponds to a simple domain discretized by a mesh consisting of 1.16 million cells. The variation of absolute permeabilities results in a discontinuous variation of the coefficients by four orders of magnitude.
(2) *Cooling jacket*. Computation of the flow through the cooling jacket of a four-cylinder engine. The underlying mesh consists of 100 000 tetrahedra cells (see Fig. 4).
(3) *Coal furnace*. Computation of the flow inside the model of a coal furnace. The underlying mesh consists of 330 000 hexahedra and a few thousand pentahedra, including many locally refined grid patches.
(4) *Underhood*. Computation of the underhood flow of a Mercedes-Benz E-class model. The mesh is highly complex and consists of 910 000 cells (see Fig. 1).
(5) *E-Class*. Computation of the exterior flow over a Mercedes-Benz E-class model (see Fig. 5). The original mesh consists of 10 million cells. Due to memory restrictions, our test runs to two reduced mesh sizes consisting of 2.23 and 2.82 million cells, respectively. (Note that the underlying mesh also includes all modeling details of the previous underhood case.)

Memory requirement is a major concern for any commercial software provider. Industrial users of commercial codes always drive their simulations to the limits of their computers, shortage of memory being a serious one. For these reasons, in a commercial environment, low-memory AMG approaches are of primary interest, even if the reduction of the memory requirement is at the expense of a (limited) increase of the total computational time. Compared to standard one-level solvers, a memory overhead of some tens of percents is certainly acceptable. In any case, however, the operator complexity c_A (11) must not be larger than 2.0 say. Therefore, in the following test runs, we employ an aggressive coarsening strategy (cf. Section 3.1) and, in order to make up for the resulting reduced convergence speed, use AMG as a pre-conditioner rather than stand-alone.

Fig. 6 shows AMG's V-cycle convergence histories for each of the above cases, based on the code RAMG05 mentioned in the introduction. The results reflect the general experience that the convergence of AMG depends, to a limited extent, on the type of elements used as well as on the type of problem, but hardly on the problem size. In particular, the three Mercedes meshes are

[3] The first case has been provided by StreamSim Technologies Inc., the other ones by Computational Dynamics Ltd.

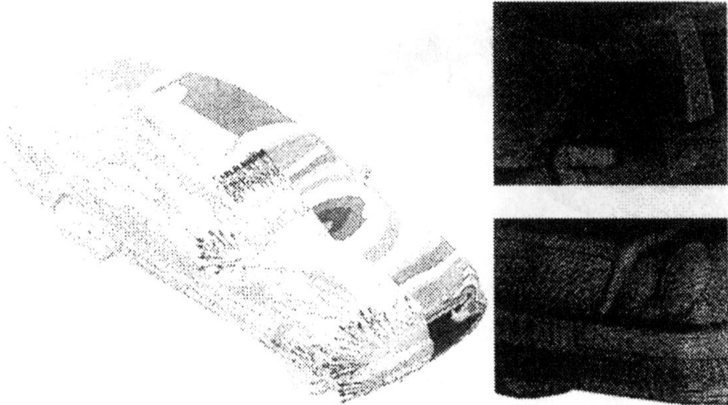

Fig. 5. Model of a Mercedes-Benz E-Class.

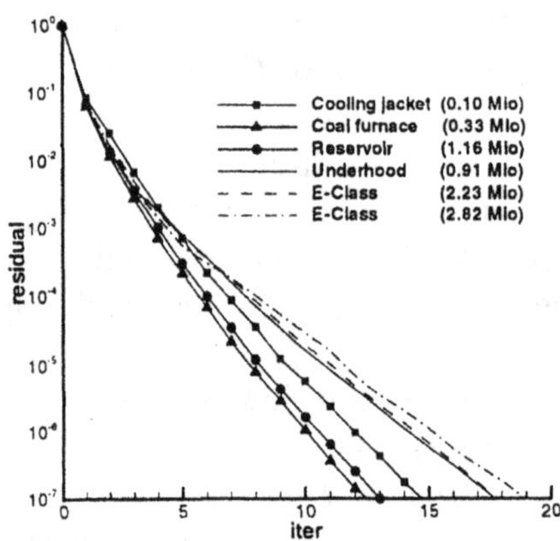

Fig. 6. RAMG05 convergence histories for various problems.

	RAMG05/cg			ILU(0)/cg		AMG1R5
problem	time	iter	c_A	time	iter	c_A
jacket	12.3	21	1.44	218.2	926	5.35
furnace	45.7	18	1.47	292.8	286	7.06
reservoir	165.0	18	1.41	2707.0	720	7.66
underhood	172.9	25	1.43	1364.0	461	5.64
eclass (2.23)	438.8	25	1.46	8282.0	1151	6.24

Fig. 7. Performance of RAMG05 vs. ILU(0)/cg.

comparable in their type but their size varies by more than a factor of three. Convergence, obviously, is influenced only marginally.

Fig. 7 compares the RAMG05 performance with that of ILU(0) pre-conditioned conjugated gradient. For both methods and for each of the above problems, the number of iterations as well as total run times (in s), required to reduce the residuals by nine digits, are shown. Compared to ILU(0)/cg.

AMG reduces the number of iterations by up to a factor of 46. In terms of run time, AMG is up to 19 times faster. The figure also shows that the industrial requirements in terms of memory, mentioned before, are fully met. In fact, the A-complexity (11) is very satisfactory for all cases, namely $c_A \approx 1.45$.

For a comparison, the last column in the figure shows the unacceptably high complexity values of RAMG05's forerunner, AMG1R5. As already mentioned in Section 3.1, AMG1R5 typically performs quite well in the case of 2D problems. In complex 3D cases as considered here, however, the results clearly demonstrate one of the advantages of the different coarsening and interpolation approaches used in RAMG05 (Fig. 7). (For more information on the differences in the two codes we refer to [51].)

5. AMG based on mere F-relaxation

In this section, we consider a very different approach [28,51] which can be used to *force* the right-hand side of (4) to become small. For a description, we assume vectors and matrices to be re-ordered so that, w.r.t a given C/F-splitting, the set of equations (1) can be written in block form

$$A_h u = \begin{pmatrix} A_{FF} & A_{FC} \\ A_{CF} & A_{CC} \end{pmatrix} \begin{pmatrix} u_F \\ u_C \end{pmatrix} = \begin{pmatrix} f_F \\ f_C \end{pmatrix} = f. \tag{13}$$

Correspondingly, the interpolation operator is re-written as $I_H^h = (I_{FC}, I_{CC})^{\mathrm{T}}$ with I_{CC} being the identity operator. Instead of $e^h = I_H^h e^H$, we simply write $e_F = I_{FC} e_C$.

5.1. The basic idea

The approach mentioned above is based on the fact that the sub-matrix A_{FF} is very well conditioned if we just select the C/F-splitting accordingly. For instance, for all problems we have in mind here, we can easily force A_{FF} to be *strongly diagonally dominant*,

$$a_{ii} - \sum_{j \in F, j \neq i} |a_{ij}| \geqslant \delta a_{ii} \quad (i \in F) \tag{14}$$

with some fixed $\delta > 0$. Assuming this to hold in the following, we can efficiently approximate the solution of the F-equations (with frozen e_C),

$$A_{FF} e_F + A_{FC} e_C = 0 \tag{15}$$

for instance, by relaxation (in the following called *F-relaxation*). Using this as the basis for both the definition of smoothing *and* interpolation, we can force the right-hand side of (4) to become as small as we wish.

To be more specific, given any e_C, interpolation is defined by applying μF-relaxation steps to approximately solve (15). In order to keep the resulting operator as "local" as possible, we only consider Jacobi-relaxation (below, we refer to this as *Jacobi-interpolation*). That is, we iteratively define a sequence of operators,

$$I_{FC}^{(\mu)} = P_{FF} I_{FC}^{(\mu-1)} - D_{FF}^{-1} A_{FC} \quad \text{where } P_{FF} = I_{FF} - D_{FF}^{-1} A_{FF}, \tag{16}$$

starting with some reasonable first-guess interpolation operator, $I_{FC}^{(0)}$. Because of (14), we have rapid convergence $(I_H^h)^{(\mu)} e_C \to \hat{e}\,(\mu \to \infty)$ at a rate which depends only on δ. Here $\hat{e} := (\hat{e}_F, e_C)^{\mathrm{T}}$ where $\hat{e}_F := -A_{FF}^{-1} A_{FC} e_C$ denotes the solution of (15).

Similarly, we also use F-relaxation for smoothing (referred to as *F-smoothing* below). That is, we define one smoothing step by $u \to \bar{u}$ where

$$Q_{FF}\bar{u}_F + (A_{FF} - Q_{FF})u_F + A_{FC}u_C = f_F, \quad \bar{u}_C = u_C. \tag{17}$$

In contrast to the interpolation, we here normally use Gauss–Seidel relaxation, i.e., Q_{FF} is the lower triangular part of A_{FF} (including the diagonal). The corresponding smoothing operator is easily seen to be

$$S_h^\nu e = \begin{pmatrix} S_{FF}^\nu(e_F - \hat{e}_F) + \hat{e}_F \\ e_C \end{pmatrix} \quad \text{where } S_{FF} = I_{FF} - Q_{FF}^{-1} A_{FF}. \tag{18}$$

As with the interpolation, for any given $e = (e_F, e_C)^{\mathrm{T}}$, we have rapid convergence $S_h^\nu e \to \hat{e}\ (\nu \to \infty)$.

5.2. Two-level convergence

For various classes of matrices A one can show that F-smoothing and Jacobi-interpolation can be used to obtain *matrix-independent* two-level convergence if the first-guess interpolation, $I_{FC}^{(0)}$, is selected properly. Moreover, two-level convergence becomes arbitrarily fast if ν, μ are chosen sufficiently large. As an example, we again consider the class of M-matrices (cf. Theorem 1).

Theorem 2. *Let A be a symmetric, weakly diagonally M-matrix. Define the interpolation by applying $\mu \geqslant 0$ Jacobi F-relaxation steps, using an interpolation as defined in Theorem* 1 *as the first-guess* (*with fixed* $0 < \tau \leqslant 1$). *Then, if $\nu \geqslant 1$ Gauss–Seidel F-relaxation steps are used for* (*pre-*) *smoothing, the following two-level convergence estimate holds,*

$$\|KS^\nu\|_A \leqslant \|S_{FF}\|_{A_{FF}}^\nu + \tilde{\tau}\|P_{FF}\|_{A_{FF}}^\mu,$$

where $\|S_{FF}\|_{A_{FF}} < 1$ and $\|P_{FF}\|_{A_{FF}} < 1$, and both norms as well as $\tilde{\tau}$ depend only on τ but not on the matrix A.

In this theorem, we have used the interpolation from Theorem 1 as a first guess. In particular, we assume the C/F-splitting to satisfy (9) which can easily be seen to ensure strong diagonal dominance (14) with $\delta = \tau$. Although one may think of various other ways to define the first-guess interpolation, we want to point out that a proper selection of the first-guess interpolation is important for obtaining matrix-independent two-level convergence (it is, for instance, not sufficient to simply select $I_{FC}^{(0)} = 0$). Generally, the first-guess interpolation has to be such that the Galerkin operator which corresponds to it, $A_H^{(0)}$, is spectrally equivalent to the Schur complement, $A_{CC} - A_{CF}A_{FF}^{-1}A_{FC}$, w.r.t. all matrices in the class under consideration. For more details and generalizations of the above theorem as well as the proofs, see [51].

Note that the AMG approach discussed here is not really in the spirit of standard multigrid since smoothing in the usual sense is not exploited. In fact, the role of F-smoothing is merely to *force* $S^\nu e \approx \hat{e}$ rather than to smooth the error of the full system. This, together with Jacobi-interpolation, is a "brute force" approach to make $\|S^\nu e - I_H^h e_C\|_A$ small for all $e = (e_F, e_C)^{\mathrm{T}}$.

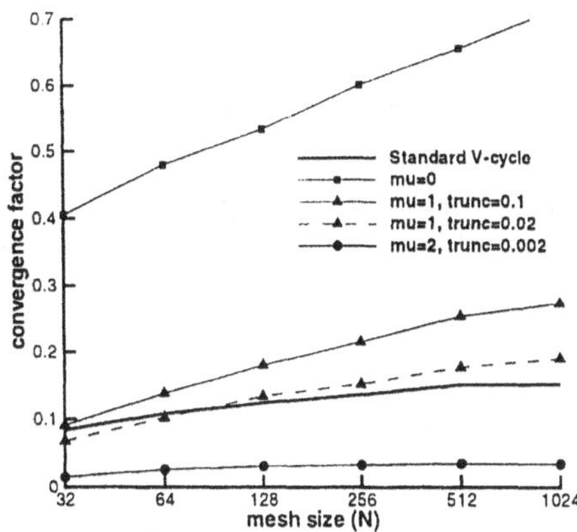

Fig. 8. Convergence factors of AMG based on F-relaxation.

5.3. Practical remarks

The mere fact that AMG can be *forced* to converge as fast as we wish, is only of little relevance in practice. Each F-relaxation step applied to the interpolation increases its "radius" by one additional layer of couplings, causing increased fill-in in the Galerkin operator. The resulting gain in convergence speed is, generally, more than eaten up for by a corresponding increase of matrix complexities towards coarser levels. Consequently, the main problem is the tradeoff between convergence and numerical work (which is directly related to the memory requirements). Note that this is, in a sense, just opposite to geometric multigrid where the numerical work per cycle is known and controllable but the convergence may not be satisfactory.

For a practical realization of Jacobi-interpolation, several things are important to observe. First, most of the new entries introduced by each additional relaxation step will be relatively small and can be truncated (before computing the Galerkin operator) without sacrificing convergence seriously (cf. also Section 3.1). Second, it is usually not necessary to perform F-relaxation with the complete matrix A_{FF}. Instead, one may well ignore all those entries of A_{FF} which are relatively small (and add them to the diagonal, say, in order to preserve the row sums of interpolation). Finally, we want to remark that, although Theorem 2 states fast convergence only if μ is sufficiently large, in practice, $\mu > 2$ is hardly ever required (at least if δ is not too small).

Fig. 8 shows some V-cycle convergence factors as a function of the mesh size for model equation

$$-((1+\sin(x+y))u_x)_x - (e^{x+y}u_y)_y = f(x,y), \tag{19}$$

discretized on the unit square with uniform mesh size $h = 1/N$. We first observe the rapid h-independent convergence of the "standard" AMG V-cycle (corresponding to the approach outlined in Section 3, using one full Gauss–Seidel relaxation step both for pre- and post-smoothing). Convergence drastically drops, and becomes strongly h-dependent, if we just replace each full smoothing step by two F-smoothing steps and leave interpolation unchanged (case $\mu = 0$). This has to be expected since the definition of interpolation in classical AMG is based on the assumption that the error after relaxation is *algebraically smooth* (cf. Section 3.1). This is, clearly, not true if only *partial* relaxation, such as F-relaxation, is performed. However, if we use just one Jacobi F-relaxation step to improve interpolation ($\mu = 1$), convergence becomes comparable to that of the standard cycle. Results are shown using two different truncation parameters, 0.1 and 0.02, respectively. Finally, the case $\mu = 2$ (and four partial relaxation steps for smoothing rather than two) gives a convergence which is about twice as fast as that of the standard cycle.

We note that, if computational time and memory requirement is taken into account in this example, the standard V-cycle is more efficient than the others. In particular, the cycle employing $\mu = 2$ is substantially inferior, mainly due to the considerably higher setup cost. This seems typical for applications for which algebraically smooth errors, in the sense of Section 3.1, can be characterized sufficiently well. The heuristic reason is that then, using *full* smoothing steps, relatively simple interpolations of the type outlined in Section 3.1 are usually sufficient to approximate algebraically smooth errors and obtain fast convergence. This is no longer true if mere F-smoothing is employed and, generally, additional effort needs to be invested to "improve" interpolation by F-relaxation in order to cope with all those error components which are not affected by mere F-smoothing. (In particular, note that all error components of the form $\hat{e}$ are not reduced at all by F-smoothing.)

In general, however, when the characterization of algebraically smooth errors is less straightforward, the use of F-relaxation provides a means to enhance convergence. Further numerical experiments employing F-smoothing and Jacobi-interpolation can be found in [28,51]. F-relaxation is a special case of a "compatible relaxation" which, in a more general context, is considered in [11].

6. Aggregation-type AMG

In the previous sections, we have considered increasingly complex interpolation approaches. In this section, we go back and consider the most simple case that each F-variable interpolates from *exactly* one C-variable only. We have already pointed out in Section 3.2 that the use of such "one-sided" interpolations is not recommendable. In fact, one important goal of the additional objective introduced in Section 3.2 was just *to avoid* such extreme interpolations. On the other hand, the resulting method is so easy to implement that it, nevertheless, has drawn some attention. We will outline the fundamental problems with this approach in Section 6.2 and summarize three possibilities of improvement in Sections 6.3–6.5. Since we just want to highlight the main ideas, we restrict our motivations to very simple but characteristic (Poisson-like) problems.

6.1. The basic approach

Let us consider C/F-splittings and interpolations (6) where, for each $i \in F$, $w_{ik} = 1$ for just one particular $k \in C$ and zero otherwise. Consequently, the total number of variables can be subdivided into "aggregates" I_k ($k \in C$) where I_k contains (apart from k itself) all indices i corresponding to F-variables which interpolate from variable k (see Fig. 9).

With this notation, the computation of the Galerkin operator becomes very simple. One easily sees that

$$I_h^H A_h I_H^h = (a_{kl}^H) \quad \text{where } a_{kl}^H = \sum_{i \in I_k} \sum_{j \in I_l} a_{ij}^h \quad (k, l \in C), \tag{20}$$

that is, the coefficient a_{kl}^H is just the sum of all cross-couplings between I_k and I_l. Obviously, regarding the coefficients a_{kl}^H, the particular role of the variables k and l (as being C-variables) is

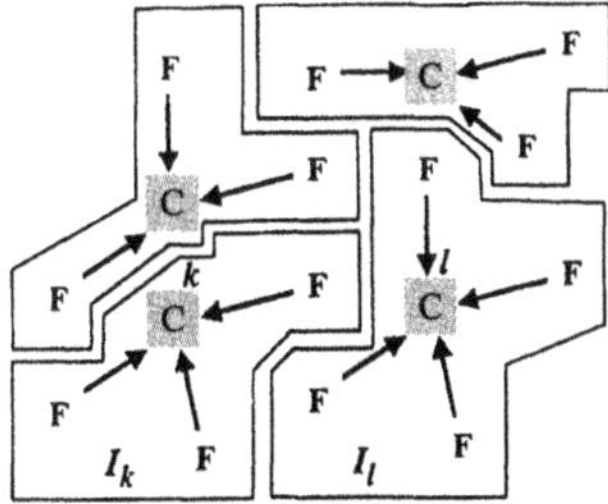

Fig. 9. Subdivision of fine-level variables into aggregates. The arrows indicate which C-variable an F-variable interpolates from.

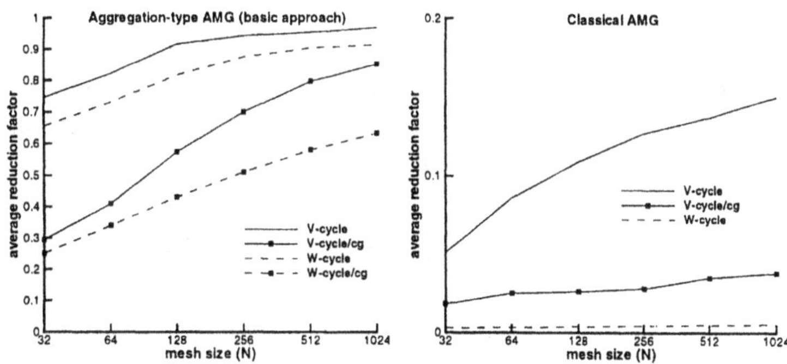

Fig. 10. Convergence of (a) aggregation-type AMG, (b) classical AMG.

not distinguished from the other variables; the Galerkin operator merely depends *on the definition of the aggregates*. Consequently, we might as well associate each aggregate I_k with some "new" coarse-level variable which has no direct relation to the C-variable k. The above interpolation then is nothing else than *piecewise constant interpolation* from these new coarse-level variables to the associated aggregates.

Originally, such *aggregation-type* AMG approaches [52,53,9] have been developed the other way around: Coarsening is defined by building aggregates (rather than constructing C/F-splittings), a new coarse-level variable is associated with each aggregate and interpolation I_H^h is defined to be piecewise constant. (That is, the set of coarse-level variables is generally *not* considered as a subset of the fine-level ones.) Clearly, for a given subdivision into aggregates to be reasonable, all variables in the same aggregate should strongly depend on each other. Otherwise, piecewise constant interpolation makes no real sense.

As expected, an immediate implementation of this simple coarsening and interpolation approach leads to very inefficient solvers, even if used only as a pre-conditioner. Fig. 10a shows the typical convergence of both the V- and the W-cycle, used as stand-alone and as pre-conditioner, in solving the model equation (19). Convergence is indeed very slow and exhibits a *strong h-dependency*. For a comparison, the much better performance of classical AMG is depicted in Fig. 10b.

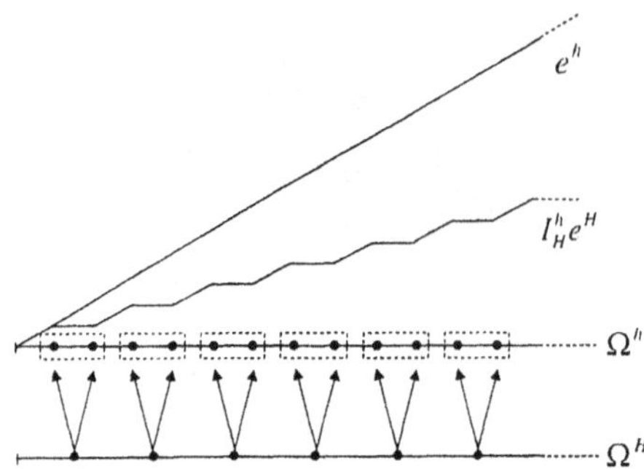

Fig. 11. Optimal approximation $I_H^h e^H$ of e^h w.r.t. the energy norm.

6.2. The reason for slow convergence

The main reason for this unsatisfactory convergence is that piecewise constant interpolation is not able to approximate the *values* of smooth error if approximation is based on the *energy norm* (cf. (4)). In fact, the approximation order becomes zero.

To illustrate this, let us consider the most simple case that A_h is derived from discretizing $-u''$ on the unit interval with meshsize h, i.e., the rows of A_h correspond to the difference stencil

$$\frac{1}{h^2}[-1 \quad 2 \quad -1]_h$$

with Dirichlet boundary conditions. Let e^h be an error satisfying the homogeneous boundary conditions. According to the variational principle, the corresponding two-level correction, $I_H^h e^H$, is optimal in the sense that it minimizes $\|e^h - I_H^h e^H\|_{A_h}$ w.r.t. all possible corrections in $\mathcal{R}(I_H^h)$. (At this point, the concrete choice of I_H^h is not relevant.) This implies that $I_H^h e^H$ minimizes

$$\|v^h\|_{A_h}^2 = (A_h v^h \cdot v^h) = \frac{1}{2h^2} \sum_{i,j}{}' (v_i^h - v_j^h)^2 + \sum_i s_i (v_i^h)^2, \tag{21}$$

where $v^h = e^h - I_H^h e^H$ and $s_i = \sum_j a_{ij}^h$. (The prime indicates that summation is only over neighboring variables i and j.) This, in turn, means that, away from the boundary (where we have $s_i \equiv 0$), the Euclidian norm of the *slope* of v^h is minimal. At the boundary itself we have $s_i \neq 0$, and v^h equals zero.

Assuming now the aggregates to be built by joining the pairs of neighboring variables, the result of this minimization is illustrated in Fig. 11 (see also [9,10]). We here consider a smooth error e^h in the neighborhood of the left boundary of the unit interval. On each aggregate, interpolation is constant and the slope of $I_H^h e^H$ necessarily vanishes. On the remaining intervals, the Euclidian norm of the slope of v^h becomes minimal if the slope of $I_H^h e^H$ equals that of e^h. Consequently, $I_H^h e^H$ has, on the average, *only half the slope of* e^h (independent of h). That is, the resulting approximation is off by a factor of approximately 0.5 if compared to the best approximation in the Euclidian sense. (Note that subsequent smoothing smooths out the "wiggles", but does not improve the quality of the correction.) Accordingly, the Galerkin operator, which can easily be computed, turns out to be too large by a factor of two compared to the "natural" $2h$-discretization of $-u''$.

If the same strategy is now used recursively to introduce coarser and coarser levels, the above arguments carry over to each of the intermediate levels and, in particular, each coarser-grid Galerkin operator is off by a factor of 2 compared to the previous one. A simple recursive argument shows that errors are accumulated from grid to grid and the asymptotic V-cycle convergence factor cannot be expected to be better than $1 - 2^{-m}$, where m denotes the number of coarser levels. That is, *the V-cycle convergence is strongly h-dependent.*

6.3. Improvement by re-scaling the Galerkin operators

The fact that piecewise constant interpolation produces badly scaled AMG components, was the basis for an improvement introduced in [9]. In that paper, it is demonstrated that convergence can be substantially improved by just multiplying corrections $I_H^h e^H$ by some suitable factor $\alpha > 1$ ("over-correction"). This is equivalent to re-scaling the Galerkin operator by $1/\alpha$

$$I_h^H A_h I_H^h \to \frac{1}{\alpha} I_h^H A_h I_H^h$$

and leaving everything else unchanged.

In case of the simple model equation $-u''$ considered in the previous section, $\alpha = 2$ would be the optimal choice. However, the main arguments carry over to the Poisson equation in 2D and 3D, assuming a uniform grid and the aggregates to be built by 2×2 and $2 \times 2 \times 2$ blocks of neighboring variables, respectively. In case of more general problems and/or different grids, the optimal weight is no longer $\alpha = 2$. Nevertheless, it has been demonstrated in [9] that a slightly reduced value of $\alpha = 1.8$ (in order to reduce the risk of "overshooting") yields substantially improved V-cycle convergence for various types of problems, *if the cycle is used as a pre-conditioner* and if the number of coarser levels is kept fixed (in [9] four levels are always used). Smoothing is done by symmetric Gauss–Seidel relaxation sweeps.

A comparison of Figs. 10a and 12a shows the convergence improvement if re-scaling by $\alpha = 1.8$ is applied to the model equation (19). (In contrast to [9], we here have not restricted the number of coarser levels.) Fig. 12a shows that there is indeed a risk of "overshooting": For larger meshes, the V-cycle starts to diverge. (Note that the above re-scaling destroys the validity of the variational principle and the iteration process may well diverge.) Using the V-cycle as a pre-conditioner, eliminates the problem.

We want to point out that the above comparison shows only the tendency of improvements due to re-scaling, the concrete gain depends on how the aggregates are chosen precisely (which is not optimized here and can certainly be improved to some extent). In any case, the gain in convergence, robustness and efficiency of this (very simple and easily programmable) approach are somewhat limited, one reason being that a good value of α depends on various aspects such as the concrete problem, the type of mesh and, in particular, the type and size of the aggregates. For instance, if the aggregates are composed of three neighboring variables (rather than two) in each spatial direction, the same arguments as in the previous section show that the best weight would be $\alpha \approx 3$ in case of Poisson's equation. If the size of the aggregates strongly varies over the domain, it becomes difficult to define a good value for α.

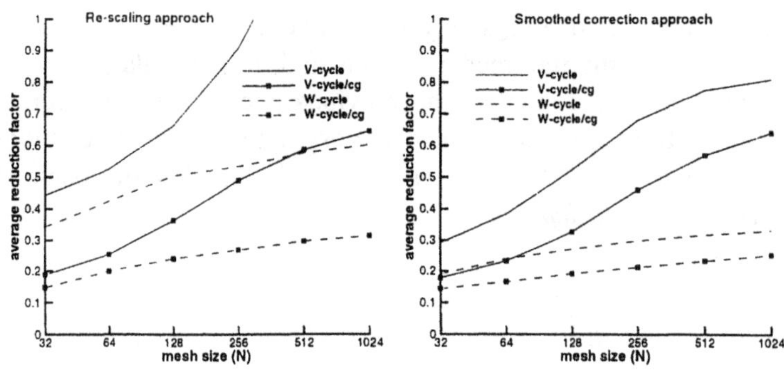

Fig. 12. (a) Re-scaling approach ($\alpha = 1.8$), (b) smoothed correction approach.

6.4. Improvement by smoothing corrections

Rather than explicitly prescribing a scaling factor α as before, a reasonable scaling can also be performed automatically. The idea is to modify the coarse-level correction step (2) by replacing the true (piece-wise constant) correction $e^h = I_H^h e^H$ by some approximation, e_0^h, and then compute u_{new}^h by

$$u_{\text{new}}^h = u_{\text{old}}^h + \alpha e_0^h \quad \text{with } \alpha = \frac{(f^h - A_h u_{\text{old}}^h, e_0^h)}{(A_h e_0^h, e_0^h)}, \tag{22}$$

instead of (2). Note that α is defined so that the energy norm of the error of u_{new}^h becomes minimal.

Clearly, for this minimization to be meaningful, the selection of e_0^h is crucial. Most importantly, e_0^h should be some sufficiently *smooth* approximation to e^h. (The choice $e_0^h = e^h$ would not give any gain: The variational principle would just imply $\alpha = 1$.) One possible selection is

$$e_0^h = S_h^\nu e^h, \tag{23}$$

which requires the application of ν smoothing steps to the *homogeneous* equations (starting with e^h). Note that, loosely speaking, this process will leave the "smooth part" of e^h essentially unchanged; only its "high-frequency part" will be reduced. Consequently, the regular smoothing steps, applied to u_{new}^h after the coarse-grid correction, will effectively correct this.

The effect of this process of *smoothing corrections*, is demonstrated in Fig. 12b (using $\nu = 2$). Apart from the fact that, compared to the re-scaling approach (see Fig. 12a), convergence is slightly better here, there is no risk of "overshooting" as before since the process "controls itself". On the other hand, the additional smoothing steps increase the overall computing time.

Although smoothing of corrections is a simple means to automatically correct wrong scalings to some extent, its possibilities are limited. In any case, the resulting overall performance is generally worse than that of classical AMG.

6.5. Improvement by smoothing the interpolation

A more sophisticated (but also more costly) way to accelerate the basic aggregation-type AMG approach is developed and analyzed in [52–54]. Here, piecewise constant interpolation is only considered as a first-guess interpolation which is improved by some *smoothing process* ("smoothed

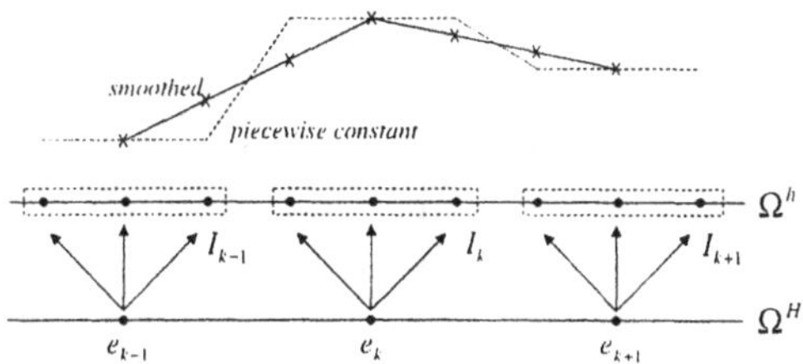

Fig. 13. Piecewise constant versus smoothed interpolation.

aggregation") before the Galerkin operator is computed. In [52,53], this smoothing is proposed to be done by applying one ω-Jacobi-relaxation step.

To be more specific, denote the operator corresponding to piecewise constant interpolation by $\tilde{I}_H^h$. Then the interpolation actually used is defined by

$$I_H^h = (I_h - \omega D_h^{-1} A_h^f)\tilde{I}_H^h,$$

where $D_h = \mathrm{diag}(A_h^f)$ and A_h^f is derived from the original matrix A_h by adding all weak connections to the diagonal ("filtered matrix"). That is, given some coarse-level vector e^H, $e^h = I_H^h e^H$ is defined by applying one ω-Jacobi relaxation step to the homogeneous equations $A_h^f v^h = 0$, starting with the piecewise constant vector $\tilde{I}_H^h e^H$. (Note that this process will increase the "radius" of interpolation and, hence, destroy the simplicity of the basic approach. Moreover, interpolation will generally *not* be of the special form (5) any more. Note also that here Jacobi relaxation serves a quite different purpose than Jacobi-F-relaxation as considered in Section 5. In particular, Jacobi-relaxation is here used as a smoother, applied to the full system of equations, which requires the use of an under-relaxation parameter, ω.)

To illustrate this process, we again consider the 1D case of $-u''$ and assume the aggregates to consist of three neighboring variables (corresponding to the typical size of aggregates used in [52,53] in each spatial direction). Note first that, since all connections are strong, we have $A_h^f = A_h$. Fig. 13 depicts both the piecewise constant interpolation (dashed line) and the smoothed interpolation obtained after the application of one Jacobi-step with $\omega = \frac{2}{3}$ (solid line). Obviously, the smoothed interpolation just corresponds to *linear* interpolation if the coarse-level variables are regarded as the fine-level analogs of those variables sitting in the center of the aggregates. Linear interpolation, however, does not exhibit a scaling problem as described in Section 6.2 for piecewise constant interpolation.

Of course, in more general situations, relaxation of piecewise constant interpolation will not give exact linear interpolation any more and a good choice of ω depends on the situation. Nevertheless, even if $\omega = \frac{2}{3}$ is kept fixed, smoothed interpolation will typically be much better than the piecewise constant one. (Actually, the real advantage of smoothed compared to piecewise constant interpolation is that errors, obtained after interpolation from the coarser level, have a much lower "energy"; see also Section 7.) This is demonstrated in [53] by means of various examples using a mixture of Gauss–Seidel and SOR sweeps for error smoothing. The tendency is to compose aggregates of three neighboring variables in each spatial direction. Note that a good value for ω depends not only on the problem and the underlying mesh, but also on the size of the aggregates. If, instead, only *two* neighbors would be aggregated in each spatial direction, one easily confirms that $\omega \approx 0.5$ should be chosen.

In general, classical AMG and AMG based on smoothed aggregation perform comparably if applied to relatively smooth (Poisson-like) problems. A certain advantage of aggregation-type AMG is that it, often, requires less memory than classical AMG (due to its very fast coarsening which causes a particularly low operator complexity c_A). On the other hand, this is payed for by a reduced robustness: The aggregation-type code seems to *require* acceleration by conjugate-gradient to maintain an acceptable efficiency and robustness in more complex situations. Since classical AMG puts more effort into the construction of interpolation and performs a slower coarsening, its performance generally depends on aspects such as strong discontinuities only to a lesser extent (for some examples, see [51]).

7. Further developments and conclusions

Algebraic multigrid provides a general approach to developing robust and efficient methods for solving large classes of matrix equations such as those typically arising in the numerical solution of elliptic PDEs, on structured as well as unstructured grids, in 2D and 3D. The construction of suitable interpolation operators (including the coarsening process) is crucial for obtaining fast and (nearly) h-independent convergence. Generally, the more effort is put into this construction, the faster the convergence can be, but, unfortunately, the required numerical work may increase even faster. That is, the main problem in designing efficient AMG algorithms is the tradeoff between convergence and numerical work, and keeping the balance between the two is the ultimate goal of any practical algorithm.

We have summarized and discussed various possibilities to realize concrete AMG algorithms. For most applications of the type discussed in this paper, methods based on the classical AMG approach still belong to the most efficient ones. An extensive list of experiments, based on the original code AMG1R5, can be found in [18]. Robustness and efficiency can substantially be improved, in particular in case of complex 3D meshes, by employing modifications as mentioned in Sections 3.1 and 5 (for more details, see [51]). AMG methods based on smoothed aggregation (see Section 6.5) are an efficient alternative to classical AMG, at least if employed as a pre-conditioner rather than stand-alone. Further developments and applications which are close to the classical AMG ideas are, for example, contained in [16,23,24,26,32,33,40,62,63]. Related approaches, but with a focus on different coarsening and interpolation strategies, are, for example, found in [22,27]. Applications of the aggregation-type approach in computational fluid dynamics are found in [30,39,46,60].

However, there are still many applications for which algebraically defined interpolation, and hence the resulting AMG performance, are not yet satisfactory. For instance, one of the major current research activities in AMG aims at its generalization to efficiently treat *systems* of PDEs such as linear elasticity problems. Although AMG has successfully been applied to various cases (see, e.g., [9,14,31,48,53]), its development has not yet reached a state where a particular approach is well settled. However, even for scalar applications, there are still questions about best ways to define coarsening and interpolation, for instance, if the given matrix is s.p.d., contains relatively large positive off-diagonal entries, and is far from being weakly diagonally dominant. In such cases, the performance of classical AMG may be only suboptimal.

It is often possible to avoid such situations by simplifying the given matrix before applying AMG [41]. One can also imagine situations where it would be advantageous (and easy) to provide AMG with some additional information on the problem at hand. For instance, information on the geometry (in terms of point locations) or more concrete descriptions on what an "algebraically smooth" error looks like (e.g. in the form of some user-provided "test-vector(s)"). This additional information can be used to fit AMG's interpolation in order to approximate certain types of error components particularly well. Straightforward possibilities have already been pointed out in [48].

In the following, we briefly summarize a few more recent approaches to define interpolation which aim at increasing the robustness in cases such as those mentioned above.

A new way to construct interpolation (AMGe, [14]) starts from the fact that an algebraically smooth error is nothing else but an error which is slow-to-converge w.r.t. the relaxation process. Hence, an algebraically smooth error, generally, corresponds to the eigenvectors of A belonging to the smallest eigen-values. Instead of defining interpolation by directly approximating (7), the goal in [14] is to define interpolation so that the smaller the associated eigenvalue is the better the eigenvectors are interpolated. To satisfy this by explicitly computing eigenvectors is, of course, much too expensive. However, in the case of finite element methods – assuming the element stiffness matrices to be known – one can derive measures (related to measures used in classical multigrid theory) whose minimization allows the determination of *local* representations of algebraically smooth error components in the above sense. The added robustness has been demonstrated in [14] by means of certain model applications. However, the approach is still in its infancy. In particular, significant development work still has to be invested to link the processes of coarsening and interpolation definition in order to obtain an optimal algorithm. In any case, it is an interesting new approach which has the potential of leading to more generally applicable AMG approaches.

Other algebraic approaches, designed for the solution of equations derived from finite-element discretizations, have been considered in [31,58]. Both approaches are aggregation based and the coarse space basis functions are defined so that their energy is minimized in some sense. (In the finite-element context it is natural to define interpolation implicitly by constructing the coarse space basis functions.) This does not require the element stiffness matrices to be known, but leads to a *global* (constraint) minimization problem the exact solution of which would be very expensive. However, iterative solution processes are proposed in both papers to obtain approximate solutions, indicating that the extra work (invested in the setup phase) is acceptable. While Wan et al. [58] concentrate on scalar applications, an extension to systems of PDEs (from linear elasticity) is one major aspect in [31]. Special attention is paid to the correct treatment of zero-energy modes (e.g. rigid-body modes in case of linear elasticity): such modes should be contained in the span of the coarse space basis functions, at least away from Dirichlet boundaries. (Note that, for typical scalar problems, this corresponds to the requirement that constants should be interpolated exactly away from Dirichlet boundaries, cf. (8).) It is interesting that the approach in [31] can be regarded as an extension of the earlier work [53] on smoothed aggregation: if only one iteration step is performed to approximately solve the energy minimization problem, the resulting method coincides with the smoothed aggregation approach. In contrast to the latter, however, further iterations will *not* increase the support of the basis functions (i.e., the radius of interpolation). Some test examples in [31] indicate the advantages of this new interpolation in terms of convergence speed. Unfortunately, however, this benefit is essentially offset by the expense of the minimization steps.

There are various other papers with focus on the development of multigrid methods to solve finite-element problems on unstructured grids. Although some of them are also based on algorithmical components which are, more or less, algebraically defined, most of them are not meant to be algebraic multigrid solvers in the sense as considered in this paper. We therefore do not want to discuss such approaches here further but rather refer, for example, to [15] and the references given therein.

In the approach of [56], A is not assumed to be asymmetric, and interpolation and restriction are constructed separately. Interpolation, for instance, is constructed so that a smooth error, $S_h e^h$, is interpolated particularly well w.r.t. the *Euclidian* norm, $\|\cdot\|_2$. More precisely, the attempt is to make

$$\|S_h e^h - I_H^h e^H\|_2,$$

where e^H denotes the straight injection of $S_h e^h$ to the coarse level, as small as possible (cf. (4)). In [56], this leads to certain local minimizations which are used to find, for each variable, pairs of neighboring variables which would allow a good interpolation in the above sense, and, at the same time, compute the corresponding weights (of both the interpolation and the restriction). Based on this information, a C/F-splitting is constructed which allows each F-variable to interpolate from one of the pairs found before. A heuristic algorithm is used to minimize the total number of C-variables.

In this context, we want to point out that, although classical AMG has been developed in the variational framework, it has successfully been applied to a large number of non-symmetric problems without any modification. This can be explained heuristically but no theoretical justification is available at this time. In the context of smoothed aggregation-based AMG, a theoretical analysis can be found in [25].

An important aspect which has not been addressed in this paper is the parallelization of AMG. An efficient parallelization of classical AMG is rather complicated and requires certain algorithmical modifications in order to limit the communication cost without sacrificing convergence significantly. Most parallelization approaches investigated up to now either refer to simple aggregation-based variants (e.g. [46]) or use straightforward domain decomposition techniques (such as Schwarz' alternating method) for parallelization. A parallelization strategy which stays very close to the classical AMG approach is presented in [29]. Results for complex 3D problems demonstrate that this approach scales reasonably well on distributed memory computers as long as the number of unknowns per processor is not too small. The method discussed in [56] is also available in parallel. There are several further ongoing parallelization activities, for instance, at the University of Bonn and the National Laboratories LLNL [17] and LANL, but no results have been published by now.

It is beyond the scope of this paper to also discuss the variety of hierarchical algebraic approaches which are not really related to the multigrid idea in the sense that these approaches are not based on the fundamental multigrid principles, smoothing and coarse-level correction. There is actually a rapid and very interesting ongoing development of such approaches. For completeness, however, we include some selected references. Various approaches based on approximate block Gauss elimination ("Schur-complement" methods) are found in [2–5,19,36–38,42,57]. Multi-level structures have also been introduced into ILU-type pre-conditioners, for example, in [49]. Very recently, some hybrid methods have been developed which use ideas from ILU and from multigrid [6–8,43–45]. For a further discussion, see also [55].

Summarizing, the development of hierarchically operating algebraic methods to efficiently tackle the solution of large sparse, unstructured systems of equations, currently belongs to one of the most active fields of research in numerical analysis. Many different methods have been investigated but, by now, none of them is really able to efficiently deal with *all* practically relevant problems. All methods seem to have their range of applicability but all of them may fail to be efficient in certain other applications. Hence, the development in this exciting area of research has to be expected to continue for the next years.

References

[1] R.E. Alcouffe, A. Brandt, J.E. Dendy, J.W. Painter, The multi-grid method for the diffusion equation with strongly discontinuous coefficients, SIAM J. Sci. Statist. Comput. 2 (1981) 430–454.

[2] O. Axelsson, The method of diagonal compensation of reduced matrix entries and multilevel iteration, J. Comput. Appl. Math. 38 (1991) 31–43.

[3] O. Axelsson, M. Neytcheva, Algebraic multilevel iteration method for Stieltjes matrices, Numer. Linear Algebra Appl. 1 (3) (1994) 213–236.

[4] O. Axelsson, P.S. Vassilevski, Algebraic multilevel preconditioning methods I, Numer. Math. 56 (1989) 157–177.

[5] O. Axelsson, P.S. Vassilevski, Algebraic multilevel preconditioning methods II, SIAM Numer. Anal. 27 (1990) 1569–1590.

[6] R.E. Bank, R.K. Smith, The incomplete factorization multigraph algorithm, SIAM J. Sci. Comput., to appear.

[7] R.E. Bank, R.K. Smith, The hierarchical basis multigraph algorithm, SIAM J. Sci. Comput., submitted.

[8] R.E. Bank, Ch. Wagner, Multilevel ILU decomposition, Numer. Math. 82 (1999) 543–576.

[9] D. Braess, Towards algebraic multigrid for elliptic problems of second order, Computing 55 (1995) 379–393.

[10] A. Brandt, Algebraic multigrid theory: the symmetric case, Appl. Math. Comput. 19 (1986) 23–56.

[11] A. Brandt, General highly accurate algebraic coarsening schemes, Proceedings of the Ninth Copper Mountain Conference on Multigrid Methods, Copper Mountain, April 11–16, 1999.

[12] A. Brandt, S.F. McCormick, J. Ruge, Algebraic multigrid (AMG) for automatic multigrid solution with application to geodetic computations, Institute for Computational Studies, POB 1852, Fort Collins, Colorado, 1982.

[13] A. Brandt, S.F. McCormick, J. Ruge, Algebraic multigrid (AMG) for sparse matrix equations, in: D.J. Evans (Ed.), Sparsity and its Applications, Cambridge University Press, Cambridge, 1984, pp. 257–284.

[14] M. Brezina, A.J. Cleary, R.D. Falgout, V.E. Henson, J.E. Jones, T.A. Manteuffel, S.F. McCormick, J.W. Ruge, Algebraic multigrid based on element interpolation (AMGe), LLNL Technical Report UCRL-JC-131752, SIAM J. Sci. Comput., to appear.

[15] T. Chan, J. Xu, L. Zikatanov, An agglomeration multigrid method for unstructured grids, Proceedings of the 10th International Conference on Domain Decomposition Methods, 1988.

[16] Q. Chang, Y.S. Wong, H. Fu, On the algebraic multigrid method, J. Comput. Phys. 125 (1996) 279–292.

[17] A.J. Cleary, R.D. Falgout, V.E. Henson, J.E. Jones, Coarse-grid selection for parallel algebraic multigrid, Proceedings of the "Fifth International Symposium on Solving Irregularly Structured Problems in Parallel", Lecture Notes in Computer Science, vol. 1457, Springer, New York, 1998, pp. 104–115.

[18] A.J. Cleary, R.D. Falgout, V.E. Henson, J.E. Jones, T.A. Manteuffel, S.F. McCormick, G.N. Miranda, J.W. Ruge, Robustness and scalability of algebraic multigrid, SIAM J. Sci. Comput., special issue on the "Fifth Copper Mountain Conference on Iterative Methods", 1998.

[19] W. Dahmen, L. Elsner, Algebraic Multigrid Methods and the Schur Complement, Notes on Numerical Fluid Mechanics, 23, Vieweg Verlag, Braunschweig, 1988.

[20] J.E. Dendy Jr., Black box multigrid, J. Comput. Phys. 48 (1982) 366–386.

[21] J.E. Dendy, S.F. McCormick, J. Ruge, T. Russell, S. Schaffer, Multigrid methods for three-dimensional petroleum reservoir simulation, Proceedings of 10th SPE Symposium on Reservoir Simulation, February 6–8, 1989.

[22] J. Fuhrmann, A modular algebraic multilevel method, Technical Report Preprint 203, Weierstrass-Institut für Angewandte Analysis und Stochastik, Berlin, 1995.

[23] T. Grauschopf, M. Griebel, H. Regler, Additive multilevel-preconditioners based on bilinear interpolation, matrix dependent geometric coarsening and algebraic multigrid coarsening for second order elliptic PDEs, Appl. Numer. Math. 23 (1997) 63–96.

[24] M. Griebel, T. Neunhoeffer, H. Regler, Algebraic multigrid methods for the solution of the Navier–Stokes equations in complicated geometries, SFB-Bericht Nr. 342/01/96 A, Institut für Informatik, Technische Universität München, 1996.

[25] H. Guillard, P. Vanek, An aggregation multigrid solver for convection-diffusion problems on unstructured meshes, Center for Computational Mathematics, University of Denver, Report 130, 1998.

[26] W.Z. Huang, Convergence of algebraic multigrid methods for symmetric positive definite matrices with weak diagonal dominance, Appl. Math. Comput. 46 (1991) 145–164.

[27] F. Kickinger, Algebraic multi-grid for discrete elliptic second order problems, Institutsbericht 513, Universität Linz, Institut für Mathematik, 1997.

[28] A. Krechel, K. Stüben, Operator dependent interpolation in algebraic multigrid, Proceedings of the Fifth European Multigrid Conference, Stuttgart, October 1–4, 1996; Lecture Notes in Computational Science and Engineering, vol. 3, Springer, Berlin, 1998.

[29] A. Krechel, K. Stüben, Parallel algebraic multigrid based on subdomain blocking, Parallel Comput., to appear.

[30] R.D. Lonsdale, An algebraic multigrid solver for the Navier–Stokes equations on unstructured meshes, Int. J. Numer. Methods Heat Fluid Flow 3 (1993) 3–14.

[31] J. Mandel, M. Brezina, P. Vanek, Energy optimization of algebraic multigrid bases, UCD/CCM Report 125, 1998.

[32] S. McCormick, J. Ruge, Algebraic multigrid methods applied to problems in computational structural mechanics, in: State-of-the-Art Surveys on Computational Mechanics, ASME, New York, 1989, pp. 237–270.

[33] R. Mertens, H. De Gersem, R. Belmans, K. Hameyer, D. Lahaye, S. Vandewalle, D. Roose, An algebraic multigrid method for solving very large electromagnetic systems, IEEE Trans. Magn. 34 (1998), to appear.

[34] W.A. Mulder, A new multigrid approach to convection problems, J. Comput. Phys. 83 (1989) 303–323.

[35] N.H. Naik, J. van Rosendale, The improved robustness of multigrid elliptic solvers based on multiple semicoarsened grids, SIAM Numer. Anal. 30 (1993) 215–229.

[36] Y. Notay, An efficient algebraic multilevel preconditioner robust with respect to anisotropies, in: O. Axelsson, B. Polman (Eds.), Algebraic Multilevel Iteration Methods with Applications, Department of Mathematics, University of Nijmegen, 1996, pp. 111–228.

[37] Y. Notay, Using approximate inverses in algebraic multilevel methods, Numer. Math. 80 (1998) 397–417.

[38] Y. Notay, Optimal V-cycle algebraic multilevel preconditioning, Numer. Linear Algebra Appl., to appear.

[39] M. Raw, A coupled algebraic multigrid method for the 3D Navier–Stokes equations, Report: Advanced Scientific Computing Ltd., 554 Parkside Drive, Waterloo, Ontario N2L 5Z4, Canada.

[40] H. Regler, Anwendungen von AMG auf das Plazierungsproblem beim Layoutentwurf und auf die numerische Simulation von Strömungen, Ph.D. thesis, TU München, 1997.

[41] S. Reitzinger, Algebraic multigrid and element preconditioning I, SFB-Report 98-15, University Linz, Austria, December 1998.

[42] A.A. Reusken, Multigrid with matrix-dependent transfer operators for a singular perturbation problem, Computing 50 (3) (1993) 199–211.

[43] A.A. Reusken, A multigrid method based on incomplete Gaussian elimination, Eindhoven University of Technology, Report RANA 95-13, ISSN 0926-4507, 1995.

[44] A.A. Reusken, On the approximate cyclic reduction preconditioner, Report 144, Institut für Geometrie und Praktische Mathematik, RWTH Aachen, 1997. SIAM J. Sci. Comput., to appear.

[45] A.A. Reusken, Approximate cyclic reduction preconditioning, in: W. Hackbusch, G. Wittum (Eds.), Multigrid Methods, Vol. 5, Lecture Notes in Computational Science and Engineering, Vol. 3, Springer, Berlin, 1998, pp. 243–259.

[46] G. Robinson, Parallel computational fluid dynamics on unstructured meshes using algebraic multigrid, in: R.B. Pelz, A. Ecer, J. Häuser (Eds.), Parallel Computational Fluid Dynamics, Vol. 92, Elsevier, Amsterdam, 1993.

[47] J.W. Ruge, K. Stüben, Efficient solution of finite difference and finite element equations by algebraic multigrid (AMG), in: D.J. Paddon, H. Holstein (Eds.), Multigrid Methods for Integral and Differential Equations, The Institute of Mathematics and its Applications Conference Series, New Series vol. 3, Clarendon Press, Oxford, 1985, pp. 169–212.

[48] J.W. Ruge, K. Stüben, Algebraic multigrid (AMG), in: S.F. McCormick (Ed.), Multigrid Methods, Frontiers in Applied Mathematics, Vol. 5, SIAM, Philadelphia, 1986.

[49] Y. Saad, ILUM: a multi-elimination ILU preconditioner for general sparse matrices, SIAM J. Sci. Comput. 17 (1996) 830–847.

[50] K. Stüben, Algebraic multigrid (AMG): Experiences and comparisons, Appl. Math. Comput. 13 (1983) 419–452.

[51] K. Stüben, Algebraic multigrid (AMG): an introduction with applications, in: U. Trottenberg, C.W. Oosterlee, A. Schüller (Eds.), Multigrid, Academic Press, New York, 2000. Also GMD Report 53, March 1999.

[52] P. Vanek, J. Mandel, M. Brezina, Algebraic multigrid on unstructured meshes, University of Colorado at Denver, UCD/CCM Report No. 34, 1994.

[53] P. Vanek, J. Mandel, M. Brezina, Algebraic multigrid by smoothed aggregation for second and fourth order elliptic problems, Computing 56 (1996) 179–196.

[54] P. Vanek, M. Brezina, J. Mandel, Convergence of algebraic multigrid based on smoothed aggregation, UCD/CCM Report 126, 1998. Numer. Math., submitted.

[55] C. Wagner, Introduction to algebraic multigrid, Course notes of an algebraic multigrid course at the University of Heidelberg in the Wintersemester 1998/99, http://www.iwr.uni-heidelberg.de/ ~ Christian.Wagner, 1999.

[56] C. Wagner, On the algebraic construction of multilevel transfer operators, IWR-Report, Universität Heidelberg, Computing (2000).

[57] C. Wagner, W. Kinzelbach, G. Wittum, Schur-complement multigrid – a robust method for groundwater flow and transport problems, Numer. Math. 75 (1997) 523–545.

[58] W.L. Wan, T.F. Chan, B. Smith, An energy minimization interpolation for robust multigrid methods, Department of Mathematics, UCLA, UCLA CAM Report 98-6, 1998.

[59] T. Washio, C.W. Oosterlee, Flexible multiple semicoarsening for 3D singularly perturbed problems, SIAM J. Sci. Comput. 19 (1998) 1646–1666.

[60] R. Webster, An algebraic multigrid solver for Navier–Stokes problems in the discrete second order approximation, Int. J. Numer. Methods Fluids 22 (1996) 1103–1123.

[61] P. Wesseling, C.W. Oosterlee, Geometric multigrid with applications to computational fluid dynamics, this volume, Comput. Appl. Math. 128 (2001) 311–334.

[62] L. Zaslavsky, An adaptive algebraic multigrid for multigroup neutron diffusion reactor core calculations, Appl. Math. Comput. 53 (1993) 13–26.

[63] L. Zaslavsky, An adaptive algebraic multigrid for reactor critically calculations, SIAM J. Sci. Comput. 16 (1995) 840–847.

ELSEVIER

Journal of Computational and Applied Mathematics 128 (2001) 311–334

JOURNAL OF COMPUTATIONAL AND APPLIED MATHEMATICS

www.elsevier.nl/locate/cam

Geometric multigrid with applications to computational fluid dynamics

P. Wesseling[a,]*, C.W. Oosterlee[b]

[a] J.M. Burgers Center and Delft University of Technology, Faculty of Technical Mathematics and Informatics, Mekelweg 4, 2628 CD Delft, The Netherlands

[b] University of Cologne and GMD, Institute for Algorithms and Scientific Computing, D-53754 Sankt Augustin, Germany

Received 5 July 1999; received in revised form 29 November 1999

Abstract

An overview is given of the development of the geometric multigrid method, with emphasis on applications in computational fluid dynamics over the last ten years. Both compressible and incompressible flow problems and their corresponding multigrid solution methods are discussed. The state of the art is described with respect to methods employed in industry as well as the multigrid efficiency obtained in academic applications. © 2001 Elsevier Science B.V. All rights reserved.

MSC: 65N55; 76Mxx

Keywords: Multigrid; Nonlinear problems; Computational fluid dynamics

1. Introduction

The multigrid or multilevel approach has the unique potential of solving many kinds of mathematical problems with N unknowns with $\mathcal{O}(N)$ work. As discussed in [15], this applies to diverse areas such as integral equations, or optimization methods in various scientific disciplines. Complexity of $\mathcal{O}(N)$ has been shown theoretically for discretizations of a large class of elliptic linear partial differential equations [38,4,43–45,149].

We can distinguish between algebraic multigrid (AMG) [121] and geometric multigrid. In algebraic multigrid no information is used concerning the grid on which the governing partial differential equations are discretized. Therefore, it might be better to speak of algebraic multilevel methods. In geometric multigrid, coarse grids are constructed from the given fine grid, and coarse grid corrections

* Corresponding author.

E-mail address: p.wesseling@math.tudelft.nl (P. Wesseling).

PII: S0377-0427(00)00517-3

are computed using discrete systems constructed on the coarse grids. Constructing coarse grids from fine grids by agglomeration of fine grid cells is easy when the fine grid is structured, but not if the fine grid is unstructured. That is where algebraic multigrid becomes useful. Unfortunately, AMG is less developed than geometric multigrid for the applications considered here. Algebraic multigrid is covered by K. Stüben in another paper in this issue. The present paper is about geometric multigrid.

For remarks on early multigrid history, see [150,152]. In the early seventies, there were methods already available with low computational complexity, such as solution methods based on fast Fourier transforms, resulting in $\mathcal{O}(N \log N)$ work. But these methods are restricted to special classes of problems, such as separable partial differential equations on cubic domains. Multigrid, however, is much more robust: it is efficient for a much wider class of problems. The interest of practitioners of large-scale scientific computing in multigrid was particularly stimulated by the 1977 paper [11] of Achi Brandt, generally regarded as a landmark in the field. Two series of conferences dedicated to multigrid were set up: the European Multigrid Conferences (EMG): Cologne (1981, 1985), Bonn (1991), Amsterdam (1993), Stuttgart (1996), Ghent (1999), and in the US the Copper Mountain Conferences on multigrid, held bi-annually from 1983 until the present. Proceedings of the European meetings have appeared in [46,48,49,54,50] and of the Copper Mountain Conferences in special issues of journals: Applied Numerical Mathematics (Vol. 13, 1983; Vol. 19, 1986), Communications in Applied Numerical Methods (Vol. 8, 1992), SIAM Journal of Numerical Analysis (Vol. 30, 1993), Electronic Transactions on Numerical Analysis (Vol. 6, 1996). Another rich source of information on multigrid is the MGNet website maintained by C.C. Douglas: `http://www.mgnet.org`.

Introductions to multigrid methods can be found in [11,17,47,134,152], to be collectively referred to in the sequel as the basic literature. A thorough introduction to multigrid methods is given in [47]. The introduction in [152] requires less mathematical background and is more oriented towards applications. In Chapter 9 of [152] the basic principles and the state of the art around 1990 of multigrid for computational fluid dynamics is described. Therefore here we will not dwell much on the basic principles, and confine ourselves mainly to developments that have taken place during the last decade.

Computational Fluid Dynamics (CFD) gives rise to very large systems requiring efficient solution methods. Not surprisingly, multigrid found applications in CFD at an early stage. The compressible potential equation was solved with multigrid in 1976 [131], the incompressible Navier–Stokes equations shortly after [12,148]. Over the years, multigrid has become closely intertwined with CFD, and has become an ingredient in major CFD codes. The viscous flow around a complete aircraft configuration can now be computed thanks to the availability of multigrid solvers [60], and also complex industrial flows in machinery are computed successfully with multigrid.

However, as remarked in [15], full textbook multigrid efficiency has not yet been achieved in realistic engineering applications in CFD in general. An important reason for this is that in CFD we often have to deal with singular perturbation problems. This gives rise to grids with cells with high aspect ratios. Another reason is that the governing equations may show elliptic or parabolic behavior in one part of the domain and hyperbolic behavior in another part of the domain. This requires careful design of both the discretization and the solver, putting a premium on robustness. With the increasing complexity of CFD applications (most of all due to the grid structures on which the equations are discretized), the demand for robustness of solution methods is increasing even more. Industrial practice is heading towards unstructured grids, more complex flow modeling, time-dependent problems and multidisciplinary applications. These developments pose new challenges for multigrid research. The potential for large further gains is there, and consequently multigrid remains an active research topic in CFD. Ways for improvement are pointed out in [15,16]. The current state of the art is surveyed below.

The field of computational fluid dynamics is too diverse and multigrid can be implemented in too many ways to make a self-contained synopsis possible within the confines of a journal article. Therefore our descriptions will be eclectic, global and fragmentary, but we will cast our net wide in referring to the literature after 1990 for further details.

2. Multigrid basics

First, we present a brief description of multigrid principles. For further information one may consult the basic literature. To establish notation and terminology, we start by formulating the basic two-grid algorithm. Let us have a system of m partial differential equations on a domain Ω, discretized on a grid $G \subset \Omega$. The resulting nonlinear algebraic system is denoted as

$$N(u) = b, \quad u \in U = \{u : G \to \mathbb{R}^m\}, \tag{1}$$

where U is the space of grid functions on G. For the moment, the problem is assumed to be independent of time. Let there also be a coarse grid $\bar{G} \subset \Omega$ with fewer nodes than G. Overbars denote coarse grid quantities.

2.1. The nonlinear multigrid method

The basic two-grid algorithm is given by

Choose u^0
Repeat until convergence:
begin
(1) $S_1(u^0, u^{1/3}, b)$; $r = b - N(u^{1/3})$;
(2) Choose $\bar{u}_\alpha, s$; $\bar{b} = \bar{N}(\bar{u}_\alpha) + \sigma R r$;
(3) $\bar{S}(\bar{u}_\alpha, \bar{u}^{2/3}, \bar{b})$;
(4) $u^{2/3} = u^{1/3} + \frac{1}{\sigma} P(\bar{u}^{2/3} - \bar{u}_\alpha)$;
(5) $S_2(u^{2/3}, u^1, b)$;
(6) $u^0 = u^1$;
end

Step (1) (pre-smoothing) consists of a few iterations with some iterative method for the fine grid problem (1) with initial iterate u^0 and result $u^{1/3}$. In step (2), $\bar{u}_\alpha$ is an approximation on the coarse grid of the exact solution, used to remain on the correct solution branch and/or to linearize the coarse grid correction problem. One may take, for example, $\bar{u}_\alpha = \tilde{R} u^{1/3}$, with $\tilde{R} : U \to \bar{U}$ a restriction operator from the fine to the coarse grid. But it may be more economical to keep $\bar{u}_\alpha$ fixed during two-grid iterations, if $\bar{N}(\bar{u}_\alpha)$ is expensive, or if $\bar{u}_\alpha$ is used to construct an expensive Jacobian. In step (2), σ is a parameter which, if chosen small enough, ensures solvability of the coarse grid problem, given by

$$\bar{N}(\bar{u}) = \bar{b} \equiv \bar{N}(\bar{u}_\alpha) + \sigma R r, \tag{2}$$

where $\bar{N}$ is a coarse grid approximation to N, obtained, for example, by discretization of the underlying system of partial differential equations on $\bar{G}$. Furthermore, $R : U \to \bar{U}$ is a fine to coarse grid restriction operator, that need not be the same as $\tilde{R}$. In step (3), $\bar{S}$ stands for solving the coarse grid problem (2) approximately by some iteration method with initial guess $\bar{u}_\alpha$ and result $\bar{u}^{2/3}$. In step (4), the coarse grid correction is added to the current fine grid iterate. Here, P is a prolongation or interpolation operator $P : \bar{U} \to U$. In step (5) post-smoothing takes place.

With $\sigma = 1$, we obtain the well-known Full Approximation Scheme (FAS) [11], which is most commonly used in CFD. The nonlinearity of the problems enters in the smoothing operators S_1, S_2 and $\bar{S}$. Global linearization is not necessary, so that there is no need to store a global Jacobian, which is, moreover, frequently very ill-conditioned in CFD. For many problems, however, the nonlinearity can also be handled globally, resulting in a sequence of linear problems that can be solved efficiently with linear multigrid. The multigrid method is obtained if solution of the coarse grid problem in step (3) is replaced by γ iterations with the two-grid method, employing a still coarser grid, and so on, until the coarsest grid is reached, where one solves more or less exactly. With $\gamma = 1$ or $\gamma = 2$, the V- or W-cycle is obtained, respectively. If N is the number of unknowns on G and N/β is the number of nodes on $\bar{G}$, then the above multigrid algorithm requires $\mathcal{O}(N)$ storage and, for accuracy commensurable with discretization accuracy, $\mathcal{O}(N \log N)$ work, if $\gamma < \beta$. To get $\mathcal{O}(N)$ work, the multigrid cycles must be preceded by nested iteration, also called full multigrid; see the basic literature. Starting on a coarse grid and refining this grid successively leads to a well-defined grid hierarchy also on unstructured grids (see, for example [7]). For a given unstructured grid it is usually not difficult to define a sequence of finer grids and a corresponding multigrid method. However, it might be difficult to define a sequence of coarser grids starting from an irregular fine grid [21]. This is where AMG [121] comes into play.

A natural generalization of multigrid is to combine it with locally refined grids. This leads to the Multilevel Adaptive Technique (MLAT) [11,5] or to the Fast Adaptive Composite Method (FAC) [90]. The locally refined regions are incorporated as extra additional levels in multigrid where all multigrid components like smoothing are defined with minor modifications, see [11,90] for details.

The favorable $\mathcal{O}(N)$ convergence behavior depends on satisfying the *smoothing property* and the *approximation property* [47]. The smoothing property requires that the smoothing processes S_1 and S_2 make the error between discrete solution and current approximation smooth. In not too difficult cases this can be checked by frozen coefficients Fourier analysis [11,134,152], leading to the determination of the smoothing factor. The approximation property says something about the accuracy of the coarse grid correction applied in step (4). Extrapolating from what is known from simple cases, this implies roughly, for example, that P and R, when applied to a given variable and its corresponding residual, should satisfy

$$m_P + m_R > M. \tag{3}$$

Here, orders m_P, m_R of P and R are defined as the highest degree plus one of polynomials that are interpolated exactly by P or $\tilde{\sigma} R^{\mathrm{T}}$, respectively, with $\tilde{\sigma}$ a scaling factor, R^{T} the transpose of R, and M the order of the highest derivative of the unknown concerned that occurs in the partial differential equation. For instance, for the Euler equations, we have $M = 1$, whereas for the Navier–Stokes equations we have $M = 1$ for the pressure but $M = 2$ for velocity components. No requirements are known for $\tilde{R}$.

2.2. The smoothing method

An important issue is the *robustness* of the smoother. This implies that the smoother should be efficient for a sufficiently large class of problems. Two model problems that represent circumstances encountered in CFD practice are the convection-diffusion equation and the rotated anisotropic diffusion equation, given by, respectively

$$cu_x + su_y - \varepsilon(u_{xx} + u_{yy}) = 0, \tag{4}$$

$$-(\varepsilon c^2 + s^2)u_{xx} - 2(\varepsilon - 1)csu_{xy} - (\varepsilon s^2 + c^2)u_{yy} = 0, \tag{5}$$

where $c = \cos\phi$, $s = \sin\phi$, and ϕ and ε are parameters. The first equation contains the effect of strong convection (if $\varepsilon \ll 1$) and is related to hyperbolic systems; a method that does not work for a scalar hyperbolic equation ($\varepsilon = 0$) is likely to fail for a system. The second equation models the effect of large mesh aspect ratios, equivalent to $\varepsilon \ll 1$ on an isotropic grid, and grid nonorthogonality ($\phi \neq 0$), giving rise to mixed derivatives in boundary-fitted coordinates. A smoothing method may be called robust if it works for all ϕ and ε. For a multigrid method that is robust for the test problems (4) and (5) one may have some hope of success in application to the governing equations of CFD. Fourier smoothing analysis for the simple test problems above is easy, and smoothing factors for many methods are reported in [151,152,104,169], and for some three-dimensional cases in [67,136]. For the two-dimensional case, robust smoothing methods exist. They are basically of line-Gauss–Seidel type and ILU type; a list is given in Section 7.12 of [152]. For the simple test problems (4) and (5), it can be said that efficient and robust multigrid smoothers and corresponding multigrid methods are available, even of black-box type for problems on structured grids, see [162]. What has kept CFD applications away from textbook multigrid efficiency is the generalization from the scalar case to the case of systems of partial differential equations, the difficulty of making the coarse grid correction sufficiently accurate, and finding efficient smoothers for higher-order upwind discretizations with limiters. Furthermore, application of multigrid to unstructured grids has started only recently, and further development is necessary for achieving textbook multigrid efficiency in this context.

The historical development of multigrid in CFD has been such that smoothing methods come in two flavors: basic iterative methods (BIMs) and multistage (MS) (Runge–Kutta) methods. Assuming instead of (1) a linear problem $Au = b$, a BIM is given by

$$Mu^{n+1} = Nu^n + b, \quad A = M - N. \tag{6}$$

Typical examples of BIMs are methods of Gauss–Seidel type and of incomplete LU factorization (ILU) type. MS smoothers for the nonlinear system (1) are obtained by artificially creating a system of ordinary differential equations:

$$\frac{\mathrm{d}u}{\mathrm{d}t} = N(u) - b. \tag{7}$$

MS methods of the following type are used:

$$u^{(0)} = u^n,$$
$$u^{(k)} = u^{(0)} - c_k \tau N(u^{(k-1)}) + c_k \tau b, \quad k = 1, \ldots, p, \tag{8}$$
$$u^{n+1} = u^{(p)},$$

where τ is a time step. The appearance of time in our algebraic problem comes as a surprise, but in fact this application of MS to solve an algebraic system is old wine in a new bag. In the linear case, where $N(u) = Au$, we obtain by elimination of $u^{(k)}$:

$$u^{n+1} = P_p(-\tau A)u^n + Q_{p-1}(-\tau A)b, \tag{9}$$

with the amplification polynomial P_p of degree p given by

$$P_p(z) = 1 + z(c_p + c_{p-1}z(1 + c_{p-2}z(\ldots(1 + c_1 z)\ldots), \tag{10}$$

and Q_{p-1} a polynomial that we will not write down. It follows that MS is an iterative method of the type

$$u^{n+1} = Su^n + Tb, \quad S = P_p(-\tau A). \tag{11}$$

Methods for which the iteration matrix S is a polynomial of the matrix of the system to be solved are well-known in numerical linear algebra, and are called semi-iterative methods; see [142] for the theory of such methods. In the multigrid context, the MS coefficients c_k are not chosen to optimize time accuracy or stability, but the smoothing performance. Because for stationary problems time is an artifact, τ need not be the same for all grid cells, but can vary, such that $c_k \tau$ is optimal for smoothing. Grids with large aspect ratios are generic for Navier–Stokes applications, due to the need to resolve thin boundary layers. Smoothing analysis reveals that when high mesh aspect ratios occur, modeled by test problem (5) by choosing $\phi = 0$, $\varepsilon \ll 1$, then point-wise smoothers such as Gauss–Seidel and MS, are not satisfactory. The unknowns in certain subsets of the grid points should be updated together, leading to line-Gauss–Seidel or ILU [162,163].

In the MS case, implicit stages could be included, but this avenue has not been followed systematically. Instead, implicit residual averaging is applied. After a MS stage, the residual

$$r^{(k)} = b - Au^{(k)}$$

is replaced by $\tilde{r}^{(k)}$ satisfying

$$B\tilde{r}^{(k)} = r^{(k)}, \tag{12}$$

with B such that B^{-1} has a smoothing effect, and such that (12) is cheap to solve. For details, see [83]; tuning parameters are involved.

These ways to obtain a robust smoother clearly have drawbacks. Extensions to systems of differential equations leads to the need to solve subsystems that are more involved and costly than tri-diagonal systems encountered in implicit residual averaging or line-Gauss–Seidel for the scalar case. Parallelizability is likely to be impaired. An alternative is to strengthen the coarse grid correction instead of the smoother. This can be done by constructing the coarse grid by semi-coarsening, i.e., the mesh-size is doubled only in selected directions, such that the coarse grid becomes more isotropic than its parent fine grid. This approach, although not novel, has only recently started to have an impact of practical CFD; we will return to semi-coarsening below.

In three dimensions, the situation is less satisfactory, see [67]. Typical reference model problems in 3D can be

$$s\tilde{c}u_x + cu_y + s\tilde{s}u_z - \varepsilon(u_{xx} + u_{yy} + u_{zz}) = 0, \tag{13}$$

where $c = \cos\phi$, $s = \sin\phi$, $\tilde{s} = \sin\theta$, $\tilde{c} = \cos\theta$ and ϕ, θ and ε are parameters, and

$$-\varepsilon_1 u_{xx} - \varepsilon_2 u_{yy} - \varepsilon_3 u_{zz} = 0. \tag{14}$$

If a strong coupling of unknowns in two space directions in (14) exists, it is not possible to use cheap basic (collective) point or line Gauss–Seidel type smoothers and still obtain a fast and efficient multigrid method. The strong coupling of unknowns should be taken into account. This means that plane smoothers, that update all unknowns in a plane simultaneously, have to be applied. However, the coupled system in a plane, even for scalar equations, does not result in a tri-diagonal system as with line smoothers, but in a more general sparse system. It is found that often, an inexact (iterative) solution method, like one multigrid V-cycle in a 2D plane, is sufficient for an overall efficient 3D solution method [136,103]. In this way a relatively cheap plane solver can be obtained for certain problems.

It helps considerably in finding efficient smoothers if the Jacobian of $N(u)$ is close to an M-matrix, as explained in Section 4.2 of [152]. Unfortunately, Godunov's order barrier theorem [42] implies that this can be the case only for first-order accurate schemes (in the linear case), so that we are restricted to first-order upwind schemes. This is unsatisfactory. Generally speaking, engineering practice requires second-order accuracy. One way around this is to use defect correction. Suppose, we have a first-order discretization $N_1(u)$ for which we have a good smoother, and a second-order scheme $N_2(u)$ that we would prefer to use. Defect correction works as follows:

```
begin Solve N1(ỹ) = b;
      for i = 1(1) n do
                Solve N1(y) = b − N2(ỹ) + N1(ỹ);
                ỹ = y;
      od
end
```

It suffices to take $n = 1$ or 2 to achieve second-order accuracy. But there are smoothers that work for second-order schemes, so that defect correction is not necessary; see [34,1,104].

2.3. The coarse grid correction

In this section, we discuss several issues related to the coarse grid correction that are relevant for CFD problems. In linear multigrid, there are two ways to approximate the fine grid operator on coarse grids, namely coarse grid discretization approximation and coarse grid Galerkin approximation. With the discretization approximation the coarse grid operator is obtained by rediscretizing the governing differential equations on the coarse grids. With the Galerkin approximation one puts

$$\bar{A} = RAP.$$

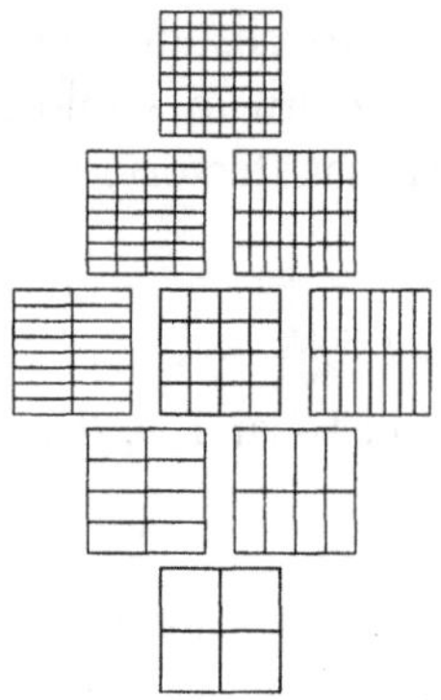

Fig. 1. Multiple semi-coarsened grids in 2D with a finest grid consisting of 9×9 points.

For more details on coarse grid Galerkin approximation, see Section 6.2 of [152]. Coarse grid construction by agglomeration of fine grid cells may lead to coarse grids that conform insufficiently to strongly curved boundaries, leading to inaccurate coarse grid corrections and degradation of multigrid efficiency. This issue is addressed in [52,128], where it is shown how to overcome this difficulty by deriving geometric information not from the coarse grid nodes but from the fine grid, effectively employing grid cells with curved boundaries. This difficulty is taken care of automatically when Galerkin coarse grid approximation is used.

Suppose the convection–diffusion equation is discretized with the first-order upwind scheme or with a second-order upwind biased scheme. In the case of dominating advection, this means that artificial viscosity is added. Standard coarse grid correction suffers from the difficulty that the artificial viscosity is implicitly multiplied by a factor 2 or more. This leads to inaccurate coarse grid correction, as explained in [13,14,161]. This difficulty is overcome in two ways. In the first place, it is often the case for convection dominated problems that the smoother reduces both rough and smooth error components so effectively, that weakness of the coarse grid correction has no consequences. Some smoothers with this property are symmetric point Gauss–Seidel, alternating symmetric line Gauss–Seidel and ILU type smoothers, (provided there is no recirculation, otherwise see [146,158]); but MS smoothers do not have this pleasant property.

In the second place, one may replace standard coarse grid approximation by something more accurate, namely by multiple semi-coarsening or by directional semi-coarsening. Instead of increasing the cost of the smoother, the coarsening strategy is changed so that cheaper smoothers (point smoothers like MS smoothers or line smoothers instead of plane smoothers) can be applied. Although in multiple semi-coarsening methods [97,98,69,147] many coarse grids exist on coarser grid levels, the work for solving a problem with N unknowns is still $\mathcal{O}(N)$, if an F-cycle is applied. Directional coarsening is applied in all directions, so that a fine grid gives rise to d coarse grids, with d the number of dimensions, making it possible to work with very simple and cheap smoothers. Some coarse grids from different finer grids coincide. In 2D a diamond-like sequence of grids is obtained (see Fig. 1). There are several options for the transfer operators between the grids [97,102,99].

The total number of cells in all grids combined is $8N$ (in three dimensions). In 3D, however, there are more possibilities for the grid coarsening. Depending on the number of coarse grids chosen, a point or a line smoother must be used to obtain a robust 3D multiple semi-coarsening-based solution method. A multiple semi-coarsening variant that uses a coarsening strategy such that line smoothers guarantee robustness for 3D reference problems is presented in [147]. A reduction of the computational overhead can be achieved by *flexible multiple semi-coarsening*: only if strong couplings in certain directions exist [147], the semi-coarsening process is continued in that direction.

The second variant is *directional coarsening*: fine grid cells are coarsened in one direction only, by removing only the largest cell faces, so that coarsening is done only in a direction roughly perpendicular to refinement zones. When the coarse grid has become sufficiently isotropic, one can switch to full coarsening. This approach is commonly used for compressible Navier–Stokes equations, to be discussed in Section 4.2. Directional semi-coarsening combined with line relaxation is presented in [27].

2.4. Structured and unstructured grids

Grid generation is one of the most time consuming activities of the CFD practitioner. It is usually the main bottleneck for project turn-around time. In comparison, solver CPU time plays a minor role. This achievement is for a large part due to multigrid. In the past, much effort has gone into development of tools for the generation of structured grids. In such grids, the number of cells that share a common vertex is uniform in the interior of the domain. The approach is to divide the domain in subdomains, each of which is mapped by a boundary-fitted coordinate mapping to a cube, in which a uniform grid is generated, the image of which gives a boundary-fitted curvilinear grid in the subdomain. In order to cope with the geometric complexity of engineering applications, the subdomain decomposition must be unstructured, leading to multiblock block-structured grids. On structured grids algorithms can be formulated that run fast on vector computers, less computer memory is required, and coarse grid generation for multigrid and the implementation of transfer operators between grids is straightforward. These are the main advantages of structured grids. Despite intensive research efforts, however, it has turned out to be not possible to automate the generation of structured grids to a sufficient extent to reduce the amount of human labor involved to a low level. In particular, the generation of the domain decomposition requires much time from the user.

As a consequence, unstructured grids are now getting more and more attention. Not only are these grids easier to generate than structured grids, they also lend themselves better to adaptive discretization, since it is relatively easy to insert and remove grid points. The development of accurate discretizations and efficient solution methods is more difficult than for structured grids. This is now an active field of research, and much remains to be done. We will give some pointers to recent work, but it is too early for a review. We will concentrate mainly on structured grids, which are still mostly used today.

A third type of grid consists of the union of locally overlapping grids that together cover the domain. The local grids are usually structured. The flexibility offered by this kind of grid is especially useful for flows in which boundary parts move relatively to each other. Furthermore, this is a way to include adaptivity in the structured grid context. The multigrid principles are the same as for structured grids. We will not discuss overlapping grids. Examples may be found in [3,55,22,138].

3. Incompressible flow

Unified methods that treat compressible and incompressible flows in a uniform way are emerging (see [9] and references quoted there). Unified methods departing from incompressible Navier–Stokes adopt corresponding solution methods, whereas low Mach equations arising from preconditioning compressible formulations are solved by compressible solution methods. Standard computing methods for compressible and incompressible flow differ substantially, and will therefore be discussed separately. We start with the incompressible case.

We will assume that discretization takes place on boundary-fitted structured grids. The flow is governed by the incompressible Navier–Stokes equations. For the high Reynolds numbers occurring in most applications thin boundary layers occur, necessitating locally refined grids, giving rise to high mesh aspect ratios (10^4 or even more). The primitive variables are commonly used. The solution of d momentum equations (d is the number of dimensions) and the continuity equation is required. The discretized continuity equation serves as an algebraic constraint, see any textbook on the incompressible Navier–Stokes equations, so that after discretization in space the incompressible Navier–Stokes equations constitute a differential-algebraic system (of index two), and the pressure acts as a Lagrangean multiplier, mathematically speaking.

Often, a semi-heuristic turbulence model is used to predict time-averaged flow variables based on a compromise between accuracy, memory requirements and computing time. More complete models such as large-eddy simulation and direct numerical simulation are time dependent and can be efficiently implemented with explicit time-stepping schemes. In these models an algebraic pressure correction equation appears that can easily be handled by classical multigrid methods, that do not need further attention here. In fact, for time-accurate solutions the main use of multigrid, for example in many engineering codes, is for some form of the elliptic equation for the pressure. It is, however, possible to benefit more from multigrid, as we will discuss below.

The spatial discretization should be such that spurious pressure modes are avoided. This has given rise to two different approaches. If grid-oriented velocity components with a staggered placement of unknowns are employed, no special measures are required, but the scheme is somewhat complicated, and must be formulated carefully to maintain accuracy on rough grids; see [153]. The alternative is to use a colocated (nonstaggered) placement of the unknowns, for which the Cartesian velocity components and pressure can be used. This makes it easier to formulate accurate schemes on curvilinear grids, but artificial stabilization measures are required. For a discussion of the relative merits of staggered and colocated schemes, see, for example, [119,153]. The two approaches will be discussed in separate subsections.

3.1. Multigrid for staggered schemes

Discretization of the stationary incompressible Navier–Stokes equations on a staggered grid results in a nonlinear algebraic system of the following form:

$$\begin{pmatrix} Q & G \\ D & 0 \end{pmatrix} \begin{pmatrix} U \\ p \end{pmatrix} = \begin{pmatrix} b_1 \\ b_2 \end{pmatrix}. \tag{15}$$

Here, U contains the velocity components, p contains the pressure values in the cell centers, b_1 and b_2 are forcing terms resulting from the boundary conditions, Q is a nonlinear algebraic operator arising from the discretization of the inertia and viscous term, and G and D are the linear discrete gradient and divergence operator, respectively.

Because of the staggered placement of the unknowns, different prolongation (P) and restriction (R) operators are required for the velocity components and pressure. This is not a big problem; see [141,152,164] for examples. Because no higher than first-order derivatives of the pressure occur, P and R can be less accurate than for the velocity, according to the accuracy rule (3). As discussed in Section 2, coarse grid approximation can be done either by discretization on the coarse grid, which is most commonly done, or by Galerkin approximation of the fine grid operators on the coarse grids.

The main issue is the design of good smoothers. Classical relaxation methods are inapplicable, due to the occurrence of a zero block on the main diagonal. Two approaches may be distinguished: box relaxation [141] and distributive iteration [12,154]. For an introduction to these methods, see Section 9.7 of [152]. A brief synopsis is given below, after which recent developments are reviewed.

In box iteration, the unknowns belonging to a cell are updated simultaneously in Gauss–Seidel fashion. These unknowns are the normal velocity components in the cell faces and the pressure in the cell center. A $(2d+1)\times(2d+1)$ system, with d the number of dimensions, has to be solved for each cell, which can be done using an explicit formula. This is the symmetric coupled Gauss–Seidel (SCGS) method, introduced in [141]. Roughly speaking, the method performs similar to the Gauss–Seidel method for a convection–diffusion equation. This implies that a symmetric version is required, in which the cells are processed both in a forward and backward (reversed) order, so as to obtain a smoother that is robust with respect to flow direction. Furthermore, on highly stretched grids a line-version is required, with lines perpendicular to the refinement zone. Line versions of SCGS are described and applied in [101,137]. Moreover, [137] presents the line smoother in an adaptive grid framework. When the grid is nonorthogonal, line-Gauss–Seidel might become less effective, as predicted by analysis of scalar model problems; see Chapter 7 of [152]. But ILU remains effective as a smoother. Further applications of multigrid with point or line SCGS are described in [66,156,168]. In [62,100] a k–ε turbulence model is included.

Distributed iteration methods are best described as follows. The system (15), denoted as $Ay=b$, is replaced by

$$AB\bar{y}=b, \quad y=B\bar{y}, \tag{16}$$

where the distribution matrix B is chosen such that AB lends itself easily to iterative solution; for example, because AB is close to a triangular M-matrix the zero block in A disappears. Let C be some approximation to AB. Then we have the following iteration method for (16):

$$\bar{y}^{m+1}=\bar{y}^{m}+C^{-1}(b-AB\bar{y}^{m}),$$

or

$$y^{m+1}=y^{m}+BC^{-1}(b-Ay^{m}). \tag{17}$$

Of course, C is chosen such that the action of C^{-1} is easily determined. It is made plain in [152], that depending on the choice of B and C, various well-known methods are obtained, the classical examples being the SIMPLE method of [109] and the distributed Gauss–Seidel method of [12]. Distributive iteration methods can be designed such that the individual velocity components and pressure are updated separately, so that for the velocity components one has essentially to deal with convection–diffusion equations, for which smoothing is relatively easy. But box variants are also possible. The box ILU distributive smoother employed in [165] is robust with respect to mesh aspect ratio and grid nonorthogonality. For accuracy, second-order schemes are required for the inertia terms; this may necessitate the use of defect correction [28].

It is found that these methods lend themselves well as smoothers in multigrid. Interesting new insights in distributive smoothers are presented in [10]. Smoothing analysis is presented in [12,123,127] (elementary) and [154,155] (advanced). The analysis in [14] is especially interesting, because the influence of boundaries is taken into account. Unfortunately, some distributed smoothing methods require problem-dependent underrelaxation parameters. In [127], the SCGS method is compared with distributed smoothers. It is found that SCGS is more efficient than distributed iteration for high Reynolds numbers, and that it is less sensitive to the value of the underrelaxation parameter. In [129], a SIMPLE type smoother is compared with point and line versions of SCGS on the basis of numerical experiments. SCGS come out best, with a line version being superior in the presence of large mesh aspect ratios. But in [107] it is found that for stratified flow problems SIMPLE is to be preferred. Both approaches are compared as single grid solvers in [51]. Recent applications using distributive smoothers are presented in [24,106,114,126,166], for unstructured grids in [64,65,57] and in combination with adaptivity in [157]. Multigrid for divergence free finite elements is described in [140].

3.2. *Multigrid for colocated schemes*

In order to rule out spurious pressure modes, usually the continuity equation is perturbed by terms involving pressure. This can be done by introducing artificial compressibility [23], or by the pressure-weighted interpolation method (PWI) of [115], or by using a Roe-type flux difference splitting [32]. The second approach is most widespread in engineering practice.

We start with methods using the PWI scheme. The stabilizing terms replace the zero block in (15) by an operator that is not given explicitly in the literature, but that is easily deduced from Eq. (49) of [92]. Nevertheless, distributive methods dominate the field, that are quite similar to SIMPLE and its sisters for staggered schemes, and carry the same names. Smoothing analysis for smoothers of SIMPLE type is given in [93,41]. Some recent publications discussing the application of multigrid methods to computation of laminar flows are [41,75,128]. Inclusion of two-equation turbulence models is discussed in [63,74]. Turbulence is included on unstructured grids in [2,57]. In [57] ILU is used as the smoothing method, and starting from a coarsest grid, adaptive unstructured grids are defined.

The numerics for artificial compressibility methods resemble closely the numerics for compressible flow methods, and will be discussed below. Laminar flow computations with the artificial compressibility method are reported in [36,76,130,144]; turbulence modeling is included in [77,124,167], adaptivity in [76]. A staggered scheme with a multigrid acceleration of a so-called fraction step solver is presented in [120], the fractional step method is employed in a finite element context in [139].

In the flux-difference splitting of [32], a stabilizing pressure term is introduced in the discrete continuity equation in a natural way. This scheme does not give an M-matrix, but a so-called vector-positive discretization. This makes Gauss–Seidel type smoothing possible in a collective or a decoupled formulation. The flux-difference scheme is first-order accurate, but second-order accuracy can be obtained in the standard way using the MUSCL approach. For details, see [32].

4. Compressible flow

One of the first applications of multigrid in CFD may be found in [131]. This work concerns the compressible potential equation. This equation is now efficiently solved routinely with multigrid in the aerospace industry, and will not be discussed here. For a survey of past multigrid developments, see Chapter 9 of [152]. A general survey of discretization schemes for compressible flow equations is given in [56].

4.1. The Euler equations

The prevalent smoothing method for multigrid solution of the Euler equations is the MS method. Since its introduction in [59], the MS smoothing method has evolved quite a bit and has become steadily more efficient. An overview is given in [61]. To begin with, efforts have been made to optimize the MS coefficients c_k to enhance smoothing. This has been done in [72,116] for the one-dimensional scalar case:

$$\frac{\partial u}{\partial t}+\lambda\frac{\partial u}{\partial x}=0.$$

Since in a system there are different eigenvalues λ, straightforward application of the optimal coefficients to the systems case is not optimal. This is shown in [72], where the straightforward approach is compared with what is called characteristic time stepping [73] (requiring the use of a flux splitting discretization scheme), for which only one effective wavespeed comes into play, so that more benefit is derived from optimization for the scalar case. The difference in efficiency between the two methods is found to be considerable. Optimal MS smoothing for central schemes with artificial dissipation, i.e., the very popular Jameson–Schmidt–Turkel scheme [59], requires the use of preconditioning. That is, the discrete scheme to be solved is given by

$$\frac{\mathrm{d}U}{\mathrm{d}t}+P(U)R(U)=0,$$

where $P(U)$ is a preconditioner. Of course, time accuracy is lost. The purpose of preconditioning is to cluster the eigenvalues, so that the coefficients obtained from scalar optimization can be more profitably applied. Preconditioning is usually done by collective Jacobi iteration [1,33,34,116,19]; this is called the multistage Jacobi method. In [34] it is shown by experiments that it is more efficient to optimize c_k for increasing the time step rather than for smoothing. We think the disparity between optimality as derived in [72] by Fourier analysis and optimality in practice is due to the influence of boundary conditions, which is accounted for faster with larger time steps.

The optimal multistage coefficients c_k as determined for the one-dimensional case can also be used in the multidimensional case, because, as remarked in [135], a two-dimensional optimization leads to optimal coefficients that are not significantly different from those obtained in the one-dimensional case. Only the optimal CFL number (which has to do with the choice of the local time step τ) differs somewhat; in [135] a recipe is given for choosing a good CFL number in more dimensions.

Implicit smoothing methods work differently. The equations are discretized in time with the implicit Euler method, giving rise to

$$\frac{1}{\Delta t}(U^{n+1}-U^{n})+R(U^{n+1})=0.$$

If steady solutions are envisaged, the time difference can be deleted. A relaxation scheme is chosen for this nonlinear system for U^{n+1}, for example the collective Gauss–Seidel (CGS) smoothing method in a FAS method. All unknowns in a cell are updated simultaneously, keeping values in neighboring cells fixed. This requires the (approximate) solution of a small nonlinear system: usually one Newton iteration suffices. This approach is followed in [53,68,96,132]. If applied in symmetric fashion (forward followed by backward ordering), CGS is a very efficient smoother for first-order schemes for hyperbolic systems, more efficient than MS, as predicted by the model problem smoothing analysis in [152]. Furthermore, CGS does not require tuning of coefficients. But, unlike MS, GS does not work for second-order schemes directly, but must be combined with defect correction, as done in [30,31,53,68,132] or line versions must be chosen [104]. According to [33,34,116] it is best to apply defect correction only on the finest grid, and to use the first-order scheme on the coarse grids; this is called mixed discretization. In [34,116] it is found that without the latest enhancements MS is less efficient than CGS, but with Jacobi preconditioning and mixed discretization it can compete with CGS and defect correction for Total Variation Diminishing (TVD) schemes. Because parallelization is easier, MS is probably to be preferred on multi-processor platforms. Adaptive grids are incorporated in compressible solvers in [8,108,145,84,39].

The multiple semi-coarsening method has been pioneered in [97,98] for the Euler equations. The combination of multistage smoothing without frills, low Mach number preconditioning and semi-coarsening is shown to be quite robust for the two-dimensional Euler equations in [25]. The efficiency is better than standard multigrid based methods; typically 500 to 1500 work units are required, with a work unit the work for a residual evaluation on the finest grid. We expect that if nested iteration (full multigrid) would have been incorporated, then optimal efficiency (100 work units, say) would not be far away. Similar performance in three dimensions and for Navier–Stokes still remains to be demonstrated.

Standard multigrid has been very successful for the Euler equations. For the Navier–Stokes equations the situation is less satisfactory, as we shall see in the next section.

4.2. The Navier–Stokes equations

As far as multigrid is concerned, the main difference with the Euler equations is the generic occurrence of highly stretched cells (with aspect ratios of up to 10^4), in order to resolve thin boundary layers. This leads to a widely observed deterioration of the standard multigrid convergence rate with MS smoothers, causing a wide gap between actual and textbook multigrid efficiency. From Fourier smoothing analysis of simple model problems (see, e.g., [152]) it is clear that, if coarse grid correction is left alone, then unknowns in columns of cells sharing large faces (more or less perpendicular to refinement zones) must be updated simultaneously, giving rise to methods such as line-Jacobi, line-Gauss–Seidel [70], ADI [20] and ILU. In the scalar case this gives rise to solving tridiagonal or similar simple systems, but in the systems case the smoother becomes more involved and computing intensive. As discussed in Section 2.3, the alternative is to leave the smoother alone, but to do something about the coarse grid correction. We will return to this shortly, but first we discuss robust smoothers for grids containing refinements zones.

The obvious extension of CGS to a line version (LU-SGS) has been undertaken in [159,160]. In [1,143] the MS scheme is made suitable for Navier–Stokes by choosing for the preconditioner $P(U)$ something similar to LU-SGS; [1] also provides Fourier smoothing analysis results, as do [133,169].

In order to take care specifically of stretched cells in a boundary layer, one may choose $P(U)$ corresponding to collective line-Jacobi iteration, with lines chosen perpendicular to the boundary layer. This is described and analyzed in [111]. In [144] the MS smoother is used with line-implicit stages to take care of high aspect ratio cells, and Fourier smoothing analysis is presented; good efficiency is obtained. More reliable than smoothing analysis for prediction of actual multigrid performance is two-grid analysis, since this gives a better representation of the influence of coarse grid correction. Two-grid analysis for compressible flow problems is presented in [58].

A semi-coarsening variant for the Navier–Stokes equations is presented in [113]. The large number of coarse grids generated makes this method impractical for industrial purposes. For Navier–Stokes, however, directional coarsening is starting to be accepted. Smoothing analysis in the presence of directional coarsening and results are given in [1,110,112]; and applications are described in [77,130]. Significant gains in efficiency over older methods are obtained by combining the MS method, point-Jacobi preconditioning and directional coarsening.

Closer towards the AMG approach is the approach followed in [85–87,95,88]. There the coarse grid operator for the compressible Navier–Stokes equations is constructed by the Galerkin coarse grid approximation. The coarsening strategy is also AMG based [95]. An efficient solver is presented in [88] with Krylov subspace acceleration of a multigrid method with directional AMG-like coarsening and "directional implicit smoothing", i.e., combining the coarsening with simultaneous smoothing of coupled unknowns. Related to the previously mentioned methods, departing from an unstructured fine mesh, is the approach presented in [94,29] and the references quoted therein.

Turbulence modeling brings in additional difficulties that are typical for reactive flows as well. The case of the k–ε model can be taken as a typical example. Stiff source terms appear, some positive, some negative. For physical as well as numerical reasons, k and ε must remain positive. It turns out to be profitable to compute the flow variables and the turbulence quantities in a coupled manner, i.e., to include the turbulence model with the flow model in the multigrid method, as discussed in [77]. This requires some delicacy in the treatment of the turbulence model. The negative part of the source terms must be treated implicitly; this can be incorporated in the preconditioner. To ensure positivity, the coarse grid corrections for k and ε must be damped or limited. For details, see [35,40,80,125,6]. A multigrid application to compressible Direct Numerical Simulation (DNS) is described in [18].

5. Multigrid and parallel computing

Selection of a good bottle of wine is trivial if the price plays no role. Similarly, design of numerical algorithms is trivial if computer resources are unlimited. In reality they are scarce. Therefore, it is not of much benefit to parallelize inefficient numerical algorithms, i.e., algorithms with computational complexity $\mathcal{O}(N^\alpha)$, $\alpha > 1$. The demands of applications in engineering and physics are such that the relevant problem size increases as much as the computer size allows. Let us ponder briefly the consequences of this fact. Assume that if p is the number of processors, then the problem size is $N = mp$. That is, we assume that the problem size is proportional to the number of processors. Let the parallel efficiency be perfect, so that the computing speed is sp flops. Then the turn-around time will be $T = \mathcal{O}(p^{\alpha-1})$. Hence, if $\alpha > 1$ the turn-around time gets worse when more processors are installed. We conclude that it makes sense if parallel computing goes hand in hand with $\mathcal{O}(N)$ numerical algorithms. Hence the special significance of multigrid for parallel computing.

An approach for the parallelization of grid-oriented problems, which is suitable for many solution methods for PDEs, is grid partitioning [82,89,78,118]. In grid partitioning, parallelization is achieved by splitting the grid into subgrids and by mapping the subgrids to different processors. This parallelization concept is reasonable for problems wherein all operations are sufficiently local. By providing "overlap" regions along all internal subgrid boundaries, local operations (for example, the operations that make up the multigrid algorithm) can be performed in parallel. The overlap regions contain the latest values of the unknowns at points belonging to neighboring blocks, thus allowing normal execution of all operations at all points, including points on internal subgrid boundaries. The latest values in the overlap regions are provided by communication between the processors. Since communication takes place on a lower dimensional subset (boundary data) than the computation (on volumes of data) and since the number of arithmetic operations per grid point is relatively large in CFD problems, the grid partitioning strategy results in a good computation/communication ratio, and hence in efficient parallel solvers, including multigrid.

As the special type of the PDE has no great influence on the parallelization, many of the following considerations carry over directly to the incompressible and the compressible equations but also to other PDE systems. A block-structured grid provides a natural basis for the parallelization of PDE solvers. If these blocks are of completely different size and if each block is handled by one processor, problems with the load balance can be expected, since the work done in a processor essentially depends on the number of grid points owned by the processor. It is therefore important to split blocks in such a way that a satisfactory load balancing is achieved. It makes no sense to spend too much time at this stage: it is harmless if one (or a few) processors have less work than the average.

In parallel multigrid algorithms based on grid partitioning, the main part of the communication time is spent in updating the overlap regions. Locality in smoothers is beneficial for parallel processing. With the explicit MS smoothers, for example, it is possible to obtain a parallel multigrid algorithm which is identical to the single processor version. The situation is somewhat different with the implicit BIM smoothers, especially if unknowns are updated in a sequential order. The easiest way to parallelize such smoothers is by adapting the partitioning such that all unknowns that need to be updated simultaneously lie within one block. In the situation that an artificial block boundary cuts a strong coupling, parallel versions of implicit smoothers, like parallel line solvers [91,71], are necessary for satisfactory convergence. Modifications in which lines within a block are handled are, for example, described in [81].

Of course, the ratio between communication and arithmetic costs on coarser grids becomes worse. An immediate response to the degradation in efficiency caused by coarse grid levels would be to use multigrid cycle types which minimize the amount of time spent on coarse grids. From this point of view the use of V- or F-cycles, which provide optimal multigrid convergence for many applications, should be preferred to W-cycles.

It might happen that on the coarse grids there are more processors available than there are grid points. An approach for treating such a coarse grid problem in parallel multigrid is found in the strategy of not going to the possible coarsest grid, but keeping all the processors busy. In this case, the parallel algorithm is different from the sequential algorithm. The efficiency of this approach depends particularly on the solution procedure on the coarsest grid.

As indicated above, an advantage of an $\mathcal{O}(N)$ method like multigrid is that the method scales well, i.e., for increasing problem sizes and for an increasing number of processors the scalability of the application is very satisfactory, if the number of grid points remains fixed per processor.

It has been found in [79] that the multigrid solution method scales well for problems from CFD applications. (The number of processors varied in that study from 1 up to 192.) In [26], a staggered incompressible Navier–Stokes solver with the SCGS smoother is parallelized with grid partitioning, a colocated incompressible solver with distributed ILU smoothing in [37]. Adaptive multigrid on parallel computers for the Euler equations with implicit smoothing methods is presented in [117]. Three-dimensional industrial codes are parallelized by a communications library CLIC (Communications Library for Industrial Codes), which also supports adaptivity in [122]. Further 3D examples are found in [105].

Also parallel multiple semi-coarsening variants are most commonly based on the grid partitioning technique [102]. This has implications for the processing of the (multiple) fine and coarse grids. A sequential processing of certain fine and coarse grids seems natural, since these parts of the semi-coarsened grids are in the same processor. It does not make much sense to solve these parts in parallel because additional wall-clock time is not gained.

6. Conclusions

We have presented an overview of the developments in geometric multigrid methods for problems from computational fluid dynamics. With many pointers to the literature of the last ten years for the compressible and the incompressible case, we hope to have given a survey helpful for many practitioners. It is also clear that the desired textbook multigrid efficiency is not yet achieved for all relevant CFD problems and that the demands of engineering applications are orienting research in interesting new directions. With the strongly anisotropic grids that are currently used, advanced multigrid features, such as semi-coarsening techniques, adaptivity and generalization to unstructured grids are becoming more important. The same holds for parallel computing. We think that there is good reason to regard the multigrid approach to be one of the most significant developments in numerical analysis in the second half of the century that now lies behind us.

References

[1] S. Allmaras, Analysis of semi-implicit preconditioners for multigrid solution of the 2-D compressible Navier–Stokes equations, AIAA Paper 95-1651-CP, 1995.

[2] W. Anderson, R. Rausch, D. Bonhaus, Implicit/multigrid algorithms for incompressible turbulent flows on unstructured grids, J. Comput. Phys. 128 (1996) 391–408.

[3] M.J.H. Anthonissen, B. van't Hof, A.A. Reusken, A finite volume scheme for solving elliptic boundary value problems on composite grids, Computing 61 (1998) 285–305.

[4] N.S. Bachvalov, On the convergence of a relaxation method with natural constraints on the elliptic operator, USSR Comput. Math. Math. Phys. 6 (1966) 101–135.

[5] D. Bai, A. Brandt, Local mesh refinement multilevel techniques, SIAM J. Sci. Comput. 8 (1987) 109–134.

[6] X.S. Bai, L. Fuchs, A multi-grid method for calculation of turbulence and combustion, in: P.W. Hemker, P. Wesseling (Eds.), Multigrid Methods IV. Proceedings of the Fourth European Multigrid Conference, Amsterdam, Birkhäuser, Basel, 1994, pp. 35–43.

[7] P. Bastian, K. Birken, K. Johannsen, S. Lang, N. Neuss, H. Rentz-Reichert, C. Wieners, UG- a flexible toolbox for solving partial differential equations, Comput. Visual. Sci. 1 (1997) 27–40.

[8] M.J. Berger, A. Jameson, Automatic adaptive grid refinement for the Euler equations, AIAA J. 23 (1985) 561–568.

[9] H. Bijl, P. Wesseling, A unified method for computing incompressible and compressible flows in boundary-fitted coordinates, J. Comput. Phys. 141 (1998) 153–173.

[10] D. Braess, R. Sarazin, An efficient smoother for the Stokes problem, Appl. Numer. Math. 23 (1997) 3–19.

[11] A. Brandt, Multi-level adaptive solutions to boundary-value problems, Math. Comput. 31 (1977) 333–390.

[12] A. Brandt, N. Dinar, Multigrid solutions to flow problems, in: S. Parter (Ed.), Numerical Methods for Partial Differential Equations, Academic Press, New York, 1979, pp. 53–147.

[13] A. Brandt, Multigrid solvers for non-elliptic and singular-perturbation steady-state problems, Report, Department of Applied Mathematics The Weizmann Institute of Science, Rehovot, Israel, 1981 unpublished.

[14] A. Brandt, I. Yavneh, On multigrid solution of high-Reynolds incompressible flows, J. Comput. Phys. 101 (1992) 151–164.

[15] A. Brandt, The Gauss Center research in multiscale scientific computing, Electron. Trans. Numer. Anal. 6 (1997) 1–34.

[16] A. Brandt, Barriers to achieving textbook multigrid efficiency (TME) in CFD, NASA Report CR-1998-207647, ICASE Hampton, 1998.

[17] W.L. Briggs, A Multigrid Tutorial, SIAM, Philadelphia, PA, 1987.

[18] J. Broeze, B. Geurts, H. Kuerten, M. Streng, Multigrid acceleration of time-accurate DNS of compressible turbulent flow, in: N.D. Melson, T.A. Manteuffel, S.F. Mc.Cormick, C.C. Douglas (Eds.), Seventh Copper Mountain Conference on Multigrid Methods, NASA Conference Publication 3339, NASA Hampton, 1996, pp. 109–121.

[19] L.A. Catalano, H. Deconinck, Two-dimensional optimization of smoothing properties of multi-stage schemes applied to hyperbolic equations, in: W. Hackbusch, U. Trottenberg (Eds.), Multigrid Methods: Special Topics and Applications II, GMD-Studien, Vol. 189, GMD St. Augustin, Germany, 1991, pp. 43–56.

[20] D.A. Caughey, Implicit multigrid techniques for compressible flows, Comput. & Fluids 22 (1993) 117–124.

[21] T.F. Chan, S. Go, L. Zikatanov, Multilevel methods for elliptic problems on unstructured grids, VKI Lecture Series 1997-02, Vol. 1, v.Karman Inst. Rhode St.G. Belgium, 1997.

[22] G. Chesshire, W.D. Henshaw, Composite overlapping meshes for the solution of partial differential equations, J. Comput. Phys. 90 (1990) 1–64.

[23] A.J. Chorin, A numerical method for solving incompressible viscous flow problems, J. Comput. Phys. 2 (1967) 12–26.

[24] W.K. Cope, G. Wang, S.P. Vanka, A staggered grid multilevel method for the simulation of fluid flow in 3-D complex geometries, AIAA Paper 94-0778, 1994.

[25] D.L. Darmofal, K. Siu, A robust multigrid algorithm for the Euler equations with local preconditioning and semi-coarsening, J. Comput. Phys. 151 (1999) 728–756.

[26] A.T. Degani, G.C. Fox, Parallel multigrid computation of the unsteady incompressible Navier–Stokes equations, J. Comput. Phys. 128 (1996) 223–236.

[27] J.E. Dendy, S.F. McCormick, J.W. Ruge, T.F. Russel, S. Schaffer, Multigrid methods for three-dimensional petroleum reservoir simulation, Paper SPE 18409, presented at the Tenth Symposium on Reservoir Simulation, Houston, TX, 1989.

[28] G.B. Deng, J. Piquet, P. Quetey, M. Visonneau, A new fully coupled solution of the Navier–Stokes equations, Internat. J. Numer. Methods Fluids 19 (1994) 605–639.

[29] A. Dervieux, C. Debiez, J. Francescatto, V. Viozat, Development of vertex-centered technologies for flow simulation, in: K.D. Papailiou et al. (Eds.), Computational Fluid Dynamics'98, Proceedings of the ECCOMAS Conference, Vol. 2, Wiley, Chichester, 1998 pp. 38–44.

[30] E. Dick, Multigrid formulation of polynomial flux-difference splitting for steady Euler equations, J. Comput. Phys. 91 (1990) 161–173.

[31] E. Dick, Multigrid solution of steady Euler equations based on polynomial flux-difference splitting, Internat. J. Numer. Methods Heat Fluid Flow 1 (1991) 51–62.

[32] E. Dick, J. Linden, A multigrid method for steady incompressible Navier–Stokes equations based on flux difference splitting, Internat. J. Numer. Methods Fluids 14 (1992) 1311–1323.

[33] E. Dick, K. Riemslagh, Multi-staging of Jacobi relaxation to improve smoothing properties of multigrid methods for steady Euler equations, J. Comput. Appl. Math. 50 (1994) 241–254.

[34] E. Dick, K. Riemslagh, Multi-staging of Jacobi relaxation to improve smoothing properties of multigrid methods for steady Euler equations, II, J. Comput. Appl. Math. 59 (1995) 339–348.

[35] E. Dick, J. Steelant, Coupled solution of the steady compressible Navier–Stokes equations and the $k-\varepsilon$ turbulence equations with a multigrid method, Appl. Numer. Math. 23 (1997) 49–61.

[36] D. Drikakis, O.P. Iliev, D.P. Vassileva, A nonlinear multigrid method for the three-dimensional incompressible Navier–Stokes equations, J. Comput. Phys. 146 (1998) 301–321.

[37] F. Durst, M. Schäfer, A parallel block-structured multigrid method for the prediction of incompressible flows, Internat. J. Numer. Methods Fluids 22 (1996) 549–565.

[38] R.P. Fedorenko, The speed of convergence of one iterative process, USSR Comput. Math. Math. Phys. 4 (3) (1964) 227–235.

[39] L. Ferm, P. Lötstedt, Blockwise adaptive grids with multigrid acceleration for compressible flow, AIAA J. 37, 1999.

[40] P. Gerlinger, D. Brüggemann, An implicit multigrid scheme for the compressible Navier–Stokes equations with low-Reynolds-number closure, Trans. ASME/J. Fluids Eng. 120 (1998) 257–262.

[41] T. Gjesdal, M.E.H. Lossius, Comparison of pressure correction smoothers for multigrid solution of incompressible flow, Internat. J. Numer. Methods Fluids 25 (1997) 393–405.

[42] S.K. Godunov, Finite difference method for numerical computation of discontinuous solutions of the equations of fluid dynamics, Mat. Sbornik 47 (1959) 271–306.

[43] W. Hackbusch, On the multi-grid method applied to difference equations, Computing 20 (1978) 291–306.

[44] W. Hackbusch, Survey of convergence proofs for multi-grid iterations, in: J. Frehse, D. Pallaschke, U. Trottenberg (Eds.), Special topics of applied mathematics, North-Holland, Amsterdam, 1980, pp. 151–164.

[45] W. Hackbusch, Convergence of multi-grid iterations applied to difference equations, Math. Comput. 34 (1980) 425–440.

[46] W. Hackbusch, U. Trottenberg (Eds.), Multigrid Methods, Lecture Notes in Mathematics, Vol. 960, Springer, Berlin, 1982.

[47] W. Hackbusch, Multi-Grid Methods and Applications, Springer, Berlin, 1985.

[48] W. Hackbusch, U. Trottenberg (Eds.), Multigrid Methods II, Lecture Notes in Mathematics, Vol. 1228, Springer, Berlin, 1986.

[49] W. Hackbusch, U. Trottenberg (Eds.), Multigrid Methods III, Proceedings of the Third International Conference on Multigrid Methods, International Series of Numerical Mathematics, Vol. 98, Birkhäuser, Basel, 1991.

[50] W. Hackbusch, G. Wittum (Eds.), Multigrid Methods V, Lecture Notes in Computational Science and Engineering, Vol. 3, Springer, Berlin, 1998.

[51] R.F. Hanby, D.J. Silvester, J.W. Chew, A comparison of coupled and segregated iterative solution techniques for incompressible swirling flow, Internat. J. Numer. Methods Fluids 22 (1996) 353–373.

[52] P. He, M. Salcudean, I.S. Gartshore, P. Nowak, Multigrid calculation of fluid flows in complex 3D geometries using curvilinear grids, Comput. Fluids 25 (1996) 395–419.

[53] P.W. Hemker, S.P. Spekreijse, Multiple grid and Osher's scheme for the efficient solution of the steady Euler equations, Appl. Numer. Math. 2 (1986) 475–493.

[54] P.W. Hemker, P. Wesseling (Eds.), Multigrid Methods IV, Proceedings of the Fourth European Multigrid Conference, Amsterdam, Birkhäuser, Basel, 1994.

[55] W.D. Henshaw, G. Chesshire, Multigrid on composite meshes, SIAM J. Sci. Comput. 8 (1987) 914–923.

[56] C. Hirsch, Numerical Computation of Internal and External Flows, Vol. 2, Wiley, Chichester, 1990.

[57] B. Huurdeman, G. Wittum, Multigrid solution for 2D incompressible turbulent flows on hybrid grids, in: K.D. Papailiou et al. (Eds.), Computational Fluid Dynamics'98, Proceedings of the ECCOMAS Conference, Vol. 2, Wiley, Chichester, 1998, pp. 266–271.

[58] S.O. Ibraheem, A.O. Demuren, On bi-grid local mode analysis of solution techniques for 3-D Euler and Navier–Stokes equations, J. Comput. Phys. 125 (1996) 354–377.

[59] A. Jameson, W. Schmidt, E. Turkel, Numerical solution of the Euler equations by finite volume methods using Runge-Kutta time stepping schemes, AIAA Paper 81-1259, 1981.

[60] A. Jameson, Transonic flow calculations for aircraft, in: F. Brezzi (Ed.), Numerical Methods in Fluid Mechanics, Lecture Notes in Mathematics, Vol. 1127, Springer, Berlin, 1985, pp. 156–242.

[61] A. Jameson, Analysis and design of numerical schemes for gas dynamics, 1: artificial diffusion, upwind biasing, limiters and their effect on accuracy and multigrid convergence, Internat. J. Comput. Fluid Dynamics 4 (1995) 171–218.

[62] D.S. Joshi, S.P. Vanka, Multigrid calculation procedure for internal flows in complex geometries, Numer. Heat Transfer B 20 (1991) 61–80.

[63] P. Johansson, L. Davidson, Modified collocated SIMPLEC algorithm applied to buoyancy-affected turbulent flow using a multigrid solution procedure, Numer. Heat Transfer Part B 28 (1995) 39–57.

[64] R. Jyotsna, S.P. Vanka, Multigrid calculation of steady, viscous flow in a triangular cavity, J. Comput. Phys. 122 (1995) 107–117.

[65] R. Jyotsna, S.P. Vanka, A pressure based multigrid procedure for the Navier–Stokes equations on unstructured grids, in: N.D. Melson, T.A. Manteuffel, S.F. Mc.Cormick, C.C. Douglas (Eds.), Seventh Copper Mountain Conference on Multigrid Methods NASA Conference Publication 3339, NASA Hampton, 1996, pp. 409–424.

[66] K.C. Karki, P.S. Sathyamurthy, S.V. Patankar, Performance of a multigrid method with an improved discretization scheme for three-dimensional fluid flow calculations, Numer. Heat Transfer B 29 (1996) 275–288.

[67] R. Kettler, P. Wesseling, Aspects of multigrid methods for problems in three dimensions, Appl. Math. Comput. 19 (1986) 159–168.

[68] B. Koren, Defect correction and multigrid for an efficient and accurate computation of airfoil flows, J. Comput. Phys. 77 (1988) 183–206.

[69] B. Koren, P.W. Hemker, C.T.H. Everaars, Multiple semi-coarsened multigrid for 3D CFD, AIAA Paper 97-2029, 1997.

[70] B. Koren, Multigrid and defect correction for the steady Navier–Stokes equations, J. Comput. Phys. 87 (1990) 25–46.

[71] A. Krechel, H-J. Plum, K. Stüben, Parallelization and vectorization aspects of the solution of tridiagonal linear systems, Parallel Comput. 14 (1990) 31–49.

[72] B. Van Leer, C.-H. Tai, K.G. Powell, Design of optimally smoothing multistage schemes for the Euler Equations, AIAA Paper 89-1933, 1989.

[73] B. van Leer, W.-T. Lee, P.L. Roe, Characteristic time-stepping or local preconditioning of the Euler equations, AIAA Paper 91-1552, 1991.

[74] F.S. Lien, M.A. Leschziner, Multigrid acceleration for recirculating laminar and turbulent flows computed with a non-orthogonal, collocated finite-volume scheme, Comput. Methods Appl. Mech. Eng. 118 (1994) 351–371.

[75] Z. Lilek, S. Muzaferija, M. Perić, Efficiency and accuracy aspects of a full-multigrid SIMPLE algorithm for three-dimensional flows, Numer. Heat Transfer B 31 (1997) 23–42.

[76] S.-Y. Lin, T.-M. Wu, An adaptive multigrid finite-volume scheme for incompressible Navier–Stokes equations, Internat. J. Numer. Methods Fluids 17 (1993) 687–710.

[77] F.B. Lin, F. Sotiropoulos, Strongly-coupled multigrid method for 3-D incompressible flows using near-wall turbulence closures, Trans. ASME - J. Fluids Eng. 119 (1997) 314–324.

[78] J. Linden, B. Steckel, K. Stüben, Parallel multigrid solution of the Navier–Stokes equations on general 2D-domains, Parallel Comput. 7 (1988) 461–475.

[79] J. Linden, G. Lonsdale, H. Ritzdorf, A. Schüller, Scalability aspects of parallel multigrid, Fut. Generation Comput. Systems 10 (1994) 429–439.

[80] F. Liu, X. Zheng, A strongly coupled time-marching method for solving the Navier–Stokes and $k-\omega$ turbulence model equations with multigrid, J. Comput. Phys. 128 (1996) 289–300.

[81] G. Lonsdale, A. Schüller, Multigrid efficiency for complex flow simulations on distributed memory machines, Parallel Comput. 19 (1993) 23–32.

[82] J.Z. Lou, R. Ferraro, A parallel incompressible flow solver package with a parallel multigrid elliptic kernel, J. Comput. Phys. 125 (1996) 225–243.

[83] L. Martinelli, A. Jameson, Validation of a multigrid method for the Reynolds averaged equations, AIAA Paper 88-0414, 1988.

[84] D.J. Mavripilis, Multigrid solution of the two-dimensional Euler equations on unstructured triangular meshes, AIAA J. 26 (1988) 824–831.

[85] D.J. Mavripilis, V. Venkatakrishnan, Agglomeration multigrid for two-dimensional viscous flows, Comput. & Fluids 24 (1995) 553–570.

[86] D.J. Mavripilis, V. Venkatakrishnan, A 3D agglomeration multigrid solver for the Reynolds averaged Navier–Stokes equations on unstructured meshes, Internat. J. Numer. Methods Fluids 23 (1996) 527–544.

[87] D.J. Mavripilis, V. Venkatakrishnan, A unified multigrid solver for the Navier–Stokes equations on mixed element meshes, Internat. J. Comput. Fluid Dynamics 8 (1998) 247–263.

[88] D.J. Mavripilis, Multigrid strategies for viscous flow solvers on anisotropic unstructured meshes, J. Comput. Phys. 145 (1998) 141–165.

[89] O.A. McBryan, P.O. Frederickson, J. Linden, A. Schüller, K. Solchenbach, K. Stüben, C.A. Thole, U. Trottenberg, Multigrid methods on parallel computers-a survey of recent developments, Impact Comput. Sci. Eng. 3 (1991) 1–75.

[90] S.F. Mc.Cormick, Multilevel Adaptive Methods for Partial Differential Equations, Frontiers in Applied Mathematics, Vol. 6, SIAM, Philadelphia, PA, 1989.

[91] P.H. Michielse, H.A. van der Vorst, Data transport in Wang's partition method, Parallel Comput. 7 (1988) 87–95.

[92] T.F. Miller, F.W. Schmidt, Use of a pressure-weighted interpolation method for the solution of the incompressible Navier–Stokes equations on a nonstaggered grid system, Numer. Heat Transfer 14 (1988) 213–233.

[93] T.F. Miller, Fourier analysis of the SIMPLE algorithm formulated on a collocated grid, Numer. Heat Transfer B 30 (1996) 45–66.

[94] E. Morano, A. Dervieux, Steady relaxation methods for unstructured multigrid Euler and Navier–Stokes solutions, Comput. Fluid Dynamics 5 (1995) 137–167.

[95] E. Morano, D.J. Mavripilis, V. Venkatakrishnan, Coarsening strategies for unstructured multigrid techniques with application to anisotropic problems, in: N.D. Melson, T.A. Manteuffel, S.F. Mc.Cormick, C.C. Douglas (Eds.), Seventh Copper Mountain Conference on Multigrid Methods NASA Conference Publication 3339, NASA Hampton, 1996, pp. 591–606.

[96] W.A. Mulder, Multigrid relaxtion for the Euler equations, J. Comput. Phys. 60 (1985) 235–252.

[97] W.A. Mulder, A new multigrid approach to convection problems, J. Comput. Phys. 83 (1989) 303–323.

[98] W.A. Mulder, A high resolution Euler solver based on multigrid, semi-coarsening, and defect correction, J. Comput. Phys. 100 (1992) 91–104.

[99] N.H. Naik, J. van Rosendale, The improved robustness of multigrid elliptic solvers based on multiple semicoarsened grids, SIAM Numer. Anal. 30 (1993) 215–229.

[100] Z.P. Nowak, M. Salcudean, Turbulent flow calculations by the nonlinear multi-grid method, Z. Angew. Math. Mech. 76 (1996) 463–469.

[101] C.W. Oosterlee, P. Wesseling, A robust multigrid method for a discretization of the incompressible Navier–Stokes equations in general coordinates, Impact Comput. Sci. Eng. 5 (1993) 128–151.

[102] C.W. Oosterlee, The convergence of parallel multiblock multigrid methods, Appl. Numer. Math. 19 (1995) 115–128.

[103] C.W. Oosterlee, A GMRES-based plane smoother in multigrid to solve 3D anisotropic fluid flow problems, J. Comput. Phys. 130 (1997) 41–53.

[104] C.W. Oosterlee, F.J. Gaspar, T. Washio, R. Wienands, Multigrid line smoothers for higher order upwind discretizations of convection-dominated problems, J. Comput. Phys. 139 (1998) 274–307.

[105] C.W. Oosterlee, F.J. Gaspar, T. Washio, Parallel adaptive multigrid with nonlinear Krylov subspace acceleration for steady 3D CFD problems In: K.D. Papailiou et al. (Eds.), Computational Fluid Dynamics'98, Proceedings of the ECCOMAS conference, Vol 2, Wiley, Chichester, 1998, 272–277.

[106] M.F. Paisley, Multigrid computation of stratified flow over two-dimensional obstacles, J. Comput. Phys. 136 (1997) 411–424.

[107] M.F. Paisley, N.M. Bhatti, Comparison of multigrid methods for neutral and stably stratified flows over two-dimensional obstacles, J. Comput. Phys. 142 (1998) 581–610.

[108] N.G. Pantelelis, A.E. Kanarachos, The parallel block adaptive multigrid method for the implicit solution of the Euler equations, Internat. J. Numer. Methods Fluids 22 (1996) 411–428.

[109] S.V. Patankar, D.B. Spalding, A calculation procedure for heat and mass transfer in three-dimensional parabolic flows, Internat. J. Heat Mass Transfer 15 (1972) 1787–1806.

[110] N.A. Pierce, M.B. Giles, Preconditioned multigrid methods for compressible flow calculations on stretched meshes, J. Comput. Phys. 136 (1997) 425–445.

[111] N.A. Pierce, M.B. Giles, A. Jameson, L. Martinelli, Accelerating three-dimensional Navier-Stokes calculations, AIAA Paper 97-1953, 1997.

[112] N.A. Pierce, J.J. Alonso, Efficient computation of unsteady viscous flows by an implicit preconditioned multigrid method, AIAA J. 36 (1998) 401–408.

[113] R. Radespiel, R.C. Swanson, Progress with multigrid schemes for hypersonic flow problems, J. Comput. Phys. 116 (1995) 103–122.

[114] D. Rayner, Multigrid flow solutions in complex two-dimensional geometries, Internat. J. Numer. Methods Fluids 13 (1991) 507–518.

[115] C.M. Rhie, W.L. Chow, Numerical study of the turbulent flow past an airfoil with trailing edge separation, AIAA J. 21 (1983) 1525–1532.

[116] K. Riemslagh, E. Dick, Multi-stage Jacobi relaxation in multigrid methods for the steady Euler equations, Internat. J. Comput. Fluid Dynamics 4 (1995) 343–361.

[117] H. Ritzdorf, K. Stüben, Adaptive multigrid on distributed memory computers, in: P.W. Hemker, P. Wesseling (Eds.), Multigrid Methods IV, Proceedings of the Fourth European Multigrid Conference, Amsterdam, Birkhäuser, Basel, 1994, pp. 77–95.

[118] H. Ritzdorf, A. Schüller, B. Steckel, K. Stüben, L_iSS- An environment for the parallel multigrid solution of partial differential equations on general domains, Parallel Comput. 20 (1994) 1559–1570.

[119] W. Rodi, S. Majumdar, B. Schönung, Finite volume methods for two-dimensional incompressible flow problems with complex boundaries, Comput. Methods Appl. Mech. Eng. 75 (1989) 369–392.

[120] M. Rosenfeld, D. Kwak, Multigrid acceleration of a fractional-step solver in generalized curvilinear coordinate systems, AIAA Paper 92-0185, 1992.

[121] J. Ruge, K. Stüben, Algebraic Multigrid (AMG), in: S.F. Mc.Cormick (Ed.), Multigrid Methods, Frontiers in Applied Mathematics, Vol. 5, SIAM, Philadelphia, PA, 1987, pp. 73–130.

[122] A. Schüller (Ed.), in: Portable Parallelization of Industrial Aerodynamic Applications (POPINDA), Results of a BMBF Project, Notes on Numerical Fluid Mechanics, Vol. 71, Vieweg Braunschweig, 1999.

[123] G.J. Shaw, S. Sivaloganathan, On the smoothing of the SIMPLE pressure correction algorithm, Internat. J. Numer. Methods Fluids 8 (1988) 441–462.

[124] C. Sheng, L.K. Taylor, D.L. Whitfield, Multigrid algorithm for three-dimensional high-Reynolds number turbulent flow, AIAA J. 33 (1995) 2073–2079.

[125] S.G. Sheffer, L. Martinelli, A. Jameson, An efficient multigrid algorithm for compressible reactive flows, J. Comput. Phys. 144 (1998) 484–516.

[126] W. Shyy, C.-S. Sun, Development of a pressure-correction/staggered-grid based multigrid solver for incompressible recirculating flows, Comput. & Fluids 22 (1993) 51–76.

[127] S. Sivaloganathan, The use of local mode analysis in the design and comparison of multigrid methods, Comput. Phys. Comm. 65 (1991) 246–252.

[128] K.M. Smith, W.K. Cope, S.P. Vanka, A multigrid procedure for three-dimensional flows on non-orthogonal collocated grids, Internat. J. Numer. Methods Fluids 17 (1993) 887–904.

[129] P. Sockol, Multigrid solution of the Navier–Stokes equations on highly stretched grids, Internat. J. Numer. Methods Fluids 17 (1993) 543–566.

[130] F. Sotiropoulos, G. Constantinescu, Pressure-based residual smoothing operators for multistage pseudocompressibility algorithms, J. Comput. Phys. 133 (1997) 129–145.

[131] J.C. South, A. Brandt, Application of a multi-level grid method to transonic flow calculations, NASA Langley Research Center, ICASE Report 76-8, Hampton, Virginia, 1976.

[132] S.P. Spekreijse, Multigrid solution of second-order discretizations of hyperbolic conservation laws, Math. Comput. 49 (1987) 135–155.

[133] J. Steelant, E. Dick, S. Pattijn, Analysis of robust multigrid methods for steady viscous low Mach number flows, J. Comput. Phys. 136 (1997) 603–628.

[134] K. Stüben, U. Trottenberg, Multigrid methods: fundamental algorithms, model problem analysis and applications, in: W. Hackbusch, U. Trottenberg (Eds.), Multigrid Methods, Lecture Notes in Mathematics, Vol. 960, Springer, Berlin, 1982, pp. 1–176.

[135] C.-H. Tai, J.-H. Sheu, B. van Leer, Optimal multistage schemes for Euler equations with residual smoothing, AIAA J. 33 (1995) 1008–1016.

[136] C.-A. Thole, U. Trottenberg, Basic smoothing procedures for the multigrid treatment of elliptic 3D-operators, Appl. Math. Comput. 19 (1986) 333–345.

[137] M.C. Thompson, J.H. Ferziger, An adaptive multigrid technique for the incompressible Navier–Stokes equations, J. Comput. Phys. 82 (1989) 94–121.

[138] J.Y. Tu, L. Fuchs, Calculation of flows using three-dimensional overlapping grids and multigrid methods, Internat. J. Numer. Methods Eng. 38 (1995) 259–282.

[139] S. Turek, A comparative study of some time-stepping techniques for the incompressible Navier–Stokes equations: from fully implicit nonlinear schemes to semi-implicit projection methods, Internat. J. Numer. Methods Fluids 22 (1996) 987–1011.

[140] S. Turek, Multigrid techniques for simple discretely divergence-free finite element spaces, in: P.W. Hemker, P. Wesseling (Eds.), Multigrid Methods IV, Proceedings of the Fourth European Multigrid Conference, Amsterdam, Birkhäuser, Basel, 1994, pp. 321–333.

[141] S.P. Vanka, Block-implicit multigrid solution of Navier–Stokes equations in primitive variables, J. Comput. Phys. 65 (1986) 138–158.

[142] R.S. Varga, Matrix Iterative Analysis, Prentice-Hall, Englewood Cliffs, NJ, 1962.

[143] V.N. Vatsa, B.W. Wedan, Development of a multigrid code for 3-D Navier–Stokes equations and its application to a grid-refinement study, Comput. & Fluids 18 (1990) 391–403.

[144] J. Vierendeels, K. Riemslagh, E. Dick, A multigrid semi-implicit line-method for viscous incompressible and low-Mach-number flows on high aspect ratio grids, J. Comput. Phys. 154 (1999) 310–341.

[145] K. Warendorf, U. Küster, R. Rühle, Multilevel methods for a highly-unstructured Euler solver, in: K.D. Papailiou, et al., (Eds.), Computational Fluid Dynamics'98, Proceedings of the ECCOMAS Conference, Vol. 1, Wiley, Chichester, 1998, pp. 1252–1257.

[146] T. Washio, C.W. Oosterlee, Krylov subspace acceleration for nonlinear multigrid schemes, Electron. Trans. Numer. Anal. 6 (1997) 271–290.

[147] T. Washio, C.W. Oosterlee, Flexible multiple semicoarsening for three-dimensional singularly perturbed problems, SIAM J. Sci. Comput. 19 (1998) 1646–1666.

[148] P. Wesseling, P. Sonneveld, Numerical experiments with a multiple grid and a preconditioned Lanczos type method, in: R. Rautmann (Ed.), Approximation Methods for Navier–Stokes Problems, Lecture Notes in Mathematics, Vol. 771, Springer, Berlin, 1980, pp. 543–562.

[149] P. Wesseling, Theoretical and practical aspects of a multigrid method, SIAM J. Sci. Comput. 3 (1982) 387–407.

[150] P. Wesseling, Multigrid methods in computational fluid dynamics, Z. Angew. Math. Mech. 70 (1990) T337–T348.

[151] P. Wesseling, A survey of Fourier smoothing analysis results, in: W. Hackbusch, U. Trottenberg (Eds.), Multigrid Methods III, Proceedings of the Third International Conference on Multigrid Methods Int. Series of Numerical Mathematics, Vol. 98, Birkhäuser, Basel, 1991, pp. 105–127.

[152] P. Wesseling, An Introduction to Multigrid Methods, Wiley, Chichester, 1992.

[153] P. Wesseling, A. Segal, C.G.M. Kassels, Computing flows on general three-dimensional nonsmooth staggered grids, J. Comput. Phys. 149 (1999) 333–362.

[154] G. Wittum, Multi-grid methods for Stokes and Navier–Stokes equations with transforming smoothers: algorithms and numerical results, Numer. Math. 54 (1989) 543–563.

[155] G. Wittum, On the convergence of multi-grid methods with transforming smoothers, Numer. Math. 57 (1990) 15–38.

[156] N.G. Wright, P.H. Gaskell, An efficient multigrid approach to solving highly recirculating flows, Comput. & Fluids 24 (1995) 63–79.

[157] J.A. Wright, W. Shyy, A pressure-based composite grid method for the Navier–Stokes equations, J. Comput. Phys. 107 (1993) 225–238.

[158] I. Yavneh, C.H. Venner, A. Brandt, Fast multigrid solution of the advection problem with closed characteristics, SIAM J. Sci. Comput. 19 (1998) 111–125.

[159] S. Yoon, A. Jameson, Lower-Upper-Symmetric-Gauss-Seidel method for the Euler and Navier–Stokes equations, AIAA J. 26 (1988) 1025–1026.

[160] S. Yoon, A. Jameson, D. Kwak, Effect of artificial diffusion schemes on multigrid convergence, AIAA Paper, 95-1670, 1995.

[161] P.M. de Zeeuw, E.J. van Asselt, The convergence rate of multi-level algorithms applied to the convection-diffusion equation, SIAM J. Sci. Comput. 6 (1985) 492–503.

[162] P.M. de Zeeuw, Matrix-dependent prolongations and restrictions in a blackbox multigrid solver, J. Comput. Appl. Math. 3 (1990) 1–27.

[163] P.M. de Zeeuw, Incomplete line LU as smoother and as preconditioner, in: W. Hackbusch, G. Wittum (Eds.), Incompete decompsitions (ILU) – algorithms, theory, and applications, Vieweg, Braunschweig, 1993, pp. 215–224.
[164] S. Zeng, P. Wesseling, Multigrid solution of the incompressible Navier–Stokes equations in general coordinates, SIAM J. Numer. Anal. 31 (1994) 1764–1784.
[165] S. Zeng, P. Wesseling, An ILU smoother for the incompressible Navier–Stokes equations in general coordinates, Internat. J. Numer Methods Fluids 20 (1995) 59–74.
[166] L. Zeng, M.D. Matovic, A. Pollard, R.W. Sellens, A distributive mass balance correction in single- and multigrid incompressible flow calculation, Comput. Fluid Dynamics 6 (1996) 125–135.
[167] X. Zheng, C. Liao, C. Liu, C.H. Sung, T.T. Huang, Multigrid computation of incompressible flows using two-equation turbulence models: Part 1 – numerical method, Trans. ASME – J. Fluids Eng. 119 (1997) 893–899.
[168] L.-B. Zhang, A multigrid solver for the steady incompressible Navier–Stokes equations on curvilinear coordinate systems, J. Comput. Phys. 113 (1994) 26–34.
[169] Z.W. Zhu, C. Lacor, C. Hirsch, A new residual smoothing method for multigrid acceleration applied to the Navier–Stokes equations, in: P.W. Hemker, P. Wesseling (Eds.), Multigrid Methods IV, Proceedings of the Fourth European Multigrid Conference, Amsterdam, Birkhäuser, Basel, 1994, pp. 345–356.

ELSEVIER

Journal of Computational and Applied Mathematics 128 (2001) 335–362

JOURNAL OF COMPUTATIONAL AND APPLIED MATHEMATICS

www.elsevier.nl/locate/cam

The method of subspace corrections ☆

Jinchao Xu

Department of Mathematics, Pennsylvania State University, 309 McAllister Bldg., University Park, PA 16802, USA

Received 28 January 2000; received in revised form 16 March 2000

Abstract

This paper gives an overview for the method of subspace corrections. The method is first motivated by a discussion on the local behavior of high-frequency components in a solution to an elliptic problem. A simple domain decomposition method is discussed as an illustrative example and multigrid methods are discussed in more detail. Brief discussions are also given to some non-linear examples including eigenvalue problems, obstacle problems and liquid crystal modelings. The relationship between the method of subspace correction and the method of alternating projects is observed and discussed.

1. Introduction

The method of subspace corrections refer to a large class of algorithms used in scientific and engineering computing. This type of method is based on an old and simple strategy: divide and conquer. Many iterative methods (simple or complicated, traditional or modern) fall into this category. Examples include the Jacobi method, Gauss–Seidel methods, point or block relaxation method, multigrid method and domain decomposition method. All these methods can be applied to both linear and nonlinear problems.

This paper is to give a glimpse of this type of method when it is applied to approximate the solutions of partial differential equations. While these methods can be applied to a large variety of problems, it is when they are applied to partial differential equations that these methods become practically most valuable and mathematically profoundly interesting. Among many such algorithms in this category, the multigrid method is certainly the most remarkable example.

As a "divide and conquer" strategy, the first question is perhaps how to "divide" namely how to divide a big (global) problem into small (local) ones. In this paper, we shall address this question by discussing the local property of high-frequency components in the solution to elliptic partial

☆ This work was partially supported by NSF DMS-9706949, NSF ACI-9800244 and NASA NAG2-1236 through Penn State university.

E-mail address: xu@math.psu.edu (J. Xu).

0377-0427/01/$ - see front matter
PII: S0377-0427(00)00518-5

differential equations. This will be discussed in Section 2. After such a discussion, an overlapping domain decomposition method is then a natural algorithm to introduce. We then proceed later on to introduce the multigrid method as a recursive application of the overlapping domain decomposition method.

Our main focus of the presentation will be on iterative methods for linear algebraic system. In Section 3, we shall give a brief introduction to basic linear iterative methods and (preconditioned) conjugate gradient methods. In Section 4, a general framework is introduced for the method of subspace corrections based on space decomposition. Basic ideas, simple examples and convergence analysis will be discussed here.

As a special example of the method of subspace corrections, the multigrid method will be discussed in some length in Section 5. Here we use a model of a simple elliptic equation discretized by linear finite elements. We introduce a simple variant of the multigrid method, \-cycle, and then sketch two different convergence proofs. We then introduce and discuss the BPX preconditioner. Finally, in this section, we give some brief discussions of algebraic multigrid methods.

The method of subspace corrections can be applied to many nonlinear problems. Several examples are given in Section 6, including an eigenvalue problem, an obstacle problem and a nonlinear nonconvex optimization problem arising from liquid crystal modeling.

There is a class of methods, called the method of alternating projections, that has been studied by many researchers in the approximation research community. In the last section, Section 7, we exam the relationship between the method of subspace corrections and the method of alternating projections. Some new observations are made here.

For convenience, following [50], the symbols $\lesssim, \gtrsim$ and $\eqsim$ will be used in this paper. That $x_1 \lesssim y_1$, $x_2 \gtrsim y_2$ and $x_3 \eqsim y_3$, mean that $x_1 \leqslant C_1 y_1$, $x_2 \geqslant c_2 y_2$ and $c_3 x_3 \leqslant y_3 \leqslant C_3 x_3$ for some constants C_1, c_2, c_3 and C_3 that are independent of mesh parameters.

2. Motivations: local behavior of high frequencies

The method of subspace corrections is based on a simple old idea: *divide and conquer*. In other words, we try to solve a big problem by breaking it apart and solving a number of smaller problems. The crucial question is then how to break a big problem into some smaller ones. In the solution of partial differential equations, in our view, the clue is in the behavior of high-frequency part of the solution.

2.1. Descriptive definition of high frequencies

High-frequency functions, loosely speaking, refer to those functions that have relatively high oscillations. Let us begin our discussion with a simple example of a differential equation. We consider the following two-point boundary value problem:

$$-u'' = f \quad x \in (0, 1), \quad u(0) = u(1) = 0. \tag{2.1}$$

It is easy to see that the solution of the above problem can be given in terms of the following

Fourier series:

$$u(x) = \sum_{k=0}^{\infty} c_k \psi_k(x), \tag{2.2}$$

where each $\psi_k(x) = \sin k\pi x$ happens to be an eigenfunction of the underlying differential operator corresponding to the eigenvalue $(k\pi)^2$. Obviously the function ψ_k oscillates more as k gets larger. Let us call k the frequency of the function ψ_k. For large k, ψ_k may be called a high-frequency function. Similarly, for small k, ψ_k may be called a low-frequency function. Apparently "high" or "low" is a relative concept.

The Fourier expansion gives a representation of the solution in terms of a linear combination of functions of different frequencies. A function is relatively smooth if its low frequency components dominate (namely the coefficients c_k are relatively large for small k's) and conversely a function is relatively rough (or nonsmooth) if its high-frequency components dominate.

The concept of frequencies described above naturally carries over to the discretized system. Consider a uniform partition of $(0,1)$ by $n+1$ equal-sized subintervals with nodal points $x_i = i/(n+1)$ $(1 \leqslant i \leqslant n)$ and discretize problem (2.1) by a linear finite element or a finite difference method. We have the following discretized system:

$$A\mu = b, \tag{2.3}$$

where $A = h^{-2}\,\mathrm{diag}(-1, 2, -1) \in R^{n\times n}$. It is easy to see that the matrix A has the eigenvalues and eigenvectors:

$$\lambda_k = (n+1)^2 \sin^2 \frac{k\pi}{n+1} \quad \text{and} \quad \xi_j^k = \sin \frac{jk\pi}{n+1},$$

which behave similar to those in the continuous case.

From an approximation point of view, higher frequencies are more difficult to resolve and they require a finer discretization scheme. But high frequencies have many very important properties that can be used advantageously to design effective numerical schemes in many situations.

2.2. Locality

One most important property of high frequencies is that they tend to behave locally in elliptic partial differential equations. Let us now use a very simple example to explain what this locality means roughly.

Let $G = (-1, 1)^n$ and $G_0 = (-\frac{1}{2}, \frac{1}{2})^n$. Then, there exists a constant c such that, for any harmonic function v on G, namely $\Delta v = 0$, the following estimate holds:

$$\| \nabla v \|_{0,G_0} \lesssim \| v \|_{0,G}\,. \tag{2.4}$$

This is, of course a well-known trivial property of harmonic functions which can be derived easily by simple integration by parts using some cut-off functions. Now, we shall use it to explain the locality of high frequencies.

Given any reasonable (say, Lipschitz) domain $\Omega \subset R^n$, let us consider the boundary value problem

$$-\Delta u = f(x) \quad \text{in } \Omega \quad \text{and} \quad u = 0 \quad \text{on } \partial\Omega. \tag{2.5}$$

We now attempt to solve this problem locally. Given any $z \in \Omega$, let $G^\rho \subset \Omega$ be an open ball of radius ρ centered at z. Let us now consider the following local problem:

$$- \Delta u_\rho = f(x), \quad x \in G^\rho \quad \text{and} \quad u_\rho = 0 \quad \text{on } \partial G^\rho. \tag{2.6}$$

We would of course not expect that this local solution u_ρ would be any good approximation to the original solution u of (2.5). But, anyway, let us look at the error $u - u_\rho$ which is obviously harmonic in G^ρ. By (2.4) and a simple scaling argument, we then have

$$\| \nabla(u - u_\rho) \|_{0,G^{\rho/2}} \lesssim \rho^{-1} \| u - u_\rho \|_{0,G^\rho} \tag{2.7}$$

where $G^{\rho/2} \subset G^\rho$ is a ball of radius $\rho/2$.

Apparently, this means that, regardless what f is, the relative frequency of $u - u_\rho$ in $G^{\rho/2}$ is, roughly speaking, at most ρ^{-1} asymptotically; in other words, the local solution u_ρ actually captures very well the frequencies in u that oscillate at distance less than or equal to ρ.

This is what we mean by saying that high frequencies behave locally in elliptic partial differential equations. Singularities, for example, are some form of high frequencies. In the finite element method, many forms of singularity can be resolved through certain local mesh refinement and the reason why this type of method works is also because of the local behavior of high frequencies.

This local property of the high frequencies is closely related to the maximum principle in elliptic equations. Many of qualitative studies in elliptic problems may be interpreted as the studies of the behavior of high frequencies.

The Poisson equation that we just discussed exhibits a pointwise locality for the high frequencies and this property is reflected from the fact that the level set of the fundamental solution of the Laplacian is an $(n-1)$-dimensional sphere.

The locality is different for an anisotropic or convection-dominated operator

$$- \partial_{xx} - \varepsilon \partial_{yy} \quad \text{or} \quad - \varepsilon \Delta + \beta \cdot \nabla, \tag{2.8}$$

where $0 < \varepsilon << 1$. The high frequencies of this equation are then local in a slightly different way. In fact, the higher frequencies exhibit (long–thin) ellipse locality. The level set of the fundamental solution associated with (2.8) is an ellipse that gets longer and thinner as ε gets smaller.

2.3. A simple domain decomposition method

After understanding the local behavior of high frequencies as discussed above, it is then rather transparent to derive a simple domain decomposition. Let us now carry out this exercise.

As discussed above, given by subdomain of size $\rho = h_0$, after solving a local problem such as (2.6), we have pretty much captured all frequencies that oscillate inside this subdomain. Thus, if we solve a number of local problems on a collection of subdomains of size, say approximately of order h_0, that actually cover the whole domain, we should then be able to capture all the frequencies that oscillate within a distance smaller than h_0. In other words, the remaining frequencies are low frequencies that oscillate at most $O(h_0)$ distance.

Let us give a slightly more precise description of this process. We again use the simple model problem (2.5). We start by assuming that we are given a set of overlapping subdomains $\{\Omega_i\}_{i=1}^J$ of Ω. One way of defining the subdomains and the associated partition is by starting with disjoint open sets $\{\Omega_i^0\}_{i=1}^J$ with $\bar{\Omega} = \bigcup_{i=1}^J \bar{\Omega}_i^0$ and $\{\Omega_i^0\}_{i=1}^J$ quasi-uniform of size h_0. The subdomain Ω_i is defined to be a subdomain containing Ω_i^0 with the distance from $\partial\Omega_i \cap \Omega$ to Ω_i^0 greater than or equal to ch_0 for some prescribed constant c.

There are different ways to proceed with the local solution with the given subdomains. Let us now describe a simple successive correction procedure:

Algorithm 2.1.

For $i = 1 : J$, *with* $u_0 = 0$,
find $e_i \in H_0^1(\Omega_i)$ *such that* $-\Delta e_i = f - (-\Delta u_{i-1})$,
set $u_i = u_{i-1} + e_i$.

Based on the discussions we had above, we see that $u - u_J$ is relatively smooth and it mainly consists of frequencies that oscillate at most $O(h_0)$ distance.

The above procedure essentially describes the main idea in a typical (overlapping) domain decomposition method. But, in practice, this type of method is often carried out on a discrete level. Let us now discuss in some detail the discrete version of this method.

We consider, for example, a finite element space $V_h \subset H_0^1(\Omega)$ consisting of piecewise linear functions on a triangulation T_h of Ω. We assume that the triangulation is compatible with subdomains Ω_i in the domain decomposition mentioned above, namely the restriction of T_h on each Ω_i is a good triangulation of Ω_i. The discrete versions of the subspaces $H_0^1(\Omega_i)$ in the above algorithm are the following finite element subspaces

$$V_i = \{v \in V : v(x) = 0, \quad \forall x \in \Omega \setminus \Omega_i\}.$$

As we have already shown, $u - u_J$ is relatively smooth and it can be well approximated by a finite element space defined on a grid of size of order h_0. Such a finite element space is called a coarse space in domain decomposition terminology and we shall denote it by $V_0 \subset V_h$.

The discrete version of our domain decomposition method is then a successive correction procedure carried out on all these finite element subspaces V_i for $i = 1 : J$ and then for $i = 0$. This is a typical subspace correction method based on a space decomposition as follows:

$$V_h = V_0 + V_1 + \cdots + V_J = \sum_{i=0}^{J} V_i. \tag{2.9}$$

In this decomposition, each subspace covers a certain range of frequencies for the space V_h. Consequently, certain uniform convergent properties can be expected from the corresponding method of subspace corrections.

2.4. *From domain decomposition to multigrid methods*

The simple domain decomposition method discussed above provides a good example on how a large global problem can be decomposed into small local problems using the local property of high frequencies. In our view, the local property of high frequencies is the key reason why the method of subspace correction works for partial differential equations.

The local subdomain solution process in the above domain decomposition method is called a smoothing process since this procedure damps out high frequencies and give rise to a much smoother component that remains to be resolved by a coarse space V_0. But the coarse space problem in V_0 may still be too large and a natural solution is then to repeat a similar procedure on V_0 and apply such a domain decomposition recursively. The resulting algorithm from this recursive procedure is nothing but a multigrid method.

The local behavior of high frequencies determines what smoothers to use in a multigrid process. For the simple Poisson equation, as illustrated above, the high frequency has a point locality and hence local relaxation such as the point Gauss–Seidel method can be used as an effective smoother. For a differential operator like (2.8), the locality of high frequency is in a long-thin region and hence line relaxation may be used as a smoother.

For any given application of multigrid methodology, in our view, the key is to understand the local property of high frequencies of the underlying linear or nonlinear (partial differential) operators. But in many applications, especially, for systems and/or nonlinear problems, how the high frequencies behave is often not clear. It appears to be necessary to have a systematic theoretical investigation on this question for various partial differential operators that are practically interesting. This is first a problem in the theory of partial differential equations, but ultimately we also need to study the same question for the discretized equations. We believe this is a research topic in partial differential equation theory that has a significant practical importance.

The domain decomposition and multigrid methods are special subspace correction methods. This type of method which is based on a proper space decomposition is a general approach to the design of iterative methods for large-scale systems arising from the discretization of partial differential equations. We shall devote the next two sections to the discussion of this type of methods in an abstract setting.

3. Elementary iterative methods

Assume $\mathscr{V}$ is a finite-dimensional vector space. The goal of this paper is to discuss iterative methods and preconditioning techniques for solving the following kind of equation:

$$Au = f. \tag{3.1}$$

Here $A : \mathscr{V} \mapsto \mathscr{V}$ is a symmetric positive-definite (SPD) linear operator over $\mathscr{V}$ and $f \in \mathscr{V}$ is given.

In this section, we discuss some basic iterative methods for solving the above system of equation.

3.1. Linear iterative methods

A single step linear iterative method which uses an old approximation, u^{old}, of the solution u of (3.1), to produce a new approximation, u^{new}, usually consists of three steps:

(1) form $r^{\text{old}} = f - Au^{\text{old}}$;
(2) solve $Ae = r^{\text{old}}$ approximately: $\hat{e} = Br^{\text{old}}$;
(3) update $u^{\text{new}} = u^{\text{old}} + \hat{e}$;

where B is a linear operator on $\mathscr{V}$ and can be thought of as an approximate inverse of A.

As a result, we have the following iterative algorithm.

Algorithm 3.1. *Given* $u^0 \in \mathscr{V}$,

$$u^{k+1} = u^k + B(f - Au^k), \quad k = 0, 1, 2, \ldots . \tag{3.2}$$

The core of the above iterative scheme is the operator B. Notice that if $B = A^{-1}$, after one iteration, u^1 is then the exact solution.

We say that an iterative scheme like (3.2) converges if $\lim_{k\to\infty} u_k = u$ for any $u_0 \in \mathscr{V}$. Assume that u and u^k are solutions of (3.1) and (3.2), respectively. Then

$$u - u^k = (I - BA)^k (u - u_0).$$

Therefore, the iterative scheme (3.2) converges if and only if $\rho(I - BA) < 1$.

Sometimes, it is more desirable to have a symmetric B. If B is not symmetric, there is a natural way to symmetrize it. The symmetrized scheme is as follows:

$$u^{k+1/2} = u^k + B(f - Au^k), \quad u^{k+1} = u^{k+1/2} + B^{\mathrm{t}}(f - Au^{k+1/2}).$$

Here and below "t" and "$*$" denote the transpositions with respect to $(\cdot,\cdot)$ and $(\cdot,\cdot)_A$ respectively. Eliminating the intermediate $u^{k+1/2}$, we have

$$u^{k+1} = u^k + \bar{B}(f - Au^k) \tag{3.3}$$

with

$$\bar{B} = B^{\mathrm{t}} + B - B^{\mathrm{t}}AB \text{ satisfying } I - \bar{B}A = (I - BA)^*(I - BA).$$

It is easy to verify the following identity:

$$\| v \|_A^2 - \| (I - \bar{B}A)v \|_A^2 = (\bar{B}Av, v)_A \quad \text{and} \quad \lambda_{\max}(\bar{B}A) \leqslant 1.$$

The above identity immediately yields a useful convergence criteria:

Scheme (3.2) *converges if* (*and only if in case* B *is symmetric*) *the symmetrized scheme* (3.3) *is convergent, namely* $\bar{B}$ *is SPD or, equivalently,* $B^{-\mathrm{t}} + B^{-1} - A$ *is SPD.*

While the symmetrized scheme is desirable when, for example, it is used with the preconditioned conjugate gradient method (see discussion below), but as a stand alone iterative method, its convergence property, as indicated by the above discussions, may not be as good as the original iterative scheme. This phenomenon has been observed and discussed by some authors (see, for example [28]).

Example 3.2. Assume $\mathscr{V} = \mathbb{R}^n$ and $A = (a_{ij}) \in \mathbb{R}^{n\times n}$ is an SPD matrix. We write $A = D - L - U$ with D being the diagonal of A and $-L$ and $-U$ the lower and upper triangular parts of A, respectively. We have the following choices of B that result in various different iterative methods:

$$B = \begin{cases} \omega & \text{Richardson,} \\ D^{-1} & \text{Jacobi,} \\ \omega D^{-1} & \text{Damped Jacobi,} \\ (D - L)^{-1} & \text{Gauss–Seidel,} \\ \omega(D - \omega L)^{-1} & \text{SOR.} \end{cases} \tag{3.4}$$

The symmetrization of the aforementioned Gauss–Seidel method is called the symmetric Gauss–Seidel method.

These simple iterative methods will serve as a basis for the more advanced iterative methods (such as multigrid methods) based on subspace corrections.

3.2. Preconditioned conjugate gradient method

The well-known conjugate gradient method is the basis of all the preconditioning techniques to be studied in this paper. The preconditioned conjugate gradient (PCG) method can be viewed as a conjugate gradient method applied to the preconditioned system:

$$BAu = Bf. \tag{3.5}$$

Here $B: V \mapsto V$ is another SPD operator and known as a preconditioner for A. Note that BA is symmetric with respect to the inner product $(B^{-1}\cdot,\cdot)$. One version of this algorithm is as follows: *Given* $u^0; r^0 = f - Au^0$; $p^0 = Br^0$; *for* $k = 1,2,\ldots,$

$$u^k = u^{k-1} + \alpha^k p^{k-1},\quad r^k = r^{k-1} - \alpha^k A p^{k-1}, p^k = Br^k + \beta^k p^{k-1},$$

$$\alpha^k = (Br^{k-1}, r^{k-1})/(Ap^{k-1}, p^{k-1}), \beta^k = (Br^k, r^k)/(Br^{k-1}, r^{k-1}).$$

It is well known that

$$\| u - u^k \|_A \leqslant 2\left(\frac{\sqrt{\kappa(BA)}-1}{\sqrt{\kappa(BA)}+1}\right)^k \| u - u^0 \|_A, \tag{3.6}$$

which implies that PCG converges faster with smaller condition number $\kappa(BA)$.

The efficiency of a PCG method depends on two main factors: the action of B and the size of $\kappa(BA)$. Hence, a good preconditioner should have two competing properties: the action of B is relatively easy to compute and that $\kappa(BA)$ is relatively small (at least smaller than $\kappa(A)$).

4. Space decomposition and subspace correction

In this section, we present a general framework for linear iterative methods and/or preconditioners using the concept of *space decomposition* and *subspace correction*. This framework will be presented here from a purely algebraic point of view. Some simple examples are given for illustration and more important applications are given in the later sections for multigrid methods.

The presentation here more or less follows Xu [50] and Bramble et al. [10,9]. For related topics, we refer to [6].

4.1. Preliminaries

A decomposition of a vector space $\mathscr{V}$ consists of a number of subspaces $\mathscr{V}_i \subset \mathscr{V}$ (for $0 \leqslant i \leqslant J$) such that

$$\mathscr{V} = \sum_{i=0}^{J} \mathscr{V}_i. \tag{4.1}$$

This means that, for each $v \in \mathscr{V}$, there exist $v_i \in \mathscr{V}_i$ $(0 \leqslant i \leqslant J)$ such that $v = \sum_{i=0}^{J} v_i$. This representation of v may not be unique in general, namely (4.1) is not necessarily a direct sum.

For each i, we define $Q_i, P_i : \mathscr{V} \mapsto \mathscr{V}_i$ and $A_i : \mathscr{V}_i \mapsto \mathscr{V}_i$ by

$$(Q_i u, v_i) = (u, v_i),\quad (P_i u, v_i)_A = (u, v_i)_A,\quad u \in \mathscr{V}, v_i \in \mathscr{V}_i \tag{4.2}$$

and

$$(A_i u_i, v_i) = (Au_i, v_i), \quad u_i, v_i \in \mathscr{V}_i. \tag{4.3}$$

Q_i and P_i are both orthogonal projections and A_i is the restriction of A on $\mathscr{V}_i$ and is SPD. Note that, $Q_i = I_i^t$, where $I_i : \mathscr{V}_i \mapsto \mathscr{V}$ is the natural inclusion. It follows from the definition that

$$A_i P_i = Q_i A. \tag{4.4}$$

This simple identity is of fundamental importance and will be used frequently in this section. A consequence of it is that, if u is the solution of (3.1), then

$$A_i u_i = f_i \tag{4.5}$$

with $u_i = P_i u$ and $f_i = Q_i f = I_i^t f$. This equation may be regarded as the restriction of (3.1) to $\mathscr{V}_i$.

We note that the solution u_i of (4.5) is the best approximation of the solution u of (3.1) in the subspace $\mathscr{V}_i$ in the sense that

$$J(u_i) = \min_{v \in \mathscr{V}_i} J(v) \quad \text{with } J(v) = \tfrac{1}{2}(Av, v) - (f, v)$$

and

$$\| u - u_i \|_A = \min_{v \in \mathscr{V}_i} \| u - v \|_A .$$

In general, the subspace equation (4.5) will be solved approximately. To describe this, we introduce, for each i, another nonsingular operator $R_i : \mathscr{V}_i \mapsto \mathscr{V}_i$ that represents an approximate inverse of A_i in a certain sense. Thus, an approximate solution of (4.5) may be given by $\hat{u}_i = R_i f_i$.

Example 4.1. Consider the space $\mathscr{V} = R^n$ and the simplest decomposition:

$$\mathbb{R}^n = \sum_{i=1}^{n} \operatorname{span}\{e^i\},$$

where e^i is the ith column of the identity matrix. For an SPD matrix $A = (a_{ij}) \in \mathbb{R}^{n \times n}$

$$A_i = a_{ii}, \quad Q_i y = y_i e^i,$$

where y_i is the ith component of $y \in \mathbb{R}^n$.

4.2. Basic algorithms

From the viewpoint of subspace correction, most linear iterative methods can be classified into two major algorithms, namely the *parallel subspace correction* (PSC) method and the *successive subspace correction* method (SSC).

PSC: Parallel subspace correction. This type of algorithm is similar to the Jacobi method. The idea is to correct the residue equation on each subspace in parallel.

Let u^{old} be a given approximation of the solution u of (3.1). The accuracy of this approximation can be measured by the residual: $r^{\text{old}} = f - Au^{\text{old}}$. If $r^{\text{old}} = 0$ or is very small, we are done. Otherwise, we consider the residual equation:

$$Ae = r^{\text{old}}.$$

Obviously, $u = u^{\text{old}} + e$ is the solution of (3.1). Instead, we solve the restricted equation on each subspace $\mathscr{V}_i$

$$A_i e_i = Q_i r^{\text{old}}.$$

It should be helpful to note that the solution e_i is the best possible correction u^{old} in the subspace $\mathscr{V}_i$ in the sense that

$$J(u^{\text{old}} + e_i) = \min_{e \in \mathscr{V}_i} J(u^{\text{old}} + e).$$

As we are only seeking a correction, we only need to solve this equation approximately using the subspace solver R_i described earlier

$$\hat{e}_i = R_i Q_i r^{\text{old}} = I_i R_i I_i^{\text{t}} r^{\text{old}}.$$

An update of the approximation of u is obtained by

$$u^{\text{new}} = u^{\text{old}} + \sum_{i=0}^{J} \hat{e}_i,$$

which can be written as

$$u^{\text{new}} = u^{\text{old}} + B(f - Au^{\text{old}}),$$

where

$$B = \sum_{i=0}^{J} R_i Q_i = \sum_{i=0}^{J} I_i R_i I_i^{\text{t}}. \tag{4.6}$$

We therefore have the following algorithm.

Algorithm 4.2. Given $u_0 \in \mathscr{V}$, apply the iterative scheme (3.2) with B given in (4.6).

Example 4.3. With $\mathscr{V} = \mathbb{R}^n$ and the decomposition given by Example 4.1, the corresponding Algorithm 4.2 is just the Jacobi iterative method.

It is well known that the Jacobi method is not convergent for all SPD problems hence Algorithm 4.2 is not always convergent. However, the preconditioner obtained from this algorithm is of great importance. We note that the operator B given by (4.6) is SPD if each $R_i : \mathscr{V}_i \to \mathscr{V}_i$ is SPD.

Algorithm 4.4. Apply the PCG method to Eq. (3.1), with B defined by (4.6) as a preconditioner.

Example 4.5. The preconditioner B corresponding to Example 4.1 is

$$B = \text{diag}(a_{11}^{-1}, a_{22}^{-1}, \ldots, a_{nn}^{-1})$$

which is the well-known diagonal preconditioner for the SPD matrix A.

SSC: Successive subspace corrections. This type of algorithm is similar to the Gauss–Seidel method.

To improve the PSC method that makes simultaneous corrections, we make the correction here in one subspace at a time by using the most updated approximation of u. More precisely, starting from $v^{-1} = u^{\text{old}}$ and correcting its residue in $\mathscr{V}_0$ gives

$$v^0 = v^{-1} + I_0 R_0 I_0^{\text{t}}(f - Av^{-1}).$$

By correcting the new approximation v^1 in the next space $\mathscr{V}_1$, we get

$$v^1 = v^0 + I_1 R_1 I_1^{\text{t}}(f - Av^0).$$

Proceeding this way successively for all $\mathscr{V}_i$ leads to the following SSC algorithm.

Algorithm 4.6. Given $u^0 \in \mathscr{V}$,

for $k = 0, 1, \ldots$ till convergence
  $v \leftarrow u^k$
  for $i = 0 : J$ $\quad v \leftarrow v + I_i R_i I_i^{\text{t}}(f - Av)$ $\quad$ endfor
  $u^{k+1} \leftarrow v$
endfor

Example 4.7. Corresponding to decomposition in Example 4.1, the Algorithm 4.6 is the Gauss–Seidel iteration.

Example 4.8. More generally, decompose $\mathbb{R}^n$ as

$$\mathbb{R}^n = \sum_{i=0}^{J} \text{span}\{e^{l_i}, e^{l_i+1}, \ldots, e^{l_{i+1}-1}\},$$

where $1 = l_0 < l_1 < \cdots < l_{J+1} = n+1$. Then Algorithms 4.2, 4.4 and 4.6 are the block Jacobi method, block diagonal preconditioner and block Gauss–Seidel methods, respectively.

Let $T_i = R_i Q_i A = I_i R_i I_i^{\text{t}} A$. By (4.4), $T_i = R_i A_i P_i$. Note that $T_i : \mathscr{V} \mapsto \mathscr{V}_i$ is symmetric with respect to $(\cdot, \cdot)_A$ and nonnegative and that $T_i = P_i$ if $R_i = A_i^{-1}$.

If u is the exact solution of (3.1), then $f = Au$. Let v^i be the ith iterate (with $v^0 = u^k$) from Algorithm 4.6. We have by definition

$$u - v^{i+1} = (I - T_i)(u - v^i), \quad i = 0, \ldots, J.$$

A successive application of this identity yields

$$u - u^{k+1} = E_J(u - u^k), \tag{4.7}$$

where

$$E_J = (I - T_J)(I - T_{J-1}) \cdots (I - T_1)(I - T_0). \tag{4.8}$$

Remark 4.9. It is interesting to look at the operator E_J in the special case that $R_i = \omega A_i^{-1}$ for all i. The corresponding SSC iteration is a generalization of the classic SOR method. In this case, we have

$$E_J = (I - \omega P_J)(I - \omega P_{J-1}) \cdots (I - \omega P_1)(I - \omega P_0).$$

One trivial fact is that E_J is invertible when $\omega \neq 1$. Following an argument in [38] for the SOR method, let us take a look at the special case $\omega=2$. Since, obviously, $(I-2P_i)^{-1}=I-2P_i$ for each i, we conclude that $E_J^{-1}=E_J^*$ where, we recall, $*$ is the adjoint with respect to the inner product $(\cdot,\cdot)_A$. This means that E_J is an orthogonal operator and, in particular, $\| E_J \|_A = 1$. As a consequence, the SSC iteration cannot converge when $\omega=2$. In fact, as we shall see below, in this special case, the SSC method converges if and only if $0<\omega<2$.

The symmetrization of Algorithm 4.6 can also be implemented as follows.

Algorithm 4.10. Given $u^0 \in \mathscr{V}$, $v \leftarrow u^0$

for $k=0,1,\ldots$ till convergence
  for $i=0:J$ and $i=J:-1:0$ $v \leftarrow v + I_iR_iI_i^t(f-Av)$ endfor
endfor

As mentioned earlier, the advantage of the symmetrized algorithm is that it can be used as a preconditioner. In fact, Algorithm 4.10 can be formulated in the form of (3.2) with operator B defined as follows: for $f \in \mathscr{V}$, let $Bf=u^1$ with u^1 obtained by Algorithm 4.10 applied to (3.1) with $u^0=0$.

Colorization and parallelization of SSC iteration: The SSC iteration is a sequential algorithm by definition, but it can often be implemented in a more parallel fashion. For example, the parallelization can be realized by coloring.

Associated with a given partition (4.1), a coloring of the set $\mathscr{J}=\{0,1,2,\ldots,J\}$ is a disjoint decomposition:

$$\mathscr{J}=\bigcup_{l=1}^{J_c} \mathscr{J}(t)$$

such that

$$P_iP_j=0 \quad \text{for any } i,j \in \mathscr{J}(l), i \neq j \ (1 \leqslant l \leqslant J_c).$$

We say that i,j have the same color if they both belong to some $\mathscr{J}(t)$.

The important property of the coloring is that the SSC iteration can be carried out in parallel in each color.

Algorithm 4.11 (Colored SSC). Given $u^0 \in \mathscr{V}$, $v \leftarrow u^0$

for $k=0,1,\ldots$ till convergence
  for $l=1:J_c$ $v \leftarrow v + \sum_{i \in \mathscr{J}(l)} I_iR_iI_i^t(f-Av)$ endfor
endfor

We note that the terms under the sum in the above algorithm can be evaluated in parallel (for each l, namely within the same color).

The best-known example of colorization is perhaps the red–black ordering for the five-point finite difference stencil (or linear finite elements) for the Poisson equation on unit square. We note that the coloring technique can be applied in very general situations.

4.3. Convergence theory

There are some very elegant convergence theories for the subspace correction methods described above. For simplicity, let us present a theory contained in [50] for a simple case, namely each subspace solver R_i is symmetric positive definite. This theory stems from Bramble et al. [9]. For a more general theory, we refer to [6,51] and more recently Xu and Zikatanov [52] (for a general sharp theory).

Our theory will be presented mainly in terms of two parameters, denoted by K_0 and K_1, defined as follows:

(1) For any $v \in \mathscr{V}$, there exists a decomposition $v = \sum_{i=0}^J v_i$ for $v_i \in \mathscr{V}_i$ such that

$$\sum_{i=0}^{J} (R_i^{-1} v_i, v_i) \leqslant K_0 (Av, v). \tag{4.9}$$

(2) For any $S \subset \{0,1,2,\dots,J\} \times \{0,1,2,\dots,J\}$ and $u_i, v_i \in \mathscr{V}$ for $i = 0,1,2,\dots,J$,

$$\sum_{(i,j)\in S} (T_i u_i, T_j v_j)_A \leqslant K_1 \left(\sum_{i=0}^{J} (T_i u_i, u_i)_A \right)^{1/2} \left(\sum_{j=0}^{J} (T_j v_j, v_j)_A \right)^{1/2}. \tag{4.10}$$

Theorem 4.12. *Assume that B is the SSC preconditioner given by* (4.6); *then*

$$\kappa(BA) \leqslant K_0 K_1.$$

In fact, we have $\lambda_{\max}(BA) \leqslant K_1$ which follows directly from the definition of K_1 and $\lambda_{\min}(BA) \geqslant K_0^{-1}$ which follows from the following identity:

$$(B^{-1}v, v) = \inf_{\substack{v_i \in \mathscr{V}_i \\ \sum v_i = v}} \sum_i (R_i^{-1} v_i, v_i). \tag{4.11}$$

This identity is implicitly contained in [50] and it may be found in [48,22]. Let us now include a proof of it in the following.

Given any decomposition $v = \sum_{i=1}^J v_i$, we have, by the Cauchy–Schwarz inequality, that

$$\begin{aligned}(v, B^{-1}v)^2 &= \left(\sum_{i=1}^{J} (v_i, Q_i B^{-1} v) \right)^2 \leqslant \left(\sum_{i=1}^{J} (R_i^{-1} v_i, v_i)^{1/2} (Q_i B^{-1} v, R_i Q_i B^{-1} v)^{1/2} \right)^2 \\ &\leqslant \sum_{i=1}^{J} (R_i^{-1} v_i, v_i) \sum_{i=1}^{J} (B^{-1} v, R_i Q_i B^{-1} v) = \sum_{i=1}^{J} (R_i^{-1} v_i, v_i)(v, B^{-1} v).\end{aligned}$$

On the other hand, for the trivial decomposition $v = \sum_{i=1}^J v_i$ with $v_i = T_i T^{-1} v$, we have $\sum_{i=1}^J (R_i^{-1} T_i T^{-1} v, T_i T^{-1} v) = (B^{-1} v, v)$. This finishes the justification of (4.11).

To present our next theorem, let us denote, for $0\leqslant i\leqslant J$, $E_i=(I-T_i)E_{i-1}$ with $E_1=I$. Then

$$I-E_i=\sum_{j=0}^{i} T_jE_{j-1} \tag{4.12}$$

and

$$(2-\omega_1)\sum_{i=0}^{J}(T_iE_{i-1}v,E_{i-1}v)_A\leqslant \| v\|_A^2-\| E_Jv\|_A^2, \quad \forall v\in\mathscr{V}. \tag{4.13}$$

The proof of this identity follows immediately from the trivial identity $E_{i-1}-E_i=T_iE_{i-1}$ and the relation that

$$\| E_{i-1}v\|_A^2-\| E_iv\|_A^2=((2I-T_i)T_iE_{i-1}v,E_{i-1}v)_A\geqslant(2-\omega_1)(T_iE_{i-1}v,E_{i-1}v)_A.$$

Theorem 4.13. *For the Algorithm* 4.6,

$$\| E_J\|_A^2\leqslant 1-\frac{2-\omega_1}{K_0(1+K_1)^2}, \tag{4.14}$$

where $\omega_1=\max_i(R_i,A_i)$.

Proof. In view of (4.13), it suffices to show that

$$\sum_{i=0}^{J}(T_iv,v)_A\leqslant(1+K_1)^2\sum_{i=0}^{J}(T_iE_{i-1}v,\ E_{i-1}v)_A, \quad \forall v\in\mathscr{V}. \tag{4.15}$$

By (4.12) $(T_iv,v)_A=(T_iv,E_{i-1}v)_A+\sum_{j=0}^{i-1}(T_iv,T_jE_{j-1}v)_A$. Applying the Cauchy–Schwarz inequality gives

$$\sum_{i=0}^{J}(T_iv,E_{i-1}v)_A\leqslant\left(\sum_{i=0}^{J}(T_iv,v)_A\right)^{1/2}\left(\sum_{i=0}^{J}(T_iE_{i-1}v,E_{i-1}v)_A\right)^{1/2}.$$

By the definition of K_1 in (4.10), we have

$$\sum_{i=0}^{J}\sum_{j=0}^{i-1}(T_iv,T_jE_{j-1}v)_A\leqslant K_1\left(\sum_{i=0}^{J}(T_iv,v)_A\right)^{1/2}\left(\sum_{j=0}^{J}(T_jE_{j-1}v,E_{j-1}v)_A\right)^{1/2}.$$

Combining these three formulae then leads to (4.15). □

This theorem shows that the SSC algorithm converges as long as $\omega_1<2$. The condition that $\omega_1<2$ is reminiscent of the restriction on the relaxation parameter in the SOR method.

Example 4.14. Let us now discuss a simple application of our theory to the overlapping domain decomposition method that we discussed in Section 3. It can be proved that for the space decomposition (2.9) based on the domain decomposition, the parameters K_0 and K_1 can be bounded uniformly with respect to h and h_0. The estimate for K_1 is straightforward and the estimate for K_0 may be obtained by a simple partition of unity. For details, we refer to Xu [50] and the reference cited there. Thus the corresponding PSU preconditioner yields a uniformly bounded condition number and SSC method has a uniform convergence rate.

5. Multigrid methods

The multigrid method is among the most efficient iterative methods for solving the algebraic system arising from the discretization of partial differential equations. In this section, we shall give a brief review of this method based on finite element discretization. For more comprehensive discussion on this topic, we refer to the research monographs of Hackbusch [26,27], McCormick [37], Wesseling [47] and Bramble [6], and to the review articles of Xu [50,51] and Yserentant [54].

5.1. A model problem and finite element discretizations

We consider the boundary value problem:

$$-\nabla \cdot a\nabla U = F \quad \text{in } \Omega, \qquad U = 0 \quad \text{on } \partial\Omega, \tag{5.1}$$

where $\Omega \subset \mathbb{R}^d$ is a polyhedral domain and a is a smooth function (or piecewise smooth) on $\bar{\Omega}$ with a positive lower bound.

Let $H^1(\Omega)$ be the standard Sobolev space consisting of square-integrable functions with square-integrable (weak) derivatives of first order, and $H_0^1(\Omega)$ the subspace of $H^1(\Omega)$ consisting of functions that vanish on $\partial\Omega$. Then $U \in H_0^1(\Omega)$ is the solution of (5.1) if and only if

$$A(U,\chi) = (F,\chi) \quad \forall \chi \in H_0^1(\Omega), \tag{5.2}$$

where

$$A(U,\chi) = \int_\Omega a\nabla U \cdot \nabla\chi \, \mathrm{d}x, \quad (F,\chi) = \int_\Omega F\chi \, \mathrm{d}x.$$

Assume that Ω is triangulated with $\Omega = \bigcup_i \tau_i$, where the τ_i are nonoverlapping simplices of size $h \in (0,1]$ and are quasi-uniform, i.e., there exist constants C_0 and C_1 not depending on h such that each simplex τ_i is contained in (contains) a ball of radius $C_1 h$ (resp. $C_0 h$). Define

$$\mathscr{V} = \{v \in H_0^1(\Omega) \colon v|_{\tau_i} \in \mathscr{P}_1(\tau_i), \ \forall \tau_i\},$$

where $\mathscr{P}_1$ is the space of linear polynomials.

The finite element approximation to the solutin of (4.1) is the function $u \in \mathscr{V}$ satisfying

$$A(u,v) = (F,v) \quad \forall v \in \mathscr{V}. \tag{5.3}$$

Define a linear operator $A : \mathscr{V} \mapsto \mathscr{V}$ by

$$(Au,v) = A(u,v), \quad u,v \in \mathscr{V}. \tag{5.4}$$

Eq. (5.3) is then equivalent to (3.1) with $f = Q_h F$. The space $\mathscr{V}$ has a natural (nodal) basis $\{\phi_i\}_{i=1}^n (n = \dim \mathscr{V})$ satisfying

$$\phi_i(x_l) = \delta_{il} \quad \forall i, \ l = 1,\dots,n,$$

where $\{x_l \colon l = 1,\dots,n\}$ is the set of all interior nodal points of $\mathscr{V}$. By means of these nodal basis functions, the solution of (5.3) is reduced to solving an algebraic system (3.1) with coefficient matrix $\mathscr{A} = ((a\nabla\phi_i, \nabla\phi_l))_{n\times n}$ and right-hand side $\eta = ((f,\phi_i)_{n\times 1})$.

Finite element spaces on multiple levels: We assume that Ω has been triangulated with a nested sequence of quasi-uniform triangulations $\mathscr{T}_k=\{\tau_k^i\}$ of size h for $k=0,\dots,j$ where the quasi-uniformity constants are independent of k. These triangulations should be nested in the sense that any triangle τ_{k-1}^l can be written as a union of triangles of $\{\tau_k^i\}$. We further assume that there is a constant $\gamma > 1$, independent of k, such that

$$h_k \eqsim \gamma^{-k}.$$

Associated with each $\mathscr{T}_k$, a finite element space $\mathscr{M}_k \subset H_0^1(\Omega)$ can be defined. One has

$$\mathscr{M}_0 \subset \mathscr{M}_1 \subset \cdots \subset \mathscr{M}_k \subset \cdots \subset \mathscr{M}_J. \tag{5.5}$$

5.2. A \ -cycle multigrid method

Let $\mathscr{T}_J$ be the finest triangulation in the multilevel structure described earlier with nodes $\{x_i\}_{i=1}^{n_J}$. With such a triangulation, a natural domain decomposition is

$$\bar{\Omega} = \bar{\Omega}_0^h \bigcup \bigcup_{i=1}^{n_J} \operatorname{supp} \phi_i,$$

where ϕ_i is the nodal basis function in $\mathscr{M}_J$ associated with the node x_i and Ω_0^h, which may be empty, is the region where all functions in $\mathscr{M}_J$ vanish.

It is easy to see that the corresponding decomposition method without a coarse space is exactly the Gauss–Seidel method which is known to be inefficient (its convergence rate is known to be $1 - \mathrm{O}(h_J^2)$). The more interesting case is when a coarse space is introduced. The choice of such a coarse space is clear here, namely $\mathscr{M}_{J-1}$. There remains to choose a solver for $\mathscr{M}_{J-1}$. To do this, we may repeat the above processs by using the space $\mathscr{M}_{J-2}$ as a "coarser" space with the supports of the nodal basis function in $\mathscr{M}_{J-1}$ as a domain decomposition. We continue in this way until we reach a coarse space $\mathscr{M}_1$ where a direct solver can be used. As a result, a multilevel algorithm based on domain decomposition is obtained. This procedure can be illustrated by the following diagram:

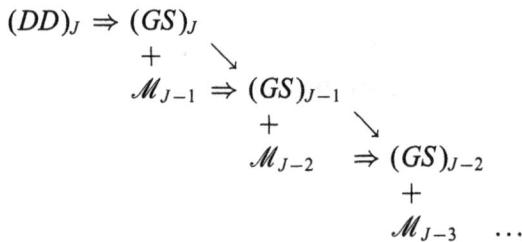

$$\begin{array}{llllll}
(DD)_J & \Rightarrow (GS)_J & & & & \\
 & \quad + & \searrow & & & \\
 & \mathscr{M}_{J-1} & \Rightarrow (GS)_{J-1} & & & \\
 & & \quad + & \searrow & & \\
 & & \mathscr{M}_{J-2} & \Rightarrow (GS)_{J-2} & & \\
 & & & \quad + & & \\
 & & & \mathscr{M}_{J-3} & \cdots &
\end{array}$$

This resulting algorithm is a very basic multigrid method cycle, which may be called a \-cycle (in comparison to the better known V- and W-cycles). We shall now give a more precise mathematical description of this multigrid method. Define $Q_k, P_k : \mathscr{M}_k \mapsto \mathscr{M}_J$ by

$$(Q_k u, v_k) = (u, v_k), \quad (\nabla P_k u, \nabla v_k) = (\nabla u, \nabla v_k). \tag{5.6}$$

Then the aforementioned multigrid method can be described by an inductive procedure in terms of a sequence of operators $B_k : \mathscr{M}_k \mapsto \mathscr{M}_k$ which are approximate inverses of A_k.

Algorithm 5.1. For $k=0$, define $B_0 = A_0^{-1}$. Assume that $B_{k-1} : \mathcal{M}_{k-1} \mapsto \mathcal{M}_{k-1}$ is defined. We shall now define $B_k : \mathcal{M}_k \mapsto \mathcal{M}_k$ which is an iterator for the equation of the form

$$A_k v = g.$$

(1) Fine grid smoothing: for $v^0 = 0$ and $l = 1,2,\ldots,m$,

$$v^l = v^{l-1} + R_k(g - A_k v^{l-1}).$$

(2) Coarse grid correction: $e_{k-1} \in \mathcal{M}_{k-1}$ is the approximate solution of the residual equation $A_{k-1}e = Q_{k-1}(g - Av^m)$ by the iterator B_{k-1}:

$$e_{k-1} = B_{k-1}Q_{k-1}(g - Av^m).$$

Define

$$B_k g = v^m + e_{k-1}.$$

In the above definition, R_k corresponds to a Gauss–Seidel iteration or a general iterative method often known as a smoother.

With the above defined B_k, there are many different ways to make use of B_k. One simple example is as follows:

$$u^{k+1} = u^k + B_J(f - Au^k). \tag{5.7}$$

We now discuss briefly the algebraic version of the above algorithm.

Let $\Phi^k = (\phi_1^k, \ldots, \phi_{n_k}^k)$ be the nodal basis vector for the space $\mathcal{M}_k$; we define the so-called prolongation matrix $\mathcal{I}_k^{k+1} \in \mathbb{R}^{n_{k+1} \times n_k}$ as follows:

$$\Phi^k = \Phi^{k+1}\mathcal{I}_k^{k+1}. \tag{5.8}$$

Let $\mathcal{A}_k = (A(\phi_i^k, \phi_j^k))$ be the stiffness matrix on level k and $\mathcal{R}_k$ is the corresponding smoother such as Gauss–Seidel or symmetric Gauss–Seidel iteration. Algorithm 5.1 is equivalent to the following algorithm that is expressed in terms of matrices and vectors.

Algorithm 5.2. Let $\mathcal{B}_0 = \mathcal{A}_0^{-1}$. Assume that $\mathcal{B}_{k-1} \in \mathbb{R}^{n_{k-1} \times n_{k-1}}$ is defined; then for $\eta \in \mathbb{R}^{n_k}, \mathcal{B}_k \in \mathbb{R}^{n_k \times n_k}$ is defined as follows.

(1) Fine grid smoothing: for $v^0 = 0$ and $l = 1,2,\ldots,m$

$$v^l = v^{l-1} + \mathcal{R}_k(\eta - \mathcal{A}_k v^{l-1}).$$

(2) Coarse grid correction: $\varepsilon_{k-1} \in \mathbb{R}^{n_{k-1}}$ is the approximate solution of the residual equation $\mathcal{A}_{k-1}\varepsilon = (\mathcal{I}_{k-1}^k)^t(\eta - \mathcal{A}_k v^m)$ by using $\mathcal{B}_{k-1}$,

$$\varepsilon_{k-1} = \mathcal{B}_{k-1}(\mathcal{I}_{k-1}^k)^t(\eta - \mathcal{A}_k v^m).$$

Define $\mathcal{B}_k\eta = v^m + \mathcal{I}_{k-1}^k \varepsilon_{k-1}$.

It is not hard to see that each iteration of multigrid cycle only requires $O(n_J)$ operations. As we shall see later, multigrid iteration converges uniformly with respect to mesh size or number of levels. Consequently, the computational complexity of a typical multigrid method is of $O(n_J)$ and at most of $O(n_J \log n_J)$. It is this optimal or nearly optimal complexity that makes the multigrid methodology one of the most powerful solution technique for solving partial differential equations.

We would like to point out that there are many variants of the \-cycle multigrid algorithm stated here. Better known examples include V-cycle, W-cycle and variable V-cycle, but these different cycles can be obtained and analyzed based on \-cycle. For example, a V-cycle can be viewed as the symmetrization of the \-cycle. For details, we refer to [51].

5.3. A convergence analysis

We shall now briefly discuss about the convergence properties of the multigrid methods. Technically speaking, there are two major approaches for multigrid convergence analysis. In this subsection, we shall discuss a more classic approach. One crucial component in this is the regularity theory of the underlying elliptic partial differential equations. Another approach will be discussed in the next section.

For simplicity of exposition, we assume that our model problem satisfies a full regularity property, namely, the solution U of (5.2) has a square-integrable second-order derivatives as long as the right-hand side F is square integrable. One sufficient condition is that either Ω has smooth boundary or it is convex with a Lipschitz continuous boundary. A direct implication of this assumption is the following error estimate which can be obtained by a well-known duality argument:

$$\| (I-P_{k-1})v \|_A^2 \leqslant c_1[\rho(A_k)]^{-1} \| A_k v \|^2 \quad \forall v \in \mathcal{M}_k. \tag{5.9}$$

Here $\rho(A_k)$ is the spectral radius of A_k.

Let us consider the case that R_k is given by symmetric Gauss–Seidel. In this case, it is easy to prove that R_k are SPD and satisfy

$$\frac{c_0}{\rho(A_k)}(v,v) \leqslant (R_k v, v) \leqslant \frac{\tilde{c}_0}{\rho(A_k)}(v,v) \quad \forall v \in \mathcal{M}_k \quad \text{and} \quad \max_{0\leqslant k\leqslant J} \rho(R_k A_k) = 1. \tag{5.10}$$

for some positive constants c_0 and $\tilde{c}_0$ independent of k.

Theorem 5.3. *For the Algorithm* 5.7, *we have*

$$\| I - B_k A_k \|_A^2 \leqslant \frac{c_1}{2c_0 + c_1}, \quad 0 \leqslant k \leqslant J.$$

Proof. Denoting $E_k = I - B_k A_k$ and $K_k = I - R_k A_k$, by definition of Algorithm 5.1, we have $E_k = (I - P_{k-1} + E_{k-1}P_{k-1})(I - R_k A_k)$ and, thus, for all $v \in \mathcal{M}_k$

$$\| E_k v \|_A^2 = \| (I - P_{k-1})K_k v \|_A^2 + \| E_{k-1}P_{k-1}K_k v \|_A^2 .$$

It follows from (5.9) and (5.10) that

$$\begin{aligned} \| (I - P_{k-1})K_k v \|_A^2 &\leqslant c_1 \lambda_k^{-1} \| A_k K_k v \|^2 = \frac{c_1}{c_0}(R_k A_k K_k v, A_k K_k v) \\ &= \frac{c_1}{c_0}((I - K_k)K_k^2 v, v)_A \leqslant \frac{c_1}{2c_0}(\| v \|_A^2 - \| K_k v \|_A^2), \end{aligned}$$

where in the last step we have used the fact that $\sigma(K_k) \subset [0,1]$ and the elementary inequality that $(1-t)t^2 \leqslant \frac{1}{2}(1-t^2)$ for $t \in [0,1]$.

Let $\delta = c_1/(2c_0 + c_1)$. We shall prove the desired estimate by induction. First it is obvious for $k = 0$. Assume it holds for $k - 1$. In the case of k, we have from the above identity that

$$\begin{aligned}\| (I - B_kA_k)v \|_A^2 &\leqslant \| (I - P_{k-1})K_kv \|_A^2 + \delta \| P_{k-1}K_kv \|_A^2\\ &\leqslant (1-\delta) \| (I - P_{k-1})K_kv \|_A^2 + \delta \| K_kv \|_A^2\\ &\leqslant (1-\delta)\frac{c_1}{2c_0}(\| v \|_A^2 - \| K_kv \|_A^2) + \delta \| K_kv \|_A^2 = \delta \| v \|_A^2 .\end{aligned}$$

This completes the proof. □

The technique used in the above analysis can be traced back to some of the earliest analysis for multigrid convergence and it has been used in most of the early theoretical papers; we refer, for example, to [3,5,7,34]. One crucial element in this proof is the elliptic regularity assumption and its resulting approximation property. While this assumption can sometimes be significantly weakened, it is an essential element that makes this kind of analysis work. Unfortunately, this assumption is not convenient in many important applications such as equations with strongly discontinuous coefficients and nonuniform finite element grids.

5.4. Application of subspace correction theory

The multigrid algorithm can also be placed in the subspace correction theoretical framework, which gives another major different approach for multigrid analysis. This relatively new approach has been successfully used to provide optimal theoretical results for many situations for which the more traditional approach fails.

Mathematically, the multigrid algorithm has many equivalent formulations. Let us describe two such formulations based on space decomposition and subspace correction.

The *first* equivalent formulation is the SSC iteration corresponding to the following space decomposition:

$$\mathscr{V} = \sum_{k=0}^{J} \mathscr{V}_k \quad \text{with } \mathscr{V}_k = \mathscr{M}_{J-k}$$

with subspace solver given by R_k.

The *second* equivalent formulation, which follows directly from the first one, is the SSC iteration corresponding to the following space decomposition:

$$\mathscr{V} = \sum_{k=0}^{J}\sum_{i=1}^{n_k} \operatorname{span}(\phi_i^k) \tag{5.11}$$

with exact subspace solvers on all *one* dimensional subspaces.

The above second formulation allows us to view a multigrid as a generalized Gauss–Seidel iteration applied to a so-called extended system given by the semi-definite stiffness matrix $(A(\phi_i^k, \phi_j^l))$ [21,23].

The above first formulation allows us to use our subspace correction convergence theory to analyze the convergence of the algorithm and furthermore it also allows us to consider the PSC variation of this method (which shall be addressed below).

To apply subspace correction convergence theory, the following norm equivalence result plays an important role:

$$\| v \|_{H^1(\Omega)}^2 \eqsim \sum_{k=0}^{\infty} \| \tilde{Q}_k v \|_{H^1\Omega}^2 \eqsim \sum_{k=0}^{\infty} h_k^{-2} \| \tilde{Q}_k v \|^2, \quad \forall v \in \mathcal{M}_J, \tag{5.12}$$

where $\tilde{Q}_k = Q_k - Q_{k-1}$. This equivalence relation is one of the most interesting result in multigrid theory. Its earliest version first appeared in [11,49] for the study of BPX preconditioner. It was then found in [39,40] to be related to certain approximation theory result based on Besov spaces. This result is also related to multi-resolution theory for wavelets; see the work in [17] and the references cited therein.

With estimates (5.12) and (5.10), it can be proved that the parameters K_0 and K_1 in our subspace correction convergence theory are bounded uniformly with respect to mesh parameters. Consequently, the multigrid method converges uniformly.

The convergence result discussed above has been for quasi-uniform grids, but we would like to remark that uniform convergence can also be obtained for locally refined grids. For details, we refer to [11,9,7,51].

A special theory: For more complicated situations, some estimates such as the norm equivalence (5.12) cannot be established. In this case, we can use weaker assumptions to derive slightly less optimal convergence results. Let us now give an example of such a theory [10] which can be derived easily from our subspace correction convergence theory presented earlier in this paper.

Theorem 5.4. *Assume that there are linear mappings* $Q_k : \mathcal{M}_J \to \mathcal{M}_J$, $Q_J = I$, *and constants* c_1 *and* c_2 *such that*

$$\| Q_k v \|_A \leqslant c_1 \| v \|_A, \quad \forall v \in \mathcal{M}_J, \ k = 1,2,\dots,J-1 \tag{5.13}$$

$$_{k-1} - Q_k v \| \leqslant \frac{c_2}{\sqrt{\rho(A_k)}} \| v \|_{A_k}, \quad \forall v \in \mathcal{M}_k, \ k = 1,2,\dots,J-1. \tag{5.14}$$

Assume that the smoothers R_k *satisfy the following estimate with constant* C_R:

$$\frac{\| u \|_k^2}{\rho(A_k)} \leqslant C_R (R_k u, u)_k \quad \textit{for all } u \in \tilde{M}_k, \tag{5.15}$$

where $\tilde{M}_k$ *is the range of* R_k, *and* $\tilde{M}_k \supset$ Range $(Q_k - Q_{k-1})$.

Then, the \\-*cycle multigrid algorithm admit the following convergence estimate*:

$$\| I - B_J A_J \|_A \leqslant 1 - \frac{1}{c_0 J} \tag{5.16}$$

with $c_0 = 1 + c_1^{1/2} + c_2^{1/2} C_R$.

The above theory can be used, for example, for analyzing multigrid method for problems with rough coefficients and it has also been used by some authors for designing and analyzing algebraic multigrid methods (see Section 5.6).

5.5. BPX multigrid preconditioners

We shall now describe a parallelized version of the multigrid method studied earlier. This method was first proposed in [11,49], and is now often known as the BPX (Bramble-Pasciak-Xu) preconditioner in the literature.

There are different ways of deriving the BPX preconditioners. The methods originally resulted from an attempt to parallelize the classical multigrid method. With the current multigrid theoretical technology, the derivation of this method is not so difficult. We shall here derive this preconditioner based on the equivalence relation (5.14).

By (5.12), we have for all $v\in\mathscr{V}$,

$$(Av,v)\eqsim\sum_{k=0}^{J} h_k^{-2}\,\|(Q_k-Q_{k-1})v\|^2=(\hat{A}v,v)\quad\text{with}\,\hat{A}=\sum_k h_k^{-2}(Q_k-Q_{k-1}).$$

Using the fact that $Q_iQ_j=Q_{\min(i,j)}$, it is easy to verify that $\hat{A}^{-1}=\sum_k h_k^2(Q_k-Q_{k-1})$. Using the fact that $h_k\approx\gamma h_{k+1}$ with $\gamma>1$, we deduce that

$$\begin{aligned}(\hat{A}^{-1}v,v)&=\sum_{k=0}^{J} h_k^2((Q_k-Q_{k-1})v,v)=\sum_{k=0}^{J} h_k^2(Q_kv,v)-\sum_{k=0}^{J-1} h_{k+1}^2(Q_kv,v)\\&\eqsim h_J^2(v,v)+\sum_{k=0}^{J-1} h_k^2(Q_kv,v)\eqsim\sum_{k=0}^{J} h_k^2(Q_kv,v)=(\tilde{B}v,v),\end{aligned}$$

where $\tilde{B}=\sum_{k=0}^{J} h_k^2Q_k$. If $R_k:\mathscr{M}_k\mapsto\mathscr{M}_k$ is given by Jacobi or symmetric Gauss–Seidel satisfying (5.10), then, for

$$B=\sum_{k=0}^{J} R_kQ_k=\sum_{k=0}^{J} I_kR_kI_k^t,\tag{5.17}$$

we have $(Bv,v)\eqsim(\tilde{B}v,v)\eqsim(A^{-1}v,v)$, namely

$$\kappa(BA)\eqsim 1.$$

As we see, this is a PSC preconditioner. Hence it is possible to use the general theory for the subspace correction method to derive optimal estimates for $\kappa(BA)$ under more general assumptions (see [51]).

5.6. Algebraic multigrid method

The multigrid methods discussed above are based on a given hierarchy of multiple levels of grids. In practice, however, such a multilevel hierarchy is not often available, which is perhaps one of the main reasons that multigrid has been difficult to popularize. It is then very desirable to design multigrid algorithms that do not depend on such a multigrid hierarchy. The algebraic multigrid method is such an algorithm and its application sometimes only need the input of the coefficient matrix and the right-hand side. This type of method can be traced back to [13,41].

Let us explain the main idea behind the algebraic multigrid method for a system of equation arising from the discretization of a two-dimensional Poisson equation discretized by linear finite elements. If we examine carefully the structure of a geometric multigrid method for this problem, we notice that we only need to use the graphic information on the underlying multilevel hierarchy. Namely, we do not need to know the actual coordinates of any nodal point and we only need to know the relevant topological location of these points (which are connected by edges). One important fact is that the graph of the stiffness matrix is more or less the same as the graph of the underlying grid. Therefore, by inspecting the sparse pattern of the stiffness matrix, we can pretty much recover the necessary topological information of the underlying grid. With such information of the finest grid, we can then proceed to obtain a sequence of coarse grids through some appropriate coarsening process.

One crucial component of the algebraic multigrid method is a proper construction of coarse grid subspaces. We shall again use the Poisson equation example to illustrate how this can be accomplished. Philosophically, it suffices to explain the two-level case.

From a subspace correction point of view, the role of the coarse grid space is to resolve those relatively low frequencies that cannot be effectively damped out by fine grid smoothings. Let V_h be the given fine grid space and V_{2h} be the coarser grid we need to construct. Let $Q_{2h}: V_h \mapsto V_{2h}$ be the L^2 projection. Following Theorem 5.4, we need the following estimates to be valid for any $v_h \in V_h$:

$$\| v_h - Q_{2h}v_h \|_{L^2} \leqslant c_0 h |v_h|_{H^1} \quad \text{and} \quad |Q_{2h}v_h|_{H^1} \leqslant c_1 |v_h|_{H^1} \tag{5.18}$$

for some positive constants c_0 and c_1.

These estimates roughly mean that a low-frequency function (which has a relatively small H^1 norm) in V_h can be well represented or approximated by functions in V_{2h}.

We shall now describe two different approaches for constructing V_{2h}. The first approach, given in [16], is to try to mimic the construction of a coarse space in the "regular" case. Since in general, we are not able to construct coarse elements (such as triangles) that are unions of fine grid elements, we try to get as close as we can. We still form patches of fine grid elements and treat them as some coarse elements. Since it is in general impossible to define continuous piecewise polynomials on these patches of elements, we take a linear combinations of fine grid basis functions that are good approximation of piecewise polynomials and that also give rise to small energy. This approach proves to be very successful and efficient multigrid codes have been developed (for details, we refer to [16]).

The second approach, due to Vanek et al. [45], is also based on patches of grids. As the first step, we use these patches to define a subspace that is more or less piecewise constant. By the continuity requirement, the piecewise constant function has to be dropped to be zero at the boundary of each patch. These sharp drops which give "big energy" are certainly not desirable. The second step is then to try to smooth out these sharp boundary drops by applying, for example, some damped Jacobi method. Under some appropriate assumptions, the resulting subspace does satisfy the approximation and stability properties (5.17).

While we have been talking about this, for convenience, in terms of grids, this can all be done in terms of graphs. The above two approaches are just two examples and many other approaches are possible. The study of effective algebraic multigrid methods for different applications is currently an active research topic.

Algebraic multigrid methods appear to have great potential for practical applications, but so far there have been no rigorous theoretical justification of these methods. The general theory developed in [10,50] has been informative both in the algorithmic design and the attempt of theoretical analysis, but there are still many gaps that need to be filled. As always, it is rather easy to establish a two level theory [16], but a truly multilevel theory is very difficult to make rigorous, although there have been many such attempts [45,44].

6. More general subspace correction methods

The subspace correction method can be generalized in a variety of different ways. In this section, we shall discuss a few such examples. We shall also discuss its relationship with another well-known class of methods, namely the method of *alternating projections.*

6.1. Nonlinear optimizations

The subspace correction method can be generalized to some nonlinear problems in some rather straightforward fashion.

For motivation, let us reformulate the subspace correction method for linear problems in a slightly different but equivalent way. Let us still consider Eq. (3.1). As we pointed out this equation is equivalent to the minimization problem

$$J(u) = \min_{v \in V} J(v) \quad \text{with } J(v) = \tfrac{1}{2}(Av, v) - (f, v).$$

Algorithm 4.6 can be formulated in the following equivalent fashion.

Algorithm 6.1. Given $u^0 \in \mathscr{V}$,

for $k = 0, 1, \ldots$ till convergence
  $v \leftarrow u^k$
  for $i = 0 : J$
    $\hat{e}_i \approx \arg\min_{e \in V_i} J(v + e)$
    $v \leftarrow v + \hat{e}_i$
  endfor
  $u^{k+1} \leftarrow v.$
endfor

Apparently, the above algorithm can be applied to more general nonlinear functional J. Such types of generalization have been studied by many authors. One important case is when J is a convex nonlinear functional. In this case, optimal convergence results may be proved under some appropriate assumptions (see [43] and the references cited therein).

An eigenvalue problem: To give a simple nonconvex optimization example, let us consider the computation of the smallest eigenvalue of the Laplacian operator with homogeneous Dirichlet boundary on a polygonal domain Ω. Let $V_h \subset H_0^1(\Omega)$ be a finite element space with a multilevel structure as described in the previous section. Then the finite element approximation of the smallest eigenvalue can be given by

$$\lambda_h = \min_{v \in V_h} R(v),$$

where R is the Rayleigh quotient:

$$R(v) = \frac{\| \nabla v \|^2}{\| v \|^2}.$$

We can apply Algorithm 6.1 with $R(\cdot)$ in place of $J(\cdot)$ using the multilevel space decomposition (5.13). This kind of algorithm has been studied in [29,15]. Under the assumption that the discrete Laplacian satisfies a discrete maximum principle, they have proved the qualitative convergence of this algorithm. But a uniformly optimal convergence, which is observed to be valid numerically, is yet to be established.

For other relevant multigrid methods for eigenvalue problems, we refer to [2,8,25,24,14,36,33,35,12].

6.2. Constrained optimizations

Subspace corrections can also be applied to constrained optimization problems. Let us illustrate such type of applications by a couple of examples.

Obstacle problem: We consider the following obstacle problem:

$$\min_{v \in H_0^1(\Omega), v(x) \geqslant 0} D(v) \quad \text{with } D(v) = \tfrac{1}{2} \| \nabla v \|^2 - (f, v).$$

This is a convex optimization problem with convex constraint. We consider applying subspace correction method for solving this problem with the multilevel space decomposition given by (5.11). The only difference here is that we need to take care of the constraint properly. Assume $\bar{u}$ is the current iterate, then the correction on the subspace span(ϕ_i^k) corresponding to the following one-dimensional constraint optimization problem:

$$\min_{\bar{u}+\alpha\phi_i^k \geqslant 0} D(\bar{u} + \alpha\phi_i^k).$$

One disadvantage of this algorithm is that it requires $O(n_J \log n_J)$ operation to finish one sweep of iteration to go through all subspaces in (5.11). Consequently, this algorithm is not quite optimal.

It is possible to modify the algorithm slightly so that each iteration has asymptotically optimal complexity, but such modifications seem to degrade the convergence properties (see [31] and references cited therein).

Algorithms of this kind have been theoretically proven to have asymptotically optimal convergence property under certain assumptions, namely the algorithms converge almost uniformly (with respect to mesh parameters) after sufficiently many iterations. Uniformly optimal convergence have been observed in numerical experiments, but this property is yet to be theoretically established.

Liquid crystal modeling: The examples given earlier fall into the category of convex optimization. The method of subspace correction can also be applied to nonconvex optimization. In [53], multilevel subspace correction method has been successfully applied to Oseen–Frank equations for liquid crystal modelings. One simple special case of Oseen–Frank equation is the so-called harmonic map problem:

$$\min \left\{ \int_\Omega |(\nabla v)(x)|^2 \, dx \colon v = (v_1, v_2, v_3) \in [H^1(\Omega)]^3, \right.$$
$$\left. \sum_i v_i(x)^2 = 1, \text{ in } \Omega \text{ and } v(x) = g(x) \text{ on } \partial\Omega \right\}.$$

This is a highly nonlinear problem and the corresponding Euler–Lagrange equation looks like

$$-\Delta u - |\nabla u|^2 u = 0.$$

Numerical experiments demonstrated that multilevel subspace correction method may be effectively applied to solve this type of problems. But theoretical analysis of this type of algorithm is still a wide open area.

7. Method of alternating projections

The method of alternating projection is, in its simplest form, due to [46]. Let us now briefly describe this method. Again, let V be a finite dimensional vector space and $M_1, M_2 \subset V$ be two subspaces. Let P_{M_1} and P_{M_2} be two orthogonal projections from V to M_1 and M_2, respectively. It is easy to see that $P_{M_1}P_{M_2} = P_{M_1 \cap M_2}$ if (and only if) P_{M_1} and P_{M_2} commutes, namely $P_{M_1}P_{M_2} = P_{M_2}P_{M_1}$. von Neumann [46] proved that, even if P_{M_1} and P_{M_2} do not commute, the following identity holds:

$$\lim_{k\to\infty} (P_{M_1}P_{M_2})^k = P_{M_1 \cap M_2}.$$

The above equation suggests an algorithm, called the method of alternating projection. It is as follows:

For any $v \in V$, set $v_0 = v$ and $v_k = P_{M_1}P_{M_2}v$ for $k = 1, 2, \ldots,$ then $v_k \to P_{M_1 \cap M_2}v$.

The following rate of convergence is known [1,30]:

$$\| (P_{M_1}P_{M_2})^k - P_{M_1 \cap M_2} \| = c^{2k-1}(M_1, M_2),$$

where $c(M_1, M_2)$ is the cosine of the angle between M_1 and M_2:

$$c(M_1, M_2) = \sup \left\{ \frac{(u, v)}{\| u \| \| v \|} : u \in M_1 \cap (M_1 \cap M_2)^\perp, \quad v \in M_2 \cap (M_1 \cap M_2)^\perp \right\}.$$

The method of alternating projections generalizes naturally to the case of more than two subspaces and similar (but less sharp) estimates for the rate of convergence have also been obtained in the literature [19,4]. It was noted in [18] that all these algorithms together with their estimates of convergence rate hold more generally when the subspaces are replaced by closed linear varieties (i.e. translates of subspaces).

The method of subspace corrections and the method of alternating projections are, not surprisingly, closely related. Let us consider a simple space decomposition

$$V = V_1 + V_2.$$

Let $M_1 = V_1^\perp$ and $M_2 = V_2^\perp$. The above identity means that $M_1 \cap M_2 = \{0\}$. Hence, we have

$$\lim_{k\to\infty} [(I - P_{V_2})(I - P_{V_1})]^k = 0.$$

This identity precisely means the convergence of the subspace correction method with subspace solvers being exact. In particular, this gives another qualitative proof of the convergence of the alternating (domain decomposition) method of Schwarz [42] (see also [32]).

The above discussion reveals a fact that some special cases of subspace correction methods (such as domain decomposition and multigrid methods) can be analyzed in the framework of method of alternating projections. This fact was correctly observed in [20]. But it was nevertheless stated in

[20] that the multigrid method, as an algorithm, would not be a method of alternating projections. Following [52], it is not difficult to verify the following statement:

The (exact) successive subspace correction method associated with the space decomposition $V = V_1 + V_2$ *for solving an SPD system* $Au = f$ *is precisely a method of alternating projection under the* $(\cdot,\cdot)_A$*-inner product associated with the following linear varieties*:

$$M_i = \{v \in V\colon (Av, \phi_i) = (f, \phi_i), \quad \forall \phi_i \in V_i\}.$$

On the other hand, the method of alternating projection method associated with the subspaces M_1 *and* M_2 *is equivalent to a method of (exact) successive subspace correction with* $V = (M_1 \cap M_2)^{\perp}$ *and* $V_i = M_i^{\perp}$ $(i = 1, 2)$ *for finding* $u \in V$ *such that*

$$(u, \phi) = (v, \phi) \quad \forall \phi \in V.$$

Furthermore, the kth iterate u_k *of the method of subspace corrections and the kth iterate* v_k *of the method of alternating projections is related by* $u_k = v - v_k$.

Hence, the method of (exact) successive subspace corrections and the method of alternating projections are mathematically equivalent. As a result, certain multigrid methods (such as those using Gauss–Seidel iterations as smoothers) can actually be viewed as a method of alternating projections.

Given the exact relationship revealed here between method of subspace corrections and method of alternating projections, it would be natural to ask if certain available error estimates in the literature for the method of alternating projections can be used to derive optimal quantitative convergence estimates for multigrid and/or domain decomposition methods. So far, we have not found this to be the case. We find that existing theories for these two classes of methods, despite their close relationship, were taken into different directions by different research communities. We are, however, able to improve some known convergence results for the method of alternating projections by using the techniques we developed for the method of subspace corrections [52].

References

[1] N. Aronszjan, Theory of reproducing kernels, Trans. Amer. Math. Soc. 68 (1950) 337–404.

[2] R.E. Bank, Analysis of a multilevel inverse iteration procedure for eigenvalue problems, SIAM J. Numer. Anal. 19 (5) (1982) 886–898.

[3] R.E. Bank, T. Dupont, An optimal order process for solving elliptic finite element equations, Math. Comput. 36 (1981) 35–51.

[4] H.H. Bauschke, J.M. Borwein, On projection algorithms for solving convex feasibility problems, SIAM Rev. 38 (3) (1996) 367–426.

[5] D. Braess, W. Hackbusch, A new convergence proof for the multigrid method including the V cycle, SIAM J. Numer. Anal. 20 (1983) 967–975.

[6] J.H. Bramble, in: Multigrid Methods, Pitman Research Notes in Mathematical Sciences, Vol. 294, Longman Scientific & Technical, Essex, England, 1993.

[7] J.H. Bramble, J.E. Pasciak, New estimates for multigrid algorithms including the V-cycle, Math. Comput. 60 (1993) 447–471.

[8] J.H. Bramble, J.E. Pasciak, A.V. Knyazev, A subspace preconditioning algorithm for eigenvector/eigenvalue computation, Adv. Comput. Math. 6 (2) (1997,1996) 159–189.

[9] J.H. Bramble, J.E. Pasciak, J. Wang, J. Xu, Convergence estimates for multigrid algorithms without regularity assumptions, Math. Comput. 57 (1991) 23–45.

[10] J.H. Bramble, J.E. Pasciak, J. Wang, J. Xu, Convergence estimates for product iterative methods with applications to domain decomposition, Math. Comput. 57 (1991) 1–21.
[11] J.H. Bramble, J.E. Pasciak, J. Xu, Parallel multilevel preconditioners, Math. Comput. 55 (1990) 1–22.
[12] A. Brandt, S. McCormick, J. Ruge, Multigrid methods for differential eigenproblems, SIAM J. Sci. Statist. Comput. 4 (2) (1983) 244–260.
[13] A. Brandt, S. McCormick, J. Ruge, Algebraic multigrid (AMG) for sparse matrix equations, in: Sparsity and its Applications, Loughborough, 1983, Cambridge University Press, Cambridge, 1985, pp. 257–284.
[14] Z. Cai, J. Mandel, S. McCormick, Multigrid methods for nearly singular linear equations and eigenvalue problems, SIAM J. Numer. Anal. 34 (1) (1997) 178–200.
[15] T.F. Chan, I. Sharapov, Subspace correction multilevel methods for elliptic eigenvalue problems, in: Proceedings of the Ninth International Conference on Domain Decomposition Methods, Wiley, New York, 1996.
[16] T.F. Chan, J. Xu, L. Zikatanov, An agglomeration multigrid method for unstructured grids, in: J. Mandel, C. Farhat and X.C. Cai (Eds.), 10th International Conference on Domain Decomposition Methods, Contemporary Mathematics, Vol. 218, American Mathematical Society, Providence, RI, 1998, pp. 67–81.
[17] W. Dahmen, Wavelet and multiscale methods for operator equations, Acta Numerica (1997) 55–228.
[18] F. Deutsch, Representers of linear functionals, norm-attaining functionals, and best approximation by cones and linear varieties in inner product spaces, J. Approx. Theory 36 (3) (1982) 226–236.
[19] F. Deutsch, The method of alternating orthogonal projections, in: S. Singh (Ed.), Approximation Theory, Spline Functions and Applications, Kluwer, Dordrecht, 1992, pp. 105–121.
[20] J. Gilbert, W.A. Light, Multigrid methods and the alternating algorithm, in: Algorithms for Approximation, Shrivenham, 1985, Oxford University Press, New York, 1987, pp. 447–458.
[21] M. Griebel, Multilevel algorithms considered as iterative methods on semidefinite systems, SIAM J. Sci. Comput. 15 (1994) 547–565.
[22] M. Griebel, P. Oswald, On additive Schwarz preconditioners for sparse grid discretizations, Numer. Math. 66 (4) (1994) 449–463.
[23] M. Griebel, P. Oswald, On the abstract theory of additive and multiplicative Schwarz algorithms, Numer. Math. 70 (2) (1995) 163–180.
[24] W. Hackbusch, On the computation of approximate eigenvalues and eigenfunctions of elliptic operators by means of a multi-grid method, SIAM J. Numer. Anal. 16 (2) (1979) 201–215.
[25] W. Hackbusch, Multigrid solutions to linear and nonlinear eigenvalue problems for integral and differential equations, Rostock. Math. Kolloq. (25) (1984) 79–98.
[26] W. Hackbusch, in: Multigrid Methods and Applications, Computational Mathematics, Vol. 4, Springer, Berlin, 1985.
[27] W. Hackbusch, Iterative Solution of Large Sparse Systems of Equations, Springer, Berlin, 1993.
[28] M. Holst, S. Vandewalle, Schwarz methods: to symmetrize or not to symmetrize, in: J. Mandel, S. McCormick (Eds.), Proceedings of the Seventh Copper Mountain Conference, Copper Mountain, NASA Langley Research Center, April 1995.
[29] M. Kaschiev, An iterative method for minimization of Rayleigh–Riesz functional, in: Computational Processes and Systems, Volume 6, Nauka, Moscow, 1988. pp. 160–170 (in Russian).
[30] S. Kayalar, H.L. Weinert, Error bounds for the method of alternating projections, Math. Control Signals Systems 1 (1) (1988) 43–59.
[31] R. Kornhuber, Adaptive Monotone Multigrid Method for Nonlinear Variational Problems, B.G. Teubner, Stuttgart, 1997.
[32] P.-L. Lions, On the Schwarz alternating method I, in: R. Glowinski, G.H. Golub, G.A. Meurant, J. Périaux (Eds.), First International Symposium on Domain Decomposition Methods for Partial Differential Equations, Philadelphia, 1988, SIAM, Philadelphia, PA, 1988, pp. 1–42.
[33] J. Mandel, S. McCormick, A multilevel variational method for $Au = \lambda Bu$ on composite grids, J. Comput. Phys. 80 (2) (1989) 442–452.
[34] J. Mandel, S.F. McCormick, R.E. Bank, Variational multigrid theory, in: S.F. McCormick (Ed.), Multigrid Methods, Frontiers in Applied Mathematics, Vol. 3, SIAM, Philadelphia, PA, 1987, pp. 131–177.
[35] S. McCormick, A mesh refinement method for $Au = \lambda Bu$ on composite grids, Math. Comput. 36 (1981) 485–498.
[36] S. McCormick, Multilevel adaptive methods for elliptic eigenproblems: a two-level convergence theory, SIAM J. Numer. Anal. 31 (6) (1994) 1731–1745.

[37] S.F. McCormick, in: Multigrid Methods, Frontiers in Applied Mathematics, Vol. 3, SIAM, Philadelphia, PA, 1987.
[38] R.A. Nicolaides, On a geometrical aspect of SOR and the theory of consistent ordering for positive definite matrices, Numer. Math. 23 (1974) 99–104.
[39] P. Oswald, On function spaces related to finite element approximation theory, Z. Anal. Anwendungen 9 (1) (1990) 43–64.
[40] P. Oswald, Multilevel Finite Element Approximation, Theory and applications, B.G. Teubner, Stuttgart, 1994.
[41] J.W. Ruge, K. Stüben, Algebraic multigrid, in: S.F. McCormick (Ed.), Multigrid Methods, Frontiers in Applied Mathematics, 1987, SIAM, Philadelphia, PA, 73–130.
[42] H.A. Schwarz, Gesammelte mathematische abhandlungen, Vierteljahrsschrift der Naturforschenden Gesellschaft 15 (1870) 272–286.
[43] X. Tai, J. Xu, Global convergence of subspace correction methods for convex optimization problems, Mathematics of Computation, to appear.
[44] P. Vanek, M. Brezina, J. Mandel, Convergence of algebraic multigrid based on smoothed aggregation, Technical Report No. 126, University of Colorado, 1998.
[45] P. Vanek, J. Mandel, M. Brezina, Algebraic multi-grid by smoothed aggregation for second and fourth order elliptic problems, Computing 56 (1996) 179–196.
[46] J. von Neumann, The geometry of orthogonal spaces, in: Functional Operators – Vol. II, Annals of Mathematical Studies, Vol. 22, Princeton University Press, Princeton, NJ, 1950 (This is a reprint of mineographed lecture notes, first distributed in 1933).
[47] P. Wesseling, An Introduction to Multigrid Methods, Wiley, Chichester, 1992.
[48] O.B. Widlund, Some Schwarz methods for symmetric and nonsymmetric elliptic problems, in: Fifth International Symposium on Domain Decomposition Methods for Partial Differential Equations, Norfolk, VA, 1991, SIAM, Philadelphia, PA, 1992, pp. 19–36.
[49] J. Xu, Theory of multilevel methods, Ph.D. Thesis, Cornell University, 1989.
[50] J. Xu, Iterative methods by space decomposition and subspace correction, SIAM Rev. 34 (1992) 581–613.
[51] J. Xu, An introduction to multilevel methods, in: M. Ainsworth, J. Levesley, W.A. Light and M. Marletta (Eds.), Wavelets, Multilevel Methods and Elliptic PDEs, Leicester, 1996, Oxford University Press, New York, 1997, pp. 213–302.
[52] J. Xu, L. Zikatanov, The method of subspace corrections and the method of alternating projections in Hilbert space, AM223, Penn state, 2000, preprint, 2000.
[53] J. Xu, L. Zikatanov, Multigrid methods for liquid crystal modeling, preprint, 2000.
[54] H. Yserentant, Old and new convergence proofs for multigrid methods, in: Acta Numerica, Cambridge University Press, Cambridge, 1993, pp. 285–326.

ELSEVIER

Journal of Computational and Applied Mathematics 128 (2001) 363–381

JOURNAL OF COMPUTATIONAL AND APPLIED MATHEMATICS

www.elsevier.nl/locate/cam

Moving finite element, least squares, and finite volume approximations of steady and time-dependent PDEs in multidimensions

M.J. Baines

Department of Mathematics, The University of Reading, P.O.Box 220, Reading RG6 6AX, UK

Received 10 August 1999; received in revised form 24 November 1999

Abstract

We review recent advances in Galerkin and least squares methods for approximating the solutions of first- and second-order PDEs with moving nodes in multidimensions. These methods use unstructured meshes and minimise the norm of the residual of the PDE over both solutions and nodal positions in a unified manner. Both finite element and finite volume schemes are considered, as are transient and steady problems. For first-order scalar time-dependent PDEs in any number of dimensions, residual minimisation always results in the methods moving the nodes with the (often inconvenient) approximate characteristic speeds. For second-order equations, however, the moving finite element (MFE) method moves the nodes usefully towards high-curvature regions. In the steady limit, for PDEs derived from a variational principle, the MFE method generates a locally optimal mesh and solution: this also applies to least squares minimisation. The corresponding moving finite volume (MFV) method, based on the l_2 norm, does not have this property however, although there does exist a finite volume method which gives an optimal mesh, both for variational principles and least squares. © 2001 Elsevier Science B.V. All rights reserved.

Keywords: Moving finite element; Least squares; Finite volume

1. Introduction

In this paper we consider standard Galerkin and least squares methods on moving meshes in multidimensions. The capabilities of mesh movement in approximating the solution of PDEs are yielding their secrets slowly, largely because there have been significant difficulties in handling the complex nonlinearities inherent in the problem and in controlling the mesh to prevent tangling. In the recent past techniques employed have included various forms of equidistribution [10,13,14], usually based

E-mail address: m.j.baines@reading.ac.uk (M.J. Baines).

PII: S0377-0427(00)00519-7

on solution shape criteria, and minimisation based on the residual of the PDE [3,12,22]. Equidistribution is a highly effective technique for the distribution of nodes in one dimension. However, there have remained question marks over how to choose the equidistribution criteria and to what purpose (although see [7]). Minimisation techniques, on the other hand, allow immediate access to multidimensions but here node distribution is less well understood. In this paper we concentrate on residual minimisation as the criterion for moving the nodes.

We begin by recalling the basis of the moving finite element (MFE) method of Miller [1,8,9,17,19] together with some of its properties. We then go on to discuss L_2 least squares methods on moving meshes, with examples, and the relationship between the two methods in the steady case.

In the second part of the paper we describe moving finite volume and discrete l_2 least squares methods are proposed using the same approach. These methods have their own character and their properties differ from the L_2 case when the nodes are allowed to move.

Finally, we discuss local optimisation methods for minimising these norms and conclude with a summary of the schemes and their properties.

2. Finite elements

The Galerkin finite element method for the generic scalar PDE

$$u_t = Lu, \tag{1}$$

where L is a second-order space operator, e.g. $Lu = -a\partial_x u + \sigma\partial_x^2 u$, is a semi-discrete method based on a weak form of the PDE. One way of deriving the weak form is by constructing the unique minimiser of the L_2 residual of the PDE with respect to *the time derivative* U_t via

$$\min_{U_t} \|U_t - LU\|_{L_2}^2, \tag{2}$$

where U is the finite-dimensional approximation to u. Differentiating (2) with respect to U_t and using an expansion of U in terms of basis functions $\phi_j(\underline{x})$ in the form

$$U = \sum_j U_j(t)\phi_j(\underline{x}) \tag{3}$$

yields the Galerkin equations

$$\langle \phi_j, U_t - LU \rangle = 0, \tag{4}$$

where the bracket notation denotes the L_2 inner product and ϕ_j is the jth basis function for the finite-dimensional subspace which contains U and therefore U_t. We take the functions U and U_t to be piecewise continuous and the basis function ϕ_j to be of local finite element type. The resulting matrix system may be solved in time using a suitable ODE package in the style of the method of lines.

2.1. Steady state

In principle the Galerkin method may be used to solve the time-*independent* equation $Lu = 0$ by driving solutions of the time-dependent equation (1) to steady state.

To reach steady state the velocity U_t in (4) may be replaced by an explicit time discretisation with index n and time step τ and the discrete equation

$$\left\langle \phi_j, \frac{U^{n+1} - U^n}{\tau} - LU^n \right\rangle = 0 \tag{5}$$

used as an iteration to drive U^n to convergence. The steady-state solution satisfies the weak form

$$\langle \phi_j, LU \rangle = 0. \tag{6}$$

Not only may the Galerkin equations (4) or (5) be used to iterate to steady state but the mass matrix may be replaced by any positive definite matrix.

2.2. *An optimal property of the steady Galerkin equations*

If the differential equation $Lu = 0$ can be derived from a variational principle, i.e. there exists a function $F(\underline{x}, u, \nabla u)$ such that

$$Lu = -\frac{\partial F}{\partial u} + \nabla \cdot \frac{\partial F}{\partial \nabla u}, \tag{7}$$

then since

$$\frac{\partial}{\partial U_j} \int F(\underline{x}, U, \nabla U)\,\mathrm{d}\underline{x} = \int \left(\frac{\partial F}{\partial U}\frac{\partial U}{\partial U_j} + \frac{\partial F}{\partial \nabla U}\frac{\partial \nabla U}{\partial U_j} \right) \mathrm{d}\underline{x}$$

$$= \int \left(\frac{\partial F}{\partial U} - \nabla \cdot \frac{\partial F}{\partial \nabla U} \right) \frac{\partial U}{\partial U_j}\,\mathrm{d}\underline{x} = -\int LU\phi_j\,\mathrm{d}\underline{x} \tag{8}$$

the Galerkin equations (6) provide an optimal U for variations of the functional

$$I(F) = \int F(\underline{x}, u, \nabla u)\,\mathrm{d}\underline{x} \tag{9}$$

in the approximation space of U. The functional (9) is minimised by solutions of the weak form (6) with LU given by (7), i.e. solutions of

$$\left\langle \phi_j, \frac{\partial F}{\partial U} \right\rangle + \left\langle \nabla \phi_j, \frac{\partial F}{\partial \nabla U} \right\rangle = 0. \tag{10}$$

In (10) integration by parts has been used over a local patch of elements surrounding node j (see Fig. 1) with the assumption that the finite element basis functions ϕ_j vanish on the boundary of the patch. Not only may the Galerkin equations (4) or (5) be used to iterate to steady state but the mass matrix may be replaced by any positive definite matrix.

In particular, if $F(\underline{x}, u, \nabla u) = \frac{1}{2}(f(\underline{x}, u, \nabla u))^2$, the functional

$$J(f) = \frac{1}{2} \int (f(\underline{x}, u, \nabla u))^2\,\mathrm{d}\underline{x} \tag{11}$$

is minimised by solutions of the weak form

$$\left\langle \phi_j, \frac{\partial f^2}{\partial U} \right\rangle + \left\langle \nabla \phi_j, \frac{\partial f^2}{\partial \nabla U} \right\rangle = 0. \tag{12}$$

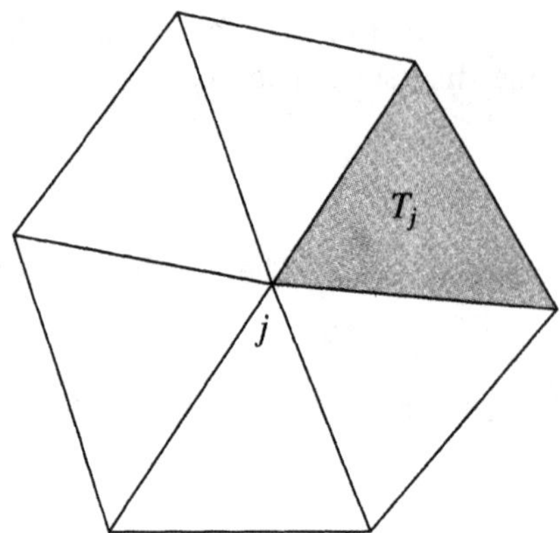

Fig. 1. A local patch of elements surrounding node j.

3. Moving finite elements

The moving finite element (MFE) procedure [1,8,9,17,19] is a semi-discrete moving mesh finite element method based on the two coupled weak forms of the PDE arising from the minimisation in Section 2 when the node locations are allowed to depend on time. Thus U becomes an explicit function of $\underline{X}_j(t)$ (the nodal positions). Then, using the result

$$\frac{\partial U}{\partial \underline{X}_j} = (-\nabla U)\phi_j \tag{13}$$

which holds if the basis functions ϕ_j are of *linear* finite element type (see [17,19] or [16]), the derivative of U with respect to t becomes

$$\begin{aligned}\frac{\partial U}{\partial t}\Big|_{\text{moving}\underline{X}} &= \frac{\partial U}{\partial t}\Big|_{\text{fixed}\underline{X}} + \sum_j \frac{\partial U}{\partial \underline{X}_j} \cdot \frac{\mathrm{d}\underline{X}_j}{\mathrm{d}t} \\ &= \frac{\partial \hat{U}}{\partial t} + \sum_j (-\nabla U)\phi_j \cdot \frac{\partial \underline{X}_j}{\mathrm{d}t} \\ &= \dot{U} - \nabla U \cdot \dot{\underline{X}}\end{aligned} \tag{14}$$

say, where U is given by (3) and $\hat{U}$ and $\underline{X}$ are independent functions of t whose time derivatives have expansions

$$\dot{U} = \frac{\partial \hat{U}}{\partial t} = \sum_j \frac{\partial U_j}{\partial t}\phi_j, \qquad \dot{\underline{X}} = \frac{\mathrm{d}\underline{X}}{\mathrm{d}t} = \sum_j \frac{\mathrm{d}\underline{X}_j}{\mathrm{d}t}\phi_j \tag{15}$$

(cf. (3)). These functions are taken to be continuous, corresponding to the evolution of a continuous linear approximation.

Using (14), minimisation of the residual in (2) over the coefficients $\dot{U}_j, \dot{\underline{X}}_j$ then takes the form

$$\min_{\dot{U}_j, \dot{X}_j} \|\dot{U} - \nabla U \cdot \dot{\underline{X}} - LU\|_{L_2}^2 \tag{16}$$

which, using (15), leads to the MFE or extended Galerkin equations

$$\langle \phi_j, \dot{U} - \nabla U \cdot \dot{\underline{X}} - LU \rangle = 0, \tag{17}$$

$$\langle (-\nabla U)\phi_j, \dot{U} - \nabla U \cdot \dot{\underline{X}} - LU \rangle = 0. \tag{18}$$

The resulting ODE system may be solved by a stiff ODE package as in the method of lines.

The method has been analysed in [1] and found to possess the following properties:

Property 1. For scalar first-order time-dependent PDEs in any number of dimensions the method is an approximate method of characteristics.

Property 2. For scalar second-order time-dependent PDEs in one dimension the method repels nodes from inflection points towards areas of high curvature. At steady state the nodes asymptotically equidistribute a power of the second derivative.

However, the method also has intrinsic singularities.

- If ∇U has a component whose values are equal in adjacent elements (dubbed parallelism by Miller [19]), the system of equations (17) and (18) becomes singular and must be regularised in some way (see [1,8,9,17,19]).
- If the area of an element vanishes, the system again becomes singular and special action is required.

Each of these singularities also leads to poor conditioning of the corresponding matrix systems near to singularity. For these reasons the method is usually regularised by adding penalties in the functional (16).

3.1. Steady state

In the same way as for fixed meshes the MFE method may in principle be used to generate weak forms for approximately solving the *steady* PDE $Lu = 0$ by driving the MFE solutions to steady state (assuming convergence). From Property 2 of Section 3 it may be expected that for scalar *second-order* PDEs in one dimension the nodes will converge towards areas of high curvature. Property 1, however, indicates that for scalar *first-order* PDEs the nodes continue to move with characteristic speeds and are not therefore expected to settle down to a steady state.

To reach a steady state we may replace the velocities $\dot{U}$ and $\dot{\underline{X}}$ by explicit time discretisations with index n and time steps τ, σ and use the resulting equations (17) and (18) to drive U^n, X^n to convergence, provided that is possible. Since we are only interested in the limit the mass matrix may be replaced by any positive definite matrix. The steady-state solution satisfies the weak forms

$$\left\langle \begin{pmatrix} 1 \\ -\nabla U \end{pmatrix} \phi_j, LU \right\rangle = 0. \tag{19}$$

We note that from (16) the MFE method in the steady case implements the minimisation

$$\min_{\dot{U}, \dot{\underline{X}}} \|LU\|_{L_2}^2. \tag{20}$$

Although $\dot{U}$ and $\dot{\underline{X}}$ no longer appear in LU, the minimisation is over all functions lying in the space spanned by $\{\phi_j, (-\nabla U)\phi_j\}$. In one dimension this space is also spanned by the discontinuous linear functions on the mesh (see [1]) (provided that U_x is not equal in adjacent elements).

3.2. The optimal property of the steady MFE equations

It has been shown in [16] that the optimal property of Section 2.2 generalises to the steady MFE equations (19). If $Lu=0$ can be derived from a variational principle then, as in Section 2.2, solutions of the weak forms (19) provide a local optimum of (9) over the approximation space spanned by the set of functions $\{\phi_j, (-\nabla U)\phi_j\}$. We shall refer to this property as the optimal property. This result essentially follows from (14) modified to apply to variations, i.e.

$$\delta U|_{\text{moving}\underline{X}} = \delta U|_{\text{fixed}\underline{X}} - \nabla U \cdot \delta \underline{X}. \tag{21}$$

The MFE method may therefore be used as an iterative procedure to generate locally optimal meshes (see [16]). If desired, the MFE mass matrix may be replaced by any positive definite matrix (see [18]).

Substituting (7) into (19), the functional (9) is minimised by solutions of the two weak forms

$$\left\langle \phi_j, \frac{\partial F}{\partial U} \right\rangle + \left\langle \nabla \phi_j, \frac{\partial F}{\partial \nabla U} \right\rangle = 0, \tag{22}$$

$$\left\langle \phi_j, \frac{\partial F}{\partial \underline{x}} \right\rangle + \left\langle \nabla \phi_j, \left(F - \nabla U \cdot \frac{\partial F}{\partial \nabla U} \right) \right\rangle = 0, \tag{23}$$

where the identity

$$\nabla \cdot \left(F - \nabla U \cdot \frac{\partial F}{\partial \nabla U} \right) = \frac{\partial F}{\partial \underline{x}} + \frac{\partial F}{\partial U} \nabla U - \left(\nabla \cdot \frac{\partial F}{\partial \nabla U} \right) \nabla U \tag{24}$$

has been used to transform the second component of (19) into the equivalent equation (23) which is formally suitable for piecewise linear approximation. In carrying out the integration by parts to arrive at (23) we have used the fact that the continuous piecewise linear finite element basis function ϕ_j vanishes on the boundary of the patch.

In particular, for the least squares functional (11) the weak forms are

$$\left\langle \phi_j, \frac{\partial f^2}{\partial U} \right\rangle + \left\langle \nabla \phi_j, \frac{\partial f^2}{\partial \nabla U} \right\rangle = 0, \tag{25}$$

$$\left\langle \phi_j, \frac{\partial f^2}{\partial \underline{x}} \right\rangle + \left\langle \nabla \phi_j, \left(f^2 - \nabla U \cdot \frac{\partial f^2}{\partial \nabla U} \right) \right\rangle = 0. \tag{26}$$

4. Least squares finite elements

Notwithstanding the use of the L_2 norm in the construction of the Galerkin and MFE methods in Sections 2 and 3, from a fully discrete point of view the procedure used there is a restricted least squares minimisation because it is carried out only over the *velocities* $\dot{U}_j$ and $\dot{\underline{X}}_j$. The variables U_j

and $\underline{X}_j$ are treated as constants, independent of $\dot{U}_j$ and $\dot{\underline{X}}_j$ and the coupling is ignored, as in the method of lines.

A fully discrete least squares approach is feasible, however, if u_t is discretised in time *before* the least squares minimisations are carried out. Minimisation is then over U_j and $\underline{X}_j$ rather than $\dot{U}_j$ and $\dot{\underline{X}}_j$ and the variational equations include additional terms that do not arise in the semi-discrete finite element formulations.

In what follows we shall restrict attention to *first-order* space operators Lu depending on $\underline{x}$, u and ∇u only.

4.1. Least squares finite elements on a fixed grid

To describe the procedure in more detail consider a one-step (explicit or implicit) time discretisation of Eq. (1) of the form

$$\frac{u^{n+1} - u^n}{\Delta t} = Lu^*, \tag{27}$$

where $*$ may denote n or $n+1$. The finite-dimensional approximation U^{n+1} at the next time step is then generated by least squares minimisation of the residual

$$R^* = \frac{U^{n+1} - U^n}{\Delta t} - LU^* \tag{28}$$

of (27) over the coefficients U_j^{n+1} via

$$\min_{U_j^{n+1}} \|R^*\|_{L_2}^2. \tag{29}$$

In the explicit case ($* = n$) the gradient of (29) with respect to U_j^{n+1} gives rise to the weak form

$$\left\langle R^n, \frac{\phi_j^{n+1}}{\Delta t} \right\rangle = 0 \tag{30}$$

which is a simple time discretisation of (4).

However, in the implicit case ($* = n+1$) the gradient of (29) with respect to U_j^{n+1} leads to

$$\left\langle R^{n+1}, \frac{\phi_j^{n+1}}{\Delta t} - \frac{\partial LU^{n+1}}{\partial U_j^{n+1}} \right\rangle = 0. \tag{31}$$

Eq. (31) is not simply an implicit time discretisation of (4) because of the additional terms in the test function.

4.2. Least squares moving finite elements (LSMFE)

Now consider minimisation of the L_2 norm in (29) over the nodal coordinates $\underline{X}_j^{n+1}$ as well as the coefficients U_j^{n+1}. This is the approach of the recent least squares moving finite element (LSMFE) method [18] which was proposed partly in an attempt to overcome the difficulties which arise with first-order PDEs when the nodes move with characteristic speeds.

By analogy with (16) consider the minimisation

$$\min_{U_j^{n+1},\underline{X}_j^{n+1}} \|R^*\|_{L_2}^2, \tag{32}$$

where R^* is the residual

$$R^* = \frac{U^{n+1}-U^n}{\Delta t} - \nabla U^* \cdot \frac{\underline{X}^{n+1}-\underline{X}^n}{\Delta t} - LU^* \tag{33}$$

(cf. (16)). In the explicit case ($*=n$) $\|R^n\|_{L_2}^2$ is quadratic in both sets of variables $U_j^{n+1},\underline{X}_j^{n+1}$ and minimisation yields

$$\left\langle R^n, \begin{pmatrix} 1 \\ -\nabla U^n \end{pmatrix} \frac{\phi_j^{n+1}}{\Delta t} \right\rangle = 0 \tag{34}$$

which is a simple time discretisation of (17).

In the implicit case ($*=n+1$), on the other hand, the gradient of $\|R^{n+1}\|_{L_2}^2$ with respect to U_j^{n+1} gives

$$\left\langle R^{n+1}, \left(\frac{\phi_j^{n+1}}{\Delta t} - \frac{\underline{X}^{n+1}-\underline{X}^n}{\Delta t} \cdot \nabla \phi_j^{n+1} - \frac{\partial}{\partial U_j^{n+1}} LU^{n+1} \right) \right\rangle = 0 \tag{35}$$

while that with respect to $\underline{X}_j^{n+1}$ gives (formally)

$$\left\langle R^{n+1}, (-\nabla U^{n+1}) \left(\frac{\phi_j^{n+1}}{\Delta t} - \frac{\underline{X}^{n+1}-\underline{X}^n}{\Delta t} \cdot \nabla \phi_j^{n+1} - \frac{\partial}{\partial U_j^{n+1}} LU^{n+1} \right) \right\rangle$$

$$+ \oint \frac{1}{2} (R^{n+1})^2 \phi_j^{n+1} \underline{\hat{n}} \, \mathrm{d}s = 0 \tag{36}$$

using (13), where the boundary integral in (36) (which appears due to variations in the mesh) is taken over the boundaries of the elements in the patch containing node j (see Fig. 1). The unit normal $\underline{\hat{n}}$ is measured inwards.

In deriving Eqs. (35) and (36) the functions U^{n+1} and $\underline{X}^{n+1}$ appearing in the time-discretised terms in (33) are regarded as independent variables, but the U^{n+1} occurring in ∇U^{n+1} and LU^{n+1} are functions of $\underline{x}$ and $\underline{X}$. We have therefore used the chain rule

$$\frac{\partial LU}{\partial U_j} = \frac{\partial U}{\partial U_j}\frac{\partial LU}{\partial U} + \frac{\partial \nabla U}{\partial U_j} \cdot \frac{\partial LU}{\partial \nabla U} = \frac{\partial LU}{\partial U}\phi_j + \frac{\partial LU}{\partial \nabla U} \cdot \nabla \phi_j \tag{37}$$

(see (3)) when differentiating ∇U^{n+1} and LU^{n+1} to obtain (36).

We shall refer to (35) and (36) as the transient LSMFE equations. These equations have been solved in [18] in one dimension and, in spite of hopes to the contrary, the method was found to still possess Property 1 of Section 3, that for scalar first-order time-dependent PDEs the method is an approximate method of characteristics. This property survives because, even though Eqs. (35) and (36) differ from (17) and (18), approximate characteristic speeds still make the residual vanish. However, as we shall see, the method does generate the optimal property of MFE at the steady state.

4.3. Steady-state least squares moving finite elements

Consider now the steady-state limit of (35) and (36) as $\Delta t \to \infty$, assuming that convergence takes place. Since $R^{n+1} \to LU$, Eqs. (35) and (36) become

$$\left\langle LU, \frac{\partial}{\partial U_j} LU \right\rangle = \left\langle LU, (-\nabla U) \frac{\partial}{\partial U_j}(LU) \right\rangle + \oint \frac{1}{2}(LU)^2 \phi_j \underline{\hat{n}} \, \mathrm{d}s = 0 \tag{38}$$

which may be written in the equivalent forms

$$\left\langle \frac{\partial (LU)^2}{\partial U}, \phi_j \right\rangle + \left\langle \frac{\partial (LU)^2}{\partial \nabla U}, \nabla \phi_j \right\rangle = 0, \tag{39}$$

$$\left\langle \frac{\partial (LU)^2}{\partial \underline{x}}, \phi_j \right\rangle + \left\langle \left((LU)^2 - (\nabla U) \cdot \frac{\partial (LU)^2}{\partial \nabla U} \right), \nabla \phi_j \right\rangle = 0, \tag{40}$$

where we have used (37) and the identity (24) with $F = (LU)^2$. Eq. (38) may be obtained by direct minimisation of $\|LU\|_{L_2}^2$ over U_j and $\underline{X}_j$.

Referring back to (25) and (26) we see that Eqs. (39) and (40) are the *steady* MFE equations for the PDE

$$-\frac{\partial (Lu)^2}{\partial u} + \nabla \cdot \frac{\partial (Lu)^2}{\partial \nabla u} = 0 \tag{41}$$

(see [16]) which corresponds to the Euler–Lagrange equation for the minimisation of the least squares functional $\|Lu\|_{L_2}^2$.

To solve the nonlinear equations (39) and (40) by iteration we may use the corresponding time-stepping method, (35) and (36), with n as the iteration parameter, or any other convenient iteration (see Section 7).

4.4. Properties of the steady LSMFE equations

(i) As we have already seen in Section 3.3, for steady problems the least squares functional $F(x, U, \nabla U) = (LU)^2$ leads to the weak forms (25) and (26) and therefore (39) and (40), and we have the optimal property, as expected.

(ii) In the LSMFE tests carried out in [18] on scalar first-order *steady-state* equations it is shown that the nodes no longer move with characteristic speeds, as in Property 1 of Section 3, but instead move to regions of high curvature as in Property 2. This is a useful property and could have been expected because the least squares procedure in effect embeds the original first-order equation in the *second-order* Eq. (41) for which the MFE steady limit yields the asymptotic equidistribution of Property 2 of Section 3.1.

(iii) In the particular case where LU takes the form of a divergence of a continuous function of U, we may apply an extension of the result in [4] which shows that, asymptotically, minimisation of $\|LU\|_{L_2}^2$ is equivalent to equidistribution of LU over all the elements in a certain sense. Thus for example in the case where

$$Lu = \nabla \cdot (\underline{a}u) \tag{42}$$

with constant $\underline{a}$, the LSMFE method asymptotically equidistributes the piecewise constant residual $LU = \nabla \cdot (\underline{a}U)$ in each element in the sense described in Section 6.1.

We now give some illustrative examples of steady LSMFE.

4.5. Examples

(i) Take

$$Lu = u - f(x) \tag{43}$$

for which the steady LSMFE weak forms, from (39) and (40), are

$$\left\langle \begin{pmatrix} 1 \\ -\nabla U \end{pmatrix} \phi_j, (U - f(\underline{x})) \right\rangle = 0, \tag{44}$$

subject to boundary conditions, which provide a local minimum for the least squares variable node approximation problem

$$\min_{U_j, \underline{X}_j} \int (U - f(\underline{x}))^2 \, d\underline{x}. \tag{45}$$

(Superior results for this problem can however be obtained by considering piecewise linear discontinuous approximation — see Section 4.6.)

(ii) Take

$$Lu = \underline{a} \cdot \nabla u = a\frac{\partial u}{\partial x} + b\frac{\partial u}{\partial y}, \tag{46}$$

where $\underline{a} = (a, b)$ is constant. In this case Eq. (41) becomes the degenerate elliptic equation

$$\nabla \cdot ((\underline{a} \cdot \nabla u)\underline{a}) = \left(a\frac{\partial}{\partial x} + b\frac{\partial}{\partial y}\right)\left(a\frac{\partial u}{\partial x} + b\frac{\partial u}{\partial y}\right) = 0. \tag{47}$$

The steady LSMFE weak forms, from (39) and (40), are

$$\langle \underline{a} \cdot \nabla U, \underline{a} \cdot \nabla \phi_j) \rangle = 0 \tag{48}$$

and

$$\left\langle \phi_j, (\underline{a} \cdot \nabla U)\frac{\partial(\underline{a} \cdot \nabla U)}{\partial \underline{x}} \right\rangle - \tfrac{1}{2}\langle \nabla \phi_j, (\underline{a} \cdot \nabla U)^2 \rangle = 0, \tag{49}$$

subject to boundary conditions, which provide a local minimum for the least squares variable node approximation problem

$$\min_{U_j, X_j} \int (\underline{a} \cdot \nabla U)^2 \, d\underline{x} \tag{50}$$

(see ([2,21])). These are also the steady MFE equations for the second-order degenerate elliptic equation

$$\nabla \cdot ((\underline{a} \cdot \nabla u)\underline{a}) = 0. \tag{51}$$

If $\underline{a}$ depends on $\underline{x}$ but is continuous and divergence-free, then $LU = \underline{a} \cdot \nabla U$ is the divergence of a continuous function and this example has the asymptotic equidistribution property referred to in

Section 4.4, in this case asymptotically equidistributing $\underline{\bar{a}} \cdot \nabla U$ where $\underline{\bar{a}}$ consists of the element averages of $a(\underline{x})$.

(iii) Take

$$Lu = |\nabla u| = \left(\left(\frac{\partial u}{\partial x} \right)^2 + \left(\frac{\partial u}{\partial y} \right)^2 \right)^{1/2} \tag{52}$$

for which Eq. (41) is Laplace's equation

$$\nabla^2 u = 0. \tag{53}$$

The steady LSMFE weak forms, from (39) and (40), are

$$\langle \nabla \phi_j, \nabla U \rangle = \left\langle \phi_j, \frac{\partial (\nabla U)^2}{\partial \underline{X}} \right\rangle - \langle \nabla \phi_j, (\nabla U)^2 \rangle = 0, \tag{54}$$

subject to boundary conditions, which provide a minimum for the variable node Dirichlet problem (see e.g. [22])

$$\min_{U_j, \underline{X}_j} \int |\nabla U|^2 \, \mathrm{d}\underline{x}. \tag{55}$$

4.6. The MBF approach

If the functional F is independent of ∇U (the best approximation problem) we may take combinations of the variations δU_j and $\delta \underline{X}_j$ to design simpler sequential algorithms. The first variation of the square of the L_2 norm of $R = LU$, using (32), (35), (36) and (38), gives

$$\delta \tfrac{1}{2} \|LU\|_{L_2}^2 = \left\langle LU, \frac{\partial LU}{\partial U} \phi_j \right\rangle (\delta U_j - \nabla U \cdot \delta \underline{X}_j) \tag{56}$$

$$- \oint \frac{1}{2} (LU)^2 \phi_j \underline{\hat{n}} \cdot \delta \underline{X}_j \, \mathrm{d}s. \tag{57}$$

Setting $\delta \underline{X}_j = 0$ gives the fixed mesh least squares equation

$$\langle LU \frac{\partial LU}{\partial U}, \phi_j \rangle = 0 \tag{58}$$

while, setting

$$\delta U_j - \nabla U \cdot \delta \underline{X}_j = 0$$

gives

$$\oint \frac{1}{2} (LU)^2 \phi_j \underline{\hat{n}} \, \mathrm{d}s = 0, \tag{59}$$

an equation for updating the nodes locally which depends only on the integral over the boundaries of the patch containing node j. The constraint (4.6), which forces δU_j to vary with $\delta \underline{X}_j$, corresponds to linear interpolation/extrapolation on the current piecewise linear U function. This approach, which depends only on local problems, is called the moving best fits (MBF) method in [1] and is the basis of best approximation algorithms in [2,21].

5. Finite volume methods

We turn now to a discussion of the use of *discrete* l_2 norms with area weighting as a basis for generating finite volume schemes, to a large part the result of a simple quadrature applied to the L_2 norm used previously.

Define the discrete l_2 norm as the weighted sum over triangles of the average residual of the PDE, viz.

$$\|R\|_{l_2}^2 = \sum_T S_T \bar{R}_T^2 \tag{60}$$

(cf. (2), (16), (29) and (32)), where the suffix T runs over all the triangles of the region, S_T is the area of triangle T and $\bar{R}_T$ is the average value of the residual R over the vertices of T.

This norm coincides with the L_2 norm in the case where R is constant on each triangle. For then

$$\|R\|_{L_2}^2 = \int R^2 \, \mathrm{d}\underline{x} = \sum_T \int_T R^2 \, \mathrm{d}\underline{x} = \sum_T \bar{R}_T^2 \int_T \mathrm{d}\underline{x} = \sum_T S_T \bar{R}_T^2 = \|R\|_{l_2}^2 \tag{61}$$

as in example (ii) of Section 4.5 where $R = LU = \underline{a} \cdot \nabla U$, the advection speed $\underline{a}$ being constant and U piecewise linear. If the area weighting in (60) is omitted this link is lost. However, one objection to the use of least squares residuals is that when triangles become degenerate the norm of the derivative is unbounded. By redefining the norm in (60) with a squared weight S_T^2 instead of S_T the norm is always well defined and still has an approximate equidistribution property (see [19]). Here we concentrate on (60); however, the modifications are straightforward.

The form (60) may be rewritten as a sum over nodes j, namely

$$\|R\|_{l_2}^2 = \frac{1}{3} \sum_j \sum_{\{T_j\}} S_{T_j} \bar{R}_{T_j}^2, \tag{62}$$

where $\{T_j\}$ runs over the patch of triangles abutting node j (see Fig. 1).

We may take $(\nabla U)_T$ to be the gradient associated with the linearly interpolated corner values of U in the triangle T, given by (see [16])

$$(\nabla U)_T = \left(\frac{-\sum^i U_i \Delta Y_i}{\sum^i X_i \Delta Y_i}, \frac{\sum^i U_i \Delta X_i}{-\sum^i Y_i \Delta X_i} \right) = \left(\frac{\sum^i Y_i \Delta U_i}{\sum^i X_i \Delta Y_i}, \frac{-\sum^i X_i \Delta U_i}{\sum^i Y_i \Delta X_i} \right), \tag{63}$$

where the sums run over the corners i of the triangle T and $\Delta X_i, \Delta Y_i, \Delta U_i$ denote the increments in the values of X, Y, U taken anticlockwise across the side of T *opposite* the corner concerned (see Fig. 1). This is of course identical to the finite element gradient with piecewise linear approximation. In the same notation the area of the triangle T is

$$S_T = \frac{1}{2} \sum_i X_i \Delta Y_i = -\frac{1}{2} \sum_i Y_i \Delta X_i \tag{64}$$

which incorporates a summation by parts.

5.1. Moving finite volumes

By analogy with the MFE method of Section 3 a moving finite volume (MFV) method may be set up by minimising the residual

$$\|\dot{U} - \nabla U \cdot \underline{\dot{X}} - LU\|_{l_2} \tag{65}$$

(see (60)) over $\dot{U}_j$ and $\underline{\dot{X}}_j$, which leads to

$$\sum_{\{T_j\}} S_{T_j}(\dot{U} - \nabla U \cdot \underline{\dot{X}} - LU)_j = 0, \tag{66}$$

$$\sum_{\{T_j\}} S_{T_j}(\dot{U} - \nabla U \cdot \underline{\dot{X}} - LU)_j(-\nabla U)_{T_j} = 0 \tag{67}$$

(cf. (17) and (18)) where $\{T_j\}$ is the set of triangles abutting node j and the suffix j indicates that terms involving $\dot{U}$ and $\underline{\dot{X}}$ are evaluated at the node j while those involving ∇U and LU are evaluated over the triangle T_j.

Property 1 of Section 3 still holds since the residual vanishes as before when $\underline{\dot{X}}$ approximates characteristic speeds. The method also has the same singularities as MFE, in particular when components of the gradients ∇U are the same in adjacent elements.

At the steady state we have the steady MFV equations

$$\sum_{\{T_j\}} S_{T_j}(LU)_j = 0, \tag{68}$$

$$\sum_{\{T_j\}} S_{T_j}(LU)_T(-\nabla U)_{T_j} = 0. \tag{69}$$

If Lu is derived from a variational principle, given by (7), these become

$$\sum_{\{T_j\}} S_{T_j}\left(-\frac{\partial F}{\partial U} + \nabla \cdot \frac{\partial F}{\partial \nabla U}\right)_j = 0, \tag{70}$$

$$\sum_{\{T_j\}} S_{T_j}\left(-\frac{\partial F}{\partial U} + \nabla \cdot \frac{\partial F}{\partial \nabla U}\right)_j (-\nabla U)_{T_j} = 0 \tag{71}$$

which, using the summation by parts implicit in (64), lead to

$$\sum_{\{T_j\}} S_{T_j}\left(-\frac{\partial F}{\partial U} - \frac{1}{2}\frac{\partial F}{\partial \nabla U} \cdot \underline{n}_j\right)_j = 0, \tag{72}$$

$$\sum_{\{T_j\}} S_{T_j}\left(-\frac{\partial F}{\partial U} - \frac{1}{2}\frac{\partial F}{\partial \nabla U} \cdot \underline{n}_j\right)_j (-\nabla U)_{T_j} = 0, \tag{73}$$

where $\underline{n}_j = (\Delta Y_j, -\Delta X_j)$ is the inward normal to the side opposite node j scaled to the length of that side (see [5] and Fig. 1).

5.2. A discrete optimisation

By analogy with (60) a discrete form of (9) is

$$I_{\rm d}(F)=\sum_T S_T\overline{F(\underline{X},U,\nabla U)}_T=\sum_T S_T\frac{1}{3}\sum_{i=1}^{3}F(\underline{X}_i,U_i,(\nabla U)_T), \tag{74}$$

where the overbar denotes the average value over the vertices of T. (When Euler first derived his variational equation he used such a finite form, although on a fixed mesh of course.)

The sum in (74) may be rewritten as a sum over nodes j as

$$I_{\rm d}(F)=\frac{1}{3}\sum_j\sum_{\{T_j\}}S_{T_j}\frac{1}{3}\sum_{i=1}^{3}F(\underline{X}_i,U_i,(\nabla U)_{T_j}), \tag{75}$$

where i runs over the corners of the triangle T_j.

A moving mesh method may be derived based on the two coupled equations which arise when (75) is optimised over both U_j and $\underline{X}_j$. Differentiating (75) with respect to U_j gives

$$\frac{\partial I_{\rm d}}{\partial U_j}=\sum_{\{T_j\}}S_{T_j}\frac{1}{3}\left(\frac{\partial F}{\partial U}+\frac{\partial F}{\partial\nabla U}\cdot\frac{\partial\nabla U}{\partial U}\right)_j \tag{76}$$

leading to the equation

$$\sum_{\{T_j\}}\left(S_T\frac{\partial F}{\partial U}+\frac{1}{2}\frac{\partial F}{\partial\nabla U}\cdot\underline{n}\right)_j=0. \tag{77}$$

This is a finite volume weak form which corresponds to the finite element weak form (10). It is also identical to (72) showing that the optimal property of Section 2.2 goes over to the steady-state finite volume case when U_j varies.

Differentiating with respect to $\underline{X}_j$ gives

$$\frac{\partial I_{\rm d}}{\partial\underline{X}_j}=\sum_{\{T_j\}}S_{T_j}\frac{1}{3}\left(\frac{\partial F}{\partial\underline{X}}+\frac{\partial\nabla U}{\partial\underline{X}}\cdot\frac{\partial F}{\partial\nabla U}\right)_j \tag{78}$$

which leads to

$$\sum_{\{T_j\}}\left(S_T\frac{\partial F}{\partial\underline{X}}-\left(\Delta U\begin{pmatrix}0&-1\\1&0\end{pmatrix}-\frac{1}{2}S_T^{-1}\nabla U_T\cdot\underline{n}\right)\frac{\partial F}{\partial\nabla U}\right)_j=0 \tag{79}$$

(cf. (73)). Eq. (79) is the companion weak form to (73), corresponding to the second finite element weak form (23). However, this differs considerably from (23) showing that the optimal property of MFE does not go over to steady state MFV when $\underline{X}$ varies. This is because differentiation of a quadrature rule with respect to $\underline{X}$ is not the same as quadrature of the derivative. In fact, it is Eqs. (77) and (79) which give the optimal property.

If $F=\frac{1}{2}(LU)^2$, Eqs. (77) and (79) can be made the basis of a least squares method (LSMFV) for steady problems associated with the PDE (41) with $u_t=0$. From (77) and (79) we have

$$\sum_{\{T_j\}}S_{T_j}\left(LU\frac{\partial LU}{\partial U}+\frac{1}{2}LU\frac{\partial LU}{\partial\nabla U}\underline{n}\right)_j=0 \tag{80}$$

and

$$\sum_{\{T_j\}} \left[S_T LU - \left(\Delta U \begin{pmatrix} 0 & -1 \\ 1 & 0 \end{pmatrix} - \frac{1}{2} S_T^{-1} \nabla U \underline{n} \right) LU \frac{\partial U}{\partial \nabla U} \right]_j = 0. \tag{81}$$

6. Time-dependent least squares moving finite volumes

As in Section 4, a fully discrete least squares finite volume method for time-dependent problems is obtained if u_t is discretised in time *before* the l_2 least squares minimisations are carried out. Minimisation is over U_j and $\underline{X}_j$ rather than $\dot{U}_j$ and $\dot{\underline{X}}_j$.

Consider again the one-step time discretisation of Eq. (1) in the form (27). Then on a moving mesh the solution at the next time step may be generated by the least squares minimisation of the implicit form of the residual over U_j^{n+1} and $\underline{X}_j^{n+1}$ via

$$\min_{U_j^{n+1}, \underline{X}_j^{n+1}} \|R^{n+1}\|_{l_2}^2, \tag{82}$$

where (cf. (60))

$$R^{n+1} = \frac{U^{n+1} - U^n}{\Delta t} - \nabla U^{n+1} \cdot \frac{\underline{X}^{n+1} - \underline{X}^n}{\Delta t} - LU^{n+1} \tag{83}$$

(with ∇U^{n+1} defined as in (63)) and

$$\|R\|_{l_2}^2 = \langle R, R \rangle_{l_2}, \qquad \langle P, Q \rangle_{l_2} = \sum_{\{T_j\}} S_T \bar{P}_T \bar{Q}_T. \tag{84}$$

Setting the gradients with respect to U_j^{n+1} and $\underline{X}_j^{n+1}$ to zero gives

$$\left\langle R^{n+1}, \frac{\partial \overline{R^{n+1}}}{\partial U_j^{n+1}} \right\rangle_{l_2} = 0 \tag{85}$$

and

$$\left\langle R^{n+1}, \frac{\partial \overline{R^{n+1}}}{\partial \underline{X}_j^{n+1}} \right\rangle_{l_2} + \frac{1}{2} \sum_{\{T_j\}} (\overline{R_{T_j}^{n+1}})^2 \frac{\partial S_{T_j}^{n+1}}{\partial \underline{X}_j^{n+1}} = 0. \tag{86}$$

We shall refer to (85) and (86) as the transient LSMFV method. For scalar first-order time-dependent PDEs the method is still an approximate method of characteristics since approximate characteristic speeds always make the residual R^{n+1} vanish. However, in the steady state it has other features.

6.1. Steady-state least squares finite volumes

Consider now the steady limit. Then from (82) we are minimising $\|LU\|_{l_2}^2$ and (85) and (86) reduce to

$$\left\langle LU, \frac{\partial (\overline{LU})}{\partial U_j} \right\rangle_{l_2} = 0 \tag{87}$$

and

$$\left\langle LU, \frac{\partial \overline{LU}}{\partial \underline{X}_j} \right\rangle_{l_2} + \frac{1}{2}\sum_{T_j} \overline{(LU)}_{T_j}^2 \frac{\partial S_{T_j}}{\partial \underline{X}_j} = 0 \tag{88}$$

which are identical to (80) and (81) (see [5,20]).

It has been shown in [4] that if LU is the divergence of a continuous function, then the optimal values of $\overline{LU}$ are equidistributed in the sense that the double sum over elements

$$\sum_e \sum_{e'} (S_e \overline{LU}_e - S_{e'} \overline{LU}_{e'})^2 \tag{89}$$

is minimised. Thus if $Lu = \underline{a} \cdot \nabla u$ with constant $\underline{a}$, as in example (ii) of Section 4.5 the piecewise constant $\overline{LU} = \underline{a} \cdot \nabla U$ is equidistributed over the elements in this sense. The same result is only asymptotically true for the LSMFE method (see Section 4.4).

6.2. Example

Consider again example (ii) of Section 4.5, for which the steady-state residual is

$$LU = \underline{a} \cdot \nabla U. \tag{90}$$

Then (85) and (86) reduce to

$$\left\langle \overline{LU}, \underline{a} \cdot \frac{\partial(\nabla U)}{\partial U_j} \right\rangle_{l_2} = \left\langle \overline{LU}, \frac{\partial(\underline{a} \cdot \nabla U)}{\partial \underline{X}_j} \right\rangle_{l_2} + \frac{1}{2}\sum_{T_j} (\overline{LU_T})^2 \frac{\partial S_{T_j}}{\partial \underline{X}_j} = 0, \tag{91}$$

subject to boundary conditions, where from (63)

$$\frac{\partial(\underline{a} \cdot \nabla U)_{T_j}}{\partial \underline{X}_j} = \Delta U_j \begin{pmatrix} -b \\ a \end{pmatrix} - \frac{1}{2} S_{T_j}^{-1} (\underline{a} \cdot \nabla U)_{T_j} \underline{n}_j. \tag{92}$$

Recall that $\underline{n}_j$ is the inward normal to the side of the triangle opposite j scaled by the length of that side and ΔU_j is the increment in U across that side, taken anticlockwise.

Eq. (91) may be written

$$\sum_{T_j} (\underline{a} \cdot \nabla U)_{T_j} (\underline{a} \cdot \underline{n}_j) = 0 \tag{93}$$

and

$$\sum_{T_j} \left[(\underline{a} \cdot \nabla U)_T \Delta U \begin{pmatrix} -b \\ a \end{pmatrix} - \frac{1}{2} (\underline{a} \cdot \nabla U)_T^2 \underline{n} \right]_j = 0. \tag{94}$$

We observe that (93) is identical to (48), noting that ∇U is constant and $\nabla \phi = S_T^{-1} \underline{n}$. However, (94) does not correspond to (49), even when $\underline{a}$ is constant, so the two methods are not identical under node movement.

7. Minimisation techniques

The fully discrete least squares methods of Sections 4 and 6, unlike the unsteady Galerkin methods of Sections 2 and 3, provide a functional to reduce and monitor. It is therefore possible to take an optimisation approach to the generation of a local minimum. Note that the time-stepping methods discussed earlier do not necessarily reduce the functional.

Descent methods are based upon the property that the first variation of a functional $\mathscr{F}$,

$$\delta\mathscr{F} = \frac{\partial\mathscr{F}}{\partial\underline{Y}} = \underline{g}^{\mathrm{T}}\delta\underline{Y} \tag{95}$$

say, is negative when

$$\delta\underline{Y} = -\tau\underline{g} = -\tau\frac{\partial\mathscr{F}}{\partial\underline{Y}}, \tag{96}$$

where τ is a sufficiently small positive relaxation parameter. For the present application the gradients $\underline{g}$ have already been evaluated in earlier sections. For example, in the LSMFE method the gradients $\underline{g}$ with respect to U_j and $\underline{X}_j$ appear on the left-hand side of (35) and (36). Thus, writing $\underline{Y}=\{\underline{Y}_j\}=\{U_j,\underline{X}_j\}$ and $\underline{g}=\{\underline{g}_j\}$ the steepest descent method applied to (32) with $* = n+1$ is

$$(\underline{Y}_j^{n+1})^{k+1} - (\underline{Y}_j^{n+1})^k = -\tau_j^k(\underline{g}_j^{n+1})^k \tag{97}$$

($k = 1,2,\ldots$) where τ_j^k is the relaxation parameter. Choice of τ_j^k is normally governed by a line search or a local quadratic model.

The left-hand side of (97) may be preconditioned by any positive definite matrix. The Hessian gives the full Newton method but may be approximated in various ways. In [18] a positive definite regularising matrix is used in place of the Hessian to generate the solution.

In the present application a local approach may be followed which consists of updating the unknowns one node at a time, using only local information. For a given j each step of the form (97) reduces the functional (32), even when the other $\underline{Y}^{n+1}$ values are kept constant. The updates may be carried out in a block (Jacobi iteration) or sequentially (Gauss–Seidel). A variation on the local approach is to update U_j and $\underline{X}_j$ sequentially, which gives greater control of the mesh. Descent methods of this type have been used by Tourigny and Baines [21] and Tourigny and Hulsemann [22] in the L_2 case and by Roe [20] and Baines and Leary [5] in the l_2 case.

8. Conclusions

We conclude with a summary of the main results.

- The MFE method is a Galerkin method extended to include node movement. Its main properties are:
  - (a) numerical imitation of the method of characteristics for first-order equations in any number of dimensions;
  - (b) repulsion of nodes from inflection points for second-order equations in one dimension;
  - (c) for

$$Lu = -\frac{\partial F}{\partial u} + \nabla\cdot\frac{\partial F}{\partial\nabla u} = 0 \tag{98}$$

the steady MFE equations provide a local optimum for the variational problem

$$\min_{U_j,\underline{X}_j} \int F(\underline{x}, U, \nabla U)\,\mathrm{d}\underline{x} \tag{99}$$

in a piecewise linear approximation space with moving nodes.

- The implicit semi-discrete in time least squares method (LSMFE) is a least squares method extended to include node movement. It differs from MFE through more complicated test functions and the extra term found in (36), although in the case of first-order equations it shares with MFE the property of being a numerical method of characteristics. However, in the steady state the LSMFE equations for $Lu = 0$ are equivalent to the steady MFE weak forms for the PDE

$$-\frac{\partial (Lu)^2}{\partial u} + \nabla \cdot \left(\frac{\partial (Lu)^2}{\partial \nabla tu} \right) = 0 \tag{100}$$

and therefore provide a local minimum for the variational problem

$$\min_{U_j,\underline{X}_j} \int (LU)^2\,\mathrm{d}\underline{x}. \tag{101}$$

Moreover, it can be shown that, if LU is the divergence of a continuous flux function then the flux across element boundaries is *asymptotically* equidistributed over the elements.

- The LSMFV method is a moving mesh method based on minimisation of a weighted l_2 norm of the residual of the semi-discrete in time PDE over the solution and the mesh. It shares with LSMFE the property of generating approximate characteristic speeds. At steady state, however, it lacks the optimal property of LSMFE but it has the more precise property that, if LU is the divergence of a continuous flux function, then the flux across element boundaries is equidistributed discretely (not just asymptotically) over the elements in the sense of (89) [4].
- Solutions may be obtained by the minimisation procedures of optimisation theory applied to the appropriate norm. A local approach to optimisation is advantageous in preserving the integrity of the mesh.

The MFE, LSMFE and LSMFV methods have been shown to be effective in generating approximate solutions to scalar differential problems in multi-dimensions which exhibit shocks and contact discontinuities [1,2,5,6,8,9,16,20]. The MFE and LSMFV methods have also been effective in obtaining approximate solutions of systems of equations [1,5,6,8,9].

Finite volume methods of the type discussed here do not give highly accurate solutions on coarse meshes. However, high accuracy is not crucial as far as the mesh is concerned. Thus it may be argued that an LSMFV method is sufficiently accurate for the mesh locations but a more sophisticated method which is robust on distorted meshes, such as high-order finite elements or multidimensional upwinding [11,15], may be required for the solution on the optimal mesh.

An outstanding problem is how to avoid the generation of characteristic speeds by the MFE and MFV methods for first-order equations. In the case of MFE a clue may be found in the LSMFE method, which embeds the original first-order equation in a second-order degenerate elliptic equation prior to moving the nodes. When solved by a relaxation method in an iterative manner, in effect applying a finite difference scheme to the associated parabolic equation, the resulting nodal speeds are not characteristic but instead move nodes from low-curvature regions to high-curvature regions, as required. Moreover, the nodes tend to align themselves with characteristic curves, although they do not actually move along them. Although the resulting nodal speeds are effective in this sense, the LSMFE does not generate the correct solution to the first-order equation since it has now been embedded a second-order equation. Thus, if these speeds are to be used it is impossible for the discrete equations to be set up from a unified approach. Instead it is necessary to generate the speeds from the LSMFE method which must then be substituted into the Lagrangian form of the first-order equation, to be solved separately using any convenient method.

References

[1] M.J. Baines, Moving Finite Elements, Oxford University Press, Oxford, 1994.

[2] M.J. Baines, Algorithms for optimal discontinuous piecewise linear and constant L_2 fits to continuous functions with adjustable nodes in one and two dimensions, Math. Comp. 62 (1994) 645–669.

[3] M.J. Baines, Grid adaptation via node movement, Appl. Numer. Math. 26 (1998) 77–96.

[4] M.J. Baines, Least squares and equidistribution in multidimensions, Numer. Methods Partial Differential Equations 15 (1999) 605–615.

[5] M.J. Baines, S.J. Leary, Fluctuations and signals for scalar hyperbolic equations on adjustable meshes, Numerical Analysis Report 6/98, Department of Mathematics, University of Reading, 1999. Comm. Numer. Methods Eng. 15 (1999) 877–886.

[6] M.J. Baines, S.J. Leary, M.E. Hubbard, A finite volume method for steady hyperbolic equations with mesh movement, in: Vilsmeier, Benkhaldoun, Hanel (Eds.), Proceedings of Conference on Finite Volumes for Complex Applications II, Hermes, Paris, 1999, pp. 787–794.

[7] C. Budd et al., Self-similar numerical solutions of the porous-medium equation using moving mesh methods, Philos. Trans. Roy. Soc. London A 357 (1999) 1047–1077.

[8] N.N. Carlson, K. Miller, Design and application of a gradient weighted moving finite element method I: in one dimension, SIAM J. Sci. Comput. 19 (1998) 728–765.

[9] N.N. Carlson, K. Miller, Design and application of a gradient weighted moving finite element method II: in two dimensions, SIAM J. Sci. Comput. 19 (1998) 766–798.

[10] C. de Boor, A Practical Guide to Splines, Springer, New York, 1978.

[11] H. Deconinck, P.L. Roe, R. Struijs, A multidimensional generalisation of Roe's flux difference splitter for the Euler equations, Comput. Fluids 22 (1993) 215.

[12] M. Delfour et al., An optimal triangulation for second order elliptic problems, Comput. Methods Appl. Mech. Eng. 50 (1985) 231–261.

[13] E.A. Dorfi, L.O'C. Drury, Simple adaptive grids for 1D initial value problems, J. Comput. Phys. 69 (1987) 175–195.

[14] W. Huang et al., Moving mesh partial differential equations, SIAM J. Numer. Anal. 31 (1994) 709–730.

[15] M.E. Hubbard, P.L. Roe, Compact high-resolution algorithms for time-dependent advection on unstructured grids, Internat. J. Numer. Methods Fluids (1999) 33 (2000) 711–736.

[16] P.K. Jimack, Local minimisation of errors and residuals using the moving finite element method. University of Leeds Report 98.17, School of Computer Science, 1998.

[17] K. Miller, Moving finite elements II, SIAM J. Numer. Anal. 18 (1981) 1033–1057.

[18] K. Miller, M.J. Baines, Least squares moving finite elements. OUCL Report 98/06, Oxford University Computing Laboratory, 1998. IMA J. Numer. Anal., submitted.

[19] K. Miller, R.N. Miller, Moving finite elements I, SIAM J. Numer. Anal. 18 (1981) 1019–1032.

[20] P.L. Roe, Compounded of many simples, In: Ventakrishnan, Salas, Chakravarthy (Eds.), Challenges in Computational Fluid Dynamics, Kluwer, Dordrecht, 1998, pp. 241–258.

[21] Y. Tourigny, M.J. Baines, Analysis of an algorithm for generating locally optimal meshes for L_2 approximation by discontinuous piecewise polynomials, Math. Comp. 66 (1997) 623–650.

[22] Y. Tourigny, F. Hulsemann, A new moving mesh algorithm for the finite element solution of variational problems, SIAM J. Numer. Anal. 34 (1998) 1416–1438.

ELSEVIER

Journal of Computational and Applied Mathematics 128 (2001) 383–398

JOURNAL OF COMPUTATIONAL AND APPLIED MATHEMATICS

www.elsevier.nl/locate/cam

Adaptive mesh movement — the MMPDE approach and its applications

Weizhang Huang[a,1], Robert D. Russell[b,*,2]

[a]Department of Mathematics, The University of Kansas, Lawrence, KS 66045, USA

[b]Department of Mathematics and Statistics, Simon Fraser University, Burnaby BC, Canada V5A 1S6

Received 25 January 2000

Abstract

A class of moving mesh algorithms based upon a so-called moving mesh partial differential equation (MMPDE) is reviewed. Various forms for the MMPDE are presented for both the simple one- and the higher-dimensional cases. Additional practical features such as mesh movement on the boundary, selection of the monitor function, and smoothing of the monitor function are addressed. The overall discretization and solution procedure, including for unstructured meshes, are briefly covered. Finally, we discuss some physical applications suitable for MMPDE techniques and some challenges facing MMPDE methods in the future. © 2001 Elsevier Science B.V. All rights reserved.

Keywords: Moving mesh algorithm; MMPDE

1. Introduction

In this paper we discuss a class of adaptive mesh algorithms for solving time-dependent partial differential equations (PDEs) whose solutions are characterized largely by varying behavior over a given physical domain. In particular, we consider a class of moving mesh methods which employ a moving mesh partial differential equation, or MMPDE, to perform mesh adaption in such a way that the mesh is moved around in an orderly fashion.

The MMPDE is formulated in terms of a coordinate transformation or mapping. For a mapping method, mesh generation is considered to be mathematically equivalent to determining a coordinate transformation from the physical domain into a computational domain. More specifically, let $\Omega \subseteq R^3$ be the physical domain on which the physical problem is defined, and let $\Omega_c \subseteq R^3$ be the computational domain chosen somewhat artificially for the purpose of mesh adaption. Denote coordinates

* Corresponding author.

E-mail address: rdr@cs.sfu.ca (R.D. Russell).

[1] The author was supported in part by the NSF (USA) Grant DMS-9626107.

[2] The author was supported in part by NSERC (Canada) Grant OGP-0008781.

0377-0427/01/$ - see front matter © 2001 Elsevier Science B.V. All rights reserved.

PII: S0377-0427(00)00520-3

in Ω and Ω_c by $\boldsymbol{x} = (x^1, x^2, x^3)^T$ and $\boldsymbol{\xi} = (\xi^1, \xi^2, \xi^3)^T$, respectively. The idea behind mapping mesh methods is to generate meshes on Ω as images of a reference mesh on Ω_c through a one-to-one coordinate transformation, say $\boldsymbol{x} = \boldsymbol{x}(\boldsymbol{\xi}) : \Omega_c \to \Omega$. It is common to define such a mapping through the so-called variational approach, i.e., to determine the mapping as the minimizer of a functional involving various properties of the mesh and the physical solution, e.g., see [33,40] and references therein. Mapping methods have traditionally been used for the generation of structured meshes, but more recently Cao et al. [13] have demonstrated how a mapping method can naturally be used for unstructured mesh generation and adaption as well.

A variety of variational methods have been developed in the past [33,40], some of the most successful being those in [41,6,22]. The mesh generation system used in [41] consists of two variable coefficient diffusion equations (the Euler–Lagrange equations for a simple functional, as we discuss below). Winslow's method is generalized in [6] by including mesh properties such as concentration, smoothness, and orthogonality into the mesh adaption functional. The method is extended further by Brackbill [5], who includes directional control. The method introduced in [22] defines the coordinate transformation as a harmonic mapping which is the extremal of a so-called action functional or energy integral. Closely related methods define the coordinate transformation in terms of an elliptic system, e.g., the Euler–Lagrange equations for the variational form. Indeed, the mesh generation method proposed in [39], which consists of solving two Poisson equations with control source terms, can be regarded as being of this form. As well, the Soviet literature is also replete with papers on adaptive methods for generating the mesh through a coordinate transformation, e.g. see the work in [24,42,34].

It is interesting to note that in order to avoid potential mesh crossings or foldings most of the mesh adaption functionals are formulated in terms of the inverse mapping $\boldsymbol{\xi} = \boldsymbol{\xi}(\boldsymbol{x})$ instead of $\boldsymbol{x} = \boldsymbol{x}(\boldsymbol{\xi})$ (e.g., see the discussion in [22]). A mesh partial differential equation is then obtained for $\boldsymbol{\xi}$ as the Euler–Lagrange equation for the functional. Often this is then transformed into an equation for $\boldsymbol{x}(\boldsymbol{\xi})$ by interchanging the roles of dependent and independent variables. The mesh is finally obtained by numerically solving this MPDE on a given computational mesh. It is necessary to interpret the physical PDE and MMPDE in both the physical and computational variables, and moreover, to do so in such a way that key properties of the transformed PDEs such as conservation are properly treated. In Section 2 we present the relevant transformation relations between the two sets of variables.

In Section 3 we present the basic MMPDEs. In the 1D (one space dimensional) case, a sufficient variety of them has been introduced within our general framework so as to incorporate virtually all the 1D moving mesh methods which have been used heretofore [27,28]. In higher dimensions, we use a gradient flow equation which arises from the variational form. A distinguishing feature of the MMPDE approach from any other moving mesh approaches is that it uses a parabolic PDE for mesh movement. In addition to the basic formulation, we discuss the issue of how to deal with the boundary and how to choose the monitor function for solution adaptivity.

In Section 4, the crucial issues of how to discretize the PDEs and solve the coupled system for the coordinate transformation and physical solution are discussed. The way in which this is extended to the unstructured mesh case is treated in Section 5. Numerical application areas which illustrate the general usefulness of the MMPDE approach are discussed in Section 6. Lastly, in Section 7 we give some conclusions and mention some outstanding issues requiring further investigation.

2. Transformation relations

Transformation relations play an essential role for the MMPDE mesh method and for other methods based on coordinate transformations. They constitute the necessary tools in use for transforming physical PDEs between physical and computational domains and formulating mesh equations. Here we derive the relations needed in this paper and refer the interested reader to [40] for a more complete list of transformation relations and their derivations.

Consider a time-dependent (invertible) coordinate transformation $\boldsymbol{x}=\boldsymbol{x}(\xi,t):\Omega_c\to\Omega$ and denote its inverse by $\xi=\xi(\boldsymbol{x},t)$. The covariant and contravariant base vectors are defined by

$$\boldsymbol{a}_i=\frac{\partial \boldsymbol{x}}{\partial \xi^i},\quad \boldsymbol{a}^i=\nabla\xi^i,\qquad i=1,2,3, \tag{1}$$

where ∇ is the gradient operator with respect to $\boldsymbol{x}$. The Jacobian matrix $\boldsymbol{J}$ of the coordinate transformation and its inverse $\tilde{\boldsymbol{J}}$ can then be expressed as

$$\boldsymbol{J}\equiv\frac{\partial(x^1,x^2,x^3)}{\partial(\xi^1,\xi^2,\xi^3)}=[\boldsymbol{a}_1,\boldsymbol{a}_2,\boldsymbol{a}_3],\quad \tilde{\boldsymbol{J}}\equiv\frac{\partial(\xi^1,\xi^2,\xi^3)}{\partial(x^1,x^2,x^3)}=\begin{bmatrix}(\boldsymbol{a}^1)^{\mathrm{T}}\\(\boldsymbol{a}^2)^{\mathrm{T}}\\(\boldsymbol{a}^3)^{\mathrm{T}}\end{bmatrix}. \tag{2}$$

Noticing that

$$\boldsymbol{J}^{-1}=\frac{1}{J}[\mathrm{cofactor}(\boldsymbol{J})]^{\mathrm{T}}=\frac{1}{J}[\boldsymbol{a}_2\times\boldsymbol{a}_3,\boldsymbol{a}_3\times\boldsymbol{a}_1,\boldsymbol{a}_1\times\boldsymbol{a}_2]^{\mathrm{T}}, \tag{3}$$

where $J=\det(\boldsymbol{J})=\boldsymbol{a}_1\cdot(\boldsymbol{a}_2\times\boldsymbol{a}_3)$ is the Jacobian (determinant), the chain rule $\boldsymbol{J}\tilde{\boldsymbol{J}}=I$ leads to the relations

$$\boldsymbol{a}^i=\frac{1}{J}\boldsymbol{a}_j\times\boldsymbol{a}_k,\quad \boldsymbol{a}_i=J\boldsymbol{a}^j\times\boldsymbol{a}^k,\quad \boldsymbol{a}_i\boldsymbol{a}^l=\delta_i^l,\ (i,j,k)\ \text{cyclic}, \tag{4}$$

where δ_i^l is the Kronecker delta function.

It is easy to verify that

$$\sum_i\frac{\partial}{\partial\xi^i}(J\boldsymbol{a}^i)=0. \tag{5}$$

This identity is important because it must be used when interchanging the conservative and non-conservative forms of many formulas.

For an arbitrary function $u=u(\boldsymbol{x},t)$, we denote its counterpart in the new coordinate set (ξ,t) by $\hat{u}$, i.e., $\hat{u}=u(\boldsymbol{x}(\xi,t),t)$. Then, the gradient operator takes the forms in ξ

$$\begin{aligned}\nabla&=\sum_i\boldsymbol{a}^i\frac{\partial}{\partial\xi^i}\quad\text{(non-conservative form)},\\&=\frac{1}{J}\sum_i\frac{\partial}{\partial\xi^i}J\boldsymbol{a}^i\quad\text{(conservative form)},\end{aligned} \tag{6}$$

which follow from

$$\nabla u = \left[\frac{\partial u}{\partial(x^1,x^2,x^3)}\right]^{\mathrm{T}}$$

$$= \left[\frac{\partial \hat{u}}{\partial(\xi^1,\xi^2,\xi^3)} \cdot \frac{\partial(\xi^1,\xi^2,\xi^3)}{\partial(x^1,x^2,x^3)}\right]^{\mathrm{T}}$$

$$= \left[\frac{\partial(\xi^1,\xi^2,\xi^3)}{\partial(x^1,x^2,x^3)}\right]^{\mathrm{T}} \cdot \left[\frac{\partial \hat{u}}{\partial(\xi^1,\xi^2,\xi^3)}\right]^{\mathrm{T}}$$

$$= \sum_i \boldsymbol{a}^i \frac{\partial \hat{u}}{\partial \xi^i}$$

and from identity (5).

The identity

$$J_t = J\nabla \cdot \boldsymbol{x}_t \tag{7}$$

follows from the fact that

$$J_t = \sum_i \frac{\partial \boldsymbol{a}_i}{\partial t} \cdot (\boldsymbol{a}_j \times \boldsymbol{a}_k) \quad (i,j,k) \text{ cyclic}$$

$$= J \sum_i \boldsymbol{a}^i \cdot \frac{\partial \boldsymbol{a}_i}{\partial t}$$

$$= J \sum_i \boldsymbol{a}^i \cdot \frac{\partial \boldsymbol{x}_t}{\partial \xi^i}.$$

Eq. (7) is frequently referred to [38] as the geometric conservation law (GCL). It relates the change rate of the volume of a cell to its surface movement.

Differentiating $\hat{u}(\boldsymbol{\xi},t) = u(\boldsymbol{x}(\boldsymbol{\xi},t),t)$ with respect to t while fixing $\boldsymbol{\xi}$ and using (7) gives

$$\begin{aligned} u_t &= \hat{u}_t - \nabla u \cdot \boldsymbol{x}_t \quad &\text{(non-conservative form)},\\ Ju_t &= (J\hat{u})_t - J\nabla \cdot (u\boldsymbol{x}_t) \quad &\text{(conservative form)}. \end{aligned} \tag{8}$$

Thus far, the transformation relations have been given in three dimensions. The corresponding two-dimensional forms can be obtained from these formulas by simply setting the third base vectors to be the unit vector, $\boldsymbol{a}_3 = \boldsymbol{a}^3 = (0,0,1)^{\mathrm{T}}$, and dropping the third component from the final results. For instance, if we denote the two-dimensional physical and computational coordinates by $\boldsymbol{x} = (x,y)^{\mathrm{T}}$ and $\boldsymbol{\xi} = (\xi,\eta)^{\mathrm{T}}$, and the base vectors by

$$\boldsymbol{a}_1 = \begin{bmatrix} x_\xi \\ y_\xi \end{bmatrix}, \quad \boldsymbol{a}_2 = \begin{bmatrix} x_\eta \\ y_\eta \end{bmatrix}, \quad \boldsymbol{a}^1 = \begin{bmatrix} \xi_x \\ \xi_y \end{bmatrix}, \quad \boldsymbol{a}^2 = \begin{bmatrix} \eta_x \\ \eta_y \end{bmatrix}, \tag{9}$$

we have

$$J = \det \begin{bmatrix} x_\xi & x_\eta & 0 \\ y_\xi & y_\eta & 0 \\ * & * & 1 \end{bmatrix} = x_\xi y_\eta - x_\eta y_\xi \tag{10}$$

and

$$\begin{aligned} \boldsymbol{a}^1 &= \frac{1}{J}\boldsymbol{a}_2 \times \boldsymbol{a}_3 \\ &= \frac{1}{J}(x_\eta \boldsymbol{i} + y_\eta \boldsymbol{j} + *\boldsymbol{k}) \times \boldsymbol{k} \\ &= \frac{1}{J}(-x_\eta \boldsymbol{j} + y_\eta \boldsymbol{i} + 0\boldsymbol{k}) \\ &= \frac{1}{J}\begin{bmatrix} y_\eta \\ -x_\eta \end{bmatrix}. \end{aligned} \tag{11}$$

Similarly, we have

$$\boldsymbol{a}^2 = \frac{1}{J}\begin{bmatrix} -y_\xi \\ x_\xi \end{bmatrix}. \tag{12}$$

We conclude this section with an illustration of how the above relations are used in practice. Consider a diffusion–convection equation in the conservative form

$$u_t + \nabla \cdot \boldsymbol{f} = \nabla \cdot (a \nabla u), \quad \text{in } \Omega, \tag{13}$$

where $\boldsymbol{f} = \boldsymbol{f}(u, \boldsymbol{x}, t)$ and $a = a(\boldsymbol{x}, t) \geqslant \alpha > 0$ are given functions. We want to transform it from the physical domain to the computational domain.

Upon replacing the inner and outer gradient operators by the non-conservative and conservative forms, respectively, the diffusion term has the conservative form

$$\nabla \cdot (a \nabla u) = \frac{1}{J} \sum_{i,j} \frac{\partial}{\partial \xi^j}\left(aJ\boldsymbol{a}^i \cdot \boldsymbol{a}^j \frac{\partial \hat{u}}{\partial \xi^i} \right). \tag{14}$$

Expanding the outer differentiation, using (5) and noticing that $\nabla^2 \xi^i = \sum_j \boldsymbol{a}^j \cdot \partial \boldsymbol{a}^i / \partial \xi^j$, we obtain

$$\nabla \cdot (a \nabla u) = \sum_{i,j} (\boldsymbol{a}^i \cdot \boldsymbol{a}^j) \frac{\partial}{\partial \xi^j}\left(a \frac{\partial \hat{u}}{\partial \xi^i} \right) + a \sum_i (\nabla^2 \xi^i) \frac{\partial \hat{u}}{\partial \xi^i}. \tag{15}$$

To find the expressions for $\nabla^2 \xi^i$, $i = 1, 2, 3$, we consider (15) with $a = 1$ and $\boldsymbol{x}$ replacing u. This results in

$$0 = \sum_{i,j} (\boldsymbol{a}^i \cdot \boldsymbol{a}^j) \frac{\partial^2 \boldsymbol{x}}{\partial \xi^i \partial \xi^j} + \sum_i (\nabla^2 \xi^i) \boldsymbol{a}_i. \tag{16}$$

Taking the inner product of the above equation with $\boldsymbol{a}^l$ and using $\boldsymbol{a}_i \cdot \boldsymbol{a}^l = \delta_i^l$ leads to

$$\nabla^2 \xi^l = -\sum_{i,j} (\boldsymbol{a}^i \cdot \boldsymbol{a}^j) \left(\boldsymbol{a}^l \cdot \frac{\partial^2 \boldsymbol{x}}{\partial \xi^i \partial \xi^j} \right). \tag{17}$$

The convection term is transformed into

$$\nabla \cdot \boldsymbol{f} = \frac{1}{J}\sum_i \frac{\partial}{\partial \xi^i}(J\boldsymbol{a}^i \cdot \hat{\boldsymbol{f}}) = \sum_i \boldsymbol{a}^i \cdot \frac{\partial \hat{\boldsymbol{f}}}{\partial \xi^i}. \tag{18}$$

Substituting (8), (14), (15), and (18) into (13), we obtain the transformed equation in the conservative form

$$(J\hat{u})_t + \sum_i \frac{\partial}{\partial \xi^i}[J\boldsymbol{a}^i \cdot (\hat{\boldsymbol{f}} - \hat{u}\boldsymbol{x}_t)] = \sum_{i,j} \frac{\partial}{\partial \xi^j}\left(aJ\boldsymbol{a}^i \cdot \boldsymbol{a}^j \frac{\partial \hat{u}}{\partial \xi^i}\right) \tag{19}$$

or in nonconservative form

$$\hat{u}_t + \sum_i \boldsymbol{a}^i \cdot \left(\frac{\partial \hat{\boldsymbol{f}}}{\partial \xi^i} - \frac{\partial \hat{u}}{\partial \xi^i}\boldsymbol{x}_t\right) = \sum_{i,j}(\boldsymbol{a}^i \cdot \boldsymbol{a}^j)\frac{\partial}{\partial \xi^j}\left(a\frac{\partial \hat{u}}{\partial \xi^i}\right) + a\sum_i(\nabla^2 \xi^i)\frac{\partial \hat{u}}{\partial \xi^i}. \tag{20}$$

Note that we have derived (20) from a conservative equation and kept the convective flux $\boldsymbol{f}$ in the divergence form $(\partial \hat{\boldsymbol{f}})/(\partial \xi^i)$. For this reason, Hindman [25] calls (20) the chain rule conservative law form (CRCLF). He also shows that (20) is able to catch shock waves.

3. Formulation of MMPDEs

For a moving mesh method, the mesh points move continuously in the space–time domain and concentrate in regions where the physical solution varies significantly. An explicit rule, called a moving mesh equation, is designed to move the mesh around. It is often very difficult to formulate a moving mesh strategy which can efficiently perform the mesh adaption while moving the mesh in an orderly fashion. In principle, the mesh equation should be designed based directly on some type of error estimates, but unfortunately, such methods often result in singular meshes due to points crossing and tangling.

We consider in this section, the MMPDE approach for formulating mesh equations. The basic strategy is to define the mesh equation as a time-dependent PDE based on the equidistribution principle or its higher-dimensional generalization. Although in the higher-dimensional case, it is usually unclear whether or not this generalization relates directly to any type of error estimates, it has proven to perform mesh adaption well and generate nondegenerate meshes.

3.1. 1D MMPDEs

MMPDEs are first introduced in [27,28] as time regularizations [2] of the differential form of the (steady-state) equidistribution principle [19]. If $g = g(x) > 0$ is the one-dimensional monitor function which controls the mesh point distribution (see below for more discussion on monitor functions), then the differential form of the equidistribution principle can be written in terms of the inverse mapping $\xi = \xi(x)$ as

$$\frac{\mathrm{d}}{\mathrm{d}x}\left(\frac{1}{g}\frac{\mathrm{d}\xi}{\mathrm{d}x}\right) = 0. \tag{21}$$

(For simplicity, we assume here that both the physical and computational domains are the unit interval (0,1).) The mesh equation for the mapping $x = x(\xi)$ can be obtained by interchanging the roles of dependent and independent variables in (21), namely,

$$\frac{\mathrm{d}}{\mathrm{d}\xi}\left(g\frac{\mathrm{d}x}{\mathrm{d}\xi}\right) = 0. \tag{22}$$

The MMPDEs are defined by adding a time derivative (or time regularization) to (22). Three popular choices are the MMPDEs

$$\text{MMPDE4:} \quad \frac{\partial}{\partial\xi}\left(g\frac{\partial x_t}{\partial\xi}\right) = -\frac{1}{\tau}\frac{\partial}{\partial\xi}\left(g\frac{\partial x}{\partial\xi}\right), \tag{23}$$

$$\text{MMPDE5:} \quad x_t = \frac{1}{\tau}\frac{\partial}{\partial\xi}\left(g\frac{\partial x}{\partial\xi}\right), \tag{24}$$

$$\text{MMPDE6:} \quad \frac{\partial^2 x_t}{\partial\xi^2} = -\frac{1}{\tau}\frac{\partial}{\partial\xi}\left(g\frac{\partial x}{\partial\xi}\right), \tag{25}$$

where $\tau > 0$ is a parameter used for adjusting the time scale of the mesh movement. These MMPDEs are analyzed numerically and theoretically in [27,28]. Their smoothed versions are discussed in [29].

The equidistribution relation (21) can also be interpreted as the Euler–Lagrange equation for the functional

$$I[\xi] = \frac{1}{2}\int_0^1 \frac{1}{g}\left(\frac{\mathrm{d}\xi}{\mathrm{d}x}\right)^2 \mathrm{d}x, \tag{26}$$

and MMPDEs can be derived using this alternative form. For example, we can define an MMPDE as a modified gradient flow equation for the functional $I[\xi]$, i.e.,

$$\frac{\partial\xi}{\partial t} = \frac{1}{\tau g}\frac{\partial}{\partial x}\left(\frac{1}{g}\frac{\partial\xi}{\partial x}\right). \tag{27}$$

Switching the roles of dependent and independent variables, (27) becomes

$$x_t = \frac{1}{\tau g^3 x_\xi^2}\frac{\partial}{\partial\xi}\left(g\frac{\partial x}{\partial\xi}\right), \tag{28}$$

which is basically MMPDE5 with a spatially varying time scaling parameter $\tau g^3 x_\xi^2$.

3.2. Higher-dimensional MMPDEs

The strategy for formulating (27) is used in [30,31] for developing higher-dimensional MMPDEs. Given a functional $I[\boldsymbol{\xi}]$ which involves various properties of the mesh and the physical solution, the MMPDE is introduced as the (modified) gradient flow equation

$$\frac{\partial\xi^i}{\partial t} = -\frac{P_i}{\tau}\frac{\delta I}{\delta\xi^i}, \quad i = 1,2,3, \tag{29}$$

where $\tau > 0$ is again a user-defined parameter used for adjusting the time scale of mesh movement and P_i, $i = 1,2,3$ are differential operators with positive spectra in a suitable function space.

In one dimension, functional (26) corresponds exactly to the equidistribution principle. In higher dimensions, the situation is more complicated, as there does not generally exist a choice for $I[\xi]$ leading to an equivalent equidistribution [32]. Nevertheless, in [30,31] the generalization of (26),

$$I[\xi] = \frac{1}{2}\int_{\Omega}\sum_{i}(\nabla\xi^{i})^{\mathrm{T}}G_{i}^{-1}\nabla\xi^{i}\,\mathrm{d}\boldsymbol{x}, \tag{30}$$

is used where the monitor functions G_i, $i=1,2,3$, are symmetric positive-definite matrices which interconnect the mesh and the physical solution. Form (30) includes several common mesh adaption functionals as examples. Indeed, with

$$G_1 = G_2 = G_3 = wI, \tag{31}$$

where w is the weight function, (30) leads to the functional corresponding to Winslow's mesh adaption method [41], while the choice

$$G_1 = G_2 = G_3 = \frac{1}{\sqrt{\det(G)}}G \tag{32}$$

gives the method based on harmonic maps [22]. Winslow's method is generalized in [6] to include terms for further mesh smoothness and orthogonality control. This has become one of the most popular methods used for steady-state mesh adaption.

Combining (29) with (30) and taking $P_i = 1/\sqrt[d]{g_i}$, where $g_i = \det(G_i)$ and d is the dimension of the spatial domain, we get

$$\frac{\partial\xi^{i}}{\partial t} = \frac{1}{\tau\sqrt[d]{g_i}}\nabla\cdot(G_i^{-1}\nabla\xi^{i}),\quad i=1,2,3. \tag{33}$$

Once the monitor function is calculated, (33) can be solved numerically for the mapping $\boldsymbol{\xi}=\boldsymbol{\xi}(\boldsymbol{x},t)$, and the physical mesh at the new time level is then obtained by interpolation.

It can be more convenient to work directly with the mapping $\boldsymbol{x}=\boldsymbol{x}(\boldsymbol{\xi},t)$ since it defines mesh point locations explicitly. The moving mesh equation for this mapping is obtained by interchanging the roles of dependent and independent variables in (33). Using (6) and

$$\boldsymbol{x}_t = -\frac{\partial\boldsymbol{x}}{\partial\boldsymbol{\xi}}\boldsymbol{\xi}_t = -\sum_{i}\boldsymbol{a}_i\xi_t^{i}, \tag{34}$$

we can rewrite (33) as

$$\boldsymbol{x}_t = -\frac{1}{\tau J}\sum_{i,j}\frac{\boldsymbol{a}_j}{\sqrt[d]{g_j}}\frac{\partial}{\partial\xi^{i}}[(\boldsymbol{a}^{i})^{\mathrm{T}}G_j^{-1}\boldsymbol{a}^{j}]. \tag{35}$$

A fully non-conservative form can sometimes be simpler and easier to solve numerically. Upon expanding the differentiation, (35) becomes

$$\boldsymbol{x}_t = \frac{1}{\tau}\sum_{i,j}A_{i,j}\frac{\partial^2\boldsymbol{x}}{\partial\xi^{i}\partial\xi^{j}} + \sum_{i}B_i\frac{\partial\boldsymbol{x}}{\partial\xi^{i}}, \tag{36}$$

where the coefficient matrices $A_{i,j}$ and B_i are functions of G_i and the transformation metrics. The expressions for the coefficients are somewhat complicated — see [30] for the two-dimensional case. For Winslow's monitor function where $G_1 = G_2 = G_3 = wI$, (36) takes the much simpler form

$$\boldsymbol{x}_t = \frac{1}{\tau w^3}\sum_{i,j}(\boldsymbol{a}^{i}\cdot\boldsymbol{a}^{j})\frac{\partial}{\partial\xi^{j}}\left(w\frac{\partial\boldsymbol{x}}{\partial\xi^{i}}\right). \tag{37}$$

The one-dimensional forms of MMPDE4 and MMPDE6 can be generalized straightforwardly for this case to

$$\text{MMPDE4:}\quad \sum_{i,j}(\boldsymbol{a}^i\cdot\boldsymbol{a}^j)\frac{\partial}{\partial\xi^j}\left(w\frac{\partial \boldsymbol{x}_t}{\partial\xi^i}\right)=-\frac{1}{\tau}\sum_{i,j}(\boldsymbol{a}^i\cdot\boldsymbol{a}^j)\frac{\partial}{\partial\xi^j}\left(w\frac{\partial \boldsymbol{x}}{\partial\xi^i}\right), \tag{38}$$

$$\text{MMPDE6:}\quad \sum_{i,j}(\boldsymbol{a}^i\cdot\boldsymbol{a}^j)\frac{\partial}{\partial\xi^j}\left(\frac{\partial \boldsymbol{x}_t}{\partial\xi^i}\right)=-\frac{1}{\tau}\sum_{i,j}(\boldsymbol{a}^i\cdot\boldsymbol{a}^j)\frac{\partial}{\partial\xi^j}\left(w\frac{\partial \boldsymbol{x}}{\partial\xi^i}\right). \tag{39}$$

3.3. Boundary treatment

To completely specify the coordinate transformation, the above MMPDEs must be supplemented with suitable boundary conditions. The simplest conditions are of Dirichlet type, with which the boundary points are fixed or their movement is prescribed. However, it is more desirable in general to move the boundary points to adapt to the physical solution. We have used two types of moving boundary conditions. One is to use orthogonality conditions which require that one set of the coordinate lines be perpendicular to the boundary. This results in mixed Dirichlet and Neumann boundary conditions. The other, proposed in [30], is to determine the boundary point distribution by using a lower-dimensional MMPDE. Since this method works better in general than using orthogonality conditions, we give it a more detailed description in the following.

Consider the two-dimensional case. For a given boundary segment Γ of $\partial\Omega$, let Γ_c be the corresponding boundary segment of $\partial\Omega_c$. Denoting by s the arc-length from a point on Γ to one of its end points and by ζ the corresponding arc-length from a point on Γ_c to one of its end points, we can identify Γ with $I=(0,\ell)$ and Γ_c with $I_c=(0,\ell_c)$. Then the arc-length coordinate $s=s(\zeta,t)$ is defined as the solution of the one-dimensional MMPDE

$$\begin{aligned}&\tau\frac{\partial s}{\partial t}=\frac{1}{M^3(\partial s/\partial\zeta)^2}\frac{\partial}{\partial\zeta}\left(M\frac{\partial s}{\partial\zeta}\right),\quad \zeta\in(0,\ell_c),\\ &s(0)=0,\quad s(\ell_c)=\ell,\end{aligned} \tag{40}$$

where M, considered as a function of s and t, is the one-dimensional monitor function. In practice, M can be defined as the projection of the two-dimensional monitor function G along the boundary, so if $\boldsymbol{t}$ is the unit tangent vector along the boundary then $M(s,t)=\boldsymbol{t}^{\mathrm{T}}G\boldsymbol{t}$. Having obtained the arc-length coordinates for the boundary points, the corresponding physical coordinates are then obtained through interpolation along the boundary.

3.4. Monitor functions

The key to the success of the MMPDE approach to mesh movement (and indeed to other mesh adaption methods based on mappings) is to define an appropriate monitor function.

This has been well-studied in one dimension. The common choice is the arc-length monitor function

$$g=\sqrt{1+u_x^2}, \tag{41}$$

where u is the adaption function. Blom and Verwer [3] give an extensive comparative study of the arc-length and curvature monitor functions. Monitor functions can also be constructed based on interpolation errors, e.g., see [16,18].

In higher dimensions, the issue becomes more difficult due to the lack of (known) direct links between functional (30) and any type of error estimates. In [13], the effect of the monitor function on the resulting mesh is analyzed qualitatively and a few guidelines for selecting monitor functions given. One special case is when $G_1 = G_2 = G_3 \equiv G$. Since G is symmetric and positive-definite, it can be decomposed into

$$G = \sum_i \lambda_i \boldsymbol{v}_i, \tag{42}$$

where $(\lambda_i, \boldsymbol{v}_i)$, $i = 1,2,3$ are (normalized) eigenpairs of G. Compression and/or expansion of coordinate lines can be shown to occur along direction $\boldsymbol{v}_i$ when the corresponding eigenvalue λ_i changes significantly in this direction. This result requires slight modification when there exist repeated eigenvalues, since eigendecomposition (42) is not unique. For the case with the triple eigenvalue $\lambda_1 = \lambda_2 = \lambda_3 = \lambda$, any three normalized orthogonal vectors will form an eigendecomposition. It is not difficult to see that in general, the coordinate line compression and/or expansion will occur mainly in one direction, the gradient direction of λ or the fastest ascent/descent direction. In the case where there is one simple eigenvalue (say λ_1) and a double eigenvalue, the mesh adaption can occur in two directions. One is along $\boldsymbol{v}_1$ when λ_1 varies and the other is the projection of the fastest ascent/descent direction of $\lambda_2 = \lambda_3$ on the orthogonal complementary subspace of $\boldsymbol{v}_1$.

Based on a qualitative analysis, Cao et al. [13] suggest that for performing mesh adaption in the gradient direction of the physical solution $u = u(\boldsymbol{x}, t)$, a class of monitor functions be constructed through (42) by taking

$$\begin{aligned} &\boldsymbol{v}_1 = \nabla u / \|u\|_2, \qquad \boldsymbol{v}_2 \perp \boldsymbol{v}_1, \qquad \boldsymbol{v}_3 \perp \boldsymbol{v}_1, \quad \boldsymbol{v}_3 \perp \boldsymbol{v}_2, \\ &\lambda_1 = \sqrt{1 + \|\nabla\|_2}, \quad \lambda_2 \text{ and } \lambda_3 \text{ are functions of } \lambda_1. \end{aligned} \tag{43}$$

For instance, the choice $\lambda_2 = \lambda_3 = \lambda_1$ leads to Winslow's monitor function (31) with $w = \sqrt{1 + \|\nabla u\|^2}$ while the choice $\lambda_2 = \lambda_3 = 1/\lambda_1$ results in (32) which corresponds to the harmonic map method. Interestingly, the intermediate choice $\lambda_2 = \lambda_3 = 1$ gives the generalization of the one-dimensional arc-length monitor function,

$$G = [I + (\nabla u)(\nabla u)^{\mathrm{T}}]^{1/2}. \tag{44}$$

As previously mentioned, additional terms can be combined into the monitor functions for mesh orthogonality and directional control, e.g., see [5,30,31]. However, in this case, the monitor functions G_i, $i = 1,2,3$ become more complicated and cannot be expressed as scalar matrices.

Monitor functions can also be constructed based on a posteriori error estimates and interpolation errors. Several such techniques are illustrated and compared numerically in [14,15]. It is shown that the monitor function based on interpolation errors can be advantageous since it is easier to compute and more precisely locates regions needing higher resolution.

In practice, the monitor function is generally smoothed before its use for the numerical solution of the MMPDEs. This is because usually the computed monitor function is changing rapidly. A smoother monitor function leads to an MMPDE which is easier to solve and has a smoother mesh solution [21,29]. Monitor functions can be smoothed in many ways, e.g., see [21,29,32]. Several cycles of low-pass filtering often suffice. The following method usually works well: For an arbitrary mesh point $\boldsymbol{x}_{\rm p}$ in Ω, let $\xi_{\rm p}$ be the corresponding mesh point in $\Omega_{\rm c}$. Then define

$$\tilde{G}_i(\boldsymbol{x}_{\rm p}) = \frac{\int_{C(\xi_{\rm p})} G_i(\boldsymbol{x}(\xi))\,{\rm d}\xi}{\int_{C(\xi_{\rm p})} {\rm d}\xi}, \tag{45}$$

where $C(\xi_{\rm p}) \subset \Omega_{\rm c}$ is a cell containing the point $\xi_{\rm p}$. In practice, $C(\xi_{\rm p})$ is normally chosen as the union of neighboring grid cells having $\xi_{\rm p}$ as one of their vertices. The integrals in (45) are evaluated by suitable quadrature formulas.

4. Discretization and solution procedures

With the moving mesh method, one must solve a coupled system consisting of physical and mesh PDEs instead of just the physical PDEs. Generally speaking, these PDEs can be solved simultaneously or alternately for the physical solution and the mesh. Simultaneous computation has commonly been used in the method of lines approach for 1D moving mesh methods (as in [28]). However, it is less straightforward in higher dimensions, where the system consisting of the physical and mesh PDEs becomes very nonlinear and its size substantially larger. The alternate solution procedure is used successfully in [12,30,31] for a variety of applications.

Once a computational mesh is given, in principle almost any spatial discretization method and time integrator can be used to solve the MMPDEs. But it is worth pointing out that very accurate time integration and spatial discretization may not be necessary. In our computations, we have used a nine-point finite-difference method (in two dimensions) or a linear finite element method in space and the backward Euler integrator in time, and have had general success in obtaining stable and sufficiently accurate meshes (see [15,31]).

The physical PDEs can be discretized in either the physical or computational domain. If this is done in the computational domain, the physical PDEs are first transformed into the computational coordinates as illustrated in Section 2. The main advantage of this approach is that one may use rectangular meshes in $\Omega_{\rm c}$ and standard finite differences. But when discretizations in the physical domain are desired or unstructured meshes are used (see next section for unstructured mesh movement), finite volume and finite element methods will generally be more appropriate. In this case of course, the physical PDEs need not be transformed into computational coordinates.

5. Unstructured mesh movement

We have seen in the previous sections that the MMPDEs are formulated in terms of the coordinate transformation $\boldsymbol{x} = \boldsymbol{x}(\xi, t)$. Thus, in order to obtain meshes in Ω using MMPDEs, the computational domain and a computational mesh must be defined initially.

Traditionally, coordinate transformation mesh methods have been used for generation of structured grids. It is a common practice to choose Ω_c to be a simple geometry, typically a square in two dimensions, and to choose the computational mesh Ω_c^h to be an orthogonal mesh. Adaptive meshes on Ω are then determined by solving an MMPDE on Ω_c^h. Of course, this structured mesh approach is too restrictive and cannot be used for very complicated domains, although the limitation can sometimes be dealt with by a multi-block approach [11].

A simple alternative approach is used in [12] for unstructured mesh movement with MMPDEs. Initially, a physical mesh $\tilde{\Omega}^h$ is generated, as is typical in finite element computations. This is usually done using one of the various mesh generators such as a Delaunay triangulation adjusted to the geometry of the physical domain. Given this geometry-oriented mesh for Ω, we can consider next the definition of Ω_c and a computational mesh Ω_c^h. Although the choice of Ω_c can be fairly arbitrary, it is recommended in [12] that Ω_c be generally chosen to be convex and to have the same number of boundary segments as Ω in order to avoid generating degenerate elements in the computational mesh. Having defined Ω_c, a computational mesh Ω_c^h can be obtained by numerically solving the boundary value problem

$$\begin{aligned} \nabla^2 \xi &= 0 \qquad \text{in } \Omega, \\ \xi(\boldsymbol{x}) &= \phi(\boldsymbol{x}), \quad \text{on } \partial\Omega. \end{aligned} \tag{46}$$

Given Ω_c^h, an adaptive initial mesh on Ω can be obtained by solving a steady-state mesh equation like (37) without the mesh speed term and then moved in time by solving an MMPDE on Ω_c.

This simple approach has been shown to work successfully in [12]. The resulting r-finite-element method is also studied in [14,15], with monitor functions being constructed using various error indicators based on solution gradients, a posteriori error estimates, and interpolation errors.

6. Some applications

Moving mesh methods have proved to be quite successful for solving 1D problems. One class of problems have been particularly amenable to solution by moving mesh techniques. These are PDEs exhibiting scaling invariance, which under suitable conditions admit self-similar solutions. Examples include: a porous medium equation (PME) [9]

$$u_t = (u u_x)_x, \tag{47}$$

where $u \geqslant 0$ taking $u(x,t) = 0$ if $|x|$ is sufficiently large; a reaction diffusion PDE with blowup [10]

$$u_t = u_{xx} + f(u), \; u(0,t) = u(1,t) = 0, \; u(0,x) = u_0(x) \tag{48}$$

where $f(u) = u^p$ or $f(u) = e^u$; and a radially symmetric nonlinear Schrödinger equation (NLS) exhibiting blowup [8]

$$\mathrm{i}\frac{\partial u}{\partial t} + \Delta u + |u|^2 u = 0, \quad t > 0 \tag{49}$$

$$u(\boldsymbol{x},0) = u_0(\boldsymbol{x}), \qquad \boldsymbol{x} \in R^d \tag{50}$$

with dimension $d > 2$.

The basic idea of the moving mesh approach for these problems is that one chooses a MMPDE that exhibits the same scaling invariance as the physical PDE. This is often straightforward to determine, e.g., equations like the above. The consequences are that these physical PDEs often admit self-similar solutions. In a general sense, a self-similar solution of the physical PDE remains attracting for the coupled system and for basically the same reasons [10], and the coordinate transformation inherits a natural spatial scaling showing the structure of the solution at the singularity. This is an ideal illustration of a situation where formal asymptotic methods and numerical methods (the moving mesh methods) work hand-in-hand to reveal the structure of self-similar solutions. Indeed, the fine resolution resulting from the numerics can even be used to predict approximately self-similar solutions (caused by boundary condition effects), and their existence can then be verified through formal arguments [7,10]. It can be important to use methods that are conservative. In particular, not all of these schemes conserve the mass of the solution, and the ones that do not can fail to admit the right similarity solutions. The momentum (and center of mass) of the solution can also have useful quantities to be preserved.

A central question is whether or not the self-similar solutions for the discrete schemes inherit stability properties of the continuous system like global attractivity. Frequently such solutions can be shown to be globally attractive for the continuous problem, with the proof breaking down for the discrete case.

There has recently been effort to generalize these results in terms of invariant spaces generated by moving mesh operators [23] applied to quasi-linear PDEs with polynomial nonlinearities. One general approach is to interpret the moving mesh methods in terms of dynamical systems based upon Lie Groups. The operators admit finite-dimensional subspaces or sets which are invariant under the corresponding nonlinear PDE operators. This situation arises in a variety of important areas such as reaction diffusion theory, combustion filtration, flame propagation, and water wave theory.

An important computational challenge will be to extend the success of moving mesh methods for these problems from 1D to higher dimensions. The choice of the MMPDE and monitor function is no longer so straightforward, partly because one is no longer using equidistribution [32], and application of the methods has been much more limited in this case.

There has been demonstrated success of our MMPDE approach to solve challenging higher-dimensional PDEs of a general nature. Notably, the 2D finite element code described in [12] is used to solve various problems: a wave equation, a convection diffusion problem for Burgers equation, a combustion problem, a porous medium, and a problem in fluid dynamics (involving flow past a cylinder). There are also efforts underway to solve other classes of problems, such as the ones exhibiting various types of singularity (including blowup), those from incompressible fluid dynamics (various Navier–Stokes problems), and from inviscid flow dynamics (airfoil analysis and wing design). Of longer term concern are multiphase problems, material manufacturing, and groundwater flow and pollution modeling. The keen interest of engineers in developing moving mesh methods for problems with moving boundaries or moving interfaces is one of the motivating factors behind designing higher-dimensional techniques. In the forefront will be the issue of determining whether interface capturing or interface tracking is preferable in the moving mesh context. Some preliminary progress on some problems in cavity flow, flow past a cylinder, and the Rayleigh–Taylor instability problem has been made in [26]. Again, other key issues are the choices of an MMPDE and a monitor function, and the design of the discretization such that underlying conservation laws are maintained by the extended system.

7. Concluding remarks

In this paper, we have given a brief review of the MMPDE approach for solving both 1D and higher-dimensional time-dependent PDEs. As we have seen, there are a variety of ways to formulate and discretize the MMPDEs, and it can be crucial to do this in such a way that key properties are conserved. Software has been written to implement a moving mesh finite-element method for both structured and unstructured meshes. Furthermore, some substantial extensions have been made, for example a general domain implementation which breaks the region into multi-blocks and solves the MMPDEs over each block separately, using overlapping Schwarz iterations and connecting the meshes on each block smoothly. While the codes have been successful for solving a reasonably large class of problems, there are many areas in which substantial improvement can be expected.

Indeed, a careful investigation is lacking for most aspects of the solution techniques for solving the physical PDE and MMPDE system, beginning with a study of whether to solve these PDEs simultaneously or alternately (the latter being the current choice). Regardless, better understanding of the types of MMPDEs, monitor functions, effects of discretizations, and solution techniques for these nonlinear equations are needed. The possibility of using a form of equidistribution effectively in higher dimensions remains attractive (e.g., see [20] for its interesting use along coordinate lines). Substantial improvements can be expected due to new preconditioners and better numerical integrators for the MMPDEs (with appropriate time integration step selection, such as investigated in [35] for PDEs with scaling invariance), as well as more sophisticated multi-level grid adaption strategies and parallelized algorithms. To obtain more robustness, an h-refinement feature is being added to our current unstructured mesh r-method finite element code. Moreover, our methods are in principle applicable to 3D problems, and extension to these problems is a goal for the future.

It is worth mentioning that there has been a substantial amount of work by others on moving mesh methods for solving PDEs. While their approaches bear some things in common with ours, there are often substantial differences. For example, the seminal early work on moving mesh methods by Miller has led the way to the development of a moving mesh finite element code [17] which uses linear elements and solves higher dimensional PDEs. However, the mesh point selection is done by minimizing a variational form involving both these points and the solution values at these points simultaneously. A similar idea has been used by Baines [1] for developing the so-called moving best fit (MBF) method. Another related class of methods based on deformation mappings has been developed in [4].

Finally, theoretical issues related to adaptivity are extremely challenging. While there has been recent progress such as the theoretical work by Qiu et al. [36,37] on adaptive meshes for singularly perturbed problems, much remains to be done even for steady-state problems in analyzing the complicated nonlinear interaction between solutions to such types of problems and the corresponding adaptive meshes.

References

[1] M.J. Baines, Moving Finite Elements, Oxford University Press, Oxford, 1994.

[2] J. Baumgarte, Stabilization of constraints and integrals of motion in dynamical systems, Comput. Methods Appl. Mech. Eng. 1 (1972) 1–16.

[3] J.G. Blom, J.G. Verwer, On the use of the arclength and curvature monitor functions in a moving-grid method which is based on the method of lines, Technical Report NM-N8902, CWI, 1989.

[4] P. Bochev, G. Liao, G.d. Pena, Analysis and computation of adaptive moving grids by deformation, Numer. Methods Partial Differential Equations 12 (1996) 489–506.

[5] J.U. Brackbill, An adaptive grid with directional control, J. Comput. Phys. 108 (1993) 38–50.

[6] J.U. Brackbill, J.S. Saltzman, Adaptive zoning for singular problems in two dimensions, J. Comput. Phys. 46 (1982) 342–368.

[7] C. Budd, J. Chen, W. Huang, R.D. Russell, Moving mesh methods with applications to blow-up problems for pdes, in: D.F. Griffiths, G.A. Watson (Eds.), Proceedings of the 1995 Biennial Conference on Numerical Analysis, Pitman Research Notes in Mathematics, Longman Scientific and Technical, New York, 1996, pp. 79–89.

[8] C. Budd, S. Chen, R.D. Russell, New self-similar solutions of the nonlinear Schrodinger equation with moving mesh methods, J. Comput. Phys. 152 (1999) 756–789.

[9] C. Budd, G. Collins, W. Huang, R.D. Russell, Self-similar numerical solutions of the porous medium equation using moving mesh methods, Phil. Trans. Roy. Soc. Lond. A 357 (1999) 1047–1078.

[10] C.J. Budd, W. Huang, R.D. Russell, Moving mesh methods for problems with blow-up, SIAM J. Sci. Comput. 17 (1996) 305–327.

[11] W. Cao, W. Huang, R.D. Russell, A moving mesh method in multi-block domains with application to a combustion problem, Numer. Methods Partial Differential Equations 15 (1999) 449–467.

[12] W. Cao, W. Huang, R.D. Russell, An r-adaptive finite element method based upon moving mesh pdes, J. Comp. Phys. 149 (1999) 221–244.

[13] W. Cao, W. Huang, R.D. Russell, A study of monitor functions for two-dimensional adaptive mesh generation, SIAM J. Sci. Comput. 20 (1999) 1978–1994.

[14] W. Cao, W. Huang, R.D. Russell, Comparison of two-dimensional r-adaptive finite element methods using various error indicators, 1999, submitted for publication.

[15] W. Cao, W. Huang, R.D. Russell, A two-dimensional r-adaptive finite element method based on a posteriori error estimates, 1999, submitted for publication.

[16] G.F. Carey, H.T. Dinh, Grading functions and mesh redistribution, SIAM J. Numer. Anal. 22 (1985) 1028–1040.

[17] N. Carlson, K. Miller, Design and application of a gradient-weighted moving finite element code. part ii in 2-d, SIAM J. Sci. Comput. 19 (1998) 766–798.

[18] K. Chen, Error equidistribution and mesh adaptation, SIAM J. Sci. Comput. 15 (1994) 798–818.

[19] C. de Boor, Good approximation by splines with variables knots ii, in: Springer Lecture Notes Series 363, Berlin, Springer, 1973.

[20] L.M. Degtyarev, T.S. Ivanova, A.A. Martynov, S.Yu. Medvedev, Generation of solution adaptive quasi orthogonal moving grids for complex structure flows. in: M. Cross, B.K. Soni, J.F. Thompson, H. Hauser, P.R. Eiseman, (Eds.), Proceedings of the sixth International Conference, University of Greenwich, 1998, pp. 99–107.

[21] E.A. Dorfi, L.O'c Drury, Simple adaptive grids for 1-d initial value problems, J. Comput. Phys. 69 (1987) 175–195.

[22] A.S. Dvinsky, Adaptive grid generation from harmonic maps on riemannian manifolds, J. Comput. Phys. 95 (1991) 450–476.

[23] V.A. Galaktionov, Invariant and positivity properties of moving mesh operators for nonlinear evolution equations, Technical Report Mathematics Preprint 99/17, University of Bath, 1999.

[24] S.K. Godunov, Prokopov, The use of moving meshes in gas-dynamical computations, U.S.S.R. Comput. Math. Math. Phys. 12 (1972) 182–195.

[25] R.G. Hindman, Generalized coordinate forms of governing fluid equations and associated geometrically induced errors, AIAA J. 20 (1982) 1359–1367.

[26] H. Huang, W. Huang, R.D. Russell, Moving finite difference solution for the Navier–Stokes equations, 1999, in preparation.

[27] W. Huang, Y. Ren, R.D. Russell, Moving mesh methods based on moving mesh partial differential equations, J. Comput. Phys. 113 (1994) 279–290.

[28] W. Huang, Y. Ren, R.D. Russell, Moving mesh partial differential equations (mmpdes) based upon the equidistribution principle, SIAM J. Numer. Anal. 31 (1994) 709–730.

[29] W. Huang, R.D. Russell, Analysis of moving mesh partial differential equations with spatial smoothing, SIAM J. Numer. Anal. 34 (1997) 1106–1126.

[30] W. Huang, R.D. Russell, A high dimensional moving mesh strategy, Appl. Numer. Math. 26 (1997) 63–76.

[31] W. Huang, R.D. Russell, Moving mesh strategy based upon a gradient flow equation for two-dimensional problems, SIAM J. Sci. Comput. 20 (1999) 998–1015.

[32] W. Huang, D.M. Sloan, A simple adaptive grid method in two dimensions, SIAM J. Sci. Comput. 15 (1994) 776–797.

[33] P. Knupp, S. Steinberg, Foundations of Grid Generation, CRC Press, Boca Raton, 1994.

[34] V.M. Kovenya, N.N. Yanenko, Numerical method for solving the viscous gas equations on moving grids, Comput. & Fluids 8 (1979) 59–70.

[35] B. Leimkuhler, C.J. Budd, M. Piggott, Scaling invariance and adaptivity, 1999 (Manuscript).

[36] Y. Qiu, D.M. Sloan, Analysis of difference approximations to a singularly perturbed two-point boundary value problem on an adaptively generated grid, J. Comput. Appl. Math. 101 (1999) 1–25.

[37] Y. Qiu, D.M. Sloan, T. Tang, Numerical solution of a singularly perturbed two-point boundary value problem using equidistribution: analysis of convergence, J. Comput. Appl. Math. (to appear).

[38] P.D. Thomas, C.K. Lombard, Geometric conservation law and its application to flow computations on moving grids, AIAA J. 17 (1979) 1030–1037.

[39] J.F. Thompson, F.C. Thames, C.W. Mastin, Automatic numerical grid generation of body fitted curvilinear coordinate system of field containing any number of arbitrary two dimensional bodies, J. Comput. Phys. 15 (1974) 299–319.

[40] J.F. Thompson, Z.A. Warsi, C.W. Mastin, Numerical Grid Generation: Foundations and Applications, North-Holland, New York, 1985.

[41] A. Winslow, Numerical solution of the quasi-linear Poisson equation in a nonuniform triangle mesh, J. Comput. Phys. 1 (1967) 149–172.

[42] N.N. Yanenko, N.T. Danaev, V.D. Liseikin, O variatsionnom metode postroieniya setok, Chislennyje metody mekhaniki sploshnoj sredy 8 (1977) 157–163.

ELSEVIER

Journal of Computational and Applied Mathematics 128 (2001) 399–422

JOURNAL OF COMPUTATIONAL AND APPLIED MATHEMATICS

www.elsevier.nl/locate/cam

The geometric integration of scale-invariant ordinary and partial differential equations

C.J. Budd *, M.D. Piggott

Department of Mathematical Sciences, University of Bath, Claverton Down, Bath, BA2 7AY, UK

Received 24 January 2000; received in revised form 24 March 2000

Abstract

This review paper examines a synthesis of adaptive mesh methods with the use of symmetry to solve ordinary and partial differential equations. It looks at the effectiveness of numerical methods in preserving geometric structures of the underlying equations such as scaling invariance, conservation laws and solution orderings. Studies are made of a series of examples including the porous medium equation and the nonlinear Schrödinger equation. © 2001 Elsevier Science B.V. All rights reserved.

Keywords: Mesh adaption; Self-similar solution; Scaling invariance; Conservation laws; Maximum principles; Equidistribution

1. Introduction

When we wish to find a numerical approximation to the solution of a partial differential equation, a natural technique is to discretise the PDE so that local truncation errors are small, and to then solve the resulting discretisation. This procedure of course underpins much of current numerical software and when linked with an effective error control strategy can lead to accurate answers. However, it does not take into account the qualitative and global features of the partial differential equation directly (although a good scheme would always aim to reproduce these in a limit). It can be argued that the global structures can often tell us more about the partial differential equation than the local information given by the expression of the equation in terms of differentials.

The recent growth in the field of geometric integration has, in contrast, led to the development of numerical methods which systematically incorporate qualitative information into their structure. Much of this work has been in the area of ordinary differential equations, and a review of this can be found

* Corresponding author.

E-mail address: cjb@maths.bath.ac.uk (C.J. Budd).

0377-0427/01/$ - see front matter
PII: S0377-0427(00)00521-5

in [10]. However, less has been done with numerical methods for partial differential equations which are often perceived as rather difficult infinite-dimensional limits of systems of ordinary differential equations. In actual fact, the strong structure imposed by a partial differential equation on the ordinary differential equations which may (for example) arise from a semi-discretisation of it, can if anything make the application of geometric ideas rather simpler.

Obviously there are many possible qualitative features that may be present in a partial differential equation and we will not attempt to list them all here. However, a possible partial listing is as follows.

1. *Symmetries.* Many partial differential equations have geometric symmetries (reflexions in space or reversibility in time), translational symmetries (in space and time) and deeper symmetries that link spatial and temporal effects (such as scaling and inversion symmetries). These symmetries are fundamental to the underlying physics that the equation represents and are often more important than the equation itself.
2. *Asymptotics.* Does the partial differential equation evolve in time so that its dynamics in some sense simplifies? For example does it ultimately evolve on a low-dimensional attractor, is its dynamics determined by an inertial manifold inside an absorbing ball? Do complex structures starting from arbitrary initial data simplify into more regular patterns? Does the partial differential equation form singularities in a finite time?
3. *Invariants.* Are quantities such as energy, momentum or potential vorticity conserved either globally or along trajectories? Do the equations have a symplectic structure? Is phase space volume conserved?
4. *Orderings.* Does the partial differential equation preserve orderings? For example if $u_0(x)$ and $v_0(x)$ are two sets of initial data leading to solutions $u(x,t)$ and $v(x,t)$ then does $u_0(x) < v_0(x)$ for all x imply that $u(x,t) < v(x,t)$? A closely related concept, important in numerical weather forcasting [14], is whether the convexity (in space) of a function $u(x,t)$ (for example pressure), is conserved during the evolution.

It is worth emphasising that these global properties are not independent and are often closely linked. For example, if the partial differential equation is derived from a variational principle linked to a Lagrangian function, then, from Noether's theorem [32], each continuous symmetry of the Lagrangian leads directly to a conservation law for the underlying equation. Symmetry when coupled with orderings frequently leads to an understanding of the asymptotic structure of the equation. Also, singularities in the equation often have more local symmetry than the general solution of the equation.

We are thus naturally drawn to ask the question of how much of this qualitative structure can be preserved by a numerical method. Of course some of these questions have been considered for some time, for example the energy preserving schemes introduced by Arakawa [1] and schemes for ordinary differential equations that conserve Lie point symmetries [25]. Furthermore, for some specific problems such as Hamiltonian ODEs or time-reversible systems, some excellent geometry preserving integrators have been developed which can be analysed using backward error analysis [36]. However, a general theory, applicable to a wide class of problems is still lacking. In this review we will look in more detail at some of the items above and then see how well numerical methods do when trying to reproduce some of the qualitative structures.

2. Symmetry

In this section we will very briefly review some of the ideas of symmetry which apply to partial differential equations in time and in one spatial dimension. We will mostly be concerned here with continuous symmetries. For a review of discrete symmetries (such as reflexional symmetries and symmetries described by finite groups) see the excellent book [21].

Suppose that $u(x,t)$ satisfies a partial differential equation

$$N(u,u_x,u_{xx},u_t,u_{tt},x,t)=0. \tag{2.1}$$

We define a *symmetry* of this equation to be any transformation of u,x and t which leaves the underlying equation unchanged.

Such symmetries can be expressed in terms of actions on the tanjent bundle TM of the solution space, or more globally as transformations of the underlying variables (t,x,u) to $(\bar{t},\bar{x},\bar{u})$ of the form

$$\bar{t}=\bar{t}(t,x,u), \quad \bar{x}=\bar{x}(t,x,u), \quad \bar{u}=\bar{u}(t,x,u), \tag{2.2}$$

so that $(\bar{t},\bar{x},\bar{u})$ satisfies the same equation as (t,x,u).

The general theory of such transformations is very rich and can be found in [32,18]. See also [15,16] for a discussion of symmetries of difference equations. Given a partial differential equation, there are systematic procedures for finding the symmetries, which generally reduce to solving an over-determined set of linear equations. Much of this can be automated, and various computer algebra packages exist to do the calculations [29]. Having found the symmetries it is then possible to go on to find exact solutions of the differential equation in many cases.

Because the theory of symmetries of partial differential equations is so large, we will not attempt to summarise all of it here, but will instead mainly look at the important sub-class of symmetries which are given by *scaling transformations.*

Typically these take the form of maps

$$t \to \lambda^{\alpha_0}t, \quad x \to \lambda^{\alpha_1}x, \quad u \to \lambda^{\alpha_2}u. \tag{2.3}$$

Here λ is considered to be an arbitrary positive quantity. Very similar transformations also apply to systems of partial and ordinary differential equations and to partial differential equations in many spatial dimensions.

The book [2] gives many examples of systems of partial differential equations with such symmetries. These arise very naturally in many problems as they express the way that a differential equation changes when the units of measurement in which it is expressed also change. For example, if u is a velocity, then a scaling of time t by λ leads to a scaling of u by $1/\lambda$.

It is an observed fact [2] that scaling (power-law) relationships have wide applications in science and in engineering. Far from being an approximation of the actual behaviour of the equations, such scalings give evidence of deep properties of the phenomena they represent, which may have no intrinsic time or length scale and which have solutions that 'reproduce themselves in time and space' under rescaling. This is an example of a covariance principle in physics that *the underlying solutions of a partial differential equation representing a physical phenomenon should not have a form which depends upon the location of the observer or the units that the observer is using to measure the system.*

Indeed, far from being a special case, scaling symmetries of the form (2.3) are universal in physics. They can be found, for example, in fluid mechanics, turbulence [27], elasticity [13], the theory of detonation and combustion [42,3,38], quantum physics, heat diffusion, convection, filtration, gas dynamics, mathematical biology [31], and structural geology [9]. Scaling invariance is also, of course, closely tied up with the theory of fractals [19], and with the general theory of dimensional analysis and renormalisation [4,20].

Motivated by definition (2.3) we introduce a vector $\alpha = (\alpha_0, \alpha_1, \alpha_2, \ldots)$ to describe the scaling group. It is evident that for any such α the vector $\mu\alpha$ also describes the same scaling transformation. It is quite possible for the same system of partial differential equations to be invariant under several such scaling transformations. It is then easy to check that the scaling operations described by two separate vectors commute. Indeed, the set of vectors corresponding to scaling transformations which leave the partial differential equation invariant form a *commutative vector space*.

2.1. Self-similar solutions

Because of the richness in the behaviour of solutions to partial differential equations, which may have arbitrary initial conditions and complex boundary conditions, it is unlikely that the general solution of the partial differential equation will itself be invariant under the action of the symmetries that leave the equation invariant. For example, the nonlinear wave equation is invariant under translations in space and time. However, only those solutions which are travelling waves retain this property.

Solutions which are so invariant are termed *self-similar solutions*. Such solutions play an important role in applied mathematics (witness the importance of travelling wave solutions). Under certain circumstances they can be attractors [26,40]. They also can differentiate between different types of initial data which lead to qualitatively different forms of solution behaviour. More significantly, they often describe the *intermediate asymptotics* of a problem [42,2]. That is, the behaviour of an evolutionary system at sufficiently long times so that the effects of initial data are not important, but before times in which the effects of boundary conditions dominate the solution. A self-similar solution also satisfies a *simpler* equation than the underlying partial differential equation. Indeed they often satisfy an ordinary differential equation. This has made them popular for computation – although they are normally singular, homoclinic or heteroclinic solutions of the ordinary differential equation and thus still remain a numerical challenge.

A most significant feature of self-similar solutions is that they need not be invariant under the full group of symmetries that leave the underlying equations invariant. In particular they may only be invariant under a particular sub-group. We see this clearly with the nonlinear wave equation

$$u_{tt} = u_{xx} + f(u).$$

For a general function $f(u)$ this equation is invariant under the two individual translation group actions $t \to t + \lambda$ and $x \to x + \mu$ as well as the combined action

$$\theta : t \to t + \lambda, \quad x \to x + c\lambda.$$

The latter group leaves invariant a travelling wave solution moving at a speed c. A solution $u(x,t)$ of the nonlinear wave equation which is invariant under the action of θ must take the form $u(x,t) = v(x - ct)$ and the function $v(y)$ then satisfies the ordinary differential equation

$$c^2 v_{yy} = v_{yy} + f(v).$$

In this equation the wave speed c is an unknown (it is a nonlinear eigenvalue) and its value must be determined as part of the solution. More generally, for a partial differential equation invariant under the action of several groups, determining the group under which the self-similar solution is invariant will form part of the solution process and any numerical method for solving them should take this into account. Problems of this kind are called *self-similar solutions of the second kind* [42,22]. Most problems are of this kind!

In certain special cases the precise group action can be determined from other considerations (such as dimensional analysis). This is of considerable advantage when computing the solution. For example if the wave speed of the travelling wave is known then the mesh can be required to move along with the solution. Such (less common) problems are called *self-similar solutions of the first kind.*

We now give three examples of partial differential equations with scaling invariance. These examples are meant to illustrate some of the range of behaviour which can occur. We consider numerical computations of each of these using invariant methods in Section 6 and will show how each such method can be derived by using geometrical methods.

Example 1. The porous medium equation.

This an example of a problem which has self-similar solutions of the first kind. The porous medium equation models the flow of a liquid through a porous medium such as fractured rock, its mathematical theory is described in [40]. The porous medium equation is given by

$$u_t = (uu_x)_x. \tag{2.4}$$

This equation admits four continuous transformation groups, the two groups of translations in time and space and the two-dimensional vector space of scaling symmetry groups spanned by the vectors

$$\alpha = (1, \tfrac{1}{2}, 0) \quad \text{and} \quad \alpha = (1, 0, -1).$$

The porous medium equation admits a family of self-similar solutions of the form

$$u(x,t) = t^{\gamma} v(x/t^{\beta}) \tag{2.5}$$

for any values of β and γ which satisfy the algebraic condition

$$2\beta - \gamma = 1.$$

Without additional conditions any such solution is possible, however, if we impose the condition that $u(x,t)$ decays as $|x| \to \infty$ then a simple calculation shows that if the mass M of the solution is given by

$$M = \int u(x,t)\,\mathrm{d}x, \tag{2.6}$$

then M is constant for all t. The only self-similar solution with this property has $\gamma = -\frac{1}{3}$ and $\beta = \frac{1}{3}$. Thus it is a self-similar solution of the first kind. Reducing the partial differential equation down to an ordinary differential equation and solving this gives a one-parameter family of compactly supported self-similar solutions of the form

$$u(x,t) = t^{-1/3}(a - x^2/t^{2/3})_+.$$

These solutions were discovered independently by Barenblatt and Pattle [2].

Example 2. The radially symmetric cubic nonlinear Schrödinger equation.

The radially symmetric solutions of the cubic nonlinear Schrödinger equation satisfy the following PDE:

$$iu_t + u_{xx} + \frac{d-1}{x}u_x + u|u|^2 = 0. \tag{2.7}$$

Where d is the dimension of the problem and x is the distance from the origin. This Hamiltonian partial differential equation models the modulational instability of water waves, plasma waves and it is important in nonlinear optics. In one dimension ($d=1$) the partial differential equation is integrable by the inverse scattering transformation and has many symmetries and invariants. Numerical methods have been designed to incorporate these symmetries and they are (in a discrete sense) also integrable. Details are given in [30]. In higher dimensions (such as $d = 2$ or 3) the equation is no longer integrable and the solutions may blow up at the point $x=0$ in a finite time T. In this case the partial differential equation has two significant symmetries. The scaling symmetry

$$(T-t) \to \lambda^2(T-t), \quad x \to \lambda x \quad u \to u/\lambda \tag{2.8}$$

and the phase symmetry

$$u \to e^{i\varphi}u, \quad \varphi \in \mathbb{R}. \tag{2.9}$$

Both symmetries act on the self-similar solutions which correspond to solutions that blow up in a finite time. These all take the singular form [41]

$$u(x,t) = \frac{1}{\sqrt{2a(T-t)}} e^{-i\log(T-t)/2a} Q(x/\sqrt{2a(T-t)}), \tag{2.10}$$

where $Q(y)$ satisfies a complex ODE. Here the scalar a represents the coupling between the amplitude and the phase of the solution and it is exactly analogous to the wave speed of the nonlinear wave equation. The value of a is a priori unknown and must be determined as part of the solution. Thus this is an example of a self-similar solution of the second kind. The Hamiltonian for this problem is given by

$$H(u) = \int_0^\infty (|u_x|^2 - \tfrac{1}{2}|u|^4)x^{d-1}\,dx \tag{2.11}$$

and is a constant of the evolution. Substituting (2.10) into (2.11), we find that

$$H(u) = (T-t)^{(d-4)/4} \int_0^\infty (|Q_y|^2 - \tfrac{1}{2}|Q|^4)y^{d-1}\,dy. \tag{2.12}$$

Thus $H(u)$ can only be constant (indeed bounded) if

$$\int_0^\infty (|Q_y|^2 - \tfrac{1}{2}|Q|^4)y^{d-1}\,dy = 0. \tag{2.13}$$

It is this additional equation which determines implicitly the value of a. Details of this problem are given in [39].

Example 3. The Heat equation.

The well-known linear heat equation

$$u_t = u_{xx} \tag{2.14}$$

is invariant under six independent symmetries [32], of which two are the groups describing translations in time and space, one is the scaling of u given by $u \to \lambda u$ so that $\alpha = (0,0,1)$, and another is the scaling given by

$$x \to \lambda x, \quad t \to \lambda^2 t,$$

so that $\alpha = (2,1,0)$. Obviously any linear combination of these gives an admitted symmetry, and we may consider a general self-similar solution to be of the form

$$u(x,t) = t^{\gamma} v((x - x_0)/t^{1/2}). \tag{2.15}$$

The value of γ is unknown a priori and thus we again have a self-similar solution of the second kind. Interestingly the value of γ may change during the evolution. Consider a bounded domain $x \in [0,1]$ with Dirichlet boundary conditions

$$u(0) = u(1) = 0.$$

If the initial conditions for $u(x,t)$ are highly localised (for example a Dirac measure centred on x_0) then initially the effect of the boundary conditions on the function $u(x,t)$ will be exponentially small. Consequently, the solution takes the approximate form of the free space fundamental solution for which $\gamma = -\frac{1}{2}$ and

$$u(x,t) = \frac{A}{\sqrt{t}} e^{-(x-x_0)^2/4t}, \tag{2.16}$$

for a suitable constant A. After a period of time proportional to the minimum of x_0^2 and $(1 - x_0)^2$ one or other of the boundaries begins to have an effect. Without loss of generality this boundary condition is at $x = 0$. In this case the boundary condition at $x = 1$ again has only exponentially small effect. The effect of this boundary at $x = 0$ is to change the value of γ so that now $\gamma = -1$. In this case there is a self-similar solution of the form

$$u(x,t) = \frac{Bx}{(t - t_0)^{3/2}} e^{-x^2/4(t-t_0)}. \tag{2.17}$$

Here B is a constant and t_0 is an effective initial time at which the boundary condition at $x = 0$ becomes important. For much greater times when the boundary condition at $x = 1$ also becomes significant, this solution is also not appropriate, and indeed the function $u(x,t)$ rapidly decays to zero. The above calculation is described in [2] where it is observed that the solution has two quite different self-similar forms for different ranges of t and between these two forms the solution has a nonself-similar transitionary form.

Examples two and three are highly instructive from the point of view of a numerical approximation to the solutions of scale invariant problems. Given an equation with a known symmetry group it is natural to exploit this when solving it numerically. For example, with the porous medium equation, we might consider rescaling the problem into new variables $v = t^{1/3}u$ and $y = x/t^{1/3}$. The advantage of doing this is that such variables will vary slowly during the course of the evolution of u and hence error estimates based upon higher derivatives in these variables will be much smaller, and indeed may be uniformly bounded. However, the two latter examples illustrate a problem with this approach. We simply may not know in advance which variables to use, or indeed the self-similar variables may only be appropriate for part of the evolution of the solution. Consequently, we must use a more general approach and consider numerical methods which themselves, rather than imposing coordinates on the method, scale in a similar way to the underlying partial differential equation. If this is achievable then many of the desirable structures of the original system will be preserved in discrete form. This approach is at the heart of the geometric method of integrating scale invariant equations and we will develop it further in the examples considered in Section 6.

3. Symmetry and the maximum principle

Before considering numerical discretisations of scale invariant problems, it is worthwhile also looking at the role played by the maximum principle as the combination of the maximum principle and scaling invariance tells us a great deal about the asymptotic behaviour of the underlying partial differential equation. Suppose that we have a partial differential equation from which may be derived a semi-group operator φ_t such that if $u(x,0)$ is some initial data then $u(x,t)=\varphi_t(u(x,0))$. Such a partial differential equation has a strong maximum principle if the ordering of solutions is preserved under the action of the semi-group [35]. Thus if $u(x,0)<v(x,0)$ for all x then $\varphi_t(u)<\varphi_t(v)$ for all x and $t>0$. Many parabolic partial differential equations (for example the nonlinear heat equation $u_t=u_{xx}+f(u)$) satisfy a strong maximum principle. Such maximum principles are invaluable when studying the dynamics of the equation. For example, if $v(x,t)$ is a known solution which is bounded above and which satisfies the partial differential equation and if $u(x,0)<v(x,0)$, then we have that $u(x,t)$ is also bounded above. Such an exact solution could easily be a self-similar solution. Furthermore, if v_1 and v_2 are two self-similar solutions such that $v_1 \to v_2$ as $t\to\infty$ then if $v_1(x,0)<u(x,0)<v_2(x,0)$ we deduce immediately that $u\to v_2$. Here we see the regularising effect of the partial differential equation. Techniques similar to this are described in [40] to prove the L_1 global attractivity of the self-similar solution of the porous medium equation – although there are considerable additional analytic difficulties due to the existence of a nonregular interface. A numerical method which has both a strong maximum principle and discrete self-similar solutions will, similarly, give the correct global asymptotic behaviour of the underlying partial differential equation. This is precisely what we seek to achieve using a geometric integration approach.

4. Scaling and adaptivity

We have shown in Section 2 that scaling invariance plays an important role in the theory and behaviour of the solutions to a partial (and indeed ordinary) differential equation. It is desirable that a numerical method to discretise such an equation should have a similar invariance principle. Ideally such a numerical method should posses *discrete self-similar solutions* which are scale invariant and which uniformly approximate the true self-similar solutions of the partial differential equation over all times. If these are global attractors (or at least have the same local stability as the underlying

PDE) then we will have a numerical method which has the correct asymptotic properties and indeed may have excellent accuracy when approximating singular solutions.

Scaling invariance of a partial or ordinary differential equation and adaptivity of the spatial and temporal meshes fit very naturally together. This is because the use of a fixed mesh in a discretisation automatically imposes an underlying spatial and temporal scale on the problem. This makes it impossible to consider scale-invariant solutions. This difficulty disappears when we introduce adaptivity as now the spatial and temporal grids become part of the solution and can easily adjust to any appropriate length and time scale consistent with the underlying problem. When considering such an approach, it is natural to look at methods of *r-adaptivity* in which spatial mesh points are moved during the solution process, rather than *h-adaptivity* in which new points are added or *p-adaptivity* in which the accuracy of the solution approximation is increased in order. The reason for doing this is that then the solution approximation, spatial and temporal mesh become one (large) dynamical system which has a lot of structure, reflecting the underlying scalings of the original problem, and this structure can then be analysed by using dynamical systems methods. A very general account of the interaction between adaptivity in space on a moving mesh for problems with a wide class of symmetries is given by the work of Dorodnitsyn and his group [15,16] etc. (See also [6]). In this paper we will look at the specific case of scaling symmetries and will call a numerical method which inherits the underlying symmetries of the system *scale invariant*.

The advantage of using an adaptive method which is invariant under the action of a group, is that such methods should, if correctly designed, admit exact discrete self-similar solutions. If we conserve maximum principles and the stability of the underlying self-similar solution, then such numerical methods will have excellent asymptotic properties. However, the discrete self-similar solutions will not be the only solutions admitted by the numerical method and thus the effect of boundary conditions and arbitrary initial conditions may be taken into account. Thus, the synthesis of adaptivity with symmetry invariance provides a flexible, general and powerful numerical tool. We will see this by looking at several examples.

To make things more precise, in this case we consider a partial differential equation of the form

$$u_t = f(u, u_x, u_{xx}). \tag{4.1}$$

We may consider a fully discrete method for solving this which gives an approximation $U_{m,n}$ on a spatial and temporal mesh $(X_{m,n}, T_m)$ so that

$$U_{m,n} \approx u(X_{m,n}, T_m).$$

Here we consider discretisations with a fixed number N of spatial mesh points $X_{m,n}$, $n=1,\dots,N$, for each time level T_m. In an adaptive scheme the values of $X_{m,n}$ and T_m are computed along with the solution $U_{m,n}$. Suppose that, in the absence of boundary conditions, the differential equation (4.1) is invariant under the action of the scaling tansformation

$$t \to \lambda^{\alpha_0} t, \quad x \to \lambda^{\alpha_1} x, \quad u \to \lambda^{\alpha_2} u. \tag{4.2}$$

Now, consider the approximation $(U_{m,n}, X_{m,n}, T_m)$ to $u(x,t)$ obtained by our method, so that $(U_{m,n}, X_{m,n}, T_m)$ is the solution of a discrete scheme. We say that the numerical method is *scale invariant* if (again in the absence of boundary conditions) the set of points

$$(\lambda^{\alpha_2} U_{m,n}, \lambda^{\alpha_1} X_{m,n}, \lambda^{\alpha_0} T_m) \tag{4.3}$$

is also a solution of the discrete system.

Condition (4.3) gives a means of defining conditions for an adaptive mesh. Observe that these are *global* conditions related to underlying scaling properties of the equation, rather than the usual *local* conditions of adaptivity in which (for example) we may choose to cluster mesh points where the local truncation error of the solution is high. The reason for this choice of condition on the mesh is that it accurately reflects the underlying geometry of the problem. For scale-invariant ODEs the two approaches (global and local) can in fact be equivalent and lead to the same time steps. Details of this equivalence are given in [11].

As a feature of this geometry, consider *discrete self-similar* solutions. A self-similar solution of (4.1) satisfies

$$u(\lambda^{\alpha_1}x, \lambda^{\alpha_0}t) = \lambda^{\alpha_2}u(x,t),$$

so that

$$u(x,t) = t^{\alpha_2/\alpha_0}v(x/t^{\alpha_1/\alpha_0}),$$

where the function $v(y)$ satisfies an ordinary differential equation. In comparison a discrete self-similar solution which has the same invariance must satisfy the condition

$$T_m = \lambda^{\alpha_0 m}T_1, \quad X_{m,n} = \lambda^{\alpha_1 m}Y_n, \quad U_{m,n} = \lambda^{\alpha_2 m}V_n. \tag{4.4}$$

The existence of such a discrete self-similar solution follows immediately from the scaling invariance condition (4.3). The vectors V_n and Y_n then satisfy an algebraic equation. Now, it is easy to verify that the two operations of rescaling a partial differential equation and discretising the same equation *commute*, with details given in [11]. It follows that if the discretisation of the PDE is consistent with the underlying PDE then the algebraic equation satisfied by Y_n and V_n is a consistent discretisation of the ordinary differential equation satisfied by $v(y)$. Hence $V_n \approx v(Y_n)$. Observe that the error implicit in this approximation *does not depend upon the value of m*. Hence we may uniformly approximate the self-similar solution over arbitrarily long times. More details of this calculation are given in [11,7] and in the following sections.

The process of introducing adaptivity is very closely linked with rescaling. Suppose that τ and ξ are computational variables in the sense of [23]. We can consider an adaptive mesh $(X_{m,n}, T_m)$ to be a *map* from the computational space (ξ,τ) to the underlying space (x,t) over which the PDE is defined, which is given in terms of the maps $X(\tau,\xi)$ and $T(\tau)$. If the computational space is covered by a uniform mesh of spacing $(\Delta\xi, \Delta\tau)$ then we have

$$T_m = T(m\Delta\tau) \quad \text{and} \quad X_{m,n} = X(m\Delta\tau, n\Delta\xi).$$

(A similar procedure can also be used in higher spatial dimensions [24].) The differential equation (4.1) when expressed in terms of the computational variables then becomes

$$u_\tau - u_x X_\tau = T_\tau f(u, u_x, u_{xx}) \quad \text{with } u_x = u_\xi/X_\xi, \text{ etc.}, \tag{4.5}$$

which retains the same invariance to scaling as (4.1). An r-adaptive approach is then equivalent to discretising Eq. (4.5) in the computational variables.

An essential part of this process is the determination of suitable functions $X(\tau,\xi)$ and $T(\tau)$. There is much arbitrariness about how this may be done, but we are guided in our choice by the scaling invariance condition (4.3). In particular, if we have a set of conditions for the mesh which lead to the solutions $T(\tau)$, $X(\tau,\xi)$, and $U(\tau,\xi)$ then these conditions should also admit a rescaled solution of the form $\lambda^{\alpha_0}T(\tau)$, $\lambda^{\alpha_1}X(\tau,\xi)$, and $\lambda^{\alpha_2}U(\tau,\xi)$.

There are various ways of obtaining such conditions. A popular choice for determing the function T is the Sundman transformation [28] in which a function $g(u)$ is introduced so that

$$\frac{\mathrm{d}T}{\mathrm{d}\tau} = g(u). \tag{4.6}$$

For scale invariance we require that $g(u)$ should satisfy the condition

$$g(\lambda^{\alpha_2} u) = \lambda^{\alpha_0} g(u). \tag{4.7}$$

In [11] a systematic method of calculating such functions is given for both single ordinary differential equations and systems of such equations.

Strategies for calculating the mesh function X vary in the literature. One direct method is to mimic the formula for T and to introduce a further function H such that

$$\frac{\partial X}{\partial \tau} = H(X,U). \tag{4.8}$$

This is a natural strategy for certain hyperbolic equations as it can correspond to advecting the mesh along the flow of the solution, see [37] where it is combined with a powerful multi-symplectic approach to study various problems in fluid mechanics. A similar strategy is also adopted by Dorodnitsyn and his co-workers for a general class of groups [15]. To give a scale invariant scheme we require that $H(X,U)$ satisfy

$$H(\lambda^{\alpha_1} X, \lambda^{\alpha_2} U) = \lambda^{\alpha_1} H(X,U) \tag{4.9}$$

or equivalently, on differentiating with respect to λ, the function H should satisfy the linear hyperbolic partial differential equation

$$\alpha_1 X H_X + \alpha_2 U H_U = \alpha_1 H. \tag{4.10}$$

A disadvantage of this approach is that it is rather local in form and it can freeze the location of the mesh, i.e. it can fail to move mesh points into regions of interest such as developing singularities.

An alternative approach is to use the ideas of equidistribution, introduced by Dorfi and Drury [17] and developed by Russell and his co-workers [23]. In this approach we introduce a monitor function $M(x,u,u_x)$ and aim to equidistribute this function over the mesh. Examples of such functions are the local truncation error of the solution [33], the commonly used arc-length

$$M(x,u,u_x) = \sqrt{1+u_x^2} \tag{4.11}$$

and more simply (and especially useful when we need to cluster points where the solution is large)

$$M(x,u,u_x) = u^{\gamma}. \tag{4.12}$$

Suppose (without much loss of generality) that the function $u(x,t)$ is defined on the interval $x \in [0,1]$ and the computational domain is $\xi \in [0,1]$ so that $X(0,t)=0$ and $X(1,t)=1$. In its simplest form, equidistribution takes the form [23]

$$\int_0^X M \,\mathrm{d}x = \xi \int_0^1 M \,\mathrm{d}x. \tag{4.13}$$

Observe that this principle is closely related to the geometric idea of conserving the function M over mesh intervals, and we can exploit this feature to help design meshes which automatically retain invariants of the evolution [7]. Differentiating (4.13) we have that X satisfies the moving mesh partial differential equation

$$(MX_\xi)_\xi = 0. \tag{4.14}$$

This problem is scale invariant if M satisfies an equation of the form

$$M(\lambda^{\alpha_1}x, \lambda^{\alpha_2}u, \lambda^{\alpha_2-\alpha_1}u_x) = \lambda^{\alpha_3}M(x,u,u_x), \tag{4.15}$$

where the value of α_3 can be very general, meaning that (4.14) can be invariant for a wide variety of different scalings.

We see significantly that the monitor function $M = u^\gamma$ satisfies this condition for any choice of α_i whereas the arc-length monitor function $\sqrt{1+u_x^2}$ only satisfies it in the very restrictive case of $\alpha_1 = \alpha_2$. Thus arc-length does not fit in well with the theory of invariant methods, although we will see in Section 6 that it can still be used for scale invariant problems. In contrast, it was shown in [11] that taking M to be relative local truncation error for a scale-invariant ordinary differential equation, often automatically ensures that Eq. (4.15) is satisfied. For this reason the choice of local truncation error for M may be a wise one!

In practice Eq. (4.14) can lead to instabilities [23], even if a scale-invariant monitor function is used [8]. Furthermore, it requires the use of a mesh which is initially equidistributed and this can be hard to achieve. To allow both for arbitrary intial meshes and to stabilise the system, a relaxed form of (4.14) is often used. One of the more common versions is the so-called MMPDE6 which takes the form

$$-\varepsilon X_{t\xi\xi} = (MX_\xi)_\xi, \tag{4.16}$$

where $\varepsilon > 0$ is a small relaxation parameter. This equation has been used with great success in many applications, see the review in [23] and a study of its applications to singular problems in [8].

The scale invariance of the equation MMPDE6 can be ensured by a suitable choice of M. This must now satisfy the more restrictive condition that

$$M(\lambda^{\alpha_1}x, \lambda^{\alpha_2}u, \lambda^{\alpha_2-\alpha_1}u_x) = \lambda^{-\alpha_0}M(x,u,u_x). \tag{4.17}$$

For example, if $\alpha_0 = 1$, $\alpha_1 = \frac{1}{2}$, $\alpha_2 = -1$ and $M = u^\gamma$ then (4.16) is scale invariant when

$$\gamma = 1, \quad M(u) = u.$$

The distinction between the two conditions (4.15) and (4.17) is not important if a single scaling group acts on the system. However in problems, such as the linear heat equation discussed in the examples of the last section, where several groups act, Eq. (4.14) is invariant under all such actions whereas (4.16) will (in general) only be invariant under the action of one scaling group.

Once a time-stepping and mesh adaption strategy has been formulated, Eqs. (4.5), (4.6) and one of (4.8), (4.14) and (4.16) can be solved using a suitable discretisation scheme. Here we discretise the problem in terms of the computational variables τ and ξ. In much work on adaptivity the underlying equation (4.5) is solved (using say a collocation method) to a higher degree of accuracy than the mesh equations (4.6), (4.8), (4.14) and (4.16). However, for scaling invariant problems this may be an unwise strategy. For such problems the close coupling of the solution with space and time means that a reduction in accuracy of the solution of the mesh equations may convert to a reduction in accuracy of the solution of the underlying partial differential equation. In particular we may wish to employ the numerical method to solve a partial differential equation from an arbitrary initial state which we believe converges toward a self-similar solution. In this case the methods described above will admit such a solution and if the mesh equations are solved accurately then the self-similar solution will be computed accurately with uniformly bounded errors.

We now consider putting this methodology into practice by looking at several case studies.

5. Scale-invariant ordinary differential equations

Allowing a little more generality than the previous section we consider systems of ordinary differential equations of the form

$$\frac{\mathrm{d}u_i}{\mathrm{d}t} = f_i(u_1, \ldots, u_N), \quad i = 1, \ldots, N, \tag{5.1}$$

for which there is an invariance of the form

$$t \to \lambda^{\alpha_0} t, \quad u_i \to \lambda^{\alpha_i} u_i. \tag{5.2}$$

Here we review some of the results in [11].

The self-similar solutions of this system (up to translations and reflexions in t) take the form

$$u_i = t^{\alpha_i/\alpha_0} v_i, \tag{5.3}$$

where the constants v_i satisfy a suitable algebraic equation. To calculate a general class of solutions of (5.1) including the self-similar solutions we introduce a computational variable τ defined through Eq. (4.5) so that we solve the transformed system

$$\frac{\mathrm{d}u_i}{\mathrm{d}\tau} = g(u_1, \ldots, u_N) f_i(u_1, \ldots, u_N), \quad \frac{\mathrm{d}T}{\mathrm{d}\tau} = g(u_1, \ldots, u_N). \tag{5.4}$$

This system is scale invariant provided that the function $g(u)$ satisfies

$$g(\lambda^{\alpha_1} u_1, \ldots, \lambda^{\alpha_N} u_N) = \lambda^{\alpha_0} g(u_1, \ldots, u_N). \tag{5.5}$$

Suppose now that we discretise the complete system (5.4) for the unknowns $u_1, \ldots, u_N, T$ using a linear multi-step method (or indeed a Runge–Kutta method) with a discrete (fixed) computational time step $\Delta\tau$. Suppose further that this method has a local truncation error of $\mathcal{O}(\Delta\tau^{p+1})$. Then, it is shown in [11] that for sufficiently small $\Delta\tau$, the resulting discrete system has discrete self-similar solutions $U_{i,n}$ taking the form

$$U_{i,n} = V_i z^{\alpha_i n}, \quad t_n = a + b z^{\alpha_0 n} \tag{5.6}$$

for suitable constants V_i, a, b and z. Significantly, the discrete self-similar solution approximates the true self-similar solution with uniform accuracy *for all times* with a relative error that does not grow with time. We state this result as follows.

Theorem. *Let $u_i(t)$ be a self-similar solution of the ordinary differential equation, then there is a discrete self-similar solution $(U_{i,n}, t_n)$ of the discrete scheme such that for all n*

$$u_i(t_n) = U_{i,n}(1 + C_{i,n}\Delta\tau^p). \tag{5.7}$$

Here $C_{i,n}$ is bounded for all n, independently of n.

Observe that this result does not depend upon the form of the self-similar solution. Thus, if this solution is developing a singularity (for example in the problem of gravitational collapse) the discrete self-similar solution continues to approximate it with uniform accuracy [11].

A proof of this result is given in [11]. We can go further with this analysis. The most interesting physical self-similar solutions are those which are attractors as they then determine the asymptotic behaviour of the solution. In [11] it is shown further that if the true self-similar solution is locally stable then so is the the discrete self-similar solution (5.6). In particular, if we introduce a small perturbation of the form

$$U_{i,n} = z^{\alpha_i n}[V_i + a_{i,n}], \qquad t_n = z^{\alpha_0 n}[T + b_n],$$

and if $a_{i,n}$ and b_n are initially small, then they remain small for all n.

We illustrate the conclusions of this theorem by considering a simple example. Suppose that $u(t)$ satisfies the ordinary differential equation

$$\mathrm{d}u/\mathrm{d}t = -u^3, \quad u(0) = u_0.$$

This is invariant under the transformation

$$t \to \lambda t, \quad u \to \lambda^{-1/2} u.$$

If we set $g(u) = 1/u^2$ in (5.4) then we have

$$\mathrm{d}u/\mathrm{d}\tau = -u, \quad \mathrm{d}t/\mathrm{d}\tau = 1/u^2 \tag{5.8}$$

with $u(0) = u_0$, $t(0) = 0$. It is easy to see that Eq. (5.8) is scale invariant, so that if $(u(\tau), t(\tau))$ is a solution then so is $(\lambda^{-1/2}u(\tau), \lambda t(\tau))$. Observe that we have linearised the equation for u. We now discretise (5.8) using (for ease of exposition) the trapezium rule. This gives

$$U_{n+1} - U_n = -\frac{\Delta\tau}{2}(U_n + U_{n+1}), \quad t_{n+1} - t_n = \frac{\Delta\tau}{2}\left(\frac{1}{U_n^2} + \frac{1}{U_{n+1}^2}\right). \tag{5.9}$$

with $U_0 = u_0$ and $t_0 = 0$. This scale-invariant discretisation admits a discrete self-similar solution of the form

$$U_n = Vy^n, \quad t_n = a + by^{-2n}, \tag{5.10}$$

where V, $y = z^{-1/2}$, a and b are to be determined. Substituting into the first equation in (5.9) (and dividing by the constant factor of Vy^n) we have

$$y - 1 = -\frac{\Delta\tau}{2}(y + 1),$$

so that

$$y = \left(1 - \frac{\Delta\tau}{2}\right) \bigg/ \left(1 + \frac{\Delta\tau}{2}\right) = \mathrm{e}^{-\Delta\tau} + \mathcal{O}(\Delta\tau^3).$$

Here we will assume that $\Delta\tau < 2$ so that $0 < y < 1$ and $u_n \to 0$ as $n \to \infty$. Similarly, from the second equation in (5.9) we have (on division by the constant factor of y^{-2n})

$$b(y^{-2} - 1) = \frac{\Delta\tau}{2} V^{-2}(y^{-2} + 1).$$

Hence, combining results gives

$$b=\frac{\Delta\tau}{2}V^{-2}\left(\frac{1+y^2}{1-y^2}\right),\quad y=\frac{1-\Delta\tau/2}{1+\Delta\tau/2}. \tag{5.11}$$

Applying the initial conditions also gives $V=u_0$ and $a+b=0$. Thus

$$b=\left[\frac{1}{2}+\frac{\Delta\tau^2}{8}+\mathcal{O}(\Delta\tau^3)\right]u_0^{-2}.$$

Hence, from the identity

$$t_n=a+bz^{-2n}=a+b(u_0/U_n)^2,$$

we have

$$t_n=\left[\frac{1}{2}+\frac{\Delta\tau^2}{8}+\mathcal{O}(\Delta\tau^3)\right](U_n^{-2}-u_0^{-2}).$$

After some manipulation we then have

$$U_n=u_0\sqrt{\frac{1+(\Delta\tau^2/4)+\mathcal{O}(\Delta\tau^3)}{2t_nu_0^2+1+(\Delta\tau^2/4)+\mathcal{O}(\Delta\tau^3)}}. \tag{5.12}$$

We now compare expression (5.12) with the exact solution which is given by

$$u(t)=u_0\sqrt{\frac{1}{2tu_0^2+1}}. \tag{5.13}$$

This is, in fact, a self-similar solution with respect to the translated time $s=t+u_0^{-2}/2$. Expanding the expression for U_n and rearranging, then gives (for the case $t_n>0$)

$$U_n-u(t_n)=\frac{1}{16u_0t_n}\Delta\tau^2+\mathcal{O}(\Delta\tau^3).$$

Observe that this error is (i) of the correct order for the trapezium rule and (ii) the error *decays* as $t_n\to\infty$.

6. Scale-invariant partial differential equations

As it is difficult to give a general theory for the convergence of numerical methods for scale invariant partial differential equations, we consider numerical methods as applied to the three examples described in Section 2. To simplify our discussions, we consider the case of semi-discretisations only. These results can then easily be extended to full discretisations by using the theory outlined in the previous section.

6.1. The porous medium equation

Some of these results can be found in the paper [7]. As described, the equation

$$u_t=(uu_x)_x=(u^2/2)_{xx},$$

has self-similar solutions of constant first integral (2.6). Both these, and the solution from general compactly supported initial data, are compactly supported and have interfaces at points s_- and s_+ which move at a finite speed. To discretise this equation we introduce an adaptive mesh $X(\xi,t)$ such that $X(0,t)=s_-$ and $X(1,t)=s_+$. To determine X we use a monitor function and a moving mesh partial differential equation. As the evolution of the porous medium equation is fairly gentle it is possible to use the mesh equation (4.13) without fear of instability in the mesh. This then allows a wide possible choice of scale-invariant monitor functions of the form $M(u)=u^{\gamma}$. A convenient function to use is

$$M(u)=u,$$

the choice of which is strongly motivated by the conservation law

$$\int_{s_-}^{s_+} u\,\mathrm{d}x = C.$$

Here C is a constant which we can take to equal 1 without loss of generality. Setting $M=u$ and $C=1$ in (4.13) gives

$$\int_{s_-}^{X} u\,\mathrm{d}x = \xi, \tag{6.1}$$

so that on differentiation with respect to ξ we have

$$uX_\xi = 1, \tag{6.2}$$

as the equation for the mesh which we will discretise. Note that this is invariant under the group action $u \to \lambda^{-1/3}u$, $X \to \lambda^{1/3}X$. Now, differentiating (6.1) with respect to t gives

$$0 = X_t u + \int_{s_-}^{X} u_t\,\mathrm{d}x = X_t u + \int_{s_-}^{X} (uu_x)_x\,\mathrm{d}x = u(X_t + u_x).$$

Thus, for the continuous problem we also have that X satisfies the equation

$$X_t = -u_x. \tag{6.3}$$

Substituting (6.2) and (6.3) into the rescaled equation (4.5) gives, after some manipulation, the following equation for u in the computational coordinates:

$$u(\xi,t)_t = \tfrac{1}{2}u^2(u^2)_{\xi\xi}. \tag{6.4}$$

Eqs. (6.2)–(6.4) have a set of self-similar solutions of the form

$$\hat{u}(\xi,t) = (t+C)^{-1/3}w(\xi), \quad \hat{X}(\xi,t) = (t-C)^{1/3}Y(\xi), \tag{6.5}$$

where C is arbitrary and the functions w and Y satisfy differential equations in ξ only. Now, consider semi-discretisations of (6.2) and (6.4) so that we introduce discrete approximations $U_i(t)$ and $X_i(t)$ to the continuous functions $u(\xi,t)$ and $X(\xi,t)$ over the computational mesh

$$\xi = i/N, \quad 1 \leqslant i \leqslant (N-1)$$

with $U_0(t)=U_N(t)=0$. A simple centred semi-discretisation of (6.4) for $1 \leqslant i \leqslant (N-1)$ is given by

$$\frac{\mathrm{d}U_i}{\mathrm{d}t} = \frac{N^2}{2}U_i^2(U_{i+1}^2 - 2U_i^2 + U_{i-1}^2). \tag{6.6}$$

To define the mesh X_i we discretise (6.2) to give the algebraic system

$$(X_{i+1} - X_i)(U_{i+1} + U_i) = 2/N, \quad 0 \leqslant i \leqslant (N-1). \tag{6.7}$$

We observe that this procedure has the geometric property of automatically conserving the discrete mass

$$\sum_{i=0}^{N-1} (X_{i+1} - X_i)(U_{i+1} + U_i). \tag{6.8}$$

An additional equation is needed to close the set of equations for the unknowns X_i and we do this by insisting that (as in the true solution) the discrete centre of mass is conserved (without loss of generality at 0), so that

$$\sum_{i=0}^{N-1} (X_{i+1}^2 - X_i^2)(U_{i+1} + U_i) = 0. \tag{6.9}$$

Observe that Eq. (6.6) for the solution and Eqs. (6.7), (6.9) for the mesh have decoupled in this system. This makes it much easier to analyse. We discuss generalisations of this principle in a forthcoming paper [12]. In particular (6.6) has two key geometrical features. *Firstly*, it is invariant under the group action

$$t \to \lambda t, \quad U_i \to \lambda^{-1/3} U_i.$$

Thus it admits a (semi) discrete self-similar solution of the form

$$\hat{U}_i(t) = (t + C)^{-1/3} W_i, \tag{6.10}$$

where $W_i > 0$ satisfies an algebraic equation approximating the differential equation for $w(\xi)$. Observe that for all C we have

$$t^{1/3} \hat{U}_i \to W_i.$$

Secondly, the discretisation satisfies a maximum principle, so that if $U_i(t)$ and $V_i(t)$ are two solutions with $U_i(0) < V_i(0)$ for all i then $U_i(t) < V_i(t)$ for all i. (See [12]).

The consequences of these two results are profound. Suppose that $U_i(0) > 0$ is a general set of initial data. By choosing values of C appropriately (say $C = t_0$ and $C = t_1$) we can find values such that

$$t_0^{-1/3} W_i < U_i(0) < t_1^{-1/3} W_i, \quad 1 \leqslant i \leqslant (N-1).$$

Consequently, applying the maximum principle to the self-similar solution we then have that for all t

$$(t + t_0)^{-1/3} W_i < U_i(t) < (t + t_1)^{-1/3} W_i, \quad 1 \leqslant i \leqslant (N-1)$$

and hence we have the convergence result

$$t^{1/3} U_i(t) \to W_i, \tag{6.11}$$

showing that the solution U_i and hence the mesh X_i converges *globally* to the self-similar solution. Hence the numerical scheme is predicting precisely the correct asymptotic behaviour. This is because the discretisation has the same underlying geometry as the partial differential equation.

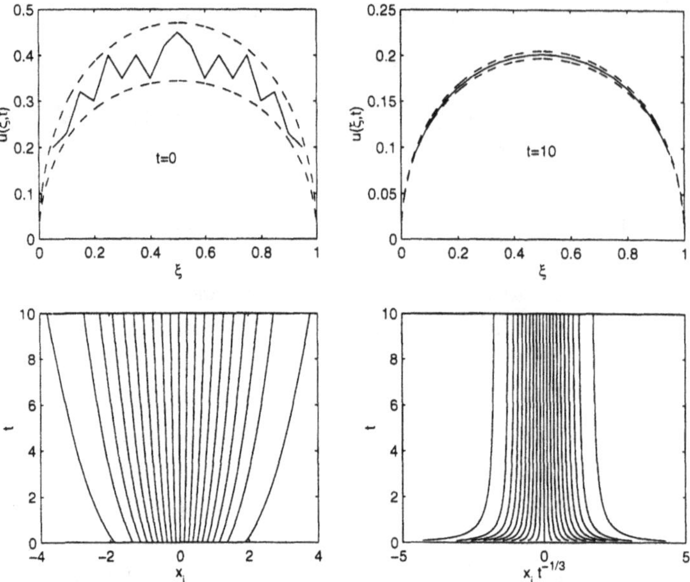

Fig. 1. Convergence of the solution in the computational domain and invariance of the computed mesh.

We illustrate this result by performing a computation. Since the equations for the solution and the equations for the mesh are decoupled, we may consider them separately. Firstly, the ODEs for the solution u in the computational domain are integrated usind a third-order BDF method. We can then simply solve for the mesh by inverting a linear system at each time level, assuming we have symmetric initial data so that the centre of mass is at $x = 0$. Fig. 1 shows a solution with arbitrary initial data being *squeezed* between two discrete self-similar solutions. The self-similar solutions shown here are those with C taking the values 0.9 and 2.3.

6.2. The nonlinear Schrödinger equation

We will summarise the method described in [5] for computing the radially symmetric blow-up solutions of the nonlinear Schrödinger equation (2.7) which become infinite at the origin in a finite time T. Previous computations [39] have used the method of *dynamic rescaling* which is closely related to the scale-invariant methods, but lacks the generality of the approach we describe. Eq. (2.7) is invariant under the action of the two groups (2.8), (2.9), with an unknown coupling constant a. It is immediate from the phase invariance (2.9) that the required monitor function $M(x, u, u_x)$ must be of the form

$$M(x, u, u_x) = M(x, |u|, |u_x|).$$

The mesh equation (4.14) applied to solutions which become infinite in a finite time leads to unstable meshes, and instead we use the stabilised equation (4.16). As a possible ansatz for a monitor function leading to solutions invariant under the scaling (2.8) we may use

$$M(x, |u|, |u_x|) = x^{\alpha}|u|^{\beta}|u_x|^{\gamma}$$

with the simplest scale-invariant choice being

$$M = |u|^2. \tag{6.12}$$

Observe that this function is now invariant under both symmetries (2.8) and (2.9) in the problem and hence under any combination of them. Thus the method does not need informing of the value of the coupling constant a. As an aside, the underlying partial differential equation is unitary and conserves the integral

$$\int_0^\infty |u|^2 x^{d-1}\,\mathrm{d}x, \tag{6.13}$$

during the evolution. Unlike the example of the porous medium equation, the monitor function (6.12) *does not* enforce this conservation law. Although this may appear to be a difficulty, in fact it is an advantage of the method. Adaptive procedures which do enforce (6.13) move points away from the developing singularity, as it is easy to show that across the developing peak of the solution, the integral of $|u|^2x^{d-1}$ tends to zero if $d > 2$, and the integral (6.13) comes from the 'nonsingular' part of the solution which does not need to be resolved with a fine mesh.

To compute the blow-up solution the rescaled equation (4.5) and the moving mesh partial differential equations (4.16) are semi-discretised by using collocation method over a large domain so that $X(0,t) = 0$ and $X(1,t) = 5$. To help with the stability of the solution the monitor function is smoothed over several adjacent mesh points. The resulting system of ordinary differential equations are then solved using DDASSL [34]. Observe that they automatically admit a self-similar solution of the form (2.10) as a special class of solutions for any value of the coupling constant a. We observe, without proof, that both the true and the (semi) discrete self-simlar solutions are stable attractors.

The resulting scheme has proved very effective at computing the singular solutions with a modest number ($N = 81$) of mesh points in the computational domain. The computations have revealed much new structure, including observations of the behaviour of multi-bump self-similar solutions when $d > 2$ [5]. We present results of a computation which evolves towards the singular blow-up self-similar solution which has a monotone profile. Fig. 2 shows two solutions taken when $d = 3$ and the estimated value of T is $T = 0.0343013614215$. These two solutions are taken close to the blow-up time when the amplitude of $|u|$ is around 10^5 and the peak has width around 10^{-5}. Observe that the resolution of the peak is very good, indicating that the mesh points are adapting correctly.

We now look at the mesh $X_i(t)$. If this is evolving in a self-similar manner then we would expect that $X_i(t) = \sqrt{T-t}Y_i$. Now, as $|u(0,t)| = Q(0)/\sqrt{(T-t)}$ then a self-similarly evolving solution and mesh should satisfy

$$X_i(t)|u(0,t)| = Q(0)Y_i.$$

Accordingly, in Fig. 3 we present a plot of $X_i(t)|u(0,t)|$ as a function of $\log(T-t)$ for a range in which u varies from 100 to 500 000. Observe that these values rapidly evolve towards constants, demonstrating that the mesh is indeed evolving in a self-similar manner.

6.3. The linear heat equation

For our final calculation we consider the linear heat equation

$$u_t = u_{xx}, \tag{6.14}$$

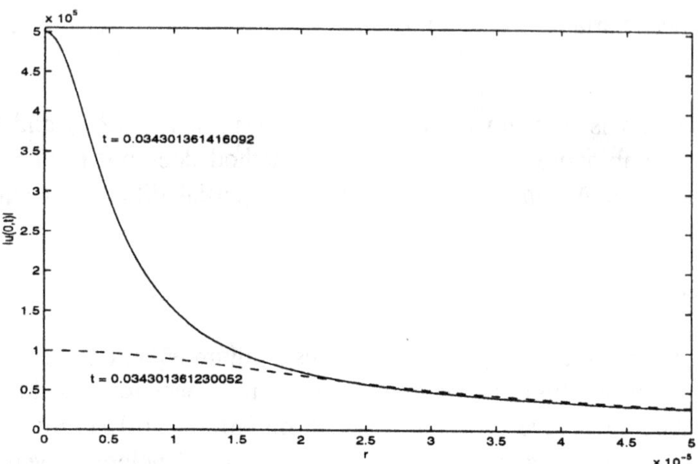

Fig. 2. The solution when $|u(0,t)| = 100\,000$ and $500\,000$.

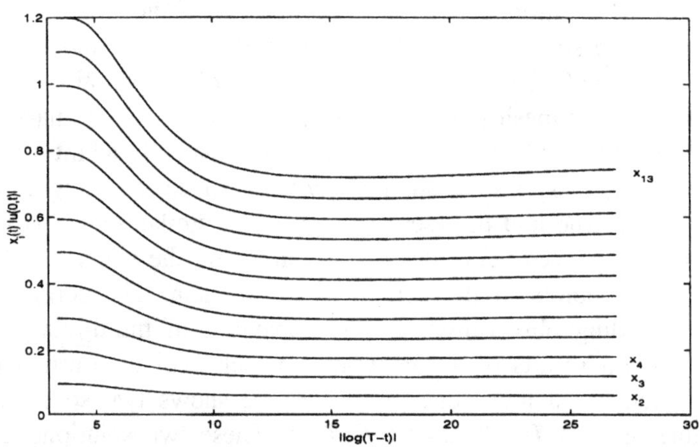

Fig. 3. The scaled mesh $X_i(t)|u(0,t)|$.

on the bounded domain $x \in [0,1]$, with Dirichlet boundary conditions

$$u(0,t) = u(1,t) = 0, \tag{6.15}$$

and initial conditions approximating a Dirac measure. As was observed in Section 2, this problem may have an infinite number of different self-similar solutions and that the precise such solution may depend upon the time t. Thus we need to construct a numerical method invariant under the action of all such groups. As the heat equation has very stable solution patterns, we may (as in the case of the porous medium equation) use the mesh equation (4.13). The advantage of this is that it permits us to consider a wide class of monitor functions. This is necessary due to the many possible self-similar solutions admitted by this problem. For our monitor function we will consider the commonly used arc-length function

$$M = \sqrt{1 + u_x^2}. \tag{6.16}$$

As was observed earlier, this function is not scale invariant in general. However, if u_x is large, then M to a first approximation is given by $|u_x|$ which is scale invariant. We can regard arc-length as a regularisation of this function. When $|u_x|$ is small then arc-length reduces to 1 which is (trivially) scale invariant and leads to a uniform mesh. This is appropriate during the final evolution of the solution. This monitor function also has the advantage that it is not too much affected by the boundary conditions. Thus the mesh $X(\xi,t)$ must satisfy

$$\int_0^X \sqrt{1+u_x^2}\,\mathrm{d}x = \xi \int_0^1 \sqrt{1+u_x^2}\,\mathrm{d}x. \tag{6.17}$$

If $|u_x|$ is large and u has a single maximum then

$$\xi \int_0^1 \sqrt{1+u_x^2}\,\mathrm{d}x \approx 2\xi \max(u). \tag{6.18}$$

We now consider how the resulting continuous mesh behaves for solutions in the two asymptotic regimes described by (2.16) and (2.17). Consider first (2.16), in which we will assume that t is sufficiently small so that this self-similar description is appropriate. In this case we have

$$\int_0^X \frac{A}{\sqrt{t}} \mathrm{e}^{-(x-x_0)^2/4t}\,\mathrm{d}x = \xi \int_0^1 \frac{A}{\sqrt{t}} \mathrm{e}^{-(x-x_0)^2/4t}\,\mathrm{d}x$$

with u having a single maximum of $A/\sqrt{t} \gg 1$ at $x = x_0$. Now, as t is small, $|u_x|$ is large, apart from the immediate neighbourhood of the point x_0 and

$$\int_0^X \sqrt{1+u_x^2}\,\mathrm{d}x \approx \int_0^X |u_x|\,\mathrm{d}x \approx u(X) \quad \text{if } X < x_0 \quad \text{or} \quad 2\max(u) - u(x) \quad \text{if } X > x_0.$$

It follows from (6.18) that if we are close to the peak of the solution then

$$\frac{A}{\sqrt{t}} \mathrm{e}^{(X-x_0)^2/4t} = \frac{2A}{\sqrt{t}}\xi \quad \text{if } X < x_0$$

and similarly

$$\frac{A}{\sqrt{t}} \mathrm{e}^{(X-x_0)^2/4t} = \frac{2A}{\sqrt{t}}(1-\xi) \quad \text{if } X > x_0.$$

The scaling invariance of the function $|u_x|$ has led to the same factor of $A/\sqrt{t}$ arising in both sides of this equation and allows us to simplify and solve it. The resulting mesh is then given by

$$X(\xi,t) \approx x_0 - 2\sqrt{t}\sqrt{\log(1/2\xi)}, \quad \xi < \tfrac{1}{2}, \tag{6.19}$$

$$X(\xi,t) \approx x_0 + 2\sqrt{t}\sqrt{\log(1/2(1-\xi))}, \quad \xi > \tfrac{1}{2}. \tag{6.20}$$

(We have ignored here the behaviour very close to x_0 where u_x vanishes and M is very locally approximated by 1.)

Observe that this mesh automatically clusters points around the maximum at $x = x_0$ of the solution of the heat equation and that these points evolve in a correct self-similar manner consistent with the underlying scaling. Indeed, $X(\xi,t)$ evolves as

$$X(\xi,t) \approx x_0 + \sqrt{t}Y(\xi)$$

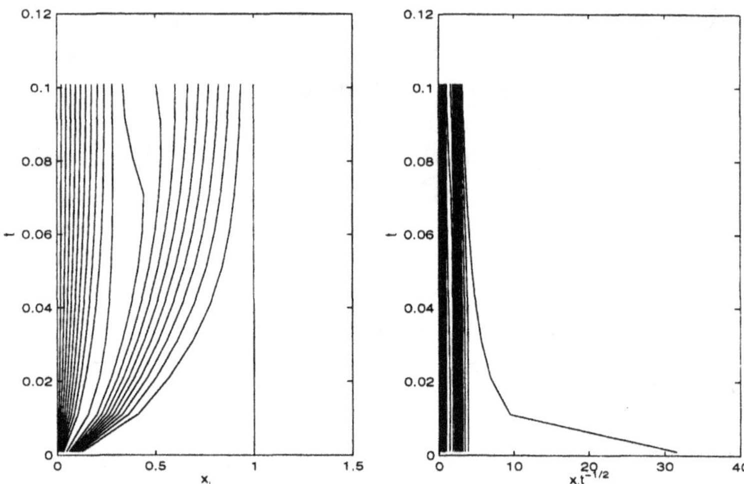

Fig. 4. The mesh and scaled mesh for the second self-similar solution to the linear heat equation.

for some function $Y(\xi)$, the form of which follows from (6.19) and (6.20). There will be a small departure from scaling invariance in the immediate neighbourhood of the peak, but this may well be within a single mesh interval and will thus not affect the computation.

Now consider the next scaling range in which the boundary at $x = 0$ is important, but for which $1/\sqrt{t}$ is still large. Now from (2.17) we have that up to a constant

$$u(x,t) = \frac{Bx}{(t-t_0)^{3/2}} e^{-x^2/4(t-t_0)}. \tag{6.21}$$

We can rewrite this as

$$u(x,t) = \frac{B}{(t-t_0)} z e^{-z^2/4}, \quad z = x/\sqrt{t-t_0},$$

taking a maximum of $B\sqrt{2}e^{-1/2}/(t-t_0)$ when $z = \sqrt{2}$. When $(t-t_0)$ is small we can analyse the motion of the mesh in a similar manner to before. In particular, if $X/\sqrt{t-t_0} < \sqrt{2}$ then we have from the integration of the approximate monitor function $|u_x|$

$$f(Z) \equiv Ze^{-Z^2/4} = 2\xi\sqrt{2}e^{-1/2} \quad \text{with} \quad Z = X/\sqrt{t-t_0}.$$

Thus

$$X(\xi,t) = \sqrt{t-t_0} f^{-1}(2\xi\sqrt{2}e^{-1/2})$$

with $\xi = \frac{1}{2}$ corresponding to $X = \sqrt{2(t-t_0)}$ and the function f^{-1} corresponding only to that part of the fuction $f(Z)$ on the interval $0 \leqslant Z \leqslant \sqrt{2}$. For $\xi > \frac{1}{2}$ we have similarly that

$$X(\xi,t) = \sqrt{t-t_0} f^{-1}(2(1-\xi)\sqrt{2}e^{-1/2}),$$

where now f^{-1} corresponds to that part of the fuction $f(Z)$ on the interval $\sqrt{2} < Z < \infty$.

We see that in both cases we have again recovered a mesh which accurately picks up the correct self-similar form of the solution, with $X(\xi,t)$ evolving like $\sqrt{t-t_0}Y(\xi)$.

Thus the use of the monitor-function-based approach for this example has allowed the mesh to evolve in the correct self-similar manner in the two cases, even though the self-similar solution is taking two rather different forms. This example demonstrates the flexibility of our approach.

We present, in Fig. 4, some calculations of the evolution of the mesh obtained by solving (6.17) in the regime for which we take u to be the self-similar solution (6.21). We see the mesh is initially scaling as $\sqrt{t}Y(\xi)$ close to $x = 0$ and eventualy becomes uniform as $u_x \to 0$ everywhere.

7. Conclusions

We have shown in this review that scale-invariant differential equations are common in many application areas and have interesting types of behaviour, with the form of the scaling underlying the solutions of the equations not always apparent at first inspection. Scale-invariant adaptive numerical methods have the virtue of preserving many of the properties of such equations which are derived from their scaling, including their asymptotic behaviour. These methods do not, however, compromise the generality required of a numerical computational technique. The next stage of the work on these ideas is to develop them for problems which have scaling invariance in several spatial dimensions. This is the subject of our future work.

References

[1] A. Arakawa, Computational design for long-term numerical integration of the equations of fluid motion: two-dimensional incompressible flow, J. Comput. Phys. 1 (1966) 119–143.
[2] G.I. Barenblatt, Scaling, self-similarity and intermediate asymptotics, Cambridge Univeristy Press, Cambridge, 1996.
[3] J. Bebernes, D. Eberly, Mathematical Problems from Combustion Theory, Applied Mathematical Sciences, Vol. 83, Springer, Berlin, 1989.
[4] P.W. Bridgman, Dimensional Analysis, Yale University Press, New York, 1931.
[5] C.J. Budd, S. Chen, R.D. Russell, New self-similar solutions of the nonlinear Schrödinger equation with moving mesh computations, J. Comput. Phys. 152 (1999) 756–789.
[6] C.J. Budd, G.J. Collins, Symmetry based numerical methods for partial differential equations, in: D. Griffiths, D. Higham, G. Watson (Eds.), Numerical Analysis 1997, Pitman Research Notes in Mathematics Series, Vol. 380, Longman, New York, 1998, pp. 16–36.
[7] C.J. Budd, G.J. Collins, W.-Z. Huang, R.D. Russell, Self-similar discrete solutions of the porous medium equation, Philos. Trans. Roy. Soc. London A 357 (1999) 1047–1078.
[8] C.J. Budd, W.-Z. Huang, R.D. Russell, Moving mesh methods for problems with blow-up, SIAM J. Sci. Computut. 17 (1996) 305–327.
[9] C.J. Budd, G.W. Hunt, M.A. Peletier, Self-similar fold evolution under prescribed end-shortening, J. Math. Geol. 31 (1999) 989–1005.
[10] C.J. Budd, A. Iserles (Eds.), Geometric integration: numerical solution of differential equations on manifolds, Philos. Trans. Roy. Soc. London A 357 (1999) 943–1133.
[11] C.J. Budd, B. Leimkuhler, M.D. Piggott, Scaling invariance and adaptivity, in: Iserles, Norsett (Eds.), Geometric Integration, 2000.
[12] C.J. Budd, M.D. Piggott, Symmetry invariance and the discrete maximum principle, in preparation.
[13] I. Budiansky, G. Carrier, The pointless wedge, SIAM J. Appl. Math. 25 (1973) 378–387.
[14] M.J.P. Cullen, J. Norbury, R.J. Purser, Generalised lagrangian solutions for atmospheric and oceanic flows, SIAM J. Appl. Math. 51 (1991) 20–31.
[15] V.A. Dorodnitsyn, Finite difference methods entirely inheriting the symmetry of the original equations, in: N. Ibragimov (Ed.), Modern Group Analysis: Advanced Analytical and Computational Methods in Mathematical Physics, 1993, pp. 191–201.

[16] V.A. Dorodnitsyn, Symmetry of finite-difference equations, in: N. Ibragimov (Ed.), CRC Handbook of Lie Group Analysis of Differential Equations, Vol. I: Symmetries, Exact Solutions and Conservation Laws, CRC Press, Boca Raton, FL, 1993, pp. 365–403.
[17] E.A. Dorfi, L. O'C. Drury, Simple adaptive grids for 1-D initial value problems, J. Comput. Phys. 69 (1987) 175–195.
[18] L. Dresner, Similarity solutions of nonlinear partial differential equations, Pitman Research Notes in Mathematics, Vol. 88, Longman, New York, 1983.
[19] K. Falconer, Fractal Geometry, Wiley, New York, 1990.
[20] N. Goldenfeld, O. Martin, Y. Oono, Intermediate asymptotics and renormalization group theory, J. Sci. Comput. 4 (1989) 355–372.
[21] M. Golubitsky, D. Schaeffer, I. Stewart, Singularities and Groups in Bifurcation Theory, Applied Mathematica Sciences, Vol. 69, Springer, Berlin, 1987.
[22] J. Hulshof, J.L. Vazquez, Self-similar solutions of the second kind for the modified porous medium equation, Eur. J. Appl. Math. 5 (1994) 391–403.
[23] W. Huang, Y. Ren, R.D. Russell, Moving mesh methods based on moving mesh partial differential equations, J. Comput. Phys. 112 (1994) 279–290.
[24] W. Huang, R.D. Russell, A high dimensional moving mesh strategy, Appl. Numer. Math. 26 (1999) 998–1015.
[25] A. Iserles, Numerical methods on (and off) manifolds, in: F. Cucker, M. Shub (Eds.), Foundations of Computational Mathematics, Springer, New York, 1997, pp. 180–189.
[26] S. Kamenomostskaya, The asymptotic behaviour of the solution of the filtration equation, Israel J. Math. 14 (1973) 279–290.
[27] A.N. Kolmogorov, The local structure of turbulence in incompressible fluids at very high Reynolds numbers, Dokl. USSR Acad. Sci. 30 (4) (1941) 299–303.
[28] B. Leimkuhler, Reversible adaptive regularisation: perturbed Kepler motion and classical atomic trajectories, Philos. Trans. Roy. Soc. 357 (1999) 1101–1133.
[29] E. Mansfield, The differential algebra package `diffgrob2`, MapleTech 3 (1996) 33–37.
[30] R. McLachlan, Symplectic integration of Hamiltonian wave equations, Numer. Math. 66 (1994) 465–492.
[31] J.D. Murray, Mathematical biology, Springer, Berlin, 1989.
[32] P. Olver, Applications of Lie Groups to Differential Equations, Springer, New York, 1986.
[33] V. Pereyra, E.G. Sewell, Mesh selection for discrete solution of boundary problems in ordinary differential equations, Numer. Math. 23 (1975) 261–268.
[34] L. Petzold, A description of DDASSL, a differential/algebraic system solver, Report SAND82-8637, Sandia National Laboratories, 1982.
[35] M.H. Protter, H.F. Weinberger, Maximum Principles in Differential Equations, Springer, New York, 1984.
[36] S. Reich, Backward error analysis for numerical integrators, SIAM J. Numer. Anal. 36 (1999) 1549–1570.
[37] S. Reich, Finite volume methods for multi-symplectic PDEs, BIT 40 (2000) 559–582.
[38] A. Samarskii, V. Galaktionov, S. Kurdyumov, A. Mikhailov, Blow-up in Quasilinear Parabolic Equations, de Gruyter, Berlin, 1995.
[39] C. Sulem, P.-L. Sulem, The Nonlinear Schrödinger Equation, Applied Mathematical Sciences, Vol. 139, Springer, Berlin, 1999.
[40] J.L. Vazquez, An introduction to the mathematical theory of the porous medium equation, in: M. Delfour (Ed.), Shape Optimisation and Free Boundaries, Kluwer Academic Publishers, Dordrecht, 1992, pp. 347–389.
[41] V.E. Zakharov, E.A. Kuznetsov, Hamiltonian formalism for systems of hydrodynamic type, Soviet Sci. Rev. Section C: Math. Phys. Rev 4 (1984) 167–220.
[42] Ya. B. Zeldovich, Yu. P. Razier, Physics of Shock Waves and High Temperature Hydrodynamic Phenomena, Academic Press, New York, 1967.

ELSEVIER

Journal of Computational and Applied Mathematics 128 (2001) 423–445

JOURNAL OF COMPUTATIONAL AND APPLIED MATHEMATICS

www.elsevier.nl/locate/cam

A summary of numerical methods for time-dependent advection-dominated partial differential equations

Richard E. Ewing[a],*, Hong Wang[b]

[a]*Institute for Scientific Computation, Texas A&M University, College Station, Texas 77843-3404, USA*

[b]*Department of Mathematics, University of South Carolina, Columbia, South Carolina 29208, USA*

Received 30 September 1999; received in revised form 8 February 2000

Abstract

We give a brief summary of numerical methods for time-dependent advection-dominated partial differential equations (PDEs), including first-order hyperbolic PDEs and nonstationary advection–diffusion PDEs. Mathematical models arising in porous medium fluid flow are presented to motivate these equations. It is understood that these PDEs also arise in many other important fields and that the numerical methods reviewed apply to general advection-dominated PDEs. We conduct a brief historical review of classical numerical methods, and a survey of the recent developments on the Eulerian and characteristic methods for time-dependent advection-dominated PDEs. The survey is not comprehensive due to the limitation of its length, and a large portion of the paper covers characteristic or Eulerian–Lagrangian methods. © 2001 Elsevier Science B.V. All rights reserved.

MSC: 65M; 65N; 76M; 76R

Keywords: Advection–diffusion equations; Characteristic methods; Eulerian methods; Numerical simulations

1. Mathematical models

We present mathematical models arising in subsurface porous medium fluid flow (e.g. subsurface contaminant transport, reservoir simulation) to motivate time-dependent advection-dominated PDEs. These types of PDEs also arise in many other important fields, such as the mathematical modeling of aerodynamics, fluid dynamics (e.g. Euler equations, Navier–Stokes equations) [70,93], meteorology [90], and semiconductor devices [72].

* Corresponding author.

E-mail address: richard-ewing@tamu.edu (R.E. Ewing).

PII: S0377-0427(00)00522-7

1.1. Miscible flows

A mathematical model used for describing fully saturated fluid flow processes through porous media is derived by using the mass balance equation for the fluid mixture [5,40]

$$\frac{\partial}{\partial t}(\phi\rho) - \nabla\cdot\left(\frac{\rho \boldsymbol{K}}{\mu}(\nabla p - \rho \boldsymbol{g})\right) = q, \quad \boldsymbol{x}\in\Omega,\ t\in[0,T]. \tag{1.1}$$

Here Ω is the physical domain, $\boldsymbol{u}$, p, and ρ are the Darcy velocity, the pressure, and the mass density of the fluid, $\boldsymbol{K}(\boldsymbol{x})$ is the absolute permeability of the medium, μ is the dynamic viscosity of the fluid, $\boldsymbol{g}$ is the acceleration vector due to gravity, and q represents the source and sink terms, which is often modeled via point or line sources and sinks.

The transport of a specific component in the fluid mixture is governed by the mass conservation for the component and is expressed as

$$\frac{\partial(\phi c)}{\partial t} + \nabla\cdot(\boldsymbol{u}c) - \nabla\cdot(\boldsymbol{D}(\boldsymbol{u})\nabla c) = \bar{c}q, \quad \boldsymbol{x}\in\Omega,\ t\in[0,T]. \tag{1.2}$$

Here c, a fraction between 0 and 1, represents the concentration of the component, ϕ is the porosity of the medium, $\bar{c}(\boldsymbol{x},t)$ is either the specified concentrations of the injected fluids at sources or the resident concentrations at sinks, and $\boldsymbol{D}(\boldsymbol{u})$ is the diffusion–dispersion tensor.

1.2. Multiphase flows

When either air or a nonaqueous-phase liquid (NAPL) contaminant is present in groundwater transport processes, this phase is immiscible with the water phase and the two phases flow simultaneously in the flow process. Likewise, in the immiscible displacement in petroleum production, the oil phase and the water phase are immiscible. In both cases, there is no mass transfer between the two phases and so the following equations hold for each phase [5,17,19,40]:

$$\frac{\partial}{\partial t}(\phi\rho_j S_j) - \nabla\cdot\left(\frac{\rho_j \boldsymbol{K} k_{rj}}{\mu_j}\nabla p_j\right) = \rho_j q_j, \quad \boldsymbol{x}\in\Omega,\ t\in[0,T]. \tag{1.3}$$

Here S_j, $\boldsymbol{u}_j$, ρ_j, p_j, k_{rj}, μ_j, and q_j are the saturation, velocity, density, pressure, relative permeability, viscosity, and source and sink terms for the phase j. The indices j=n and w stand for the nonwetting and wetting phases, respectively. The saturations S_{n} and S_{w} satisfy the relation $S_{\mathrm{n}} + S_{\mathrm{w}} = 1$.

Eqs. (1.3) may be rearranged in a form that resembles Eqs. (1.1) and (1.2) by letting $S_{\mathrm{n}} = 1 - S_{\mathrm{w}}$. The pressure between the two phases is described by the capillary pressure $p_{\mathrm{c}}(S_{\mathrm{w}}) = p_{\mathrm{n}} - p_{\mathrm{w}}$. The global pressure p and total velocity $\boldsymbol{u}$ of a two-phase flow model is given by the following equations [21]:

$$S_{\mathrm{n}}\mathrm{C}_{\mathrm{n}}\frac{Dp}{Dt} - \nabla\cdot(\boldsymbol{K}\lambda\nabla p) = q(\boldsymbol{x}, S_{\mathrm{w}}, p), \quad \boldsymbol{x}\in\Omega,\ t\in[0,T], \tag{1.4}$$

where $(D/Dt) = \phi(\partial/\partial t) + (\boldsymbol{u}_{\mathrm{n}}/S_{\mathrm{n}})\cdot\nabla$ and $p = \frac{1}{2}(p_{\mathrm{n}} + p_{\mathrm{w}}) + \frac{1}{2}\int_{S_{\mathrm{c}}}^{S}((\lambda_{\mathrm{n}} - \lambda_{\mathrm{w}})/\lambda)(\mathrm{d}p_{\mathrm{c}}/\mathrm{d}\xi)\,\mathrm{d}\xi$ with $p_{\mathrm{c}}(S_{\mathrm{c}}) = 0$. The total mobility $\lambda = \lambda_{\mathrm{n}} + \lambda_{\mathrm{w}}$, the phase mobility $\lambda_j = k_{rj}/\mu_j$, and the compressibility $\mathrm{C}_j = (1/\rho_j)(\mathrm{d}\rho_j/\mathrm{d}p_j)$ are functions of time and space.

The governing equation for the wetting phase now has a form

$$\phi\frac{\partial S_{\mathrm{w}}}{\partial t}+\nabla\cdot(f(S_{\mathrm{w}})\boldsymbol{u}-\boldsymbol{D}(S_{\mathrm{w}})\nabla S_{\mathrm{w}})=q_{\mathrm{w}},\quad \boldsymbol{x}\in\Omega,\ t\in[0,T], \tag{1.5}$$

where $\rho_{\mathrm{c}}=\rho_{\mathrm{n}}-\rho_{\mathrm{w}}$. The capillary diffusion $\boldsymbol{D}(S_{\mathrm{w}})=-\boldsymbol{K}\lambda_{\mathrm{n}}f_{\mathrm{w}}(\mathrm{d}p_{\mathrm{c}}/\mathrm{d}S_{\mathrm{w}})$ and the fractional flow functions $f_j=\lambda_j/\lambda$.

In practice, the diffusion term in Eq. (1.2) or (1.5) is often a small phenomenon relative to advection. Hence, these equations are time-dependent advection–diffusion partial differential equations (PDEs) in terms of the concentration c or the saturation S. In particular, Eq. (1.5) has an S-shaped nonlinear flux function f and a degenerate capillary diffusion term [19,40]. Sometimes the diffusion phenomenon is so small that its effect is neglected. In this case, Eq. (1.2) or (1.5) is reduced to a first-order hyperbolic PDE. Finally, initial and boundary conditions also need to be specified to close the system (1.1)–(1.2) or (1.4)–(1.5).

2. Conventional finite difference and finite element methods

We carry out a brief historical review of classical numerical methods in this section and a survey of the recent developments on the Eulerian and characteristic methods in the next section primarily for time-dependent advection-dominated PDEs, including first-order hyperbolic PDEs and nonstationary advection–diffusion PDEs. Because of the extensive research carried out in these areas, it is impossible to describe adequately all these developments in the space available. Hence, this review is not comprehensive in that we try to describe and review only some representatives of the huge amount of works in the literature. Notice that since relatively more references and survey papers can be found on the Eulerian methods for unsteady state advection-dominated PDEs, we intend to use a relatively large portion to cover characteristic or Eulerian–Lagrangian methods for advection-dominated PDEs. Finally, we refer interested readers to the works of Morton [73] and Roos et al. [86] for detailed descriptions on the recent developments for the numerical methods for stationary advection–diffusion PDEs.

It is well known that advection-dominated PDEs present serious numerical difficulties due to the moving steep fronts present in the solutions of advection–diffusion transport PDEs or shock discontinuities in the solutions of pure advection PDEs or advection–diffusion PDEs with degenerate diffusion. Additional difficulties include the strong couplings and nonlinearities of advection-dominated PDE systems, the effect of the singularities at point/line sources and sinks, the strong heterogeneity of the coefficients, anisotropic diffusion–dispersion in tensor form, and the enormous sizes of field-scale applications.

2.1. Finite difference methods (FDMs)

Due to their simplicity, FDMs were first used in solving advection-dominated PDEs. For convenience, of presentation, we consider the one-dimensional constant-coefficient analogue of Eq. (1.2)

$$\frac{\partial c}{\partial t}+V\frac{\partial c}{\partial x}-D\frac{\partial^2 c}{\partial x^2}=0,\quad x\in(a,b),\ t\in[0,T] \tag{2.1}$$

and assume a uniform spatial and temporal partition $x_i = a + i\Delta x$ for $i = 0, 1, \ldots, I$ with $\Delta x = (b-a)/I$ and $t^m = m\Delta t$ for $m = 0, 1, \ldots, M$ with $\Delta t = T/M$.

We define the Courant number $\mathrm{Cr} = V\Delta t/\Delta x$ and the Peclet number $\mathrm{Pe} = V\Delta x/D$. It is known that the solution to the space-centered explicit scheme

$$\frac{c_i^{m+1} - c_i^m}{\Delta t} + V\frac{c_{i+1}^m - c_{i-1}^m}{2\Delta x} - D\frac{c_{i+1}^m - 2c_i^m + c_{i-1}^m}{(\Delta x)^2} = 0 \tag{2.2}$$

does not oscillate only when the Peclet number $\mathrm{Pe} \leqslant 2$ and the CFL condition ($\mathrm{Cr} \leqslant 1$) is satisfied [30,81]. For $\mathrm{Pe} > 2$, damped oscillations occur with nonreal eigenvalues [48,81]. Furthermore, for the linear hyperbolic PDE

$$\frac{\partial c}{\partial t} + V\frac{\partial c}{\partial x} = 0, \quad x \in (a,b), \ t \in [0,T], \tag{2.3}$$

which can be viewed as a limiting case of $D \to 0$ in Eq. (2.1), the corresponding scheme to scheme (2.2)

$$\frac{c_i^{m+1} - c_i^m}{\Delta t} + V\frac{c_{i+1}^m - c_{i-1}^m}{2\Delta x} = 0 \tag{2.4}$$

is unconditionally unstable [48,94].

The upwind FDM (UFDM) uses a one-sided finite difference in the upstream direction to approximate the advection term in the transport PDE (2.1) and can be expressed as follows (assuming $V > 0$):

$$\frac{c_i^{m+1} - c_i^m}{\Delta t} + V\frac{c_i^m - c_{i-1}^m}{2\Delta x} - D\frac{c_{i+1}^m - 2c_i^m + c_{i-1}^m}{(\Delta x)^2} = 0. \tag{2.5}$$

The Lax–Friedrichs scheme

$$\frac{c_i^{m+1} - (c_{i+1}^m + c_{i-1}^m)/2}{\Delta t} + V\frac{c_{i+1}^m - c_{i-1}^m}{2\Delta x} - D\frac{c_{i+1}^m - 2c_i^m + c_{i-1}^m}{(\Delta x)^2} = 0 \tag{2.6}$$

is obtained by replacing c_i^m in the first term in Eq. (2.2) by its mean value $(c_{i+1}^m + c_{i-1}^m)/2$.

Remark 1. Schemes (2.5) and (2.6) eliminate the nonphysical oscillations present in Scheme (2.2), and generate stable solutions even for very complicated multiphase and multicomponent flows. It can be shown that the UFDM scheme is actually a second-order approximation to Eq. (2.1) with a modified diffusion $D(1 + (\mathrm{Pe}/2)(1 - \mathrm{Cr}))$, while the Lax–Friedrichs scheme is a second-order approximation to Eq. (2.1) with an extra numerical diffusion $((\Delta x)^2/2\Delta t)(1 - \mathrm{Cr}^2)$ [40,59,70]. Hence, these methods introduce excessive numerical diffusion and the numerical solutions are dependent upon grid orientation. Detailed description on the theory and the use of modified equations can be found in [59,70,110].

The Lax–Wendroff scheme is based on the Taylor series expansion and Eq. (2.3)

$$\begin{aligned} c(x,t^{m+1}) &= c(x,t^m) + \Delta t\frac{\partial c(x,t^m)}{\partial t} + \frac{(\Delta t)^2}{2}\frac{\partial^2 c(x,t^m)}{\partial t^2} + \mathrm{O}((\Delta t)^3) \\ &= c(x,t^m) - V\Delta t\frac{\partial c(x,t^m)}{\partial x} + \frac{(V\Delta t)^2}{2}\frac{\partial^2 c(x,t^m)}{\partial x^2} + \mathrm{O}((\Delta t)^3). \end{aligned} \tag{2.7}$$

Dropping the $O((\Delta t)^3)$ term in Eq. (2.7) and using centered differences to approximate the spatial derivatives yields the Lax–Wendroff scheme

$$c_i^{m+1} = c_i^m - \frac{\mathrm{Cr}}{2}(c_{i+1}^m - c_{i-1}^m) + \frac{\mathrm{Cr}^2}{2}(c_{i+1}^m - 2c_i^m + c_{i-1}^m), \tag{2.8}$$

which is a second-order scheme.

The Beam–Warming scheme is a one-sided version of the Lax–Wendroff scheme. It uses second-order accurate one-sided differences to approximate the spatial derivatives in Eq. (2.7)

$$c_i^{m+1} = c_i^m - \frac{\mathrm{Cr}}{2}(3c_i^m - 4c_{i-1}^m + c_{i-2}^m) + \frac{\mathrm{Cr}^2}{2}(c_i^m - 2c_{i-1}^m + c_{i-2}^m). \tag{2.9}$$

Remark 2. The Lax–Wendroff scheme and the Beam–Warming scheme give third-order approximations to the modified advection–dispersion equation

$$\frac{\partial c}{\partial t} + V\frac{\partial c}{\partial x} - \beta\frac{\partial^3 c}{\partial x^3} = 0, \quad x \in (a,b),\ t \in [0,T]$$

with $\beta = (V(\Delta x)^2/6)(\mathrm{Cr}^2 - 1)$ for (2.8) and $(V(\Delta x)^2/6)(2 - 3\mathrm{Cr} + \mathrm{Cr}^2)$ for (2.9). The theory of dispersive waves and its utility in the study of numerical methods are covered in [97,112], which show that the Lax–Wendroff scheme tends to develop oscillations behind shock fronts while the Beam–Warming scheme tends to develop oscillations in front of shock fronts.

Remark 3. Solving Eq. (2.3) yields $c(x,t^{m+1}) = c(x - V\Delta t, t^m)$. When the CFL condition is satisfied, the UPFD or the Lax–Friedrichs scheme can be viewed as an linear interpolation of $c(x - V\Delta t, t^m)$ by the nodal values $c(x_{i-1},t^m)$ and $c(x_i,t^m)$, or $c(x_{i-1},t^m)$ and $c(x_{i+1},t^m)$, respectively. This explains why these schemes are free of oscillations and introduce smearing from another point of view [70]. Second, because UFDM takes advantage of upstream information, it is slightly more accurate than the Lax–Friedrichs scheme. On the other hand, the latter is symmetric and can be easily implemented, which is an important feature for nonlinear hyperbolic conservation laws. In contrast, the Lax–Wendroff scheme (2.8) or Beam–Warming scheme (2.9) can be viewed as a quadratic interpolation of $c(x - V\Delta t, t^m)$ by the nodal values $c(x_{i-1},t^m)$, $c(x_i,t^m)$, and $c(x_{i+1},t^m)$, or $c(x_i,t^m)$, $c(x_{i+1},t^m)$, and $c(x_{i+2},t^m)$. This is why they introduce oscillations across shock discontinuities.

The leap-frog scheme for Eq. (2.3) is obtained by replacing the forward difference in time in (2.4) by a centered difference

$$\frac{c_i^{m+1} - c_i^{m-1}}{\Delta t} + V\frac{c_{i+1}^m - c_{i-1}^m}{2\Delta x} = 0. \tag{2.10}$$

Scheme (2.10) has an improved truncation error of $O((\Delta x)^2 + (\Delta t)^2)$, but it is a multi-level scheme. This leads to increased computational storage, a particular disadvantage for large multi-dimensional nonlinear systems.

These methods can be extended to solve nonlinear hyperbolic conservation laws

$$\frac{\partial c}{\partial t} + \frac{\partial f(c)}{\partial x} = 0, \quad x \in (a,b),\ t \in [0,T] \tag{2.11}$$

and their viscous analogue [48,70]. For example, a large class of upwind schemes have been developed, based on the Godunov scheme [50]; they have often been presented in terms of Riemann solvers. The Lax–Friedrichs scheme is a basis for the development of nonoscillatory central schemes (see e.g. [75]).

2.2. Galerkin and Petrov–Galerkin finite element methods (FEMs)

Many FEM schemes have been developed in parallel. For instance, the Galerkin and Petrov–Galerkin FEMs that are analogues to Scheme (2.2) and the UFDM (2.5) for Eq. (2.1) can be uniformly written as follows:

$$\begin{aligned}
&\int_a^b c(x,t^{m+1})w_i(x)\,\mathrm{d}x - \int_a^b c(x,t^m)w_i(x)\,\mathrm{d}x \\
&\quad + \gamma\Delta t\left[\int_a^b D\frac{\partial c(x,t^{m+1})}{\partial x}\frac{\partial w_i(x)}{\partial x}\,\mathrm{d}x + \int_a^b V\frac{\partial c(x,t^{m+1})}{\partial x}w_i(x)\,\mathrm{d}x\right] \\
&\quad = -(1-\gamma)\Delta t\left[\int_a^b D\frac{\partial c(x,t^m)}{\partial x}\frac{\partial w_i(x)}{\partial x}\,\mathrm{d}x + \int_a^b V\frac{\partial c(x,t^m)}{\partial x}w_i(x)\,\mathrm{d}x\right].
\end{aligned} \tag{2.12}$$

Here $c(x,t^{m+1})$ is a piecewise-linear trial function. In the linear Galerkin FEM, the test functions $w_i(x)$ are standard hat functions centered at the node x_i and correspond to the space-centered scheme (2.2) (see e.g. [40,48]).

In the quadratic Petrov–Galerkin FEM (QPG), the test functions are constructed by adding an asymmetric perturbation to the original piecewise-linear hat functions [4,15,22]

$$w_i(x) = \begin{cases} \dfrac{x - x_{i-1}}{\Delta x} + v\dfrac{(x-x_{i-1})(x_i - x)}{\Delta x^2}, & x \in [x_{i-1}, x_i], \\ \dfrac{x_{i+1} - x}{\Delta x} - v\dfrac{(x-x_i)(x_{i+1} - x)}{\Delta x^2}, & x \in [x_i, x_{i+1}], \\ 0 & \text{otherwise.} \end{cases}$$

With a choice of v=3, the QPG reproduces the UFDM. With an optimal choice of $v=3[\coth(\mathrm{Pe}/2)-2/\mathrm{Pe}]$, the QPG is reduced to the optimal FDM of Allen and Southwell [1]. For a stationary analogue of Eq. (2.1), the QPG method yields solutions that coincide with the exact solution at the nodal points, and minimizes the errors in approximating spatial derivatives [4,15]. However, the QPG is susceptible to strong time truncation errors that introduce numerical diffusion and the restrictions on the size of the Courant number, and hence tends to be ineffective for transient advection-dominated PDEs.

In the cubic Petrov–Galerkin FEM (CPG), the test functions are defined as the original piecewise-linear hat functions with a symmetric cubic perturbation added to each nonzero piece [10,111]

$$w_i(x) = \begin{cases} \dfrac{x - x_{i-1}}{\Delta x} + v\dfrac{(x - x_{i-1})(x_i - x)(x_{i-1} + x_i - 2x)}{\Delta x^3}, & x \in [x_{i-1}, x_i], \\ \dfrac{x_{i+1} - x}{\Delta x} - v\dfrac{(x - x_i)(x_{i+1} - x)(x_i + x_{i+1} - 2x)}{\Delta x^3}, & x \in [x_i, x_{i+1}], \\ 0 & \text{otherwise.} \end{cases}$$

Here $v = 5\mathrm{Cr}^2$. The CPG intends to use nonzero spatial error to cancel the temporal error to improve the overall accuracy. In these treatments the effects on mass balance come from spatial dependence of test functions in the first terms on both the sides of Eq. (2.12). Detailed descriptions of the FDMs and FEMs that have been used in the petroleum industry can be found in [40,89].

Corresponding to the Lax–Wendroff scheme (2.8) and the leap-frog scheme (2.10) is the Taylor–Galerkin scheme

$$\frac{c_{i+1}^{m+1} - c_{i+1}^{m-1}}{6} + \frac{2(c_i^{m+1} - c_i^{m-1})}{3} + \frac{c_{i-1}^{m+1} - c_{i-1}^{m-1}}{6}$$
$$= -\frac{\mathrm{Cr}}{2}(c_{i+1}^m - c_{i-1}^m) + \frac{\mathrm{Cr}^2}{2}(c_{i+1}^m - 2c_i^m + c_{i-1}^m),$$

and the leap-frog Galerkin scheme

$$\frac{c_{i+1}^{m+1} - c_{i+1}^{m-1}}{6\Delta t} + \frac{2(c_i^{m+1} - c_i^{m-1})}{3\Delta t} + \frac{c_{i-1}^{m+1} - c_{i-1}^{m-1}}{6\Delta t} + V\frac{c_{i+1}^m - c_{i-1}^m}{2\Delta x} = 0.$$

In addition, a wide variety of other methods can be devised for advection-dominated transport PDEs by using different FDM and FEM approximations, or Taylor expansions. Many large-scale simulators use fully implicit discretization so that large time steps can be allowed. However, in implicit methods, the temporal error and the spatial error add together. Hence, increasing the size of time steps can significantly reduce the accuracy of the solutions [40]. This is also observed computationally [106]. In contrast, in explicit schemes the temporal error and the spatial error cancel each other. Hence, reducing the time step size further with fixed spatial step size will actually reduce the accuracy of the numerical solutions. The sizes of spatial grids and temporal steps have to be reduced simultaneously to improve the accuracy of the solutions, leading to significantly increased overall computational and storage cost [106].

3. Recent developments for advection–diffusion PDEs

Recent developments in effectively solving advection–diffusion PDEs have generally been along one of two approaches: Eulerian or characteristic Lagrangian methods. Eulerian methods use the standard temporal discretization, while the main distinguishing feature of characteristic methods is the use of characteristics to carry out the discretization in time.

3.1. Eulerian methods for advection–diffusion PDEs

Many methods directly apply to a nonconservative analogue of Eq. (1.2)

$$\phi\frac{\partial c}{\partial t} + \boldsymbol{u}\cdot\nabla c - \nabla\cdot(\boldsymbol{D}\nabla c) = \bar{c}q, \quad \boldsymbol{x}\in\Omega,\ t\in[0,T]. \tag{3.1}$$

3.1.1. The streamline diffusion finite element method (SDFEM)

The SDFEM directly applies to Eq. (3.1). It is based on the framework of space–time FEMs on the space–time strip $\bar{\Omega}\times[t^m,t^{m+1}]$, and uses continuous and piecewise polynomial trial and test functions in space as standard FEM but a discontinuous Galerkin approximation in time at time level t^m and t^{m+1} such that

$$\begin{aligned}&\int_{t^m}^{t^{m+1}}\int_\Omega\left[\phi\frac{\partial c}{\partial t} + \boldsymbol{u}\cdot\nabla c - \nabla\cdot(\boldsymbol{D}\nabla c)\right]\left[w + \delta\left(\phi\frac{\partial w}{\partial t} + \boldsymbol{u}\cdot\nabla w\right)\right]\mathrm{d}\boldsymbol{x}\,\mathrm{d}t\\ &\quad + \int_{t^m}^{t^{m+1}}\int_\Omega \nabla w\cdot(\boldsymbol{D}\nabla c)\,\mathrm{d}\boldsymbol{x}\,\mathrm{d}t + \int_\Omega c(\boldsymbol{x},t^m_+)w(\boldsymbol{x},t^m_+)\,\mathrm{d}\boldsymbol{x}\\ &\quad = \int_{t^m}^{t^{m+1}}\int_\Omega \bar{c}q\left[w + \delta\left(\phi\frac{\partial w}{\partial t} + \boldsymbol{u}\cdot\nabla w\right)\right]\mathrm{d}\boldsymbol{x}\,\mathrm{d}t + \int_\Omega c(\boldsymbol{x},t^m_-)w(\boldsymbol{x},t^m_+)\,\mathrm{d}\boldsymbol{x}.\end{aligned}$$

Here $w(\boldsymbol{x},t^m_+)=\lim_{t\to t^m,t>t^m} w(\boldsymbol{x},t)$ and $w(\boldsymbol{x},t^m_-)=\lim_{t\to t^m,t<t^m} w(\boldsymbol{x},t)$. At the initial time step, $c(\boldsymbol{x},t^0_-)=c_0(\boldsymbol{x})$ is the prescribed initial condition. The second term on the left-hand side is carried out elementwise, since it is not well defined for continuous and piecewise polynomials. The parameter δ, which determines the amount of numerical diffusion introduced, is typically chosen to be of order $O(\sqrt{(\Delta x)^2+(\Delta t)^2})$.

The SDFEM was first proposed by Hughes and Brooks [62]. Since then, various SDFEM schemes have been developed and studied extensively by Brooks and Hughes and Hughes [11,61] and Johnson et al. [52,65]. The SDFEM adds a numerical diffusion only in the direction of streamlines to suppress the oscillation and does not introduce any crosswind diffusion. However, the undetermined parameter δ in the SDFEM scheme needs to be chosen very carefully in order to obtain accurate numerical results. An optimal choice of the parameter is heavily problem-dependent. We refer readers to the work of Shih and Elman on the study of the choice δ in the SDFEM formulation and the related numerical experiments [91].

While the SDFEM can capture a jump discontinuity of the exact solution in a thin region, the numerical solution may develop over- and under-shoots about the exact solution within this layer. A modified SDFEM with improved shock-capturing properties was proposed [63,66], which consists of adding a "shock-capturing" term to the diffusion by introducing a "crosswind" control that is close to the steep fronts or "shocks". This modified SDFEM performs much better in terms of catching the steep fronts or the jump discontinuities of the exact solutions. However, the modified SDFEM is a nonlinear scheme and involves another undetermined parameter.

3.1.2. Total variation diminishing (TVD) methods

Notice that when oscillations arise, the numerical solutions will have larger total variation. TVD methods are designed to yield well-resolved, nonoscillatory shock discontinuities by enforcing that the numerical schemes generate solutions with nonincreasing total variations. One approach is to take a high-order method and add an additional numerical diffusion term to it. Since this numerical diffusion is needed only near discontinuities, one wants it to vanish sufficiently quickly so that the order of accuracy of the method on smooth regions of the solutions is retained. Hence, the numerical diffusion should depend on the behavior of the solutions, being larger near shock regions than in smooth regions. This leads to a nonlinear method even for the linear advection equation (2.3). The idea of adding a variable amount of numerical diffusion dates back to some of the earliest work on the numerical solution of fluid dynamics [31,68,100]. The difficulty with this approach is that it is hard to determine an appropriate amount of numerical diffusion that introduces just enough dissipation without causing excessive smearing.

For this reason, the high-resolution methods developed more recently are based on fairly different approaches, including flux- and slope-limiter approaches that impose the nonoscillatory requirement more directly. In the flux-limiter approach, one first chooses a high-order numerical flux $F_{\mathrm{H}}(c,i) = F_{\mathrm{H}}(c_{i-l_{\mathrm{H}}}, c_{i-l+1}, \ldots, c_{i+r_{\mathrm{H}}})$ that generates accurate approximations in smooth regions and a low-order numerical flux $F_{\mathrm{L}}(c,i) = F_{\mathrm{L}}(c_{i-l_{\mathrm{L}}}, c_{i-l+1}, \ldots, c_{i+r_{\mathrm{L}}})$ that yields nonoscillatory solutions near shock discontinuities. One then combines F_{H} and F_{L} into a single numerical flux F, e.g. in the form of

$$F(c,i) = F_{\mathrm{L}}(c,i) + \Psi(c,i)(F_{\mathrm{H}}(c,i) - F_{\mathrm{L}}(c,i)), \tag{3.2}$$

such that F reduces to F_{H} in smooth regions and to F_{L} in shock regions. Here $\Psi(c,i)$, the flux limiter, should be near one in smooth regions and close to zero near shock discontinuities.

The flux-corrected transport (FCT) method of Boris and Book can be viewed as one of the earliest flux limiter methods [8,9,114]. In the FCT method, an anti-diffusive term (i.e., the correction term in (3.2)) is added to reduce the excessive numerical diffusion introduced by the lower-order flux F_{L} as much as possible without increasing the total variation of the solution.

Sweby studied a family of flux-limiter methods in [95]. By choosing

$$F_{\mathrm{L}}(c^m,i) = Vc_i^m \quad \text{and} \quad F_{\mathrm{H}}(c^m,i) = Vc_i^m + \tfrac{1}{2}V(1-\mathrm{Cr})(c_{i+1}^m - c_i^m),$$

to be the first-order upwind flux in (2.5) and the Lax–Wendroff flux in (2.8) and using (3.2), a family of flux-limiter methods can be defined

$$c_i^{m+1} = c_i^m - \frac{\Delta t}{\Delta x}[F(c^m,i) - F(c^m,i-1)] \tag{3.3}$$

with the flux $F(c^m,i)$ being given by

$$F(c^m,i) = Vc_i^m + \frac{\Psi(c^m,i)}{2}V(1-\mathrm{Cr})(c_{i+1}^m - c_i^m).$$

One way to measure the smoothness of the solution is to look at the ratio of consecutive gradients and to define the flux limiter Ψ accordingly

$$\Psi(c^m,i) = \psi(\theta_i) \quad \text{with } \theta_i = \frac{c_i^m - c_{i-1}^m}{c_{i+1}^m - c_i^m}.$$

Sweby obtained algebraic conditions on the limiter functions that guarantee second-order accuracy and the TVD property of the derived methods [95]. Harten proved a sufficient condition on ψ that can be used to impose constraints on ψ [53,55]. Among the different choices of limiters are the "superbee" limiter of Roe [85]

$$\psi(\theta) = \max\{0, \min\{1, 2\theta\}, \min\{\theta, 2\}\}$$

and a smoother limiter by van Leer [98]

$$\psi(\theta) = \frac{|\theta| + \theta}{1 + |\theta|}.$$

The extension of flux limiter methods to nonlinear conservation laws and numerical comparisons can be found in [29,95,115].

Another approach is to use slope limiters. These intend to replace the piecewise-constant representation of the solutions in Godunov's method by more accurate representations, and can be expressed in the following steps for Eq. (2.11):

(i) Given the piecewise-constant cell-average representation $\{\bar{c}_i^m\}_{i=-\infty}^{i=+\infty}$ of the solution at time level t^m, define a (e.g., piecewise-linear) reconstruction at time t^m by

$$\hat{c}(x, t^m) = \bar{c}_i^m + \sigma_i^m (x - x_i) \tag{3.4}$$

for x on the cell $[x_{i-1/2}, x_{i+1/2}]$. Here σ_i^m is a slope on the ith cell that is based on the data $\{\bar{c}_i^m\}$.

(ii) Solve Eq. (2.11) with the data $\hat{c}(x, t^m)$ at time t^m to obtain the solution $c(x, t^{m+1})$ at time t^{m+1}.

(iii) Compute the cell average $\{\bar{c}_i^{m+1}\}_{i=-\infty}^{i=+\infty}$ of the solution $c(x, t^{m+1})$ at time t^{m+1}.

Note that the cell average of reconstruction (3.4) is equal to $\bar{c}_i^m$ on the cell $[x_{i-1/2}, x_{i+1/2}]$ for any choice of σ_i^m. Since Steps 2 and 3 are also conservative, the methods with slope limiters are conservative. Secondly, the choice of $\sigma_i^m = 0$ in Eq. (3.4) recovers Godunov's method. It is well known that Godunov's method generates solutions with excessive numerical diffusion. More accurate reconstructions, such as Eq. (3.4), could be used to reduce the numerical diffusion and to improve the accuracy of the numerical solutions.

In the context of the linear advection PDE (2.3), the solution of Step (ii) is simply $c(x, t^{m+1}) = \hat{c}(x - V\Delta t, t^m)$. Computing the cell average of $c(x, t^{m+1})$ in Step (iii) leads to the following expression:

$$\bar{c}_i^{m+1} = \bar{c}_i^m - \mathrm{Cr}(\bar{c}_i^m - \bar{c}_{i-1}^m) - \frac{\Delta x}{2}\mathrm{Cr}(1 - \mathrm{Cr})(\sigma_i^m - \sigma_{i-1}^m). \tag{3.5}$$

A natural choice of $\sigma_i^m = (\bar{c}_{i+1}^m - \bar{c}_i^m)/\Delta x$ in Eq. (3.5) leads to the Lax–Wendroff method. Thus, it is possible to obtain second-order accuracy by this approach. Secondly, the slope-limiter methods could generate oscillatory solutions (since the Lax–Wendroff method could do so), if the slope limiters σ_i^m are not chosen properly. Geometrically, the oscillations are due to a poor choice of slopes, which leads to a piecewise-linear reconstruction $\hat{c}(x, t^m)$ with much larger total variation than the given data $\{\bar{c}_i^m\}_{i=-\infty}^{i=+\infty}$ [70].

Hence, because of their importance, extensive research has been conducted on how to choose the slope σ_i^m in Eq. (3.4) to ensure the resulting methods to be total variation diminishing (TVD). These methods include the monotonic upstream-centered scheme for conservation laws (MUSCL) of van Leer and Minmod methods among others [28,51,98,99]. The simplest choice of the slope is probably the minmod slope defined by

$$\sigma_i^m = \mathrm{minmod}\left\{\frac{\bar{c}_{i+1}^m - \bar{c}_i^m}{\Delta x}, \frac{\bar{c}_i^m - \bar{c}_{i-1}^m}{\Delta x}\right\}$$

with $\mathrm{minmod}(a, b) = \frac{1}{2}(\mathrm{sgn}(a) + \mathrm{sgn}(b))\min(|a|, |b|)$.

In concluding this part, we notice the connection between the flux- and slope-limiter methods. Using formulation (3.3), we see that the numerical flux for the slope-limiter method (3.5) is

$$F(c^m, i) = Vc_i^m + \frac{\Delta x}{2} V(1 - \mathrm{Cr})\sigma_i^m,$$

which is of the same form as the flux-limiter method (3.3) if the slope-limiter σ_i^m is related to the flux-limiter $\Psi(c^m, i)$ by $\sigma_i^m = [(c_{i+1}^m - c_i^m)/\Delta x]\Psi(c^m, i)$.

3.1.3. Essentially nonoscillatory (ENO) schemes and weighted essentially nonoscillatory (WENO) schemes

Traditional finite difference methods are based on fixed stencil interpolations of discrete data using polynomials. The resulting scheme is linear for linear PDEs. However, fixed stencil interpolation of second- or higher-order accuracy is necessarily oscillatory across a discontinuity; this is why the Lax–Wendroff scheme (2.8) and the Beam–Warming scheme (2.9) introduce oscillations across shock discontinuities (see Remark 3). One common approach to eliminate or reduce spurious oscillations near discontinuities is to add a numerical diffusion as in the SDFEM presented earlier. The numerical diffusion should be tuned so that it is large enough near discontinuities but is small enough elsewhere to maintain high-order accuracy. One disadvantage of this approach is that it is hard to determine an appropriate amount of numerical diffusion that introduces just enough dissipation without causing excessive smearing. Another approach is to apply (flux or slope) limiters to eliminate the oscillations. By carefully designing such limiters (e.g., reducing the slope of a linear interpolant or using a linear rather than a quadratic interpolant near shock discontinuities), the TVD property could be achieved for some numerical schemes for nonlinear scalar conservation laws in one space dimension. Unfortunately, Osher and Chakravarthy proved that TVD methods must degenerate to first-order accuracy at local maximum or minimum points [78].

The ENO and WENO schemes are high-order accurate finite difference/volume schemes designed for nonlinear hyperbolic conservation laws with piecewise smooth solutions containing discontinuities [54,56,64,71]. By delicately defining a nonlinear adaptive procedure to automatically choose the locally smooth stencil, the ENO and WENO schemes avoid crossing discontinuities in the interpolation procedure and thus generate uniformly high-order accurate, yet essentially nonoscillatory solutions. These schemes have been quite successful in applications, especially for problems containing both shock discontinuities and complicated smooth solution structures [92].

3.1.4. The discontinuous Galerkin (DG) method

The original discontinuous Galerkin finite element method was introduced by Reed and Hill for solving a linear neutron transport equation [84], in which the method can be carried out element by element when the elements are suitably ordered according to the characteristic directions. Lesaint and Raviart [69] carried out the first analysis for this method and proved a convergence rate of $(\Delta x)^k$ for general triangular partitions and $(\Delta x)^{k+1}$ for Cartesian grids. Johnson and Pitkäranta [65] obtained an improved estimate of $(\Delta x)^{k+1/2}$ for general triangulations, which is confirmed to be optimal by Peterson [79]. Chavent and Salzano [20] constructed an explicit DG method for Eq. (2.11), in which piecewise linear FEM is used in space and an explicit Euler approximation is used in time. Unfortunately, the scheme is stable only if the Courant number Cr $=\mathrm{O}(\sqrt{\Delta x})$. Chavent and Cockburn [18] modified the scheme by introducing a slope limiter, and proved the scheme to be total variation bounded (TVB) when Cr $\leqslant \frac{1}{2}$. However, the slope limiter introduced compromises the accuracy of the approximation in smooth regions. Cockburn and Shu [26] introduced the first Runge–Kutta DG (RKDG) method, which uses an explicit TVD second-order Runge–Kutta discretization and modifies the slope limiter to maintain the formal accuracy of the scheme at the extrema. The same authors then extended this approach to construct higher-order RKDG methods [25], to multidimensional scalar conservation laws [24,27], and to multidimensional systems. We refer interested readers to the survey article [23] in this volume for detailed discussions on the DG methods.

3.2. Characteristic methods

Because of the hyperbolic nature of advective transport, characteristic methods have been investigated extensively for the solution of advection–diffusion PDEs. In a characteristic (or Lagrangian) method, the transport of the fluid is referred to a Lagrangian coordinate system that moves with the fluid velocity. One tracks the movement of a fluid particle and the coordinate system follows the movement of the fluid. The time derivative along the characteristics of the advection–diffusion PDE (3.1) is expressed as

$$\frac{Dc}{Dt} = \frac{\partial c}{\partial t} + \frac{\boldsymbol{u}}{\phi} \cdot \nabla c. \tag{3.6}$$

Consequently, the advection–diffusion PDE (3.1) is rewritten as the following parabolic diffusion–reaction PDE in a Lagrangian system:

$$\phi\frac{Dc}{Dt} - \nabla \cdot (\boldsymbol{D}\nabla c) = \bar{c}q \tag{3.7}$$

and the advection has seemingly disappeared. In other words, in a Lagrangian coordinate system (that moves with the flow) one would only see the effect of the diffusion, reaction, and the the right-hand side source terms but not the effect of the advection or moving steep fronts. Hence, the solutions of the advection–diffusion PDEs are much smoother along the characteristics than they are in the time direction. This explains why characteristic methods usually allow large time steps to be used in a numerical simulation while still maintaining its stability and accuracy. Unfortunately, Eq. (3.7) is written in a Lagrangian coordinate system, which is constantly moving in time. Consequently, the corresponding characteristic or Lagrangian methods often raise extra and nontrivial analytical, numerical, and implementational difficulties, which require very careful treatment. In contrast, Eq. (1.2) or (3.1) is written in an Eulerian system which is fixed in space. Hence, Eulerian methods are relatively easy to formulate and to implement.

3.2.1. Classical characteristic or Eulerian–Lagrangian methods

The classical Eulerian–Lagrangian method is a finite difference method based on the forward tracking of particles in cells. In this method, the spatial domain is divided into a collection of elements or cells and a number of particles are placed within each cell. Then the governing PDE is used to determine the movement of the particles from cell to cell. In this algorithm, the solution is determined by the number of particles within a cell at any given time. Related works can be found in [47,49,96]. In these methods, the diffusion occurs at the time step t^m and the solution is advected forward in time to the time step t^{m+1}, leading to the following scheme for Eq. (2.1):

$$\frac{\tilde{c}_i^{m+1} - c_i^m}{\Delta t} - D\frac{c_{i+1}^m - 2c_i^m + c_{i-1}^m}{(\Delta x)^2} = 0.$$

Here $\tilde{c}_i^{m+1} = c(\tilde{x}_i, t^{m+1})$ with $\tilde{x}_i = x_i + V\Delta t$. Because the advected nodes $\tilde{x}_i$ need not be nodes at time t^{m+1}, they are irregular, in general.

Neuman developed an Eulerian–Lagrangian finite element method using a combination of forward and backward tracking algorithms [76,77]. Near a steep front, a forward tracking algorithm is used to move a cloud of particles from time t^m to new positions at time t^{m+1} according to the advection, as done by Garder et al. [49]. An implicit scheme is then used to treat the diffusion at time t^{m+1}. Away from a front, a backward tracking algorithm is used, in which one finds a point that ends up at position x at time t^{m+1}.

Eulerian methods carry out the temporal discretization in the time direction, so they cannot accurately simulate all of the wave interactions that take place if the information propagates more than one cell per time step (i.e., if the CFL condition is violated), either for the reason of stability (for explicit methods) or for the reason of accuracy (for implicit methods). By using characteristic tracking, characteristic methods follow the movement of information or particles as well as their interactions. However, forward tracked characteristic methods often distort the evolving grids severely and greatly complicate the solution procedures, especially for multi-dimensional problems.

3.2.2. The modified method of characteristics (MMOC)

In this part we briefly review the MMOC, which was proposed by Douglas and Russell for solving advection–diffusion PDEs in a nonconservative form [37] and can be viewed as a representative of the Eulerian–Lagrangian methods developed during the same time period [6,80,82]. Using the Lagrangian form (3.7), we can combine the first two terms on the left-hand side of (3.1) to form one term through characteristic tracking (3.6) (see, e.g. [37])

$$\phi\frac{Dc(\boldsymbol{x}, t^{m+1})}{Dt} \approx \phi(\boldsymbol{x})\frac{c(\boldsymbol{x}, t^{m+1}) - c(\boldsymbol{x}^*, t^m)}{\Delta t} \tag{3.8}$$

with $\boldsymbol{x}^* = \boldsymbol{x} - \boldsymbol{u}(\boldsymbol{x}, t^{m+1})\Delta t/\phi(\boldsymbol{x})$.

Substituting (3.8) for the first two terms on the left-hand side of Eq. (3.1) and integrating the resulting equation against any finite element test functions $w(\boldsymbol{x})$, one obtains the following MMOC scheme [37,43] for Eq. (3.1):

$$\int_\Omega \phi(\boldsymbol{x})\frac{c(\boldsymbol{x}, t^{m+1}) - c(\boldsymbol{x}^*, t^m)}{\Delta t}w(\boldsymbol{x})\,\mathrm{d}\boldsymbol{x} + \int_\Omega \nabla w(\boldsymbol{x}) \cdot \boldsymbol{D}\nabla c(\boldsymbol{x}, t^{m+1})\,\mathrm{d}\boldsymbol{x}$$
$$= \int_\Omega \bar{c}q(\boldsymbol{x}, t^{m+1})w(\boldsymbol{x})\,\mathrm{d}\boldsymbol{x}. \tag{3.9}$$

Eq. (3.9) follows the flow by tracking the characteristics backward from a point $\boldsymbol{x}$ in a fixed grid at the time step t^{m+1} to a point $\boldsymbol{x}^*$ at time t^m. Hence, the MMOC avoids the grid distortion problems present in forward tracking methods. Moreover, MMOC symmetrizes and stabilizes the transport

PDEs, greatly reducing temporal errors; therefore MMOC allows for large time steps in a simulation without loss of accuracy and eliminates the excessive numerical dispersion and grid orientation effects present in many Eulerian methods [36,40,89]. However, the MMOC and the characteristic methods presented earlier have the following drawbacks:

Remark 4. In the context of the MMOC and other characteristic methods using a backtracking algorithm, the $\int_\Omega \phi(\boldsymbol{x})c(\boldsymbol{x}^*,t^m)w(\boldsymbol{x})\,\mathrm{d}\boldsymbol{x}$ term in Eq. (3.9) is defined on the domain at time t^{m+1}. In this term, the test functions $w(\boldsymbol{x})$ are standard FEM basis functions on Ω at time t^{m+1}, but the value of $c(\boldsymbol{x}^*,t^m)$ has to be evaluated by a backtracking method where $\boldsymbol{x}^* = \boldsymbol{r}(t^m;\boldsymbol{x},t^{m+1})$ is the point at the foot corresponding to $\boldsymbol{x}$ at the head [37,43]. For multidimensional problems, the evaluation of this term with a backtracking algorithm requires significant effort, due to the need to define the geometry at time t^m that requires mapping of points along the boundary of the element and subsequent interpolation and mapping onto the fixed spatial grid at the previous time t^m [7,74]. This procedure introduces a mass balance error and leads to schemes that fail to conserve mass [15,74,107]. Moreover, in these methods it is not clear how to treat flux boundary conditions in a mass-conservative manner without compromising the accuracy, when the characteristics track to the boundary of the domain [15,88,104,106,107].

3.2.3. The modified method of characteristics with adjusted advection (MMOCAA)

Recently, Douglas et al. proposed an MMOCAA scheme to correct the mass error of the MMOC by perturbing the foot of the characteristics slightly [34,35]. For Eq. (3.1) with a no-flow or periodic boundary condition, the summation of Eq. (3.9) for all the test functions (that add exactly to one) yields the following equation:

$$\int_\Omega \phi(\boldsymbol{x})c(\boldsymbol{x},t^{m+1})\,\mathrm{d}\boldsymbol{x} - \int_\Omega \phi(\boldsymbol{x})c(\boldsymbol{x}^*,t^m)\,\mathrm{d}\boldsymbol{x} = \Delta t\int_\Omega \bar{c}q(\boldsymbol{x},t^{m+1})\,\mathrm{d}\boldsymbol{x}.$$

Recall that the term on the right-hand side of this equation is obtained by an Euler approximation to the temporal integral in this term. On the other hand, integrating the original PDE (1.2) on the domain $\Omega\times[t^m,t^{m+1}]$ yields the following equation:

$$\int_\Omega \phi(\boldsymbol{x})c(\boldsymbol{x},t^{m+1})\,\mathrm{d}\boldsymbol{x} - \int_\Omega \phi(\boldsymbol{x})c(\boldsymbol{x},t^m)\,\mathrm{d}\boldsymbol{x} = \int_{t^m}^{t^{m+1}}\int_\Omega \bar{c}q\,\mathrm{d}\boldsymbol{x}\,\mathrm{d}t.$$

Therefore, to maintain mass balance, we must have

$$\int_\Omega \phi(\boldsymbol{x})c(\boldsymbol{x},t^m)\,\mathrm{d}\boldsymbol{x} \equiv Q^m = Q^m_* \equiv \int_\Omega \phi(\boldsymbol{x})c(\boldsymbol{x}^*,t^m)\,\mathrm{d}\boldsymbol{x}.$$

For some fixed constant $\kappa > 0$, we define

$$\boldsymbol{x}^*_+ = \boldsymbol{x} - \frac{\boldsymbol{u}(\boldsymbol{x},t^{m+1})}{\phi(\boldsymbol{x})}\Delta t + \kappa\frac{\boldsymbol{u}(\boldsymbol{x},t^{m+1})}{\phi(\boldsymbol{x})}(\Delta t)^2,$$

$$\boldsymbol{x}^*_- = \boldsymbol{x} - \frac{\boldsymbol{u}(\boldsymbol{x},t^{m+1})}{\phi(\boldsymbol{x})}\Delta t - \kappa\frac{\boldsymbol{u}(\boldsymbol{x},t^{m+1})}{\phi(\boldsymbol{x})}(\Delta t)^2.$$

We also define

$$c_{\#}(\boldsymbol{x}^*,t^m)=\begin{cases}\max\{c(\boldsymbol{x}^*_+,t^m),c(\boldsymbol{x}^*_-,t^m)\}, & \text{if } Q^m_* \leqslant Q^m,\\ \min\{c(\boldsymbol{x}^*_+,t^m),c(\boldsymbol{x}^*_-,t^m)\}, & \text{if } Q^m_* > Q^m.\end{cases}$$

Because $c(\boldsymbol{x},t^{m+1})$ is unknown in the evaluation of Q^m, an extrapolation of $2c(\boldsymbol{x},t^m)-c(\boldsymbol{x},t^{m-1})$ is used. We set

$$Q^m_{\#}=\int_\Omega \phi(\boldsymbol{x})c_{\#}(\boldsymbol{x}^*,t^m)\,\mathrm{d}\boldsymbol{x}.$$

If $Q^m_{\#}=Q^m_*$, we let $\check{c}(\boldsymbol{x}^*,t^m)=c(\boldsymbol{x}^*,t^m)$. In this case, the mass is not conserved. Otherwise, find θ^m such that $Q^m=\theta^m Q^m_*+(1-\theta^m)Q^m_{\#}$ and let $\check{c}(\boldsymbol{x}^*,t^m)=\theta^m c(\boldsymbol{x}^*,t^m)+(1-\theta)c_{\#}(\boldsymbol{x}^*,t^m)$. In latter case, one has

$$\int_\Omega \phi(\boldsymbol{x})\check{c}(\boldsymbol{x}^*,t^m)\,\mathrm{d}\boldsymbol{x}=Q^m.$$

Hence, mass is conserved globally. In the MMOCAA procedure one replaces $c(\boldsymbol{x}^*,t^m)$ in (3.8) and (3.9) by $\check{c}(\boldsymbol{x}^*,t^m)$.

3.2.4. The Eulerian–Lagrangian localized adjoint method (ELLAM)

The ELLAM formalism was introduced by Celia et al. for the solution of one-dimensional advection–diffusion PDEs [16,60]. It provides a general characteristic solution procedure for advection-dominated PDEs, and it presents a consistent framework for treating general boundary conditions and maintaining mass conservation. The ELLAM formulation directly applies to Eq. (1.2) in a conservative form. Multiplying Eq. (1.2) with space–time test functions w that vanish outside $\Omega\times(t^m,t^{m+1}]$ and are discontinuous in time at time t^m, and integrating the resulting equation over the space–time domain $\Omega\times(t^m,t^{m+1}]$, we obtain a space–time weak formulation for Eq. (1.2) with a noflow boundary condition

$$\begin{aligned}&\int_\Omega \phi(\boldsymbol{x})c(\boldsymbol{x},t^{m+1})w(\boldsymbol{x},t^{m+1})\,\mathrm{d}\boldsymbol{x}+\int_{t^m}^{t^{m+1}}\int_\Omega \nabla w\cdot(\boldsymbol{D}\nabla c)\,\mathrm{d}\boldsymbol{x}\,\mathrm{d}t\\&\quad-\int_{t^m}^{t^{m+1}}\int_\Omega c\rho(\phi w_t+\boldsymbol{u}\cdot\nabla w)\,\mathrm{d}\boldsymbol{x}\,\mathrm{d}t\\&\quad=\int_\Omega \phi(\boldsymbol{x})c(\boldsymbol{x},t^m)w(\boldsymbol{x},t^m_+)\,\mathrm{d}\boldsymbol{x}+\int_{t^m}^{t^{m+1}}\int_\Omega \bar{c}qw\,\mathrm{d}\boldsymbol{x}\,\mathrm{d}t,\end{aligned}\tag{3.10}$$

where $w(\boldsymbol{x},t^m_+)=\lim_{t\to t^m_+}w(\boldsymbol{x},t)$ takes into account that $w(x,t)$ is discontinuous in time at time t^m.

Motivated by the localized adjoint method, the ELLAM formalism chooses the test functions from the solution space of the homogeneous adjoint equation of Eq. (1.2) (e.g. see [16,60])

$$-\phi(\boldsymbol{x})\frac{\partial w}{\partial t}-\boldsymbol{u}\cdot\nabla w-\nabla\cdot(\boldsymbol{D}\nabla w)=0.\tag{3.11}$$

Because the solution space for Eq. (3.11) is infinite dimensional and only a finite number of test functions should be used, an operator splitting technique is applied to Eq. (3.11) to define the test functions.

(i) In the first splitting, the two terms involving spatial derivatives are grouped together, leading to the following system of equations:

$$-\phi(\boldsymbol{x})\frac{\partial w}{\partial t}=0,$$

$$-\boldsymbol{u}\cdot\nabla w-\nabla\cdot(\boldsymbol{D}\nabla w)=0.$$

This splitting leads to a class of optimal test function methods involving upstream weighting in space [4,15,22], which yield solutions with significant temporal errors and numerical diffusion.

(ii) In the second splitting, the terms involving first-order derivatives are grouped together, leading to the following system of equations:

$$\begin{aligned}&-\phi(\boldsymbol{x})\frac{\partial w}{\partial t}-\boldsymbol{u}\cdot\nabla w=0,\\&-\nabla\cdot(\boldsymbol{D}\nabla w)=0.\end{aligned}\tag{3.12}$$

The first equation in (3.12) implies that the test functions should be constant along the characteristics defined by

$$\frac{\mathrm{d}\boldsymbol{r}}{\mathrm{d}t}=\frac{\boldsymbol{u}(\boldsymbol{r},\theta)}{\phi(\boldsymbol{r})},\tag{3.13}$$

which reflects the hyperbolic nature of Eq. (1.2) and assures Lagrangian treatment of advection. The second equation in (3.13) is an elliptic PDE, so standard FEM approximations would be a natural choice for the spatial configuration of the test functions.

Using splitting (3.12), we define the test functions to be standard FEM basis functions on the spatial domain $\bar{\Omega}$ at time t^{m+1} and extend them by a constant into the space–time strip $\bar{\Omega}\times[t^m,t^{m+1}]$ along the characteristics defined by (3.13). Incorporating these test functions into the reference equation (3.10), we obtain an ELLAM scheme as follows:

$$\begin{aligned}&\int_{\Omega}\phi(\boldsymbol{x})c(\boldsymbol{x},t^{m+1})\,\mathrm{d}\boldsymbol{x}+\Delta t\int_{\Omega}(\nabla w\cdot(\boldsymbol{D}\nabla c))(\boldsymbol{x},t^{m+1})\,\mathrm{d}\boldsymbol{x}\\&\quad=\int_{\Omega}\phi(\boldsymbol{x})c(\boldsymbol{x},t^m)w(\boldsymbol{x},t^m_+)\,\mathrm{d}\boldsymbol{x}+\Delta t\int_{\Omega}(\bar{c}qw)(\boldsymbol{x},t^{m+1})\,\mathrm{d}\boldsymbol{x}.\end{aligned}\tag{3.14}$$

Remark 5. The ELLAM scheme (3.14) symmetrizes the transport PDE (1.2), and generates accurate numerical solutions without excessive numerical diffusion or nonphysical oscillation even if coarse spatial grids and large time steps are used [87,104,106]. Second, it is proved that the ELLAM scheme conserves mass [16,88]. Third, in contrast to the MMOC and many other characteristic methods that treat general boundary conditions in an ad hoc manner, the ELLAM formulation can treat any combinations of boundary conditions and provides a systematic way to calculate the boundary conditions accurately [13,16,104,106]. Thus, the ELLAM formulation overcomes the drawbacks of many previous characteristic methods while maintaining their numerical advantages.

Remark 6. Most integrals in the ELLAM scheme (3.14) are standard in FEMs and can be evaluated in a straightforward manner. The only exception is the $\int_{\Omega}\phi(\boldsymbol{x})c(\boldsymbol{x},t^m)w(\boldsymbol{x},t^m_+)\,\mathrm{d}\boldsymbol{x}$ term on the right-hand side of Eq. (3.14). This term corresponds to the $\int_{\Omega}\phi(\boldsymbol{x})c(\boldsymbol{x}^*,t^m)w(\boldsymbol{x})\,\mathrm{d}\boldsymbol{x}$ term in the

MMOC scheme (3.9). As discussed in Remark 4, in the MMOC and other characteristic methods using a backtracking algorithm the evaluation of the $\int_\Omega \phi(\boldsymbol{x})c(\boldsymbol{x}^*, t^m)w(\boldsymbol{x})\,\mathrm{d}\boldsymbol{x}$ term requires significant effort and introduces mass balance error [7,74]. In the ELLAM scheme (3.14), the term $\int_\Omega \phi(\boldsymbol{x})c(\boldsymbol{x}, t^m)w(\boldsymbol{x}, t^m_+)\,\mathrm{d}\boldsymbol{x}$ is evaluated by a forward tracking algorithm that was proposed by Russell and Trujillo [88]. In this approach, an integration quadrature would be enforced on $\bar{\Omega}$ at time t^m with respect to a fixed spatial grid on which $c(\boldsymbol{x}, t^m)$ is defined. The difficult evaluation of $w(\boldsymbol{x}, t^m_+) = \lim_{t \to t^m_+} w(\boldsymbol{x}, t^m) = w(\tilde{\boldsymbol{x}}, t^{m+1})$ is carried out by a forward tracking algorithm from $\boldsymbol{x}$ at time t^m to $\tilde{\boldsymbol{x}} = \boldsymbol{r}(t^{m+1}; \boldsymbol{x}, t^m)$ at time t^{m+1}. Because this forward tracking is used only in the evaluation of the right-hand side of (3.14), it has no effect on the solution grid or the data structure of the discrete system. Therefore, the forward tracking algorithm used here does not suffer from the complication of distorted grids, which complicates many classical forward tracking algorithms.

In the past few years, Wang et al. developed ELLAM schemes for multidimensional advection–diffusion PDEs [101,104,109]; Ewing and Wang [45] and Wang et al. [106] also developed ELLAM schemes for multidimensional advection–reaction PDEs; Celia and Ferrand [14] and Healy and Russell [57,58] developed ELLAM schemes in a finite-volume setting. Dahle et al. developed ELLAM for two-phase flow [33,42]. The computational experiments carried out in [104,106] showed that the ELLAM schemes often outperform many widely used and well-received numerical methods in the context of linear advection–diffusion or advection–reaction PDEs. In addition, Binning and Celia developed a backtracking finite-volume ELLAM scheme for unsaturated flow [7], Wang et al. developed an ELLAM-MFEM solution technique for porous medium flows with point sources and sinks [108]. These works illustrate the strength of the ELLAM schemes in solving the coupled systems of advection–diffusion PDEs. From a viewpoint of analysis, ELLAM methods introduce further difficulties and complexities to the already complicated analyses of characteristic methods. We refer readers to the works of Wang et al. for the convergence analysis and optimal-order error estimates for the ELLAM schemes for advection–diffusion or advection–diffusion–reaction PDEs [102,103,105,107], and the corresponding analysis of Ewing and Wang for the ELLAM schemes for advection–reaction PDEs [44–46].

3.2.5. The characteristic mixed finite element method (CMFEM)

The CMFEM was presented by Arbogast et al. in [2,3] and Yang in [113], and can be viewed as a procedure of ELLAM type [3]. It is also based on the space–time weak formulation (3.10), but uses a mixed finite element approach by introducing the diffusive flux $\boldsymbol{z} = -\boldsymbol{D}\nabla$ as a new variable. Let $\boldsymbol{V}_h \times W_h$ be the lowest-order Raviart–Thomas spaces [83], and $\hat{W}_h$ be the space of discontinuous piecewise-linear functions on the same partition. Then, the CMFEM scheme can be formulated as follows: find $c(\boldsymbol{x}, t^{m+1}) \in W_h$ and $\boldsymbol{z}(\boldsymbol{x}, t^{m+1}) \in \mathscr{V}_h$ such that

$$\int_\Omega \phi(\boldsymbol{x}) \frac{c(\boldsymbol{x}, t^{m+1}) - \hat{c}(\boldsymbol{x}^*, t^m)}{\Delta t} w(\boldsymbol{x})\,\mathrm{d}\boldsymbol{x} + \int_\Omega \nabla \cdot \boldsymbol{z}(\boldsymbol{x}, t^{m+1}) w(\boldsymbol{x})\,\mathrm{d}\boldsymbol{x}$$

$$= \int_\Omega (\bar{c} - c) q(\boldsymbol{x}, t^{m+1}) w(\boldsymbol{x})\,\mathrm{d}\boldsymbol{x}, \quad \forall w \in W_h,$$

$$\int_\Omega \boldsymbol{D}^{-1} \boldsymbol{z}(\boldsymbol{x}, t^{m+1})\,\mathrm{d}\boldsymbol{x} - \int_\Omega c(\boldsymbol{x}, t^{m+1}) \nabla \cdot \boldsymbol{v}\,\mathrm{d}\boldsymbol{x} = 0, \quad \forall \boldsymbol{v} \in \boldsymbol{V}_h,$$

where $\boldsymbol{x}^*$ is defined in (3.8) and $\hat{c}(\boldsymbol{x},t^m) \in \hat{W}_h$ is a post-processing of $c(\boldsymbol{x},t^m)$ and $z(\boldsymbol{x},t^m)$ defined by

$$\int_\Omega \phi(\boldsymbol{x})\hat{c}(\boldsymbol{x},t^m)w(x)\,\mathrm{d}\boldsymbol{x} = \int_\Omega c(\boldsymbol{x},t^m)w(\boldsymbol{x})\,\mathrm{d}\boldsymbol{x}, \quad \forall w \in W_h,$$

$$\int_\Omega \nabla\hat{w}(\boldsymbol{x})\cdot(\boldsymbol{D}\nabla\hat{c})(\boldsymbol{x},t^m)\,\mathrm{d}\boldsymbol{x} = -\int_\Omega z(\boldsymbol{x},t^m)\hat{w}(\boldsymbol{x})\,\mathrm{d}\boldsymbol{x}, \quad \forall\hat{w} \in \hat{W}_h.$$

It is well known that in the mixed method, the scalar variable $c(\boldsymbol{x},t^{m+1})$ is of first order accuracy in space. This post-processing procedure is used to improve the accuracy to the order of $O((\Delta x)^{3/2})$ [3].

Remark 7. Theoretically the CMFEM is locally mass conservative. The situation might not be so clear numerically due to the following reasons: (i) The post-processing procedure is anti-diffusive and, hence, could yield $\hat{c}$ with undershoot or overshoot. A slope limiter has been used in the implementation of CMFEM to overcome this problem [2]. It is not clear how the local mass conservation is achieved in this case. (ii) The CMFEM inherently requires a backtracking procedure and thus has to exactly determine the backtracked image at the previous time step t^m of each cell at the future time step t^{m+1} in order to conserve mass. Since the backtracked image of each cell typically has curved boundaries in general, it is not clear how to trace these cell boundaries exactly to conserve mass numerically. Finally, the theoretically proved error estimate for the CMFEM is obtained only for Eq. (1.2) with a periodic boundary condition and is of $O((\Delta x)^{3/2})$ which is suboptimal by a factor $O((\Delta x)^{1/2})$.

3.2.6. Characteristic methods for immiscible fluid flows, operator splitting techniques

In the governing equation (1.5) for immiscible flows, the hyperbolic part is given by Eq. (2.11) with a typically S-shaped function of the unknown, while the unknown function is a decreasing function in space. Hence, Eq. (2.11) could develop a non-unique solution [12,48,67,70]. Thus, characteristic methods do not apply directly. Espedal and Ewing [38] presented an operator-splitting technique to overcome this difficulty. The fractional flow function $f(c)$ is split into an advective concave hull $\bar{f}(c)$ of $f(c)$, which is linear in what would be the shock region of Eq. (2.11), and a residual anti-diffusive part. The modified advection PDE

$$\frac{\partial c}{\partial t} + \frac{\partial \bar{f}(c)}{\partial x} = 0, \quad \boldsymbol{x} \in \Omega,\ t \in [0,T]$$

yields the same entropy solution as the PDE (2.11), and thus defines characteristic directions uniquely. The residual anti-diffusive advection term is grouped with the diffusion term in the governing PDE so that correct balance between nonlinear advection and diffusion is obtained. Numerically, the PDE is solved by a quadratic Petrov–Galerkin FEM. This technique has been applied in numerical simulation for immiscible flow by Espedal, Ewing, and their collaborators [32,39]. Subsequently, Ewing [41] and Dahle et al. have applied the operator-splitting technique to develop an ELLAM scheme for nonlinear advection–diffusion PDEs, which has shown very promising results.

References

[1] D. Allen, R. Southwell, Relaxation methods applied to determining the motion in two dimensions of a fluid past a fixed cylinder, Quart. J. Mech. Appl. Math. 8 (1955) 129–145.

[2] T. Arbogast, A. Chilakapati, M.F. Wheeler, A characteristic-mixed method for contaminant transport and miscible displacement, in: T.F. Russell, R.E. Ewing, C.A. Brebbia, W.G. Gray and G.F. Pinder (Eds.), Computational Methods in Water Resources IX, Vol. 1, Computational Mechanics Publications and Elsevier Applied Science, London and New York, 1992, pp. 77–84.

[3] T. Arbogast, M.F. Wheeler, A characteristic-mixed finite element method for advection-dominated transport problems, SIAM J. Numer. Anal. 32 (1995) 404–424.

[4] J.W. Barrett, K.W. Morton, Approximate symmetrization and Petrov–Galerkin methods for diffusion-convection problems, Comput. Methods Appl. Mech. Eng. 45 (1984) 97–122.

[5] J. Bear, Hydraulics of Groundwater, McGraw-Hill, New York, 1979.

[6] J.P. Benque, J. Ronat, Quelques difficulties des modeles numeriques en hydraulique, in: R. Glowinski, J.-L. Lions (Eds.), Computer Methods in Applied Mechanics and Engineering, North-Holland, Amsterdam, 1982, pp. 471–494.

[7] P.J. Binning, M.A. Celia, A finite volume Eulerian–Lagrangian localized adjoint method for solution of the contaminant transport equations in two-dimensional multi-phase flow systems, Water Resour. Res. 32 (1996) 103–114.

[8] J.P. Boris, D.L. Book, Flux-corrected transport I, SHASTA, a fluid transport algorithm that works, J. Comput. Phys. 11 (1973) 38–69.

[9] J.P. Boris, D.L. Book, Flux-corrected transport, III, Minimal-error FCT algorithms, J. Comput. Phys. 20 (1976) 397–431.

[10] E.T. Bouloutas, M.A. Celia, An improved cubic Petrov–Galerkin method for simulation of transient advection–diffusion processes in rectangularly decomposable domains, Comput. Meth. Appl. Mech. Eng. 91 (1991) 289–308.

[11] A. Brooks, T.J.R. Hughes, Streamline upwind Petrov–Galerkin formulations for convection dominated flows with particular emphasis on the incompressible Navier–Stokes equations, Comput. Meth. Appl. Mech. Eng. 32 (1982) 199–259.

[12] S.E. Buckley, M.C. Leverret, Mechanism of fluid displacement in sands, Trans. AIME 146 (1942) 107–116.

[13] M.A. Celia, Eulerian–Lagrangian localized adjoint methods for contaminant transport simulations, in: Peters et al. (Eds.), Computational Methods in Water Resources X, Vol. 1, Water Science and Technology Library, Vol. 12, Kluwer Academic Publishers, Dordrecht, Netherlands, 1994, pp. 207–216.

[14] M.A. Celia, L.A. Ferrand, A comparison of ELLAM formulations for simulation of reactive transport in groundwater, in: Wang (Ed.), Advances in Hydro-Science and Engineering, Vol. 1(B), University of Mississippi Press, Mississippi, 1993, pp. 1829–1836.

[15] M.A. Celia, I. Herrera, E.T. Bouloutas, J.S. Kindred, A new numerical approach for the advective-diffusive transport equation, Numer. Meth. PDEs 5 (1989) 203–226.

[16] M.A. Celia, T.F. Russell, I. Herrera, R.E. Ewing, An Eulerian–Lagrangian localized adjoint method for the advection–diffusion equation, Adv. Water Resour. 13 (1990) 187–206.

[17] G. Chavent, A new formulation of diphasic incompressible flows in porous media, in: Lecture Notes in Mathematics, Vol. 503, Springer, Berlin, 1976.

[18] G. Chavent, B. Cockburn, The local projection P^0, P^1-discontinuous Galerkin finite element method for scalar conservation laws, M^2AN 23 (1989) 565–592.

[19] G. Chavent, J. Jaffré, Mathematical Models and Finite Elements for Reservoir Simulation, North-Holland, Amsterdam, 1986.

[20] G. Chavent, G. Salzano, A finite element method for the 1D water flooding problem with gravity, J. Comput. Phys. 45 (1982) 307–344.

[21] Z. Chen, R.E. Ewing, Comparison of various formulations of three-phase flow in porous media, J. Comput. Phys. 132 (1997) 362–373.

[22] I. Christie, D.F. Griffiths, A.R. Mitchell, O.C. Zienkiewicz, Finite element methods for second order differential equations with significant first derivatives, Int. J. Numer. Eng. 10 (1976) 1389–1396.

[23] B. Cockburn, Finite element methods for conservation laws, this volume, J. Comput. Appl. Math. 128 (2001) 187–204.

[24] B. Cockburn, S. Hou, C.-W. Shu, TVB Runge–Kutta local projection discontinuous Galerkin finite element method for conservation laws IV: the multidimensional case, Math. Comp. 54 (1990) 545–590.

[25] B. Cockburn, C.-W. Shu, The Runge–Kutta local projection P^1-discontinuous Galerkin finite element method for scalar conservation laws II: general framework, Math. Comp. 52 (1989) 411–435.

[26] B. Cockburn, C.-W. Shu, The Runge–Kutta local projection P^1-discontinuous Galerkin finite element method for scalar conservation laws, M^2AN 25 (1991) 337–361.

[27] B. Cockburn, C.-W. Shu, The Runge–Kutta discontinuous Galerkin finite element method for conservation laws V: multidimensional systems, J. Comput. Phys. 141 (1998) 199–224.

[28] P. Colella, A direct Eulerian MUSCL scheme for gas dynamics, SIAM J. Sci. Statist. Comput. 6 (1985) 104–117.

[29] P. Colella, P. Woodward, The piecewise-parabolic method (PPM) for gasdynamical simulations, J. Comput. Phys. 54 (1984) 174–201.

[30] R. Courant, K.O. Friedrichs, H. Lewy, Uber die partiellen differenzen-gleichungen der mathematisches physik, Math. Ann. 100 (1928) 32–74.

[31] R. Courant, E. Isaacson, M. Rees, On the solution of nonlinear hyperbolic differential equations by finite differences, Comm. Pure. Appl. Math. 5 (1952) 243–255.

[32] H.K. Dahle, M.S. Espedal, R.E. Ewing, O. Sævareid, Characteristic adaptive sub-domain methods for reservoir flow problems, Numer. Methods PDE's 6 (1990) 279–309.

[33] H.K. Dahle, R.E. Ewing, T.F. Russell, Eulerian–Lagrangian localized adjoint methods for a nonlinear convection-diffusion equation, Comput. Methods Appl. Mech. Eng. 122 (1995) 223–250.

[34] J. Douglas Jr., F. Furtado, F. Pereira, On the numerical simulation of waterflooding of heterogeneous petroleum reservoirs, Comput. Geosci. 1 (1997) 155–190.

[35] J. Douglas Jr., C.-S. Huang, F. Pereira, The modified method of characteristics with adjusted advection, Technical Report, No. 298, Center for Applied Mathematics, Purdue University, 1997.

[36] J. Douglas Jr., I. Martinéz-Gamba, M.C.J. Squeff, Simulation of the transient behavior of a one-dimensional semiconductor device, Mat. Apl. Comput. 5 (1986) 103–122.

[37] J. Douglas Jr., T.F. Russell, Numerical methods for convection-dominated diffusion problems based on combining the method of characteristics with finite element or finite difference procedures, SIAM J. Numer. Anal. 19 (1982) 871–885.

[38] M.S. Espedal, R.E. Ewing, Characteristic Petrov–Galerkin sub-domain methods for two-phase immiscible flow, Comput. Methods Appl. Mech. Eng. 64 (1987) 113–135.

[39] M.S. Espedal, R.E. Ewing, J.A. Puckett, R.S. Schmidt, Simulation techniques for multiphase and multicomponent flows, Comm. Appl. Numer. Methods 4 (1988) 335–342.

[40] R.E. Ewing (Ed.), The Mathematics of Reservoir Simulation, Research Frontiers in Applied Mathematics, Vol. 1, SIAM, Philadelphia, PA, 1984.

[41] R.E. Ewing, Operator splitting and Eulerian–Lagrangian localized adjoint methods for multi-phase flow, in: Whiteman (Ed.), The Mathematics of Finite Elements and Applications VII, MAFELAP, 1990, Academic Press, San Diego, CA, 1991, pp. 215–232.

[42] R.E. Ewing, Simulation of multiphase flows in porous media, Transport in Porous Media 6 (1991) 479–499.

[43] R.E. Ewing, T.F. Russell, M.F. Wheeler, Simulation of miscible displacement using mixed methods and a modified method of characteristics, SPE 12241 1983 (71–81).

[44] R.E. Ewing, H. Wang, Eulerian–Lagrangian localized adjoint methods for linear advection equations, in: Computational Mechanics '91, Springer, Berlin, 1991, pp. 245–250.

[45] R.E. Ewing, H. Wang, Eulerian–Lagrangian localized adjoint methods for linear advection or advection–reaction equations and their convergence analysis, Comput. Mech. 12 (1993) 97–121.

[46] R.E. Ewing, H. Wang, An optimal-order error estimate to Eulerian–Lagrangian localized adjoint method for variable-coefficient advection–reaction problems, SIAM Numer. Anal. 33 (1996) 318–348.

[47] C.L. Farmer, A moving point method for arbitrary Peclet number multi-dimensional convection-diffusion equations, IMA J. Numer. Anal. 5 (1980) 465–480.

[48] B.A. Finlayson, Numerical Methods for Problems with Moving Fronts, Ravenna Park Publishing, Seattle, 1992.

[49] A.O. Garder, D.W. Peaceman, A.L. Pozzi, Numerical calculations of multidimensional miscible displacement by the method of characteristics, Soc. Pet. Eng. J. 4 (1964) 26–36.

[50] S.K. Godunov, A difference scheme for numerical computation of discontinuous solutions of fluid dynamics, Mat. Sb. 47 (1959) 271–306.
[51] J.B. Goodman, R.J. LeVeque, A geometric approach to high-resolution TVD schemes, SIAM J. Numer. Anal. 25 (1988) 268–284.
[52] P. Hansbo, A. Szepessy, A velocity-pressure streamline diffusion finite element method for the incompressible Navier–Stokes equations, Comput. Methods Appl. Mech. Eng. 84 (1990) 107–129.
[53] A. Harten, High resolution schemes for hyperbolic conservation laws, J. Comput. Phys. 49 (1983) 357–393.
[54] A. Harten, B. Engquist, S. Osher, S. Chakravarthy, Uniformly high-order accurate essentially nonoscillatory schemes, III, J. Comput. Phys. 71 (1987) 231–241.
[55] A. Harten, J.M. Hyman, P.D. Lax, On finite-difference approximations and entropy conditions for shocks, Comm. Pure Appl. Math. 29 (1976) 297–322.
[56] A. Harten, S. Osher, Uniformly high-order accurate nonoscillatory schemes, I, SIAM J. Numer. Anal. 24 (1987) 279–309.
[57] R.W. Healy, T.F. Russell, A finite-volume Eulerian–Lagrangian localized adjoint method for solution of the advection-dispersion equation, Water Resour. Res. 29 (1993) 2399–2413.
[58] R.W. Healy, T.F. Russell, Solution of the advection-dispersion equation in two dimensions by a finite-volume Eulerian–Lagrangian localized adjoint method, Adv. Water Res. 21 (1998) 11–26.
[59] G. Hedstrom, Models of difference schemes for $u_t + u_x = 0$ by partial differential equations, Math. Comp. 29 (1975) 969–977.
[60] I. Herrera, R.E. Ewing, M.A. Celia, T.F. Russell, Eulerian–Lagrangian localized adjoint methods: the theoretical framework, Numer. Methods PDEs 9 (1993) 431–458.
[61] T.J.R. Hughes, Multiscale phenomena: Green functions, the Dirichlet-to-Neumann formulation, subgrid scale models, bubbles, and the origins of stabilized methods, Comput. Methods Appl. Mech. Eng. 127 (1995) 387–401.
[62] T.J.R. Hughes, A.N. Brooks, A multidimensional upwind scheme with no crosswind diffusion, in: T.J.R. Hughes (Ed.), Finite Element Methods for Convection Dominated Flows, Vol. 34, ASME, New York, 1979, pp. 19–35.
[63] T.J.R. Hughes, M. Mallet, A new finite element formulation for computational fluid dynamics: III, The general streamline operator for multidimensional advective–diffusive systems, Comput. Methods Appl. Mech. Eng. 58 (1986) 305–328.
[64] G. Jiang, C.-W. Shu, Efficient implementation of weighted ENO schemes, J. Comput. Phys. 126 (1996) 202–228.
[65] C. Johnson, J. Pitkäranta, An analysis of discontinuous Galerkin methods for a scalar hyperbolic equation, Math. Comp. 46 (1986) 1–26.
[66] C. Johnson, A. Szepessy, P. Hansbo, On the convergence of shock-capturing streamline diffusion finite element methods for hyperbolic conservation laws, Math. Comp. 54 (1990) 107–129.
[67] P.D. Lax, Hyperbolic Systems of Conservation Laws and the Mathematical Theory of Shock Waves, SIAM, Philadelphia, PA, 1973.
[68] P.D. Lax, B. Wendroff, Systems of conservation laws, Comm. Pure Appl. Math. 13 (1960) 217–237.
[69] P. Lesaint, P.A. Raviart, On a finite element method for solving the neutron transport equation, in: DeBoor (Ed.), Mathematics Aspects of Finite Elements in Partial Differential Equations, Academic Press, New York, 1974, pp. 89–123.
[70] R.J. LeVeque, Numerical Methods for Conservation Laws, Birkhäuser, Basel, 1992.
[71] X.-D. Liu, S. Osher, T. Chan, Weighted essentially nonoscillatory schemes, J. Comput. Phys. 115 (1994) 200–212.
[72] P.A. Markowich, C. Ringhofer, S. Schmeiser, Semiconductor Equations, Springer, Vienna, 1990.
[73] K.W. Morton, Numerical Solution of Convection-Diffusion Problems, Applied Mathematics and Computation, Vol. 12, Chapman & Hall, London, 1996.
[74] K.W. Morton, A. Priestley, E. Süli, Stability of the Lagrangian-Galerkin method with nonexact integration, RAIRO M^2AN 22 (1988) 123–151.
[75] H. Nessyahu, E. Tadmor, Non-oscillatory central differencing for hyperbolic conservation laws, J. Comput. Phys. 87 (1990) 408–463.
[76] S.P. Neuman, An Eulerian–Lagrangian numerical scheme for the dispersion-convection equation using conjugate space–time grids, J. Comput. Phys. 41 (1981) 270–294.
[77] S.P. Neuman, Adaptive Eulerian–Lagrangian finite element method for advection-dispersion equation, Int. J. Numer. Methods Eng. 20 (1984) 321–337.

[78] S. Osherm S. Chakravarthy, High resolution schemes and the entropy condition, SIAM J. Numer. Anal. 21 (1984) 984–995.

[79] T. Peterson, A note on the convergence of the discontinuous Galerkin method for a scalar hyperbolic equation, SIAM J. Numer. Anal. 28 (1991) 133–140.

[80] O. Pironneau, On the transport-diffusion algorithm and its application to the Navier–Stokes equations, Numer. Math. 38 (1982) 309–332.

[81] H.S. Price, R.S. Varga, J.E. Warren, Applications of oscillation matrices to diffusion-convection equations, J. Math. Phys. 45 (1966) 301–311.

[82] J. Pudykiewicz, A. Staniforth, Some properties and comparative performance of the semi-Lagrangian method of robert in the solution of the advection–diffusion equation, Atmosphere-Ocean 22 (1984) 283–308.

[83] P.-A. Raviart, J.-M. Thomas, A mixed finite element method for second order elliptic problems, in: I. Galligani, E. Magenes (Eds.), Mathematical Aspects of Finite Element Methods, Lecture Notes in Mathematics, Vol. 606, Springer, Berlin, 1977, pp. 292–315.

[84] W.H. Reed, T.R. Hill, Triangular mesh methods for the neutron transport equation, Los Alamos Scientific Laboratory Report LA-UR-73-479, 1973.

[85] P.L. Roe, Some contributions to the modeling of discontinuous flows, Lecture Notes in Applied Mathematics, Vol. 22, 1985, pp. 163–193.

[86] H.G. Roos, M. Stynes, L. Tobiska, Numerical Methods for Singular Perturbed Differential Equations, Convection-Diffusion and Flow Problems, Computational Mathematics, Vol. 24, Springer, Berlin, 1996.

[87] T.F. Russell, Eulerian–Lagrangian localized adjoint methods for advection-dominated problems, in: Griffiths, Watson (Eds.), Proceedings of the 13th Dundee Conference on Numerical Analysis, Pitmann Research Notes in Mathematics Series, Vol. 228, Longman Scientific & Technical, Harlow, United Kingdom, 1990, pp. 206–228.

[88] T.F. Russell, R.V. Trujillo, Eulerian–Lagrangian localized adjoint methods with variable coefficients in multiple dimensions, Computational Methods in Surface Hydrology, Proceedings of the Eighth International Conference on Computational Methods in Water Resources, Venice, Italy, 1990, pp. 357–363.

[89] T.F. Russell, M.F. Wheeler, Finite element and finite difference methods for continuous flows in porous media, in: R.E. Ewing (Ed.), The Mathematics of Reservoir Simulation, SIAM, Philadelphia, PA, 1984, pp. 35–106.

[90] R. Salmon, Lectures on Geophysical Fluid Dynamics, Oxford University Press, New York, 1998.

[91] Y.T. Shih, H.C. Elman, Modified streamline diffusion schemes for convection-diffusion problems, Comput. Methods Appl. Mech. Eng. 174 (1999) 137–151.

[92] C.-W. Shu, Essentially nonoscillatory (ENO) and weighted essentially nonoscillatory (WENO) schemes for hyperbolic conservation laws, in: Quarteroni (Ed.), Advanced Numerical Approximation of Nonlinear Hyperbolic Equations, Lecture Notes in Mathematics, Vol. 1697, Springer, New York, 1997, pp. 325–432.

[93] J. Smoller, Shock Waves and Reaction-Diffusion Equations, 2nd Edition, Fundamental Principles of Mathematical Sciences, Vol. 258, Springer, New York, 1994.

[94] J.C. Strikwerda, Finite Difference Schemes and Partial Differential Equations, Wadsworth and Brooks/Cole, Pacific Grove, CA, 1989.

[95] P.K. Sweby, High resolution schemes using flux limiters for hyperbolic conservation laws, SIAM J. Numer. Anal. 28 (1991) 891–906.

[96] G. Thomaidis, K. Zygourakis, M.F. Wheeler, An explicit finite difference scheme based on the modified method of characteristics for solving diffusion-convection problems in one space dimension, Numer. Methods Partial Differential Equations 4 (1988) 119–138.

[97] L.N. Trefethen, Group velocity in finite difference schemes, SIAM Rev. 24 (1982) 113–136.

[98] B. van Leer, Towards the ultimate conservative difference scheme II, Monotonicity and conservation combined in a second order scheme, J. Comput. Phys. 14 (1974) 361–370.

[99] B. van Leer, On the relation between the upwind-differencing schemes of Godunov, Engquist-Osher, and Roe, SIAM J. Sci. Statist. Comput. 5 (1984) 1–20.

[100] J. Von Neumann, R.D. Richtmyer, A method for the numerical calculation of hydrodynamic shocks, J. Appl. Phys. 21 (1950) 232–237.

[101] H. Wang, Eulerian–Lagrangian localized adjoint methods: analyses, implementations, and applications, Ph.D. Thesis, Department of Mathematics, University of Wyoming, 1992.

[102] H. Wang, A family of ELLAM schemes for advection–diffusion–reaction equations and their convergence analyses, Numer. Meth. PDEs 14 (1998) 739–780.

[103] H. Wang, An optimal-order error estimate for an ELLAM scheme for two-dimensional linear advection–diffusion equations, SIAM J. Numer. Anal. 37 (2000) 1338–1368.

[104] H. Wang, H.K. Dahle, R.E. Ewing, M.S. Espedal, R.C. Sharpley, S. Man, An ELLAM Scheme for advection–diffusion equations in two dimensions, SIAM J. Sci. Comput. 20 (1999) 2160–2194.

[105] H. Wang, R.E. Ewing, Optimal-order convergence rates for Eulerian–Lagrangian localized adjoint methods for reactive transport and contamination in groundwater, Numer. Methods PDEs 11 (1995) 1–31.

[106] H. Wang, R.E. Ewing, G. Qin, S.L. Lyons, M. Al-Lawatia, S. Man, A family of Eulerian–Lagrangian localized adjoint methods for multi-dimensional advection–reaction equations, J. Comput. Phys. 152 (1999) 120–163.

[107] H. Wang, R.E. Ewing, T.F. Russell, Eulerian–Lagrangian localized methods for convection-diffusion equations and their convergence analysis, IMA J. Numer. Anal. 15 (1995) 405–459.

[108] H. Wang, D. Liang, R.E. Ewing, S.L. Lyons, G. Qin, An ELLAM-MFEM solution technique for compressible fluid flows in porous media with point sources and sinks, J. Comput. Phys. 159 (2000) 344–376.

[109] H. Wang, R.C. Sharpley, S. Man, An ELLAM scheme for advection–diffusion equations in multi-dimensions, in: Aldama et al. (Eds.) Computational Methods in Water Resources XI, Vol. 2, Computational Mechanics Publications, Southampton, 1996, pp. 99–106.

[110] R. Warming, B.J. Hyett, The modified equation approach to the stability and accuracy analysis of finite difference methods. J. Comput. Phys. 14 (1974) 159–179.

[111] J.J. Westerink, D. Shea, Consider higher degree Petrov–Galerkin methods for the solution of the transient convection-diffusion equation, Int. J. Numer. Methods Eng. 28 (1989) 1077–1101.

[112] G. Whitham, Linear and Nonlinear Waves, Wiley, New York, 1974.

[113] D. Yang, A characteristic-mixed method with dynamic finite element space for convection-dominated diffusion problems, J. Comput. Appl. Math. 43 (1992) 343–353.

[114] S.T. Zalesak, Fully multidimensional flux-corrected transport algorithms for fluids, J. Comput. Phys. 31 (1978) 335–362.

[115] S.T. Zalesak, A preliminary comparison of modern shock-capturing schemes: linear advection equation, in: R. Vichnevetsky, R.S. Stepleman (Eds.) Advances in Computer Methods for PDEs, Vol. 6, IMACS, Amsterdam, 1987, pp. 15–22.

ELSEVIER Journal of Computational and Applied Mathematics 128 (2001) 447–466

JOURNAL OF COMPUTATIONAL AND APPLIED MATHEMATICS

www.elsevier.nl/locate/cam

Approximate factorization for time-dependent partial differential equations ☆

P.J. van der Houwen, B.P. Sommeijer *

CWI, P.O. Box 94079, 1090 GB Amsterdam, The Netherlands

Received 1 June 1999; received in revised form 28 October 1999

Abstract

The first application of approximate factorization in the numerical solution of time-dependent partial differential equations (PDEs) can be traced back to the celebrated papers of Peaceman and Rachford and of Douglas of 1955. For linear problems, the Peaceman–Rachford–Douglas method can be derived from the Crank–Nicolson method by the approximate factorization of the system matrix in the linear system to be solved. This factorization is based on a splitting of the system matrix. In the numerical solution of time-dependent PDEs we often encounter linear systems whose system matrix has a complicated structure, but can be split into a sum of matrices with a simple structure. In such cases, it is attractive to replace the system matrix by an approximate factorization based on this splitting. This contribution surveys various possibilities for applying approximate factorization to PDEs and presents a number of new stability results for the resulting integration methods. © 2001 Elsevier Science B.V. All rights reserved.

MSC: 65L06; 65L20; 65M12; 65M20

Keywords: Numerical analysis; Partial differential equations; Approximate factorization; Stability

1. Introduction

The first application of approximate factorization in the numerical solution of time-dependent partial differential equations (PDEs) can be traced back to the celebrated papers of Peaceman and Rachford [20] and of Douglas [5] of 1955. More explicitly, approximate factorization was formulated by Beam and Warming [1] in 1976.

☆ Work carried out under project MAS 1.2 - 'Numerical Algorithms for Surface Water Quality Modelling'

* Corresponding author. Tel.: +31-20-592-41-92; fax: +31-20-592-41-99.

E-mail address: bsom@cwi.nl (B.P. Sommeijer).

0377-0427/01/$ - see front matter © 2001 Elsevier Science B.V. All rights reserved.

PII: S0377-0427(00)00523-9

In order to illustrate the idea of approximate factorization, consider the initial-boundary value problem for the two-dimensional diffusion equation

$$\frac{\partial u(t,x,y)}{\partial t} = \frac{\partial^2 u(t,x,y)}{\partial x^2} + \frac{\partial^2 u(t,x,y)}{\partial y^2}$$

and let this problem be discretized in space by finite differences. Then, we obtain an initial-value problem (IVP) for a system of ordinary differential equations (ODEs)

$$\frac{\mathrm{d}\boldsymbol{y}(t)}{\mathrm{d}t} = J_1\boldsymbol{y} + J_2\boldsymbol{y}, \tag{1.1}$$

where $\boldsymbol{y}(t)$ contains approximations to $u(t,x,y)$ at the grid points and J_1 and J_2 are matrices representing finite-difference approximations to $\partial^2/\partial x^2$ and $\partial^2/\partial y^2$. System (1.1) can be integrated by, e.g., the second-order trapezoidal rule, yielding the well-known Crank–Nicolson method [3]

$$(I - \tfrac{1}{2}\Delta t(J_1 + J_2))\boldsymbol{y}_{n+1} = \boldsymbol{y}_n + \tfrac{1}{2}\Delta t(J_1 + J_2)\boldsymbol{y}_n. \tag{1.2}$$

Here, I denotes the identity matrix, Δt is the timestep and $\boldsymbol{y}_n$ represents a numerical approximation to $\boldsymbol{y}(t_n)$. Each step requires the solution of a linear system with system matrix $I - \frac{1}{2}\Delta t(J_1 + J_2)$. Due to the relatively large bandwidth, the solution of this system by a *direct* factorization of the system matrix is quite expensive. Following Beam and Warming [1], (1.2) is written in the equivalent form

$$(I - \tfrac{1}{2}\Delta t(J_1 + J_2))(\boldsymbol{y}_{n+1} - \boldsymbol{y}_n) = \Delta t(J_1 + J_2)\boldsymbol{y}_n, \tag{1.2a}$$

and the system matrix is replaced by an *approximate* factorization, to obtain

$$(I - \tfrac{1}{2}\Delta tJ_1)(I - \tfrac{1}{2}\Delta tJ_2)(\boldsymbol{y}_{n+1} - \boldsymbol{y}_n) = \Delta t(J_1 + J_2)\boldsymbol{y}_n. \tag{1.3}$$

This method is easily verified to be identical with the alternating direction implicit method (ADI method) of Peaceman–Rachford and Douglas, usually represented in the form

$$\boldsymbol{y}_{n+1/2} = \boldsymbol{y}_n + \tfrac{1}{2}\Delta t(J_1\boldsymbol{y}_{n+1/2} + J_2\boldsymbol{y}_n), \quad \boldsymbol{y}_{n+1} = \boldsymbol{y}_{n+1/2} + \tfrac{1}{2}\Delta t(J_1\boldsymbol{y}_{n+1/2} + J_2\boldsymbol{y}_{n+1}). \tag{1.3a}$$

Although we now have to solve two linear systems, the small bandwidth of the matrices $I - \frac{1}{2}\Delta tJ_k$ causes that direct solution methods are not costly. Since the factorized system matrix in (1.3) is a second-order approximation to the system matrix in (1.2a), the ADI method is a third-order perturbation of (1.2a), and hence of (1.2), so that it is second-order accurate. Note that directly applying approximate factorization to the system matrix in (1.2) would yield a first-order-accurate method. Hence, the intermediate step which replaces (1.2) by (1.2a) is essential.

The application of approximate factorization is not restricted to schemes resulting from time discretizations by the trapezoidal rule. For example, one may replace the trapezoidal rule (1.2) by a second-order linear multistep method and proceed as described above. In fact, approximate factorization can be applied in many more cases where linear or nonlinear time-dependent PDEs are solved numerically. We mention (i) the linear multistep approach of Warming and Beam [28] described in Section 2.1, (ii) linearly implicit integration methods like Rosenbrock methods (see Section 2.2), (iii) linearization of a nonlinear method (Section 2.3), and (iv) iterative application of approximate factorization for solving linear systems (Section 3). In all these cases, we are faced with linear systems whose system matrix has the form $I - \Delta tM$, where the matrix M itself has a complicated structure, but can be split into a sum $\sum M_k$ with matrices M_k possessing a simple structure. This leads us to replace $I - \Delta tM$ by the approximate factorization $\Pi(I - \Delta tM_k)$.

In this paper, we discuss the application of the approximate factorization technique to the four cases mentioned above and we present stability theorems for the resulting integration methods, many of which are new results. One of the results is that in the case of three-component splittings $M = \sum M_k$, where the M_k have purely imaginary eigenvalues, iterative approximate factorization leads to methods with substantial stability boundaries. Such methods are required in the numerical solution of three dimensional, convection-dominated transport problems.

2. Noniterative factorized methods

Consider an initial-boundary-value problem for the PDE

$$\frac{\partial u(t,\boldsymbol{x})}{\partial t} = L(t,\boldsymbol{x},u(t,\boldsymbol{x})), \tag{2.1}$$

where L is a differential operator in the d-dimensional space variable $\boldsymbol{x} = (x_1,\dots,x_d)$. Spatial discretization yields an IVP for a system of ODEs

$$\frac{\mathrm{d}\boldsymbol{y}(t)}{\mathrm{d}t} = \boldsymbol{f}(t,\boldsymbol{y}), \quad \boldsymbol{y}(t_0) = \boldsymbol{y}_0. \tag{2.2}$$

In order to simplify the notations, we shall assume that (2.2) is rewritten in autonomous form. Furthermore, it will be assumed that the Jacobian matrix $J(\boldsymbol{y}) := \partial \boldsymbol{f}(\boldsymbol{y})/\partial \boldsymbol{y}$ can be split into a sum of m matrices, i.e., $J(\boldsymbol{y}) = \sum J_k$, where the splitting is either according to the spatial dimensions (as in the early papers on splitting methods), or to the physical terms in the PDE (2.1), or according to any other partition leading to matrices J_k with a convenient structure. In this paper, we only use splittings of the Jacobian and not of the right-hand side function $\boldsymbol{f}(\boldsymbol{y})$. This is often convenient in the case of nonlinear PDEs.

We discuss three options for applying noniterative approximate factorization techniques, viz. (i) the ADI method of Warming and Beam, (ii) approximate factorization of linearly implicit integration methods and (iii) approximate factorization in the linearization of nonlinear methods.

2.1. The method of Warming and Beam

Consider the linear multistep method (LM method)

$$\rho(E)\boldsymbol{y}_{n-\mu+1} = \Delta t\sigma(E)\boldsymbol{f}(\boldsymbol{y}_{n-\mu+1}), \quad \rho(z) := \sum_{i=0}^{\mu} a_i z^{\mu-i}, \quad \sigma(z) := \sum_{i=0}^{\mu} b_i z^{\mu-i}, \quad a_0 = 1, \tag{2.3}$$

where E is the forward shift operator and $\mu \geqslant 1$. Warming and Beam [28] rewrite (2.3) in the form

$$\rho(E)(\boldsymbol{y}_{n-\mu+1} - b_0\Delta t\boldsymbol{f}(\boldsymbol{y}_{n-\mu+1})) = \Delta t(\sigma(E) - b_0\rho(E))\boldsymbol{f}(\boldsymbol{y}_{n-\mu+1}). \tag{2.3a}$$

Since the degree of ρ is larger than that of $\sigma - b_0\rho$, the right-hand side does not depend on $\boldsymbol{y}_{n+1}$. In [28] it is assumed that $\boldsymbol{f}$ is linear, i.e., $\boldsymbol{f}(\boldsymbol{y}) = J\boldsymbol{y}$, so that (2.3a) becomes a linear system for $\rho(E)\boldsymbol{y}_{n-\mu+1}$. However, by replacing (2.3a) with

$$\rho(E)(\boldsymbol{y}_{n-\mu+1} - b_0\Delta tJ\boldsymbol{y}_{n-\mu+1}) = \Delta t(\sigma(E) - b_0\rho(E))\boldsymbol{f}(\boldsymbol{y}_{n-\mu+1}), \tag{2.3b}$$

we can also deal with ODE systems where $\boldsymbol{f}$ is nonlinear (see [2]). Assuming that (2.3) is consistent, so that $\rho(1) = 0$, it can be shown that (2.3b) is an $O((\Delta t)^3)$ perturbation of (2.3a), and hence of (2.3). Method (2.3b) is linearly implicit in the quantity $\boldsymbol{q}_n := \rho(E)\boldsymbol{y}_{n-\mu+1}$ with system matrix $I - b_0\Delta tJ = I - b_0\Delta t\sum J_k$, where I denotes the identity matrix (in the following, the identity matrix will always be denoted by I without specifying its order, which will be clear from the context). Approximate factorization of this system matrix leads to the method of Warming and Beam:

$$\begin{aligned} \Pi\boldsymbol{q}_n &= \Delta t(\sigma(E) - b_0\rho(E))\boldsymbol{f}(\boldsymbol{y}_{n-\mu+1}), \quad \Pi := \prod_{k=1}^{m}(I - b_0\Delta tJ_k), \\ \boldsymbol{y}_{n+1} &= \boldsymbol{q}_n - (\rho(E) - E^{\mu})\boldsymbol{y}_{n-\mu+1}. \end{aligned} \tag{2.4}$$

Since $\boldsymbol{q}_n = \mathrm{O}(\Delta t)$ it follows that (2.4) is an $\mathrm{O}((\Delta t)^3)$ perturbation of (2.3b) which was itself an $\mathrm{O}((\Delta t)^3)$ perturbation of (2.3). Thus, if (2.3) is at least second-order accurate, then (2.4) is also second-order accurate. Since the LM method (2.3) cannot be A-stable if its order is higher than two and because A-stability of (2.3) will turn out to be a necessary condition for (2.4) to be A-stable (see Section 2.4), this order limitation is not restrictive.

If the PDE is linear and if (2.3) is defined by the trapezoidal rule, then (2.4) is identical to the Peaceman–Rachford method (1.3) for $m = 2$. Hence, (2.4) might be considered as an extension of the Peaceman–Rachford method (1.3) (or (1.3a)) to nonlinear PDEs with multicomponent splittings.

The computational efficiency of (2.4) depends on the structure of the successive system matrices $I - b_0\Delta tJ_k$. Let us consider the case of an m-dimensional convection-dominated problem where the convection terms are discretized by third-order upwind formulas. Using dimension splitting, the J_k become block-diagonal whose blocks are penta-diagonal matrices. The LU-decomposition of $I - b_0\Delta tJ_k$ and the forward/backward substitution each requires about $8N$ flops for large N, N denoting the dimension of J_k (see, e.g., [9, p. 150]). Hence, the total costs are only proportional to N, viz. $8mN$ flops per step and an additional $8mN$ flops if the LU-decompositions are recomputed. Moreover, there is scope for a lot of vectorization, so that on vector computers the solution of the linear systems in (2.4) is extremely fast. Furthermore, there is a lot of intrinsic parallelism, because of the block structure of J_k. However, the crucial point is the magnitude of the stepsize for which the method is stable. This will be the subject of Section 2.4.

Finally, we remark that the Warming–Beam method (2.4) was originally designed as an ADI method based on dimension splitting, but it can of course be applied to any Jacobian splitting $J = \sum J_k$.

2.2. Factorized linearly implicit methods

In the literature, various families of linearly implicit methods have been proposed. The first methods of this type are the Rosenbrock methods, proposed in 1962 by Rosenbrock [22]. A more general family contains the linearly implicit Runge–Kutta methods developed by Strehmel and Weiner [26]. Here, we illustrate the factorization for Rosenbrock methods which are defined by (cf. [11, p. 111])

$$\begin{aligned} \boldsymbol{y}_{n+1} &= \boldsymbol{y}_n + (\boldsymbol{b}^{\mathrm{T}} \otimes I)\boldsymbol{K}, \quad \boldsymbol{K} := (\boldsymbol{k}_i), \quad i = 1,\dots,s, \; n \geqslant 0, \\ (I - T \otimes \Delta tJ)\boldsymbol{K} &= \Delta t\boldsymbol{F}(\boldsymbol{e} \otimes \boldsymbol{y}_n + (L \otimes I)\boldsymbol{K}), \quad J \approx J(\boldsymbol{y}_n) := \frac{\partial \boldsymbol{f}(\boldsymbol{y}_n)}{\partial \boldsymbol{y}}, \end{aligned} \tag{2.5}$$

where $\boldsymbol{b}$ and $\boldsymbol{e}$ are s-dimensional vectors, $\boldsymbol{e}$ has unit entries, T is an $s \times s$ diagonal or lower triangular matrix, L is a strictly lower triangular $s \times s$ matrix, and $\otimes$ denotes the Kronecker or direct matrix product. Furthermore, for any vector $\boldsymbol{V} = (\boldsymbol{v}_i), \boldsymbol{F}(\boldsymbol{V})$ is defined by $(\boldsymbol{f}(\boldsymbol{v}_i))$. If the order of the method (2.5) is independent of the choice of the Jacobian approximation J, then (2.5) is called a Rosenbrock-W method [25]. Note that the steppoint formula in (2.5) is explicit, so that the main computational effort goes into the computation of the implicitly defined vector $\boldsymbol{K}$. Since T is lower triangular and L is strictly lower triangular, the s subsystems for $\boldsymbol{k}_i$ can be solved successively. Moreover, although the system for $\boldsymbol{K}$ is nonlinear, these subsystems are linear.

Let us rewrite the system for $\boldsymbol{K}$ in the equivalent form

$$(I - D \otimes \Delta tJ)\boldsymbol{K} = \Delta t\boldsymbol{F}(\boldsymbol{e} \otimes \boldsymbol{y}_n + (L \otimes I)\boldsymbol{K}) + ((T - D) \otimes \Delta tJ)\boldsymbol{K},$$

where D is a diagonal matrix whose diagonal equals that of T. Then, approximately factorizing the block-diagonal system matrix $I - D \otimes \Delta tJ = I - D \otimes \Delta t \sum J_k$ leads to the *factorized Rosenbrock method*

$$\begin{aligned} \boldsymbol{y}_{n+1} &= \boldsymbol{y}_n + (\boldsymbol{b}^T \otimes I)\boldsymbol{K}, \\ \Pi\boldsymbol{K} &= \Delta t\boldsymbol{F}(\boldsymbol{e} \otimes \boldsymbol{y}_n + (L \otimes I)\boldsymbol{K}) + ((T - D) \otimes \Delta tJ)\boldsymbol{K}, \quad \Pi := \prod_{k=1}^{m} (I - D \otimes \Delta tJ_k). \end{aligned} \tag{2.6}$$

If the Rosenbrock method (2.5) is at least second-order accurate and if $J = J(\boldsymbol{y}_n) + O(\Delta t)$, then (2.6) is also at least second-order accurate. However, as observed in [27], if (2.5) is a Rosenbrock-W method with a diagonal matrix T with constant diagonal entries κ, then the approximate factorization does not affect the order of accuracy. This follows from the fact that for $T = \kappa I$ we can write $\Pi = I - \kappa I \otimes \Delta tJ^*$. Hence, we may consider the factorized Rosenbrock method (2.6) as the original Rosenbrock-W method with $J = J^*$. Since in Rosenbrock-W methods the Jacobian can be freely chosen, Rosenbrock-W methods and their factorized versions have the same order of accuracy.

As to the computational efficiency of factorized Rosenbrock methods, we observe that if in the underlying Rosenbrock method $T = D$ and $L = O$, then the s subsystems for $\boldsymbol{k}_i$ in (2.6) can be solved concurrently. These subsystems have the same structure as in the Warming–Beam method (2.4), so that the computational efficiency is comparable on a parallel computer system. As an example of such a parallel Rosenbrock method, we have

$$\boldsymbol{b} = \frac{1}{2(\kappa_2 - \kappa_1)} \begin{pmatrix} 2\kappa_2 - 1 \\ -2\kappa_1 + 1 \end{pmatrix}, \quad T = \begin{pmatrix} \kappa_1 & 0 \\ 0 & \kappa_2 \end{pmatrix}, \quad L = O, \quad \kappa_1 \neq \kappa_2, \tag{2.7}$$

which is second-order accurate if $J = J(\boldsymbol{y}_n) + O(\Delta t)$.

However, if either $T \neq D$ or $L \neq O$, then the s subsystems in (2.6) have to be solved sequentially.

2.3. Approximate factorization of linearized methods

Instead of starting with a linearly implicit integration method, we may also linearize a nonlinear method. In fact, the Rosenbrock methods of the preceding section can be introduced by linearizing diagonally implicit Runge–Kutta (DIRK) methods (cf. [11, p. 111]). In the literature, many other examples of linearization can be found. For instance, the linearization of the θ-method applied to the porus media equation (in [21, p. 203]), the linearization of the Crank–Nicolson method for hyperbolic conservation laws (in [1]) and the linearization of LM methods for the compressible Navier–Stokes equations (in [2]). In this paper, we consider the linearization of a class of methods which contains most methods from the literature:

$$\begin{aligned} &y_{n+1} = (a^{\mathrm{T}} \otimes I)Y_{n+1} + g_n, \quad Y_{n+1} := (y_{n+c_i}), \\ &Y_{n+1} - \Delta t(T \otimes I)F(Y_{n+1}) = G_n, \quad F(Y_{n+1}) := (f(y_{n+c_i})), \end{aligned} \qquad i = 1,\ldots,s, \; n \geqslant 0. \tag{2.8}$$

Here, a is an s-dimensional vector and T is again an $s \times s$ matrix. The steppoint value y_{n+1} and the components y_{n+c_i} of Y_{n+1} represent numerical approximations to the exact solution values $y(t_n + \Delta t)$ and $y(t_n + c_i\Delta t)$, where the c_i are given abscissae. Y_{n+1} is called the *stage vector*, its components y_{n+c_i} the *stage values*. G_n and g_n are assumed to be defined by preceding steppoint values y_n, y_{n-1}, etc. and by the preceding stage vectors Y_n, Y_{n-1}, etc. and their derivatives. Again, the steppoint formula is explicit, so that the main computational effort goes into the solution of the stage vector Y_{n+1}.

Let us linearize the stage vector equation in (2.8) to obtain for Y_{n+1} the linear system

$$Y_{n+1} - \Delta t(T \otimes I)(F(Y^0) + (I \otimes J)(Y_{n+1} - Y^0)) = G_n, \quad J \approx J(y_n) := \frac{\partial f(y_n)}{\partial y}. \tag{2.9}$$

Here, Y^0 is an approximation to Y_{n+1}, for example, $Y^0 = Y_n$ or $Y^0 = e \otimes y_n$. However, with this simple choice, the order of the linearized method is not necessarily the same as the original method (2.8). For instance, if (2.8) has order $p \geqslant 2$ and if $J = J(y_n) + O(\Delta t)$, then the order of the linearized method is in general not higher than two. If (2.8) has order $p \geqslant 3$, then higher-order formulas for Y^0 should be used. Of course, if the ODE system (2.2) is already linear, i.e., $y' = Jy$, then Y^0 does not play a role, because (2.9) is identical with the stage vector equation in (2.8) for all Y^0. Note that this also implies that the *linear* stability properties of (2.8) and its linearization are identical for all Y^0.

It turns out that approximate factorization of linear systems of the type (2.9) is most effective if T is either diagonal or (lower) triangular as in the case of the Rosenbrock method (2.5). Therefore, from now on, we impose this condition on T. Furthermore, instead of directly applying approximate factorization to the linear system (2.9), we first rewrite it into the equivalent form (compare (1.2a))

$$\begin{aligned} (I - \Delta t D \otimes J)(Y_{n+1} - Y^0) = {} & G_n - Y^0 + \Delta t(T \otimes I)F(Y^0) \\ & + \Delta t((T - D) \otimes J)(Y_{n+1} - Y^0), \end{aligned} \tag{2.9a}$$

where again $D = \mathrm{diag}(T)$. Proceeding as in the preceding section leads to the factorized method

$$\begin{aligned} &y_{n+1} = (a^{\mathrm{T}} \otimes I)Y_{n+1} + g_n, \\ &\Pi(Y_{n+1} - Y^0) = G_n - Y^0 + \Delta t(T \otimes I)F(Y^0) + \Delta t((T - D) \otimes J)(Y_{n+1} - Y^0) \end{aligned} \tag{2.10}$$

with Π defined as in (2.6). If $Y_{n+1} - Y^0 = O(\Delta t)$, then (2.10) presents a third-order perturbation of (2.9). Hence, by setting $Y^0 = Y_n$ or $Y^0 = e \otimes y_n$, the resulting method is second-order accurate provided that (2.8) is also (at least) second-order accurate. We shall refer to the approximately factorized, linearized method (2.10) as the *AFL method.*

If T is diagonal, then the subsystems for the components of $Y_{n+1} - Y^0$ can be solved concurrently, and if T is lower triangular, then these subsystems should be solved successively (note that $T - D$

is strictly lower triangular). The computational efficiency of solving the linear systems in (2.10) is comparable with that of (2.6).

2.4. Stability

As already remarked, the crucial point is the stability of the factorized methods. We shall discuss stability with respect to the model problem $y' = Jy = \sum J_k y$, where the matrices J_k commute. Application of the factorized methods to this model problem leads to linear recursions. The roots ζ of the corresponding characteristic equations define the *amplification factors* of the method. These amplification factors are functions of the vector $z=(z_1,\ldots,z_m)^{\mathrm{T}}$, where z_k runs through the eigenvalues of $\Delta t J_k$. We call a method *stable* at the point z if its amplification factor $\zeta(z)$ is on the unit disk. Likewise, we shall call a function $R(z)$ *stable* at z if $R(z)$ is on the unit disk. In the stability definitions and stability theorems given below, we shall use the notation

$$\mathbb{W}(\alpha) := \{w \in \mathbb{C}\colon |\arg(-w)| \leqslant \alpha\},$$

$$\mathbb{R}(\beta) := (-\beta, 0],$$

$$\mathbb{I}(\beta) := \left\{w \in \mathbb{C}\colon \arg(w) = \pm\frac{\pi}{2},\ |w| < \beta\right\}.$$

Definition 2.1. A method or a function is called

- *$A(\alpha)$-stable* if it is stable for $z_k \in \mathbb{W}(\alpha)$, $k = 1,\ldots,m$,
- *A-stable* if it is stable for $z_k \in \mathbb{W}(\pi/2)$, $k = 1,\ldots,m$,
- *$A_r(\alpha)$-stable* if it is stable for $z_1,\ldots,z_r \in \mathbb{R}(\infty) \wedge z_{r+1},\ldots,z_m \in \mathbb{W}(\alpha)$.

The first two definitions of stability are in analogy with the definitions in numerical ODE theory. The third type of stability was introduced by Hundsdorfer [17] and will be referred to as $A_r(\alpha)$-stability. This type of stability is relevant in the case of convection–diffusion–reaction equations. For example, for *systems* of two-dimensional convection–diffusion–reaction equations in which the Jacobian of the reaction terms has real, stiff eigenvalues, we would like to have $A_1(\pi/2)$-stability for $m = 3$, that is, stability in the region $\mathbb{R}(\infty) \times \mathbb{W}(\pi/2) \times \mathbb{W}(\pi/2)$. Then, by choosing the splitting such that J_1 corresponds to the reaction terms, and J_2 and J_3 to the convection–diffusion terms in the two spatial directions, we achieve unconditional stability. We remark that in the case of a *single* two-dimensional convection–diffusion–reaction equation, we need only A-stability for $m = 2$, because we can choose the splitting such that J_1 corresponds to the reaction term and the convection–diffusion in one spatial direction, and J_2 to convection–diffusion in the other spatial direction. Note that in this splitting, the matrices J_1 and J_2 both have a band structure with small band width.

In the following, we shall often encounter stability regions containing subregions of the form $\mathbb{S}_1 \times \mathbb{S}_2 \times \mathbb{S}_3$. In the case of approximate factorizations that are *symmetric* with respect to the Jacobians J_1, J_2 and J_3, as in the methods (2.4), (2.6) and (2.10), this means that the stability region also contains the subregions $\mathbb{S}_1 \times \mathbb{S}_3 \times \mathbb{S}_2$, $\mathbb{S}_2 \times \mathbb{S}_1 \times \mathbb{S}_3, \ldots$.

In the next sections, we give stability theorems for the method of Warming and Beam, and a few AFL and factorized Rosenbrock methods.

2.5. Method of Warming and Beam

Applying the Warming–Beam method (2.4) to the stability test problem yields the following characteristic equation for the amplification factor $\zeta = \zeta(z)$ of the method:

$$\rho(\zeta) - \psi(z)\sigma(\zeta) = 0, \quad \psi(z) := e^{\mathrm{T}} z \left[b_0 e^{\mathrm{T}} z + \prod_{k=1}^{m} (1 - b_0 z_k) \right]^{-1}. \tag{2.11}$$

Stability properties of (2.4) can be derived by using the following lemma of Hundsdorfer [17]:

Lemma 2.2. *Let H_m be the function defined by*

$$H_m(w) := 1 + e^{\mathrm{T}} w \prod_{k=1}^{m} (1 - \tfrac{1}{2} w_k)^{-1}, \quad w = (w_1, \ldots, w_m)^{\mathrm{T}}, \quad m \geqslant 2.$$

H_m is $A(\alpha)$-stable if and only if $\alpha \leqslant \frac{1}{2}\pi(m-1)^{-1}$ and $A_r(\alpha)$-stable if and only if $\alpha \leqslant \frac{1}{2}\pi(m-r)^{-1}$.

Theorem 2.3. *Let the LM method* (2.3) *be A-stable. Then the Warming–Beam method* (2.4) *is*
(a) *$A(\alpha)$-stable for $m \geqslant 2$ if and only if $\alpha \leqslant \frac{1}{2}\pi(m-1)^{-1}$.*
(b) *$A_r(\alpha)$-stable for $m \geqslant 2$ and $r \geqslant 1$ if and only if $\alpha \leqslant \frac{1}{2}\pi(m-r)^{-1}$.*
(c) *Stable in the region $\mathbb{I}(\beta_1) \times \mathbb{I}(\beta_1) \times \mathbb{R}(\beta_2)$ for $m = 3$ if $b_0^2\beta_1^2(b_0\beta_2 - 3) = 1$.*

Proof. If the method (2.3) is A-stable, then (2.4) is stable at the point z if $\mathrm{Re}(\psi(z)) \leqslant 0$, or equivalently, if $|(1 + c\psi(z))(1 - c\psi(z))^{-1}| \leqslant 1$ for some positive constant c. Let us choose $c = b_0$ (the A-stability of (2.3) implies that $b_0 > 0$). Then, it follows from (2.11) that

$$\frac{1 + b_0\psi(z)}{1 - b_0\psi(z)} = H_m(2b_0 z), \tag{2.12}$$

where H_m is defined in Lemma 2.2. Applying this lemma with $w = 2b_0 z$ proves part (a) and (b). Part (c) is proved by analysing the inequality $|H(2b_0 z)| \leqslant 1$ for $z = (\mathrm{i}y_1, \mathrm{i}y_2, x_3)$. For $m = 3$ this leads to

$$((1 - b_0^2 y_1 y_2)(1 - b_0 x_3) + 2b_0 x_3)^2 + (b_0(y_1 + y_2)(1 + b_0 x_3))^2$$
$$\leqslant (1 + b_0^2 y_1^2)(1 + b_0^2 y_2^2)(1 - b_0 x_3)^2.$$

The most critical situation is obtained if y_1 and y_2 assume their maximal value. Setting $y_1 = y_2 = \beta_1$ and taking into account that $x_3 \leqslant 0$, the inequality reduces to $x_3 \geqslant -(1 + 3b_0^2\beta_1^2)/b_0^3\beta_1^2$ from which assertion (c) is immediate. □

This theorem implies A-stability for $m = 2$, a result already obtained by Warming and Beam [28]. Furthermore, the theorem implies $A(0)$-stability for all $m \geqslant 2$, and $A(\alpha)$-stability with $\alpha \leqslant \pi/4$ for $m \geqslant 3$. Hence, we do not have unconditional stability in the case where all Jacobians J_k have eigenvalues close to the imaginary axis. We even do not have stability in regions of the form $\mathbb{I}(\beta) \times \mathbb{I}(\beta) \times \mathbb{I}(\beta)$ or $\mathbb{I}(\beta) \times \mathbb{I}(\beta) \times \mathbb{R}(\infty)$ with $\beta > 0$ (see also [12]). However, part (c) of the theorem implies for $m = 3$ stability in $\mathbb{W}(\pi/2) \times \mathbb{W}(\pi/2) \times \mathbb{R}(3/b_0)$. Such regions are suitable for *systems* of two-dimensional convection–diffusion–reaction equations with real, *nonstiff* eigenvalues in the reaction part. Note that it is advantageous to use small b_0-values, whereas L-stability of the underlying LM method does not lead to better stability properties.

Remark 2.4. The amplification factors of the stabilizing corrections method of Douglas (cf. [6,7]) are given by $\zeta = H(z)$, so that it has similar stability properties as the Warming–Beam method.

Remark 2.5. As already remarked in Section 2.1, (2.4) can be seen as a generalization of the Peaceman–Rachford method (1.3) to nonlinear PDEs with multicomponent splittings. In the literature, a second, direct generalization of (1.3a) is known, however its stability is less satisfactory. For the definition of this generalization, let $\boldsymbol{F}$ be a splitting function with m arguments satisfying the relation $\boldsymbol{F}(\boldsymbol{y},\dots,\boldsymbol{y})=\boldsymbol{f}(\boldsymbol{y})$ and define $\boldsymbol{F}_k$ by setting the kth argument of $\boldsymbol{F}$ equal to $\boldsymbol{y}^k$ and all other arguments equal to $\boldsymbol{y}^{k-1}$. Then, the direct generalization of (1.3a) reads

$$\boldsymbol{y}^0 = \boldsymbol{y}_n, \quad \boldsymbol{y}^k = \boldsymbol{y}^{k-1} + \frac{\Delta t}{m}\boldsymbol{F}_k, \quad \boldsymbol{y}_{n+1} = \boldsymbol{y}^m, \quad k = 1,\dots,m$$

(cf., e.g., [19, p. 278] and [16]). This scheme is second-order accurate for all $\boldsymbol{F}$. Evidently, it reduces to (1.3a) for linear problems and $m = 2$. Its amplification factor is given by

$$\zeta(z) = \prod_{k=1}^{m} \frac{m + \boldsymbol{e}^{\mathrm{T}} z - z_k}{m - z_k},$$

showing that unlike the Warming–Beam method, it is not even $A(0)$-stable for $m \geqslant 3$ (e.g., $\zeta(\boldsymbol{e} z_0) \approx (1-m)^m$ as $z_0 \to \infty$), so that it is only of use for $m = 2$.

2.6. AFL–LM methods

We start with AFL methods based on the class of LM methods (2.3). Writing (2.3) in the form (2.8) and applying the AFL method (2.10) to the stability test problem leads to the characteristic equation

$$\rho(\zeta) - \boldsymbol{e}^{\mathrm{T}} z \sigma(\zeta) = \left[1 - b_0 \boldsymbol{e}^{\mathrm{T}} z - \prod_{k=1}^{m}(1 - b_0 z_k)\right](\zeta - 1)\zeta^{\mu-1}. \tag{2.13}$$

This equation does not allow such a general stability analysis as in the case (2.11). Therefore, we confine our considerations to two particular cases, viz. the AFL methods based on the trapezoidal rule and the BDF method.

2.6.1. The trapezoidal rule

The trapezoidal rule is defined by $\rho(\zeta) = \zeta - 1$, $\sigma(\zeta) = \frac{1}{2}(\zeta + 1)$. This leads to the characteristic equation $\zeta = H(z)$, where H is defined in Lemma 2.2. Hence, according to the proof of Theorem 2.3 the AFL-trapezoidal rule and the Warming–Beam method with $b_0 = \frac{1}{2}$ possess the same stability region, so that Theorem 2.3 applies (with $b_0 = \frac{1}{2}$ in part (c)).

2.6.2. The BDF method

For the BDF with $\rho(\zeta)=\zeta^2-\frac{4}{3}\zeta+\frac{1}{3}$, $\sigma(\zeta)=\frac{2}{3}\zeta^2$ the characteristic equation (2.13) assumes the form

$$\zeta^2 - C_1\zeta + C_2 = 0,$$

$$C_1 = \frac{P(z)}{Q(z)}, \quad C_2 = \frac{1}{Q(z)}, \quad P(z) := Q(z) + 1 + 2e^{\mathrm{T}}z, \quad Q(z) := 3\prod_{k=1}^{m}(1-\tfrac{2}{3}z_k). \tag{2.14}$$

In order to find the stability region, we use Schur's criterion stating that the amplification factors are on the unit disk if $|C_2|^2 + |C_1 - C_1^* C_2| \leqslant 1$.

Theorem 2.6. *The AFL–BDF method is A-stable for $m = 2$ and $A(\pi/4)$-stable for $m = 3$.*

Proof. Let P^* and Q^* denote the complex conjugates of P and Q. Then, in terms of P and Q, the Schur criterion requires that the function $R(z) := |(P(z)Q^*(z) - P^*(z))/(Q(z)Q^*(z) - 1)|$ is bounded by one in the product space defined by $\operatorname{Re} z_k < 0$, $k = 1,\ldots,m$. Since $|1-\frac{2}{3}z_k|^2 > 1$ for z_k in the left half-plane, we easily see that $R(z)$ is analytic in this product space. Hence, the maximum principle reveals that we need to require $|R(\mathrm{i}y_1,\ldots,\mathrm{i}y_m)| \leqslant 1$, where we have written $z_k = \mathrm{i}y_k$, with y_k real. This condition on its own is equivalent to requiring that the polynomial $E(\mathrm{i}y) := (|Q(\mathrm{i}y)|^2 - 1)^2 - |P(\mathrm{i}y)Q^*(\mathrm{i}y) - P^*(\mathrm{i}y)|^2$ with $y := (y_1,\ldots,y_m)^{\mathrm{T}}$ is nonnegative. For $m = 2$ we straightforwardly find that

$$E(\mathrm{i}y_1,\mathrm{i}y_2) = \tfrac{16}{9}(y_1+y_2)^2(9y_1^2 + 9y_2^2 + 4y_1^2y_2^2 + 6y_1y_2).$$

It is easily seen that $E(\mathrm{i}y_1,\mathrm{i}y_2) \geqslant 0$ for all y_1 and y_2, proving the A-stability for $m = 2$.

For $m = 3$ we set $z_k = x_k - \mathrm{i}x_k$, $k = 1,2,3$, with $x_k \leqslant 0$ and derived an expression for $E(z)$ with the help of Maple. This expression has the form $-e^{\mathrm{T}}xs(x)$, where $x = (x_1,x_2,x_3)^{\mathrm{T}}$ and $s(x)$ consists of a sum of terms each term being of the form $x_1^p x_2^q x_3^r$, where p,q and r are nonnegative integers. We verified that the coefficients of these terms are all positive if $p+q+r$ is even and negative otherwise (the length of the formulas prevents us from presenting $s(x)$ here). Hence, $E(z) \geqslant 0$ for all $z_k = x_k - \mathrm{i}x_k$ with $x_k \leqslant 0$. Likewise, it can be shown that $E(z) \geqslant 0$ for all $z_k = x_k + \mathrm{i}x_k$ with $x_k \leqslant 0$, proving the $A(\pi/4)$-stability for $m = 3$. □

In addition, we determined stability regions of the form $\mathbb{I}(\beta_1) \times \mathbb{I}(\beta_1) \times \mathbb{R}(\beta_2)$ by analysing the stability boundary curve $E(\mathrm{i}y_1,\mathrm{i}y_2,x_3) = 0$ with the help of Maple. In particular, we found that in the region $\mathbb{W}(\pi/2) \times \mathbb{W}(\pi/2) \times \mathbb{R}(\beta)$ the value of β is determined by the equation $E(\mathrm{i}\infty,\mathrm{i}\infty,\beta) = 0$ and in the region $\mathbb{I}(\beta) \times \mathbb{I}(\beta) \times \mathbb{R}(\infty)$ by the equation $E(\mathrm{i}\beta,\mathrm{i}\beta,\infty) = 0$. This leads to $\beta = (9+3\sqrt{17})/4$ and $\beta = \frac{3}{4}\sqrt{2}$, respectively. For the sake of easy comparison, we have listed a number of stability results derived in this paper in Table 1. This table shows that the AFL–BDF regions $\mathbb{I}(\beta) \times \mathbb{I}(\beta) \times \mathbb{R}(\infty)$ and $\mathbb{W}(\pi/2) \times \mathbb{W}(\pi/2) \times \mathbb{R}(\beta)$ are larger than the corresponding stability regions of the Warming–Beam method generated by the BDF ($b_0 = \frac{2}{3}$). We now even have stability in $\mathbb{I}(\beta) \times \mathbb{I}(\beta) \times \mathbb{I}(\beta)$ with nonzero imaginary stability boundary β, however these boundaries are quite small ($\beta < 1/10$).

2.7. AFL–DIRK methods

If we define in (2.8) $g_n = (1 - a^{\mathrm{T}}e)y_n$ and $G_n = e \otimes y_n$, then (2.8) becomes a diagonally implicit Runge–Kutta (DIRK) method. We define an AFL–DIRK method by approximating Y_{n+1} by means of (2.10) with $Y^0 = e \otimes y_n$. The amplification factor with respect to the stability test model becomes

$$\zeta(z) = 1 + e^{\mathrm{T}} z a^{\mathrm{T}} \left(\prod_{k=1}^{m} (I - z_k D) - e^{\mathrm{T}} z (T - D) \right)^{-1} Te. \tag{2.15}$$

Let us consider the second-order, L-stable DIRK methods

$$\begin{aligned} &y_{n+1} = (e_2^{\mathrm{T}} \otimes I) Y_{n+1}, \quad Y_{n+1} = \begin{pmatrix} y_{n+\kappa} \\ y_{n+1} \end{pmatrix}, \\ &Y_{n+1} - \Delta t (T \otimes I) F(Y_{n+1}) = e \otimes y_n, \quad T = \begin{pmatrix} \kappa & 0 \\ 1-\kappa & \kappa \end{pmatrix} \end{aligned} \tag{2.16a}$$

and

$$\begin{aligned} &y_{n+1} = (1 - a^{\mathrm{T}} e) y_n + a^{\mathrm{T}} Y_{n+1}, \quad Y_{n+1} = \begin{pmatrix} y_{n+\kappa} \\ y_{n+1-\kappa} \end{pmatrix}, \quad a = \frac{1}{2\kappa^2} \begin{pmatrix} 3\kappa - 1 \\ \kappa \end{pmatrix}, \\ &Y_{n+1} - \Delta t (T \otimes I) F(Y_{n+1}) = e \otimes y_n, \quad T = \begin{pmatrix} \kappa & 0 \\ 1-2\kappa & \kappa \end{pmatrix}, \end{aligned} \tag{2.16b}$$

where $e_2 = (0,1)^{\mathrm{T}}$ and $\kappa = 1 \pm \frac{1}{2}\sqrt{2}$. The amplification factor (2.15) becomes in both cases

$$\zeta(z) = 1 + \frac{e^{\mathrm{T}} z}{\pi(z)} + \frac{\kappa(1-\kappa)(e^{\mathrm{T}} z)^2}{\pi^2(z)}, \quad \pi(z) := \prod_{k=1}^{m} (1 - \kappa z_k), \quad \kappa = 1 \pm \tfrac{1}{2}\sqrt{2}. \tag{2.17}$$

Theorem 2.7. *The AFL versions of* (2.16) *are A-stable for* $m = 2$ *and* $A(\pi/4)$*-stable for* $m = 3$.

Proof. Writing $\zeta(z) = P(z) Q^{-1}(z)$, where P and Q are polynomials in z_1, z_2 and z_3, it follows that we have A-stability if the E-polynomial $E(z) := |Q(z)|^2 - |P(z)|^2$ is nonnegative for all purely imaginary z_k. Using Maple, we found for $\kappa = 1 \pm \frac{1}{2}\sqrt{2}$

$$E(\mathrm{i}y_1, \mathrm{i}y_2, 0) = \tfrac{1}{4}(17 \pm 12\sqrt{2})(y_1 + y_2)^4,$$

which proves the A-stability for $m = 2$. Similarly, the $A(\pi/4)$ stability can be shown for $m = 3$. □

Thus, the A-stability and $A(\alpha)$-stability properties of the Warming–Beam, AFL-trapezoidal, AFL–BDF, and the above AFL–DIRK methods are comparable for $m \leqslant 3$. However, for the AFL–DIRK methods we found (numerically) the stability regions $\mathbb{I}(\beta 1) \times \mathbb{I}(\beta_1) \times \mathbb{R}(\infty)$ and $\mathbb{W}(\pi/2) \times \mathbb{W}(\pi/2) \times \mathbb{R}(\beta_2)$ with $\beta_1 \approx 1.26$ and $\beta_2 \approx 10.2$ for $\kappa = 1 - \frac{1}{2}\sqrt{2}$ and with $\beta_1 \approx 0.28$ and $\beta_2 \approx 1.75$ for $\kappa = 1 + \frac{1}{2}\sqrt{2}$. Hence, choosing $\kappa = 1 - \frac{1}{2}\sqrt{2}$ we have larger stability regions than the corresponding stability regions of the other methods (see Table 1). We also have stability in regions of the type $\mathbb{I}(\beta) \times \mathbb{I}(\beta) \times \mathbb{I}(\beta)$ with $\beta > 0$, but β is uselessly small.

2.8. Factorized Rosenbrock methods

Finally, we consider the factorized Rosenbrock method (2.6). With respect to the stability test model its amplification factor is given by

$$\zeta(z) = 1 + e^{\mathrm{T}} z b^{\mathrm{T}} \left(\prod_{k=1}^{m} (I - z_k D) - e^{\mathrm{T}} z (L + T - D) \right)^{-1} e. \tag{2.18}$$

We consider the original, second-order, L-stable Rosenbrock method [20] defined by (2.5) with

$$b = \begin{pmatrix} 0 \\ 1 \end{pmatrix}, \quad T = \begin{pmatrix} \kappa & 0 \\ 0 & \kappa \end{pmatrix}, \quad L = \tfrac{1}{2} \begin{pmatrix} 0 & 0 \\ 1 - 2\kappa & 0 \end{pmatrix}, \quad \kappa = 1 \pm \tfrac{1}{2}\sqrt{2}, \tag{2.19a}$$

and the second-order, L-stable Rosenbrock-W method (see [4, p. 233]) with

$$b = \tfrac{1}{2} \begin{pmatrix} 1 \\ 1 \end{pmatrix}, \quad T = \begin{pmatrix} \kappa & 0 \\ -2\kappa & \kappa \end{pmatrix}, \quad L = \begin{pmatrix} 0 & 0 \\ 1 & 0 \end{pmatrix}, \quad \kappa = 1 \pm \tfrac{1}{2}\sqrt{2}, \tag{2.19b}$$

The amplification factor (2.18) is for both methods (2.19) identical with the amplification factor (2.17) of the DIRK methods (2.16), so that all results of the preceding section apply to (2.19). The factorization of the Rosenbrock-W method (2.19b) in transformed form has successfully been used by Verwer et al. [27] for the solution of large scale air pollution problems. See also Sandu [23] for a discussion of this method.

3. Factorized iteration

Except for the factorized Rosenbrock-W methods, the factorized methods discussed in the preceding section are at most second-order accurate. As already observed by Beam and Warming [2], a simple way to arrive at higher-order methods that are still computationally efficient, is factorized iteration of higher-order integration methods. Evidently, if the iteration method converges, then we retain the order of accuracy of the underlying integration method (to be referred to as the corrector). Likewise, if the convergence conditions are satisfied, then the stability properties of the iterated method are the same as those of the corrector. Hence, the stability region of the iterated method is the intersection of the convergence region of the iteration method and the stability region of the corrector. Thus, if we restrict our considerations to A-stable, preferably L-stable correctors, then the stability region of the iterated method is the same as the convergence region of the iteration method.

Perhaps, even more important than the possibility of constructing higher-order methods is the increased robustness of the iterative approach. The reason is that the stability problem for the noniterative approach is replaced by a convergence problem for the iterative approach. However, unlike stability, which concerns accumulation of perturbations through a large number of integration steps, convergence can be controlled in each single step.

In Section 3.1, we discuss (i) *AFN iteration*, that is, approximately factorized Newton iteration of the nonlinear stage vector equation in (2.8), and (ii) *AF iteration*, that is, approximately factorized iteration of the linearized stage vector equation (2.9). AFN and AF iteration enables us to achieve stability in regions of the form $\mathbb{I}(\beta) \times \mathbb{I}(\beta) \times \mathbb{W}(\pi/2)$.

The AFN and AF methods treat all terms in the ODE system implicitly. In the case where the ODE system contains terms that are nonstiff or mildly stiff with respect to the other terms, it may be advantageous to treat these terms explicitly. This will be illustrated in Section 3.2.

Finally, in Section 3.3 we show how A-stability for three-component Jacobian splittings can be obtained, albeit at the cost of an increase of the computational complexity.

3.1. The AFN and AF iteration methods

Applying Newton iteration to the stage vector equation in (2.8) yields the linear Newton systems

$$(I - \Delta t T \otimes J)(\boldsymbol{Y}^j - \boldsymbol{Y}^{j-1}) = \boldsymbol{G}_n - \boldsymbol{Y}^{j-1} + \Delta t(T \otimes I)\boldsymbol{F}(\boldsymbol{Y}^{j-1}), \quad j \geqslant 1. \tag{3.1}$$

Next, we apply approximate factorization to obtain the *AFN iteration method*

$$\Pi(\boldsymbol{Y}^j - \boldsymbol{Y}^{j-1}) = \boldsymbol{G}_n - \boldsymbol{Y}^{j-1} + \Delta t(T \otimes I)\boldsymbol{F}(\boldsymbol{Y}^{j-1}) + \theta\Delta t((T - D) \otimes J)(\boldsymbol{Y}^j - \boldsymbol{Y}^{j-1}), \quad j \geqslant 1, \tag{3.2}$$

where Π is defined as before, $\boldsymbol{Y}^0$ is a suitable initial approximation to $\boldsymbol{Y}_{n+1}$, and where θ is a free parameter to be explained later. Note that after one iteration the AFN process is identical with (2.10) if we set $\theta = 1$ and if (2.10) and (3.2) use the same approximation $\boldsymbol{Y}^0$.

In the case of the linear system (2.9), we apply the *AF iteration method*

$$\Pi(\boldsymbol{Y}^j - \boldsymbol{Y}^{j-1}) = \boldsymbol{G}_n - \boldsymbol{Y}^{j-1} + \Delta t(T \otimes I)\boldsymbol{F}(\boldsymbol{Y}^0) + \Delta t((T \otimes J)(\boldsymbol{Y}^{j-1} - \boldsymbol{Y}^0), \tag{3.3}$$

$$+ \theta\Delta t((T - D) \otimes J)(\boldsymbol{Y}^j - \boldsymbol{Y}^{j-1}), \quad j \geqslant 1 \tag{3.3}$$

which is of course just the linearization of (3.2). The AFN and AF processes are consistent for all θ, that is, if the iterates $\boldsymbol{Y}^j$ converge, then they converge to the solutions $\boldsymbol{Y}_{n+1}$ of (2.8) and (2.9), respectively. Since the formulas (3.2) and (3.3) have the same structure as the AFL method (2.10), we conclude that, given the LU-decompositions of the factor matrices in Π, the costs of performing one iteration are comparable with those of applying (2.10). Hence, the efficiency of the AFN and AF processes is largely determined by the number of iterations needed to more or less solve the implicit system. The large-scale 3D shallow water transport experiments reported in [15,24,12] indicate that two or three iterations suffice. Also note that for $\theta = 0$ the subsystems in the resulting iteration processes can be solved in parallel, even if T is a triangular matrix.

AFN iteration can also be applied for solving *simultaneously* the subsystems for the components $\boldsymbol{k}_i$ of $\boldsymbol{K}$ from the Rosenbrock method (2.5). Similarly, AF iteration can be applied *successively* to these (linear) subsystems. Here, we shall concentrate on the iteration of (2.8) and (2.9). For details on the AFN and AF iteration of Rosenbrock methods we refer to [13].

3.1.1. The iteration error

Let us consider the recursions for the error $\varepsilon^j := \boldsymbol{Y}^j - \boldsymbol{Y}_{n+1}$. From (2.8) and (3.2) it follows that the AFN error satisfies the *nonlinear* recursion

$$\begin{aligned}
&\varepsilon^j = Z\varepsilon^{j-1} + \Delta t\Phi(\varepsilon^{j-1}), \quad j \geqslant 1,\\
&Z := I - (\Pi - \theta\Delta t((T - D) \otimes J))^{-1}(I - \Delta t T \otimes J),\\
&\Phi(\varepsilon) := (\Pi - \theta\Delta t((T - D) \otimes J))^{-1}(T \otimes I)(\boldsymbol{F}(\boldsymbol{Y}_{n+1} + \varepsilon) - \boldsymbol{F}(\boldsymbol{Y}_{n+1}) - (I \otimes J)\varepsilon).
\end{aligned} \tag{3.4}$$

Similarly, we deduce from (2.9) and (3.3) for the AF error the *linear* recursion

$$\varepsilon^j = Z\varepsilon^{j-1}, \quad j \geqslant 1. \tag{3.5}$$

It is difficult to decide which of the two iteration processes has a better rate of convergence. However, in a first approximation, the rates of convergence are comparable, because in the neighbourhood of the origin the Lipschitz constant of the function Φ is quite small, provided that J is a close approximation to $J(y_n)$. Therefore, we will concentrate on the amplification matrix Z.

First of all, we consider the convergence for small Δt. Since $Z=(1-\theta)\Delta t(T-D)\otimes J+\mathrm{O}((\Delta t)^2)$, the following theorem is easily proved (cf. [13]):

Theorem 3.1. *The iteration errors of the AFN and AF iteration processes* (3.2) *and* (3.3) *satisfy*

$$\varepsilon^j = \mathrm{O}((\Delta t)^{2j})\varepsilon^0, \quad j\geqslant 1 \quad \text{if } T \text{ is diagonal or if } \theta = 1,$$

$$\varepsilon^j = \begin{cases} \mathrm{O}((\Delta t)^j)\varepsilon^0 & \text{for } 1\leqslant j\leqslant s-1 \\ \mathrm{O}((\Delta t)^{2j+1-s})\varepsilon^0 & \text{for } j\geqslant s \end{cases} \quad \text{if } T \text{ is lower triangular and } \theta \neq 1.$$

This theorem shows that we always have convergence if Δt is sufficiently small. It also indicates that the nonstiff error components (corresponding with eigenvalues of J_k of modest magnitude) are rapidly removed from the iteration error. Furthermore, we now see the price to be paid if we set $\theta = 0$, while T is lower triangular (and not diagonal). In such cases, the subsystems in (3.2) and (3.3) can still be solved in parallel, however, at the cost of a lower order of convergence.

3.1.2. Convergence and stability regions

The eigenvalues $\lambda(Z)$ of the amplification matrix Z will be called the *amplification factors* in the iteration process. As in the stability analysis, we consider the test equation where the Jacobian matrices J_k commute. For this model problem, they are given by the eigenvalues of the matrix $I - \Pi^{-1}(I - \Delta t T \otimes J)$, so that

$$\lambda(Z) = 1 - (1 - \lambda(T)e^{\mathrm{T}}z)\prod_{k=1}^{m}(1 - \lambda(T)z_k)^{-1}.$$

Note that $\lambda(Z)$ does not depend on the parameter θ. We shall call a method convergent at z if $\lambda(Z)$ is within the unit circle at z. This leads us to the following analogue of Definition 2.1.

Definition 3.2. The iteration method is called

- *A(α)-convergent* if it is convergent for $z_k \in \mathbb{W}(\alpha)$, $k = 1,\ldots,m$,
- *A-convergent* if it is convergent for $z_k \in \mathbb{W}(\pi/2)$, $k = 1,\ldots,m$,
- *$A_r(\alpha)$-convergent* if it is convergent for $z_1,\ldots,z_r \in \mathbb{R}(\infty) \wedge z_{r+1},\ldots,z_m \in \mathbb{W}(\alpha)$.

From now on, we shall explicitly assume that

the corrector method is A-stable or L-stable,
the matrix T has nonnegative eigenvalues, (3.6)
the iteration process is performed until convergence.

These assumptions imply that the region of stability equals the region of convergence. The following theorem provides information on the $A(\alpha)$-stability characteristics [8].

Theorem 3.3. *Let the conditions* (3.6) *be satisfied. Then, AFN and AF iteration is $A(0)$-stable for $m \geqslant 2$, A-stable for $m=2$, and $A(\pi/4)$-stable for $m=3$.*

A comparison with the Theorems 2.3, 2.6 and 2.7 reveals that for $m \leqslant 3$ AFN and AF iteration have the same $A(\alpha)$-stability characteristics as obtained for the noniterative methods discussed in this paper. However, the stability results of Theorem 3.3 apply to any A-stable or L-stable integration method of the form (2.8) or (2.9) with $\lambda(T) \geqslant 0$, so that the order of accuracy can be raised beyond 2.

Furthermore, we found the stability region $\mathbb{W}(\pi/2) \times \mathbb{W}(\pi/2) \times \mathbb{R}(\beta)$ with $\beta = (1+\sqrt{2})\rho^{-1}(T)$, where $\rho(T)$ denotes spectral radius of T (do not confuse $\rho(T)$ with the Dahlquist polynomial $\rho(\zeta)$ used in (2.3)). Hence, this region can be made greater than the corresponding stability regions of all preceding noniterative methods (see Table 1) by choosing corrector methods such that $\rho(T)$ is sufficiently small. In Section 4, we give methods with $\rho(T)$ in the range [0.13, 0.5].

An even greater advantage is that factorized iteration leads to stability in regions of the form $\mathbb{I}(\beta_1) \times \mathbb{I}(\beta_1) \times \mathbb{I}(\beta_2)$ with substantial values of β_1 and β_2. In [13] it was shown that

$$\beta_1 = \min_{\lambda \in \Lambda(T)} \min_{0 \leqslant x \leqslant \lambda\beta_2} \frac{g(x)}{\lambda},$$

where $\Lambda(T)$ denotes the spectrum of T and g is defined by $4xg^3 + 2(x^2-1)g^2 - x^2 - 1 = 0$. Thus, if we choose β_1 not larger than the minimal value of $g(x)\rho^{-1}(T)$ in the interval $[0,\infty]$, then we have stability in the region $\mathbb{I}(\beta_1) \times \mathbb{I}(\beta_1) \times \mathbb{W}(\pi/2)$. This optimal value of β_1 is given by

$$\beta_1 = \frac{1}{6\rho(T)}(2 + (26+6\sqrt{33})^{1/3} - 8(26+6\sqrt{33})^{-1/3}) \approx \frac{0.65}{\rho(T)}. \tag{3.7}$$

Since usually $\rho(T)$ is less than 1, we obtain quite substantial values for β_1. This makes the iterative approach superior to the noniterative approach, where we found stability regions of the form $\mathbb{I}(\beta) \times \mathbb{I}(\beta) \times \mathbb{I}(\beta)$ with at best quite small β. The stability region $\mathbb{I}(\beta_1) \times \mathbb{I}(\beta_1) \times \mathbb{W}(\pi/2)$ enables us to integrate shallow water problems where we need unconditional stability in the vertical direction (because of the usually fine vertical resolutions) and substantial imaginary stability boundaries in the horizontal directions (because of the convection terms). The AFN–BDF method was successfully used in [15,24] for the solution of large-scale, three-dimensional shallow water transport problems.

3.2. Partially implicit iteration methods

The AFN and AF iteration methods (3.2) and (3.3) are implicit with respect to all Jacobians J_k in the splitting $J(y) = \sum J_k$. However, Table 1 frequently shows finite values for the stability boundaries. This raises the question whether it is necessary to treat all terms in the corresponding splitting implicitly. Afterall, when applying the standard, explicit, fourth-order Runge–Kutta method, we have real and imaginary stability boundaries of comparable size, viz. $\beta \approx 2.8$ and $2\sqrt{2}$, respectively.

In [13] this question is addressed and preliminary results are reported for iteration methods where Π does not contain all Jacobians J_k. In this approach, the iteration method can be fully tuned to the problem at hand. In this paper, we illustrate the partially implicit approach for transport problems in three-dimensional air pollution, where the horizontal spatial derivatives are often treated explicitly. In such problems, the Jacobian matrix $J(y)$ can be split into three matrices where J_1 corresponds with the convection terms and the two horizontal diffusion terms, J_2 corresponds with the vertical diffusion term, and J_3 corresponds with the chemical reaction terms. It is typical for air pollution terms that J_2 and J_3 are extremely stiff (that is, possess eigenvalues of large magnitude), and that J_1 is moderately stiff in comparison with J_2 and J_3 (see, e.g., [23,27]). This leads us to apply (3.2) or (3.3) with Π replaced with $\Pi_1 := (I - \Delta t D \otimes J_2)(I - \Delta t D \otimes J_3)$. Thus, only the vertical diffusion and the chemical interactions are treated implicitly. In the error recursions (3.4) and (3.5), the amplification matrix Z should be replaced by

$$Z_1 := I - (\Pi_1 - \theta\Delta t((T-D)\otimes J))^{-1}(I - \Delta t T \otimes J), \quad \Pi_1 := (I - \Delta t D \otimes J_2)(I - \Delta t D \otimes J_3).$$

Since $Z_1 = O(\Delta t)$, the nonstiff components in the iteration error are less strongly damped than by the AFN and AF processes (see Theorem 3.1). This is partly compensated by the lower iteration costs when using Π_1 instead of Π.

Let us assume that the eigenvalues z_2 of $\Delta t J_2$ are negative (vertical diffusion) and the eigenvalues z_3 of J_3 are in the left halfplane (chemical reactions). We are now interested to what region we should restrict the eigenvalues z_1 of $\Delta t J_1$ in order to have convergence. For the model problem this region is determined by the intersection of the domains bounded by the curve

$$|\lambda z_1|^2 + 2\lambda^3 z_2 \operatorname{Im}(z_3)\operatorname{Im}(z_1) = (1+\lambda^2|z_3|^2)(1-\lambda z_2)^2 - \lambda^4 z_2^2|z_3|^2, \tag{3.8}$$

where $\lambda \in \Lambda(T)$, $z_2 \in \mathbb{R}(\infty)$ and $z_3 \in \mathbb{W}(\pi/2)$. It can be verified that this intersection is given by the points $|z_1| < \rho^{-1}(T)$. Thus, we have proved:

Theorem 3.4. *Let the conditions* (3.6) *be satisfied, let* $m=3$, *let* Π *be replaced by* Π_1 *in the AFN and AF iteration methods, and define the disk* $\mathbb{D}(\beta) := \{w \in C: |w| < \beta\}$. *Then, the stability region contains the region* $\mathbb{D}(\rho^{-1}(T)) \times \mathbb{R}(\infty) \times \mathbb{W}(\pi/2)$.

We remark that the approximate factorization operator Π_1 is not symmetric with respect to all three Jacobians. This means that the stability region of the methods of Theorem 3.4 also contain the region $\mathbb{D}(\rho^{-1}(T)) \times \mathbb{W}(\pi/2) \times \mathbb{R}(\infty)$, but not, e.g., the region $\mathbb{R}(\infty) \times \mathbb{D}(\rho^{-1}(T)) \times \mathbb{W}(\pi/2)$.

3.3. A-stability for three-component Jacobian splitting

So far, the approximate factorization methods constructed in this paper are not A-stable for three-component splittings. However, all these methods can be modified such that they become A-stable for $m = 3$. The idea is to start with, e.g., a two-component splitting $J = J_1 + J^*$, where J_1 has the desired simple structure, but J^* has not, and to solve the linear system containing J^* iteratively with approximate factorization iteration. We illustrate this for the AFN process (3.2). Consider the process

$$(I - \Delta t\, D \otimes J_1)\tilde{\Delta}^j = \boldsymbol{G}_n - \boldsymbol{Y}^{j-1} + \Delta t(T \otimes I)\boldsymbol{F}(\boldsymbol{Y}^{j-1}) + \theta\Delta t((T-D)\otimes J)(\boldsymbol{Y}^j - \boldsymbol{Y}^{j-1}), \tag{3.9a}$$

$$(I - \Delta t\, D \otimes (J_2 + \cdots + J_m))\Delta^j = \tilde{\Delta}^j, \quad \boldsymbol{Y}^j - \boldsymbol{Y}^{j-1} = \Delta^j, \tag{3.9b}$$

where $j \geqslant 1$. This method can be interpreted as the AFN method (3.2) with m replaced by 2 and J_2 replaced by $J_2 + \cdots + J_m$. Hence, Theorem 3.3 implies that we have A-stability with respect to the eigenvalues of J_1 and $J_2 + \cdots + J_m$ (assuming that the corrector is A-stable). If the matrix $J_2 + \cdots + J_m$ does not have a 'convenient' structure (e.g., a small band width), then the system for Δ^j cannot be solved efficiently. In such cases, we may solve this system by an AFN (inner) iteration process:

$$\Delta^{j,0} = \tilde{\Delta}^j,$$

$$\Pi_1(\Delta^{j,i} - \Delta^{j,i-1}) = \tilde{\Delta}^j - (I - \Delta t\, D \otimes (J_2 + \cdots + J_m))\Delta^{j,i-1}, \quad i = 1, \ldots, r, \tag{3.9c}$$

$$Y^j = Y^{j-1} + \Delta^{j,r}, \quad \Pi_1 := \prod_{k=2}^{m} (I - \Delta t\, D \otimes J_k).$$

Since the stability theory of Section 3.1 applies to this process, we can apply the stability results from this section. The following theorem summarizes the main results for {(3.9a),(3.9c)}.

Theorem 3.5. *Let* (3.6) *be satisfied. Then, the inner-outer AFN process* {(3.9a),(3.9c)} *is*

(a) *$A(0)$-stable for $m \geqslant 2$, A-stable for $m = 2$ and $m = 3$, and $A(\pi/4)$-stable for $m = 4$,*
(d) *Stable in $\mathbb{W}(\pi/2) \times \mathbb{I}(\beta) \times \mathbb{I}(\beta) \times \mathbb{W}(\pi/2)$ with $\beta \approx 0.65\, \rho^{-1}(T)$* for $m = 4$,
(e) *Stable in $\mathbb{W}(\pi/2) \times \mathbb{W}(\pi/2) \times \mathbb{W}(\pi/2) \times \mathbb{R}(\beta)$ with $\beta = (1 + \sqrt{2})\, \rho^{-1}(T)$ for $m = 4$.*

Thus the process {(3.9a),(3.9c)} has excellent stability properties, but its computational complexity is considerably larger than that of (3.2). In order to compare this, let the number of outer iterations be denoted by q. Then (3.2) requires the solution of qm linear systems, whereas {(3.9a),(3.9c)} requires $q(rm + 1 - r)$ linear system solutions. For example, if $m = 3$, then we need $3q$ and $(2r + 1)q$ linear system solutions, respectively, so that for $r \geqslant 2$ the nested approach is more expensive.

Evidently, the above approach can also be applied to the AF method (3.3), but also to the noniterative methods (2.4), (2.6), and (2.10). In the noniterative methods, the computational complexity increases from m to $rm + 1 - r$ linear system solutions, i.e., by the same factor as in the iterative case. It should be remarked that there exist several splitting methods, not based on approximate factorization, that are also A-stable for three-component Jacobian splittings. We mention the ADI method of Gourlay and Mitchell [10] and the trapezoidal and midpoint splitting methods of Hundsdorfer [18]. These methods are second-order accurate and possess the same amplification factor $\zeta(z) = \prod_{k=1}^{3}(1 + \frac{1}{2}z_k)(1 - \frac{1}{2}z_k)^{-1}$ from which the A-stability is immediate. However, the internal stages of these methods are not consistent, that is, in a steady state the internal stage values are not stationary points of the method. This leads to loss of accuracy (cf. [17]).

4. Methods with minimal $\rho(T)$

The stability regions in Table 1 and the Theorems 3.4 and 3.5 indicate that small values of $\rho(T)$ increase the stability regions of the iterated methods. Similarly, Theorem 2.3 shows that small values of b_0 increases the stability region $\mathbb{I}(\beta_1) \times \mathbb{I}(\beta_1) \times \mathbb{R}(\beta_2)$ of the Warming–Beam method (note that for LM methods $b_0 = \rho(T)$). Therefore, it is relevant to look for methods with small $\rho(T)$.

Let us first consider the two-parameter family of all second-order, A-stable linear two-step methods (cf. [28, Fig. 2]). Taking b_0 and a_2 as the free parameters (see (2.3)), this family is defined by

$$\rho(\zeta) = \zeta^2 - (a_2 + 1)\zeta + a_2, \quad \sigma(\zeta) = b_0\zeta^2 + \tfrac{1}{2}(3 - a_2 - 4b_0)\zeta + b_0 - \tfrac{1}{2}(1 + a_2), \tag{4.1}$$

where $-1 \leqslant a_2 < 1$ and $b_0 \geqslant \frac{1}{2}$. Hence, the smallest value of b_0 is $\frac{1}{2}$. Moreover, from an implementation point of view, the trapezoidal rule choice $a_2 = 0$ is attractive.

Next, we consider the family of DIRK methods. We recall that they are defined by (2.8) with $\boldsymbol{g}_n = (1 - \boldsymbol{a}^{\mathrm{T}}\boldsymbol{e})\boldsymbol{y}_n$ and $\boldsymbol{G}_n = \boldsymbol{Y}^0 = \boldsymbol{e} \otimes \boldsymbol{y}_n$. In [14] a number of methods with minimal $\rho(T)$, relative to the number of stages, have been derived. Here, we confine ourselves to presenting a few second-order, L-stable methods by specifying the matrix T (in all cases $\boldsymbol{a} = \boldsymbol{e}_s$ and $\rho(T) = \kappa$):

$$T = \begin{pmatrix} \kappa & 0 \\ 1-\kappa & \kappa \end{pmatrix}, \quad \kappa = 1 - \tfrac{1}{2}\sqrt{2} \approx 0.29, \tag{4.2}$$

$$T = \begin{pmatrix} \kappa & 0 & 0 \\ \dfrac{1 - 4\kappa + 2\kappa^2}{2(1-\kappa)} & \kappa & 0 \\ 0 & 1-\kappa & \kappa \end{pmatrix}, \quad \kappa = \tfrac{1}{12}\left(9 + 3\sqrt{3} - \sqrt{72 + 42\sqrt{3}}\right) \approx 0.18, \tag{4.3}$$

$$T = \frac{1}{4}\begin{pmatrix} 4\kappa & 0 & 0 & 0 \\ \dfrac{1 - 8\kappa + 16\kappa^2 + 8\kappa^3}{1 - 4\kappa + 2\kappa^2} & 4\kappa & 0 & 0 \\ 0 & \dfrac{2 - 8\kappa + 4\kappa^2}{1-\kappa} & 4\kappa & 0 \\ 0 & 0 & 4(1-\kappa) & 4\kappa \end{pmatrix},$$

$$\kappa = \frac{4 + 2\sqrt{2} - \sqrt{20 + 14\sqrt{2}}}{4} \approx 0.13. \tag{4.4}$$

5. Summary of stability results

We conclude this paper with Table 1 which compares a number of stability results for various factorized methods based on three-component splittings.

Table 1
Stability regions of approximate factorization methods with three-component splittings

Methods	Stability region	Stability boundaries
Warming–Beam (2.4) and AFL-trapezoidal ($b_0 = \frac{1}{2}$)	$\mathbb{W}(\pi/4) \times \mathbb{W}(\pi/4) \times \mathbb{W}(\pi/4)$	
	$\mathbb{W}(\pi/2) \times \mathbb{W}(\pi/2) \times \mathbb{R}(\beta)$	$\beta = 3/b_0$
	$\mathbb{I}(\beta) \times \mathbb{I}(\beta) \times \mathbb{R}(\infty)$,	$\beta = 0$
AFL–BDF	$\mathbb{W}(\pi/4) \times \mathbb{W}(\pi/4) \times \mathbb{W}(\pi/4)$	
	$\mathbb{W}(\pi/2) \times \mathbb{W}(\pi/2) \times \mathbb{R}(\beta)$,	$\beta = (9 + 3\sqrt{17})/4 \approx 5.34$
	$\mathbb{I}(\beta) \times \mathbb{I}(\beta) \times \mathbb{R}(\infty)$,	$\beta = \frac{3}{4}\sqrt{2} \approx 1.06$
AFL–DIRK (2.16) and Factorized Rosenbrock (2.19) with $\kappa = 1 - \frac{1}{2}\sqrt{2}$	$\mathbb{W}(\pi/4) \times \mathbb{W}(\pi/4) \times \mathbb{W}(\pi/4)$	
	$\mathbb{W}(\pi/2) \times \mathbb{W}(\pi/2) \times \mathbb{R}(\beta)$	$\beta \approx 10.2$
	$\mathbb{I}(\beta) \times \mathbb{I}(\beta) \times \mathbb{R}(\infty)$,	$\beta \approx 1.26$
AFN/AF iteration (3.2)/(3.3)	$\mathbb{W}(\pi/4) \times \mathbb{W}(\pi/4) \times \mathbb{W}(\pi/4)$	
	$\mathbb{W}(\pi/2) \times \mathbb{W}(\pi/2) \times \mathbb{R}(\beta)$	$\beta = (1 + \sqrt{2})\rho^{-1}(T) \approx 2.41\ \rho^{-1}(T)$
	$\mathbb{I}(\beta) \times \mathbb{I}(\beta) \times \mathbb{W}(\pi/2)$	$\beta \approx 0.65\ \rho^{-1}(T)$
AFN/AF iteration (3.2)/(3.3) with Π_1 (Section 3.2)	$\mathbb{D}(\beta) \times \mathbb{R}(\infty) \times \mathbb{W}(\pi/2)$	$\beta = \rho^{-1}(T)$
Nested AFN {(3.9a),(3.9c)}	$\mathbb{W}(\pi/2) \times \mathbb{W}(\pi/2) \times \mathbb{W}(\pi/2)$	

References

[1] M. Beam, R.F. Warming, An implicit finite-difference algorithm for hyperbolic systems in conservation-law form, J. Comput. Phys. 22 (1976) 87–110.

[2] M. Beam, R.F. Warming, An implicit factored scheme for the compressible Navier–Stokes equations II, The numerical ODE connection, Paper No. 79-1446, Proceedings of AIAA 4th Computational Fluid Dynamics Conference, Williamsburg, VA, 1979.

[3] J. Crank, P. Nicolson, A practical method for numerical integration of solutions of partial differential equations of heat-conduction type, Proc. Cambridge Philos. Soc. 43 (1947) 50–67.

[4] K. Dekker, J.G. Verwer, Stability of Runge–Kutta Methods for Stiff Nonlinear Differential Equations, North-Holland, Amsterdam, 1984.

[5] J. Douglas Jr., On the numerical integration of $u_{xx} + u_{yy} = u_t$, J. Soc. Ind. Appl. Math. 3 (1955) 42–65.

[6] J. Douglas Jr., Alternating direction methods for three space variables, Numer. Math. 4 (1962) 41–63.

[7] J. Douglas Jr., J.E. Gunn, A general formulation of alternating direction methods. Part I. Parabolic and hyperbolic problems, Numer. Math. 6 (1964) 428–453.

[8] C. Eichler-Liebenow, P.J. van der Houwen, B.P. Sommeijer, Analysis of approximate factorization in iteration methods, Appl. Numer. Math. 28 (1998) 245–258.

[9] G.H. Golub, C.F. Van Loan, Matrix Computations, 2nd Edition, The Johns Hopkins University Press, Baltimore, MD, 1989.

[10] A.R. Gourlay, A.R. Mitchell, On the structure of alternating direction implicit (A.D.I.) and locally one dimensional (L.O.D.) difference methods, J. Inst. Math. Appl. 9 (1972) 80–90.

[11] E. Hairer, G. Wanner, Solving Ordinary Differential Equations II. Stiff and Differential–Algebraic Problems, Springer, Berlin, 1991.

[12] P.J. van der Houwen, B.P. Sommeijer, Approximate factorization in shallow water applications, Report MAS R9835, CWI, Amsterdam, 1998.

[13] P.J. van der Houwen, B.P. Sommeijer, Factorization in block-triangularly implicit methods for shallow water applications, Appl. Numer. Math. 36 (2001) 113–128.

[14] P.J. van der Houwen, B.P. Sommeijer, Diagonally implicit Runge–Kutta methods for 3D shallow water applications, J. Adv. Comput. Math. 12 (2000) 229–250.

[15] P.J. van der Houwen, B.P. Sommeijer, J. Kok, The iterative solution of fully implicit discretizations of three-dimensional transport models, Appl. Numer. Math. 25 (1997) 243–256.

[16] P.J. van der Houwen, J.G. Verwer, One-step splitting methods for semi-discrete parabolic equations, Computing 22 (1979) 291–309.

[17] W. Hundsdorfer, A note on the stability of the Douglas splitting method, Math. Comp. 67 (1998) 183–190.

[18] W. Hundsdorfer, Trapezoidal and midpoint splittings for initial-boundary value problems, Math. Comp. 67 (1998) 1047–1062.

[19] G.I. Marchuk, Splitting methods and alternating directions methods, in: P.G. Ciarlet, J.L Lions (Eds.), Handbook of Numerical Analysis, Vol. I, Part I, North-Holland, 1990, pp. 197–459.

[20] D.W. Peaceman, H.H. Rachford Jr., The numerical solution of parabolic and elliptic differential equations, J. Soc. Indust. Appl. Math. 3 (1955) 28–41.

[21] R.D. Richtmyer, K.W. Morton, Difference Methods for Initial-Value Problems, Wiley, New York, 1967.

[22] H.H. Rosenbrock, Some general implicit processes for the numerical solution of differential equations, Comput. J. 5 (1962–1963) 329–330.

[23] A. Sandu, Numerical aspects of air quality modelling, Ph.D. Thesis, University of Iowa, 1997.

[24] B.P. Sommeijer, The iterative solution of fully implicit discretizations of three-dimensional transport models, in: C.A. Lin, A. Ecer, J. Periaux, N. Satofuka (Eds.), Parallel Computational Fluid Dynamics — Developments and Applications of Parallel Technology, Proceedings of the 10th International Conference on Parallel CFD, May 1998, Hsinchu, Taiwan, Elsevier, Amsterdam, 1999, pp. 67–74.

[25] T. Steihaug, A. Wolfbrandt, An attempt to avoid exact Jacobian and nonlinear equations in the numerical solution of stiff differential equations, Math. Comp. 33 (1979) 521–534.

[26] K. Strehmel, R. Weiner, Linearly Implicit Runge–Kutta Methods and their Application, Teubner, Stuttgart, 1992. (in German).

[27] J.G. Verwer, W.H. Hundsdorfer, J.G. Blom, Numerical time integration for air pollution models, Report MAS R9825, CWI, Amsterdam, 1998, Surveys for Mathematics in Industry, to appear.

[28] R.F. Warming, M. Beam, An extension of A-stability to alternating direction implicit methods, BIT 19 (1979) 395–417.

ELSEVIER Journal of Computational and Applied Mathematics 128 (2001) 467

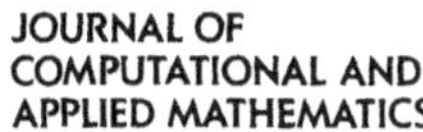

JOURNAL OF COMPUTATIONAL AND APPLIED MATHEMATICS

www.elsevier.nl/locate/cam

Author Index Volume 128 (2001)

PII: S0377-0427(01)00352-1

www.ingramcontent.com/pod-product-compliance
Lightning Source LLC
Chambersburg PA
CBHW080127220326
41598CB00032B/4982
9780444506160